Decoding Neural Circuit Structure and Function

Arzu Çelik · Mathias F. Wernet
Editors

Decoding Neural Circuit Structure and Function

Cellular Dissection Using Genetic Model Organisms

Springer

Editors
Arzu Çelik
Department of Molecular Biology and Genetics
Boğaziçi University
Bebek, Istanbul
Turkey

Mathias F. Wernet
Fachbereich Biologie, Chemie, Pharmazie, Institut für Biologie—Neurobiologie
Freie Universität Berlin
Berlin
Germany

ISBN 978-3-319-57362-5 ISBN 978-3-319-57363-2 (eBook)
DOI 10.1007/978-3-319-57363-2

Library of Congress Control Number: 2017940534

Printed on acid-free paper

This Springer imprint is published by Springer Nature
The registered company is Springer International Publishing AG
The registered company address is: Gewerbestrasse 11, 6330 Cham, Switzerland

Foreword

We Live in Exciting Times

We live in exciting times. Blocking or activating single cell types and recording from them in the awake behaving animal has so much become standard that one could almost forget that it has not always been this way. In fact, it is only about 20 years ago that the usual way to establish structure–function relationship, e.g., in *Drosophila* was to screen randomly generated mutants. If the screen was a structural one, it was followed by behavioral tests, if the screen was a behavioral one, it was followed by a structural study. Much brain power went into ingenious design of these screens as well as in the elegant behavioral setups, many of them are still in use today. While the results obtained this way were certainly an interesting first step, they suffered from important limitations insurmountable at that time. First of all, the resolution one obtained was a course one, far away from the single cell level: a certain brain structure was found to be involved in a certain behavior, but which type of neuron was doing what job and how could not be answered. Another almost lethal disadvantage of *Drosophila* these days was the fact that the nervous system of *Drosophila* was silent for the investigator: except for the giant descending neuron, no neuron in the CNS could be recorded from. So how could one go further to disentangle a neural circuit, even when given a complete catalogue of neuron types from Golgi staining?

Those were the days when the leech, lobster, and stick insects dominated the field of neural oscillators, when honeybees, silk moths, and cockroaches were model organisms for insect olfaction, precise timing was studied in barn owls, bats, and weakly electric fish and *Aplysia* was the hot spot for synaptic plasticity. One just needs to open a textbook on Neuroethology from the 1980s in order to see that. Interestingly, mammals seemed to be almost entirely excluded from this field since, except for place cells, their brain was studied mostly in anesthetized animals. As a systems neuroscientist working with *Drosophila* and submitting a paper to a major journal, one had to be prepared to receive reviews that started with "A beautiful study …", and ended with a rejection by saying "if only it had been done in the right animal." By now, many of the aforementioned model organisms have been ousted by *Drosophila*, the mouse and the zebrafish, as exemplified in this book.

But not only have the favorite animal species changed and new tools have been developed: the 'genetic revolution' also led to a merger of different fields and cultures, bringing together people with different backgrounds. Nowadays, an advanced neuroscience student might have done some PCR in the morning, while after lunch, he/she goes on and pulls electrodes or prepares for calcium recording under the 2-Photon microscope. If I should name a few areas in the fly brain where this has led to unforeseen progress in our understanding, I would list olfactory processing and memory formation in the mushroom bodies, neural circuit for motion vision and the fly's inner compass within the central complex. What made the change? Genetically addressable neurons are the first and most important prerequisite, whether it be through the cell-specific promotors available for olfactory receptor neurons or via cell-specific enhancers further refined by intersectional expression strategies like split-Gal4. Next are genetically encoded calcium indicators, kicked off by Roger Tsien with the famous paper 'Miyawaki et al, 1997' that opened the door to optical recording from every neuron you wanted. Proteins to block or activate identified neurons, like Shibire and Channelrhodopsin, represent other key ingredients to the success of the genetic approach. Of course, the combination of driver and effector flies mostly rests on the binary expression system Gal4-UAS, introduced by Brand and Perrimon in 1993.

This book represents a beautiful collection of studies where the neural circuits for different types of behavior are dissected in genetic model organism using state-of-the-art tools from genetics, physiology, and computational analysis. Now that we have everything we ever dreamed of, we are mostly limited by our imagination and intellect—if we fail, we have no excuse.

Alexander Borst
Max-Planck-Institute of Neurobiology
Martinsried, Germany

Contents

Notes from the Editors

The idea for this book project was born after a very successful EMBO workshop of almost the same name: "Decoding Neural Circuit Structure and Function." Organized by Arzu Çelik (Boğaziçi University), Nilay Yapici (then Rockefeller University, now Cornell University), and Hernan Lopez-Schier (Helmholtz Zentrum Muenchen) this workshop took place from September 26 to 28, 2014 on the wonderful campus of Boğaziçi University right in the center of the buzzing metropolis Istanbul.

What makes meetings like this one so exciting is the interdisciplinary nature of the presentations, the very different backgrounds of the participating people, and the broad range of topics and techniques covered. Also, while teaching courses on neural circuit structure and function both in Istanbul and Berlin, we noticed that teaching material about this exciting and fast evolving area of research is under-represented. This book now intends to give a glimpse of what is currently possible. At the center of many questions lies the concept of a neuronal cell type and the question about its role within the brain in general, and within a more restricted neural circuit linked to a particular function. The progress toward reproducibly labeling neuronal cell types using molecular genetic tools in different genetic model organisms has been a major step forward over the last two decades. We are very happy that Alexander Borst has agreed to contribute a preface to this book project, in which he describes the impact that this truly historic ascension of genetic model organisms has had on a field that suffered from numerous technical challenges. The model systems we ended up focusing on in this book are the fruit fly, zebrafish, and the mouse. We apologize that the worm *C. elegans* ended up somewhat under-represented in our project—that was not at all intended, but instead resulted from a series of unfortunate coincidences, when several authors had to drop out at the last minute. So, how does one organize 20 Chapters on 'The Structure and Function of Neural Circuits?' In this book project, we decided to cover four large areas, which we think together represent the Neural Circuit field using genetic model organisms: Anatomy, Behavior, Physiology, and cell type specification (Development).

In Part I (Anatomy), the authors cover an amazing range of topics, beginning with an introduction to the complexity of molecular genetic toolkits, which can certainly seem intimidating (Chap. 1). Two chapters then present different efforts to reconstruct neural circuits in 3D using electron microscopy as well as light microscopy (Chaps. 2 and 3). Additional aspects of connectivity are also addressed: Trans-synaptic tracing using neurotropic viruses (Chap. 4), as well as live imaging of neural circuit formation (Chap. 5).

Part II (Behavior) then moves on to describe the manipulation of cellular elements and the resulting consequences on animal behavior, both in fly larvae (Chap. 6), as well as adults (Chap. 7). The quantitative study of vertebrate behavior is also covered by discussing the visual responses of zebrafish (Chap. 8). In the following chapter, the important link between genetic manipulations, behavioral responses, and the computations executed by identified neural circuits are discussed (Chap. 9). The final two chapters remind us that nothing is set in stone and remains unchanged forever and describe the modulation of neural circuitry (Chap. 10), as well as the important question about what kind of problems can arise from drawing conclusions about the necessity or sufficiency of identified circuit elements purely from genetic manipulations in behaving animals (Chap. 11).

Part III (Physiology) then presents important examples for how genetically encoded indicators of neuronal activity can be used to probe the nervous system of flies (Chap. 12), as well as the entire zebrafish brain (Chap. 13). This is followed by a chapter describing how the neural circuits affected in mood disorders can be studied in the mouse, using a combination of electrophysiology and optogenetics (Chap. 14). Finally, the fly visual system serves as an elegant example for how many of the so far covered techniques can be combined to reveal the cellular basis for motion computation in the invertebrate brain (Chap. 15).

The question how the vast diversity in neuronal cell types is generated will be addressed in Part IV (Development). This Part starts with two examples for how neuronal cell fate decisions are regulated in different sensory systems of fruit flies (Chaps. 16 and 17). The investigation of neuronal cell types using transcriptomic approaches is addressed in two complementary chapters: One covers single cell transcriptomics in the mouse (Chap. 18), and the other one covers different transcriptomic approaches in invertebrates (Chap. 19).

In summary, the four parts aim toward providing a 'holistic' description of the role of identified cell types within neural circuits. It must be pointed out that many of the techniques and approaches described above become increasingly intertwined: many laboratories are implementing them simultaneously, resulting in impressive datasets of unprecedented precision. Nevertheless, the need for such synergy also presents important challenges: how can one single laboratory unite all the necessary expertise, the required funds, and equipment? It is not surprising that several large research centers have been established, where such environments are being provided: Janelia Research Campus, The Allen Brain Institute, The Champalimaud Center for the Unknown, and the Max-Planck Institute for Neurobiology, just to name a few prominent examples. We are particularly happy to unite authors from all these institutions within this book project.

Where do we go from here? The complexity of neural circuit architecture and function that is arising from the efforts presented in this book is truly mind-blowing. The amount of data provided from the 3D reconstruction efforts can now be overlaid with the functional properties of identified cell types, resulting in a (synaptic) resolution previously deemed impossible. Given the number of different connectomic endeavors that are currently going on, we are very happy that Ian Meinertzhagen (Dalhousie University) agreed to discuss the future of 'Comparative Connectomics' in his perspective piece.

Going through the exciting chapters in this book, one wonders whether our limitation on genetic model organisms (worm, fruit fly, zebrafish, mouse) will disappear in the future? Recent progress in genome editing via Crispr/Cas9 could provide everyone with his or her 'model organism of choice' in a relatively straightforward manner. This truly feels like the 'golden age' for Neural Circuit Science. The possibilities seem limitless, and one can hardly wait to see where we will be taken next.

In conclusion, we would like to thank all the contributing authors (many of them being up-and-coming junior Scientists), as well as the participants of the 2014 EMBO workshop, and our contacts at Springer for making this project possible. In theory, the speed at which the field of Neural Circuit Science progresses already seems to make it necessary to start planning a new follow-up book project, which would almost certainly look entirely different. But let us put first things first. For now, we look forward to more workshops, seminars, and meetings on this fundamental topic. The year 2014 and our peaceful gathering in Istanbul seems very far away right now in this rapidly changing world. We can only hope that the desire of people across the globe (Scientists and non-Scientists alike) to come together and discuss on the basis of facts and data will prevail. In this particular case, we are proud to present the very tangible result of such interactions that we hope any interested reader will enjoy as much as we did: "Decoding Neural Circuit Structure and Function (Cellular Dissection Using Genetic Model Organisms)".

Berlin, Germany
January 2017

Arzu Çelik
Mathias F. Wernet

Part I
Anatomy: High-Resolution Neuroanatomy Using Molecular-Genetic Tools

Chapter 1
The Current State of the Neuroanatomy Toolkit in the Fruit Fly *Drosophila melanogaster*

Daryl M. Gohl, Javier Morante and Koen J.T. Venken

Abstract The fruit fly *Drosophila melanogaster* is a popular workhorse model organism that has tremendously contributed to our understanding of the nervous system across eukaryotic multicellular species. Through molecular, developmental, histochemical, anatomical, and physiological experimentation, studies that incorporate fruit flies have had immediate biomedical impact relevant to neurobiology and neuropathology. *D. melanogaster* is one of the most well-established eukaryotic multicellular model organisms, largely due to its sophisticated and expanding in vivo targeted neurogenetic manipulations. Here, we summarize the current status of techniques for precisely targeted spatiotemporal manipulation of the fly's nervous system, focused on the most recent developments within the field.

1.1 Introduction

The fruit fly *Drosophila melanogaster* is one of the very few model organisms with a large arsenal of genetic tools for highly sophisticated genetic manipulation (Venken and Bellen 2012, 2014; Venken et al. 2011a, 2016). Through this toolbox, fruit flies have significantly contributed to our understanding of developmental, biological, physiological, and behavioral aspects of the developing and adult

D.M. Gohl (✉)
University of Minnesota Genomics Center, Minneapolis, MN 55455, USA
e-mail: dmgohl@umn.edu

J. Morante (✉)
Instituto de Neurociencias, Consejo Superior de Investigaciones Científicas, Universidad Miguel Hernández, Campus de Sant JoanAvenida Ramon y Cajal s/n, 03550 Sant Joan, Alicante, Spain
e-mail: j.morante@umh.es

K.J.T. Venken (✉)
Verna and Marrs McLean Department of Biochemistry and Molecular Biology, Department of Pharmacology, Dan L. Duncan Cancer Center, Baylor College of Medicine, Houston, TX 77030, USA
e-mail: Koen.Venken@bcm.edu

A. Çelik and M.F. Wernet (eds.), *Decoding Neural Circuit Structure and Function*, DOI 10.1007/978-3-319-57363-2_1

nervous system (Bellen et al. 2010), including basic neurobiological and behavioral functions such as vision (Silies et al. 2014), olfaction (Wilson 2013), taste (Freeman and Dahanukar 2015), circadian rhythm (Tataroglu and Emery 2014; Allada and Chung 2010), sleep (Donelson and Sanyal 2015), memory (Keene and Waddell 2007), pain (Leung et al. 2013; Tracey et al. 2003) and courtship (Dickson 2008), to name just a few. In addition, since the early 2000s, work in flies has contributed more and more to our understanding of the mechanisms and neuropathological characteristics associated with neurodevelopmental diseases that have their origins during development (Gatto and Broadie 2011), and neurodegenerative (McGurk et al. 2015), neurological (Shulman 2015), and neuropsychiatric disorders (van Alphen and van Swinderen 2013), which often occur much later in life. This toolbox comprises a large number of different molecular players that are used for a variety of purposes (Venken and Bellen 2012, 2014; Venken et al. 2011a, 2016; del Valle Rodríguez et al. 2012). The goal of this chapter is to provide a summary of the available genetic reagents that are used to spatiotemporally manipulate neurons.

1.2 Binary Activation Systems

Cellular manipulation in *D. melanogaster* is almost exclusively performed through binary activation systems (Fig. 1.1). A prototypical binary activation system has two parts: a transactivator and an effector. The transactivator is a heterologous transcription factor with a DNA-binding domain (DBD) and an activation domain (AD), typically expressed from a regulatory element, such as enhancer or promoter that will direct expression in a specific tissue or subset of cells (see Sect. 1.4). This artificial transcription factor can then drive the expression of any effector through binding at a synthetic promoter with specific multimerized binding sites (see Sect. 1.3). To further refine overexpression, the effector can be preceded or followed by additional RNA or protein regulatory elements that will tune expression levels in a negative or positive fashion, e.g., minimal promoters required for transcriptional initiation (Ni et al. 2009), translational enhancers (Pfeiffer et al. 2012), introns (Ni et al. 2008, 2009; Pfeiffer et al. 2010, 2012), RNA stabilizing elements (Pfeiffer et al. 2010, 2012), transcriptional termination signals (Pfeiffer et al. 2010, 2012; Brand and Perrimon 1993; Shearin et al. 2013), or protein-destabilizing domains (Nern et al. 2011). The different binary expression systems have incorporated the expression patterns of many 1000s of simple regulatory elements to efficiently and systematically drive expression of arbitrary effectors, e.g., fluorescent markers for cell labeling (see Sect. 1.3.1), or neuromodulators to influence neuronal physiological activity (see Chap. 7). Currently, there are three binary activation systems that are commonly used in *Drosophila*: the GAL4, LexA, and Q systems.

(a)

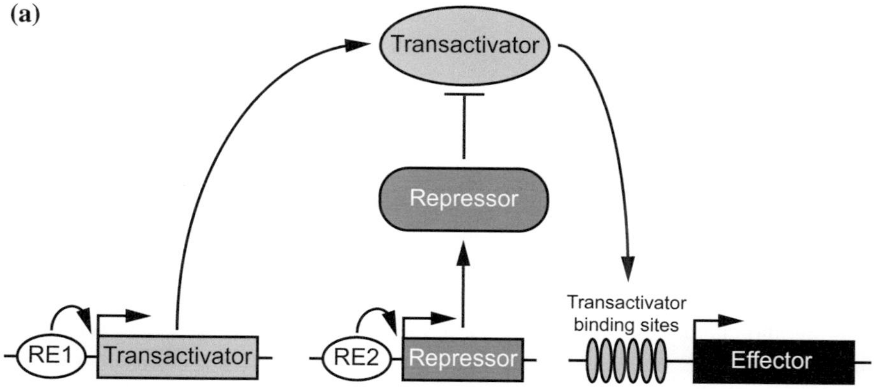

(b)

Transactivator	Repressor	Transactivator binding sites
GAL4	GAL80	*UAS*
LexA	--	*LexAop*
QF	QS	*QUAS*

Fig. 1.1 Binary transcriptional activation systems commonly used for neuroanatomical intersectional analysis in *Drosophila melanogaster* are GAL4, LexA, and QF. GAL4, LexA, and QF encode transactivators that can capture the expression patterns of regulatory elements (RE) either through random transposon integration or using cloned fragment fusions. The transactivator can activate expression of an effector gene construct that contains a multimerized copy of the transactivator's DNA binding site. The GAL4 and QF systems also have repressor proteins, i.e., GAL80 and QS, respectively, that target the activation domain of the transactivator protein, resulting in inhibited transactivation activity. The repressor's expression is regulated by a second regulatory element, i.e., same or different than the regulatory element driving the transactivator

1.2.1 The GAL4 System

The GAL4 system uses the heterologous GAL4 transcription factor, a regulator of galactose-induced genes in *Saccharomyces cerevisiae*, and an effector construct containing the GAL4 recognition site, the upstream activating sequence (*UAS*) (Giniger et al. 1985; Johnston and Hopper 1982; Laughon and Gesteland 1982). GAL4 was initially shown capable of inducing reporter gene expression outside of *S. cerevisiae*, including *Drosophila* (Fischer et al. 1988; Kakidani and Ptashne 1988). Following this demonstration, a two-part GAL4/*UAS Drosophila* toolkit was developed (Brand and Perrimon 1993). In contrast to the enhancer fusion approach that preceded it (Fischer et al. 1988; Kakidani and Ptashne 1988), the modular nature of the GAL4/*UAS*-system allowed construction of both driver lines (see

Sect. 1.4) and *UAS*-linked effector lines (see Sect. 1.3) that could be used in any combination by crossing a driver line with a desired expression pattern (i.e., determined by enhancer and/or promoter) to a fly carrying the desired effector construct. Since the introduction of the GAL4 system, collections of thousands of GAL4 lines and *UAS* effector lines have been generated (Venken et al. 2011a; Duffy 2002; Hayashi et al. 2002; Jenett et al. 2012; Jory et al. 2012; Li et al. 2014; Manning et al. 2012). We will discuss different types of GAL4 lines in detail below: some lines are enhancer traps or fusions (see Sect. 1.4.1), some are promoter traps or fusions (see Sect. 1.4.2), and some are protein traps (see Sect. 1.4.3). The value of these collections has been significantly enhanced through a number of large-scale imaging and characterization projects that have generated annotated and searchable databases of expression data throughout the nervous system as well as other tissues (Hayashi et al. 2002; Jenett et al. 2012; Jory et al. 2012; Li et al. 2014; Manning et al. 2012; Chiang et al. 2011; Peng et al. 2011).

In *S. cerevisiae*, GAL4-mediated expression is repressed by GAL80, which binds to the GAL4 AD and prevents transcriptional activation in the absence of galactose (Ma and Ptashne 1987a). Interestingly enough, GAL80 can also function as a negative regulator of GAL4 in a heterologous model system (e.g., *Drosophila*), a function that was first exploited in the context of the elegant MARCM (i.e., mosaic analysis with a repressible cell marker) system to positively mark clones in mosaic mitotic analysis (Lee and Luo 1999) (see Sect. 1.6). Subsequently, GAL80 has also become an important tool for intersectional refinement of enhancer trap expression patterns (see Sect. 1.5) (Pfeiffer et al. 2010; Suster et al. 2004).

While the GAL4 system provides spatial control of gene expression, for many experiments it is desirable to also have temporal control. For instance, many genes have dual roles in development as well as adult nervous system function, and phenotypes resulting from GAL4-mediated expression at both stages may obscure these dual roles of the gene. Variants of GAL4 have been developed that provide temporal control over gene expression. A hormone-inducible derivative of GAL4, i.e., the GAL4 DBD fused to the estrogen receptor domain, was shown to function in *Drosophila* oocytes (Han et al. 2000). Similarly, the more widely adopted GeneSwitch system utilizes a synthetic protein fusion consisting of a GAL4 DBD, progesterone receptor ligand binding domain and p65 activation domain, which can be induced in a dose-dependent manner with RU486 (mifepristone) (Nicholson et al. 2008; Osterwalder et al. 2001; Roman and Davis 2002; Roman et al. 2001).

Another mechanism that adds temporal control to the GAL4 system is the use of a temperature-sensitive mutation in GAL80 (Matsumoto et al. 1978). Shifting flies to the nonpermissive temperature abolishes repression of GAL4-mediated expression by GAL80^{ts} (McGuire et al. 2003). One limitation of these tools is their timescale of induction (and hence their temporal resolution), which ranges from 6 h to achieve steady-state expression and 36 h to return to baseline expression with GAL80^{ts}, and about 24 to 48 h to achieve maximal expression with GeneSwitch (McGuire et al. 2003). Finally, temperature itself strongly influences GAL4-mediated expression because most enhancer trap lines use promoter elements from the temperature-sensitive Hsp70 promoter (Brand and Perrimon 1993;

Mondal et al. 2007), although GAL4's transcriptional activity itself is temperature-independent (Mondal et al. 2007). Temperature shifts can cause unexpected physiological responses that influence many behavioral phenotypes (Kuo et al. 2012). Thus, it is important to control for this additional experimental variable in any experiments involving $GAL80^{ts}$.

1.2.2 The LexA System

In its native context in regulating the *Escherichia coli* SOS stress response, LexA functions as a transcriptional repressor (Walker 1984). However, when LexA is fused to a heterologous transcriptional AD, it can activate transcription from transgenes containing LexA operator (LexAop) sites in heterologous systems, including *D. melanogaster* (Lai and Lee 2006; Szuts and Bienz 2000). Development of the LexA system as a second binary expression system in *Drosophila* made orthogonal expression of multiple transgenes in the same animal possible. In addition, fusing LexA to the GAL4 or VP16 ADs generated GAL80-sensitive and -insensitive versions of LexA (Lai and Lee 2006), which have applications in intersectional targeting (see Sect. 1.5), and also enabled more sophisticated versions of the MARCM technology, e.g., dual-expression-control MARCM (Lai and Lee 2006) (see Sect. 1.6). Fusing estrogen and progesterone receptor domains to the LexA DBD provides spatiotemporal control by β-estradiol and RU486 respectively (Kuo et al. 2012).

1.2.3 The Q System

The Q system is based on a transcription factor, QF, from the *Neurospora crassa qa* gene cluster (Geever et al. 1989), which regulates quinic acid metabolism in its native context by binding to the so-called QUAS sites (Potter et al. 2010). Like the LexA system, the Q system provides an orthogonal system for labeling or manipulating specific populations of cells and also enables "coupled MARCM" experiments in which two-cell populations arising from a single-cell division can be independently labeled using the Q and GAL4 systems (see Sect. 1.6) (Potter et al. 2010).

Like GAL4, QF is targeted by a negative regulator, QS, which can repress QF-mediated expression (Huiet and Giles 1986). Repression by QS can also be disrupted in a dose-dependent manner by feeding flies quinic acid, providing a means to temporally control transgene expression (Potter et al. 2010; Potter and Luo 2011).

The initial QF construct exhibited some toxicity in *D. melanogaster*, precluding the establishment of pan-neuronal or pan-organismal driver lines. However, subsequent protein engineering efforts have yielded nontoxic variants, QF2 and QF2w, as well as chimeric GAL4QF (i.e., a protein fusion between the GAL4 DBD and the QF AD) and LexAQF (i.e., a protein fusion between the LexA DBD and the QF AD) transactivators, which are QS suppressible and quinic acid inducible (Riabinina et al. 2015).

1.3 Neurogenetic Labeling

The three binary expression systems, GAL4, LexA, and Q, can be used to drive expression of genetically encoded reporters to label the entire cytoplasm or subcompartments of neurons, i.e., cell compartments and organelles common to most cells (e.g., nucleus, mitochondria, endoplasmatic reticulum, and Golgi), or cellular compartments exclusive to neurons (e.g., synaptic vesicles, active zones, and dendrites) (Fig. 1.2). Fluorescent reporters can be used for live imaging or analysis of fixed specimens (i.e., directly or after immunohistochemistry using antibodies), and non-fluorescent reporter proteins can be used for immunohistochemistry. Alternatively, neuronal modulators or activity sensors can be targeted to a subset of neurons to affect or measure neuronal physiology respectively.

1.3.1 Fluorescent Protein Reporters

Currently, most existing fluorescent reporters are only GAL4 compatible. Expression of fluorescent reporters without an organelle—or compartment-targeting peptide label the entire cytoplasm and provide a full internal labeling of the host neuron (Pfeiffer et al. 2010; Halfon et al. 2002; Shearin et al. 2014; Yeh et al. 1995). While some earlier reporters inefficiently labeled the cytoplasm of entire neurons, codon optimization (Pfeiffer et al. 2010), or multimerization (Shearin et al. 2014) of the reporters has significantly improved labeling. On the other hand, fluorescent markers fused to membrane targeting motifs or membrane targeted domains solely label the cellular outline and their enrichment in membranes provides intricate detail about neuronal morphology (Pfeiffer et al. 2010; Lee and Luo 1999; Ritzenthaler et al. 2000; Ye et al. 2007; Yu et al. 2009). Protein fusions between synaptic vesicle proteins and reporters predominantly label synaptic vesicles and the presynaptic portion of the synaptic contact (Estes et al. 2000; Rolls et al. 2007; Zhang et al. 2002). A fluorescent protein fused to the active zone localized proteins *bruchpilot* (*brp*) (Wagh et al. 2006) or *cacophony* (*cac*) (Kawasaki et al. 2004) labels active zones. Dendrites are preferentially labeled by a synthetic fusion protein between a fluorescent reporter and the mouse protein ICAM5/Telencephalin (i.e., Denmark) (Nicolai et al. 2010) or an exon encoding a specific membrane targeting domain of Down syndrome cell adhesion molecule (Dscam) (i.e., Dscam[exon 17.1]) (Wang et al. 2004). A fluorescent protein fusion to the neurotransmitter

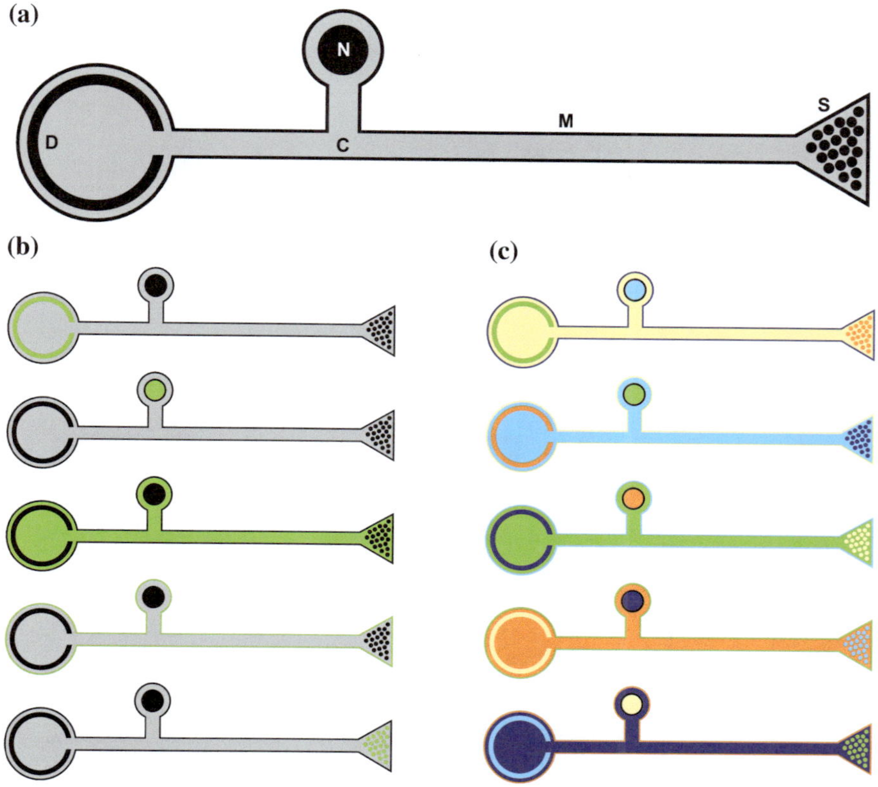

Fig. 1.2 Reporters useful for neuroanatomical analysis in *Drosophila melanogaster*. **a** Schematic of a typical *Drosophila melanogaster* neuron indicating five neuronal compartments most relevant to neuroanatomical analysis: dendrites (D), nucleus (N), cytoplasm (C), membrane (M), and synapse (S). **b** Classically, individual reporters are targeted to unique compartments and analyzed separately. **c** Future analysis may incorporate combinatorial reporters that could be activated through Brainbow strategies (see Sect. 1.6 and Fig. 1.12), i.e., each of the five Brainbow cassettes encodes five fluorescent proteins each labeling a separate neuronal compartment

receptor proteins resistant to dieldrin (Rdl) and nicotinic acetylcholine Receptor α7 (nAChRα7) can also be used to identify synapses (Leiss et al. 2009; Sanchez-Soriano et al. 2005). Protein fusions between fluorescent proteins and targeting elements specific for nuclei (Yasunaga et al. 2006), mitochondria (LaJeunesse et al. 2004), endoplasmatic reticulum (LaJeunesse et al. 2004), and Golgi (LaJeunesse et al. 2004) result in subcellular labeling enriched for the targeted organelle. Only recently fluorescent reporters as described above for the GAL4 system have also been generated for the LexA and Q systems, including several markers that label cytoplasm (Shearin et al. 2014; Yagi et al. 2010), membrane (Pfeiffer et al. 2010; Lai and Lee 2006; Potter et al. 2010; Diegelmann et al. 2008; Petersen and Stowers 2011), and synaptic vesicles (Shearin et al. 2013; Petersen and Stowers 2011).

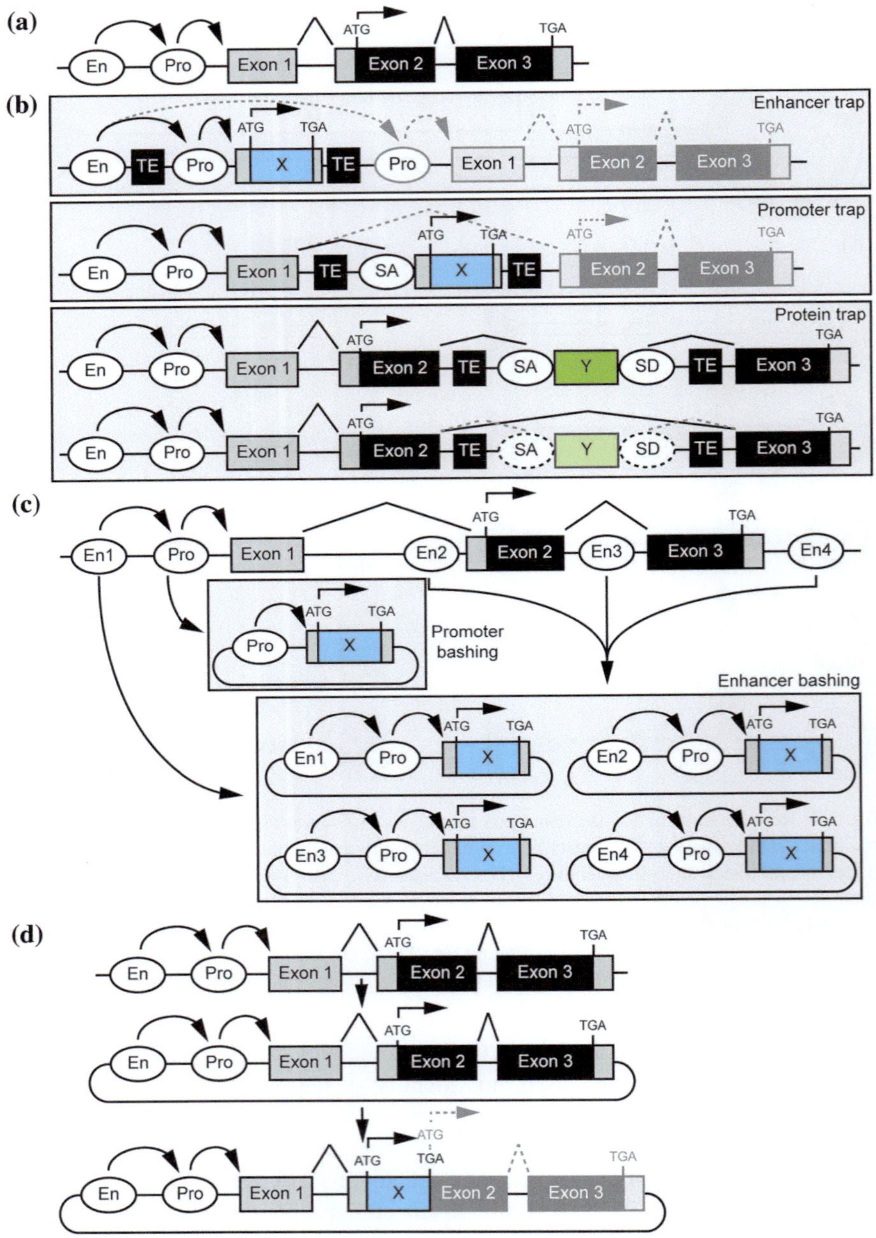
(a)
En
Pro
Exon 1
ATG
Exon 2
Exon 3
TGA
(b)
Enhancer trap
TE
X
Promoter trap
SA
Protein trap
Y
SD
(c)
En1
En2
En3
En4
Promoter bashing
Enhancer bashing
(d)

◂**Fig. 1.3** Overview of regulatory element trapping by transposons versus regulatory element cloning. **a** Diagram of a gene indicating enhancer (En), promoter (Pro), and exon/intron structure (i.e., 5′ and 3′ UTRs are indicated in *gray* while coding portions are indicated in *black*), and splicing patterns. **b** Diagrams of typical enhancer- (*top*), promoter- (*middle*), and protein-transposon trap (bottom) insertions containing genetically encoded cargo, i.e., intersectional tool "X" for enhancer- and promoter trap insertions, or protein tag "Y" for protein trap insertions. While theoretically genome regulation is thought to be absent downstream of the trap, leaky "read through" regulation may escape the trap (indicated by *gray dashed lines*, *grayly colored exons*, and *light green* "Y" encoded exon). **c** Diagram of enhancer and promoter bashing. Individual enhancers (En1, En2, En3, and En4) and promoters (Pro) are subcloned in appropriate plasmids, i.e., upstream of a binary activator for promoter bashing, or upstream of a minimal promoter and binary activator for enhancer bashing. While a promoter can be defined as a piece of DNA of arbitrary length upstream of the 5′ UTR that directly influences transcriptional initiation, enhancers can be located at different positions near or within the gene. **d** Diagram of the generation of a genomic clone encompassing all regulatory elements that can be upgraded to include an intersectional tool "X" by DNA cloning methods, e.g., recombineering (see Sect. 1.4.2 and Fig. 1.6b). While theoretically genome regulation is thought to be absent downstream of the TGA stop codon, leaky "read through" regulation may escape the event (indicated by *gray dashed lines* and *gray colored exons*). Enhancer (En), promoter (Pro), start codon (ATG), stop codon (TGA), intersectional tool (X), protein tag (Y), transposon end (TE), splice acceptor site (SA), splice donor site (SD)

1.3.2 Non-fluorescent Protein Reporters

Besides fluorescent markers some non-fluorescent reporters are useful as well. A fusion with horseradish peroxidase is useful for transmission electron microscopy (Larsen et al. 2003; Watts et al. 2004). Recently, a family of highly antigenic molecules was engineered combining the advantages of both fluorescent proteins (i.e., high solubility and stability, and well tolerated by cells) and peptide epitope tags (i.e., small size and readily available, well validated, and reliable primary antibodies) (Viswanathan et al. 2015). The GFP protein backbone was used as a scaffold for numerous copies (i.e., 10–15) of single peptide epitope tags. Each of these epitope tags can bind many primary antibodies significantly amplifying the signal. The resulting tags were dubbed 'spaghetti monster' fluorescent proteins. Spaghetti monsters were generated for several commonly used peptide tags, i.e., HA, Myc, V5, Flag, OLLAS, and strep II. Orthogonal spaghetti monsters were used to reveal stereotyped cell arrangements in the fly visual system through multicolor stochastic labeling (Nern et al. 2015) (see Sect. 1.6).

1.4 Regulating Binary Activators

The expression pattern of a binary transcriptional activator depends on the regulatory elements that drive its expression (Fig. 1.3a). Regulatory elements can be connected to binary transcriptional activators through random transposition of mobile elements with a "trap" that encodes a synthetic piece of DNA that captures genomic regulatory

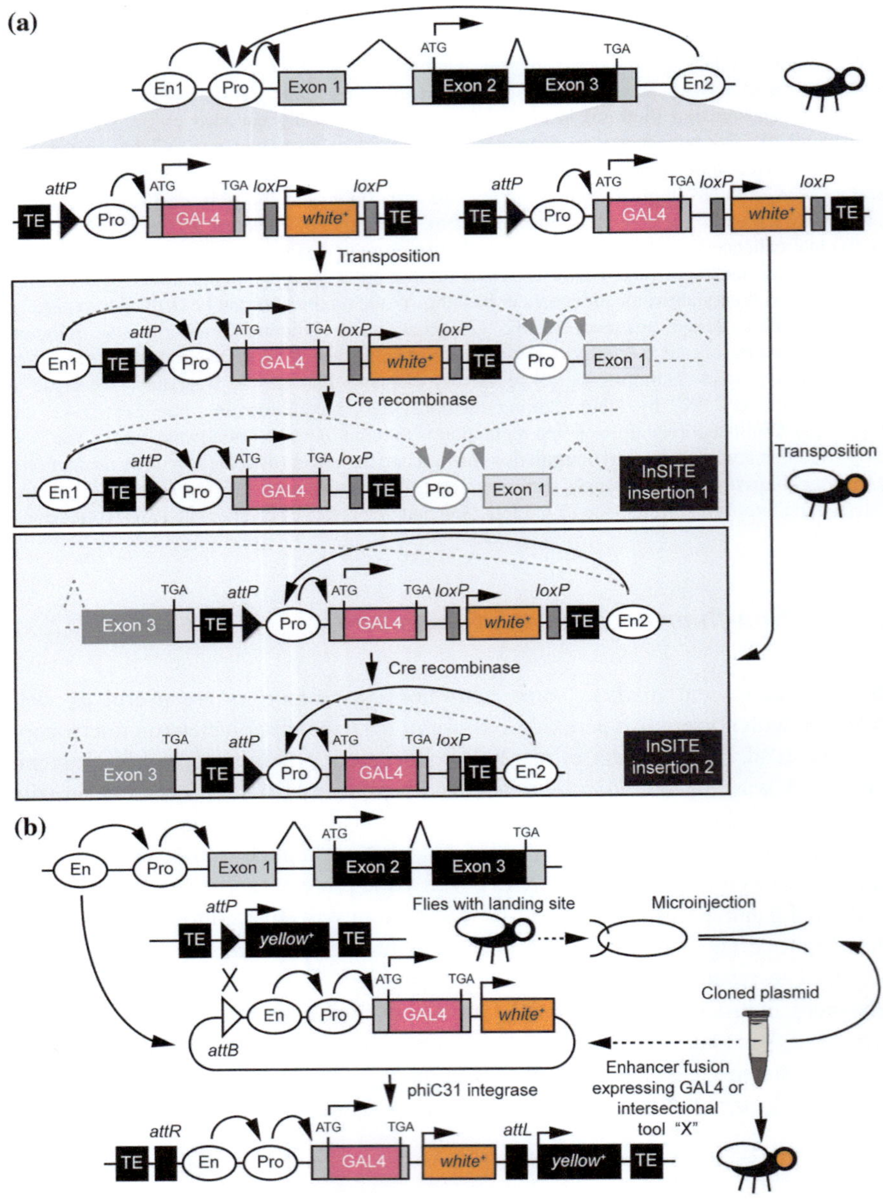
(a)
En1
Pro
Exon 1
ATG
Exon 2
Exon 3
TGA
En2
attP
TE
GAL4
loxP
white+
Transposition
Cre recombinase
InSITE insertion 1
InSITE insertion 2
(b)
En
yellow+
Flies with landing site
Microinjection
attB
Cloned plasmid
phiC31 integrase
Enhancer fusion expressing GAL4 or intersectional tool "X"
attR
attL

◂**Fig. 1.4** Enhancer analysis by enhancer element cloning and enhancer trapping by transposons. **a** Enhancer trapping illustrated by InSITE transposons. The same transposon can integrate at different locations in the genome. After isolation of individual insertion events, the transformation marker, flanked by *loxP* sites, can be removed by Cre recombinase. **b** In the pipeline, developed by Janelia Farm Research Campus, a cloned donor plasmid containing *attB* attachment site for phiC31-mediated site-specific integration, an enhancer, a minimal promoter, an intersectional genetic tool (e.g., GAL4 binary transactivator), and a transformation marker (the dominant color marker *white*$^{+}$) is microinjected into embryos, and integrates into an *attP* attachment-containing landing site. Using this system, enhancer expression patterns can be repurposed by cloning a different intersectional tool downstream of the enhancer and generating a new transgenic line by microinjection and phiC31-mediated integration. Transposon end (TE), enhancer (En), promoter or minimal promoter (Pro), intersectional tool (X), locus of crossover in P1 (*loxP*), attachment phage site (*attP*), attachment bacteria site (*attB*), attachment *left* site (*attL*), attachment *right* site (*attR*)

information surrounding the transposon insertion site (Fig. 1.3b). Traps come in different flavors: enhancer, promoter, or protein trap. An enhancer trap captures cumulative regulatory information from surrounding enhancers and silencers. Promoter and protein traps, on the other hand, capture all regulatory information of the host gene in which the transposon is integrated. Alternatively, a regulatory element and binary transcriptional activator are coupled together by bacterial cloning in a plasmid that is used for fly transgenesis. For historical reasons, as DNA elements regulating a specific gene or developmental process were often screened for by the tedious brute force process of random cloning, this method is called “bashing” (Fig. 1.3c). Finally, genomic DNA clones (Ejsmont et al. 2009; Venken et al. 2006, 2009), each potentially encompassing the majority if not all of a gene’s entire regulatory repertoire can be used as starting material to dissect regulatory elements (Fig. 1.3d). To no one’s surprise, each of these methods has advantages and disadvantages. Determining the expression behavior of enhancers requires empirical experimentation (Arnold et al. 2013; Kvon 2015), while the expression of genes can be deduced from large-scale RNA sequencing efforts (Graveley et al. 2011). The latter category is particularly useful to probe regulatory intersection between expression domains of previously characterized neuronal enhancers and genes that have important function during synaptic communication within the nervous system, e.g., neurotransmitters and neuropeptides (see Sect. 1.4.3).

1.4.1 Enhancer Trapping and Bashing

The concept of enhancer trapping was first demonstrated in *E. coli* by integrating transposons containing reporter genes near regulatory elements, upstream or downstream of the transposon insertion site, in order to study endogenous expression patterns (Casadaban and Cohen 1979). The development of *P* element-mediated transgenesis (Rubin and Spradling 1982) opened the door to enhancer trapping in *Drosophila* (O’Kane and Gehring 1987). The first *Drosophila* enhancer traps contained the *E. coli LacZ* gene, which enabled the visualization of expression patterns, but only after a colorimetric X-gal staining (O’Kane and

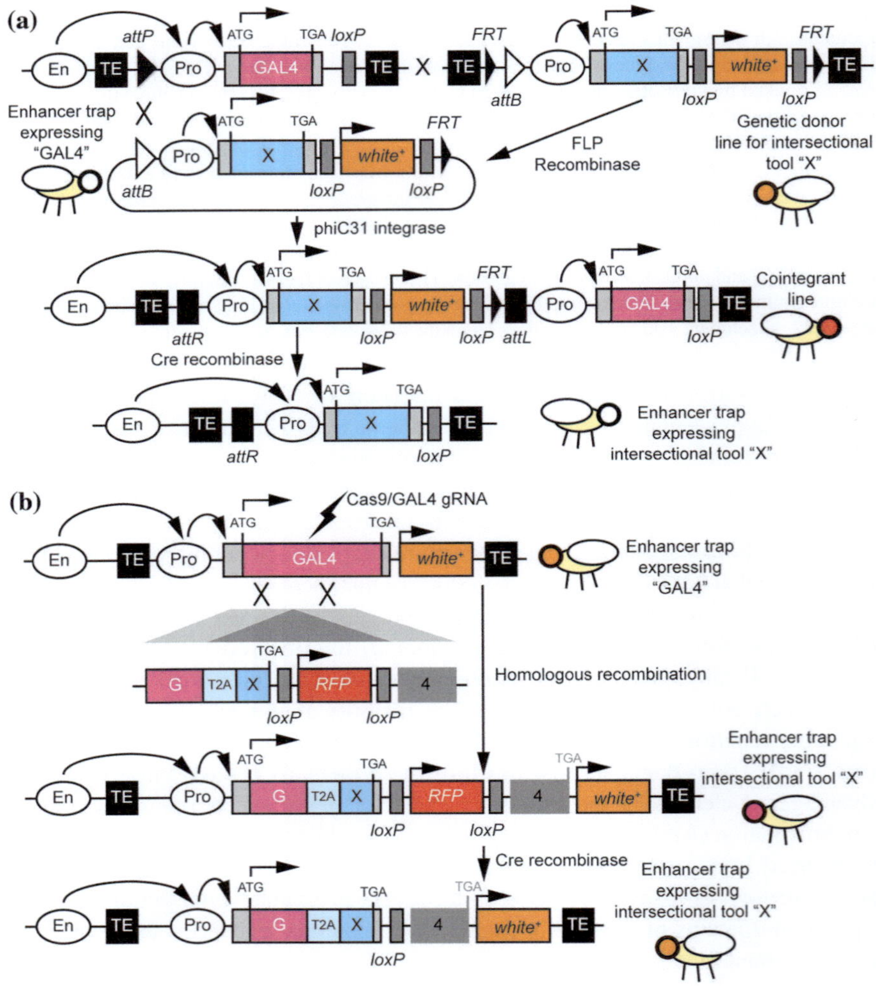

Gehring 1987). Replacing the LacZ gene with the GAL4 transcriptional activator provided limitless opportunities to express any reporter or modulator by binary activation regulated by any regulatory element (Brand and Perrimon 1993). Similar enhancer trap collections were made for GAL80 (Suster et al. 2004), GeneSwitch (Nicholson et al. 2008) and LexA (Miyazaki and Ito 2010), and expanded for GAL4 to accommodate the InSITE system (Gohl et al. 2011) (Fig. 1.4a).

To generate binary drivers with more restricted expression patterns, genomic DNA pieces encompassing putative enhancers are subcloned into transgenesis-competent plasmids upstream of a minimal promoter and GAL4 transcriptional activator (Fig. 1.4b). This method is generally known as enhancer bashing. The resulting plasmids are then integrated by *P* element-mediated transposition (Rubin and Spradling 1982), or at a specific docking sites in the fly genome

◂**Fig. 1.5** In vivo repurposing of enhancer traps by genetic crosses using InSITE and HACK methods. **a** The InSITE system allows the GAL4 transactivator in an enhancer trap transposon to be swapped for another intersectional genetic tool "X" using genetic crosses. An *attB*-containing genetic donor construct, flanked by *FRT* recombination sites, is liberated with FLP recombinase and is integrated by phiC31 integrase into the *attP* site of the enhancer trap construct. The cointegration event (i.e., donor construct integrated into acceptor enhancer trap construct), which is tracked by an eye color marker (*white*$^{+}$), is then treated with Cre recombinase in a second genetic cross to remove all genetic material located between *loxP* sites, i.e., GAL4 and the eye color marker, completing the swap of GAL4 to the new intersectional genetic tool "X". **b** Overview of the HACK method. An enhancer-driven GAL4 "acceptor" line, e.g., enhancer trap, can be converted to a new binary activator line through gene conversion after a Cas9/gRNA-mediated cut at GAL4. The "donor" element, located at a different genomic location has two targeting arms homologous to GAL4 flanking a T2A translational self-cleaving peptide fused to a novel binary activator, as well as the dominant eye color marker, red fluorescent protein (*RFP*) flanked by *loxP* sites, which can be removed by Cre recombinase in a second genetic cross. A similar strategy can be employed for transgenics generated by the JFRC pipeline. Transposon end (TE), enhancer (En), promoter or minimal promoter (Pro), intersectional tool (X), FLP recognition target (*FRT*), locus of crossover in P1 (*loxP*), attachment phage site (*attP*), attachment bacteria site (*attB*), attachment *left* site (*attL*), attachment *right* site (*attR*), 2A peptide of *Thosea asigna virus* (T2A), red fluorescent protein (RFP)

(Pfeiffer et al. 2010; Venken et al. 2006; Bischof et al. 2007; Groth et al. 2004; Knapp et al. 2015; Markstein et al. 2008), using the phiC31 integrase (Bischof et al. 2007; Groth et al. 2004), followed by extensive expression analysis (Jenett et al. 2012; Jory et al. 2012; Manning et al. 2012; Pfeiffer et al. 2008). Due to variable position effects that occur between different transposon insertion sites (Levis et al. 1985), site-specific integration is preferred; since transgenes with different regulatory elements can be integrated at the same docking site, position effects are mostly neutralized (Pfeiffer et al. 2008). Plasmids for enhancer bashing are available for fusions with GAL4 (Chiang et al. 2011; Pfeiffer et al. 2008; Apitz et al. 2004; Sharma et al. 2002).

At Janelia Farm Research Campus (JFRC), a collection of 7000 transgenic lines was generated and the expression patterns have been characterized in the adult brain and ventral nerve cord (Jenett et al. 2012), the embryonic central nervous system (Manning et al. 2012), and in larval imaginal discs (Jory et al. 2012). Since the JFRC collection is based upon cloned GAL4 enhancer fusions, repurposing an enhancer pattern using another binary system transactivator or intersectional tool can be accomplished by cloning the enhancer fragment upstream of the gene of interest and establishing a new transgenic line by microinjection (Pfeiffer et al. 2008, 2010). While creating a large number of lines by injection is labor-intensive, this system has the long-term advantage that once characterized, a large collection of enhancer trap lines does not need to be maintained in continuous culture (as is the case with other enhancer trap lines, since *Drosophila* cannot be readily cryopreserved), as any given driver line can be readily regenerated by microinjection when desired.

To simplify these labor-intensive strategies, a number of methods now unify all the tools for binary activation of gene expression under the same umbrella. These extensible genetic toolkits are all based on in vivo or in vitro exchange of the genes being driven by a captured regulatory element. One of these systems, Integrase

Swappable In vivo Targeting Element (InSITE) uses a two-step cassette exchange strategy with phiC31 integrase and Cre recombinase (Bischof et al. 2007) to convert a binary transactivator into another intersectional genetic tool (Fig. 1.5a) (Gohl et al. 2011). In the InSITE system, an enhancer trap line containing GAL4 and an appropriately positioned *attP* and *loxP* site serves as a target or "landing site" for phiC31-mediated integration of an *attB* and *loxP*-containing donor plasmid. The donor plasmid can be used to introduce other binary transactivators, hemidrivers, binary system repressors, or any other effector of interest. Once an integrant has been isolated, germline treatment with Cre recombinase can be used to remove GAL4 and to generate a cleanly swapped enhancer trap line. An independently developed method, G-MARET, is very similar to InSITE (Yagi et al. 2010).

One key design advantage of the InSITE system is that it can be carried out in vivo purely through genetic crosses, obviating the need to inject embryos with the plasmids necessary to generate the swaps (Gohl et al. 2011). To facilitate this process, chromosomally integrated *FRT*-flanked *attB* donor lines for commonly used intersectional tools have been established. Activating FLP recombinase liberates a circular episome from the chromosome analogous to an injected *attB* donor plasmid that can integrate into the *attP* site in an InSITE enhancer trap line (Fig. 1.5a). Because the recombinase and integrase reactions are very robust, generation of swaps is highly efficient either by injecting a donor plasmid or through genetic crosses only (Gohl et al. 2011).

A collection of more than 1000 InSITE GAL4 enhancer trap lines in an isogenic genetic background has been generated (Gohl et al. 2011; Silies et al. 2013). The chromosomal insertion sites of this collection have been mapped using a novel next-generation sequencing (NGS)-based strategy in which line identity was encoded in a small number of pools using digital error-correcting (Hamming) codes, and NGS libraries were prepared, enriched for *piggyBac* transposon ends using PCR, and sequenced (Gohl et al. 2014). Using this approach, the pattern of appearance of a transposon-adjacent sequence in the pools could then be used to determine the association between insertion site and line identity.

Most recently, a method was developed to convert any existing GAL4 line to a QF2 line using injections or genetic crosses, similar to InSITE. This method called Homology Assisted CRISPR Knock-in (HACK), utilizes the CRISPR/Cas9 system to induce double stranded breaks in a GAL4 transgene, followed by gene conversion at a QF2 donor transgene (Lin and Potter 2016) (Fig. 1.5b). While the method was demonstrated for conversion of GAL4 to QF2, it should be fairly straightforward to implement other binary activators and repressors in the pipeline.

1.4.2 Promoter Bashing and Trapping

To generate binary drivers with expression patterns closely representing endogenous genes, genomic DNA pieces encompassing promoters are subcloned into transgenesis-competent plasmids upstream of the GAL4 transcriptional activator

(Fig. 1.6a). This method is generally known as promoter bashing. The resulting plasmids can then be integrated by transposition (Osterwalder et al. 2001; Roman et al. 2001), or at a specific docking site in the fly genome (Pfeiffer et al. 2010; Venken et al. 2006; Bischof et al. 2007; Groth et al. 2004; Knapp et al. 2015; Markstein et al. 2008), using the phiC31 integrase (Bischof et al. 2007; Groth et al. 2004). Again, site-specific integration is preferred over transposition since the latter results in variable position effects between different insertions (Levis et al. 1985). Plasmids for promoter bashing are available for fusions with GAL4 (Petersen and Stowers 2011; Pfeiffer et al. 2008), GeneSwitch (Osterwalder et al. 2001; Roman et al. 2001), LexA (Shearin et al. 2013; Petersen and Stowers 2011), and QF (Petersen and Stowers 2011). Such cloned promoters do not always accurately reflect endogenous expression of a gene, primarily because the cloned fragment may lack enhancer and/or repressor elements necessary for appropriate regulation (Gnerer et al. 2015).

A valuable alternative strategy is to use recombineering to integrate binary transcriptional activators in large genomic DNA clones that presumably cover the entire regulatory repertoire (Ejsmont et al. 2009; Venken et al. 2006, 2009; Sharan et al. 2009) (Fig. 1.6b). Binary transcriptional activators, such as GAL4 (Chan et al. 2011; Jin et al. 2012; Stowers 2011) and QF (Stowers 2011), can be amplified by PCR and readily introduced in the genomic locus through recombineering. Subsequently, recombineered plasmids are integrated in specific *attP* docking sites in the fly genome to neutralize genomic position effects (Pfeiffer et al. 2010; Venken et al. 2006; Bischof et al. 2007; Groth et al. 2004; Knapp et al. 2015; Markstein et al. 2008). Another approach to capturing and dissecting the entire regulatory region of a gene is through in situ enhancer bashing. This has been accomplished by introducing an *attP* landing site into a locus by gene conversion and using FRT mediated recombination to delete regulatory elements (Bieli et al. 2015a). Cloned rescue constructs containing full length, partial, or modified fragments of the deleted regulatory domain can be introduced to parse the functional elements of the regulatory domain (Bieli et al. 2015b).

To ensure full capture of all regulatory information acting on a gene, *Minos*-Mediated Integration Cassette (MiMIC) provides a trapping alternative for catching promoters (Venken et al. 2011b) (Fig. 1.7a). MiMIC is a *Minos*-based transposon with two inverted phiC31 *attP* sites flanking a marker that can be swapped with a replacement cassette using recombinase-mediated cassette exchange (RMCE) (Bateman et al. 2006). MiMIC insertions that are located in a 5′ UTR non-coding intron of a gene can be replaced with a splice acceptor site followed by a binary factor revealing the expression pattern of the gene. This was illustrated for GAL4 (Gnerer et al. 2015; Venken et al. 2011b), LexA (Gnerer et al. 2015), and QF (Venken et al. 2011b). This strategy is feasible for ~13% of MiMIC insertions (Venken et al. 2011b; Nagarkar-Jaiswal et al. 2015). Genes without a 5′ UTR non-coding intronic MiMIC insertion can easily be modified using CRISPR/Cas9-stimulated gene targeting (Gratz et al. 2014), and an ectopic targeting template accommodating promoter trapping (Fig. 1.7b).

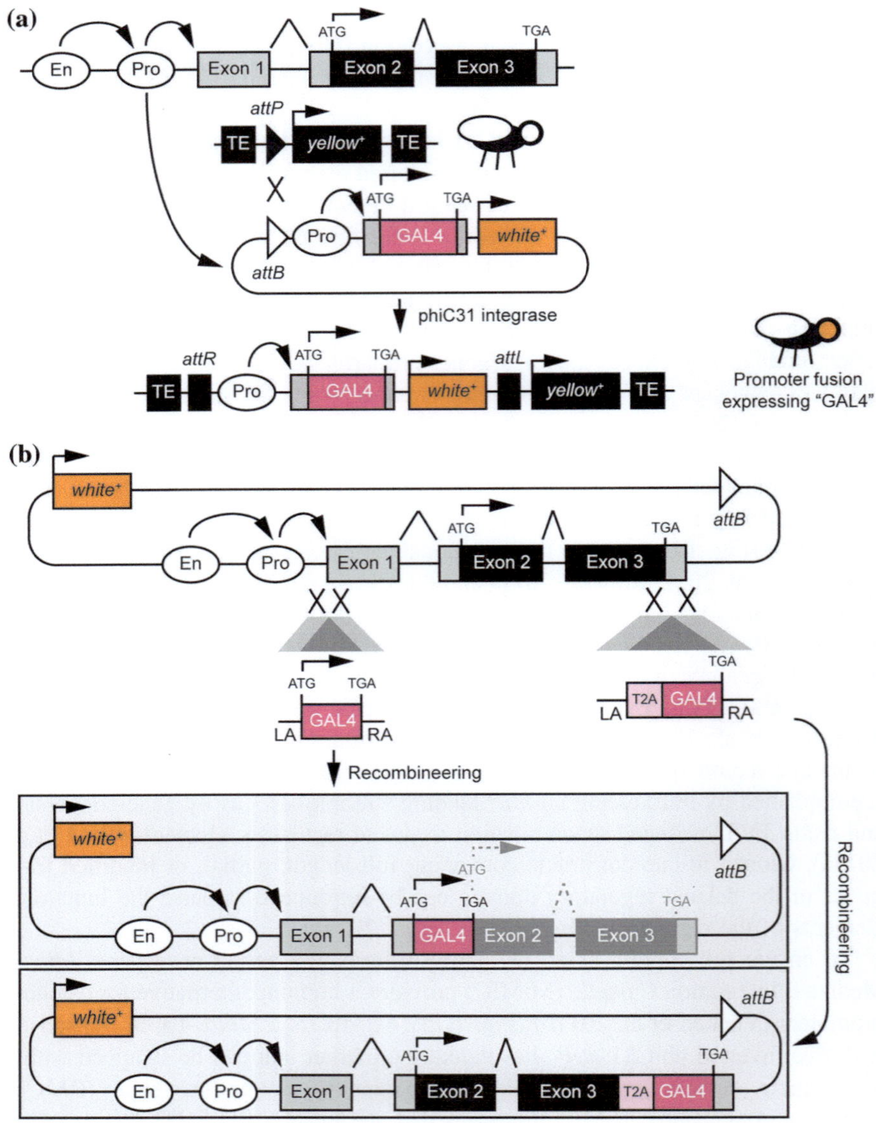

1.4.3 *Protein Trapping*

An alternative strategy to generate promoter traps is through protein trapping using the MiMIC system, introduced in the previous section. Under normal circumstances, a protein trap is made by converting a MiMIC transposon insertion in a coding intron into an artificial exon encoding a genetically encoded protein tag (e.g., superfolder GFP) to visualize endogenous protein localization. However, each

◂**Fig. 1.6** Promoter analysis by promoter bashing and recombineering in genomic DNA clones. **a** Similar to enhancer bashing (see Fig. 1.4b), promoter pieces can be cloned into a plasmid containing *attB* attachment site for phiC31-mediated site-specific integration, the GAL4 binary transactivator, and a transformation marker (the dominant color marker *white*$^{+}$). After microinjection, plasmids can integrate into an *attP* attachment-containing landing site. Using this system, promoter expression patterns can be repurposed by cloning a different intersectional tool (i.e., "X") downstream of the promoter and generating a new transgenic line by microinjection and phiC31-mediated integration. **b** The integration of the GAL4 binary transactivator by recombineering into genomic DNA clones encompassing all regulatory elements. Starting clone and recombineered clone can be integrated in docking sites as described previously (see Figs. 1.4b and 1.6a). While theoretically translation is thought to be absent downstream of the recombineering event, leaky "read through" regulation may escape the event (indicated by *gray dashed lines* and *gray colored exons*). Alternatively, the inclusion of a T2A peptide strategy can generate a bicistronic transcript resulting in the expression of both the GAL4 binary transactivator and the host gene product (see also Figs. 1.5b and 1.8a, b). Using this system, promoter expression patterns can be repurposed by recombineering a different intersectional tool into the genomic clone and generating a new transgenic line by microinjection and phiC31-mediated integration. Enhancer (En), promoter (Pro), transposon end (TE), attachment phage (*attP*), attachment bacteria (*attB*), attachment *left* (*attL*), attachment *right* (*attR*), *left* homology arm for recombineering (LA), *right* homology arm for recombineering (RA), 2A peptide of *Thosea asigna virus* (T2A), splice acceptor site (SA), enhanced green fluorescent protein (EGFP)

of these intragenic intronic insertions can be converted into gene-specific binary factors, through the use of novel exchange cassettes containing a splice acceptor followed by a self-cleaving T2A peptide sequence fused to the coding sequence of the transcriptional activator followed by a 3′ UTR (Fig. 1.8a). This method was illustrated for GAL4 (Gnerer et al. 2015; Diao et al. 2015), LexA (Diao et al. 2015), QF2 (Diao et al. 2015), split GAL4 (Diao et al. 2015), and GAL80 (Gnerer et al. 2015; Diao et al. 2015). Similar to InSITE, this method also works through genetic crosses (Diao et al. 2015). This strategy is feasible for $\sim$18% of all MiMIC insertions (Venken et al. 2011b; Nagarkar-Jaiswal et al. 2015). When a MiMIC insertion is not available in a gene, MiMIC-like elements compatible with phiC31-catalyzed RMCE can be integrated using CRISPR/Cas9 at any location in the fly genome (Diao et al. 2015) (Fig. 1.8b).

1.5 Refining Genetic Targeting by Intersectional Perturbations

Enhancer traps are rarely expressed in a single cell or cell type, or at a single stage in development. To further refine enhancer trap expression patterns, a number of intersectional genetic targeting approaches have been developed. These approaches effectively implement Boolean logic gates within cells, i.e., the integration of regulatory information coming from multiple expression patterns (Fig. 1.9a). Given their utility, modularity, and widespread adoption, binary systems form the basis of most intersectional methods. In addition, the FLP/*FRT* recombinase system provides another useful tool (Golic and Lindquist 1989), which can be used in

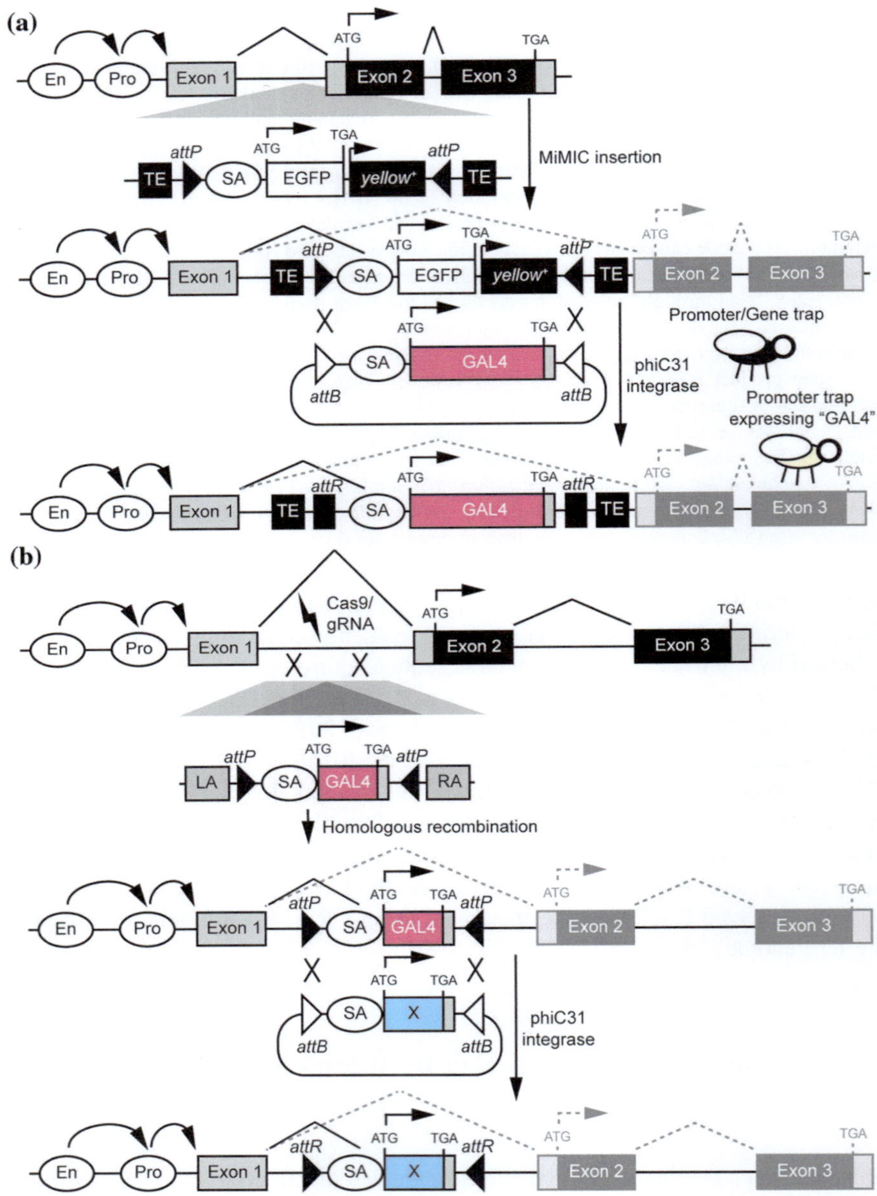
(a)
En
Pro
Exon 1
ATG
Exon 2
Exon 3
TGA
MiMIC insertion
TE
attP
SA
ATG
EGFP
TGA
yellow+
attP
TE
Promoter/Gene trap
phiC31
integrase
GAL4
attB
attB
Promoter trap
expressing "GAL4"
attR
attR
(b)
Cas9/
gRNA
LA
RA
Homologous recombination
X

◂**Fig. 1.7** Promoter analysis by promoter trapping using MiMIC transposons integrated in 5′ UTR non-coding introns. **a** Promoter trapping of a 5′ UTR non-coding intronic MiMIC transposon insertion by recombinase-mediated cassette exchange between two pairs of *attP* and *attB* sites. The exchanged cassette encodes a splice acceptor site followed by the GAL4 binary transactivator. Using this system, promoter expression patterns can be repurposed by cloning a different intersectional tool downstream of the splice acceptor and generating a new *transgenic line* by microinjection and phiC31-driven recombinase-mediated cassette exchange. **b** Genes that do not have a 5′ UTR non-coding intronic insertion can be trapped by CRISPR/Cas9 targeting of a promoter trap element, similar to what has been reported for protein traps within a MiMIC-style exchange element (see Fig. 1.8b). Subsequent recombinase-mediated cassette exchange catalyzed by phiC31 integrase can replace the GAL4 binary transactivator with a novel intersectional tool "X". In both cases, while theoretically genome regulation is thought to be absent downstream of the promoter trap, leaky "read through" regulation may escape the trap (indicated by *gray dashed lines* and *gray colored exons*). Enhancer (En), promoter (Pro), splice acceptor site (SA), enhanced green fluorescent protein (EGFP), transposon end (TE), attachment phage (*attP*), attachment bacteria (*attB*), attachment *left* (*attL*), *left* homology arm for homologous recombination (LA), *right* homology arm for homologous recombination (RA)

conjunction with binary expression systems for intersectional refinement of expression patterns (Bohm et al. 2010). While there are too many possible intersectional strategies to enumerate, with the tools currently available, essentially any desired basic logic gate can be implemented to refine overlapping expression patterns (Fig. 1.9b). Below we provide some examples and also highlight additional tools that enable specific intersectional operations.

1.5.1 OR Gates

OR gates, which combine the expression patterns of two separate transgenes, can be simply constructed by co-expressing two drivers from a single binary system, or by using two driver lines from different binary systems along with appropriate effector lines expressing the same gene. Since enhancer trap patterns are typically broader than desired, the use of a true OR gate is limited in practice as the goal is typically to refine rather than combine expression patterns. More commonly, two orthogonal driver lines will be used to drive expression independently in two distinct tissues, in order to, for instance, manipulate or monitor neighboring cell populations (Potter et al. 2010).

1.5.2 AND Gates

AND gates (Fig. 1.9c), and NOT gates (see Sect. 1.5.3) are the most useful operations in order to combinatorially refine expression patterns. There are multiple ways in which an AND gate (i.e., expressing an effector gene only in the cells that overlap between two expression patterns) can be constructed. One common way is to use two independent binary system drivers in conjunction with the FLP/*FRT*

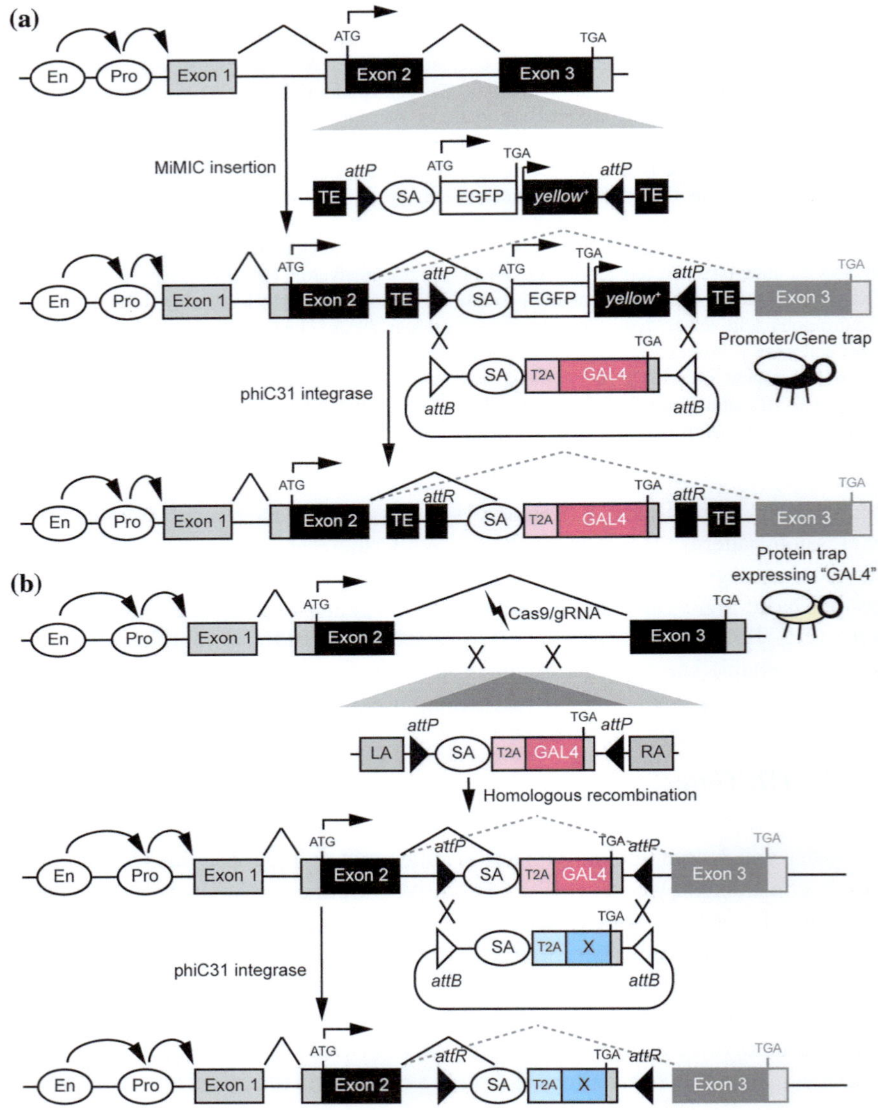

system. The FLP gene encodes a site-specific recombinase protein that catalyzes the recombination of a pair of *FRT* sites (Golic and Lindquist 1989). Depending on the relative orientation of the two *FRT* sites on a linear chromosome, recombination can result in either a deletion or an inversion of the intervening sequence. For use in intersectional targeting, two basic strategies, FLP-in (Fig. 1.10a) (Basler and Struhl 1994; Struhl and Basler 1993), and FLP-out (Fig. 1.10b) (Bohm et al. 2010), are used. In the FLP-in approach, an *FRT*-flanked "stop" cassette containing transcriptional or translational terminators is inserted between a ubiquitous or

◂**Fig. 1.8** Protein trapping by transposons and gene targeted protein traps. **a** Protein trapping of a coding intronic MiMIC transposon insertion by recombinase-mediated cassette exchange between two pairs of *attP* and *attB* sites. The exchanged cassette encodes a splice acceptor site followed by a T2A self-cleaving peptide fused to the GAL4 binary transactivator. *Note* that this is not a classical protein trap resulting in the incorporation of a protein tag in the middle of the host gene product (see Fig. 1.3b, *bottom*); this is rather a hybrid protein/gene trap resulting in C-terminal truncated gene product followed by intersectional tool "X". While theoretically genome regulation is thought to be absent downstream of the transposon trap, leaky "read through" regulation may escape the trap (indicated by *gray dashed lines* and *gray colored exons*). Using this system, gene expression patterns can be repurposed by cloning a different intersectional tool downstream of the splice acceptor site and generating a new transgenic line by microinjection and phiC31-driven recombinase-mediated cassette exchange. **b** Simplified overview of CRISPR/Cas9 targeting of a protein trap element in a coding intron of a target gene. The protein trap (embedded in a MiMIC-style exchange element cassette) encodes a splice acceptor site followed by a T2A self-cleaving peptide fused to the GAL4 binary transactivator. Subsequent recombinase-mediated cassette exchange catalyzed by phiC31 integrase can replace the GAL4 binary transactivator with a novel intersectional tool "X". While theoretically genome regulation is thought to be absent downstream of the Crispr/Cas9 t targeted trap, leaky "read through" regulation may escape the trap (indicated by *gray dashed lines* and *gray colored exons*). Enhancer (En), promoter (Pro), transposon end (TE), splice acceptor (SA), enhanced green fluorescent protein (EGFP), 2A peptide of *Thosea asigna virus* (T2A), intersectional tool (X), attachment phage site (*attP*), attachment bacteria site (*attB*), attachment *left* site (*attL*)

transactivator-inducible promoter and an effector gene of interest. Upon exposure to FLP recombinase, the stop cassette is removed, resulting in expression of the transgene in tissues where the promoter is active. In the FLP-out approach, the effector gene itself is flanked by *FRT* sites, resulting in deletion of the effector gene upon FLP expression. An AND gate can be created by using a FLP-in cassette expressed under the control of one binary transactivator and using a second transactivator to drive FLP expression (Fig. 1.10c).

Another elegant method of constructing an AND gate is to use a split hemidriver system (Fig. 1.10d). Since the AD of GAL4 is genetically separable from the DBD (Ma and Ptashne 1987b), it is possible to generate split-GAL4 proteins where the AD and DBD are each fused to heterodimerizing leucine zippers. When the two split-GAL4 proteins are expressed in the same cell, they are able to dimerize and reconstitute functional GAL4 activity (Pfeiffer et al. 2010; Luan et al. 2006). Since the LexA system uses a transactivator that is by design a chimeric fusion of a DBD and an AD, split-LexA hemidrivers can also be made (Ting et al. 2011). Notably, the establishment of *UAS*-LexA-DBD lines allows GAL4 lines to be intersected with the split-LexA system. The DBD and AD of QF are also separable and functional GAL4: QF and LexA:QF chimeras have been made (Riabinina et al. 2015). However, no split-QF hemidriver lines currently exist in *Drosophila*, though this system has been generated in the nematode worm *Caenorhabditis elegans* (Wei et al. 2012). The level of expression driven by hemidriver AND gates can also be controlled by using ADs of different strengths. In addition to the originally reported GAL4-AD hemidriver (which drives weak expression) and VP16AD hemidriver (which drives stronger expression) (Luan et al. 2006), a p65AD hemidriver has been generated which drives still stronger levels of effector expression (Pfeiffer et al. 2010).

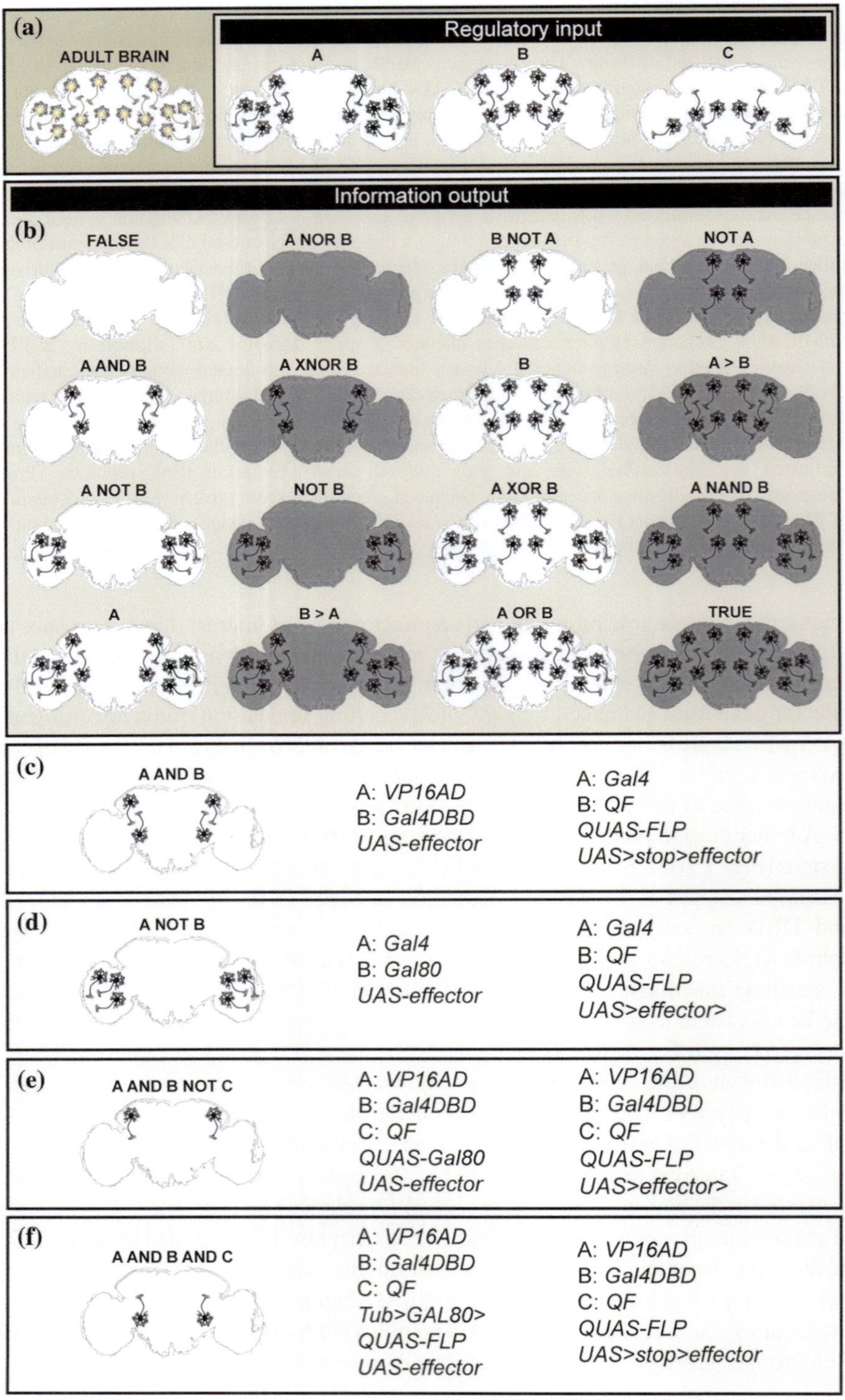
(a)
ADULT BRAIN
Regulatory input
A
B
C
Information output
(b)
FALSE
A NOR B
B NOT A
NOT A
A AND B
A XNOR B
B
A > B
A NOT B
NOT B
A XOR B
A NAND B
A
B > A
A OR B
TRUE
(c)
A AND B
A: VP16AD
B: Gal4DBD
UAS-effector
A: Gal4
B: QF
QUAS-FLP
UAS>stop>effector
(d)
A NOT B
A: Gal4
B: Gal80
UAS-effector
A: Gal4
B: QF
QUAS-FLP
UAS>effector>
(e)
A AND B NOT C
A: VP16AD
B: Gal4DBD
C: QF
QUAS-Gal80
UAS-effector
A: VP16AD
B: Gal4DBD
C: QF
QUAS-FLP
UAS>effector>
(f)
A AND B AND C
A: VP16AD
B: Gal4DBD
C: QF
Tub>GAL80>
QUAS-FLP
UAS-effector
A: VP16AD
B: Gal4DBD
C: QF
QUAS-FLP
UAS>stop>effector

◂**Fig. 1.9** Logic gates for intersectional refinement of expression patterns in *Drosophila melanogaster*. **a** Schematic of the integration of regulatory information coming from three expression domains. Seven hypothetical neurons in each half brain are illustrated in the adult brain that can be activated by regulatory input, i.e., "A", "B", or "C" regulation. **b** Schematic diagrams of all sixteen logic gates, just coming from two regulatory inputs, i.e., "A" and "B". *Dark* coloring indicates active regulation. (**c**, **d**) Most useful logic gates and representative transgenic implementations of intersectional logic gates. **c** A two-input AND gate, and two examples with requirements at the level of genetic components. **d** A two-input NOT gate, and two examples with requirements at the level of genetic components. **e**, **f** Examples of three-input combinatorial gating. **e** A three-input AND/NOT gate, and two examples with requirements at the level of genetic components. **f** A three-input AND gate, and two examples with requirements at the level of genetic components

A specialized case of an AND gate is the GRASP (i.e., GFP reconstitution across synaptic partners) system (Feinberg et al. 2008; Gordon and Scott 2009), recently expanded toward multicoloring (Macpherson et al. 2015). In this system, the goal is not to refine expression of a binary system driven effector gene, but rather to report on synaptic interactions between adjacent cells. This is accomplished by expressing non-fluorescent split GFP proteins under the control of two orthogonal binary systems, GAL4 and LexA. When the two-cell populations targeted by these binary systems are in close proximity, which occurs at synapses, GFP function is reconstituted.

1.5.3 NOT Gates

The natural repressor proteins of GAL4 and QF (GAL80 and QS, respectively) provide a convenient means for generating NOT gates (i.e., subtracting one expression pattern from a second expression pattern) (Fig. 1.9d) (Lee and Luo 1999; Potter et al. 2010). NOT gates can also be implemented using FLP-out constructs, where the FLP-out construct is driven by one binary transactivator, and FLP is expressed under the control of another transactivator (Fig. 1.10e). An important consideration in using GAL80 or QS is that in order to get effective disruption of transactivator activity, the repressor protein must be expressed at a comparable level to that of the transactivator. Optimizing the transcriptional and translational regulatory elements associated with the effector or enhancer trap construct can help achieve the necessary high levels of repressor expression for a functional NOT gate (Pfeiffer et al. 2010, 2012).

1.5.4 Combinatorial AND/NOT Gating

While more esoteric logic gates can be devised, e.g., NAND, NOR, XOR, and XNOR gates (Fig. 1.9b), their practical utility is limited. A more useful

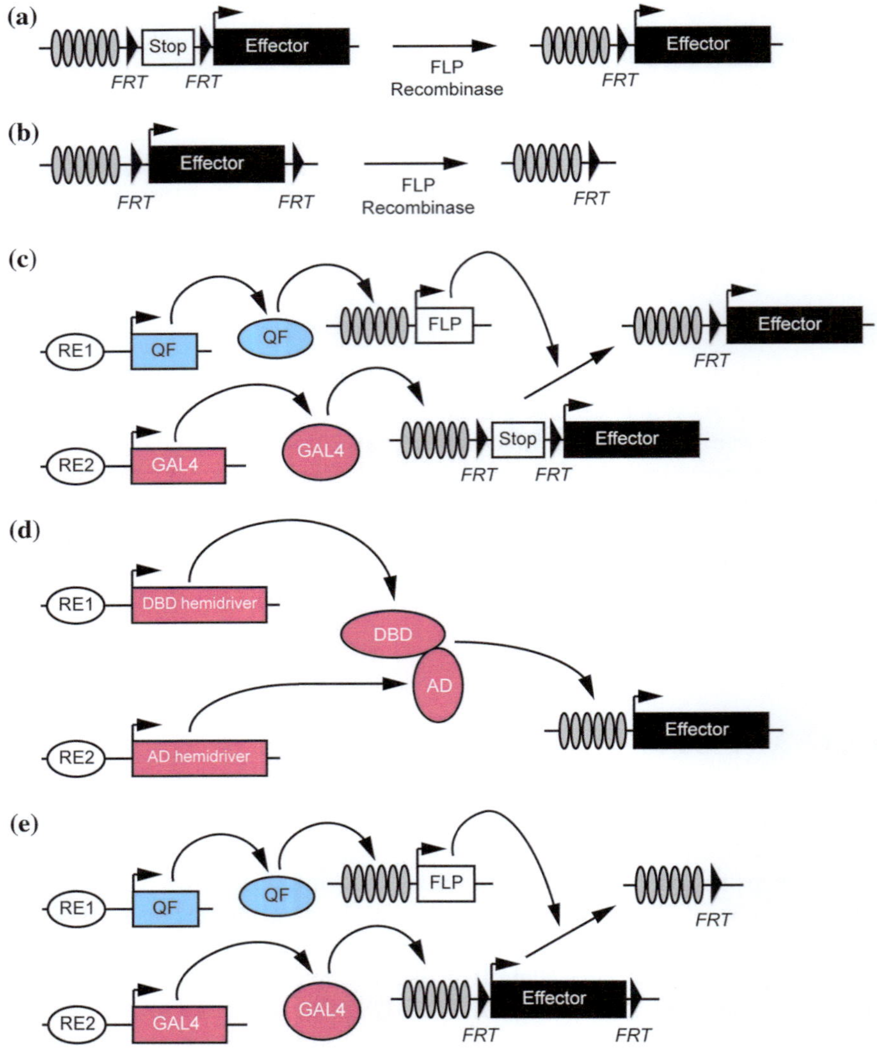

experimental application is the layering of multiple AND and/or NOT gates to dramatically refine an expression pattern. Examples of such layered logic gates are illustrated in Fig. 1.9e (i.e., A AND B NOT C) and Fig. 1.9f (i.e., A AND B AND C). The development of new binary activation systems and recombinases together with orthogonal recombination target sites will further broaden the combinatorial palette (Nern et al. 2011; Hadjieconomou et al. 2011).

◄**Fig. 1.10** Experimental intersectional paradigms commonly used in *Drosophila melanogaster*. **a** The FLP-in approach uses an effector transgene that is interrupted by an *FRT*-flanked stop cassette. In the presence of FLP recombinase, the stop cassette is excised, allowing expression of the effector gene. **b** In the FLP-out approach, the effector gene itself is flanked by *FRT* sites and is excised in the presence of FLP recombinase, preventing effector gene expression. **c** An AND gate based on FLP-in technology. The FLP-in is controlled by one binary transactivator (e.g., GAL4 driven by one regulatory input) while a second transactivator (e.g., QF driven by a second regulatory input) activates FLP expression. **d** An AND gate base on a split binary system. Split hemidrivers are constructed by fusing heterodimerizing leucine zippers to the modular DNA binding domain (DBD) or activation domain (AD) of transactivator proteins. Typically, each hemidriver half is expressed from regulatory elements (RE1 and RE2) and transactivator function is reconstituted only in cells that express both hemidriver halves. DBD hemidriver can be GAL4DBD or LexADBD, while AD hemidrivers can be GAL4AD, VP16AD, or p65AD. **e** A NOT gate based on implementing FLP-out technology. The FLP-out construct is driven by one binary transactivator (e.g., GAL4 driven by one regulatory input), while FLP is expressed by an orthogonal transactivator system (e.g., QF driven by a second regulatory input). Flp recognition target (*FRT*), regulatory element (RE), DNA binding domain (DBD), activation domain (AD)

1.6 Mitotic Analysis and Multicolor Stochastic Labeling Strategies in *D. melanogaster*

Comprehensive brain wiring maps are needed to understand how information flows in order to orchestrate behaviors. The ability to label isolated single neurons and to examine their entire projection patterns (dendritic and axonal processes) has been a critical anatomical limitation to study how neural circuits are organized.

Visual information is processed in the adult *Drosophila* optic lobe which contains ~60,000 neurons (Hofbauer and Campos-Ortega 1990), whose cell bodies are found in the outer brain surface and their projections cluster in internal neuropil structures. Classical studies using Golgi impregnations have allowed the study of the organization of the fly optic lobe and revealed the enormous cell diversity, with over 100 morphologically distinct cell types (Cajal and Sanchez 1915; Fischbach and Dittrich 1989; Strausfeld 1976). More recent studies (Gao et al. 2008; Morante and Desplan 2008) have made use of modern genetic tools to label individual neurons or groups with a Golgi staining-like resolution using stochastic recombination events with MARCM (Lee and Luo 1999) (Fig. 1.11), MARCM derivatives (Potter et al. 2010), or FLP-in techniques (Fig. 1.10a), especially Brainbow technologies (Fig. 1.12). Although it has been possible to collect large neural datasets with single-color MARCM labeling (Chiang et al. 2011; Costa et al. 2016) (Fig. 1.13a), a major limitation of these labeling studies has been the inability to resolve the morphology of individual cells when cells are in close proximity with one another (Fig. 1.13b). Thus, visualization of the neuronal morphology and the spatial arrangements to discriminate adjacent neurons and visualize cellular interactions within the same brain requires methods to distinguish the processes of multiple individual neurons in different colors (compare Fig. 1.13c, d), as discussed below (Fig. 1.12). While full electron microscopic reconstruction can be used to track and determine the connectivity of groups of neighboring neurons (Takemura

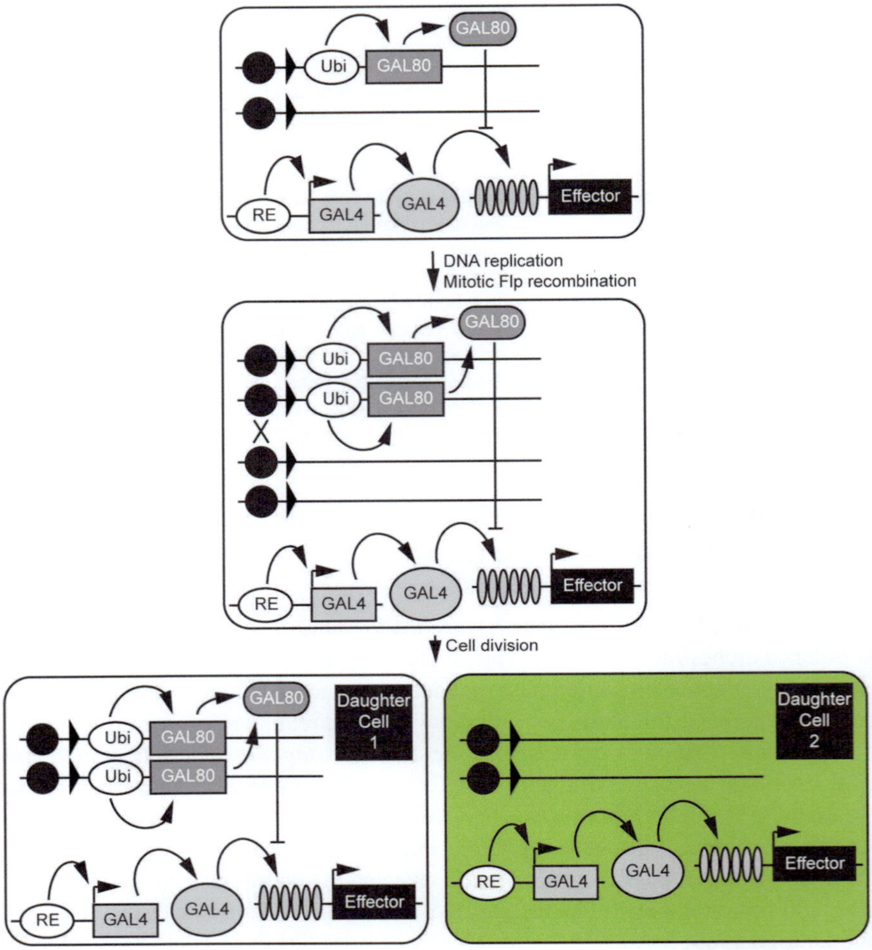

Fig. 1.11 Neuronal labeling strategies in *Drosophila melanogaster* using mitotic analysis. Schematic of typical mitotic analysis illustrated by mitotic analysis using a cell repressible marker (MARCM). Before cell division, all cells are unlabeled (i.e., the GAL80 repressor inhibits GAL4-mediated activation of effector). After cell division, one daughter cell remains unlabeled (i.e., twice the amount of the GAL80 repressor maintains inhibition of GAL4-mediated activation of effector), while the second cell becomes *labeled* or "*marked*" (i.e., lack of the GAL80 repressor ensures GAL4-mediated activation of effector). Ubiquitous promoter (Ubi), regulatory element (RE)

et al. 2013, 2015), such approaches are extremely labor-intensive and not practical for large-scale characterization of complex brain regions.

In this regard, the visualization of multiple individual neurons from defined cell populations has been greatly enriched by the development of methods for combinatorial multicolor stochastic labeling inspired by the mouse Brainbow technique

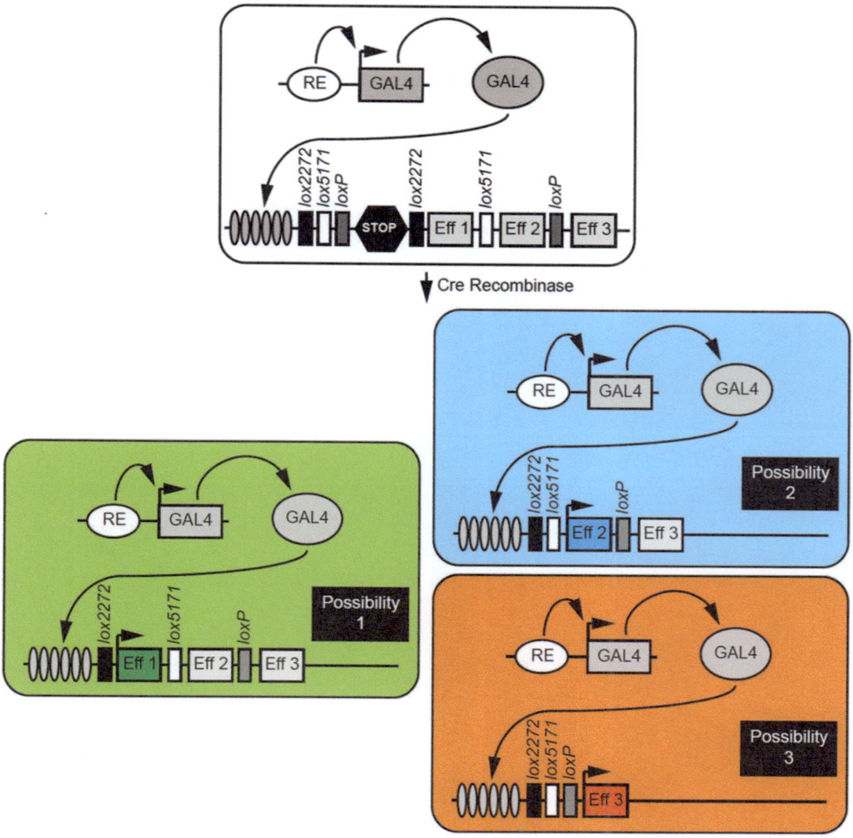

Fig. 1.12 Neuronal labeling strategies in *Drosophila melanogaster* using postmitotic single choice stochastic analysis. Schematic of typical postmitotic single choice stochastic analysis illustrated by dBrainbow. After Cre-mediated stochastic recombination, cells have activation of one effector (i.e., possibility 1), a second effector (i.e., possibility 2), or a third effector (i.e., possibility 3). Regulatory element (RE), locus of crossover in P1 (*loxP*), orthogonal *loxP* sites (*lox2272* and *lox5171*)

(Livet et al. 2007), that uses Cre/*LoxP*-mediated site-specific recombination to drive stochastic and combinatorial expression of multiple fluorescent proteins within a cell population. The use of multiple copies of this construct allowed 90 different distinguishable hues, enabling many individual neurons to be simultaneously identified (Livet et al. 2007).

Several adaptations of Brainbow are available in *Drosophila* (Hadjieconomou et al. 2011; Hampel et al. 2011) that combine the power to regulate transgene expression specifically in different neural populations using the GAL4/*UAS*-system (Brand and Perrimon 1993) with the label diversity provided by stochastic color choice. dBrainbow (Hampel et al. 2011) consists of a single *UAS*-reporter construct

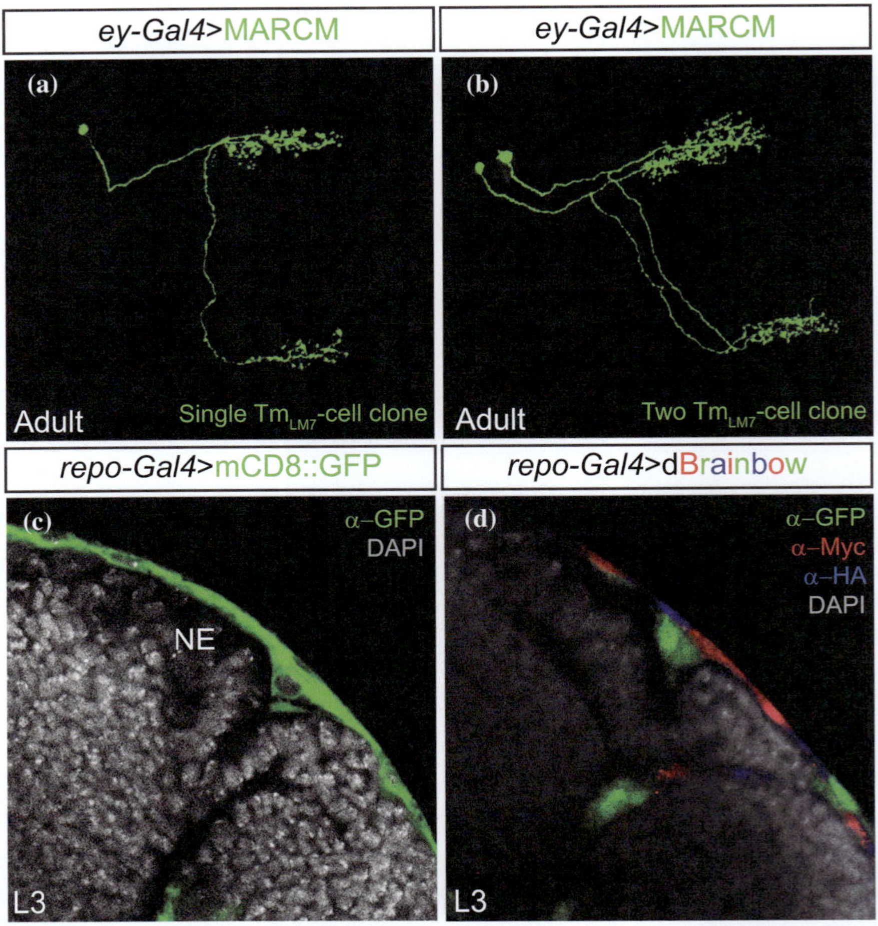

Fig. 1.13 Comparison between mitotic and postmitotic single choice stochastic neuronal labeling strategies in *Drosophila melanogaster*. **a** Single MARCM clones in adult TmLM7 neurons using *ey-GAL4*. **b** Two-cell MARCM clones in adult TmLM7 neurons using *ey-GAL4*. **c** Surface-associated glial cells are labeled by the pan-glial driver *repo-GAL4* driving *UAS*-mCD8::GFP (*green*) in a third instar larval optic lobe. DAPI counterstaining highlight all nuclei. *NE* neuroepithelial cells. **d** Surface-associated glial cells: expression of *UAS*-dBrainbow in all glia by *repo*-GAL4 reveals perineurial (*blue*), subperineurial (*red*) and optic-lobe-associated cortex (*green*) glia in a third instar larval optic-lobe. DAPI counterstaining highlight all nuclei. *NE* neuroepithelial cells

containing a transcriptional stop sequence followed by genes encoding three cytoplasmic fluorescent proteins, that are flanked by incompatible Cre recombinase recognition sites (*LoxP* sites) and thus allows three recombination outcomes. In the presence of a Cre recombinase, recombination between one of the identical pairs of lox sites will lead to the random and permanent selection of one of the three

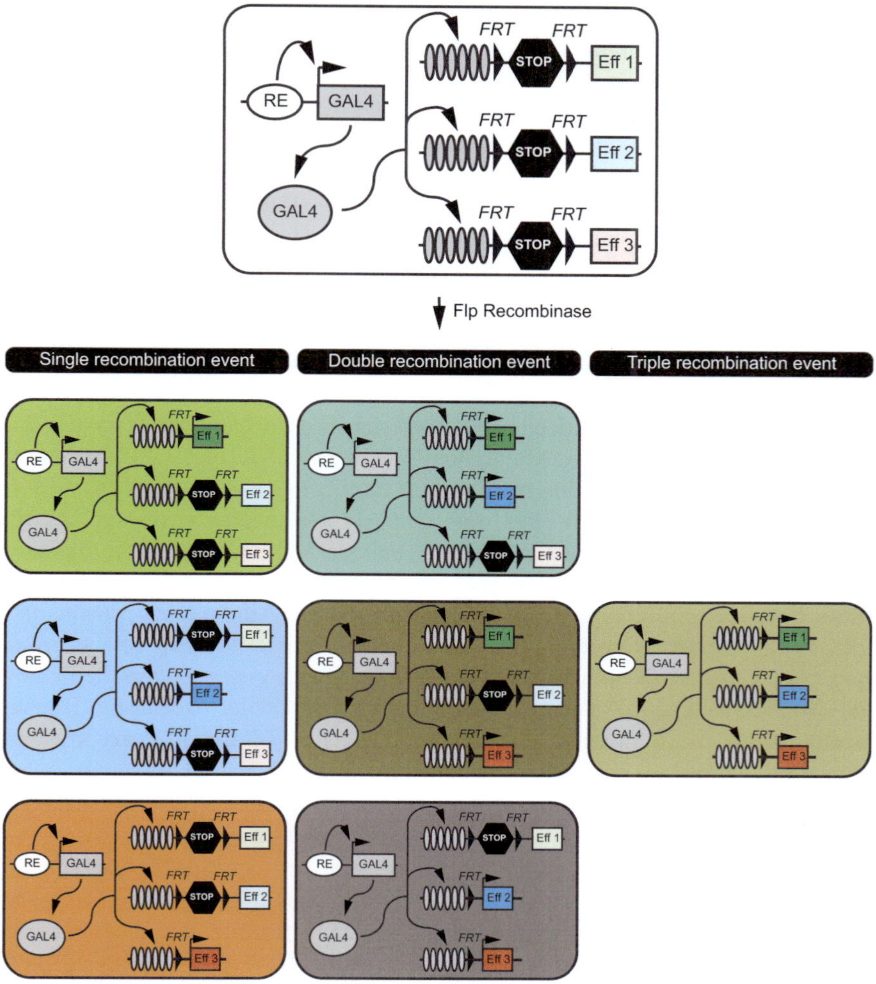

Fig. 1.14 Neuronal labeling strategies in *Drosophila melanogaster* using postmitotic multiple choice stochastic analysis. Schematic of multiple choice stochastic analysis illustrated by Multi-color FLP Out (MCFO) technology. After FLP-mediated stochastic removal of a transcriptional terminator present in each of three orthogonal labeling units (i.e., one unit controlling effector 1, a second unit controlling effector 2, and a third unit controlling effector 3), all combinatorial events are possible ($2^3 - 1$). All events indicating single recombination events (i.e., 3 events), two recombination events (i.e., 3 events), and three recombination events (i.e., 1 event) are shown. Regulatory element (RE), FLP recognition target (*FRT*), effector (Eff)

fluorescent proteins (Fig. 1.12). However, the use of Cre recombinase poses two problems in flies. First, it is potentially toxic when expressed at high levels (Heidmann and Lehner 2001). Second, there is a lack of efficient inducible Cre-lines (Siegal and Hartl 1996), resulting in labeling of clonal groups of cells

(Hampel et al. 2011). The LOLLIbow (live imaging optimized multicolor labeling by light-inducible Brainbow) method uses a photo-inducible form of Cre (split-Cre) to activate recombination in vivo after the illumination with a blue light (Boulina et al. 2013), allowing the acquisition of data at multiple times from the same sample. Thus, this method is ideal to analyze morphogenesis with single-cell resolution at embryonic, larval and pupal stages.

To bypass the limitations of Cre in flies, Flybow (Hadjieconomou et al. 2011) instead uses Flp recombinase to rearrange a single *UAS*-construct with two cassettes, each encoding two fluorescent proteins in opposing orientations flanked by *FRT* sites. Both Flybow and dBrainbow have been applied to study embryonic, larval, pupal, and adult nervous systems in fixed tissues, but also can be used for live imaging of endogenous proteins (Hadjieconomou et al. 2011; Hampel et al. 2011).

A novel strategy for multicolor stochastic labeling in the nervous system has been recently developed called Multicolor Flip-Out (MCFO) (Fig. 1.14) (Nern et al. 2015; Wolff et al. 2015). This method is a multicolor adaptation of the FLP-in technique (Basler and Struhl 1994) and employs a combination of stop cassette constructs with multiple copies of different epitope tags inserted into a myristoylated non-fluorescent GFP backbone (Pfeiffer et al. 2010; Viswanathan et al. 2015). These novel protein reporters called "spaghetti monster" fluorescent proteins (see Sect. 1.3.2) improve targeting the plasma membrane to identify fine neuronal processes. The heat shock inducible modified FLP/*FRT* system used in the Flybow and MCFO techniques allow dense or sparse multicolor stochastic labeling depending on the duration and timing of induction of the FLP recombinase expression (Nern et al. 2015; Hadjieconomou et al. 2011). Early heat shocks (when neuroblasts are dividing) favor labeling clonal groups of cells, while late heat shocks facilitate the identification of single postmitotic cells.

The cell labeling tools described so far are very useful for determining single-cell properties, by labeling a small percentage of cells within a tissue. However, multicolor labeling methods like TIE-DYE (three independent excisions dye) (Worley et al. 2013) and Raeppli (Kanca et al. 2014) have been designed to mark multiple cell lineages and allow whole-tissue labeling in fixed and live animals, respectively, aiming to distinguish the contribution of each of those lineages to the adult structure. Moreover, both systems allow simultaneous multicolor lineage analysis with overexpression or knockdown of *UAS*-constructs in a subset of the marked clones, allowing the comparison with control clones.

In summary, Brainbow-derived technologies offer an unprecedented opportunity to map cell diversity and arrangement of cells in developing and mature circuits, but also to dissect the mechanisms that contribute to organ morphogenesis.

1.7 Conclusions and Future Directions

Given the breadth of genetic tools available for precise, targeted control of gene expression, it is clear that the fruit fly, *D. melanogaster*, will continue to remain a major model system for the functional dissection of the nervous system. In the future, new genetic tools such as the CRISPR/Cas9 system (Gasiunas et al. 2012; Jinek et al. 2012), will enable increasingly precise manipulation of the nervous system. For instance, CRISPR can be used to introduce transgenic constructs at specific loci via homologous recombination, to repurpose binary system transgenic components (Lin and Potter 2016), to carry out high-throughput lineage tracing (McKenna et al. 2016), and in the future could potentially be used as another means of creating intersectional logic gates (for instance, by using sgRNAs targeting GAL4 or other binary system components to disrupt their expression in specific tissues).

Acknowledgements We apologize to those whose work we did not cite due to focus and space limitations. We thank Herman Dierick for critical comments on the chapter. JM is supported by the Ramon y Cajal Program (RyC-2010-07155) and grants from the Ministerio de Economia y Competitividad (SAF2012-31467 and BFU2016-76295-R), co-financed by the European Regional Development Fund (ERDF) and the "Severo Ochoa" Program for Centers of Excellence in R&D (SEV-2013-0317). KV is supported by startup funds kindly provided by Baylor College of Medicine, the Albert and Margaret Alkek Foundation, and the McNair Medical Institute, as well as grants from the March of Dimes Foundation (#1-FY14-315), the Cancer Prevention and Research Institute of Texas (R1313), and the National Institutes of Health (1R21HG006726, 1R21GM110190, 1R21OD022981, and R01GM109938).

References

Allada R, Chung BY (2010) Circadian organization of behavior and physiology in Drosophila. Annu Rev Physiol 72:605–624

Apitz H, Kambacheld M, Hohne M et al (2004) Identification of regulatory modules mediating specific expression of the roughest gene in *Drosophila melanogaster*. Dev Genes Evol 214:453–459

Arnold CD, Gerlach D, Stelzer C, Boryń ŁM, Rath M, Stark A (2013) Genome-wide quantitative enhancer activity maps identified by STARR-seq. Science 339:1074–1077

Basler K, Struhl G (1994) Compartment boundaries and the control of Drosophila limb pattern by hedgehog protein. Nature 368:208–214

Bateman JR, Lee AM, Wu CT (2006) Site-specific transformation of Drosophila via phiC31 integrase-mediated cassette exchange. Genetics 173:769–777

Bellen HJ, Tong C, Tsuda H (2010) 100 years of Drosophila research and its impact on vertebrate neuroscience: a history lesson for the future. Nat Rev Neurosci 11:514–522

Bieli D, Kanca O, Gohl D et al (2015a) The *Drosophila melanogaster* Mutants apblot and apXasta Affect an Essential apterous Wing Enhancer. G3: Genes|Genomes|Genetics 5:1129–1143

Bieli D, Kanca O, Requena D et al (2015b) Establishment of a developmental compartment requires interactions between three synergistic Cis-regulatory modules. PLoS Genet 11: e1005376

Bischof J, Maeda RK, Hediger M et al (2007) An optimized transgenesis system for Drosophila using germ-line-specific phiC31 integrases. Proc Natl Acad Sci USA 104:3312–3317

Bohm RA, Welch WP, Goodnight LK et al (2010) A genetic mosaic approach for neural circuit mapping in Drosophila. Proc Natl Acad Sci USA 107:16378–16383
Boulina M, Samarajeewa H, Baker JD et al (2013) Live imaging of multicolor-labeled cells in Drosophila. Development 140:1605–1613
Brand AH, Perrimon N (1993) Targeted gene expression as a means of altering cell fates and generating dominant phenotypes. Development 118:401–415
Cajal SR, Sanchez D (1915) Contribucion al conocimiento de los centros nerviosos de los insectos. Trab Lab Invest Biol XIII:1–167
Casadaban MJ, Cohen SN (1979) Lactose genes fused to exogenous promoters in one step using a Mu-lac bacteriophage: in vivo probe for transcriptional control sequences. Proc Natl Acad Sci USA 76:4530–4533
Chan CC, Scoggin S, Wang D et al (2011) Systematic discovery of Rab GTPases with synaptic functions in Drosophila. Curr Biol 21:1704–1715
Chiang AS, Lin CY, Chuang CC et al (2011) Three-dimensional reconstruction of brain-wide wiring networks in Drosophila at single-cell resolution. Curr Biol 21:1–11
Costa M, Manton JD, Ostrovsky AD et al (2016) NBLAST: rapid, sensitive comparison of neuronal structure and construction of neuron family databases. Neuron 91:293–311
del Valle Rodríguez A, Didiano D, Desplan C (2012) Power tools for gene expression and clonal analysis in Drosophila. Nat Methods 9:47–55
Diao F, Ironfield H, Luan H et al (2015) Plug-and-play genetic access to drosophila cell types using exchangeable exon cassettes. Cell Rep 10:1410–1421
Dickson BJ (2008) Wired for sex: the neurobiology of Drosophila mating decisions. Science 322:904–909
Diegelmann S, Bate M, Landgraf M (2008) Gateway cloning vectors for the LexA-based binary expression system in Drosophila. Fly 2:236–239
Donelson NC, Sanyal S (2015) Use of Drosophila in the investigation of sleep disorders. Exp Neurol 274:72–79
Duffy JB (2002) GAL4 system in Drosophila: a fly geneticist's Swiss army knife. Genesis 34:1–15
Ejsmont RK, Sarov M, Winkler S et al (2009) A toolkit for high-throughput, cross-species gene engineering in Drosophila. Nat Methods 6:435–437
Estes PS, Ho GL, Narayanan R et al (2000) Synaptic localization and restricted diffusion of a Drosophila neuronal synaptobrevin-green fluorescent protein chimera in vivo. J Neurogenet 13:233–255
Feinberg EH, Vanhoven MK, Bendesky A et al (2008) GFP reconstitution across synaptic partners (GRASP) defines cell contacts and synapses in living nervous systems. Neuron 57:353–363
Fischbach KF, Dittrich AP (1989) The optic lobe of *Drosophila melanogaster*. I. A Golgi analysis of wild-type structure. Cell Tissue Res 441–475
Fischer JA, Giniger E, Maniatis T et al (1988) GAL4 activates transcription in Drosophila. Nature 332:853–856
Freeman EG, Dahanukar A (2015) Molecular neurobiology of Drosophila taste. Curr Opin Neurobiol 34:140–148
Gao S, Takemura S, Ting C-Y et al (2008) The neural substrate of spectral preference in Drosophila. Neuron 60:328–342
Gasiunas G, Barrangou R, Horvath P et al (2012) Cas9-crRNA ribonucleoprotein complex mediates specific DNA cleavage for adaptive immunity in bacteria. Proc Natl Acad Sci USA 109:E2579–E2586
Gatto CL, Broadie K (2011) Drosophila modeling of heritable neurodevelopmental disorders. Curr Opin Neurobiol 21:834–841
Geever RF, Huiet L, Baum JA et al (1989) DNA sequence, organization and regulation of the qa gene cluster of *Neurospora crassa*. J Mol Biol 207:15–34
Giniger E, Varnum SM, Ptashne M (1985) Specific DNA binding of GAL4, a positive regulatory protein of yeast. Cell 40:767–774
Gnerer JP, Venken KJ, Dierick HA (2015) Gene-specific cell labeling using MiMIC transposons. Nucleic Acids Res 43:e56

Gohl DM, Silies MA, Gao XJ et al (2011) A versatile in vivo system for directed dissection of gene expression patterns. Nat Methods 8:231–237
Gohl DM, Freifeld L, Silies M et al (2014) Large-scale mapping of transposable element insertion sites using digital encoding of sample identity. Genetics 196:615–623
Golic KG, Lindquist S (1989) The FLP recombinase of yeast catalyzes site-specific recombination in the Drosophila genome. Cell 59:499–509
Gordon MD, Scott K (2009) Motor control in a Drosophila taste circuit. Neuron 61:373–384
Gratz SJ, Ukken FP, Rubinstein CD et al (2014) Highly specific and efficient CRISPR/Cas9-catalyzed homology-directed repair in Drosophila. Genetics 196:961–971
Graveley BR, Brooks AN, Carlson JW et al (2011) The developmental transcriptome of *Drosophila melanogaster*. Nature 471:473–479
Groth AC, Fish M, Nusse R et al (2004) Construction of transgenic Drosophila by using the site-specific integrase from phage phiC31. Genetics 166:1775–1782
Hadjieconomou D, Rotkopf S, Alexandre C et al (2011) Flybow: genetic multicolor cell labeling for neural circuit analysis in *Drosophila melanogaster*. Nat Methods 8:260–266
Halfon MS, Gisselbrecht S, Lu J et al (2002) New fluorescent protein reporters for use with the Drosophila Gal4 expression system and for vital detection of balancer chromosomes. Genesis 34:135–138
Hampel S, Chung P, McKellar CE et al (2011) Drosophila Brainbow: a recombinase-based fluorescence labeling technique to subdivide neural expression patterns. Nat Methods 8: 253–259
Han DD, Stein D, Stevens LM (2000) Investigating the function of follicular subpopulations during *Drosophila oogenesis* through hormone-dependent enhancer-targeted cell ablation. Development 127:573–583
Hayashi S, Ito K, Sado Y et al (2002) GETDB, a database compiling expression patterns and molecular locations of a collection of Gal4 enhancer traps. Genesis. 34:58–61
Heidmann D, Lehner CF (2001) Reduction of Cre recombinase toxicity in proliferating Drosophila cells by estrogen-dependent activity regulation. Dev Genes Evol 211:458–465
Hofbauer A, Campos-Ortega JA (1990) Proliferation pattern and early differentiation of the optic lobes in *Drosophila melanogaster*. Roux's Arch Dev Biol 198:264–274
Huiet L, Giles NH (1986) The qa repressor gene of *Neurospora crassa*: wild-type and mutant nucleotide sequences. Proc Natl Acad Sci USA 83:3381–3385
Jenett A, Rubin GM, Ngo TT et al (2012) A GAL4-driver line resource for Drosophila neurobiology. Cell Rep 2:991–1001
Jin EJ, Chan CC, Agi E et al (2012) Similarities of Drosophila rab GTPases based on expression profiling: completion and analysis of the rab-Gal4 kit. PLoS One 7:e40912
Jinek M, Chylinski K, Fonfara I et al (2012) A programmable dual-RNA-guided DNA endonuclease in adaptive bacterial immunity. Science 337:816–821
Johnston SA, Hopper JE (1982) Isolation of the yeast regulatory gene GAL4 and analysis of its dosage effects on the galactose/melibiose regulon. Proc Natl Acad Sci USA 79:6971
Jory A, Estella C, Giorgianni MW et al (2012) A survey of 6300 genomic fragments for cis-regulatory activity in the imaginal discs of *Drosophila melanogaster*. Cell Rep 2: 1014–1024
Kakidani H, Ptashne M (1988) GAL4 activates gene expression in mammalian cells. Cell 52: 161–167
Kanca O, Caussinus E, Denes AS et al (2014) Raeppli: a whole-tissue labeling tool for live imaging of Drosophila development. Development. 141:472–480
Kawasaki F, Zou B, Xu X et al (2004) Active zone localization of presynaptic calcium channels encoded by the cacophony locus of Drosophila. J Neurosci 24:282–285
Keene AC, Waddell S (2007) Drosophila olfactory memory: single genes to complex neural circuits. Nat Rev Neurosci 8:341–354
Knapp JM, Chung P, Simpson JH (2015) Generating customized transgene landing sites and multi-transgene arrays in Drosophila using phiC31 integrase. Genetics 199:919–934

Kuo SY, Tu CH, Hsu YT et al (2012) A hormone receptor-based transactivator bridges different binary systems to precisely control spatial-temporal gene expression in Drosophila, PLoS One 7:e50855
Kvon EZ (2015) Using transgenic reporter assays to functionally characterize enhancers in animals. Genomics 106:185–192
Lai SL, Lee T (2006) Genetic mosaic with dual binary transcriptional systems in Drosophila. Nat Neurosci 9:703–709
LaJeunesse DR, Buckner SM, Lake J et al (2004) Three new Drosophila markers of intracellular membranes. Biotechniques 36:790
Larsen CW, Hirst E, Alexandre C et al (2003) Segment boundary formation in Drosophila embryos. Development 130:5625–5635
Laughon A, Gesteland RF (1982) Isolation and preliminary characterization of the GAL4 gene, a positive regulator of transcription in yeast. Proc Natl Acad Sci USA 79:6827–6831
Lee T, Luo L (1999) Mosaic analysis with a repressible cell marker for studies of gene function in neuronal morphogenesis. Neuron 22:451–461
Leiss F, Koper E, Hein I et al (2009) Characterization of dendritic spines in the Drosophila central nervous system. Dev Neurobiol 69:221–234
Leung C, Wilson Y, Khuong TM et al (2013) Fruit flies as a powerful model to drive or validate pain genomics efforts. Pharmacogenomics. 14:1879–1887
Levis R, Hazelrigg T, Rubin GM (1985) Effects of genomic position on the expression of transduced copies of the white gene of Drosophila. Science 229:558–561
Li H-H, Kroll JR, Lennox SM et al (2014) A GAL4 driver resource for developmental and behavioral studies on the larval CNS of Drosophila. Cell Rep 8:897–908
Lin C-C, Potter CJ (2016) Editing transgenic DNA components by inducible gene replacement in *Drosophila melanogaster*. Genetics 203:1613–1628
Livet J, Weissman TA, Kang H et al (2007) Transgenic strategies for combinatorial expression of fluorescent proteins in the nervous system. Nature 450:56–62
Luan H, Peabody NC, Vinson CR et al (2006) Refined spatial manipulation of neuronal function by combinatorial restriction of transgene expression. Neuron 52:425–436
Ma J, Ptashne M (1987a) The carboxy-terminal 30 amino acids of GAL4 are recognized by GAL80. Cell 50:137–142
Ma J, Ptashne M (1987b) Deletion analysis of GAL4 defines two transcriptional activating segments. Cell 48:847–853
Macpherson LJ, Zaharieva EE, Kearney PJ et al (2015) Dynamic labelling of neural connections in multiple colours by trans-synaptic fluorescence complementation. Nat Commun 6:10024
Manning L, Heckscher ES, Purice MD et al (2012) A resource for manipulating gene expression and analyzing cis-regulatory modules in the Drosophila CNS. Cell Rep 2:1002–1013
Markstein M, Pitsouli C, Villalta C et al (2008) Exploiting position effects and the gypsy retrovirus insulator to engineer precisely expressed transgenes. Nat Genet 40:476–483
Matsumoto K, Toh-e A, Oshima Y (1978) Genetic control of galactokinase synthesis in Saccharomyces cerevisiae: evidence for constitutive expression of the positive regulatory gene gal4. J Bacteriol 134:446–457
McGuire SE, Le PT, Osborn AJ et al (2003) Spatiotemporal rescue of memory dysfunction in Drosophila. Science 302:1765–1768
McGurk L, Berson A, Bonini NM (2015) Drosophila as an in vivo model for human neurodegenerative disease. Genetics 201:377–402
McKenna A, Findlay GM, Gagnon JA et al (2016) Whole-organism lineage tracing by combinatorial and cumulative genome editing. Science 353:aaf7907
Miyazaki T, Ito K (2010) Neural architecture of the primary gustatory center of *Drosophila melanogaster* visualized with GAL4 and LexA enhancer-trap systems. J Comp Neurol 518:4147–4181
Mondal K, Dastidar AG, Singh G et al (2007) Design and isolation of temperature-sensitive mutants of Gal4 in yeast and Drosophila. J Mol Biol 370:939–950

Morante J, Desplan C (2008) The color-vision circuit in the Medulla of Drosophila. Curr Biol 18:553–565

Nagarkar-Jaiswal S, Lee P-T, Campbell ME et al (2015) A library of MiMICs allows tagging of genes and reversible, spatial and temporal knockdown of proteins in Drosophila. eLife 4: e05338

Nern A, Pfeiffer BD, Svoboda K et al (2011) Multiple new site-specific recombinases for use in manipulating animal genomes. Proc Natl Acad Sci USA 108:14198–14203

Nern A, Pfeiffer BD, Rubin GM (2015) Optimized tools for multicolor stochastic labeling reveal diverse stereotyped cell arrangements in the fly visual system. Proc Natl Acad Sci USA 112: E2967–E2976

Ni JQ, Markstein M, Binari R et al (2008) Vector and parameters for targeted transgenic RNA interference in *Drosophila melanogaster*. Nat Methods 5:49–51

Ni JQ, Liu LP, Binari R et al (2009) A Drosophila resource of transgenic RNAi lines for neurogenetics. Genetics 182:1089–1100

Nicholson L, Singh GK, Osterwalder T et al (2008) Spatial and temporal control of gene expression in Drosophila using the inducible GeneSwitch GAL4 system. I. Screen for larval nervous system drivers. Genetics 178:215–234

Nicolai LJ, Ramaekers A, Raemaekers T et al (2010) Genetically encoded dendritic marker sheds light on neuronal connectivity in Drosophila. Proc Natl Acad Sci USA 107:20553–20558

O'Kane CJ, Gehring WJ (1987) Detection in situ of genomic regulatory elements in Drosophila. Proc Natl Acad Sci USA 84:9123–9127

Osterwalder T, Yoon KS, White BH et al (2001) A conditional tissue-specific transgene expression system using inducible GAL4. Proc Natl Acad Sci USA 98:12596–12601

Peng H, Chung P, Long F et al (2011) BrainAligner: 3D registration atlases of Drosophila brains. Nat Methods 8:493–498

Petersen LK, Stowers RS (2011) A gateway MultiSite recombination cloning toolkit. PLoS One 6: e24531

Pfeiffer BD, Jenett A, Hammonds AS et al (2008) Tools for neuroanatomy and neurogenetics in Drosophila. Proc Natl Acad Sci USA 105:9715–9720

Pfeiffer BD, Ngo TT, Hibbard KL et al (2010) Refinement of tools for targeted gene expression in Drosophila. Genetics 186:735–755

Pfeiffer BD, Truman JW, Rubin GM (2012) Using translational enhancers to increase transgene expression in Drosophila. Proc Natl Acad Sci 109:6626–6631

Potter CJ, Luo L (2011) Using the Q system in *Drosophila melanogaster*. Nat Protoc 6:1105–1120

Potter CJ, Tasic B, Russler EV et al (2010) The Q system: a repressible binary system for transgene expression, lineage tracing, and mosaic analysis. Cell 141:536–548

Riabinina O, Luginbuhl D, Marr E et al (2015) Improved and expanded Q-system reagents for genetic manipulations. Nat Methods 12:219–222

Ritzenthaler S, Suzuki E, Chiba A (2000) Postsynaptic filopodia in muscle cells interact with innervating motoneuron axons. Nat Neurosci 3:1012–1017

Rolls MM, Satoh D, Clyne PJ et al (2007) Polarity and intracellular compartmentalization of Drosophila neurons. Neural Dev 2:7

Roman G, Davis RL (2002) Conditional expression of UAS-transgenes in the adult eye with a new gene-switch vector system. Genesis. 34:127–131

Roman G, Endo K, Zong L et al (2001) P[Switch], a system for spatial and temporal control of gene expression in *Drosophila melanogaster*. Proc Natl Acad Sci USA 98:12602–12607

Rubin GM, Spradling AC (1982) Genetic transformation of Drosophila with transposable element vectors. Science 218:348–353

Sanchez-Soriano N, Bottenberg W, Fiala A et al (2005) Are dendrites in Drosophila homologous to vertebrate dendrites? Dev Biol 288:126–138

Sharan SK, Thomason LC, Kuznetsov SG et al (2009) Recombineering: a homologous recombination-based method of genetic engineering. Nat Protoc 4:206–223

Sharma Y, Cheung U, Larsen EW et al (2002) PPTGAL, a convenient Gal4 P-element vector for testing expression of enhancer fragments in drosophila. Genesis 34:115–118

Shearin HK, Dvarishkis AR, Kozeluh CD et al (2013) Expansion of the gateway multisite recombination cloning toolkit. PLoS One 8:e77724
Shearin HK, Macdonald IS, Spector LP et al (2014) Hexameric GFP and mCherry reporters for the Drosophila GAL4, Q, and LexA transcription systems. Genetics 196:951–960
Shulman JM (2015) Drosophila and experimental neurology in the post-genomic era. Exp Neurol 274:4–13
Siegal ML, Hartl DL (1996) Transgene Coplacement and high efficiency site-specific recombination with the Cre/loxP system in Drosophila. Genetics 144:715–726
Silies M, Gohl DM, Fisher YE et al (2013) Modular use of peripheral input channels tunes motion-detecting circuitry. Neuron 79:111–127
Silies M, Gohl DM, Clandinin TR (2014) Motion-detecting circuits in flies: coming into view. Annu Rev Neurosci 37:307–327
Stowers RS (2011) An efficient method for recombineering GAL4 and QF drivers. Fly 5:371–378
Strausfeld NJ (1976) Atlas of an insect brain. Springer-Verlag, Berlin, Heidelberg
Struhl G, Basler K (1993) Organizing activity of wingless protein in Drosophila. Cell 72:527–540
Suster ML, Seugnet L, Bate M et al (2004) Refining GAL4-driven transgene expression in Drosophila with a GAL80 enhancer-trap. Genesis 39:240–245
Szuts D, Bienz M (2000) LexA chimeras reveal the function of Drosophila Fos as a context-dependent transcriptional activator. Proc Natl Acad Sci USA 97:5351–5356
Takemura S, Bharioke A, Lu Z et al (2013) A visual motion detection circuit suggested by Drosophila connectomics. Nature 500:175–181
Takemura S, Xu CS, Lu Z et al (2015) Synaptic circuits and their variations within different columns in the visual system of Drosophila. Proc Natl Acad Sci USA 112:13711–13716
Tataroglu O, Emery P (2014) Studying circadian rhythms in *Drosophila melanogaster*. Methods 68:140–150
Ting CY, Gu S, Guttikonda S et al (2011) Focusing transgene expression in Drosophila by coupling Gal4 with a novel split-LexA expression system. Genetics 188:229–233
Tracey WD, Wilson RI, Laurent G et al (2003) Painless, a Drosophila gene essential for nociception. Cell 113:261–273
van Alphen B, van Swinderen B (2013) Drosophila strategies to study psychiatric disorders. Brain Res Bull 92:1–11
Venken KJT, Bellen HJ (2012) Genome-wide manipulations of *Drosophila melanogaster* with transposons, Flp recombinase, and ΦC31 integrase. Methods Mol Biol 859:203–228
Venken KJT, Bellen HJ (2014) Chemical mutagens, transposons, and transgenes to interrogate gene function in *Drosophila melanogaster*. Methods 68:15–28
Venken KJT, He Y, Hoskins RA et al (2006) P[acman]: a BAC transgenic platform for targeted insertion of large DNA fragments in D. *melanogaster*. Science 314:1747–1751
Venken KJT, Carlson JW, Schulze KL et al (2009) Versatile P[acman] BAC libraries for transgenesis studies in *Drosophila melanogaster*. Nat Methods 6:431–434
Venken KJT, Simpson JH, Bellen HJ (2011a) Genetic manipulation of genes and cells in the nervous system of the fruit fly. Neuron 72:202–230
Venken KJT, Schulze KL, Haelterman NA et al (2011b) MiMIC: a highly versatile transposon insertion resource for engineering *Drosophila melanogaster* genes. Nat Methods 8:737–743
Venken KJT, Sarrion-Perdigones A, Vandeventer PJ et al (2016) Genome engineering: *Drosophila melanogaster* and beyond. Wiley Interdisc Rev. Dev Biol 5:233–267
Viswanathan S, Williams ME, Bloss EB et al (2015) High-performance probes for light and electron microscopy. Nat Methods 12:568–576
Wagh DA, Rasse TM, Asan E et al (2006) Bruchpilot, a protein with homology to ELKS/CAST, is required for structural integrity and function of synaptic active zones in Drosophila. Neuron 49:833–844
Walker GC (1984) Mutagenesis and inducible responses to deoxyribonucleic acid damage in *Escherichia coli*. Microbiol Rev 48:60–93
Wang J, Ma X, Yang JS et al (2004) Transmembrane/juxtamembrane domain-dependent Dscam distribution and function during mushroom body neuronal morphogenesis. Neuron 43:663–672

Watts RJ, Schuldiner O, Perrino J et al (2004) Glia engulf degenerating axons during developmental axon pruning. Curr Biol 14:678–684

Wei X, Potter CJ, Luo L et al (2012) Controlling gene expression with the Q repressible binary expression system in *Caenorhabditis elegans*. Nat Methods 9:391–395

Wilson RI (2013) Early olfactory processing in Drosophila: mechanisms and principles. Annu Rev Neurosci 36:217–241

Wolff T, Iyer NA, Rubin GM (2015) Neuroarchitecture and neuroanatomy of the Drosophila central complex: a GAL4-based dissection of protocerebral bridge neurons and circuits. J Comp Neurol 523:997–1037

Worley MI, Setiawan L, Hariharan IK (2013) TIE-DYE: a combinatorial marking system to visualize and genetically manipulate clones during development in *Drosophila melanogaster*. Development 140:3275–3284

Yagi R, Mayer F, Basler K (2010) Refined LexA transactivators and their use in combination with the Drosophila Gal4 system. Proc Natl Acad Sci USA 107:16166–16171

Yasunaga K, Saigo K, Kojima T (2006) Fate map of the distal portion of *Drosophila proboscis* as inferred from the expression and mutations of basic patterning genes. Mech Dev 123:893–906

Ye B, Zhang Y, Song W et al (2007) Growing dendrites and axons differ in their reliance on the secretory pathway. Cell 130:717–729

Yeh E, Gustafson K, Boulianne GL (1995) Green fluorescent protein as a vital marker and reporter of gene expression in Drosophila. Proc Natl Acad Sci USA 92:7036–7040

Yu HH, Chen CH, Shi L et al (2009) Twin-spot MARCM to reveal the developmental origin and identity of neurons. Nat Neurosci 12:947–953

Zhang YQ, Rodesch CK, Broadie K (2002) Living synaptic vesicle marker: synaptotagmin-GFP. Genesis. 34:142–145

Chapter 2
Retinal Connectomics

Kevin L. Briggman

Abstract The use of electron microscopy (EM) to describe the detailed synaptic connectivity in the retina has a rich history. Recent technological advances in serial electron microscopy (EM) have placed a complete description of the synaptic connectivity of the mammalian retina within reach. These new tools have recently been used to densely reconstruct the largest piece of a mammalian retina to date. Connectivity mapping has also revealed an unprecedented degree of wiring specificity in retinal circuitry. This is only the beginning of what can be learned from the comprehensive mapping of retinal circuits. The field of retinal connectomics is rapidly contributing to the functional understanding of retinal computation and enabling wiring comparisons both within and across species.

2.1 Introduction

The circuitry of the mammalian retina is primarily responsible for decomposing (or filtering) the visual world into a series of parallel representations (Gollisch and Meister 2010). While the purpose of these filters are intuitive to describe such as the detection of edges or motion or the adaptation to changes in color or luminance, the synaptic circuitry that underlies the majority of retinal computations remains to be discovered. In other words, we know a lot about what the retina can do, but much less about how it accomplishes specific computations. Indeed, even the absolute number of output channels in the mammalian retina—defined as distinct retinal ganglion cell (GC) types—continues to increase with each new functional or anatomical survey of cell types (Kong et al. 2005; Helmstaedter et al. 2013; Sanes and Masland 2015; Baden et al. 2016). For the few retinal circuits in which detailed synaptic wiring has been described, it is clear that circuit mapping is one essential

K.L. Briggman (✉)
Circuit Dynamics and Connectivity Unit, National Institute
of Neurological Disorders and Stroke (NINDS), Bethesda
MD 20892, USA
e-mail: kevin.briggman@nih.gov

A. Çelik and M.F. Wernet (eds.), *Decoding Neural Circuit Structure and Function*, DOI 10.1007/978-3-319-57363-2_2

component for understanding how functional responses in the retina arise (Briggman et al. 2011; Kim et al. 2014; Marc et al. 2014).

This chapter explores both the history of electron microscopy-based circuit mapping in the mammalian retina and the current state of large connectomic reconstruction efforts. In particular, modern electron microscopy (EM) tools and analysis algorithms have transformed the experimental landscape, enabling the collection of sufficiently large data volumes, at unprecedented spatial resolutions, to analyze complete retinal microcircuits. When combined with functional recordings from retinal neurons—ideally in the same piece of tissue—these methods now allow direct correlations between structure and function in the mammalian retina and provide crucial information for anatomically accurate computational models of retinal circuits.

2.2 Some Basic Retinal Anatomy

Unlike nearly all other regions of the central nervous system, the mammalian retina is essentially a standalone computational device. It consists of highly ordered layers of neuronal cell bodies [the outer and inner nuclear layers (ONL, INL) and the ganglion cell layer (GCL)] interconnected by dense synaptic neuropil layers [the outer and inner plexiform layers (OPL, IPL)] (Fig. 2.1). Five general cell classes populate the nuclear layers: photoreceptors, horizontal cells, bipolar cells (BCs), amacrine cells (ACs), and ganglion cells (GCs). The basic wiring plan of the retina was described by Dowling and Boycott by examining electron micrographs of the primate retina (Dowling and Boycott 1966). Their plan outlines a forward propagation of information from photoreceptors to ganglion cells with extensive lateral interactions implemented by horizontal and amacrine cells (Fig. 2.2a). Two ultrastructurally distinct classes of chemical synapses—ribbon synapses formed by photoreceptors and BCs, and more classical synapses formed by ACs—were also identified (Kidd 1962; Dowling and Boycott 1966) (Fig. 2.2b).

The major cell classes have been further subdivided based on distinct morphological differences and, for BCs, ACs, and GCs, stratification level within the IPL (MacNeil and Masland 1998; Masland 2012; Sanes and Masland 2015). While the diversity in cell-type morphologies was already appreciated by Cajal (Ramón y Cajal 1972) and early Golgi impregnation studies (Kolb et al. 1981), counting the actual number of subtypes has proven difficult. A historical timeline of anatomical cell-type surveys demonstrates both the upward trend in the number of defined subtypes and the increasing frequency of attempted surveys (Fig. 2.3). Part of the reason for the increased interest in defining cell types was the introduction of new experimental tools to label neuronal morphologies, including the stochastic expression of genetic reporters (Badea and Nathans 2004), photofilling (MacNeil et al. 1999), DIOlistic labeling (Gan et al. 2000), and the availability of mouse lines labeling genetically distinct subtypes (Sümbül et al. 2014). Each of these methods, however, comes with intrinsic biases that influence which neurons will be labeled.

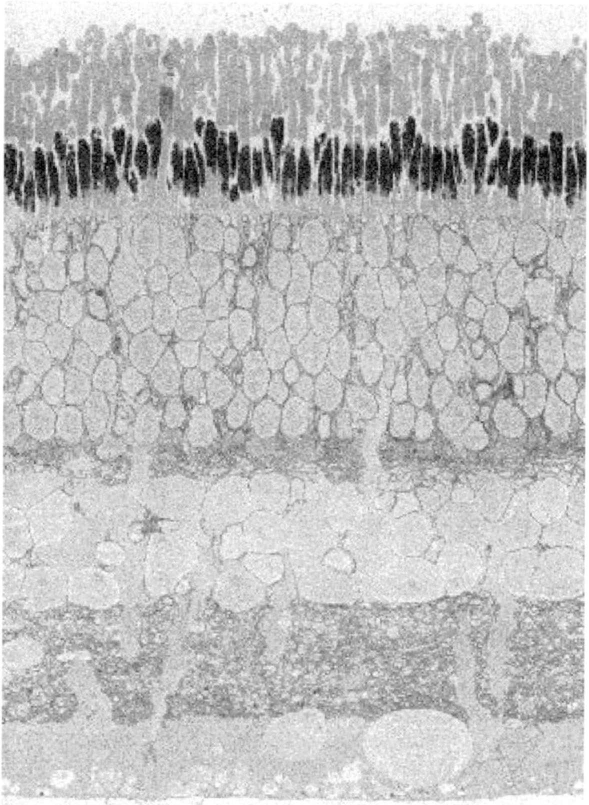

Fig. 2.1 A low power electron micrograph of a cross section through a rabbit retina. Retinal ganglion cell somas reside in the GCL. Bipolar, amacrine, and horizontal cell somas in the INL. Rod and cone photoreceptors in the ONL. Synapses between photoreceptors and bipolar cell dendrites occur in the OPL. Synapses between bipolar cell axons, amacrine cells, and ganglion cells are found in the IPL. *Scale bar* 50 μm

For example, any filling technique that depends on randomly sampling a grid placed over neuronal cell bodies will bias the filling of neurons with large somas. The only truly unbiased approach is a method in which every neuron within a given volume can be comprehensively reconstructed. This is perhaps a trivial point but one that is often underappreciated and is a clear advantage for EM-based dense volume reconstructions (Helmstaedter et al. 2013; Kasthuri et al. 2015).

As of 2016, the current cell-type counts in the mouse retina are 10 BC types, 45 AC types, and 30+ GC types (Helmstaedter et al. 2013; Sanes and Masland 2015). These numbers could easily increase as larger EM volumes become available that encompass sparse and/or large, wide-field cell types. Moreover, it is not clear how representative these numbers are for other species in which large-scale surveys have not yet been undertaken. In other words, the most basic prerequisite

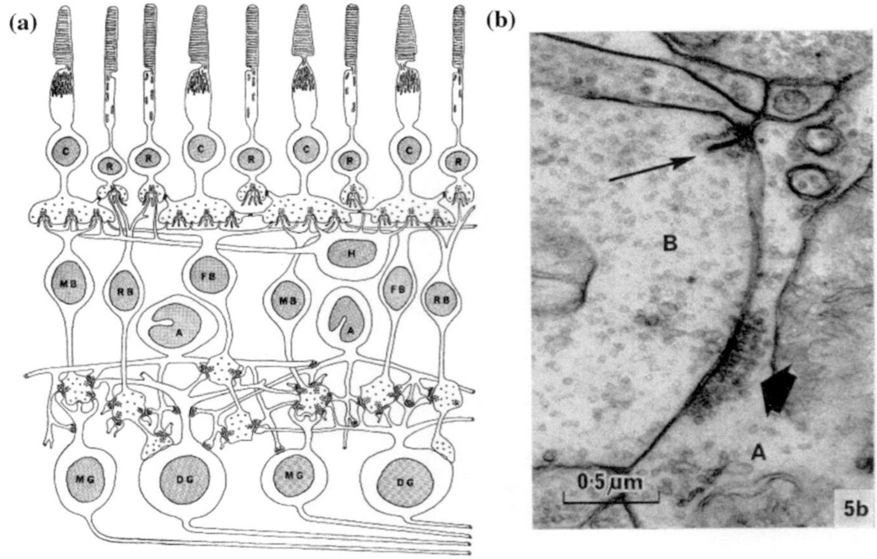

Fig. 2.2 **a** Basic synaptic organization of the mammalian retina based on early EM reconstructions by Dowling and Boycott in the central primate retina. This classic picture of the synaptic organization of the retina has not fundamentally changed in the last 50 years. During that time period, however, the number of unique cell types has continuously risen and the complexity of the wiring diagram is still under investigation. **b** A classic example of the two types of chemical synapses encountered in the ribbon; ribbon synapses found in photoreceptors and bipolar cells and classical chemical synapses formed by amacrine cells. Reproduced from Dowling and Boycott (1966)

for mapping neuronal circuitry—knowing how many distinct cell types exist in the retina—has only very recently reached some semblance of completeness and, at that, in only one mammalian species.

2.3 Why Is Retinal Connectomics Difficult?

The use of EM to describe the wiring of the retina has waxed and waned since the original images of synapses acquired by Dowling and Boycott (1966). Early efforts to identify synaptic contacts onto defined cell types typically used Golgi impregnation or horseradish peroxidase (HRP) to first fill a neuron for viewing under a light microscope and then collecting small, local EM volumes around a stained dendrite (Kolb 1970). The first serious attempts to collect and annotate volumes sufficiently large to trace dendritic trees were undertaken by Peter Sterling and colleagues (Stevens et al. 1980a, b). Their data volumes were typically around

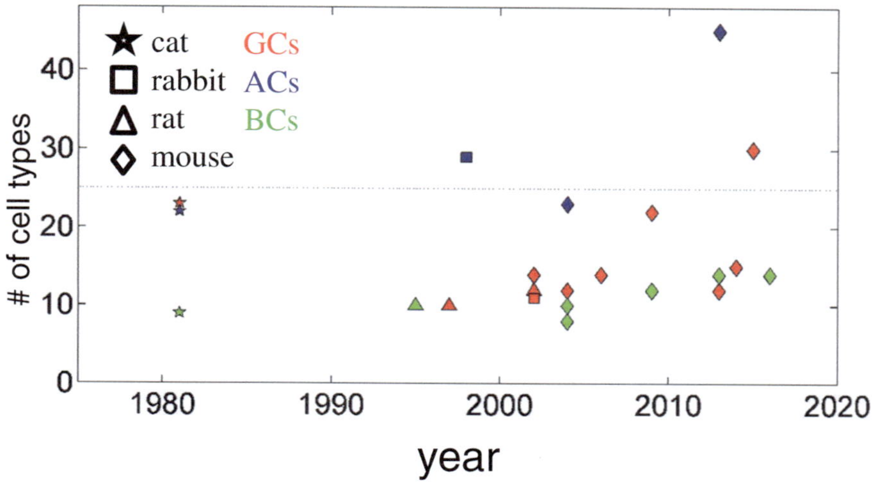

Fig. 2.3 Historical view of the number of ganglion cell (*GC*), amacrine cell (*AC*), and bipolar cell (*BC*) types estimated from the cat, rabbit, rat, and mouse retinas. Early estimates were based on the stochastic Golgi method. More recent estimates used random filling of neurons with fluorescent dyes, gene expression, or dense reconstruction of SBEM datasets. Some variability is due to differences in counting conventions, but note the general upward trend and increasing frequency of estimates. Data points derived from Kolb et al. (1981), Euler and Wassle (1995), Huxlin and Goodchild (1997), MacNeil et al. (1999), Rockhill et al. (2002), Sun et al. (2002a), Sun et al. (2002b), Badea and Nathans (2004), Ghosh et al. (2004), Coombs et al. (2006), Volgyi et al. (2009), Wassle et al. (2009), Helmstaedter et al. (2013), Sümbül et al. (2014), Sanes and Masland (2015), and Greene et al. (2016)

10^4 μm^3 (Fig. 2.4) and were focused on mapping the bipolar cell input pathways to several GC types in the cat and primate retinas (Stevens et al. 1980a, b; McGuire et al. 1984; Freed and Sterling 1988; Sterling et al. 1988; Cohen and Sterling 1990; Calkins et al. 1998).

Given the importance of accurately describing cell types and mapping the connectivity among all cell types, why have large-scale efforts equivalent to the comprehensive mapping of the *C. elegans* nervous system (White et al. 1986) only recently been undertaken? What took so long? First, the necessity to manually collect and analyze thousands of sections to reconstruct a sufficiently large volume to contain complete circuits was daunting (and tedious) (Stevens et al. 1980a, b). Second, many of the neuronal structures in the retina require higher resolutions than had been achievable. The limiting resolution in serial section electron microscopy is not the imaging resolution of an electron microscope, but rather, the actual section thickness. For decades, section thicknesses of 50–80 nm were considered state-of-the-art (Fig. 2.4). Are 50–80 nm sections sufficient to unambiguously reconstruct all arbitrarily oriented thin neurites and synaptic structures in a volume? The answer is clearly no, see discussion below. Third, the computing requirements to both manage and analyze large-scale datasets have only recently become readily available (Jain et al. 2010; Turaga et al. 2010; Helmstaedter et al. 2011; Berning et al. 2015).

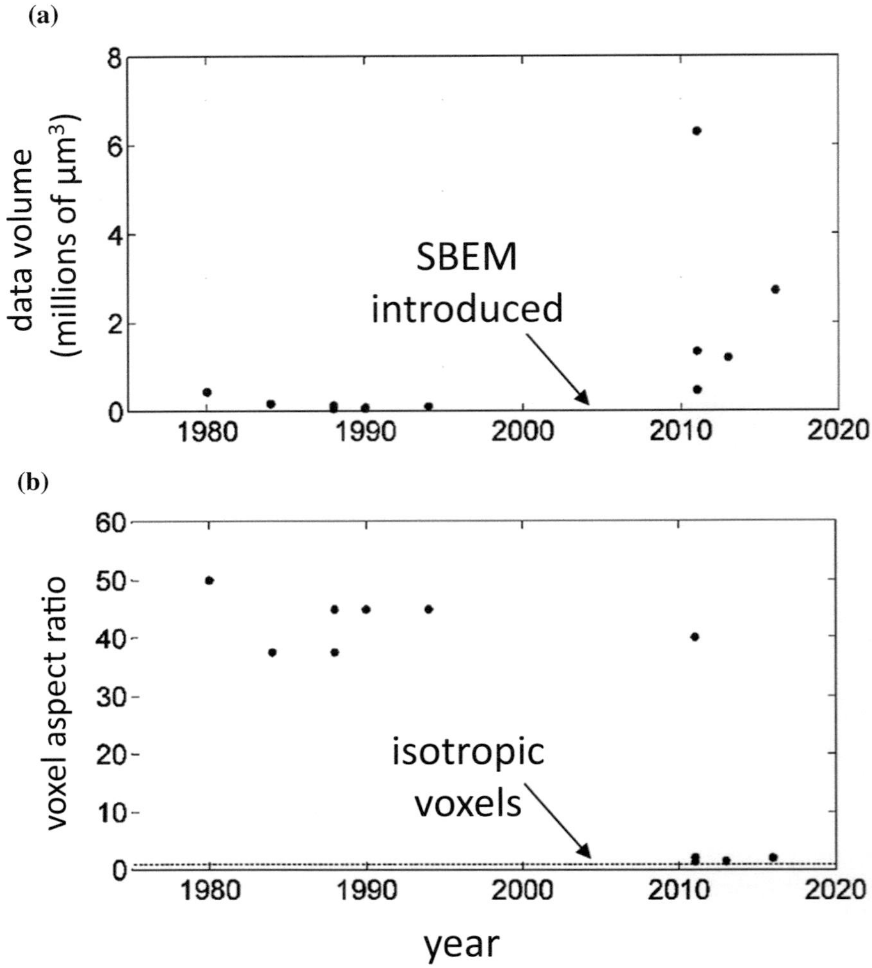

Fig. 2.4 **a** Historical view of the volume encompassed by a selection of mammalian retina EM datasets. Studies not reporting section thicknesses and/or imaged area dimensions are not shown. Since the introduction of automated acquisition techniques, dataset volumes have grown, in one case by an order of magnitude over what was collected be ssTEM. **b** Automated sectioning methods have allowed section thicknesses to dramatically decrease from 70–90 nm to 25–30 nm. Consequently, the aspect ratio (defined as the ratio of section thickness over lateral pixel resolution) of voxels has decreased. Current methods are approaching isotropic voxel resolutions. For earlier studies that used photographic film rather than cameras, a lateral resolution of 2 nm was assumed. Data points derived from Stevens et al. (1980a, b), McGuire et al. (1984), Freed and Sterling (1988), Cohen and Sterling (1990), Strettoi et al. (1990), Calkins et al. (1998), Anderson et al. (2011), Helmstaedter et al. (2011), Helmstaedter et al. (2013), and Ding et al. (2016)

2.4 The Long Overdue Automation of Electron Microscopy

For decades 3D EM data, including retinal EM data, were collected by serial section transmission electron microscopy (ssTEM, Fig. 2.5a). ssTEM involves cutting ultrathin sections with an ultramicrotome and the manual collection of the sections from a water boat. This is a tedious and inherently error-prone process that can result in a number of artifacts such as lost or folded sections. The sections are then sequentially imaged with a TEM and then aligned and assembled into a 3D volume. There was remarkably little innovation in this process until the early 2000s, when two complementary approaches were introduced to automate the sectioning process. One was termed the automated tape-collecting ultramicrotome (ATUM) (Schalek et al. 2011). The idea was to remove human interaction from the collection of sections by automatically feeding sections onto a support film like a conveyor belt. The support film is subsequently cut into strips, mounted on a metal substrate and imaged in a scanning electron microscope (SEM). A second approach was named serial block-face scanning electron microscopy (SBEM) (Denk and Horstmann 2004). SBEM removes the requirement of pre-sectioning tissue by mounting a custom microtome within the vacuum chamber of a SEM. Images are formed from the block face of a sample and tissue sections are then shaved off the block face.

Both of these advances benefited from the relatively recent development of high-resolution scanning electron microscopes and efficient electron detectors. Indeed, block-face imaging had been attempted decades earlier (Leighton 1981), but did not catch on with the SEMs available at that time. The automation afforded by both ATUM and SBEM is a clear advantage when collecting tens of thousands of sections. Perhaps more important is the reduction in section thickness that automation enabled. The typical slice thickness for ssTEM is still around 50–70 nm. ssTEM voxels are therefore extremely anisotropic (assuming a lateral TEM resolution of 2 nm) (Fig. 2.4b). Even 50-nm-thick sections are clearly inadequate when the diameters of neurites approach 100 nm (of which several examples exist in the retina, such as the dendrites of starburst amacrine cells). One of the few published accounts of the difficulties encountered while tracing neurons was reported by Freed and Sterling describing their experience tracing bipolar cells using 75 nm sections: "We could not trace the processes back to an axon stalk because the connections between these varicosities were extremely fine and tortuous" (Freed and Sterling 1988). Both ATUM and SBEM allow substantially thinner sections to be reliably cut down to a range of 20–30 nm (approaching isotropic voxels, Fig. 2.4) and miss far fewer sections than ssTEM. However, even using modern automated EM approaches with section thicknesses of ~25 nm, inter-annotator tracing discrepancies still persist (Helmstaedter et al. 2011).

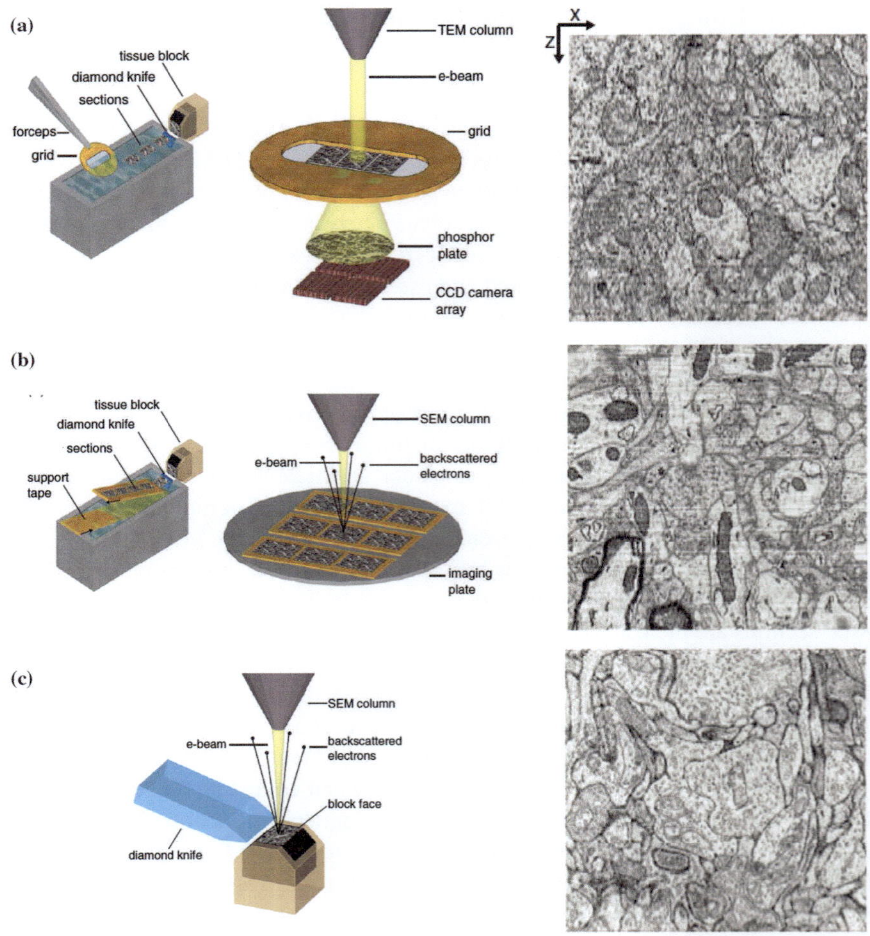

Fig. 2.5 Automated methods to collect serial EM datasets. **a** The classical ssTEM approach was used until the advent of automated sectioning methods. One limitation of this approach is the poor *Z* axis resolution due to typical section thicknesses of 70–90 nm (*right panel*). **b** The automated tape-collecting ultramicrotome approach developed by K. Hayworth and J. Lichtman automates the collection of serial sections from a water boat. The automation improves the reliability of sectioning and minimal slice thicknesses of 29 nm have been reported (Kasthuri et al. 2015). The improvement in *Z* axis resolution is evident (*right panel*). **c** The serial block-face scanning electron microscopy (*SBEM*) approach developed by W. Denk. Both automated sectioning and imaging are performed within a SEM and repeatable section thicknesses down to 23 nm have been reported (Briggman et al. 2011). The improvement in *Z* axis resolution compared to ssTEM is evident (*right panel*). *Scale bar* 1 μm. Figures reproduced from Briggman and Bock (2012)

2.5 The Real Bottleneck—Data Analysis

As automated EM methods have significantly eased the acquisition of high-quality 3D EM data, the size of collectable volumes has dramatically risen (Fig. 2.4a). The largest retinal volume to date spanned 10^6 μm^3, nearly two orders of magnitude larger than previous volumes, at a voxel resolution of $16.5 \times 16.5 \times 23$ nm^3 (Briggman et al. 2011). Such data volumes now exceed terabytes worth of raw grayscale image data. The bottleneck is therefore now the analysis of the data. The goal is twofold: first, neuronal morphologies must be reconstructed ('traced') by linking together all the neurites belonging to each neuron in the volume and second, the synaptic connections between neurons must be identified. Of the two goals, the accurate reconstruction of neuronal morphologies is by far the more difficult problem. The difficulty stems from the arborized nature of most neurons. A tracing error early on in a dendritic tree can lead to the misassignment of all downstream synapses. Synapse identification, on the other hand, is a local judgement and synapse annotation errors do not typically compound.

The first serious effort to quantify error rates when reconstructing neurons was performed in an SBEM block of mouse retina (Helmstaedter et al. 2011). This study demonstrated that annotating neurons by 'skeletonization', delineating the midline of neurons with nodes and edges, is significantly faster than full volume reconstructions (Fig. 2.6a, b). More importantly, it was found that even experts can disagree at multiple locations when tasked with reconstructed complete morphologies. It was shown that redundant tracing of neurons by multiple independent humans can substantially reduce error rates and allow hundreds of microns of path length to be accurately traced with a small group of human annotators (Fig. 2.6c, d). This strategy was ultimately applied to reconstruct about 1000 neurons from a volume of mouse retina (Helmstaedter et al. 2013).

2.6 A Retinal 'Contactome'

The first dense reconstruction all of the cells within a piece of mammalian retina was recently published (Fig. 2.7a) (Helmstaedter et al. 2013). A $80 \times 114 \times 132$ μm^3 volume of mouse retina was collected using SBEM, spanning the PRL to the GCL. The analysis of this dataset yielded more than 1000 neuronal morphologies that included BCs, ACs, and GCs. One caveat of this reconstruction effort was the unconventional way in which the tissue was stained. Prior to collecting the dataset it was clear that the analysis, rather than the actual data acquisition, would be the rate-limiting step. At the time, however, it was unclear how well machine learning-based analysis algorithms (Turaga et al. 2010) would be able to automatically segment conventionally stained tissue in which, in addition to the plasma membrane, many intracellular structures are stained (such as organelles, vesicles, filaments, etc.). The retina was therefore processed with a protocol in which the surfaces of cells were

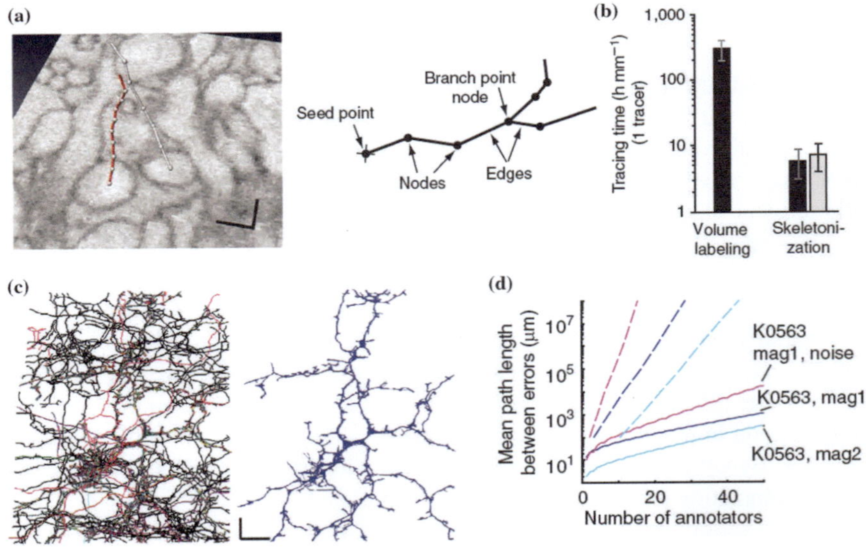

Fig. 2.6 Annotation accuracy of mouse retina EM data. **a** Neuronal morphologies can be described by tracing the midline 'skeletons', consisting of a series of nodes and edges. **b** Skeletonization in the retina speeds up morphology annotation by a factor of 50 compared to 3D volume labeling. **c** The overlay of 50 human annotated skeletons of one amacrine cell demonstrates the unreliability of human annotation (*left*). A weighted consensus of the skeletons yields a more accurate representation of the cell's morphology (*right*). **d** A calibration of the number of human annotators required to achieve a particular error rate. *Solid lines* indicate the actual performance of 50 annotators, the *dashed lines* indicate simulated error rates if the tracings were focused on locations of disagreement between annotators. Figure reproduced from Helmstaedter et al. (2011)

selectively stained and intracellular structures remained unstained (Briggman et al. 2011). The assumption was that segmentation algorithms would have an easier time delineating neurons without intracellular 'clutter'. Qualitatively, it was far easier to follow thin neurites over hundreds of microns in such cell-surface labeled tissue compared to conventionally stained tissue. However, this approach comes with the obvious downside that vesicles remain unstained and, with them, the absolute positions of synapses are unknown.

Because the locations of synapses could not be absolutely identified, the contact area between pairs of neurons was used as a proxy for actual synaptic connectivity (Fig. 2.7b). The pairwise contact areas between all neurons thus yielded a retinal 'contactome' rather than a true retinal 'connectome' (Fig. 2.7c). Was this tradeoff worth it? Probably not, since a fully automated analysis was not ultimately error-free and multiple research groups have subsequently shown that conventionally stained tissue can be reasonably segmented to a similar degree of accuracy as the cell-surface labeled tissue (Berning et al. 2015; Pallotto 2015). In fact, the analysis of the contactome relied on a hybrid approach in which redundant

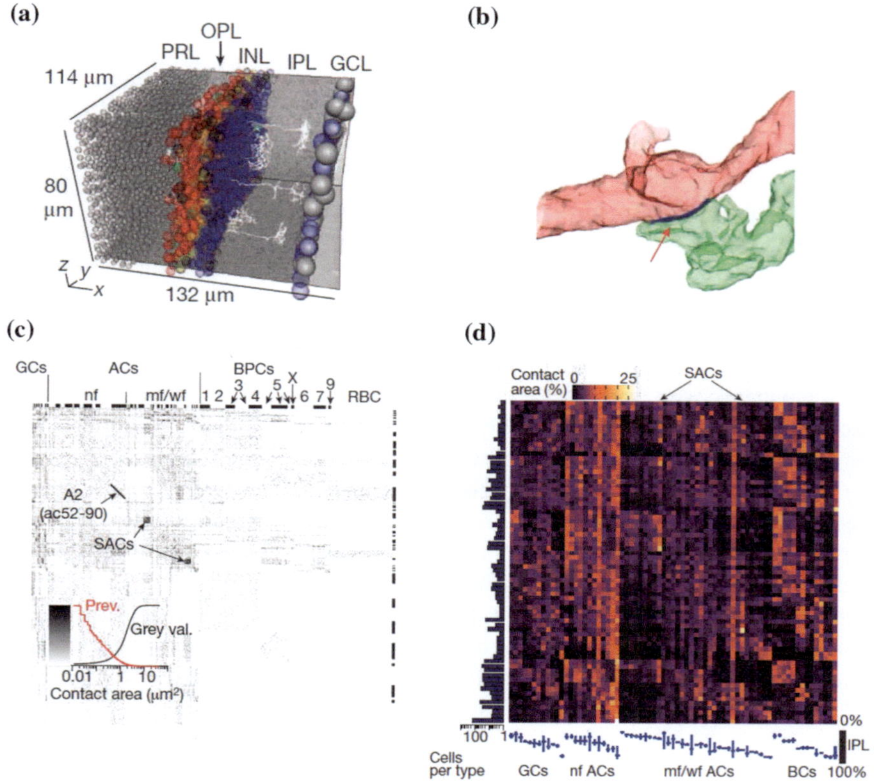

Fig. 2.7 A contactome of the mouse retina. **a** A block of mouse retina collected by SBEM. Soma locations indicated by *spheres* and *color coded* by cell class: GCs *large gray spheres*; ACs *blue*; BCs *red*; glia *yellow*; horizontal cells *green*; photoreceptors *small gray spheres*. **b** An example of the contact area that was quantified between every pair of touching cells. **c** A cell-to-cell contact matrix quantifying that shared contact area between 950 neurons, sorted by class. The intensity of the *grayscale* in each matrix entry is proportional to the contact area. **d** Cells were grouped by type and the contact matrix averaged by cell type to generate a cell-type contact matrix among 71 cell types. Figures reproduced from Helmstaedter et al. (2013)

skeletonization was used to first define morphologies and automated segmentations were then used to fill out the 3D morphologies around skeletons.

Caveats aside, the contactome provided at least three crucial insights into retinal connectivity: (1) the dense reconstruction effort provided a definitive accounting of all the neurons that were in the volume, (2) any cell types that did not touch each other cannot form circuits, and (3) an analysis of shared contact area between cell types generated testable circuit hypotheses.

Quantifying how many cell types reside in the retina is perhaps the most trivial result of a dense reconstruction, but arguably one of the most crucial outcomes. The data volume was only 80 × 114 μm^2 in the plane of the retina, precluding the reconstruction of wide-field ACs and GCs, but was sufficiently large to reconstruct all BC types and probably the majority of AC types. Only partial GC reconstructions were obtained, but cell typing was accomplished based on IPL stratification depth. A major benefit of the retina (compared to other brain regions) is its repeating mosaic organization. Any cell typing based on morphological criterion such as stratification depth or axonal spread can therefore be verified against the mosaic formed by cells classified as a single type (with a few notable exceptions such as starburst amacrine cells that form a highly overlapping mosaic).

A novel sparse BC type was discovered (BC type X) with this approach. Given the relatively low density of this cell type, it is perhaps not surprising that this cell was missed in previous surveys (Ghosh et al. 2004). This underscores the benefit of completeness enabled by EM-based dense reconstructions. Furthermore, type 5 BCs which are morphologically difficult to separate were subtyped based on their contact patterns with specific GCs. This is an example of using connectivity to discriminate cell types and is arguably, along with gene expression profiling (Sanes and Masland 2015), one of the best ways to define distinct cell types.

Once all cell types had been identified, the shared contact area between each cell was used a proxy for synaptic 'weight' in the contact matrix (Fig. 2.7c). That is, a pair of cells that share more contact area are more likely to form a stronger synaptic relationship than a pair of cells with less shared contact area. This assumption probably breaks down at some point, particularly for the special case of the ribbon synapses of bipolar cells. But the contact weight metric proved to be useful for validating existing circuit motifs and proposing new ones. Stratification between cell types has commonly been used to infer potential connectivity at the light microscopic level (Famiglietti 1992; Brown and Masland 1999). The contactome can readily rule out connectivity based on a lack of shared contact area even for cell types that extensively costratify. For example, Type 7 BCs costratify with direction selective ganglion cells (DSGCs), but were not found to share a large contact area with DSGCs. The novel BC type X was similarly found to avoid forming contacts with a GC with which it costratifies.

In addition to ruling out connectivity patterns, several circuits were proposed based on the analysis of the dominant contact areas between cell types (Fig. 2.7d). For example, the GC thought to be responsible for local-edge detection (GC type W3) (Zhang et al. 2012) was found to receive its strongest input from BC type 5R and from an ON AC, suggesting a selective role for ON feedforward inhibition onto the W3 cell. Such findings significantly narrow the number of possible cell types presynaptic to the W3 GC that need to be examined physiologically. The retinal contactome is therefore perhaps best utilized as a hypothesis generator to guide physiological studies, especially for the less well-studied circuits in the mammalian retina.

2.7 Correlating Retinal Structure with Function

While the retinal contactome provides an overview of possible connectivity motifs, a major goal of connectomics is to relate detailed synaptic connectivity to the function of neuronal circuits. One of the most direct demonstrations of how detailed retinal connectivity mapping can enhance the understanding of a particular computation is the analysis of the direction selective (DS) circuitry. The cell types that play a role in the computation of DS are some of the most well known and frequently studied in the mammalian retina. The uniquely shaped GABAergic starburst amacrine cells (SACs) are at the core of the circuit and their dendrites are responsible for computing direction in a radial manner relative to their central somas (Tauchi and Masland 1984; Famiglietti 1991; Euler et al. 2002). SACs violate 'normal' retinal mosaic principles, at least in terms of their dendritic spread. The dendrites from at least 80 or so individual SAC can be found at any given location on the retina (Ding et al. 2016). This over-representation is seemingly required by the number of postsynaptic cells with which SACs synapse—at least 7 GCs (3 subtypes of ON DSGCs and 4 subtypes of ON–OFF DSGCs) and the extensive lateral inhibition among SACs themselves.

A long-standing question regarding the computation of retinal DS was how the radial direction preference of SACs is converted to the rectilinear direction preference of DSGCs (Barlow et al. 1964; Barlow and Levick 1965). It was known from electrophysiology and synaptic pharmacology that directionally selective inhibition onto DSGCs reduces spiking along one axis and the absence of inhibition along opposing axes allow DSGCs to spike (Fried et al. 2002; Taylor and Vaney 2002). The question was how a single cell type, the SAC, could provide such directionally specific inhibition if there does not exist a dedicated SAC for each DSGC subtype?

Vaney et al. (1989) proposed a wiring principle that could explain the observed physiology. They suggested an unprecedented level of wiring specificity in which radial dendritic sectors of a single presynaptic cell selectively synapse onto different postsynaptic subtypes. In their model, the direction of presynaptic inhibitory SAC dendrites defines the null direction of a postsynaptic DSGC. How could this hypothesis be tested anatomically? Several groups analyzed proximities between SAC dendrites and DSGCs at the light microscopic level but came to differing conclusions about whether an asymmetry in SAC-to-DSGC wiring existed (Famiglietti 2002; Fried et al. 2002; Dong et al. 2004; Chen and Chiao 2008). The dense plexus of SACs makes it very difficult to accurately infer connectivity onto DSGCs with the diffraction limited optics of light microscopes. This was therefore an ideal application for modern 3D EM to test a long-standing hypothesis about retinal connectivity.

If, in fact, an asymmetric SAC-to-DSGC wiring pattern existed, the first question would be whether the pattern related to the functional properties of DSGCs. That is, to make the strongest link between structure and function, the coding properties of retinal neurons (in this case the preferred direction of DSGCs) should be characterized in the same piece of tissue used for EM analysis. This was achieved using 2P calcium imaging of hundreds of GCs covering a $300 \times 350\ \mu m^2$ patch of mouse retina in response to moving bars (Briggman and Euler 2011; Briggman et al. 2011). The relatively unbiased imaging approach used provided the locations of dozens of ON–OFF DSGC somas. A SBEM volume was then collected from the functionally imaged area that spanned the IPL (Fig. 2.8a). Using a sparse skeletonization approach, 6 ON–OFF DSGCs and 24 SACs were reconstructed and SAC–DSGC synapses were annotated (Fig. 2.8). The analysis clearly demonstrated that the asymmetric wiring hypothesis was correct. There is a strong directional asymmetry of SAC dendrites that provide inputs to each ON–OFF DSGC subclass. Moreover, because the functional tuning curves from the reconstructed DSGCs were known, the anatomical asymmetry could be confirmed to be antiparallel to the functionally measured preferred direction of each DSGC (Fig. 2.8c). This is a clear example that mapping the structure of a circuit can actually predict its function. In other words, if one collected a new retinal EM dataset that was not functionally characterized, a key functional property of DSGCs—their preferred directions—could be with near certainty predicted solely from examining their presynaptic SAC wiring patterns.

The precision of SAC-to-DSGC connectivity is also an example of an unprecedented form of wiring specificity in the nervous system. The targeting of synapses between particular cell types is known to exist throughout the brain. The targeting of BC axons to specific sublaminae in the IPL that form synapses onto specific GCs is just one example of this. A higher degree of specificity can be found in circuits in which the axons of a given cell type (such as cortical interneurons) target particular subcellular structures (such as basal versus apical dendrites) of postsynaptic cell types (Burkhalter 2008). The SAC-to-DSGC circuit represents an even more elaborate degree of specificity in which the different dendrites of a *single* cell target different postsynaptic cell types (Fig. 2.8b).

The large SBEM volume has proven to be useful for additional studies related to the computation of DS in SAC dendrites. A massive crowd-sourced effort to reconstruct BCs in the volume revealed a spatial offset in the BC types that provide inputs to SAC dendrites (Kim et al. 2014). This spatial offset, combined with potential differences in the release kinetics of the BCs, formed the basis of a space-time wiring hypothesis that potentially contributes to the centrifugal preference of SAC dendrites.

2.8 Species-Dependent Differences in Retinal Wiring

While the mapping of the SAC–DSGC circuit was a definitive demonstration of the anatomical underpinnings of the DS computation, it essentially confirmed a long-standing hypothesis that was likely to be true. The broader promise of comprehensive EM-based connectomics is the ability to incorporate synaptic wiring details into novel computational models that extend our current understanding of

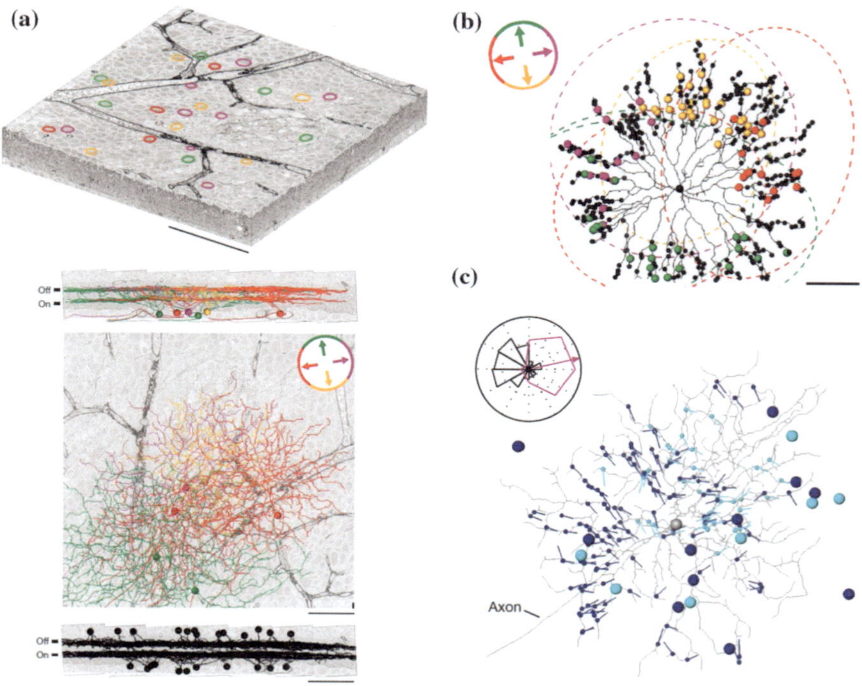

Fig. 2.8 Wiring specificity in the direction selectivity circuit. **a** A block of mouse retina spanning the IPL collected by SBEM. GCs in the retina were functionally characterized using two-photon calcium imaging prior to EM processing. Direction selective GCs are *circled* and *color coded* by their preferred directions. The morphologies of 6 reconstructed DSGCs and 24 SACs are shown in the *lower panel*. **b** The dendrites of SACs form synapses to DSGCs with antiparallel preferred directions. Synapses color coded by the preferred direction of the postsynaptic cell (see *inset*). **c** DSGCs collect synapses from SAC dendrites oriented along their null directions. *Vectors* indicate the direction of presynaptic SAC dendrites. *Inset* The directional distribution of presynaptic SAC dendrites (*black histrogram*) aligns with the null direction of a DSGC, antiparallel to the preferred direction (*magenta tuning curve*). Figure reproduced from Briggman et al. (2011)

retinal processing. A recent example of this involves the identification of species differences in the detailed wiring of the SAC-to-SAC network (Ding et al. 2016).

There are many obvious differences between the retinas of mammalian species, including the variety of high acuity regions found in different animals that are adapted to their particular environments (Linberg et al. 2001). Relatively few studies, however, have focused on species differences in the detailed synaptic wiring among cell types (Chun et al. 1993). To explore whether species differences exist in the DS circuit, a large conventionally stained SBEM dataset spanning the mouse IPL was collected that allows the positive identification of synapse locations (Fig. 2.9a). During the course of mapping the distribution of excitatory (BC synapses) and inhibitory (AC synapses) inputs onto mouse SACs, a segregation of the inputs was found that was substantially different compared to earlier EM reconstructions of rabbit SACs (Famiglietti 1991). Excitatory BC ribbon inputs were clustered along the proximal two-thirds of SAC dendrites and were largely nonoverlapping with output synapses clustered along the outer third of the dendrites (Fig. 2.9b, c). In addition, inhibitory inputs originating primarily from neighboring SACs were located exclusively along the proximal third of SAC dendrites (Fig. 2.9c, d). By comparison, SAC-to-SAC synapses in the rabbit had been found to occur along the distal dendrites based on EM reconstruction (Famiglietti 1991) and physiological mapping of inhibitory currents (Lee and Zhou 2006).

It is important to note that the basic properties of SACs are quite similar between mouse and rabbit, including average dendritic arbor diameter, soma density, and coverage factor (Tauchi and Masland 1984; Vaney 1984; Keeley et al. 2007). Therefore, in principle, mouse SACs could have been wired together similar to rabbit SACs and vice versa. However, the most obvious difference between the two species is something much simpler than any anatomical details of the retina—their

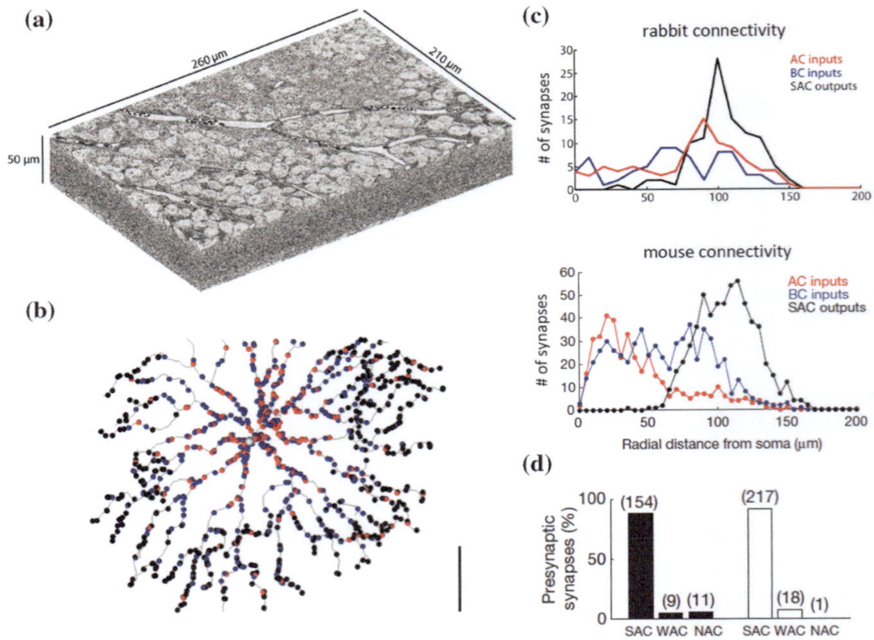

Fig. 2.9 Species-specific wiring in the direction selectivity circuit. **a** A block of mouse retina spanning the IPL collected by SBEM. This tissue block was conventionally stained so that synapses could be positively identified. **b** A reconstruction of a SAC and annotation of excitatory (*blue dots*) and inhibitory (*red dots*) input synapses and output synapses (*black dots*). **c** The radial distribution of synapse types measured from rabbit SACs (*upper panel*) and mouse SACs (*lower panel*). The locus of inhibitory inputs is notably different between the two species. **d** A quantification of the types of neurons providing inhibition to mouse SACs. The majority of inputs are from neighboring SACs. Figures reproduced from Ding et al. (2016)

eye diameters. A rabbit eye is approximately 5 times larger in diameter than a mouse eye, which means an object in visual space cover a 5 times larger area on the retinas of rabbits compared to mice. This is relevant for motion processing because it also implies an object moving at constant angular velocity in visual space moves 5 times slower across the mouse retina compared to the rabbit retina. If we assume mice and rabbits experience similar objects that move at similar angular velocities during their lives, how can their retinas ensure these velocities are accurately encoded?

To explore why the SAC circuitry of the two species would be wired differently, two computational models were constructed that incorporated the detailed anatomical data derived from the EM reconstructions (Fig. 2.10a). The major difference between the two models was the separation between connected SACs. Both the rabbit and mouse models were able to generate a preference for centrifugal motion in SAC dendrites (Fig. 2.10b). However, the difference in the spatial offset of inter-SAC lateral inhibition shifted the velocity-tuning curve to prefer lower velocities in mice (Fig. 2.10c). The shift in the tuning curve was roughly by a factor of five, in remarkable agreement with the difference in eye diameters between the two species. These results imply that the rabbit and mouse retinas have adapted the SAC wiring pattern to compensate for the constraints imposed by something as simple as eye size. The detailed mapping of synaptic connectivity was essential for recognizing the difference in the first place and provided the data needed to construct anatomically constrained computational models.

2.9 Future of Retinal Connectomics

It has been just the few years since EM datasets large enough to encompass entire retinal circuits have become available. Even in this brief span, these datasets have been mined for multiple purposes by multiple research groups demonstrating the power of the availability of high-quality EM data to the retina research community. Both dense complete reconstructions and more limited sparse reconstructions have proven useful and both analysis approaches have identified novel circuit motifs that would have been far more difficult to identify by light microscopy. Analyzing terabyte scale EM datasets remains the rate-limiting step for generating full retinal wiring diagrams, but impressive advances in computer science promise to further decrease automated segmentation error rates in the near future. In addition to algorithm improvements, altering tissue preparation techniques, such as preserving extracellular space, can further ease automated analysis (Pallotto et al. 2015).

Will EM always be needed for connectivity mapping? Recent advances in super-resolution light microscopy have broken the diffraction limit of light and enabled proteins distributions to be mapped at nanometer scale resolutions (Rust et al. 2006). This method was recently applied to map the distribution of subunit-specific inhibitory inputs to ON–OFF DSGCs (Sigal et al. 2015). This method therefore provides multiplexed molecular information that would be

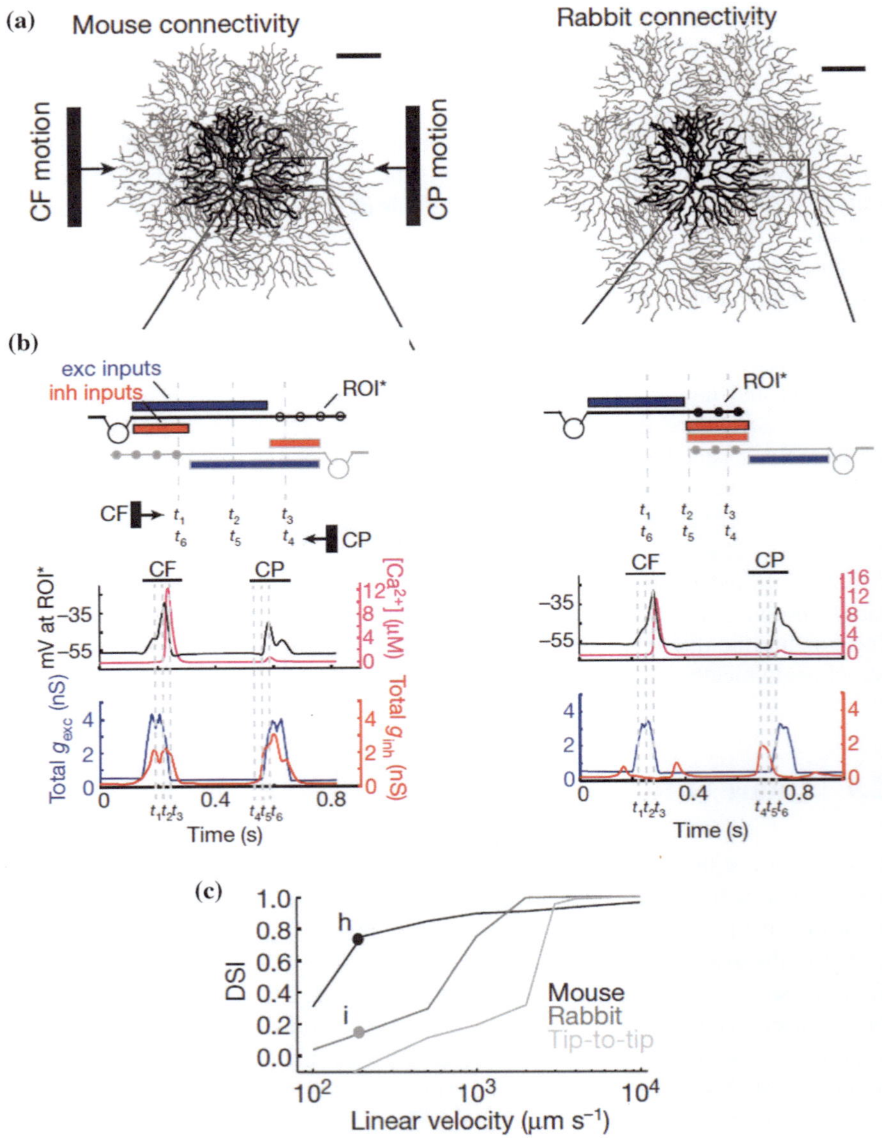

Fig. 2.10 Functional consequences of species specific wiring. **a** Anatomically constrained network models of mouse (*left*) and rabbit (*right*) network connectivity. **b** Both network models are able to generate DS responses in which both the membrane potential and intracellular calcium concentration at dendrite tips (ROI*) is greater for CF compared to CP motion. **c** A key difference of the two models is the velocity-tuning profiles. The models predict mouse SACs prefer slower velocities compared to rabbit SACs, consistent with the difference in eye diameter between the two species. Figures reproduced from Ding et al. (2016)

difficult to achieve with electron microscopy. Whether super-resolution methods can be extended to accurately trace fine neuronal processes in dense neuropil remains to be seen. Another recent alternative to EM is an innovative method to isotropically expand tissue before fluorescence imaging, essentially allowing proteins to be imaged at the molecular scale using diffraction limited optics (Chen et al. 2015). Again, the challenge will be to extend this technique to provide the same level of detail afforded by EM to map ultrastructure.

Now that modern automated EM methods have reached some degree of maturity, perhaps the most exciting next steps will be to compare the retinal wiring diagrams of different species in the phylogenetic tree. Most of the recent EM studies have focused on mouse, but similar large-scale efforts are underway in the rabbit retina (Anderson et al. 2011). Mapping the similarities and differences between evolutionarily distant retinas will likely provide key information to determining which parts of a circuit are essential for a computation and which are the results of species specializations.

References

Anderson JR, Jones BW et al (2011) Exploring the retinal connectome. Mol Vis 17:355–379

Badea TC, Nathans J (2004) Quantitative analysis of neuronal morphologies in the mouse retina visualized by using a genetically directed reporter. J Comp Neurol 480(4):331–351

Baden T, Berens P et al (2016) The functional diversity of retinal ganglion cells in the mouse. Nature 529(7586):345–350

Barlow HB, Levick WR (1965) The mechanism of directionally selective units in rabbit's retina. J Physiol 178(3):477–504

Barlow HB, Hill RM et al (1964) Retinal ganglion cells responding selectively to direction and speed of image motion in the rabbit. J Physiol 173:377–407

Berning M, Boergens KM et al (2015) SegEM: efficient image analysis for high-resolution connectomics. Neuron 87(6):1193–1206

Briggman KL, Bock DD (2012) Volume electron microscopy for neuronal circuit reconstruction. Curr Opin Neurobiol 22(1):154–161

Briggman KL, Euler T (2011) Bulk electroporation and population calcium imaging in the adult mammalian retina. J Neurophysiol 105(5):2601–2609

Briggman KL, Helmstaedter M et al (2011) Wiring specificity in the direction-selectivity circuit of the retina. Nature 471(7337):183–188

Brown SP, Masland RH (1999) Costratification of a population of bipolar cells with the direction-selective circuitry of the rabbit retina. J Comp Neurol 408(1):97–106

Burkhalter A (2008) Many specialists for suppressing cortical excitation. Front Neurosci 2(2):155–167

Calkins DJ, Tsukamoto Y et al (1998) Microcircuitry and mosaic of a blue-yellow ganglion cell in the primate retina. J Neurosci 18(9):3373–3385

Chen YC, Chiao CC (2008) Symmetric synaptic patterns between starburst amacrine cells and direction selective ganglion cells in the rabbit retina. J Comp Neurol 508(1):175–183

Chen F, Tillberg PW et al (2015) Expansion microscopy. Science 347(6221):543

Chun MH, Han SH et al (1993) Electron microscopic analysis of the rod pathway of the rat retina. J Comp Neurol 332(4):421–432

Cohen E, Sterling P (1990) Demonstration of cell types among cone bipolar neurons of cat retina. Philos Trans R Soc Lond B Biol Sci 330(1258):305–321

Coombs J, van der List D et al (2006) Morphological properties of mouse retinal ganglion cells. Neuroscience 140(1):123–136
Denk W, Horstmann H (2004) Serial block-face scanning electron microscopy to reconstruct three-dimensional tissue nanostructure. PLoS Biol 2(11):e329
Ding H, Smith RG et al (2016) Species-specific wiring for direction selectivity in the mammalian retina. Nature 535(7610):105–110
Dong W, Sun W et al (2004) Dendritic relationship between starburst amacrine cells and direction-selective ganglion cells in the rabbit retina. J Physiol 556(Pt 1):11–17
Dowling JE, Boycott BB (1966) Organization of the primate retina: electron microscopy. Proc R Soc Lond B Biol Sci 166(1002):80–111
Euler T, Wassle H (1995) Immunocytochemical identification of cone bipolar cells in the rat retina. J Comp Neurol 361(3):461–478
Euler T, Detwiler PB et al (2002) Directionally selective calcium signals in dendrites of starburst amacrine cells. Nature 418(6900):845–852
Famiglietti EV (1991) Synaptic organization of starburst amacrine cells in rabbit retina: analysis of serial thin sections by electron microscopy and graphic reconstruction. J Comp Neurol 309 (1):40–70
Famiglietti EV (1992) Dendritic co-stratification of ON and ON-OFF directionally selective ganglion cells with starburst amacrine cells in rabbit retina. J Comp Neurol 324(3):322–335
Famiglietti EV (2002) A structural basis for omnidirectional connections between starburst amacrine cells and directionally selective ganglion cells in rabbit retina, with associated bipolar cells. Vis Neurosci 19(2):145–162
Freed MA, Sterling P (1988) The on-alpha ganglion cell of the cat retina and its presynaptic cell types. J Neurosci 8(7):2303–2320
Fried SI, Munch TA et al (2002) Mechanisms and circuitry underlying directional selectivity in the retina. Nature 420(6914):411–414
Gan W-B, Grutzendler J et al (2000) Multicolor "diolistic" labeling of the nervous system using lipophilic dye combinations. Neuron 27(2):219–225
Ghosh KK, Bujan S et al (2004) Types of bipolar cells in the mouse retina. J Comp Neurol 469(1):70–82
Gollisch T, Meister M (2010) Eye smarter than scientists believed: Neural computations in circuits of the retina. Neuron 65(2):150–164
Greene MJ, Kim JS et al (2016) Analogous convergence of sustained and transient inputs in parallel on and off pathways for retinal motion computation. Cell Rep 14(8):1892–1900
Helmstaedter M, Briggman KL et al (2011) High-accuracy neurite reconstruction for high-throughput neuroanatomy. Nat Neurosci 14(8):1081–1088
Helmstaedter M, Briggman KL et al (2013) Connectomic reconstruction of the inner plexiform layer in the mouse retina. Nature 500(7461):168–174
Huxlin KR, Goodchild AK (1997) Retinal ganglion cells in the albino rat: revised morphological classification. J Comp Neurol 385(2):309–323
Jain V, Seung HS et al (2010) Machines that learn to segment images: a crucial technology for connectomics. Curr Opin Neurobiol 20(5):653–666
Kasthuri N, Hayworth KJ et al (2015) Saturated reconstruction of a volume of neocortex. Cell 162 (3):648–661
Keeley PW, Whitney IE et al (2007) Dendritic spread and functional coverage of starburst amacrine cells. J Comp Neurol 505(5):539–546
Kidd M (1962) Electron microscopy of the inner plexiform layer of the retina in the cat and the pigeon. J Anat 96:179–187
Kim JS, Greene MJ et al (2014) Space-time wiring specificity supports direction selectivity in the retina. Nature 509(7500):331–336
Kolb H (1970) Organization of the outer plexiform layer of the primate retina: electron microscopy of Golgi-impregnated cells. Philos Trans R Soc Lond B Biol Sci 258(823):261–283
Kolb H, Nelson R et al (1981) Amacrine cells, bipolar cells and ganglion cells of the cat retina: a Golgi study. Vision Res 21(7):1081–1114

Kong JH, Fish DR et al (2005) Diversity of ganglion cells in the mouse retina: unsupervised morphological classification and its limits. J Comp Neurol 489(3):293–310
Lee S, Zhou ZJ (2006) The synaptic mechanism of direction selectivity in distal processes of starburst amacrine cells. Neuron 51(6):787–799
Leighton SB (1981) SEM images of block faces, cut by a miniature microtome within the SEM—a technical note. Scan Electron Microsc 2(Pt 2):73–76
Linberg K, Cuenca N et al (2001) Comparative anatomy of major retinal pathways in the eyes of nocturnal and diurnal mammals. Prog Brain Res 131:27–52
MacNeil MA, Masland RH (1998) Extreme diversity among amacrine cells: implications for function. Neuron 20(5):971–982
MacNeil MA, Heussy JK et al (1999) The shapes and numbers of amacrine cells: matching of photofilled with Golgi-stained cells in the rabbit retina and comparison with other mammalian species. J Comp Neurol 413(2):305–326
Marc RE, Anderson JR et al (2014) The AII amacrine cell connectome: a dense network hub. Front Neural Circuits 8:104
Masland RH (2012) The neuronal organization of the retina. Neuron 76(2):266–280
McGuire BA, Stevens JK et al (1984) Microcircuitry of bipolar cells in cat retina. J Neurosci 4(12):2920–2938
Pallotto M, Watkins PV et al. (2015) Extracellular space preservation aids the connectomic analysis of neural circuits. elife 4
Ramón y Cajal S (1972) The structure of the retina. C. C. Thomas, Springfield
Rockhill RL, Daly FJ et al (2002) The diversity of ganglion cells in a mammalian retina. J Neurosci 22(9):3831–3843
Rust MJ, Bates M et al (2006) Sub-diffraction-limit imaging by stochastic optical reconstruction microscopy (STORM). Nat Meth 3(10):793–796
Sanes JR, Masland RH (2015) The types of retinal ganglion cells: current status and implications for neuronal classification. Annu Rev Neurosci 38:221–246
Schalek R, Kasthuri N et al (2011) Development of high-throughput, high-resolution 3d reconstruction of large-volume biological tissue using automated tape collection ultramicrotomy and scanning electron microscopy. Microsc Microanal 17(S2):966–967
Sigal Yaron M, Speer Colenso M et al (2015) Mapping synaptic input fields of neurons with super-resolution imaging. Cell 163(2):493–505
Sterling P, Freed MA et al (1988) Architecture of rod and cone circuits to the on-beta ganglion cell. J Neurosci 8(2):623–642
Stevens JK, Davis TL et al (1980a) A systematic approach to reconstructing microcircuitry by electron microscopy of serial sections. Brain Res 2(3):265–293
Stevens JK, McGuire BA et al (1980b) Toward a functional architecture of the retina: serial reconstruction of adjacent ganglion cells. Science 207(4428):317–319
Strettoi E, Dacheux RF et al (1990) Synaptic connections of rod bipolar cells in the inner plexiform layer of the rabbit retina. J Comp Neurol 295(3):449–466
Sümbül U, Song S et al. (2014) A genetic and computational approach to structurally classify neuronal types. Nat Commun 5:3512
Sun W, Li N et al (2002a) Large-scale morphological survey of rat retinal ganglion cells. Vis Neurosci 19(4):483–493
Sun W, Li N et al (2002b) Large-scale morphological survey of mouse retinal ganglion cells. J Comp Neurol 451(2):115–126
Tauchi M, Masland RH (1984) The shape and arrangement of the cholinergic neurons in the rabbit retina. Proc R Soc Lond B Biol Sci 223(1230):101–119
Taylor WR, Vaney DI (2002) Diverse synaptic mechanisms generate direction selectivity in the rabbit retina. J Neurosci 22(17):7712–7720
Turaga SC, Murray JF et al (2010) Convolutional networks can learn to generate affinity graphs for image segmentation. Neural Comput 22(2):511–538
Vaney DI (1984) 'Coronate' amacrine cells in the rabbit retina have the 'starburst' dendritic morphology. Proc R Soc Lond B Biol Sci 220(1221):501–508

Vaney DI, Collin SP et al. (1989) Dendritic relationships between cholinergic amacrine cells and direction-selective retinal ganglion cells. Neurobiology of the inner retina, Springer, pp 157–168

Volgyi B, Chheda S et al (2009) Tracer coupling patterns of the ganglion cell subtypes in the mouse retina. J Comp Neurol 512(5):664–687

Wassle H, Puller C et al (2009) Cone contacts, mosaics, and territories of bipolar cells in the mouse retina. J Neurosci 29(1):106–117

White JG, Southgate E et al (1986) The structure of the nervous system of the nematode caenorhabditis elegans. Philos Trans R Soc Lond B Biol Sci 314(1165):1–340

Zhang Y, Kim I-J et al (2012) The most numerous ganglion cell type of the mouse retina is a selective feature detector. Proc Natl Acad Sci 109(36):E2391–E2398

Chapter 3
Recent Progress in the 3D Reconstruction of *Drosophila* Neural Circuits

Kazunori Shinomiya and Masayoshi Ito

Abstract The brain of fruit fly *Drosophila melanogaster* has been used as a model system for functional analysis of neuronal circuits, including connectomics research, due to its modest size (~700 μm) and availability of abundant molecular genetics tools for visualizing neurons. Three-dimensional (3D) reconstruction of high-resolution images of neurons or circuits visualized with appropriate methods is a critical step for obtaining information such as morphology and connectivity patterns of neuronal circuits. In this chapter, we introduce methods for generating 3D reconstructed images with data acquired from confocal laser scanning microscopy (CLSM) or electron microscopy (EM) to analyze neuronal circuits found in the central nervous system (CNS) of the fruit fly. Comparisons of different algorithms and strategies for reconstructing neuronal circuits, using actual studies as references, will be discussed within this chapter.

3.1 Introduction

3.1.1 The Fly CNS as a Model System of Connectomics Study

The central nervous system (CNS) of the fruit fly *Drosophila melanogaster* has served as an important model system in neuronal circuit analysis. The fruit fly brain has advantages in its small size, the variety of available cell visualization methods, and the abundance of anatomical or developmental knowledge gained through intensive research over the past century. Research in this area significantly expanded after the development of the GAL4/UAS enhancer-trap technique, which enabled target neurons/pathways/circuits or substructures of cells to be easily visualized under light microscopy (Brand and Perrimon 1993). Beginning with the sensory and motor systems, now neural circuits in the higher order neuropils that

K. Shinomiya (✉) · M. Ito
HHMI/Janelia Research Campus, Ashburn, VA 20147, USA
e-mail: shinomiyak@janelia.hhmi.org

A. Çelik and M.F. Wernet (eds.), *Decoding Neural Circuit Structure and Function*, DOI 10.1007/978-3-319-57363-2_3

integrate multiple modalities have been intensively studied (Otsuna and Ito 2006; Kamikouchi et al. 2006; Aso et al. 2014; Wolff et al. 2015; Kim et al. 2009).

Drosophila CNS has also been an object of neuronal circuit investigation for electron microscopy (EM). Different areas of the brain have been imaged using electron microscopy to identify synaptic connections of neurons; neurons are visualized using light microscopy resolution with a higher degree of precision (Meinertzhagen and O'Neil 1991; Meinertzhagen 1996; Yasuyama et al. 2002; Marin et al. 2005). In recent years, the fly CNS has been regarded as a prospective target of local and global connectomics research, with the aim of visualizing and comprehensively reconstructing neuronal circuits in different scales. Here, we introduce the recent progress in fly CNS connectomics, focusing on both methods of visualization, light and electron microscopy, as well as on the fundamentals of reconstruction procedures.

3.1.2 Anatomy of the Drosophila CNS

The central nervous system of the fruit fly comprises the brain and the thoracico-abdominal ganglia (TAG). In the adult fly, the brain and the TAG are separated by a pair of fiber bundles called the cervical connective, whereas they are severely fused in the larval stage. The brain and TAG appear to be a single neuro-mass in the larval stage, where the TAG is more commonly referred as the ventral nerve cord (VNC) (Ito et al. 1995; Richter et al. 2010).

The brain can be further divided into the cerebral ganglia (CRG) and the gnathal ganglia (GNG) from developmental aspects. The CRG, which lies dorsally to, and around, the esophagus foramen, comprises three severely fused neuromeres; proto-, deuto-, and tritocerebra (Ito et al. 2014). The protocerebrum (PR) occupies most of the dorsal area of the brain, including synaptic neuropils such as the optic lobe (OL), neuropils, and the mushroom body (MB). The antennal lobe (AL), the primary olfactory sensory center, is the only neuropil which is unambiguously ascribed to the deutocerebrum (DE). The extent of the tritocerebrum (TR) is ambiguous, so are the boundaries between each of the neuromeres. Synaptic subcompartments of neuromeres are called neuropils, which are uniquely identified and reconstructed using the surface rendering algorithm (discussed in Sect. 3.2.2.1) with the software Amira (Mercury Inc.) (Fig. 3.1). As large-scale connectomics projects have been launched, demand to unambiguously identify projection sites of various types of neurons in the brain has led to the re-defining and reestablishment of the nomenclature system and the 3D map of the brain (Ito et al. 2014).

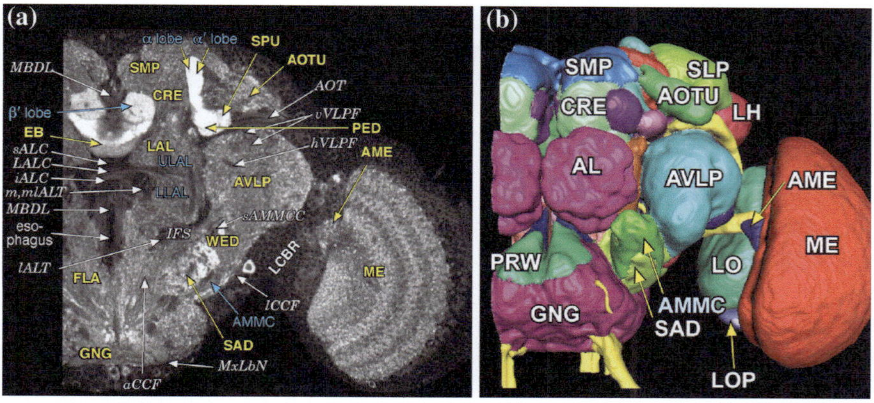

Fig. 3.1 Neuropils of the fruit fly brain. The brain neuropils of the fly were visualized with multiple methods including immunostaining, enhancer-trap, and silver staining. The names of the neuropils and their boundaries are uniquely defined and they serve as a scaffold of connectomics researches on the fly brain. **a** A frontal section of a brain hemisphere labeled with a presynaptic active zone protein conjugated with GFP (n-synaptobrevin::GFP) induced by a pan-neuronal driver elav^{C155}-GAL4. Synaptic areas (neuropils) throughout the brain are visualized in a high contrast. **b** A frontal view of a brain hemisphere, with neuropils and their boundaries visualized by the surface rendering software Amira (FEI Co., USA). The images are modified from Ito et al. (2014)

3.2 Reconstruction of Neuronal Circuits and Analysis Using CLSM Image Data

3.2.1 Visualizing Neurons with Confocal Laser Microscopy

Visualization of neurons using light microscopy (LM), in particular confocal laser scanning microscopy (CLSM), and three-dimensional reconstruction is a mainstream, and also serves as an indispensable part in recent neurobiology research. CLSM is the most widely used method to observe various types of neurons in the whole brain and has a reasonably high spatial resolution (up to several hundreds of nanometers).

In general, specifically labeling neuronal fibers is the first step in studying neurobiology, as well as connectomics. In the fruit fly, neurons can be specifically inhibited or excited to perform behavioral analyses. In addition to the aforementioned, neuronal activity can be observed by live imaging, thanks to highly sophisticated gene expression induction systems, including the GAL4/UAS enhancer-trap system (Brand and Perrimon 1993), the GAL4/UAS enhancer-fragment system (Jenett et al. 2012) and the split GAL4 system (Pfeiffer et al. 2010). Using the GAL4/UAS system, multiple reporter genes can be specifically expressed in a single neuron, or in a subset of neurons. Now tens of thousands of GAL4 lines, including strains of MZ, NP, VT and GMR libraries, are publicly available through stock centers.

3.2.2 Three-Dimensional Reconstruction of Neurons in the Fly Brain

3.2.2.1 Algorithms for Three-Dimensional Reconstruction

It is difficult and also unpractical to observe neurons and neuronal circuits in a three-dimensional image stack per se. To observe neurons imaged with CLSM in a single picture, signals on an image stack need to be projected or reconstructed so that the entire structure can be seen on a 2D plane. While there is more than one method to perform projection or reconstruction, not all of them can effectively visualize different structures labeled by reporter fluorescent proteins, or immunostaining. 3D reconstruction algorithms widely used for visualization of CLSM volume data include:

– Maximum intensity projection (MIP)

The MIP is the simplest volume rendering method which is widely used by biologists for visualizing three-dimensional image data. In the MIP algorithm, a reconstructed image is generated by selecting the maximum value ("the brightest point") through the volume data projected to a 2D plane with a parallel ray from a certain visualization angle. Many existing image processing programs use the MIP as the default projection algorithm, because the calculation amount can be minimized, and can effectively visualize bright objects. However, by using this projection, weak signals in a three-dimensionally dense volume data tend to be buried among stronger signals, and the entire image can become as if everything is uniformly painted out (Fig. 3.2a, d).

– Average intensity projection (AIP)

The AIP algorithm averages intensity of voxels of an image stack by projecting it to a 2D plane with a parallel ray. It is suitable for visualizing dense structures, such as a brain labeled with a presynaptic active zone marker (e.g., immunostaining with nc82 antibody). The AIP algorithm has the ability to observe internal structures without being obscured by bright objects (Fig. 3.2b). This algorithm is also frequently used for visualizing 3D image data by computer tomography (CT) or magnetic resonance imaging (MRI) in the medical imaging field. AIP is not used for reconstructing sparse objects such as neuronal fibers in CLSM data, as they become significantly blurred by averaging (Fig. 3.2e).

– Volume ray tracing

In volume ray tracing, a volume object is rendered by extracting information of density (and color) on a ray cast from a certain point of view. The object is projected on a 2D plane by summing intensity of voxels on the ray, while calculating transparency of each voxel at the same time. It is suitable for visualizing spatial density or concentration of an object. Since the volume ray tracing can reconstruct objects while retaining the spatial (front and back) information, which is

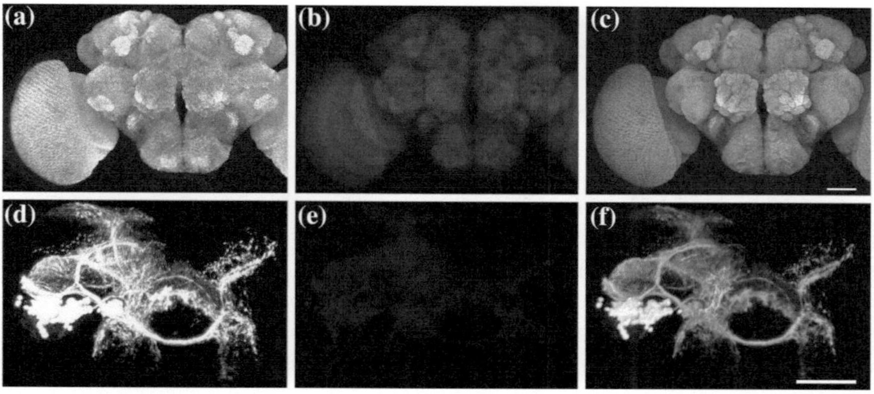

Fig. 3.2 Three different ways of 3D reconstruction. **a–c** Brain neuropils immunolabeled with the nc82 antibody. The same sample was rendered with **a** maximum intensity projection, **b** average intensity projection, **c** volume ray tracing. **d–f** Neurons of a clonal unit rendered with **d** maximum intensity projection, **e** average intensity projection, **f** volume ray tracing. *Scale bar* 50 μm

lost in the MIP, multiple neuronal fibers traversing a volume three-dimensionally can be faithfully reproduced. It also can visualize internal structures and the external surface of an object simultaneously (Fig. 3.2c, f).

– Surface rendering

The surface of an object in a volume data is extracted by thresholding signals, and is displayed as a polygonal surface with shading (Fig. 3.1b). The surface rendering is suitable for visualizing solid objects with a high contrast, such as neuropils labeled with a synaptic marker, but is not appropriate for visualizing objects with an ambiguous surface texture, like neuronal processes. Since the surface of objects is approximated by generating polygons, fine structures may not be appropriately visualized with surface rendering. In volumetric EM methods with segmentation (see Sect. 3.3.2), the surface rendering is often used to reconstruct neurons.

3.2.2.2 FluoRender: A Volume Rendering Tool Optimized for CLSM Volume Data

Most commercial 3D reconstruction software used in neuroscience research is developed to reconstruct CT/MRI image data, where different organs have their own absolute grayscale value. Neuron data imaged with CLSM usually includes weaker signals, which may be buried in fluorescent noise and may not be properly visualized if inappropriate reconstruction procedures are used.

FluoRender (http://www.sci.utah.edu/software/fluorender.html) is a 3D volume rendering program which is optimized to reconstruct multi-channel grayscale data imaged with CLSM, especially neuronal tissues, including the fruit fly brain (Wan et al. 2009). FluoRender performs graphics processing unit (GPU)-based volume rendering, which allows real-time reconstruction using the volume ray tracing or MIP algorithms. It also has a function to adjust image quality of each channel independently and interactively. It can visualize both volumetric and polygonal data on a single platform.

One of the useful functions of FluoRender is direct neuron segmentation on 3D image space. While segmentation of neurons, along with 3D reconstruction, is an essential part in light microscopy-based connectome research, segmentation of neurons of interest could only be done by tracing neuronal fibers section by section on 2D image planes (before reconstruction), or by simply cropping an image volume stack. In case of FluoRender, reconstructed single neurons or groups of neurons, such as clonal units, can be segmented directly and interactively on the rendering view using the paint tool, drastically reducing time and workload required for neuron segmentation. Segmented neurons become the basis of various subsequent analyses, especially combined with brain registration, such as synaptic site prediction (discussed in the Sect. 3.2.4.)

3.2.3 3D Registration of Brains

In connectomics research, hundreds of, or sometimes more, brains are routinely scanned to compare innervation patterns of different neuronal pathways. It is extremely difficult for a researcher to find neurons of interest among a large number of 3D brain image stacks from different animals by scanning them only by eye. In some cases, the positional relationship of multiple neurons, visualized with different GAL4 lines (e.g., if they are identical cells, or if they have direct synaptic contacts), need to be confirmed. It is theoretically possible to analyze expression patterns of multiple enhancer-trap lines in a single brain space by directly crossing them along with a reporter line, though crossing all possible combinations of hundreds of GAL4 lines is not realistic.

Expression patterns of GAL4 or other genes among multiple brains cannot be overlapped because of individual variability. Neurons from different brain samples can be directly compared in a single brain space when the brains are three-dimensionally aligned using registration algorithms. Humans can easily figure out positional relationship of neuronal fibers from multiple brains by performing registration and displaying them in a single image canvas. When neurons from multiple brains are handled in the same space, overlap between multiple neurons, indicating potential synaptic contacts, can also be detected.

In fruit fly neurobiology, a tool called Computational Morphometry Toolkit (CMTK; https://www.nitrc.org/projects/cmtk) is often used to align multiple brains (Rohlfing and Maurer 2003). CMTK aligns sample brains to a template brain, by morphing volume data of the sample brains using affine and nonrigid transformations. While the program was originally developed for aligning MRI volume data,

CLSM data, with a high contrast, can also be aligned using the same algorithm. The tool can handle single- and multi-channel CLSM data, with at least one reference channel with high contrast. The other channels are sample channels which may contain any type of signal, such as neuronal fibers, synapses, or immunostaining. The reference channel is used for positioning alignment with the template brain, and the sample channels are morphed in accordance with the reference channel.

Immunostaining, using the nc82 antibody, is frequently used as a reference channel for the fruit fly brain. nc82 is an antibody for the presynaptic active zone protein Bruchpilot, and is universally used for background staining due to its ability to label neuropils with high contrast (Wagh et al. 2006; Wichmann and Sigrist 2010).

In most cases, it is not appropriate to attempt to register brain volume data imaged with CLSM without first processing them. Misalignment can be caused by brain angle variation, variation in the signal intensity of the reference channel and from debris being present in the reference channel. If the brains are significantly deformed during dissection or image processing, they cannot be used for registration. Such variability should be removed as much as possible before alignment, using other image processing programs. Image intensity should be averaged amongst the template and sample brains. Debris, other than the brain, should be removed by cropping or masking. Once these preprocesses are complete, each channel containing brain images is saved as a separate image file and aligned to the template brain. An example of brain registration using CMTK is shown in Fig. 3.3.

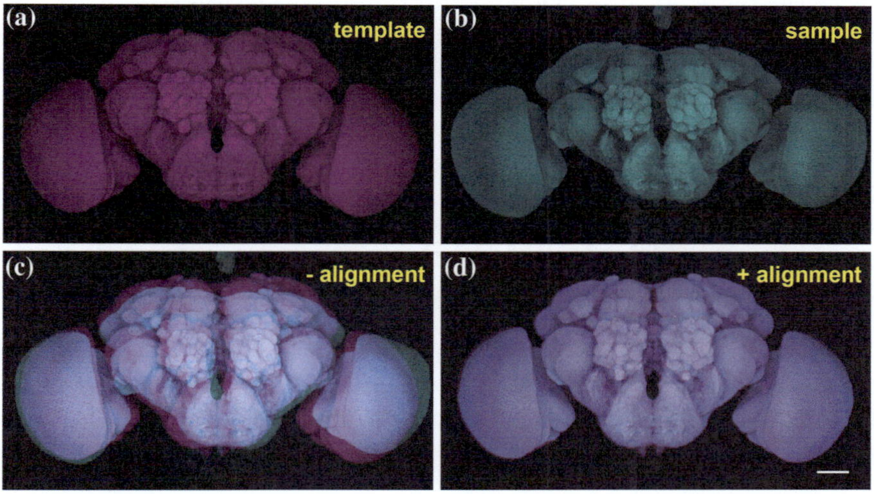

Fig. 3.3 Brain samples before and after registration. **a** The template brain and **b** a sample brain are immunolabeled with the nc82 antibody. **c** Overlay of (**a**) and (**b**) without registration. **d** The sample brain was aligned to the template brain, and overlaid on the template brain

3.2.4 Application of 3D Reconstruction and Registration

3.2.4.1 Clonal Unit Analysis

3D reconstruction and brain registration are indispensable parts of neurobiology research because of the ever-present need to compare neurons imaged from multiple brain samples. Here we provide an example study of neuroblast lineages which were three-dimensionally reconstructed after being aligned and segmented. A neural stem cell of the brain divides asymmetrically to give birth to a family of clonally associated neurons during development. A set of clonally related neurons with their specific arborization pattern is called a clonal unit, which serves as a building block in neural networks in neuropils (Ito and Awasaki 2008). The fruit fly brain consists of about 100,000 neurons and the central part of the brain, the cerebrum (which excludes the optic lobe and the gnathal ganglia), is constructed from 114 neuroblasts (Urbach and Technau 2003). These neuroblasts repeat cell divisions in two distinct developmental stages; cell divisions of the primary lineage occur in the embryonic stage, generating the larval nervous system, and the secondary lineage is generated during the late first instar larva to the early pupal stages (Ito and Hotta 1992).

Therefore, the fruit fly brain is constructed from 114 groups of progeny neurons generated from the neuroblast (through cell divisions). Morphology of all clonal units in the adult brain was identified by two independent groups through the recruitment of the *m*osaic *a*nalysis with a *r*epressible *c*ell *m*arker (MARCM) technique with a pan-neuronal GAL4 driver to label all progeny neurons for each neuroblast (Ito et al. 2013; Yu et al. 2013; Lee and Luo 1999). Chromosomal recombination, mediated by the flippase protein (FLP) induced by a heat-shock promoter (hs-FLP), was induced during the late embryonic or the early first instar larval stage to label all neurons in the secondary lineage.

114 different types of clonal units were imaged by repeating the process. 3D image data of the brains were first aligned to a standard brain template using CMTK, recruiting the nc82 channel as a reference. Neurons of interest were then segmented on FluoRender, while unnecessary signals, including single cell clones that were stochastically labeled, were removed at the same time. Finally, the clonal units were reconstructed three-dimensionally and piled up on a single template brain (Fig. 3.4).

After aligning sample neurons to the standard brain, location of the cell bodies and trajectories of neural fibers of a sample brain can be directly compared with other aligned sample brains. In the fruit fly brain, 43 neuropil regions, which were identified using various labeling methods, were mapped on the standard brain (Ito et al. 2014). Although the boundaries of the neuropils were uniquely determined by recruiting different brain structures as landmarks, it is sometimes difficult to specifically map projection sites of neurons on a volume data, especially for people who are not familiar

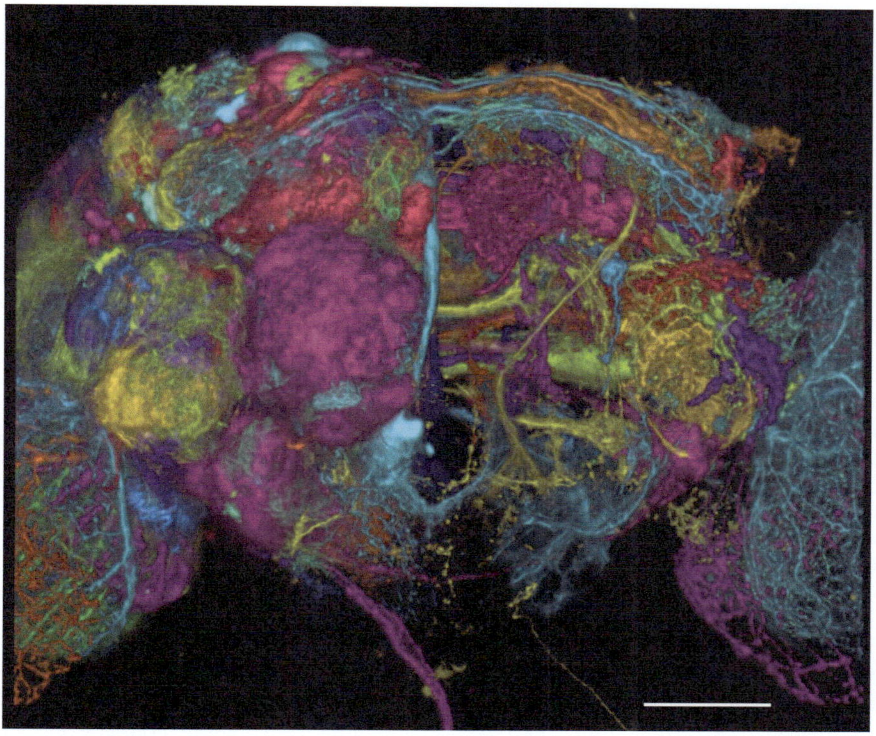

Fig. 3.4 Assembly of all clonal units in the fruit fly brain. Image stacks of brains with labeled clonal units were aligned to a standard brain template based on the neuropil structure using CMTK. Each clonal unit was carefully segmented on the 3D rendering view of FluoRender. The assembled image of clonal units was reconstructed three-dimensionally on FluoRender after optimizing graphical parameters for each sample channel. *Scale bar* 50 μm

with neuroanatomy. By aligning the clonal units to the standard brain, their projection sites can be unambiguously determined with the boundaries of the neuropils.

3.2.4.2 Prediction of Synaptic Partners

Light microscopy-based connectome research greatly depends on analysis using large-scale enhancer-trap strain libraries. In recent studies, functional neuronal circuits such as sensory information pathways have been systematically identified using a large number of GAL4 lines. However, a single GAL4 line only labels a part of the circuit in most cases, and expression patterns of multiple lines must be consolidated to visualize an entire circuit that is comprised of multiple neurons. While it is difficult to identify candidate neurons that comprise a specific neuronal circuit by comparing them using the human eye, they can easily be identified by combining registration and 3D reconstruction of neurons.

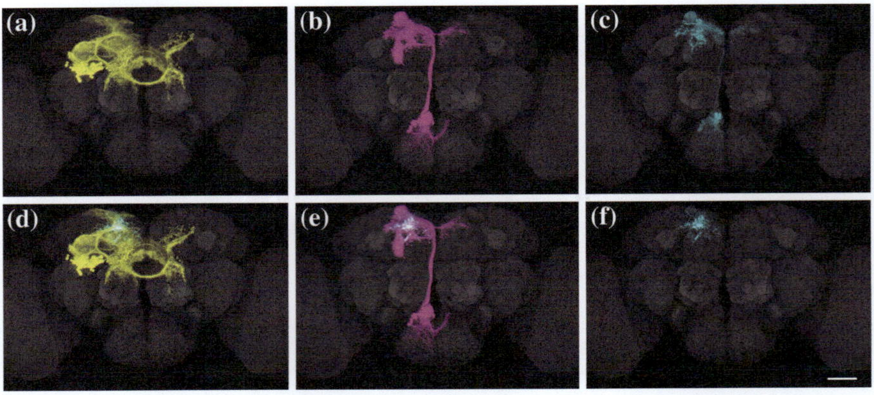

Fig. 3.5 Detecting possible synaptic contact between neurons. **a** Clonal unit innervating the anterior optic tubercle (AOTU) (*yellow*). **b** Clonal unit innervating the mushroom body (MB) (*magenta*). **c** Presynapse localization of the MB clone, visualized with n-synaptobrevin::GFP (*cyan*). **d** Possible synaptic sites with the MB clone in the AOTU clone (*yellow*: **a**, *cyan* **f**). **e** Possible synaptic sites with the AOTU clone in the MB clone (*magenta* **b**, *cyan* **f**). **f** Overlap between the AOTU clone and presynaptic sites of the MB clone (*cyan*). *Scale bar* 50 μm

When synaptic contacts exist between neurons aligned on a template brain, the signal overlap of the neurons can be observed with CLSM (Fig. 3.5). Although false-positives can happen because individual differences cannot be completely compensated by registration, large-scale screening of synaptic contacts among multiple neurons can be performed efficiently by using the aforementioned method.

3.3 Electron Microscopy-Based Reconstruction and Connectomics

3.3.1 Ultrastructures Visualized with EM

Some subcellular ultrastructure is clearly and more easily visualized using standard EM protocols than LM methods. Included are the presynaptic active zone and the postsynaptic densities (PSDs) at dendritic terminals, where neurotransmitter receptors are localized. Most presynaptic active zones of *Drosophila* are accompanied by a structure commonly referred as the “T-bar”, since it displays a T-shaped electron density that is comprised of a “pedestal” and a “platform”, when visualized with EM (Meinertzhagen 1996; Shaw and Meinertzhagen 1986; Wichmann and Sigrist 2010). Identifying both T-bars and PSDs is critically important to identify synaptic connections of neurons, as well as to acquire a synapse-level connectome. Other structures such as synaptic vesicles and cytoskeletons, in particular microtubules, are also clearly visible in most of EM methods used for connectomics, being good indicators of chemical synapses and axonal fibers, respectively.

In most large-scale connectomics projects, wild-type flies have been used for specimens, therefore neurons are more or less equally labeled. In wild-type EM samples, however, locating specific neurons in grayscale images is extremely difficult and time-consuming, as their morphology and connectivity need to first be verified by 3D reconstruction. Specific subsets of neurons can be labeled by expressing peroxidase proteins such as horseradish peroxidase (HRP) using the enhancer or promotor-trap method. Neurons expressing peroxidase are treated with diaminobenzidine (DAB), which is colorized by oxidization. Besides the plasma membrane, other organelles such as the mitochondria can be labeled in specific cell populations by expressing peroxidases conjugated with target peptides (Martell et al. 2012; Lin et al. 2016).

3.3.2 *Strategies of 3D Reconstruction*

In general, EM methods can equally visualize all neurons within an imaging range, as compared to common LM methods (using genetically induced fluorescent proteins). Therefore, compared to LM, EM methods are advantageous when identifying synaptic connections within a certain system on a large scale. For this purpose, neurons need to be reconstructed three-dimensionally from a grayscale 3D image volume, but not from several image sections. To perform 3D reconstruction from a 3D volume, proper strategies need to be selected depending on the nature of the project. Provided below are various reconstruction strategies that can be utilized.

3.3.2.1 Dense (Saturated) Reconstruction/Sparse Reconstruction

Dense (or saturated) reconstruction means reconstructing all neurons within a region of interest (ROI) (Kasthuri et al. 2015; Takemura et al. 2013). Although it usually requires a great amount of time and effort, since neuronal connections are identified completely and exhaustively. In addition to this, the number of reconstructed neurons can range from several hundred to thousands, depending on the size of the ROI. A comprehensive map of the entire region can be made once it is finished, enabling a wide-range of statistical analysis on the connectivity. Sparse reconstruction, in contrast, indicates a strategy to only reconstruct selected neurons, leaving most neurons in the ROI untraced (Hu et al. 2013; Shinomiya et al. 2014). When certain neuronal pathways or neuronal functions are the focus and the morphology of the target neurons is already known, sparse reconstruction methodology can reduce the amount of time and effort needed to complete circuit reconstruction.

3.3.2.2 Volumetric Reconstruction/Skeletonized Reconstruction

To perform volumetric reconstruction of neurons, grayscale images need to be segmented into individual profiles, using the plasma membrane as the boundary. Neurons are reconstructed by merging segmented profiles through 3D image stacks. Segmenting grayscale images, regardless if done manually or (semi-) automatically, may be time-consuming, but morphology of neurons or thickness of neuronal fibers can be the most realistically illustrated by the volumetric reconstruction.

Skeletonized reconstruction characterizes branching points or synaptic terminals as "nodes", and reconstructs the entire neuron by connecting the nodes. As neuronal fibers are typically represented as lines, thickness information may be omitted and thus the reconstructed cells will appear less realistic compared to volumetric reconstruction. The skeletonized method is more suitable for sparse reconstruction, as usually the tracing is carried out manually. Tracing speed may be faster in the skeletonized method than the volumetric, but if the grayscale images are segmented automatically and accurately, many neurons can be reconstructed volumetrically in relatively short time with less effort.

3.3.2.3 Manual Tracing/Automatic Tracing

Neuron tracing and segmentation can be done either manually or by algorithms. In volumetric reconstruction, profiles may be manually "painted" section by section, or segmented using a segmentation algorithm. Since manual tracing requires more time and effort per neuron, especially when a neuron is large and branchy, it is suitable for relatively small projects, such as targeting a smaller numbers of neurons. Automatic segmentation requires special software, and classifiers for segmentation are usually customized for each dataset. When using 3D EM techniques including FIB-SEM or SBF-SEM (see the next section), segmentation can also be performed three-dimensionally (3D segmentation), which may significantly reduce the effort for neuron tracing.

Identifying synaptic sites may also be done either manually or automatically. Synapses can be manually annotated using presynaptic T-bars and postsynaptic densities (PSDs) as landmarks. Automated synapse detection is most efficiently performed in FIB-SEM volume datasets with nearly isotropic voxels (Kreshuk et al. 2011; Huang and Plaza 2014). As the precision of prediction and the high recall of synapses cannot be fulfilled at the same time, false-positive and false-negative synapses need to be corrected after the prediction (Huang and Plaza 2014).

3.3.3 3D EM Methods in Connectomics Studies of *Drosophila* CNS

Several different 3D EM methods have been used for connectomics research on the fly CNS (Fig. 3.6). Observable structures, the difficulty of sample preparation and reconstruction of neurons, the quality of reconstructed images, and so on and so forth, may differ significantly based on the method. All of the methods are being improved in order to automate image acquisition processes.

3.3.3.1 Serial-Section Transmission EM (ssTEM)

ssTEM is the most traditional method to perform 3D reconstruction in EM (Fig. 3.6a). Ultrathin serial sections, with a thickness of 30–50 nm are prepared and scanned using TEM. After stitching the section images two-dimensionally and aligning them three-dimensionally, neurons or structures of interest are traced section by section, either manually or automatically. ssTEM is suitable for imaging relatively large areas (several tens of microns to hundreds of microns), as imaging a wider area using TEM is relatively easy. Manually cutting serial sections that contain more than several thousand sections (equivalent to tens of microns in thickness) requires a high degree of skill and is technically challenging. This problem can be solved by introducing automated sectioning systems, such as the automated tape-collecting ultramicrotome (ATUM) (Schalek et al. 2011; Hayworth et al. 2014), which allows the cutting and collecting of long serial section series relatively quickly.

Artifacts, such as folds and stains of ultrathin sections, may occur during sample preparation, and they may make tracing of neurons difficult. Such artifacts can be minimalized by employing automated section preparation techniques, but they cannot be completely avoided since each section needs to be processed individually. While samples can only be scanned once in SBF-SEM or FIB-SEM, methods which are discussed later, SBF-SEM or FIB-SEM prepared sections can be observed for multiple times in ssTEM as necessary.

The two-dimensional resolution in the *XY* plane can be 2–3 nm per pixel or higher using the ssTEM method, with which ultrastructures, such as synaptic vesicles and pre- and postsynaptic densities, can be clearly visualized and identified (Figs. 3.7a, c and 3.8). The resolution along the *Z* axis, on the other hand, is restricted by the section thickness and therefore significantly worse than the two-dimensional resolution. Because of this constraint, tracing and reconstruction of neuronal processes thinner than the section thickness running in parallel to the section planes may be difficult, as some dendritic processes of *Drosophila* neurons are as thin as 20–30 nm in diameter and therefore they can be lost during tracing. As relative positions of T-bars and corresponding dendritic terminals may also be difficult to be figured out, some PSDs may be missed due to the limitation of the spatial resolution.

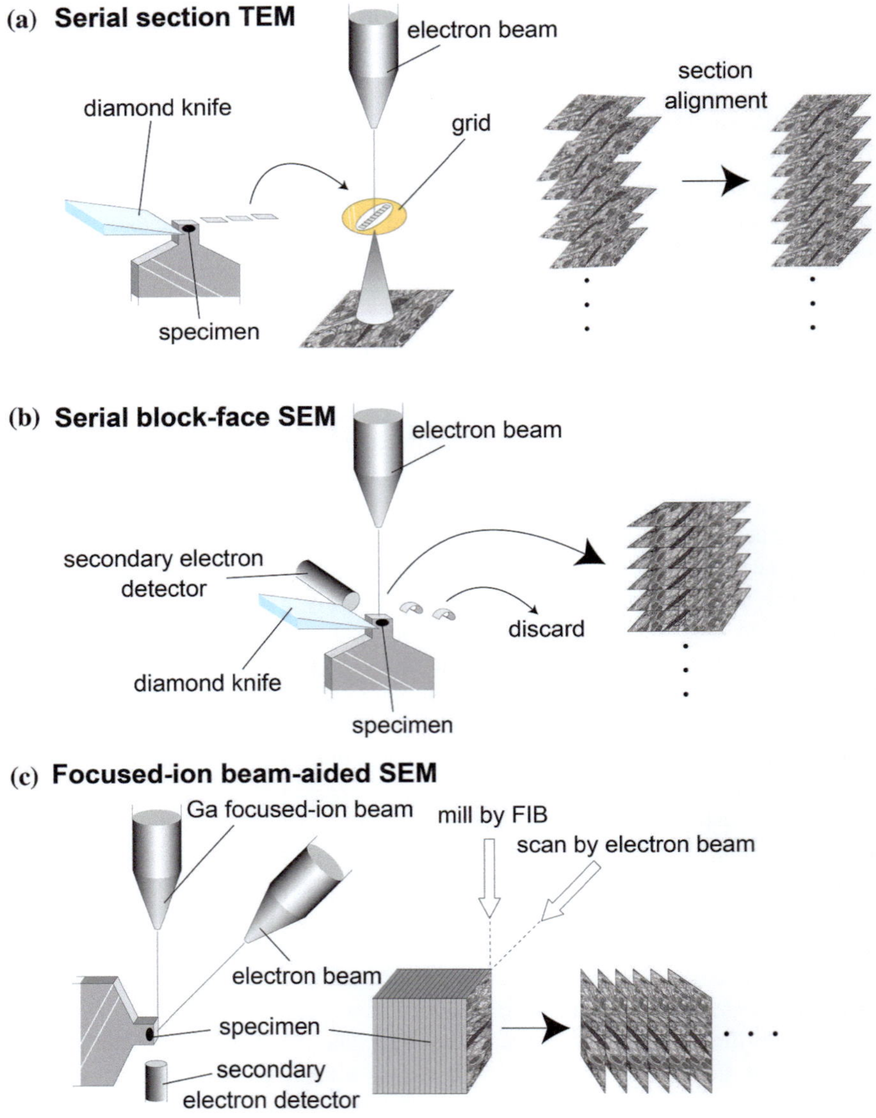

Fig. 3.6 TEM and SEM methods for 3D reconstruction. **a** Serial section transmission EM (ssTEM) requires ultrathin sectioning of a plastic-embedded specimen with a diamond knife. Sections are collected on a grid and imaged. Acquired images are computationally stitched and aligned before further analysis. **b** Specimen is continuously sectioned with a diamond knife in serial block-face scanning EM (SBF-SEM). The block-face is then scanned with an electron beam, and emission of electrons from the sample surface is detected by a secondary electron detector for imaging. Sectioned pieces are discarded. **c** Focused-ion beam-aided scanning EM (FIB-SEM) uses both an electron beam for scanning and a gallium (Ga) focused-ion beam for milling. Surface of the specimen is continuously scraped off by the FIB, and then scanned with the electron beam

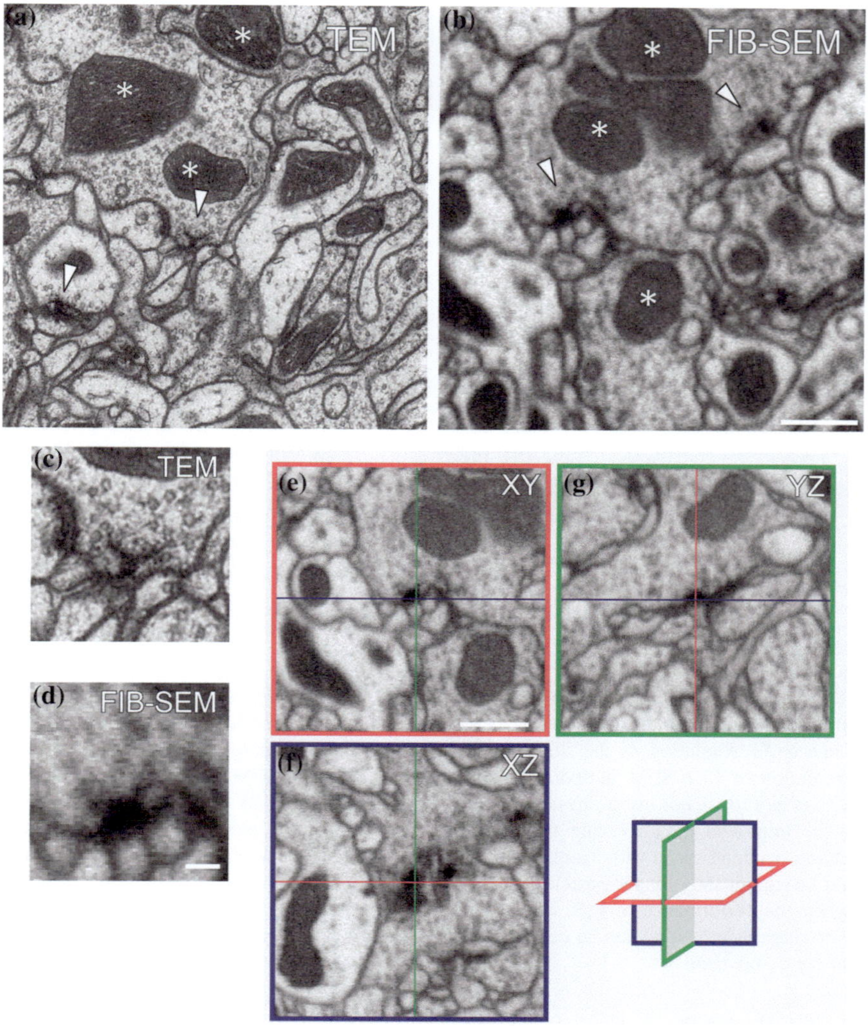

Fig. 3.7 Comparison of TEM and FIB-SEM images. **a, b** Pre- and postsynaptic terminals in the layer M10 of the medulla imaged using ssTEM (**a**) and FIB-SEM (**b**). Besides mitochondria (*asterisks*), T-bars (*arrowheads*) are sufficiently visible in the both methods as landmarks of presynaptic sites. *XY* resolution is 3.1 nm/pixel for TEM, and 8 nm/pixel for FIB-SEM. **c, d** Magnified images of T-bars. Synaptic vesicles are clearly visualized in the TEM image, while they are not very distinctive in the FIB-SEM image. **e–g** Orthogonal views of a FIB-SEM image around a T-bar. *Red frame* (e) and *lines* **(f, g)** indicate an original *XY* plane. Likewise, *blue* and *green,* respectively, indicate resliced *XZ* and *YZ* planes. In **f** and **g**, image quality is almost identical with that of the original plane regardless of reslicing. Structures of complex synapses can be figured out with multiple views. *XY* and *Z* resolutions are 8 nm/pixel. *Scale bar* **a**, **b,** and **e–g**: 500 nm, **c** and **d**: 100 nm

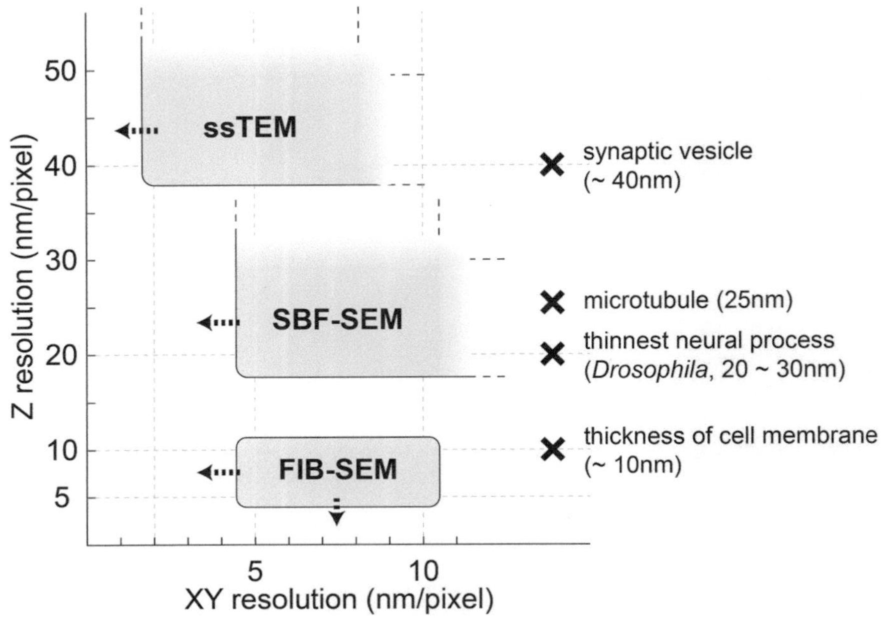

Fig. 3.8 Comparison of *XY* and *Z* resolution of different 3D EM imaging methods. ssTEM can yield better *XY* resolution compared to the SEM methods (SBF-SEM and FIB-SEM), up to around 2 to 3 nm/pixel, while its *Z* resolution is significantly inferior due to the limitation of section thickness. *XY* resolution of the SEM methods is usually 5–10 nm/pixel or less. The minimal *Z* resolution of SBF-SEM is about 20 nm/pixel, whereas FIB-SEM can typically yield 5–10 nm/pixel as it does not use a physical knife for milling. *Note* that *XY* resolution of each method may be improved by increasing the accelerating voltage of the electron beam (TEM) or by reducing scanning speed with a smaller spot (SEM) (*arrows* in the *X* axis), however, imaging time could be significantly longer especially in case of SEM. In FIB-SEM, the *Z* resolution is restricted only by the performance of FIB, and can be further improved up to several nanometers/pixel. Major ultrastructures that can be observed with EM and their approximate sizes are listed on the *right* for comparison

3.3.3.2 Serial Block-Face Scanning EM (SBF-SEM)

The processes of sectioning and image scanning are fully automated in the serial block-face SEM (SBF-SEM) method (Fig. 3.6b). A microtome is installed in a SEM chamber, with which a plastic-embedded specimen, which is continuously sectioned with a diamond knife, while the sectioned surface of the specimen is scanned with the electron beam of the SEM. Due to scattering of secondary electrons, the resolution in the *XY* plane is typically under 10 nm per pixel, which is inferior to that of TEM. However, this is good enough to recognize most of the ultrastructure needed to reconstruct connectomes, such as T-bars or synaptic vesicles. The *Z* resolution can be as good as 20–30 nm, which is limited by section thickness, reducing the chances of losing thin fibers during tracing (Fig. 3.8) (Denk and Horstmann 2004).

Since images are less distorted in the *X* and *Y* directions in the SEM methods, image alignment can be more accurately and easily done compared to that of ssTEM, and therefore volumetric reconstruction of neurons can be performed more precisely, reflecting the structures of living tissue. As the method does not directly require the handling of fragile, ultrathin sections, the risk of image loss by section damage or contamination is lower than the case of conventional ssTEM.

3.3.3.3 Focused-Ion Beam-Aided Scanning EM (FIB-SEM)

The focused-ion beam-aided SEM (FIB-SEM) method shares similar techniques with SBF-SEM (Fig. 3.6c). Instead of a diamond knife, FIB-SEM employs a focused gallium ion beam to scrape off ("mill") the surface of the embedded sample, typically yielding a *Z* resolution of 5–10 nm or higher. As the *XY* resolution is also generally 5–10 nm in SEM methods, image voxels can be isotropic using FIB-SEM, therefore the three-dimensionally reconstructed images can be observed with a constant image resolution from any angle (Fig. 3.7e–g). Complex structures such as spatial distribution of multiple T-bars and their partner PSDs can be precisely figured out (Fig. 3.7b, d). Similarly to SBF-SEM, imaging and milling processes are fully automated in FIB-SEM.

Generally, imaging of the same volume area with FIB-SEM takes longer than with ssTEM, since it requires more scans in the same thickness as the *Z* resolution becomes better. When the length of the milled sample is longer than ~ 100 μm, in the direction of the ion beam, the acquired images accumulate artifacts, such as streaks and waves of thickness variation, caused by ion polishing. Hence, constraining the size of scanning areas smaller than those found in ssTEM or SBF-SEM. For imaging and reconstructing larger areas, embedded specimens need to be cut into a smaller volume using the "hot knife" technique (discussed in Sect. 3.3.5.2) before being imaged using FIB-SEM.

3.3.4 EM Connectomics Studies in the Fly CNS

Here are several examples of different EM connectomics projects on the fly's brain or the ventral nerve cord (VNC). A general workflow of EM-based neuron or circuit reconstruction is shown in Fig. 3.9.

3.3.4.1 Connectivity Analysis in the Optic Lobe Neuropils Using ssTEM and FIB-SEM Methods

The optic lobe, situated on both sides of the brain and comprising four neuropils, called the lamina (LA), medulla (ME), lobula (LO), and the lobula plate (LOP), is a region that processes visual inputs from the photoreceptor cells located at the

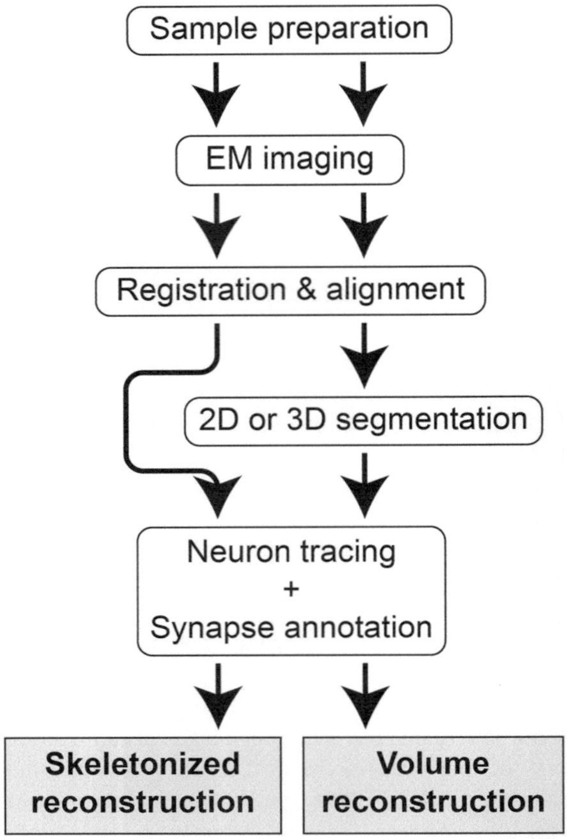

Fig. 3.9 General procedures of EM-based reconstruction. The procedures are largely shared among the ssTEM, SBF-SEM, and FIB-SEM imaging methods, although detailed processes may vary depending on actual preparation/reconstruction methods. Volume reconstruction of neurons require 2D or 3D segmentation of grayscale images, often as a preprocessing of tracing, though in cases like reconstruction using TrakEM2, neurons are segmented at the time of tracing, mainly sparsely. The steps of neuron tracing and synapse annotation, which are sometimes collectively referred as "proofreading" of neurons, commonly require a considerable amount of human effort

compound eyes. A number of different types of neurons have been identified in the optic lobe through observation using light microscopy. As the optic lobe neuropils are characterized by both tangential layered patterns and repetitive columnar units, identifying neurons and observing the synaptic contacts for these neurons is relatively easy when using EM. Also, as the thickness of each neuropil is less than 80 μm, connectivity within each neuropil can be sufficiently traced with any of the existing volumetric EM methods. Because of these advantages, the optic lobe neuropils have been investigated as a model system in EM connectomics studies.

Takemura et al. identified neurons in the medulla columns, as well as analyzed connections between these neurons using the ssTEM and FIB-SEM methods (Takemura et al. 2013, 2015). In the ssTEM project, the medulla was scanned as 1729 serial sections (each 40 nm thick), and then the acquired images were stitched and aligned to form an image stack (Chklovskii et al. 2010; Takemura et al. 2013). A region of interest (ROI) of 37 × 37 × 69 μm covering 7 medulla columns (one column and its surrounding 6 columns) was set, and neurons in the central reference column, as well as those in the surrounding columns having connections with the reference column, were densely reconstructed. As a result, 379 neurons and 8637 chemical synaptic connections between them were identified. In total, 10,093 presynaptic sites and 38,465 associated postsynaptic sites were annotated within the 7 columns, including synapses of unannotated neurons.

The acquired grayscale images were two-dimensionally segmented using an automatic segmentation algorithm, which used the cell membranes as borders. As automatic segmentation still needs to be corrected by humans, these over-segmented areas ("superpixels") were then manually agglomerated into individual neurons. Both pre- and postsynaptic densities (T-bars and PSDs), were manually identified. These processes were collectively called "proofreading". Proofreading of neurons was done manually using a custom software tool called Raveler (Olbris and Plaza 2016). The entire proofreading processes required ~14,400 person-hours (including ~2500 expert hours).

In the other project, using FIB-SEM, an area of 40 × 40 × 80 μm was imaged in the medulla, and the area of dense reconstruction was expanded to 7 columns (Fig. 3.10a). A series of 32,000 images was acquired at 10 nm per pixel in this area. The block-face of the specimen was milled by 2.5 nm with FIB in-between imaging. The images were aligned using affine transformation, and consecutive sets of four images were summed to generate the final image stack (Takemura et al. 2015). As a result, ~900 reconstructed cells, ~53,500 presynaptic sites, and ~315,500 postsynaptic sites have been identified in the 7 column area. In this project, about 1.5 more T-bars and PSDs, as compared to the ssTEM project, were identified per column, mainly due to the improvement of the *Z* axis resolution. Also, as the entire volume is segmented three-dimensionally using FIB-SEM, unlike ssTEM, time and effort for reconstructing a neuron, by connecting superpixels, were significantly reduced.

In the two projects, the authors primarily focused on a motion detection circuit in the medulla, which processes information of moving light edge, or ON-edge, and sends output to the lobula plate. By comprehensively identifying synaptic inputs to T4 cells, which integrate inputs from the medulla neurons, and by plotting the synaptic sites on a two-dimensional coordinate, they revealed that synaptic inputs to T4 from different medulla-intrinsic (Mi) and transmedulla (Tm) cells were spatially displaced, suggesting that the T4 cell, together with its input neurons, is part of an elementary motion detector, which was predicted earlier either as theoretical model or by experiment (Hassenstein and Reichardt 1956; Barlow and Levick 1965).

Other than the medulla neurons discussed above, the lamina and a part of the lobula have been so far imaged using the ssTEM technique as connectomics

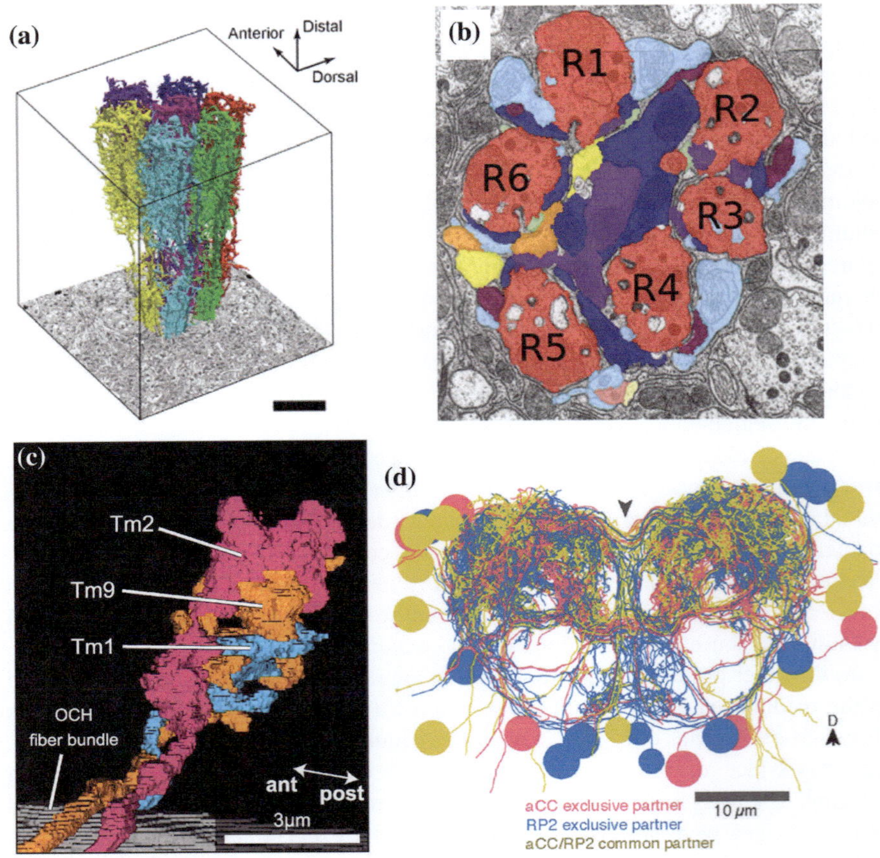

Fig. 3.10 Examples of EM-based reconstruction. **a** Three-dimensional volume reconstruction of columnar neurons in seven medulla columns from a FIB-SEM dataset. **b** Single TEM image of a lamina cartridge in which the cell profiles have been segmented and labeled with different *colors*, including the six photoreceptor cell terminals (R1–R6). **c** Volume reconstruction of three Tm cell terminals in the lobula from a ssTEM dataset. **d** Skeletonized neurons in the A3 segment of the larval VNC. Neurons are color-coded according to their presynaptic partner motor neuron cell types. Panels are adopted and rearranged from **a** Takemura et al. (2015), **b** Rivera-Alba et al. (2011), **c** Shinomiya et al. (2014), **d** Schneider-Mizell et al. (2016)

approaches with a smaller scale. Rivera-Alba et al. (2011) densely reconstructed neurons in lamina cartridges, including terminals of the photoreceptor cells, lamina neurons and three different types of glial cells, using Raveler (Fig. 3.10b). By annotating the cells and synaptic connections, they completed a comprehensive connectivity diagram of the lamina, revealing that relative placement of terminals for each cell in the cartridge is determined by the wiring economy of neurons together with the volume exclusion of synaptic terminals. In the lobula, neuronal terminals comprising a part of the moving dark edge, or OFF-edge, motion detection circuit were sparsely reconstructed using TrakEM2, a neural circuit reconstruction/visualization plugin of ImageJ/Fiji (Shinomiya et al. 2014; Schindelin et al. 2012; Cardona et al. 2012). They identified inputs from four different types of Tm cells to the T5 cell dendrites in the lobula layer 1 (Lo1), and revealed special displacement of the receptive field of the Tm cell inputs onto the T5 cells that is necessary for detecting movement, suggesting the cells comprise an elementary motion detector (EMD) circuit in parallel to the T4 pathway (Fig. 3.10c) (Shinomiya et al. 2014).

3.3.4.2 Imaging the Blowfly Brain with SBF-SEM

The SBF-SEM method has been widely used especially in connectomics research of the vertebrate CNS, such as mouse and zebrafish (Helmstaedter et al. 2011, 2013; Friedrich et al. 2013).

In Dipteran species, the brain of the blowfly, *Calliphora vicina*, has been imaged using the SBF-SEM method (Macke et al. 2008; Maack 2008). They developed a semi-automated segmentation algorithm and software packages called Neuron 2D and Neuron 3D, which bundles functions of image registration, image preprocessing (to enhance the signal/noise ratio), segmentation and 3D visualization together. Neuron 2D and Neuron 3D were used to reconstruct the first (outer) optic chiasm which connects the lamina and the medulla. With a resolution of 26.40 nm/pixel for X and Y direction and 50 nm/pixel for Z thickness, major intracellular components such as synaptic structures were distinguished. Although the resolution of the dataset was not good enough to trace the thinnest processes in synaptic neuropils, which can be less than 30 nm, the method is shown to be applicable to reconstruct tissues composed of thicker axons, such as fiber bundles. This method can be further used for reconstructing fine structures as well by enhancing the resolution.

3.3.4.3 Reconstruction of the Larval CNS with ssTEM

The central nervous system (CNS) of the *Drosophila* larva consists of the brain and the VNC, which are severely fused altogether. As the first instar larval CNS is small ($<$1 mm in the longitudinal axis) and contains a relatively small number of neurons ($\sim$10,000 cells), it has been frequently used for circuit mapping, both with electron and light microscopy.

A project to reconstruct the entire larval CNS is ongoing, and some functional pathways have been reconstructed with sparse reconstruction. Ohyama et al. (2015) investigated multimodal sensory integration pathways in the first instar larval CNS using two ssTEM series, one covering the 1.5 abdominal segments and the other covering the entire CNS, with 45–50 nm of section thickness. They identified neuronal circuits integrating mechanosensory and nociceptive inputs through sensory cells situated at the VNC. Combining the anatomical results of behavioral and physiological analyses, the circuits turned out to trigger escape locomotion in larvae, integrating the two different sensory modalities.

Several other pathways have been reported by analyzing the same ssTEM datasets. A neuronal circuit that maintains bilateral symmetrical motor patterns was identified in the VNC (Heckscher et al. 2015). A group of interneurons expressing the transcription factor Even-skipped (Eve) was identified as neurons necessary for symmetric larval locomotion. The neurons were screened by expressing the warmth-activated cation channel TRPA1 in a collection of Gal4 lines, and then their pre-and postsynaptic partners were identified using ssTEM images. In another study, an intersegmental circuit in the larval VNC which underlies the sequential muscle contraction mechanism was identified using the whole CNS ssTEM sample (Fushiki et al. 2016).

In these ssTEM projects, the web-based large image data viewer CATMAID (http://catmaid.readthedocs.io/en/stable/) was used for tracing neurons and 3D reconstruction (Fig. 3.10d) (Saalfeld et al. 2009; Schneider-Mizell et al. 2016). Neurons of interest were sparsely traced as skeletonized models, and their pre- and postsynapses were annotated. By sequentially identifying the pre- and postsynaptic neurons by the synaptic sites, entire circuits are thoroughly reconstructed. Neurons to be traced were initially identified by comparing roughly traced low-order branches of neurons in EM data with expression patterns of specific Gal4 lines in light microscopy data. Dendritic processes with postsynaptic sites are thin and usually lack microtubule cytoskeleton (thus they are called "twigs"), whereas in thicker fibers one or more microtubules are found, and therefore they are regarded as "backbones". Tracing was initiated from such thicker parts, and once the neurons were identified as the target neurons by confirming synaptic connections, they were completely traced and reconstructed up to the synaptic terminals on twigs (Fushiki et al. 2016; Schneider-Mizell et al. 2016).

3.3.5 Other Imaging Techniques for Large-Scale Connectomics

To reconstruct whole neurons or circuits composed of multiple neurons, but not only local synaptic connections, acquiring three-dimensional EM images of a relatively large (larger than roughly 100 × 100 μm) area may be required. In the case of ssTEM, imaging large areas is relatively easy, however, the Z resolution may not be good enough for tracing fine fibers and synapse identification. While the SEM methods can yield sufficient resolution, since samples embedded in the plastic are scanned *en bloc*, cracks or flaws in the block, which affect data quality, can be hardly noticed from outside before imaging. For FIB-SEM, the scanning area cannot be longer than 100 μm in the direction of the ion beam to avoid artifacts caused by ion polishing. To solve these problems, the following methods may be applied.

3.3.5.1 X-ray Tomography for Detecting Defects in Sample Blocks

Plastic-embedded specimens may include cracks or flaws during fixation and embedding. These artifacts damage tissue and interfere with 3D reconstruction of neurons when imaged using SEM methods, however, as the sample block is opaque, confirming these defects from the outside before scanning is not easy. By non-invasively checking the condition of the inside of the block, using X-ray microtomography, or "micro-CT", samples without defects can be selected in advance (Landis and Keane 2010; Mizutani et al. 2013). X-ray microtomography can yield micrometer to submicrometer resolutions, good enough for detecting cracks that may affect image quality. Since scanning of a specimen with several tens of microns of thickness takes a long time (several days to weeks) in general, the amount of time needed to get a good 3D image can be drastically reduced by carefully selecting the sample to scan before starting.

3.3.5.2 "Hot-Knife" Sectioning of Resin-Embedded Specimens for Large-Scale FIB-SEM Imaging

To avoid image disruption caused by ion polishing in FIB-SEM, large embedded samples can be divided into smaller (<100 μm) chunks before being scanned. Hayworth et al. (2015) developed a method to cut plastic-embedded samples into smaller blocks of 20 μm thickness with a diamond knife heated to 60 °C, then re-embed each slab and scanned them individually using FIB-SEM. While scanned grayscale image data is distorted due to sectioning and re-embedding, neuronal fibers running across multiple slabs can be traced with practical accuracy by correcting the distortion using a computer and stitching together the slabs of cutting planes. The process is called "hot-knife sectioning", which is expected to be applied to connectomics research. Hot-knife sectioning will be used to target larger samples, such as multiple neuropils or whole *Drosophila* brain. This technique may be utilized to subdivide very large (millimeter order) samples into smaller blocks of several tens of microns which can be imaged with current technology.

References

Aso Y, Hattori D, Yu Y, Johnston RM, Iyer NA, Ngo TT, Dionne H, Abbott LF, Axel R, Tanimoto H, Rubin GM (2014) The neuronal architecture of the mushroom body provides a logic for associative learning. Elife 3:e04577. doi:10.7554/eLife.04577

Barlow HB, Levick WR (1965) The mechanism of directionally selective units in rabbit's retina. J Physiol 178(3):477–504

Brand AH, Perrimon N (1993) Targeted gene expression as a means of altering cell fates and generating dominant phenotypes. Development 118(2):401–415

Cardona A, Saalfeld S, Schindelin J, Arganda-Carreras I, Preibisch S, Longair M, Tomancak P, Hartenstein V, Douglas RJ (2012) TrakEM2 software for neural circuit reconstruction. PLoS One 7(6):e38011. doi:10.1371/journal.pone.0038011

Chklovskii DB, Vitaladevuni S, Scheffer LK (2010) Semi-automated reconstruction of neural circuits using electron microscopy. Curr Opin Neurobiol 20(5):667–675. doi:10.1016/j.conb.2010.08.002

Denk W, Horstmann H (2004) Serial block-face scanning electron microscopy to reconstruct three-dimensional tissue nanostructure. PLoS Biol 2(11):e329. doi:10.1371/journal.pbio.0020329

Friedrich RW, Genoud C, Wanner AA (2013) Analyzing the structure and function of neuronal circuits in zebrafish. Front Neural Circuits 7:71. doi:10.3389/fncir.2013.00071

Fushiki A, Zwart MF, Kohsaka H, Fetter RD, Cardona A, Nose A (2016) A circuit mechanism for the propagation of waves of muscle contraction in *Drosophila*. eLife 5:e13253

Hassenstein B, Reichardt W (1956) Systemtheoretische analyse der zeit-, reihenfolgen-und vorzeichenauswertung bei der bewegungsperzeption des rüsselkäfers chlorophanus. Zeitschrift für Naturforschung B 11(9–10):513–524

Hayworth KJ, Morgan JL, Schalek R, Berger DR, Hildebrand DG, Lichtman JW (2014) Imaging ATUM ultrathin section libraries with WaferMapper: a multi-scale approach to EM reconstruction of neural circuits. Front Neural Circ 8:68. doi:10.3389/fncir.2014.00068

Hayworth KJ, Xu CS, Lu Z, Knott GW, Fetter RD, Tapia JC, Lichtman JW, Hess HF (2015) Ultrastructurally smooth thick partitioning and volume stitching for large-scale connectomics. Nat Methods 12(4):319–322. doi:10.1038/nmeth.3292

Heckscher ES, Zarin AA, Faumont S, Clark MQ, Manning L, Fushiki A, Schneider-Mizell CM, Fetter RD, Truman JW, Zwart MF, Landgraf M, Cardona A, Lockery SR, Doe CQ (2015) Even-skipped(+) interneurons are core components of a sensorimotor circuit that maintains left-right symmetric muscle contraction amplitude. Neuron 88(2):314–329. doi:10.1016/j.neuron.2015.09.009

Helmstaedter M, Briggman KL, Denk W (2011) High-accuracy neurite reconstruction for high-throughput neuroanatomy. Nat Neurosci 14(8):1081–1088. doi:10.1038/nn.2868

Helmstaedter M, Briggman KL, Turaga SC, Jain V, Seung HS, Denk W (2013) Connectomic reconstruction of the inner plexiform layer in the mouse retina. Nature 500(7461):168–174. doi:10.1038/nature12346

Hu T, Nunez-Iglesias J, Vitaladevuni S, Scheffer L, Xu S, Bolorizadeh M, Hess H, Fetter R, Chklovskii D (2013) Electron microscopy reconstruction of brain structure using sparse representations over learned dictionaries. IEEE Trans Med Imaging. doi:10.1109/TMI.2013.2276018

Huang GB, Plaza S (2014) Identifying synapses using deep and wide multiscale recursive networks. *arXiv preprint* arXiv:14091789

Ito K, Awasaki T (2008) Clonal unit architecture of the adult fly brain. Adv Exp Med Biol 628:137–158. doi:10.1007/978-0-387-78261-4_9

Ito K, Hotta Y (1992) Proliferation pattern of postembryonic neuroblasts in the brain of *Drosophila melanogaster*. Dev Biol 149(1):134–148

Ito K, Shinomiya K, Ito M, Armstrong JD, Boyan G, Hartenstein V, Harzsch S, Heisenberg M, Homberg U, Jenett A, Keshishian H, Restifo LL, Rossler W, Simpson JH, Strausfeld NJ, Strauss R, Vosshall LB, Insect Brain Name Working G (2014) A systematic nomenclature for the insect brain. Neuron 81(4):755–765. doi:10.1016/j.neuron.2013.12.017

Ito K, Urban J, Technau G (1995) Distribution, classification, and development of *Drosophila* glial cells in the late embryonic and early larval ventral nerve cord. Roux's Arch Dev Biol 204(5):284–307. doi:10.1007/BF02179499

Ito M, Masuda N, Shinomiya K, Endo K, Ito K (2013) Systematic analysis of neural projections reveals clonal composition of the *Drosophila* brain. Curr Biol 23(8):644–655. doi:10.1016/j.cub.2013.03.015

Jenett A, Rubin GM, Ngo TT, Shepherd D, Murphy C, Dionne H, Pfeiffer BD, Cavallaro A, Hall D, Jeter J, Iyer N, Fetter D, Hausenfluck JH, Peng H, Trautman ET, Svirskas RR,

Myers EW, Iwinski ZR, Aso Y, DePasquale GM, Enos A, Hulamm P, Lam SC, Li HH, Laverty TR, Long F, Qu L, Murphy SD, Rokicki K, Safford T, Shaw K, Simpson JH, Sowell A, Tae S, Yu Y, Zugates CT (2012) A GAL4-driver line resource for Drosophila neurobiology. Cell Rep 2(4):991–1001. doi:10.1016/j.celrep.2012.09.011

Kamikouchi A, Shimada T, Ito K (2006) Comprehensive classification of the auditory sensory projections in the brain of the fruit fly *Drosophila melanogaster*. J Comp Neurol 499(3):317–356. doi:10.1002/cne.21075

Kasthuri N, Hayworth KJ, Berger DR, Schalek RL, Conchello JA, Knowles-Barley S, Lee D, Vazquez-Reina A, Kaynig V, Jones TR, Roberts M, Morgan JL, Tapia JC, Seung HS, Roncal WG, Vogelstein JT, Burns R, Sussman DL, Priebe CE, Pfister H, Lichtman JW (2015) Saturated reconstruction of a volume of neocortex. Cell 162(3):648–661. doi:10.1016/j.cell.2015.06.054

Kim MD, Wen Y, Jan YN (2009) Patterning and organization of motor neuron dendrites in the Drosophila larva. Dev Biol 336(2):213–221. doi:10.1016/j.ydbio.2009.09.041

Kreshuk A, Straehle CN, Sommer C, Koethe U, Cantoni M, Knott G, Hamprecht FA (2011) Automated detection and segmentation of synaptic contacts in nearly isotropic serial electron microscopy images. PLoS One 6(10):e24899. doi:10.1371/journal.pone.0024899

Landis EN, Keane DT (2010) X-ray microtomography. Mater Charact 61(12):1305–1316

Lee T, Luo L (1999) Mosaic analysis with a repressible cell marker for studies of gene function in neuronal morphogenesis. Neuron 22(3):451–461

Lin TY, Luo J, Shinomiya K, Ting CY, Lu Z, Meinertzhagen IA, Lee CH (2016) Mapping chromatic pathways in the *Drosophila* visual system. J Comp Neurol 524(2):213–227. doi:10.1002/cne.23857

Maack N (2008) 3D reconstruction of neural circuits from serial EM images. Dissertation, LMU Munich, Munich, Germany

Macke JH, Maack N, Gupta R, Denk W, Scholkopf B, Borst A (2008) Contour-propagation algorithms for semi-automated reconstruction of neural processes. J Neurosci Methods 167(2):349–357. doi:10.1016/j.jneumeth.2007.07.021

Marin EC, Watts RJ, Tanaka NK, Ito K, Luo L (2005) Developmentally programmed remodeling of the Drosophila olfactory circuit. Development 132(4):725–737. doi:10.1242/dev.01614

Martell JD, Deerinck TJ, Sancak Y, Poulos TL, Mootha VK, Sosinsky GE, Ellisman MH, Ting AY (2012) Engineered ascorbate peroxidase as a genetically encoded reporter for electron microscopy. Nat Biotechnol 30(11):1143–1148. doi:10.1038/nbt.2375

Meinertzhagen IA (1996) Ultrastructure and quantification of synapses in the insect nervous system. J Neurosci Methods 69(1):59–73. doi:10.1016/S0165-0270(96)00021-0

Meinertzhagen IA, O'Neil SD (1991) Synaptic organization of columnar elements in the lamina of the wild type in *Drosophila melanogaster*. J Comp Neurol 305(2):232–263. doi:10.1002/cne.903050206

Mizutani R, Saiga R, Takeuchi A, Uesugi K, Suzuki Y (2013) Three-dimensional network of Drosophila brain hemisphere. J Struct Biol 184(2):271–279. doi:10.1016/j.jsb.2013.08.012

Ohyama T, Schneider-Mizell CM, Fetter RD, Aleman JV, Franconville R, Rivera-Alba M, Mensh BD, Branson KM, Simpson JH, Truman JW, Cardona A, Zlatic M (2015) A multilevel multimodal circuit enhances action selection in *Drosophila*. Nature 520(7549):633–639. doi:10.1038/nature14297

Olbris DJ, Plaza SM (2016) Fly EM: Raveler, https://openwiki.janelia.org/wiki/display/flyem/Raveler

Otsuna H, Ito K (2006) Systematic analysis of the visual projection neurons of *Drosophila melanogaster*. I. Lobula-specific pathways. J Comp Neurol 497(6):928–958. doi:10.1002/cne.21015

Pfeiffer BD, Ngo TT, Hibbard KL, Murphy C, Jenett A, Truman JW, Rubin GM (2010) Refinement of tools for targeted gene expression in Drosophila. Genetics 186(2):735–755. doi:10.1534/genetics.110.119917

Richter S, Loesel R, Purschke G, Schmidt-Rhaesa A, Scholtz G, Stach T, Vogt L, Wanninger A, Brenneis G, Doring C, Faller S, Fritsch M, Grobe P, Heuer CM, Kaul S, Moller OS,

Muller CH, Rieger V, Rothe BH, Stegner ME, Harzsch S (2010) Invertebrate neurophylogeny: suggested terms and definitions for a neuroanatomical glossary. Front Zool 7:29. doi:10.1186/1742-9994-7-29

Rivera-Alba M, Vitaladevuni SN, Mishchenko Y, Lu Z, Takemura SY, Scheffer L, Meinertzhagen IA, Chklovskii DB, de Polavieja GG (2011) Wiring economy and volume exclusion determine neuronal placement in the *Drosophila* brain. Curr Biol 21(23):2000–2005. doi:10.1016/j.cub.2011.10.022

Rohlfing T, Maurer CR Jr (2003) Nonrigid image registration in shared-memory multiprocessor environments with application to brains, breasts, and bees. IEEE Trans Inf Technol Biomed 7 (1):16–25

Saalfeld S, Cardona A, Hartenstein V, Tomancak P (2009) CATMAID: collaborative annotation toolkit for massive amounts of image data. Bioinformatics 25(15):1984–1986. doi:10.1093/bioinformatics/btp266

Schalek R, Kasthuri N, Hayworth K, Berger D, Tapia J, Morgan J, Turaga S, Fagerholm E, Seung H, Lichtman J (2011) Development of high-throughput, high-resolution 3D reconstruction of large-volume biological tissue using automated tape collection ultramicrotomy and scanning electron microscopy. Microsc Microanal 17(S2):966–967

Schindelin J, Arganda-Carreras I, Frise E, Kaynig V, Longair M, Pietzsch T, Preibisch S, Rueden C, Saalfeld S, Schmid B, Tinevez JY, White DJ, Hartenstein V, Eliceiri K, Tomancak P, Cardona A (2012) Fiji: an open-source platform for biological-image analysis. Nat Methods 9(7):676–682. doi:10.1038/nmeth.2019

Schneider-Mizell CM, Gerhard S, Longair M, Kazimiers T, Li F, Zwart MF, Champion A, Midgley FM, Fetter RD, Saalfeld S, Cardona A (2016) Quantitative neuroanatomy for connectomics in Drosophila. eLife 5:e12059. doi:10.7554/eLife.12059

Shaw SR, Meinertzhagen IA (1986) Evolutionary progression at synaptic connections made by identified homologous neurones. Proc Natl Acad Sci U S A 83(20):7961–7965

Shinomiya K, Karuppudurai T, Lin TY, Lu Z, Lee CH, Meinertzhagen IA (2014) Candidate neural substrates for off-edge motion detection in *Drosophila*. Curr Biol 24(10):1062–1070. doi:10.1016/j.cub.2014.03.051

Takemura S-y XuCS, Lu Z, Rivlin PK, Parag T, Olbris DJ, Plaza S, Zhao T, Katz WT, Umayam L, Weaver C, Hess HF, Horne JA, Nunez-Iglesias J, Aniceto R, Chang L-A, Lauchie S, Nasca A, Ogundeyi O, Sigmund C, Takemura S, Tran J, Langille C, Le Lacheur K, McLin S, Shinomiya A, Chklovskii DB, Meinertzhagen IA, Scheffer LK (2015) Synaptic circuits and their variations within different columns in the visual system of *Drosophila*. Proc Natl Acad Sci 112(44):13711–13716. doi:10.1073/pnas.1509820112

Takemura SY, Bharioke A, Lu Z, Nern A, Vitaladevuni S, Rivlin PK, Katz WT, Olbris DJ, Plaza SM, Winston P, Zhao T, Horne JA, Fetter RD, Takemura S, Blazek K, Chang LA, Ogundeyi O, Saunders MA, Shapiro V, Sigmund C, Rubin GM, Scheffer LK, Meinertzhagen IA, Chklovskii DB (2013) A visual motion detection circuit suggested by *Drosophila* connectomics. Nature 500(7461):175–181. doi:10.1038/nature12450

Urbach R, Technau GM (2003) Molecular markers for identified neuroblasts in the developing brain of Drosophila. Development 130(16):3621–3637

Wagh DA, Rasse TM, Asan E, Hofbauer A, Schwenkert I, Durrbeck H, Buchner S, Dabauvalle MC, Schmidt M, Qin G, Wichmann C, Kittel R, Sigrist SJ, Buchner E (2006) Bruchpilot, a protein with homology to ELKS/CAST, is required for structural integrity and function of synaptic active zones in Drosophila. Neuron 49(6):833–844. doi:10.1016/j.neuron.2006.02.008

Wan Y, Otsuna H, Chien CB, Hansen C (2009) An interactive visualization tool for multi-channel confocal microscopy data in neurobiology research. IEEE Trans Vis Comput Graph 15(6):1489–1496. doi:10.1109/TVCG.2009.118

Wichmann C, Sigrist SJ (2010) The active zone T-bar–a plasticity module? J Neurogenet 24(3):133–145. doi:10.3109/01677063.2010.489626

Wolff T, Iyer NA, Rubin GM (2015) Neuroarchitecture and neuroanatomy of the Drosophila central complex: a GAL4-based dissection of protocerebral bridge neurons and circuits. J Comp Neurol 523(7):997–1037. doi:10.1002/cne.23705

Yasuyama K, Meinertzhagen IA, Schurmann FW (2002) Synaptic organization of the mushroom body calyx in *Drosophila melanogaster*. J Comp Neurol 445(3):211–226

Yu HH, Awasaki T, Schroeder MD, Long F, Yang JS, He Y, Ding P, Kao JC, Wu GY, Peng H, Myers G, Lee T (2013) Clonal development and organization of the adult Drosophila central brain. Curr Biol 23(8):633–643. doi:10.1016/j.cub.2013.02.057

Chapter 4
Connectivity and Circuit Architecture Using Transsynaptic Tracing in Vertebrates

Kazunari Miyamichi and Lindsay A. Schwarz

Abstract The functions of the brain—such as sensory perception, memory formation, and behavioral responses—are based on the activity patterns of large numbers of interconnected neurons that form information-processing neuronal circuits. Most brain areas contain diverse types of neurons with specific morphology, gene expression profiles, input/output connectivity, and physiological response profiles. One major goal of neuroscience is to decipher connection patterns among different brain regions and cell types at the scale of the entire brain while keeping synaptic resolution. In this chapter, we first review various circuit tracing methods, and then introduce rabies virus (RV)-mediated transsynaptic tracing methods, which allow one to identify presynaptic neurons of genetically, anatomically, or functionally defined target neurons in a given brain area. This is achieved by genetic control of 'starter' cells, from which retrograde transsynaptic spread of RV occurs for only a single synaptic step. We will detail diverse methods that have been developed to restrict starter cells to a unique neuronal type. Following an introduction of RV transsynaptic tracing, the applications of these tools to three diverse biological systems in mice will be discussed: olfaction, neuromodulation, and motor control. From these examples, we will review how RV-mediated transsynaptic tracing has begun to decipher complex circuit architectures throughout the brain and spinal cord, and provides an important link between neuronal connections and circuit function.

K. Miyamichi (✉)
ERATO Touhara Chemosensory Signal Project Group,
The University of Tokyo, Tokyo, Japan
e-mail: amiyami@mail.ecc.u-tokyo.ac.jp

L.A. Schwarz
Department of Developmental Neurobiology,
St. Jude Children's Research Hospital, Memphis, TN, USA

A. Çelik and M.F. Wernet (eds.), *Decoding Neural Circuit Structure and Function*, DOI 10.1007/978-3-319-57363-2_4

4.1 Introduction

In the mammalian brain, billions of neurons with highly diversified cell types form neuronal circuits by making trillions of synapses. One major goal of neuroscience is to decipher the basic organization and connection patterns among different brain regions and cell types. Diverse methods have been developed to contribute to this goal. Figure 4.1 shows a schematic comparison of multiple methods in terms of the limit of resolution in the presynaptic side (*x*-axis) and the postsynaptic side (*y*-axis).

4.1.1 Methods for Mapping of Neuronal Circuits at Region-to-Region Resolution

The first category (purple in Fig. 4.1) is characterized by region-to-region resolution in mapping neuronal circuits. For example, axon degeneration [also known as Wallerian degeneration after Waller's histological finding that sectioning the nerve caused degeneration of axons distal to the injured site (Waller 1850)] has greatly contributed to our understanding of coarse brain organization. It cannot distinguish, however, multiple intermingled cell types on the presynaptic side or pinpoint the postsynaptic partners that the degenerating nerves innervated before induction of the injury.

The second category (cyan) is classical neuronal tracers, which can be injected locally into a brain region, without causing damage, to visualize region-to-region connections in an intact brain (for review, Vercelli et al. 2000; Nassi et al. 2015).

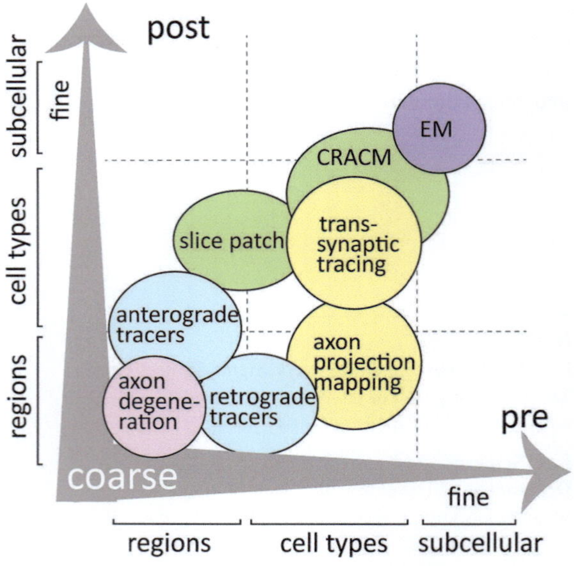

Fig. 4.1 Schematic comparison of various mapping methods. The resolutions of each mapping method on the presynaptic side (*x*-axis) and postsynaptic side (*y*-axis) are mapped. For details, see Sects. 4.1.1–4.1.5

Importantly, some of these tracers exhibit direction selectivity: retrograde tracers, such as Fluoro-gold (Schmued and Fallon 1986) and Retrobeads (Waselus et al. 2006), can be preferentially taken up by axons at the injection site and transported back to the cell bodies of these neurons, which may be located in distant brain areas. In contrast, anterograde tracers, represented by biotinylated dextran amine (Veenman et al. 1992) and isotope labeled amino acids (e.g., ^{3}H-leucline; Cowan et al. 1972), can be taken up by cell bodies and dendrites at the injection site and then spread through the neuron to label their axons. These methods do not distinguish between different cell types intermingled at the injection site and therefore are primarily categorized as region-to-region resolution tracers. When combined with other histochemical and electrophysiological methods, however, these tracers can provide additional characterization of labeled neurons. For example, neurons labeled with a retrograde tracer can then be stained with cell type specific markers or have their electrophysiological properties determined, in order to better characterize cell types (for example, see Fig. 4.2a).

4.1.2 Viral and Genetic Approaches for Axon Mapping

The third category in Fig. 4.1 (yellow) represents viral and genetic methods for mapping neuronal circuits that provide genetic control for cell types from which the tracing is initiated. One method in this category is axon projection mapping, which allows the labeling of entire axonal projections from genetically defined neurons located in a specific brain area. An example shown in Fig. 4.2b represents the axon projection mapping from serotonin neurons located in the dorsal raphe nucleus. GFP-labeled axons from these neurons are visualized throughout the brain, but are enriched in selected nuclei such as the central amygdala (CeA) (for details of serotonin circuit organization, see Sect. 4.4.1).

Utilizing differences in gene expression between cells is a powerful way to define cells types in the brain. To gain access to specific types of neurons expressing a *gene X*, researchers have generated a large number of transgenic and knock-in mouse lines where expression of a fluorescent protein gene is under the control of the promoter region of *gene X* (Fig. 4.3a). For example, to visualize inhibitory GABAergic neurons in the brain, *GAD2-GFP* mice were generated (Tamamaki et al. 2003), where *GAD2* stands for glutamate decarboxylase 2 gene, which encodes the specific enzyme that synthesizes GABA. Although powerful for in vivo identification of GABAergic neurons, this mouse alone cannot be used to precisely map axons of individual GABAergic neurons, as it labels millions of neurons throughout the brain. To provide greater spatial resolution, a better approach utilizes local infection by viral vectors (Fig. 4.3b). Tables 4.1 and 4.2 summarize the viral vectors that are often used in circuit mapping. Among them, adeno-associated virus (AAV) is particularly useful, as it can be easily constructed, is safe to handle, and stably drives transgene expression in cells, without apparent cytotoxicity, for months to years. The major drawback of the AAV vector is its

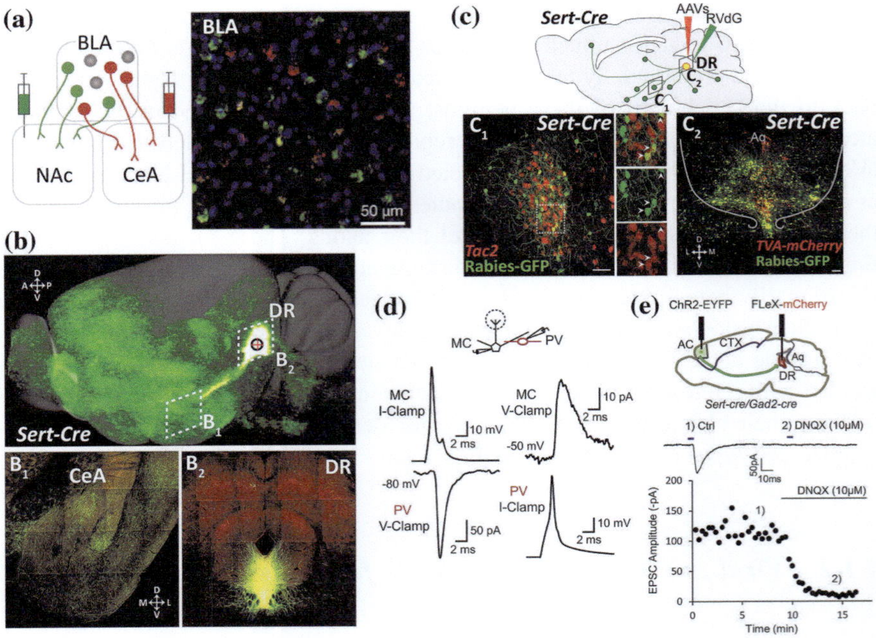

Fig. 4.2 Various examples of mapping methods. **a** Retrograde labeling of neurons in the basolateral amygdala (*BLA*) projecting to either nucleus of accumbens (*NAc*, *green*) or central amygdala (*CeA*, *red*). Two different colors of Retrobeads were used in this study. Researchers analyzed electrophysiological properties of these labeled neurons following fear or reward conditioning. Adapted with permission from Namburi et al. (2015). **b** Axon projection mapping (Sect. 4.1.2) from serotonin neurons in the dorsal raphe nucleus (*DR*, $\mathbf{B_2}$). In this sample, a Cre-dependent AAV vector (Fig. 4.3c–d) expressing GFP was injected into the DR of *Sert-Cre* mice in which Cre is expressed in serotonin neurons. Dense axon arborization was visualized throughout the brain, including central amygdala (*CeA*, $\mathbf{B_1}$). Images were taken with permission from Allen Mouse Connectivity Atlas at http://connectivity.brain-map.org/ (sample#114155190). Also see Oh et al. (2014). **c** An example of rabies virus (*RV*)-mediated transsynaptic tracing (Sects. 4.1.3 and 4.2) starting from the serotonin neurons in the DR using *Sert-Cre* mice ($\mathbf{C_2}$). Presynaptic neurons were labeled throughout the brain, including *Tac2* positive neurons in the CeA ($\mathbf{C_1}$). Adapted with permission from Weissbourd et al. (2014). **d** An example of paired recordings (Sect. 4.1.4) from a synaptically connected mitral cell (*MC*) and parvalbumin neuron (*PVN*) in an acute slice of the olfactory bulb. *Left* an action potential in the MC evokes excitatory postsynaptic current (*EPSC*) in the PVN. *Right* in the same pair of cells, an action potential in the PVN evokes an inhibitory postsynaptic current in the MC. This data unambiguously determines that these cells are monosynaptically connected. Adapted with permission from Kato et al. (2013). See Sect. 4.3.1. for details. **e** An example of ChR2-assisted circuit mapping (*CRACM*, Sect. 4.1.4) that shows direct monosynaptic excitatory input from the anterior cortical (*AC*) areas to serotonin and GABAergic neurons in the DR. EPSCs, generated by photostimulation, (*1*, *blue bar*) are abolished by application of DNQX (*2*), an AMPA receptor agonist. Top traces are the average of six trials from the same serotonin neurons. *Bottom graph* shows the change in EPSC amplitude over time. As exemplified in this case, CRACM determines monosynaptically connected neurons over a distance, with the ability to characterize their synaptic properties. Adapted with permission from Weissbourd et al. (2014). *A* anterior; *P* posterior; *D* dorsal; *V* ventral; *L* lateral; *M* medial. Scale bar in panel $\mathbf{C_1}$ corresponds to 100 µm

limited capacity; an AAV can only accommodate up to 4.7 kb of transgene, including the promoter and transcriptional stop cassette, into its genome. Therefore, in many cases, the limited AAV capacity does not allow inclusion of the full promoter region that is necessary for restricting transgene expression in the defined cell type. How can this problem be solved?

Researchers found an innovative solution by combining mouse genetics and viral vectors, to create an approach now recognized as viral and genetic technology (Fig. 4.3c). In 2003, the first Flip-Flop Excision (*FLEx*) switch (Fig. 4.3d) was invented (Schnutgen et al. 2003), where a transgene is initially placed in the

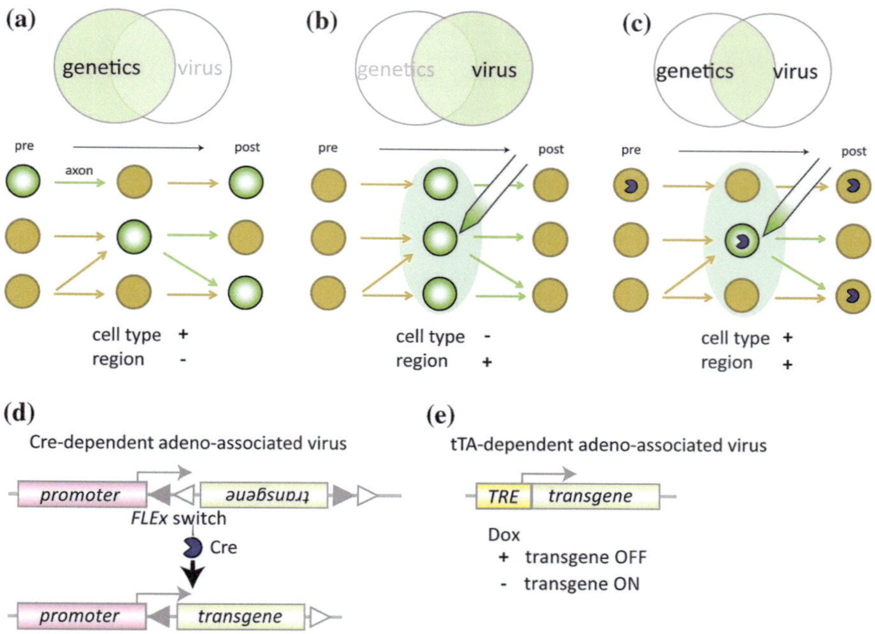

Fig. 4.3 Schematic representation of viral and genetic technology and Cre or tTA dependent AAV vectors. **a–c** In this simplified neuronal circuit, three brain regions (*columns*) each containing three neurons are connected by axons (*arrows*) as indicated. Genetically targeted cells (for example, labeled by a fluorescent protein) are represented in *green*. Genetic methods can label a specific neuronal type in each brain area, but labeling is often widely distributed in the brain (**a**). Virus injection can provide better regional resolution, but it is difficult to regulate which cell types in the injection site express the transgene (**b**). In viral and genetic technology, Cre-dependent adeno-associated virus (*AAV*) is injected into the target brain region of a transgenic mouse line expressing Cre (shown as *blue PacMan symbols*). Local injection of AAV provides regional regulation and Cre expression provides the control of cell type-specific labeling. **d** Cre-dependent AAV using the Flip-Flop Excision (*FLEx*) switch. *Gray and white triangles* represent *loxP* and its mutant *lox2272*. Cre-mediated recombination between two *loxP* sites and between two *lox2272* sites guarantees unidirectional inversion of a transgene. **e** Tetracycline transactivator (*tTA*)-dependent AAV driven by tetracycline responsive element (*TRE*) promoter. A transgene under the control of *TRE* promoter is activated by tTA, whose activity is blocked in the presence of doxycycline (*Dox*)

Table 4.1 Viruses that are replication-deficient (Viral vectors that are unable to replicate at all are called replication-deficient. Unlike their parental strains, replication-deficient viral vectors are usually nontoxic to the infected cells, and therefore can stably transduce transgene expression.) and do not spread across cells

Species	Genetic materials	Typical applications	Form of transduction	Cytotoxicity
Lentivirus	Positive single-strand RNA	Stable transgene introduction into host genome	Mostly local by using VSV glycoprotein	–
Gammaretrovirus		Transgene introduction into dividing cells	Mostly local	–
Adeno-associated virus (AAV)	Single-strand DNA	Stable transgene expression (without viral integration into host genome)	Mostly local infection but some serotypes support retrograde transduction	–
Canine adenovirus type 2 (CAV2)[a]	Double strand DNA	Retrograde transgene expression from axons	From axons (retrograde)	–
Human adenovirus 5		Stable transgene expression (without viral integration into host genome)	From axons (retrograde)	–
Helps simplex virus (HSV1)[a]	Double strand DNA	Retrograde transgene expression from axons	From axons (retrograde)	–
Pseudorabies virus (PRV)[a]				

[a]Replication-deficient forms of CAV2 (Soudais et al. 2001), HSV1 (Spaete and Frenkel 1982) and PRV (Oyibo et al. 2014) are useful retrograde vectors; their particles are taken up by axons and retrogradely transported to the cell bodies of neurons, allowing transgenes to be introduced into neurons from their projection targets. We will discuss a specific application of CAV2 vector in Sect. 4.2.4

opposite direction relative to the promoter activity, rendering it transcriptionally inactive. Only after Cre recombinase irreversibly inverts the *FLEx* switch can the transgene be expressed. This compact and efficient Cre-dependent switch has been widely used with great success in many applications of AAV vectors.

Typically, for virus-mediated axon projection mapping (Figs. 4.2b, and 4.3c, d), an AAV vector expressing a fluorescent marker protein (e.g., GFP) in a Cre-dependent manner is injected into a target brain area of a transgenic mouse, where Cre is expressed under a specific genetic promoter. Thus, the fluorescent marker will only be expressed in the specific cell types that also contain Cre. In this way, axons of the defined cell type in the defined brain region can be precisely traced throughout the brain. Many brain areas have been mapped in this way using a large number of different Cre lines (for example, see Allen Mouse Brain Connectivity Atlas at http://connectivity.brain-map.org/). Although this method

Table 4.2 Viruses that spread across cells

Species	Genetic materials	Synapse specificity	Typical applications	Direction	Cytotoxicity
Herpes simplex virus (HSV1)	Double strand DNA	Some data show nonspecific spill over[a]	Multistep tracing; Cre-dependent version can be used to control starter cells of multistep tracing	Mixed. H129 strain supports anterograde spread	+++
Pseudorabies virus (PRV)		Not determined		Mixed. Bartha strain supports retrograde spread	+++
Rabies virus (RV)	Negative single-strand RNA	Supported by both in vivo[b] and ex vivo[c] experiments	Replication-competent version is used for multistep retrograde tracing. Replication-conditional version offers monosynaptic tracing	Retrograde spread (except sensory neurons)	+[d]
Vesicular stomatitis virus (VSV)		Supported by ex vivo[c] experiments	Multistep or monosynaptic tracing to either anterograde or retrograde, depending on the type of glycoprotein. Cytotoxicity of VSV is severer than RV in rodents, but may be milder in other species	Mixed, but can be controlled by the type of glycoprotein	+++

[a]For nonspecific spillover of HSV1, see Sect. 4.2.1 and Ugolini et al. (1987), Ugolini (2011). [b]RV spreads to synaptic partners (Ugolini 2011; Callaway and Luo 2015) without ‘spillover’ to nonconnecting structures (Ugolini 1995) or cell types (Reardon et al. 2016; Miyamichi et al. 2011; see also Sects. 4.2.1 and 4.5.1). [c]Paired recordings from virally labeled cultured neurons support synapse specificity of virus spread in the case of RV (Wickersham et al. 2007a) and VSV (Beier et al. 2011; see also Sect. 4.2.2). [d]Different strains of RV exhibit a range of cytotoxicity, with a strain called CVS being less toxic (Reardon et al. 2016)

provides specificity for the cell types that are labeled on the presynaptic side of a circuit, it cannot pinpoint the postsynaptic cell types to which the labeled axons connect. How might one further improve this spatial resolution?

4.1.3 Transsynaptic Tracing with Genetic Control

Transsynaptic tracing offers genetic control of the cells from which the tracing is initiated (called starter cells), and it can label synaptically connected cells throughout the nervous system. Labeled neurons can then be characterized by location, morphology, gene expression patterns, electrophysiological properties, and receptive fields. Figure 4.2c shows an example of rabies virus (RV)-mediated transsynaptic tracing, which will be discussed in greater detail in Sect. 4.2.

Transsynaptic tracing can be classified into two types: protein-based and virus-based methods. Protein-based tracers, represented by Wheat Germ Agglutinin (WGA) (Schwab et al. 1978), can be targeted into genetically defined cells, from which it is transported both to upstream (retrograde) and downstream (anterograde) neurons, presumably via synaptic connections (Yoshihara et al. 1999). As WGA is not toxic to cells, this method allows one to map neuronal circuits in intact animals in vivo with the potential to perform histochemical analyses on traced cell types. The major drawback of this strategy, however, is the fact that the tracer protein is easily diffused below a detection threshold because it is only generated in the starter cells. Also, this method cannot control the direction of labeling (upstream versus downstream). Finally, it is not fully established if the transneuronal transfer of WGA is indeed restricted to the synaptic connections.

The second category of transsynaptic tracing includes viral tracers. Table 4.2 summarizes the most commonly used vectors. Compared with protein-based tracers, viral tracers can be replicated in each step of cell-to-cell spread and therefore do not suffer from decreased labeling intensity due to diffusion. There are five basic properties to be considered when comparing viral tracers: (a) direction preference, (b) synapse specificity, (c) tropism, (d) cytotoxicity, and (e) availability of replication-conditional virus. Similar to classical chemical tracers, some viral tracers exhibit direction preference, that is, they dominantly move in either the anterograde or retrograde direction. For example, RV is a typical retrograde tracer (Ugolini 1995), whose direction preference is determined by the glycoprotein on the surface of the viral particles (Beier et al. 2011). A specific strain of pseudorabies virus (PRV) called Bartha strain also exhibits retrograde direction preference (Card et al. 1992). Although herpes simplex virus type 1 (HSV1) spreads both in retro- and anterograde directions, a specific strain of HSV1, named HSV-H129, is known to travel primarily in the anterograde direction (Zemanick et al. 1991; Sun et al. 1996). When viruses have clear direction specificity, researchers can easily map the labeled cells relative to the starter cells to create precise connection diagrams.

Although viral tracers travel from neuron to neuron (referred to as 'transneuronal'), this alone does not guarantee synapse specificity of the viral spread, meaning that neurons can be labeled that are not directly connected to starter cells. Several viral vectors have been characterized that vary in terms of their synapse specificities for transneuronal spread. Thus far, synapse specificity of RV spread has been the most intensively characterized in various contexts, including in vivo circuit mapping (see Sects. 4.2.1 and 4.5.1) and ex vivo slice cultures (see Sect. 4.2.2).

When using viral tracers, it is also important to keep in mind that not all neurons in the brain are equally infected by a given viral tracer. The preference of viruses for certain cell types is called viral tropism. Currently the full tropism of transsynaptic viral tracers is not established and therefore caution is required when comparing labeling of different cell types in the brain. Furthermore, unlike AAV, most transsynaptic viral tracers that have been characterized so far are toxic to the infected cells. The cytotoxicity of viral tracers is an important issue if one wants to perform further analyses on infected neurons, such as electrophysiological recording or gene expression analysis, or use infected animals for behavioral experiments. Different viral tracers show a range of cytotoxicity (see Table 4.2). For example, cytotoxicity of RV is low to modest and therefore researchers can conduct Ca^{2+} imaging of RV-positive neurons several weeks after the infection (Osakada et al. 2011; Reardon et al. 2016). In contrast, HSV or PRV is usually more toxic, with signs of toxicity observed in the infected cells within a few days (Ugolini et al. 1987; Ugolini 2011).

Lastly, replication-conditional viral tracers are very useful for controlling viral spread. Wild-type viral tracers have full capacity to replicate (referred to as replication-competent). For some viral tracers, a single gene has been identified that is necessary for viral replication. This has enabled researchers to delete or transcriptionally inactivate that gene to generate a crippled virus that cannot replicate or propagate. However, replication of the virus can still occur if the missing gene is supplied in *trans*, or transcriptional inactivation of the gene is removed. This modified virus is called a replication-conditional virus. For example, replication-conditional HSV-H129 (Lo and Anderson 2011) and Bartha strain of PRV (DeFalco et al. 2001) were generated by transcriptional silencing of an essential thymidine kinase (*tk*) gene. Once the transcription of *tk* gene from the virus genome is restored by Cre-mediated recombination, the functional virus particles are generated in the Cre+ starter cells and spread in the anterograde (HSV-H129) or retrograde (PRV-Bartha) direction. As we will discuss in Sect. 4.2, replication-conditional RV is generated by deleting a single essential glycoprotein gene from the RV genome. By supplying rabies glycoprotein in *trans* only in the starter cells, viral replication is restricted to the starter cells, and therefore allows monosynaptic spread of the virus from these cells. Replication-conditional viral tracers offer higher resolution and precision in mapping neuronal connections.

4.1.4 *Electrophysiological Methods for Mapping Neuronal Circuits*

In Fig. 4.1, the fourth category (green) represents electrophysiological methods to map circuits. For local circuits, applying patch-clamp methods in brain slice preparations, combined with genetic labeling of neurons, can identify synaptically connected pairs of cells and also provides information regarding their cellular identity. Figure 4.2d shows a paired recording from a parvalbumin positive interneuron that is reciprocally connected to a mitral cell in the mouse olfactory bulb (OB) (Kato et al. 2013). This method allows highly precise mapping of connected neurons, although it can only be applied to a local circuit. In principle, bulk stimulation of axon bundles while recording from the postsynaptic neuron can map long-range connections with region-to-cell resolution. However, this approach cannot distinguish between intermingled axons from different cell types in the presynaptic structure that may be innervating the postsynaptic cell. The development of optogenetics (Deisseroth 2015) has greatly improved the resolution on the presynaptic side. Channelrhodopsin-2 (ChR2), a light activated cation channel isolated from a green alga, can be targeted into genetically defined neurons, for example by injecting Cre-dependent AAVs expressing ChR2 into a transgenic animal where Cre is expressed in specific classes of cells (Fig. 4.3c, d). In this way, the cell bodies and axons of a defined cell type, located in a defined brain region, can be activated by light. Simultaneously, one can monitor electrical responses in the presumed postsynaptic neurons with patch electrodes. If there is a monosynaptic connection between the ChR2-expressing neurons and the neuron being recorded from, an excitatory or inhibitory postsynaptic current should be observed immediately (within several milliseconds) after the onset of the light stimulation. By observing how long it takes the postsynaptic cell to respond after ChR2 activation, researchers can distinguish between monosynaptic and polysynaptic connections. This method, referred to as ChR2-Assisted Circuit Mapping (CRACM) (Petreanu et al. 2007), can identify pre- and postsynaptic partners, and provide additional information regarding their synaptic properties and cell types (defined by gene expression patterns). As an example, Fig. 4.2e shows a direct excitatory monosynaptic connection from frontal cortex pyramidal cells to serotonin neurons in dorsal raphe nucleus (Weissbourd et al. 2014). In this case, the distance between the connected cells is >7 mm. CRACM can also characterize which subcellular structures on the postsynaptic side are targeted by the ChR2-expressing axons, if light activation is restricted to a very small brain region. Given these advantages, CRACM is a standard way to map synaptic connections in many brain regions. One major drawback of CRACM is its low-throughput nature: in a single recording experiment, only one type of neuron can be labeled on the presynaptic side, while electrophysiological methods limit the number of neurons sampled on the postsynaptic side.

4.1.5 *Ultrastructural Method to Map Synaptic Connections*

The last categories of Fig. 4.1 (blue) are ultrastructural methods, represented by electron microscopy (EM). EM is a highly reliable method to determine if two cells form synaptic connections. In principle, complete reconstructions of serial EM images can reveal neuronal circuits with synaptic resolution throughout the brain, a method that has been established in nematode *Caenorhabditis elegans* (White et al. 1986). To generate a 3D reconstruction connection map, thin sections (usually less than 50 nm) are collected and imaged. Then, neuronal structures are extracted from individual images, consecutive images are aligned, and individual neuronal segments across many sections are reconstructed into a 3D volume. Finally, each synaptic contact is assigned to a defined pair of neurons. For large nervous systems such as entire mouse and human brain, these procedures are not only labor intensive, but also require very high precision in each step, as small errors in stacking many thousands of images could accumulate and lead to incorrect connection diagrams. Although this approach is currently only applied to a small piece of brain tissue (up to a few hundred microns, for examples, Helmstaedter et al. 2013; Kasthuri et al. 2015), given rapid advances in computer science, most of the above procedures and quality controls could be fully automated in the future.

4.2 Rabies Virus-Mediated Transsynaptic Tracing with Genetic Control

In this section, we will focus on transsynaptic tracing methods introduced in Sect. 4.1.3, starting with a brief history. Then we will introduce the principle of monosynaptic restriction of RV tracing, followed by various ways to apply this method to in vivo tracing experiments.

4.2.1 *Development of Replication-Conditional Rabies Tracer*

Early in the twentieth century, a limited number of studies suggested that certain types of neurotrophic viruses can travel across neuronal pathways, e.g., Goodpasture and Teague (1923). In the 1980s, researchers started to exploit replication-competent viral tracers to map neuronal pathways from peripheral body tissues to the central brain (for review Vercelli et al. 2000; Nassi et al. 2015; Ugolini 2011). In these studies, HSV1 and PRV (Table 4.2) were widely used in rodents and non-human primates. However, major pitfalls of these viral tracers are that they induce neuronal degeneration and possibly spread via nonspecific spillover of viral particles from infected neurons to nonconnected nearby cells (Ugolini et al. 1987). An important advancement was made in 1995 when a detailed time-course for RV spread was characterized (Ugolini 1995). In this study, retrograde tracing by RV (challenge virus standard, or CVS strain) was initiated from hypoglossal motor neurons (MNs) that control tongue movement in rat, and viral spread was followed to the brainstem second-order neurons and higher order motor control regions. The results established two important properties of RV: low toxicity and no 'suspicious' labeling that would indicate nonsynaptic infection. Initially infected MNs were totally intact in their size and morphology at later stages when third-order neurons were infected in the brain. Also, RV-infected hypoglossal MNs did not spread virus to the inferior olive, which the RV-infected MNs passed through without forming synaptic connections. By contrast, an equivalent experiment using HSV1 did intensively label the inferior olive, presumably through nonspecific spillover of the virus. Together, this pioneering study highlighted that RV is a specific retrograde tracer with a long asymptomatic period.

The next landmark event for transsynaptic tracing techniques was the development of replication-conditional RV, which later allowed genetic control of starter cells and monosynaptic tracing (Wickersham et al. 2007a; Sect. 4.2.2). To understand how this was achieved, it is useful to cover some basic facts of RV virology (Fig. 4.4a). RV is a single-strand negative RNA virus (meaning the genome of RV is complementary to its mRNA) with a small genome (about 12 kb) encoding only five proteins: a nucleoprotein (N), a matrix protein (M), an RNA-dependent RNA polymerase (L), a polymerase cofactor phosphorylated protein (P), and a rabies glycoprotein (RG). The core of the virion consists of helically arranged genomic RNA associated with N, P, and L. These proteins support initial replication of the RV genome in the host cell before RV's own gene transcription. The core is surrounded by M and lipid bilayers originated from the host cell, on which RG is anchored. Thanks to this simple organization, the functional viral particles can be de novo generated in cultured cells by introducing plasmids containing the RV genome and some of the RV genes listed above (Schnell et al. 1994; Fig. 4.4b, c). This made it possible to genetically manipulate the viral genome: for example, inserting a marker transgene (Mebatsion et al. 1996a) or deleting a rabies gene (Mebatsion et al. 1996b). Using this technique, a complete loss-of-function mutant was generated from an attenuated RV strain called Street Alabama Dufferin (SAD) B19 by deleting the *RG* gene. Because RG is necessary for efficient budding of RV (Mebatsion et al. 1996b), simple deletion of *RG* from the RV genome made it difficult to recover viral particles (with naked envelope) from the cultured cells. Researchers, therefore, added back the missing *RG* in *trans* via a plasmid transfected into the cultured cells, which allowed the mutant RV genome to be packed into an intact viral envelope (Fig. 4.4d). Hereafter, we will refer to this virus as RV*dRG*+RG, where the italic *dRG* represents the genome of the RV with *d*eletion of *RG*, and +RG indicates the envelope protein supplied in *trans*. When RV*dRG*+RG was injected into rat and mouse brains, initial infection occurred normally, since the virus was coated with normal RG, allowing the virus to interact with

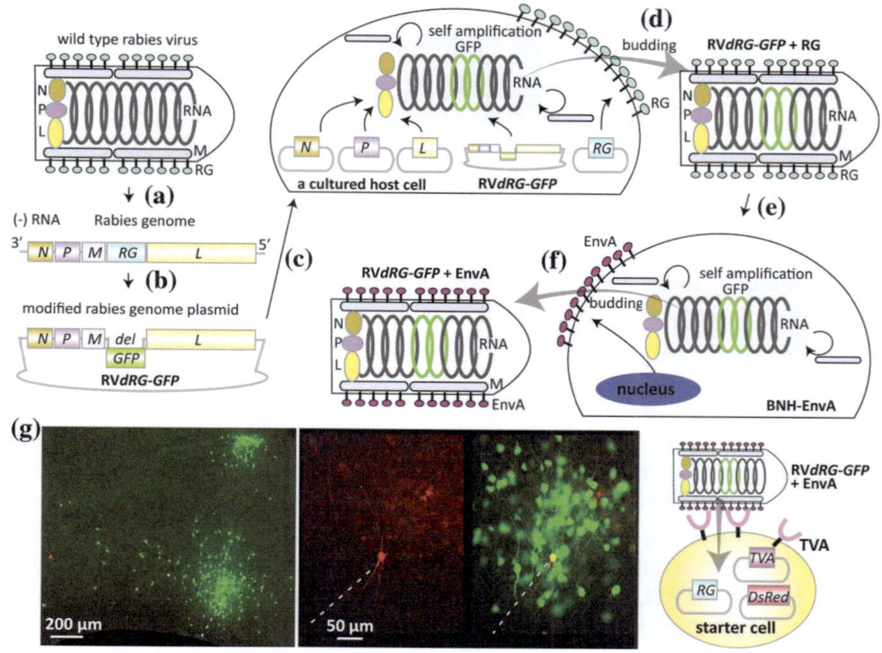

Fig. 4.4 Generation of replication-conditional RV and the monosynaptic tracing technique. **a–f** Schematic flow showing how to generate RV mutant virus that is used for monosynaptic tracing. For details, see text. A wild type RV particle is shown in the *top left*. **a** Cloning the RV genome into a plasmid. **b** Genetic modification of the RV genome to delete (*del*) *RG* and insert *GFP*. **c** Transfection of the mutant RV genome plasmid, as well as plasmids for *N*, *P*, *L*, and *RG* into a cultured host cell. **d** Budding of the mutant particle RV*dRG-GFP*+RG. **e** Infection of RV*dRG-GFP*+RG into a BHK-EnvA cell that stably expresses EnvA. **f** Budding of the pseudotyped RV*dRG-GFP*+EnvA. This particle is used for monosynaptic tracing. **g** An example of RV-mediated monosynaptic tracing using *ex vivo* slice cultures. In this sample, TVA, RG and DsRed2-expressing red neurons in hippocampal slice culture are selectively infected by RV*dRG-GFP*+EnvA (shown in *yellow*) to become a starter cell. Many putative presynaptic neurons of starter cells were labeled with GFP. *Right* schematic of a starter cell. Images are taken with permission from Wickersham et al. (2007a)

rabies receptors expressed in the mammalian nervous system (Lafon 2005). However, transsynaptic spread was completely abolished and infected animals were still healthy 11 days after infection, while control animals that had received the same amount of wild-type RV were killed by rabies (Etessami et al. 2000). This experiment clearly demonstrated that RG is necessary for transsynaptic spread of RV and that RV*dRG*+RG is a valuable nonpathogenic, replication-conditional tracer for in vivo applications.

4.2.2 Development of Monosynaptic Tracing Technique

The key idea behind introducing RV as a monosynaptic tracing technique is to provide in vivo *trans*-complementation of replication-conditional RV*dRG,* by supplying the missing *RG* as a transgene. If RG is expressed in neurons of interest, and these neurons are then infected with RV*dRG*, then they can produce RV*dRG* +RG particles by *trans*-complementation (as in cultured cells, see Fig. 4.4d). According to the natural tropism of RG, these particles can spread retrogradely and transsynaptically to the presynaptic neurons, where proliferation of RV*dRG* occurs. However, if these second order presynaptic neurons do not express RG, RV*dRG* cannot spread to third-order neurons, because RG is essential for viral spread from neuron to neuron. In this way, this approach can identify monosynaptic connections that are presynaptic to the initially infected neurons of interest.

To make this technique widely applicable, two problems must be solved: unequivocal visualization of RV*dRG*-infected neurons, and specific introduction of RV*dRG* into RG-expressing neurons of interest. The first problem was solved by introducing a fluorescent protein gene (e.g., GFP) into the RV*dRG* genome plasmid, resulting in RV*dRG-GFP* (Wickersham et al. 2007b; Fig. 4.4b). When coated with RG, these viral particles can efficiently label neurons, including their fine subcellular structures, such as dendritic spines. This allows unambiguous detection of RV-infected neurons in fixed sections, in time-lapse imaging of cultured neurons, or even in vivo via two-photon microscopy. The second problem was also elegantly solved by using a virology technique called pseudotyping, that is, altering the envelope proteins of a virus to change its tropism. The envelope protein of avian sarcoma and leucosis virus, called EnvA, restricts viral infection to cells expressing the corresponding receptor, TVA, a protein which is found in birds but not in mammals (Bates et al. 1993; Young et al. 1993). When a cell line constitutively expressing EnvA is infected with RV*dRG-GFP*, mutant RV particles coated with EnvA (referred to as RV*dRG-GFP*+EnvA) are released into the culture medium (Fig. 4.4f). Alone, this virus cannot infect mammalian cells, since they do not express TVA. However, it can infect genetically modified target neurons where expression of a TVA transgene has been transduced.

As a proof-of-principle, transduction of TVA and RG was introduced into a very small number of neurons in hippocampus slice culture, along with a red fluorescent marker, DsRed2. When RV*dRG-GFP*+EnvA was applied to the culture medium, starter cells that expressed both DsRed2 (a marker for co-expression of RG and TVA) and GFP (from RV) were generated. These neurons, labeled yellow in the culture, specifically promoted de novo production of RV*dRG-GFP*+RG. Dozens of neurons labeled with GFP alone surrounded the starter cells, suggesting that they were transsynaptically labeled from at least one of the starter cells (Fig. 4.4g; Wickersham et al. 2007a). Along with negative control experiments, these experiments in slice culture neurons established that *RVdRG-GFP*+EnvA infection is specific to TVA expressing neurons, and that RG is necessary and sufficient for transsynaptic spread of RV. Importantly, monosynaptic restriction of RV spread could be validated electrophysiologically. Using a slice with very sparse starter cells (yellow) and putative presynaptic partners (green), paired recording between these labeled neurons validated that 9 out of the 11 recorded pairs were indeed direct synaptic partners.

Before discussing how this monosynaptic RV tracing technique has been applied to in vivo circuit tracing, let us consider the advantages of monosynaptic tracing over classical multistep tracing. First, monosynaptic tracing allows genetic control of starter cells by selectively expressing TVA receptor and RG in a defined cell type (see Sect. 4.2.3). This specificity is impossible to achieve when tracing with wild type RV, as RV infects multiple types of neurons indiscriminately. Second, monosynaptic tracing technique allows unequivocal identification of direct synaptic partners, whereas classical multistep tracing can only infer the order of connectivity by the timing of the viral spread (Ugolini 1995). Although informative, variability in viral replication in different host cells, the speed and distance of retrograde transport, and the type of synapses that the virus crosses may create timing differences in transsynaptic labeling, compromising the accuracy in which the synaptic order of connected neurons can be identified. Third, RV*dRG* is less toxic to infected animals compared to classical RV tracing, and therefore allows one to visualize neuronal circuits in relatively healthy animals. This also allows monosynaptic tracing to be combined with other physiological and behavioral experiments that require live animals (for example see Sect. 4.2.5).

4.2.3 Cre-dependent Transsynaptic Tracing

A significant advance of monosynaptic RV tracing lies in the ability to genetically control the generation of starter cells. This can be done by selective targeting of RG and TVA to defined types of neurons. Generally, cell types can be characterized by a combination of the following criteria: stereotyped locations, unique gene expression, unique morphology, axonal projection patterns, developmental history, electrophysiological properties, and unique receptive fields. Currently, not all of these criteria can be readily utilized to generate starter cells, but some are available. In this section, we will discuss the methodology behind Cre-dependent tracing methods. Other methods will be discussed in the following sections.

Numerous mouse lines are available that express Cre recombinase in a selective cell type. The viral and genetic technology (see Sect. 4.1.3 and Fig. 4.3c) is a powerful way to generate starter cells of a defined cell type in a defined brain region. Let us use a specific example, such as parvalbumin-positive interneurons (PVNs) in the left primary motor cortex in mice (Wall et al. 2010; Miyamichi et al. 2013). To transsynaptically trace from these cells, we first need a transgenic mouse line where Cre is specifically expressed in the PVNs. Next, we stereotactically inject a small volume (usually 50–300 nl) of a mixture of two Cre-dependent AAVs, expressing RG and TVA fused with a red fluorescent marker mCherry (TVA-mCherry), into the left motor cortex of a *PV-Cre* mouse (Fig. 4.5a). In 2

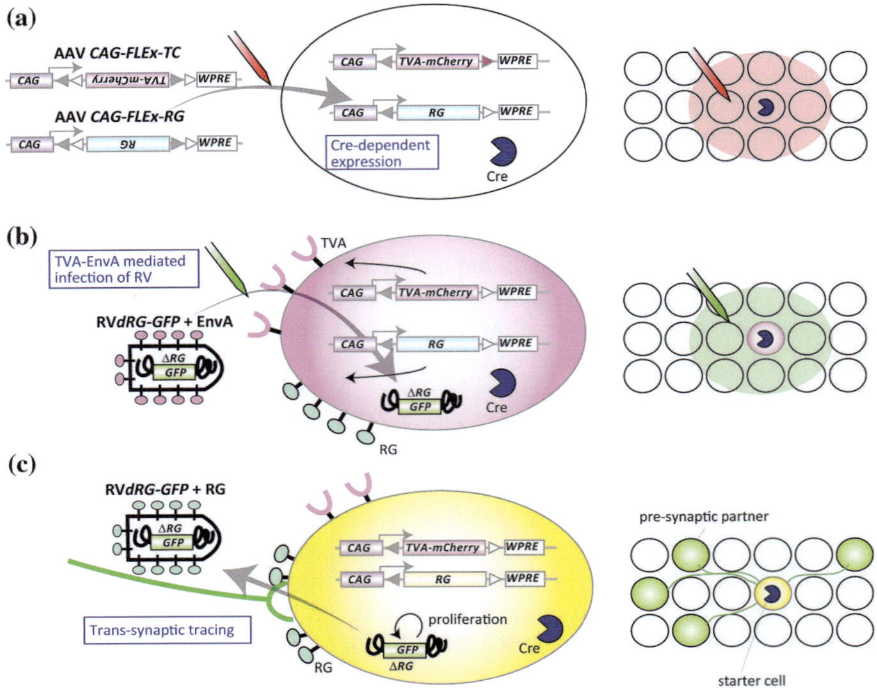

Fig. 4.5 Cre-dependent RV transsynaptic tracing. **a** *Left* constructs of two AAV vectors, one expressing TVA-mCherry and one expressing RG in a Cre-dependent manner using *FLEx* switch (see Fig. 4.3d). WPRE, woodchuck hepatitis virus posttranscriptional regulatory element, which can enhance transgene expression from an AAV vector. The *oval in the middle* represents a cell expressing Cre (*blue PacMan*) in which TVA-mCherry and RG are also expressed. *Circles on the right* represent neurons, one of which corresponds to a Cre-positive cell (with *blue PacMan*). **b** RV*dRG-GFP*+EnvA infects a Cre-positive cell that is expressing TVA and RG, resulting in a *yellow starter cell* (due to TVA-mCherry from the AAV and GFP from the RV*dRG*). Note that by mixing two separate AAVs before injection, >80% neurons at the injection site co-express both transgenes (Miyamichi et al. 2013). **c** If RG is present at the plasma membrane of starter cells (*yellow*), RV*dRG-GFP*+RG can transsynaptically spread to presynaptic partners and label them with GFP

weeks, hundreds of PVNs in the left MC will express RG and TVA-mCherry. There are millions of PVNs throughout the mouse brain, but focal injection of the AAVs restricts the location of the starter cells. In the motor cortex, >95% neurons are non-PVNs, but they do not express RG or TVA-mCherry because they do not contain Cre. We then stereotactically deliver a small amount of RV*dRG-GFP* +EnvA into the same left motor cortex (Fig. 4.5b) to initiate transsynaptic tracing. Finally, 4–7 days after RV infection, we harvest the brain and detect neurons labeled with both GFP and mCherry (starter cells, labeled yellow, Fig. 4.5c) and those labeled with GFP alone (presynaptic partners of the starter cells). In one example, 1012 starter cells were generated close to the injection site, and 8656 GFP

+ cells were detected far away from the starter cells in the contralateral motor cortex, ipsilateral somatosensory cortex, and motor-related thalamus. In addition, many thousands of locally infected neurons labeled with GFP were detected near the starter cells (Miyamichi et al. 2013). This tracing strategy has been used with great success in a wide range of brain regions including the olfactory system (Miyamichi et al. 2011, 2013), neuromodulatory systems (Weissbourd et al. 2014; Lerner et al. 2015; Schwarz et al. 2015; Ogawa et al. 2014; Watabe-Uchida et al. 2012; Menegas et al. 2015; Pollak Dorocic et al. 2014; Beier et al. 2015), amygdala (Haubensak et al. 2010), hypothalamus (Krashes et al. 2014), hippocampus (Sun et al. 2014; Kohara et al. 2014), neocortex (Fu et al. 2014; Zhang et al. 2014; Adelson et al. 2014; DeNardo et al. 2015; Kim et al. 2015), striatum (Reardon et al. 2016; Wall et al. 2013), cerebellum (Wall et al. 2010), and spinal cord (Reardon et al. 2016; Ni et al. 2014).

At a glance, this method seems complex. Instead of serially injecting AAVs and RV, one could drive expression of RG and TVA via Cre-dependent transgenes introduced into the genome of the animal. In this scenario, all Cre-expressing neurons throughout the brain (for instance all PVNs in a *PV-Cre* animal) will express RG and TVA. Stereotactic injection of RV*dRG-GFP*+EnvA into the left motor cortex of these animals will convert PVNs at the injection site into starter cells, and transsynaptic tracing will occur. However, if there are any PVNs located presynaptically to the starter PVNs, they can become 'secondary' starter cells because they will also express RG, and their presynaptic partners will also be labeled. Thus, this process can significantly compromise the accuracy of restricted, monosynaptic RV tracing. It is therefore advantageous to spatially restrict the expression of RG using an AAV.

For successful Cre-dependent transsynaptic tracing, two important points should be considered. First, starter cells should be unequivocally labeled to validate their numbers and cell types. Second, Cre-independent, nonspecific labeling of RV should be monitored and controlled. Ideally, Cre-dependent tracing should only originate from Cre-positive starter cells. However, in many cases, it is possible that a small amount of 'leaky' expression of TVA and RG can occur. In the example mentioned above (*PV-Cre* tracing in the motor cortex), negative control experiments omitting Cre still labeled ~70 GFP+ neurons close to the injection site. This Cre-independent GFP expression is likely due to the extremely efficient interaction between TVA receptor and RV*dRG-GFP*+EnvA. Because these nonspecific GFP+ neurons are indistinguishable from real presynaptic neurons of the starter cells, they compromise the accuracy of tracing within local circuits near the injection site, although usually less than 5% of local GFP+ neurons arise from Cre-independent tracing (Weissbourd et al. 2014; Miyamichi et al. 2013; Schwarz et al. 2015). Note that long-range tracing outside of the injection site is less affected by nonspecific labeling, since leaky expression of RG is not strong enough to support *trans*-complementation of RV*dRG*.

To eliminate Cre-independent local labeling, TVA-EnvA interactions can be reduced without affecting TVA-mCherry expression levels. To achieve this, a TVA mutant with 10% affinity to the EnvA-pseudotyped virus (Rong et al. 1998),

TVA_{66T}, was used instead of wild-type TVA. The resulting tracing with TVA_{66T} showed no Cre-independent GFP labeling within the brain (Weissbourd et al. 2014; Miyamichi et al. 2013). Although the tracing efficiency of long-range inputs were also reduced (due to a decrease in the number of RV*dRG-GFP*+EnvA viral particles that initially infect the starter cells), this technique is suitable for analyzing local neuronal circuits. We will see applications of this technique in the OB in Sect. 4.3.

Approaching the end of this section, let us consider a way to control the number of starter cells, in addition to specifying cell types as discussed above. Generally, reducing the titer of AAVs would decrease the number of starter cells. To more empirically control the number of starter cells, a tamoxifen-inducible Cre (CreER) transgenic mouse was crossed with a mouse line conditionally expressing tetracycline transactivator (tTA2) in a Cre-dependent manner, to obtain double transgenic mice. Small amounts of tamoxifen were injected into these mice, resulting in a sparse induction of neurons expressing tTA2 throughout the brain. Instead of using Cre-dependent AAVs, an AAV vector expressing RG and TVA (with a fluorescent marker), driven by tTA2, was injected into a defined brain area (Fig. 4.3e). In the olfactory cortex (Miyamichi et al. 2011), this strategy generated sparse starter cells (samples varied from 4 to 105 starter cells). A 'lucky' example in the somatosensory cortex contained a single starter cell, which is useful for analyzing how individual neurons integrate information from different brain regions. Generation of sparse starter cells can also be achieved by single-cell electroporation method (see Sect. 4.2.5).

4.2.4 Tracing the Relationship Between Input and Output (TRIO)

Axonal projection patterns can be used to define different classes of neurons in the brain. For example, pyramidal neurons of motor cortex located in layer 5 (one of the major output layers of the cortex) can be classified into two major types based on their projection patterns: callosal projection neurons (CPNs) and subcerebral projection neurons (SCPNs) (Greig et al. 2013). CPNs provide callosal projections to the contralateral motor cortex, in addition to bilateral projections to somatosensory cortex and striatum, while SCPNs provide long descending axons to the ipsilateral striatum, medulla and pons, on their way to the spinal cord. Importantly, these two types of neurons are highly intermingled in the cortex, and therefore can potentially be contacted by the same presynaptic neurons that send axons into the motor cortex. Given that their target areas are mostly nonoverlapping, a question arises: do CPNs and SCPNs receive the same input? And if not, what is unique about their presynaptic inputs, and can differences in their input patterns be quantitatively compared? This question can be generalized to many different brain regions, and thus can provide valuable insight into different circuit logics that underlie specific information flow.

Axonal projection patterns have been successfully used to define starter cells in specific cases, such as MNs in the spinal cord and retinal ganglion cells (RGCs) in the eye. Let us analyze these cases, and discuss how to generalize the concept of input–output connectivity for applications in the central brain. In the spinal cord, a group of MNs (a motor pool) innervating the same muscle can be selectively labeled by RV injected into the target muscle. Injecting RV*dRG-GFP*+RG, along with AAV expressing RG, into the target muscle can convert the MNs of a defined motor pool to starter cells. The RV can then transsynaptically spread to premotor neurons that are presynaptic to that motor pool (for details, see Sect. 4.5.1; Stepien et al. 2010; Tripodi et al. 2011). Note that in this particular case, AAV serotype 6 can efficiently infect MNs retrogradely from their axons at the target muscle only during the neonatal stage. Retrograde infection of AAV in the central brain at adult stages is usually inefficient, though efficiency varies considerably by AAV serotype, titer, lot, and the target brain regions.

In the retina, a specific type of RGCs that convey direction-selective motion were converted to starter cells (Yonehara et al. 2011; Yonehara et al. 2013). This was possible because these RGCs selectively project to the medial terminal nucleus (MTN) in the midbrain. By injecting RV*dRG-GFP*+RG together with a helper virus (HSV1 or AAV) expressing RG into the MTN of neonatal mice, researchers successfully detected amacrine cells and bipolar cells connecting to the starter RGCs in the retina. These methods that deliver RV*dRG-GFP*+RG by using the outputs of a neuronal population, however, cannot easily select specific starter cells in the central brain, because many different brain regions and cell types often send convergent projections to the same target. Going back to the motor cortex example, layer 5 SCPNs can be targeted retrogradely from the medulla, but there are millions of neurons throughout the brain that also project to the medulla, including cells in other parts of the cortex, hypothalamus, cerebellum, and spinal cord. To make spatially restricted starter cells, one needs a method to generate starter cells based both on brain region and projection type.

Given that Cre-dependent RV tracing with AAV helper viruses can restrict the location (by AAV injection site) and cell type (by Cre expression) of starter cells (Sect. 4.2.3, Figs. 4.5 and 4.6a), a natural extension of this technique is to supply Cre via the axons of starter cells (Fig. 4.6b). A canine adenovirus 2 (CAV2) driving Cre is well suited for this purpose, as it can efficiently and retrogradely introduce Cre into many types of neurons in the mouse brain (Table 4.1; Soudais et al. 2001; Hnasko et al. 2006). By pairing Cre-dependent RV tracing with *CAV2-Cre*, (a method called tracing the relationship between input and output, or TRIO), one can determine and compare synaptic inputs to starter cells that have been defined by their location in the brain and output projection patterns (Schwarz et al. 2015). This system was applied to quantitatively analyze presynaptic partners of CPNs and SCPNs in motor cortex layer 5, by injecting *CAV2-Cre* into contralateral motor cortex versus ipsilateral medulla, respectively. However, a problem with this method is that CPNs are distributed broadly from layer 2/3 to layer 6; the projection pattern alone is not specific enough to restrict starter cells to CPNs only in the layer 5. Note that this is a general problem for many brain regions and cell types.

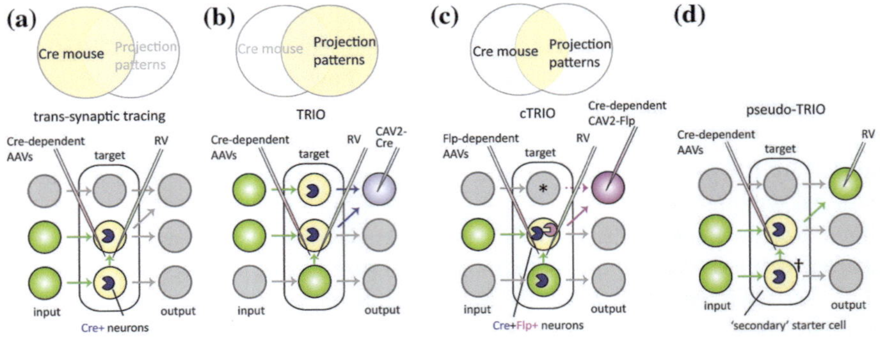

Fig. 4.6 Schematic representations of trans-synaptic tracing, TRIO, cTRIO and pseudo-TRIO. In this simplified neuronal circuit, three brain regions (*columns of circles*), each containing three neurons, are connected by axons (*arrows*) as indicated. **a** In Cre-dependent RV trans-synaptic tracing (Miyamichi et al. 2013; Watabe-Uchida et al. 2012; Sect. 4.2.3), starter cells (*yellow*) are generated in the neurons expressing Cre (represented by *blue PacMan symbols*) in the target brain region (*middle column*) regardless of their projection patterns. **b** In TRIO (Schwarz et al. 2015), *CAV2-Cre* is injected (*blue needle*) into one of the output regions to retrogradely supply Cre to the target region. Then, Cre-dependent RV tracing is conducted as in (**a**). In this case, different cell types within the target regions are not distinguished. **c** In cTRIO (Schwarz et al. 2015), a TRIO experiment is conducted in transgenic mice in which Cre recombinase is expressed in a specific cell type. The *purple PacMan* represents Flp recombinase. Note that the cell indicated by the *asterisk* cannot become a starter cell because it does not express Cre, even though it projects to the same output region. **d** Schematic representation of pseudo-TRIO. In this experiment using a Cre transgenic mouse, RV*dRG-GFP*+EnvA is directly injected into one of the output regions, instead of Cre-dependent *CAV2-Flp* in (**c**). Cre-dependent AAVs expressing RG and TVA are injected into the target region as in (**a**). A starter cell, defined by Cre expression and its projection, is generated in the middle of the target region. However, there is another Cre-expressing cell, indicated by a *dagger symbol* (†) in the target region that also receives the AAV expressing RG. This cell does not project to the RV injection site, but does connect to the starter cell. This cell can become a secondary starter cell, and its presynaptic partners will also be labeled by RV particles (in *green*). This causes pseudo-positive labeling of input cells (compare input patterns in panel **c** and **d**)

For example, targeting dopamine neurons in the ventral tegmental area (VTA) that project to the striatum would also label GABAergic neurons in the VTA that also project to the same striatum region, ‘contaminating’ the specificity of the starter cell type (see Sect. 4.4.3). What we need here is an additional restriction of starter cells based on the cell type (defined by a genetic marker) in the TRIO method.

This led the development of cell-type-specific TRIO, or cTRIO (Fig. 4.6c), which allows input mapping to genetically defined neuronal populations based on their output patterns (Schwarz et al. 2015). In cTRIO, a transgenic mouse line is used in which Cre is expressed in a specific cell type. Cre-dependent CAV2 driving Flp recombinase is injected into an output region to which the Cre-expressing neurons send their axons. This results in expression of FLP only in the neurons of a defined cell type (by Cre) with a defined projection pattern (by CAV2), which will become the starter cells. Next, Flp-dependent AAVs driving RG and TVA-mCherry

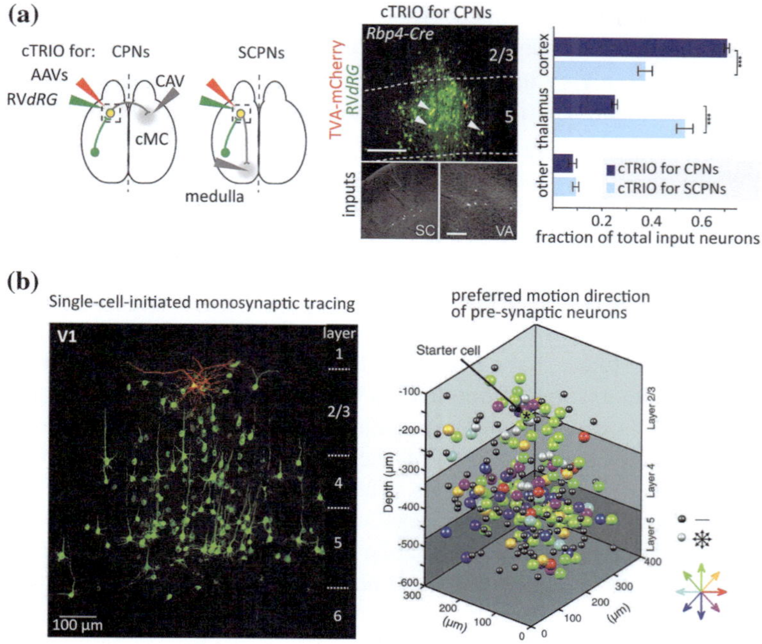

Fig. 4.7 Example of cTRIO and single-cell-initiated monosynaptic tracing. **a** Schematic of cTRIO in mouse motor cortex. Cre-dependent *CAV2-Flp* was injected into contralateral motor cortex (cMC, for CPNs) or medulla (for SCPNs) along with AAVs expressing Flp-dependent TVA-mCherry and RG into the motor cortex, followed by RV*dRG-GFP*+EnvA. Example coronal sections of motor cortex CPNs starter cells in cTRIO are shown in the *middle*. Labeled cells include starter cells (*yellow*, a subset indicated by *arrowheads*) and local input cells (*green*), as well as input neurons from the somatosensory cortex (*SC*) and ventral anterior thalamus (*VA*). Average fraction of total input neurons in cTRIO is shown in the *right graph*. *Values* represent the average fraction of input in each category. *Error bars* s.e.m. ***, $p < 0.001$. Scale, 250 µm (*middle row*), 100 µm (*bottom row*). Adapted with permission from Schwarz et al. (2015). **b** Single cell-initiated RV tracing expressing GCaMP6s in the primary visual cortex (*V1*). *Left image* represents a 300-µm thick slice containing an electroporated layer 2/3 pyramidal starter cell (*yellow*) and its local presynaptic neurons (*green*). *Right image* shows an example of 3D-reconstruction of the location of a starter cell and its presynaptic neurons. Each *filled circle* represents a neuron and is colored according to the preferred motion direction (color code is shown at *bottom right*). Cells that did not respond to the visual stimuli are represented by *small black circles*. Cells that responded to motion equally in all directions are represented by *small gray circles*. Images are taken with permission from Wertz et al. (2015)

are injected directly into the brain region where these neurons are located, followed by RV*dRG-GFP*+EnvA to initiate transsynaptic tracing. As a proof-of-principle, *retinol binding protein 4* (*Rbp4*)*-Cre* mice were used to restrict starter cells in layer 5 of cortex (Gerfen et al. 2013), with *CAV2-FLEx-Flp* injected into contralateral motor cortex or ipsilateral medulla. This time, CPNs and SCPNs in the motor cortex layer 5 were selectively converted to starter cells, and GFP+ input neurons were quantified in the cortex and thalamus. Interestingly, CPNs received proportionally more input from the cortex, while SCPNs obtained more input from the thalamus (Fig. 4.7a). Thus, CPNs and SCPNs in the motor cortex layer 5, albeit with highly intermingled dendritic trees, receive differential cortical versus thalamic input. The functional significance of this organization, as well as the local circuit connectivity of CPNs and SCPNs (Kiritani et al. 2012) remains to be explored.

At first glance, TRIO/cTRIO looks complicated. Indeed, it requires precise targeting of three different viruses into the correct stereotactic coordinates. One may want to eliminate the CAV2 injection and instead generate starter cells in a pathway-selective manner using the fact that RV*dRG-GFP* can retrogradely transduce starter cells from axons (Fig. 4.6d). This strategy, although used in several influential studies, has a major pitfall regarding input–output specificity. Let us use a specific example from the motor cortex again. In this hypothetical experiment, AAVs conditionally expressing RG and TVA-mCherry are injected into the motor cortex of *Rbp4-Cre* mice, such that layer 5 CPNs and SCPNs express TVA in their cell bodies and axons. Thanks to the sensitivity of TVA-EnvA interactions, RV*dRG-GFP*+EnvA injected into the medulla could convert SCPNs in motor cortex layer 5 into starter cells (although in reality this long-range retrograde transduction of EnvA-pseudotyped RV is not very efficient). Presynaptic partners of SCPNs are then labeled by transsynaptic RV spread, resembling cTRIO. However, if any CPNs in motor cortex layer 5 are presynaptic partners to any of the starter SCPNs, they can also become starter cells, because they contain Cre, and therefore can express RG, if infected by AAV (see Fig. 4.6d, 'secondary' starter cell indicated by a dagger symbol). This would result in transsynaptic labeling of SCPN and CPN inputs. Therefore, this strategy does not guarantee that starter cells are specific for a defined projection target. Introducing RG based on the projection pattern is crucial for the specificity achieved by TRIO/cTRIO, accomplished through coordinated injections of CAV2 and AAVs.

Finally, a few basic properties of CAV2 that are useful to know for interpreting tracing results must be discussed. The degree of local CAV2 spread at the injection site is important for precisely targeting neurons based on their projections. The olfactory system was used to assess this (Schwarz et al. 2015), thanks to the well-understood organization of mitral cell axonal projections. Mitral cells (MCs) are the projection neurons of the OB, and their axons form a bundle from which thin collaterals innervate cortical pyramidal cells in layer 1a of the piriform cortex. *CAV2-Cre* was injected into Cre-reporter mice at varied distances from layer 1a in the piriform cortex and labeled MCs in the OB were quantified. This analysis revealed that CAV2 local spread is mostly limited to within 200 μm from the injection needle. However, it was also observed that CAV2 has potential to infect axons in passage, as MCs in the accessory OB, which send myelinated axon bundles through the piriform cortex without making synapses (Shepherd 2004), were also labeled. Careful design of the CAV2-injection site is needed to avoid unwanted labeling from passing axons. Although CAV2 can transduce diverse cell types throughout the brain, the labeling efficiency may vary depending on the brain regions, type of neurons, and distance of retrograde transport. Keeping these cautions in mind, CAV2-based tracing tools are broadly applicable to many brain regions. As no genetically engineered animal is required for TRIO, it can be used in wild type mammals beyond mice. As a proof-of-principle demonstration, presynaptic partners of motor cortex neurons that project to striatum or contralateral motor cortex were visualized in rat (Schwarz et al. 2015). We will also discuss applications of TRIO/cTRIO in neuromodulatory systems (Lerner et al. 2015; Schwarz et al. 2015; Beier et al. 2015) in Sect. 4.4.

4.2.5 Tracing from Defined Neurons by Developmental History

Utilizing developmental history is a powerful way to define cell types. A salient example is cell birth date, which can be used to target developmentally defined types of neurons associated with a specific brain region, layer, projection pattern, and function. Generation of starter cells based on the birth date of neurons has been used with great success in the research field of 'adult-born' neurons. Although most neurons in the brain are generated in a short time window during embryonic development, neurogenesis continues throughout life at two specific locations: the subventricular zone (SVZ) of the lateral ventricles and the subgranular zone (SGZ) of the dentate gyrus (DG) in the hippocampus (Zhao et al. 2008). The SVZ generates OB interneurons that migrate a great distance through the rostral migratory stream, while the SGZ generates local granule cells (GCs) of the DG. To analyze how these neurons are integrated into existing circuits, researchers selectively converted adult-born neurons to starter cells. This was achieved by using a gammaretrovirus that selectively infects dividing cells (Table 4.1), which made it possible to infect neuronal progenitors, but not postmitotic neurons. Gammaretrovirus-expressing TVA and RG constitutively (Deshpande et al. 2013; Vivar et al. 2012) or Cre-dependently (Nakashiba et al. 2012) was injected into the DG of postnatal mice to generate starter cells, and RV tracing labeled their presynaptic neurons. The Cre-dependent gammaretrovirus can be used to further refine the starter cells based on expression of a specific gene. Another strategy is to utilize transgenic mouse lines expressing TVA and RG in a Cre-dependent manner, in conjunction with gammaretrovirus-expressing Cre (Li et al. 2013). Collectively, these studies revealed the timecourse of progressive integration: adult-born neurons receive local connections from multiple types of interneurons before long-range projections are established.

In the case of the OB interneurons, transient induction of a transgene into the SVG can target neurons of a defined birthdate. Several methods have been successfully used to provide TVA and RG to the SVG including electroporation of a plasmid (Arenkiel et al. 2011), infection of gammaretrovirus (Deshpande et al. 2013), or infection of lentivirus-expressing Cre into a transgenic mouse that conditionally expresses TVA and RG (Garcia et al. 2014). These studies revealed previously uncharacterized connectivity in the local OB circuits (see Sect. 4.3), as well as a temporal sequence for the integration of input to the adult-born GCs in the OB. Some basic properties of RV spread were also reported, which have relevancy beyond OB circuits. For example, researchers found that enriched olfactory experience induced a 3-fold increase in the number of RV-labeled local presynaptic partners of newborn GCs (Arenkiel et al. 2011), a phenomenon that was also observed for newborn GCs in the DG after exercise (Deshpande et al. 2013). This can be explained by two scenarios: (1) neuronal activity increased the number of synaptic connections from local neurons to the starter GCs, or (2) RV particles more efficiently spread across active synapses. To distinguish between these two possibilities, neuronal activity was manipulated in cultured OB explants during RV tracing. It was found that blocking SNARE-dependent neurotransmitter release, action potentials, or fast glutamatergic neurotransmission had no significant effect on the number of transsynaptically labeled cells (Arenkiel et al. 2011). Thus, transsynaptic spread of RV is insensitive to changes in neuronal activity, a phenomenon also suggested from in vivo tracing experiments of NMDA receptor knockout starter cells in the neocortex (DeNardo et al. 2015).

Neuronal birth date is also useful to define starter cells beyond just adult-born neurons. In utero electroporation of a Cre-expressing plasmid during specific developmental time windows successfully targeted neurons of defined layers in the neocortex (DeNardo et al. 2015). This method, combined with Cre-dependent RV tracing (Sect. 4.2.3), allowed researchers to map local and long-distance input to the layer 2/3 neurons in the somatosensory cortex, which were compared to inputs of layer 5 starter cells (generated using an *Rbp4-Cre* mouse). In principle, *in utero* electroporation-based methods can be used in many animals beyond mice to define cell types or layers of starter cells.

4.2.6 Tracing from Defined Neurons by Electrophysiological Properties

Other useful characteristics for defining starter cells include their electrophysiological properties and their receptive fields. To achieve this, electroporation of plasmids into a single cell in vivo was performed after whole-cell patch-clamp recording (Rancz et al. 2011), which promoted transgene expression in the recorded cells after patching. In this way, TVA and RG were introduced into layer 5 pyramidal cells with defined orientation selectivity in mouse primary visual cortex. RV transsynaptic tracing resulted in the generation of a single starter cell and labeled on average 346 presynaptic neurons. A follow-up study (Velez-Fort et al. 2014) demonstrated that corticocortical projection neurons in the visual cortex layer 6 are broadly tuned in their orientation selectivity and predominantly receive input from deep layers of local visual cortex, whereas cortical-thalamic projection neurons in the same area are sharply tuned to orientation and direction information, and receive more long-range input from higher cortical areas. Thus, similar to CPNs and SCPNs in the motor cortex layer 5 analyzed by cTRIO (Sect. 4.2.4), visual cortex layer 6 projection neurons with distinct output specificities integrate different contextual and stimulus-related information within and outside of the cortical network. Thanks to advancements in engineering-modified RV variants (Osakada et al. 2011; Reardon et al. 2016), RV-expressing Ca^{2+}-indicator GCaMP6 was used in single cell-initiated monosynaptic tracing in visual cortex (Wertz et al. 2015). Two-photon Ca^{2+} imaging revealed motion direction preferences of individual presynaptic neurons labeled from a single visual cortex layer 2/3 neuron of a defined direction preference (Fig. 4.7b). This allowed a systematic comparison of the direction preference of each presynaptic neuron with its postsynaptic partner across all layers of cortex. The result revealed rather unexpected properties: (1) Presynaptically labeled neurons within each layer exhibited similar direction preference; and (2) only one-third of visual cortex layer 2/3 neurons received presynaptic input with similar preferred direction selectivity. In other words, two-thirds of visual cortex layer 2/3 neurons receive 'untuned' presynaptic input that cannot be distinguished from a pool of neurons with random direction selectivity. How the postsynaptic receptive field is correctly tuned, as well as the significance of such an organization, remains to be explored. Collectively, these pioneering studies uncovered a link between the neuronal connectivity, electrophysiological properties, and receptive fields of each neuron at single-cell resolution in vivo.

In addition to technical advances and biological findings, these studies also provide a useful insight with regards to the number of presynaptic neurons that are labeled by RV from a single starter cell. An important limitation of RV tracing is that it only labels a subset of inputs, presumably due to the limited number of RV particles that each starter cell can generate. The labeling efficiency of RV can be enhanced by increasing the amount of RG expressed in the starter cells for *trans*-complementation and increasing the number of RV*dRG* particles entering the starter cells (Miyamichi et al. 2013). The strain of RV also has an influence on efficiency (CVS strain is more efficient than SAD) (Reardon et al. 2016). As the actual number of inputs per cell in the brain is unknown, it is impossible to determine the exact efficiency of RV tracing. However, based on a few simple assumptions (Miyamichi et al. 2011), cortical pyramidal cells in mice are estimated to collect input from 400–1600 presynaptic neurons. The number of RV-labeled cells per a single starter cell observed in the single-cell tracing sample was about 250 [somatosensory cortex layer 5 neuron (Miyamichi et al. 2011)], about 350 [V1 layer 5 neuron (Rancz et al. 2011)], about 380 [cortical-striatal projection neurons in V1 layer 6 (Velez-Fort et al. 2014)] and about 420 [V1 layer 2/3 neuron (Wertz et al. 2015)]. Thus, the RV tracing is estimated to visualize 20–100% of putative presynaptic partners in these cases. With improvements to the RV genome, strains of RV, and methods to drive RG, the efficiency of RV tracing will be further increased in the future.

4.3 Transsynaptic Tracing in the Olfactory System

In this section, we will discuss how RV tracing has been used to identify new components of a neuronal circuit, and to analyze the topography of the neuronal map. The mouse olfactory system provides examples for these applications of RV tracing. First, let us summarize the basic knowledge of the organization of the mouse olfactory system (Mori and Sakano 2011). The sense of smell is the primary means for most animals to find food, mates, and predators from a distance. Airborne odorants in the environment are initially detected by odorant receptors (ORs), a diverse family of G-protein-coupled receptors expressed in the olfactory epithelium. The olfactory coding in the epithelium is combinatorial: each OR is activated by multiple odorants, and each odorant is recognized by a unique combination of ORs. This enables the olfactory system to distinguish many more odorants than the number of OR genes. Two important rules underlie the transformation of this combinatorial OR code in the nose into neuronal activity patterns by which the brain can interpret the identity, concentration, and in some cases, the meanings of odorants. (1) One neuron–one receptor rule, that is, individual olfactory receptor neurons (ORNs), the sensory neurons in the olfactory epithelium, express a single type of OR. By this rule, the combinatorial code of OR activation is converted into population neuronal activities of ORNs on a one-to-one basis. (2) One OR–one projection site rule, that is, thousands of ORNs expressing the same OR-type project their axons to a common projection site, called a glomerulus, in the OB (Fig. 4.8a). In mice, there are ~1000 intact OR genes, and therefore ~1000 ORN types. Their projections form ~2000 glomeruli, with each ORN type corresponding to approximately two glomeruli on the surface of the OB. This organization allows the transformation of ORNs' population activities in the epithelium into discrete patterns of glomerular activities (an odorant map) in the OB. How does the odorant map inform higher olfactory centers? RV transsynaptic tracing has been used in several efforts to answer this question.

4.3.1 Transsynaptic Tracing in the Olfactory Bulb

Within each glomerulus in the OB, ORN axons form glutamatergic synapses with the primary dendrites of MCs and tufted cells (TCs), the two types of projection neurons of the OB (Fig. 4.8a; Shepherd 2004). Hereafter, these two types of projection neurons are collectively referred to as MTCs. Individual MTCs send a single primary dendrite to a single glomerulus. Importantly, MTCs are not just relaying olfactory information from the epithelium to the cortex, as there are diverse local interneurons in the OB. The precise function of these local networks remains elusive because of the heterogeneity of these interneurons, their diverse physiological properties and their complex synaptic connectivity. By using Cre-dependent RV transsynaptic tracing with reduced nonspecific local background (Sect. 4.2.3)

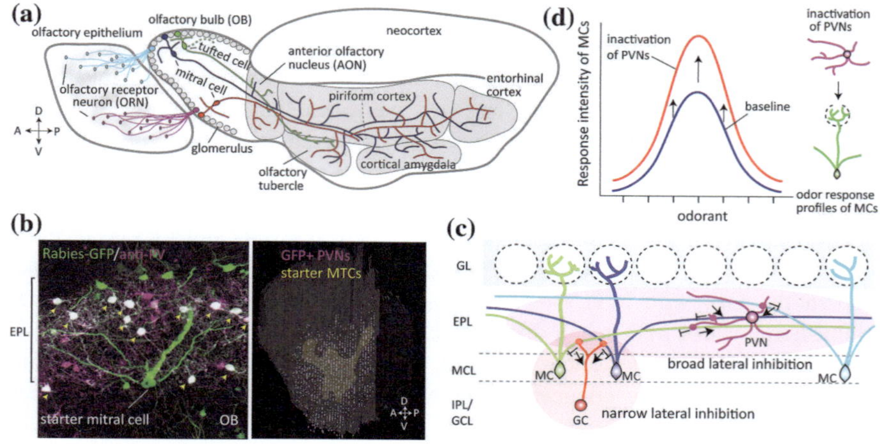

Fig. 4.8 Basic organization of mouse olfactory system and RV trans-synaptic tracing in the OB. **a** Side view of the mouse brain highlighting the organization of the olfactory system. Two types of ORNs (out of ~1000 types) are shown in two colors. ORNs of the same type send convergent axonal projections to the target glomerulus. In the OB, each mitral (*blue*, *red*) and tufted (*green*) cell sends an apical dendrite to a single glomerulus, where it receives input from a single type of ORN. MCs project long-distance axons to many olfactory cortical areas, whereas tufted cells innervate only anterior parts of the olfactory cortices (Igarashi et al. 2012). © 2015 from Principles of Neurobiology by Luo. Reproduced by permission of Garland Science/Taylor & Francis Group LLC. **b** *Left* a coronal section of the OB showing one of starter MCs and presynaptically labeled cells with GFP in the OB. PVNs are visualized by anti-PV antibodies in magenta. Note that most GFP+ cells in the external plexiform layer (*EPL*) are PV+ (indicated by *yellow arrowheads*). *Right* a 3D-reconstructed OB sample (lateral view) showing the spatial distribution of starter MTCs (*yellow*) and their presynaptic PVNs located in the EPL (*magenta*). Images are obtained with permission from Miyamichi et al. (2013). **c** Top, schematic drawing of OB local circuits. Three MCs are shown in *three colors*. Each PVN in the EPL (represented by a *magenta cell*) receives input from, and sends output to, widely distributed MTCs (for example, *green* and *cyan* MCs) for broad lateral inhibition of MTC output. In contrast, the feedback loop mediated by GCs (represented by a *red cell*) connects nearby MCs (*blue* and *cyan* MCs) for local lateral inhibition. *GL* glomerular layer; *MCL* mitral cell layer; *IPL* internal plexiform layer; *GCL* granule cell layer. **d** Schematic summary of pharmacogenetic inactivation of PVNs to analyze odorant response profiles of MCs. The *graph* represents tuning curves of MCs to seven different odorants before (*blue*) and after (*red*) inactivation of nearby PVNs. Adapted with permission from Kato et al. (2013). *A* anterior; *P* posterior; *D* dorsal; *V* ventral; *L* lateral; *M* medial

and a transgenic mouse line driving Cre exclusively in the MTCs (*Pcdh21-Cre*), the presynaptic neurons to MTCs within the local OB circuits were determined (Miyamichi et al. 2013). In addition to GCs, which are known to provide feedback inhibition to MCs (Shepherd 2004), an unexpected presynaptic partner of MTCs were labeled via these tracing studies: cells within the external plexiform layer. About 40% of GFP+ neurons presynaptic to the starter MTCs were located in this layer, ~90% of which were parvalbumin positive interneurons (PVNs, Fig. 4.8b). 3D-reconstruction of labeled GFP+ cells showed that MTCs receive input from widely distributed PVNs (~300 μm in distance), which was in sharp contrast to the narrowly organized input from GCs (<100 μm in distance). To characterize the electrophysiological properties and odorant response profiles of these PVNs in the OB, individual PVNs in the dorsal surface of the OB were recorded by in vivo two-photon targeted patch-clamp method, while the animals received odorant stimulations. This revealed that PVNs were activated less-selectively by many odorants. Because PVNs do not directly receive input from the ORNs, the source of this broad odorant receptive range was unclear. RV transsynaptic tracing was therefore applied to PVNs (using *PV-Cre* mice), which revealed that PVNs received broad MTC input. Together, these data uncovered a previously unknown feedback loop in the OB, with MTCs → PVNs → MTCs, by which a single MTC can inhibit hundreds of MTCs at a distance (Fig. 4.8c).

Before dissecting the function of this circuit, let us confirm that RV tracing reports precise synaptic connectivity between neurons. Although synapse specificity of RV spread has been supported in various contexts (Ugolini 1995; Reardon et al. 2016; Miyamichi et al. 2011; Wickersham et al. 2007a), OB interneurons are unique in that they form dendrodendritic synapses: PVNs and GCs send dendrites to the EPL where they receive excitatory input from the secondary dendrites of MTCs, and send inhibitory output to the dendrites of the same or other MTCs. It was not clear if RV transsynaptic tracing works in such connections. RV tracing data in the OB was therefore compared with the connection diagram obtained from an independent mapping method, paired recording in the OB slice (Kato et al. 2013). Paired recording revealed that MTC → PVN connections are about tenfold more frequent than MTC → CG connections. This ratio is very similar to what was observed in the RV tracing, supporting the precision of these methods. Thus, RV tracing can map synaptic partners connected by dendrodendritic synapses.

What would be the function of PVNs in processing olfactory information in the OB? To answer this question, researchers combined in vivo Ca^{2+} imaging of MCs (Kato et al. 2012) with pharmacogenetic silencing (Magnus et al. 2011) of PVNs. Upon inactivation of local PVNs, the odorant responses of MTCs were elevated almost linearly without changing their tuning curve (Fig. 4.8d; Kato et al. 2013). This linear transformation of odorant tuning is consistent with the model that PVNs provide output gain control to MTCs, that is, scaling odorant response intensities of individual MTCs to allow intensity-invariant information processing.

Another example of using RV tracing to find an uncharacterized circuit component is the identification of local corticotropin releasing hormone (CRH)-expressing inhibitory interneurons in the EPL of the OB that connect to adult born GCs (Garcia et al. 2014; also see Sect. 4.2.5). Although CRH-expressing neurons in the paraventricular nucleus of the hypothalamus (PVN) are well known for mediating systemic stress response, the function of CRH in the OB was completely unknown. Therefore researchers analyzed gene knockout mice that lack CRH, and found that these animals had a reduced number of surviving adult born GCs. In addition, conditional loss of the CRH receptor in the adult-born GCs decreased their survival. Overexpression of a constitutively active form of the CRH receptor in the adult born GCs induced an increase in dendritic arborization and expression of synaptic proteins. Together, these data suggest that EPL interneurons provide CRH to newborn GCs and facilitate synaptogenesis in the OB. As CRH+ neurons in the EPL significantly overlap with PVNs, which are tuned to a wild range of odorants, this CRH-mediated assembly of newborn GCs can partially explain why an odorant-enriched environment increases the number of GCs (Arenkiel et al. 2011) in the OB. As highlighted in these studies, RV tracing can identify a previously uncharacterized circuit component and provide anatomical basis for dissecting its function. In the OB local circuits, many types of neurons remain to be functionally analyzed in future investigations.

4.3.2 Transsynaptic Tracing from the Olfactory Cortex

Olfactory information, after being processed in the OB local circuits, is then transferred to various regions of the olfactory cortex via long axonal projections of MCs and TCs (Fig. 4.8a). How the spatial glomerular map in the OB is anatomically represented in the olfactory cortex is a fundamental question that would help us to ultimately understand the nature of olfactory perception. Therefore, RV tracing was applied to generate a small number of starter cells in the different parts of the olfactory cortex, and the spatial organization of RV-labeled MCs and their corresponding glomeruli were analyzed in a 3D-reconstructed OB model (Miyamichi et al. 2011). This study revealed three principles that define cortical representations of the OB input.

First, individual cortical neurons receive direct input from MCs representing at least four glomeruli. This convergence of multiple MC inputs enables cortical neurons to integrate information from discrete olfactory channels. In addition, layer 1 GABAergic neurons in the piriform cortex receive input from a greater number of glomeruli than layer 2/3 pyramidal cells, consistent with slice physiology data (Poo and Isaacson 2009; Stokes and Isaacson 2010). This organization allows layer 1 GABAergic neurons to integrate broad olfactory information and provide global feedforward inhibition to the cortical pyramidal neurons. Second, neurons restricted to small olfactory cortical regions (as few as 2–4 neurons that are less than 100 μm apart) receive input from glomeruli that are broadly distributed in the OB (Fig. 4.9). This organization suggests that, in contrast to visual and somatosensory systems where cortical neurons maintain a peripheral topographic map, olfactory cortex discards the fine spatial map used in the OB. Third, different cortical areas receive differentially organized OB input. For example, the cortical amygdala (CoA) preferentially receives dorsal OB input, whereas the piriform cortex samples the whole OB without obvious bias. The lack of spatial organization in MC → piriform cortex connections revealed by RV tracing is also supported by fine axon tracing experiments from defined glomeruli (Sosulski et al. 2011) and individual MCs (Igarashi et al. 2012), supporting the reliability of RV tracing in the olfactory cortex.

In vivo Ca^{2+} imaging experiments demonstrated that individual piriform cortex neurons activated by specific odorants were distributed broadly across the piriform

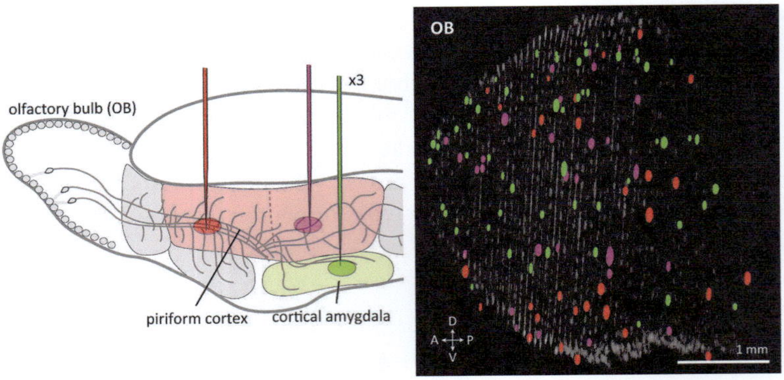

Fig. 4.9 Organization and function of olfactory cortical areas. Patterns of glomeruli retrogradely labeled by RV tracing from piriform cortex (*red* anterior; *magenta* posterior) and cortical amygdala (three samples combined, *green*). *Left* schematic of injection sites. *Right* standard OB glomeruli model and superimposed glomeruli map visualized by RV tracing after reconstruction into 3D space based on serial sections. *A* anterior; *P* posterior; *D* dorsal; *V* ventral. Adapted from with permission from Miyamichi et al. (2011). © 2015 from Principles of Neurobiology by Luo. Reproduced by permission of Garland Science/Taylor & Francis Group LLC

cortex without obvious spatial patterns (Stettler and Axel 2009). Seemingly random input from MCs to the piriform cortex provides an anatomical basis for this physiological data, together with intensive association fibers that connect broadly distributed piriform neurons (Franks et al. 2011). This organization allows piriform cortex neurons to receive a large number of combinations of different olfactory channels, and can serve as a basis on which olfactory experiences plastically adjust the strength of connections to create odorant representations based on an individual's experience. In contrast, MC axons originating from defined glomeruli exhibited stereotyped broad patches of innervation in the CoA (Sosulski et al. 2011), and many neurons in this structure received biased input that favors glomeruli in the dorsal OB (Miyamichi et al. 2011), which is known to mediate odorants with innately defined meanings (Kobayakawa et al. 2007).

To decipher the principles of information processing in each olfactory cortical area, more integrated connection diagrams are needed among olfactory cortical areas, which receive input from higher brain centers and also send outputs to various brain areas in the hypothalamus, frontal cortex, hippocampus, and ventral striatum. These maps should include detailed information about the types of pre- and postsynaptic neurons that are connected, and their synaptic properties. RV tracing and TRIO will be valuable tools to draw such connection diagrams in the future.

4.4 Transsynaptic Tracing in Neuromodulatory Systems

So far, the RV tracing studies we have described were performed on neuronal circuits with fairly specific functions in the nervous system. For instance, to relay detection of a specific odorant, MCs in the OB project to brain areas that are highly relevant for olfaction (see Sect. 4.3). But not all neuronal circuits in the brain have such a narrow function with regards to the information they relay, the brain areas they project to, or the behaviors they regulate. Frequently, it is important for the brain to adjust the activity of many neuronal circuits simultaneously, depending on internal and external cues, to modify behaviors accordingly. The process it uses to accomplish this is called neuromodulation (Marder 2012).

The specific cells that mediate this process are called neuromodulatory neurons, and they differ in several important ways from other types of neurons. First of all, whereas most neurons in the brain communicate via the rapid release of neurotransmitters, such as glutamate or GABA, at specific synaptic sites, neuromodulatory neurons can release their neurotransmitters more diffusely into the extracellular space. This bulk release of neurotransmitter allows neuromodulatory neurons to broadly activate both synaptic and extra-synaptic receptors (most commonly G-protein coupled receptors) that are located on target cells. This method of release allows neuromodulatory neurons to signal over longer distances, and slower and longer time courses, than synaptic communication. Another key difference is that classes of neuromodulatory neurons release different types of neurotransmitter, including monoamines [dopamine, serotonin, histamine, and norepinephrine (NE)], acetylcholine, and a variety of neuropeptides. Of note, it was long thought that different classes of neuromodulatory neurons release distinct neurotransmitters, but recent studies indicate that the neurotransmitter profiles of certain neuromodulatory neurons are actually quite diverse, and can contain combinations of fast neurotransmitters and neuromodulators (Stuber et al. 2010; Liu et al. 2013; Hnasko et al. 2010; Saunders et al. 2015). Overall though, the majority of neurons that express and secrete different types of neuromodulators can be found in largely segregated areas of the brain (Fig. 4.10a).

Now that the concept of neuromodulation has been introduced, let us quickly discuss the unique challenges that researchers face when studying the organization and function of these systems. Unlike many of the circuits that we have discussed so far, which have specialized roles, neuromodulatory neurons participate in a huge assortment of complex behaviors related to motivation, mood, attention, movement, sleep, and memory (to name a few!). Thus, these neurons must communicate with many anatomically and functionally distinct areas of the brain and spinal cord, making their anatomical organization exceedingly complex. Additionally, the location of the neuromodulatory neurons themselves presents several challenges. As mentioned above, the majority of neurons for each of the neuromodulatory systems can be found in specific regions of the brain (Fig. 4.10a). However, these regions are located far below the surface of the brain, and they often lack clear anatomical landmarks or boundaries, making them difficult to target with high

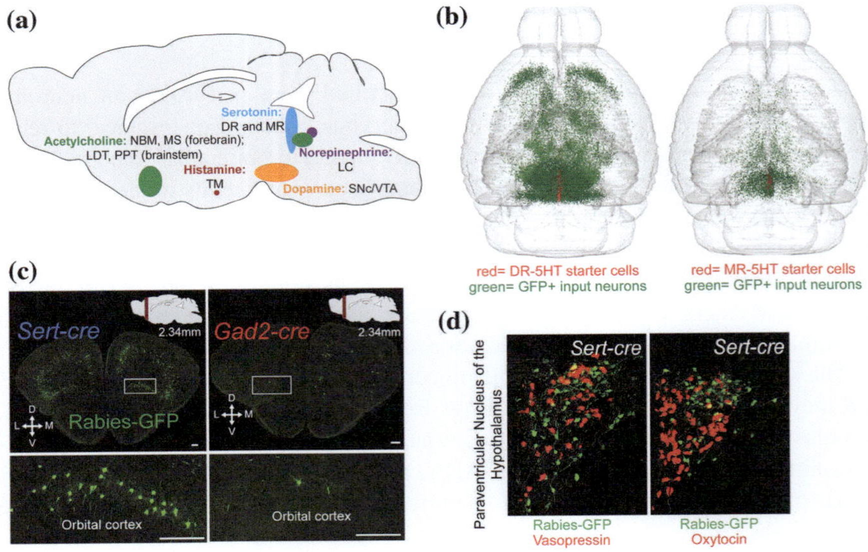

Fig. 4.10 Transsynaptic tracing studies in the serotonin system. **a** Schematic of cell body locations for the major neuromodulatory systems in the brain. **b** 3D reconstructions of whole brain monosynaptic inputs to 5-HT neurons in the DR (*left*) or MR (*right*). Starter cells generated in the DR or MR of *Sert-Cre* mice are labeled in *red*. RV*dRG-GFP* infected input neurons are labeled in *green*. Adapted with permission from Pollak Dorocic et al. (2014). **c** Examples of RV transsynaptic spread from either 5-HT+ or Gad2+ starter cells within the DR. DR-5HT neurons receive much more input from prefrontal cortical areas than DR-GABA neurons. *D* dorsal; *V* ventral; *L* lateral; *M* medial. Adapted with permission from Weissbourd et al. (2014). **d** RV tracing (*green*) was combined with in situ hybridization labeling (*red*) to identify specific cell types within the PVN that provide input to DR-5HT neurons. Adapted with permission from Weissbourd et al. (2014). Abbreviations: *DR* dorsal raphe; *LC* locus coeruleus; *LDT* laterodorsal tegmental nucleus; *MR* median raphe; *MS* medial septal nucleus; *NBM* nucleus basalis of Meynert; *PVN* paraventricular nucleus of the hypothalamus; *PPT* pedunculopontine nucleus; *TM* tuberomammilary nucleus; *SNc* substantia nigra compacta; *VTA* ventral tegmental nucleus. Scale bars: 250 μm (**c**)

specificity. In addition, neuromodulatory neurons within these brain areas are often interspersed with other cell types, such as excitatory or inhibitory neurons. This heterogeneity limits the accuracy with which neuromodulatory neurons can be specifically manipulated for experiments, while not affecting neighboring cells.

Thus, genetically regulated RV tracing provides several advantages for studying the anatomical organization of neuromodulatory systems. The robustness of the RV spread allows inputs to be visualized throughout the entire brain, which is ideal for neuromodulatory neurons that likely receive inputs from diverse brain regions. Also, the ability to precisely control which neurons become starter cells is beneficial for neuromodulatory circuit tracing, due to the challenges discussed above. Because of these advantages, RV tracing of neuromodulatory systems has produced many new insights into how the anatomy of these systems contributes to their functional diversity in the brain.

4.4.1 *Transsynaptic Tracing in the Serotonin System*

The main source of serotonin (5-HT) in the brain comes from two adjacent areas at the midline of the midbrain and brainstem, called the dorsal and median raphe (DR and MR, Fig. 4.10a). Neurons that express and release 5-HT ($\sim$20,000 in the rat brain) are heterogeneously distributed throughout these structures, along with large populations of glutamatergic, GABAergic, and peptidergic neurons (Bang and Commons 2012; Hioki et al. 2010). In fact, it is estimated that only a quarter of the neurons within the DR and MR express 5-HT (Steinbusch et al. 1980)! Release of 5-HT from DR and MR neurons can regulate mood, arousal, food seeking, sexual behavior, pain modulation, and aggression. Dysfunction in 5-HT release is also strongly correlated with several severe neurological disorders, such as depression and anxiety, and drugs that change brain levels of 5-HT are the most commonly prescribed treatment for these diseases (Müller and Jacobs 2010). Therefore, identifying the organizational principles controlling delivery of 5-HT throughout the brain has both physiological and clinical implications.

While inputs to the DR and MR regions of the brain have been classically characterized (Aghajanian and Wang 1977), the variability of cell types within these structures made it impossible to discretely determine which inputs were connected to which cell types. Thus, the ability to restrict the transsynaptic spread of RV from specific starter cells was a huge advancement for determining input connectivity within the 5-HT system. Utilizing the Cre-dependent transsynaptic tracing system (Sect. 4.2.3), several groups recently characterized the inputs received by specific populations of neurons within the DR and MR, shedding light on how these neurons may achieve their functional diversity (Weissbourd et al. 2014; Ogawa et al. 2014; Pollak Dorocic et al. 2014). Overall, the DR received more inputs than the MR, but the densest inputs into both regions came from neurons within the somatomotor cortices, amygdala, lateral hypothalamus, striatum, pallidum, inferior colliculus, and sensory and motor-related areas of the pons and medulla (Fig. 4.10b). MR-5HT neurons received less input from the prefrontal cortex, but had a preference for inputs from the hypothalamus and brainstem, compared to inputs targeting DR-5HT neurons. Meanwhile, DR-5HT neurons received more input from the CeA. Of importance, RV tracing illuminated several inputs onto DR-5HT neurons that had previously been disputed in the field; inputs arising from the prefrontal cortex and lateral habenula (LHb) (Fig. 4.10c). Using a combination of electrophysiology and optogenetics, the authors were able to functionally verify these anatomical projections, supporting the ability of RV reagents to accurately identify specific connections throughout the brain (Weissbourd et al. 2014; Pollak Dorocic et al. 2014; see also Fig. 4.2e).

While these studies comprehensively described the brain-wide inputs received by 5-HT neurons in the DR and MR, it is important to remember that extensive diversity also exists within these structures, which could contribute to differences in connectivity. To address this question, RV tracing was used to characterize the inputs received by two different cell types (5-HT and GABA) within the DR

(Weissbourd et al. 2014). Overall, these neuronal subsets had similar input profiles, with the densest number of inputs to both populations contributed by the hypothalamus, amygdala, medulla, and cortex. However, differences were also observed, such as the DR-5HT neurons receive twice as much input from the cortex, while DR-GABA neurons receive greater input from the CeA and BNST. Cortical to DR connections were also functionally verified using CRACM strategies (see Sect. 4.1.4). Another important advance of broad interest from this work was the pairing of RV tracing with in situ labeling methods to validate the cell type identity (glutamatergic, GABAergic, or peptidergic) for many of the inputs received by DR neuron populations (Fig. 4.10d). These experiments highlighted that differences in neuronal circuit anatomy exist on multiple levels: starter cells can receive input from different brain areas, and also receive input from different cell types within the same brain area.

4.4.2 Transsynaptic Tracing in the Norepinephrine System

Next, we will discuss the organization of another neuromodulatory system with significant functional overlap to the 5-HT system, specifically, the NE system. Structurally, the NE system is very different from many of the other neuromodulatory systems, in that the neurons supplying almost all of the brain's NE are restricted to a very tiny nucleus in the hindbrain, called the Locus Coeruleus (LC) (Fig. 4.10a). Also unlike other neuromodulatory structures, the LC is thought to be homogeneous in that all cells contained within it express NE ($\sim$1500 neurons per LC) (Swanson and Hartman 1975). This small population of neurons sends projections to almost all areas of the brain and spinal cord to release NE. LC-NE neurons are activated by a huge range of diverse behaviors broadly related to arousal, such as wakefulness, attention, memory formation, and stress response (Berridge and Waterhouse 2003). Thus, similar to the 5-HT system, an important question for LC-NE neurons is how they might be organized to mediate these different behavioral responses. One hypothesis is that subsets of LC neurons may be connected with distinct brain areas that participate in different arousal behaviors. Traditionally, this has been especially challenging to address for the NE system, as the small size of the LC makes it difficult to specifically target for anatomical studies. However, RV tracing and TRIO methodologies (see Sect. 4.2.4) have revealed new organizational properties regarding the brain-wide connectivity of NE neurons. RV tracing indicates that LC neurons receive inputs from over one hundred different brain areas (Schwarz et al. 2015), which is a much more diverse input profile than previously reported (Aston-Jones et al. 1986; Cedarbaum and Aghajanian 1978). However, TRIO and cTRIO studies illuminated that the NE system is highly integrative. Specifically, LC neurons projecting to diverse areas of the brain (OB, auditory cortex, hippocampus, cerebellum, or medulla) received similar distributions of inputs from widespread areas of the brain (Fig. 4.11a). Additionally, LC neurons were found to collateralize their axons extensively between many brain regions, further emphasizing a broad anatomical organization (Schwarz et al. 2015; Nagai et al. 1981; Room et al. 1981). This is in contrast to the dopamine system, where several specificities with regards to the input–output connectivity of these neurons have been observed (see Sect. 4.4.3 and Fig. 4.11c, d). The functional significance of this broad organization, and how LC neurons promote distinct arousal behaviors, have yet to be determined, but the pairing of viral-genetic tools with functional manipulations should provide additional insight.

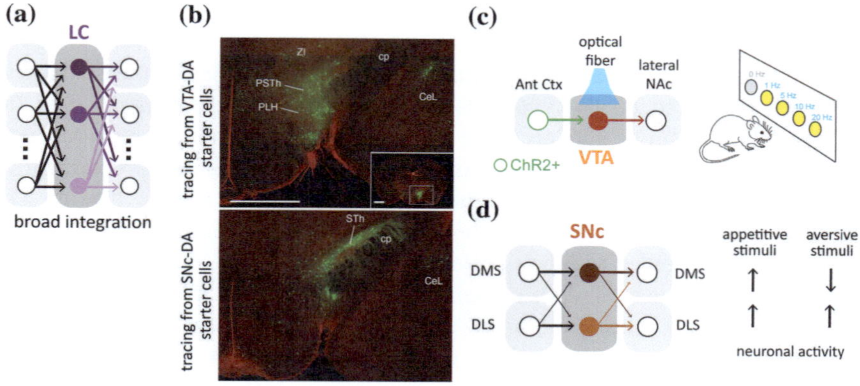

Fig. 4.11 Transsynaptic tracing studies in the Catecholamine systems. **a** Schematic of TRIO results from subpopulations of LC neurons projecting to different regions of the brain. Overall, it was observed that populations of LC neurons received similar input regardless of the brain regions that they projected to. Adapted with permission from Schwarz et al. (2015). **b** Unlike the norepinephrine system, differences were observed in the amount of RV-labeled input neurons for different populations of DA neurons. For instance, VTA-DA neurons received more input from the PLH (*top*), while SNc-DA neurons received more input from the STh (*bottom*). Scale bar = 1 mm. Adapted with permission from Watabe-Uchida et al. (2012). **c** TRIO identified a specific population of VTA-DA neurons that receive increased input from the Ant Ctx, and send specific projections to the lateral NAc. To identify a role of this circuit, researchers expressed ChR2 in Ant Ctx neurons while implanting an optical fiber over the VTA. Mice were subjected to an intracranial self-stimulation procedure, where nosepokes into different ports induced different frequencies of photostimulation. Experimental mice developed a strong preference for the 20 Hz stimulation, suggesting that Ant Ctx to VTA projections are reinforcing. Adapted with permission from Beier et al. (2015). **d** TRIO performed on populations of SNc-DA neurons also uncovered specificities. Anatomically, DMS- and DLS-projecting neurons make strong reciprocal connections (*left*). These populations of SNc-DA neurons also responded differently to appetitive (sucrose) or aversive (shock) stimuli (*right*). Adapted with permission from Lerner et al. (2015). Abbreviations: *Ant Ctx* anterior prefrontal cortex; *CeL* lateral division of the central amygdala; *cp* cortical peduncle; *DLS* dorsolateral striatum; *DMS* dorsomedial striatum; *DR* dorsal raphe; *DS* dorsal striatum; *LC* locus coeruleus; *NAc* nucleus accumbens; *PLH* peduncular part of the lateral hypothalamus; *PSTh* parasubthalamic nucleus; *SNc* substantia nigra compacta; *STh* subthalamic nucleus; *VTA* ventral tegmental nucleus; *ZI* zona incerta

4.4.3 Transsynaptic Tracing and Functional Analysis in the Dopamine System

Unlike the 5-HT and NE systems discussed above, dopaminergic (DA) neurons have traditionally been thought to have a more specific role in the brain: modulation of goal-directed behavior. DA neurons in the VTA and substantia nigra compacta (SNc) are activated by rewards, or cues indicating a reward is coming (Mirenowicz and Schultz 1996). This is supported by classic neuroanatomical data showing that midbrain DA neurons send strong projections to brain areas involved in goal-directed behavior, such as the striatum, nucleus accumbens (NAc), and frontal cortex (Beckstead et al. 1979). Studies have also indicated that several drugs of abuse increase DA signaling in these output areas to enhance their addictive properties (Bassareo et al. 1996). However, other studies demonstrate that the function of DA neurons is likely more complex, as they can also encode saliency, valence, aversion, memory formation, and movement coordination (Bromberg-Martin et al. 2010). Further supporting a more diverse role for DA neurons, subsets of DA neurons within the VTA and SNc differ in their electrophysiological properties and their projection targets (Lammel et al. 2011; Lammel et al. 2008; Margolis et al. 2008). Therefore, while it is increasingly clear that DA neurons have many roles in the brain, their organizational properties (e.g., who do they receive input from, and where do they project to?) are less obvious. To comprehensively address this question, several groups have utilized RV strategies.

The first study to do this performed transsynaptic RV tracing from DA neurons in the VTA and SNc to visualize the brain-wide inputs received by these neurons (Watabe-Uchida et al. 2012). Several important advances arose from this work. First of all, the authors introduced Cre-mediated RV transsynaptic tracing (see Sect. 4.2.3). Furthermore, the authors performed a brain-wide analysis of RV-labeled inputs to VTA and SNc-DA neurons to compare the synaptic input profiles received by these two populations. Their comprehensive analysis revealed that DA neurons in both the VTA and SNc receive direct synaptic inputs from many more brain areas than previously thought (such as many areas of the cortex), emphasizing the improved sensitivity of RV tracing for detecting anatomical connectivity of neurons over traditional tracing reagents (discussed in Sect. 4.1). While most of the input received by these two DA populations arose from similar brain areas, specific differences were also observed. For instance, DA neurons in the VTA received more input from the lateral hypothalamus, while SNc-DA neurons received preferential input from the subthalamic nucleus and somatosensory and motor cortices (Fig. 4.11B). The authors hypothesized that these input differences may underlie the preference of different DA neurons to encode saliency (cortical/thalamic inputs) versus value (hypothalamic inputs). RV tracing also resolved an ongoing debate in the field regarding which neurons in the SNc receive input from the striatum, as input tracing from DA neurons (using *DAT-Cre*) or GABA neurons (using *Vgat-Cre*) labeled distinct subsets of neurons in the striatum (Chuhma et al. 2011; Xia et al. 2011).

Thus, this study lays an important anatomical groundwork for how differences in the anatomical connectivity of DA circuits may affect their function. However, it focused on identifying inputs received by DA neurons, but it is also important to consider where DA neurons send their projections to. Also, to make the leap between anatomy and function, it is necessary to expand upon the observations made from anatomical studies with other technologies, such as electrophysiology, optogenetics, or pharmacology. As a first step towards this, non-pseudotyped RV labeling was paired with other labeling and electrophysiology methods to make several findings about connectivity differences for subsets of DA neurons. It was found that DA neurons located in the lateral VTA preferentially project to the NAc and receive input from the laterodorsal tegmentum (LDT) (Lammel et al. 2012). Meanwhile, medially located VTA-DA neurons project strongly to the prefrontal cortex (PFC) and receive input from the LHb. To explore this functionally, RV expressing channelrhodopsin-2 (RV-ChR2) was injected into the VTA, to retrogradely infect neurons in the LDT or LHb, and an optical fiber was implanted into the LDT or LHb. Phasic stimulation of LDT → VTA neurons in vivo induced robust conditioned place preference (CPP), meaning that the mice strongly preferred the side of a behavioral arena that was paired with light stimulation. Remarkably, activation of the LHb → VTA neurons had the opposite effect: strong conditioned place aversion (CPA) was induced in these animals. Overall, this study nicely highlights the power of combining RV tracing with other methods to uncover important specificities within seemingly similar neuronal populations (DA neurons of the VTA). However, experiments were focused on a small number of brain areas connected to the VTA. Furthermore, tracing was limited to DA neurons within the VTA. Yet, neurons within the VTA are known to receive input from many brain areas, and are molecularly heterogeneous. Therefore, it is likely that greater diversity exists within the DA system beyond the specific circuits characterized here (Watabe-Uchida et al. 2012; Margolis et al. 2012; Hnasko et al. 2012).

Toward this, two groups recently performed input–output anatomical and connectivity studies on a greater scale (Fig. 4.11c, d). Specifically, one study used RV tracing and TRIO methodologies to compare the inputs received by populations of VTA-DA neurons projecting to the lateral or medial shell of the NAc, the medial prefrontal cortex (mPFC), or the amygdala (Beier et al. 2015). The largest differences in connectivity were observed for the NAc-projecting populations, with VTA-DA neurons projecting to the lateral NAc receiving more input from the anterior cortex, dorsal striatum, and NAc, but less input from the dorsal raphe. Meanwhile, VTA-DA neurons projecting to the medial shell receive more medial shell input (forming a strong reciprocal connection), but much less striatal input. But how would the activation of these specific circuits affect the behavior of the mouse? Focusing on the anterior cortex → VTA → lateral NAc, the authors introduced ChR2 into the anterior cortex while placing an optical fiber over the VTA (Fig. 4.11c). The animals were then trained on intracranial self-stimulation protocol, where nose pokes induced stimulation of different frequencies (1–20 Hz). Overall, mice had a strong preference for the 20 Hz stimulation, indicating that anterior cortex → VTA connections could be strongly rewarding.

Finally, could there also be differences in the connectivity of SNc-DA neurons that further explain the diverse roles of DA release in the brain? Such a hypothesis seems plausible since it was observed that DA subtypes had differences in their anatomical connectivity (Watabe-Uchida et al. 2012). In addition, subsets of SNc-DA neurons vary in their firing rates (Schiemann et al. 2012). However, others have argued that SNc neurons may be functionally similar, as manipulating SNc-DA neurons projecting to either the dorsal medial striatum (DMS) or dorsal lateral striatum (DLS) similarly affects cognitive processes (Darvas and Palmiter 2010; Darvas and Palmiter 2009). To explore these potentially disparate results, TRIO was applied to trace the brain-wide inputs of either DMS or DLS-projecting SNc-DA neurons (Lerner et al. 2015). Whereas DMS-projecting SNc-DA neurons received strong input from NAc and DMS, DLS-projecting SNc-DA neurons received strong input from DLS, indicating a robust reciprocal connectivity for these SNc sub-circuits (Fig. 4.11d). But how might these anatomical differences relate to the animal's behavior? To address this, GCaMP6 was expressed in either DMS or DLS-projecting SNc neurons, and an optical fiber was implanted into SNc. Using fiber photometry, the activity of these neuronal subsets was monitored while the mice received either appetitive or aversive stimuli. DMS and DLS-projecting SNc-DA neurons showed similar activation patterns when the outcome was rewarding (sucrose delivery), but did not respond if sucrose was withheld. However, these neuron populations had opposing responses to aversive stimuli (foot shock), with DMS-projecting neurons showing a dip in their activity during the shock, while DLS-projecting neurons increased their activity (Fig. 4.11d). Together, these experiments emphasize that discrete circuits exist through the SNc to relay a variety of information related not only to reward, but also for aversive cues.

An independent study of the input–output organization of DA circuits additionally observed that a subset of VTA-DA neurons, projecting to the very posterior end of the striatum, had the greatest differences from other VTA-DA neurons with regards to the inputs they received (Menegas et al. 2015). Of note, this study failed to observe some of the differences in input–output connectivity reported in previous papers, though this could be explained by differences in technical aspects of the tracing methodologies. Still, this collection of papers nicely emphasize that RV-based anatomical studies provide important insight to guide more functional studies of relevant brain circuits.

4.4.4 Transsynaptic Tracing in Peptidergic Circuits

Beyond the main neuromodulatory systems discussed above, RV tracing has also been useful for determining the connectivity of complex peptidergic circuits in the brain. Many of these circuits are contained within the hypothalamus, which is an incredibly diverse brain region, both in terms of the many different cell types that are present there, and the many brain regions that it connects to. Thus, RV tracing has allowed researchers to target several intermingled peptidergic circuits within the hypothalamus with an accuracy and resolution that has not been possible with classic anatomical tools. For instance, neurons within the arcuate nucleus (ARC) of the hypothalamus express the neuropeptide called agouti-related peptide (AgRP) and regulate feeding behaviors (Fig. 4.12a). These neurons become activated by caloric deficiency, and their stimulation induces hunger and an increase in food intake (Atasoy et al. 2012; Krashes et al. 2011; Aponte et al. 2011). However, knowing which neurons upstream to AgRP cells are driving their activation is important for understanding how hunger and appetite are regulated in the brain. RV tracing from the ARC in *AgRP-Cre* animals showed that these neurons receive most of their input from neurons locally in the ARC, as well as from the PVN and dorsal medial hypothalamus (DMH) (Fig. 4.12b; Krashes et al. 2014). These anatomical connections were verified using CRACM methods. Furthermore, researchers were able to functionally link specific subsets of AgRP inputs to feeding behavior using Designer Receptor Exclusively Activated by Designer Drugs (DREADD) methodologies (Roth 2016). DREADD activation of PVN subpopulations (Pacap- or TRH-expressing neurons) induced robust feeding in animals, an effect that could be blocked if AgRP neurons in the ARC were simultaneously silenced (Fig. 4.12c; Krashes et al. 2014).

Additional studies have further characterized the projections of the ARC-AgRP neurons (Betley et al. 2013). Optogenetic stimulation of AgRP+ terminals (expressing ChR2) in different brain areas led to the interesting observation that activation of only some of the AgRP projections caused increased feeding. Anatomical experiments using RV to infect ARC-AgRP+ neurons from their axonal terminals suggested that the projections to different brain areas were discrete, with minimal collateralization between regions, implying that subsets of ARC-AgRP neurons have different roles in the brain.

More recently, RV has also been used to explore the connectivity of another cell type important for feeding behavior, the pro-opiomelanocortin (POMC) neurons. These cells are located both in the ARC (along with the AgRP neurons) (Fig. 4.12a), as well as in the nucleus tractus solitarius (NTS) (Cowley et al. 2001). Unlike AgRP neurons, stimulation of POMC neurons in the ARC or NTS suppresses feeding (Aponte et al. 2011; Zhan et al. 2013). Thus, comparing the circuit organization of these two cell types, especially within the ARC where their cell bodies are intermingled, greatly enhances our understanding of how these systems are coordinated to regulate feeding behaviors. Overall, RV tracing revealed that input patterns for POMC and AgRP neurons in the ARC were largely similar, which is perhaps surprising since these neurons have opposing functions (Wang et al. 2015). Such a finding might suggest that the anatomical organization of these circuits plays a minor role in mediating their activity. Instead, their function may be more directly controlled by other means, such as neuromodulation or hormonal signaling. Alternatively, while the distribution of GFP+ neurons received by ARC-POMC and ARC-AgRP neurons was similar, it could be that differences exist in the cell type identity of these inputs. In contrast, the inputs received by POMC neurons in the NTS were very different, originating from several nuclei in the pons

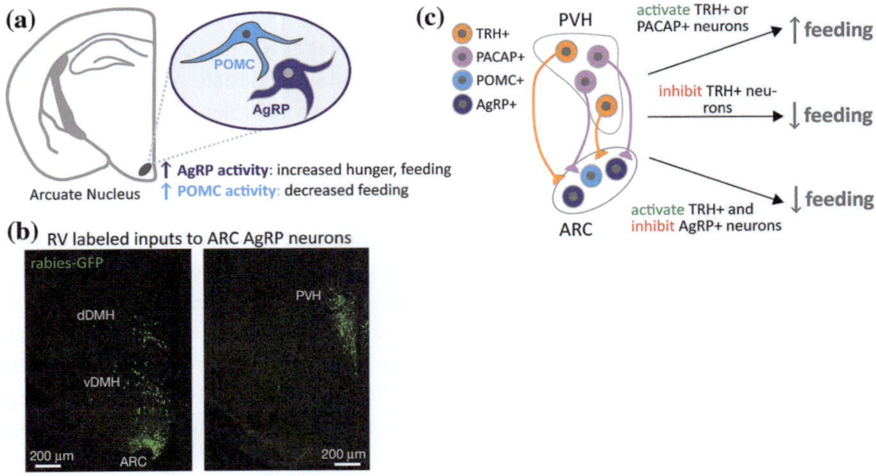

Fig. 4.12 Anatomical and functional characterization of peptidergic feeding circuits in the hypothalamus. **a** The ARC contains a diverse population of cell types, including POMC+ and AgRP+ neurons, whose activity have opposing effects on feeding behaviors (Atasoy et al. 2012; Krashes et al. 2011; Aponte et al. 2011; Zhan et al. 2013). **b** Monosynaptic spread of RV*dRG-GFP* from AgRP+ starter cells in the ARC identified inputs locally within the ARC, as well as inputs in distinct hypothalamic nuclei, such as the DMH and PVH. Adapted with permission from Krashes et al. (2014). **c** Using DREADD technology, it was found that specific cell types within the PVH send projections to AgRP+ neurons in ARC to mediate feeding. hM3Dq-mediated activation of PVH-TRH or PVH-PACAP neurons increased feeding behaviors in mice, an effect that could be blocked by simultaneous inhibition of ARC-AgRP neurons. Inhibition of PVH-TRH neurons also reduced normal food intake. Adapted with permission from Krashes et al. (2014). Abbreviations: *AgRP* agouti-related peptide; *ARC* arcuate nucleus; *DMH* dorsal medial hypothalamus; *PACAP* pituitary adenylate cyclase-activating polypeptide; *POMC* pro-opiomelanocortin; *PVH* paraventricular nucleus of the hypothalamus; *TRH* thyrotropin-releasing hormone

and brainstem. This result indicates that distinct circuits exist in the brain (ARC-POMC, NTS-POMC) to suppress feeding behaviors.

Together, the studies described in this section highlight the huge advances that RV reagents have provided for uncovering the anatomical connectivity of complex circuits in the brain. In addition, we hope that we have emphasized how viral tracing can be used in parallel with other methods to discern the function of specific neuronal populations to regulate diverse behaviors.

4.5 Transsynaptic Tracing in the Spinal Cord

In the motor system, each skeletal muscle is controlled by a pool of MNs, the cell bodies of which are clustered in stereotypic positions in the ventral horn of the spinal cord (Fig. 4.13a). MNs integrate information from multiple sources, including input from diverse spinal premotor neurons, descending commands from

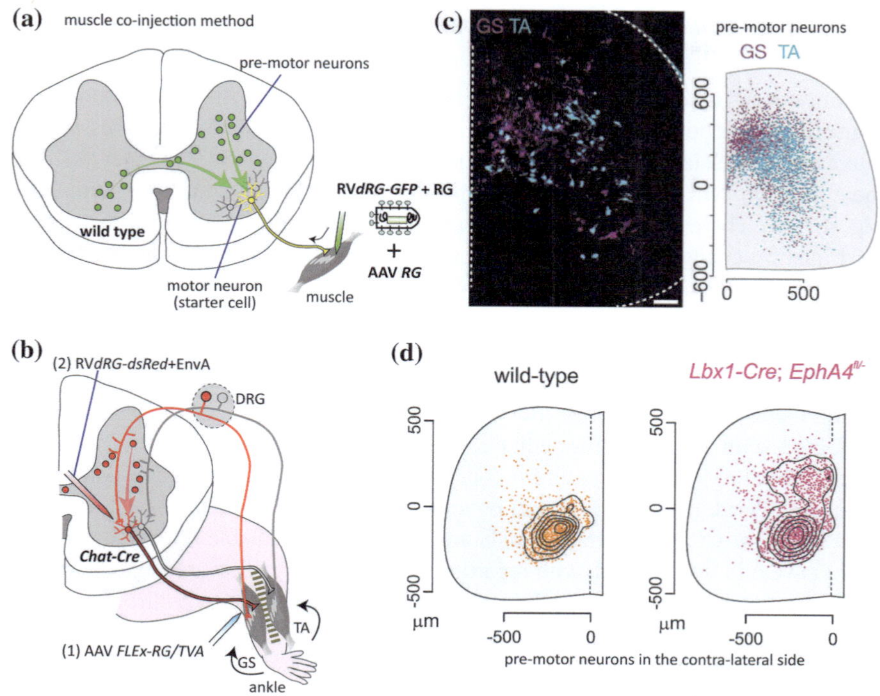

Fig. 4.13 Transsynaptic tracing studies in the spinal cord. **a** Schematic representation of the muscle coinjection method. Coinjection of RV*dRG-GFP*+RG and a helper AAV expressing RG into the target muscle can retrogradely label MNs and convert them to starter cells (one of which is shown in *yellow*). Their presynaptic partners are labeled in *green*. Adapted with permission from Stepien et al. (2010). **b** Schematic demonstration of synapse specificity of RV transsynaptic spread using sensory-motor circuits. Starter cells that are genetically restricted to the ankle muscle had transsynaptically labeled proprioceptive neurons in the dorsal root ganglion (*DRG*) connecting to the same GS (but not antagonistic TA) muscles. Note that in this experiment dsRed expressing RV was used. Adapted with permission from Reardon et al. (2016). **c** *Left* premotor neurons labeled by two distinct MN pools that regulate antagonistic muscles in the ankle, TA and GS. *Right* transverse projections of digital reconstructions showing distribution of GS and TA premotor neurons in the spinal cord. *Dots* indicate labeled neurons. Scale bar, 70 µm. Images are taken with permission from Tripodi et al. (2011). **d** Premotor neurons labeled from the contralateral MN pools innervating TA muscles in wild type (*left*) and in *Lbx1*-lineage specific *EphA4* conditional knockout mouse (*right*). Images are taken with permission from Satoh et al. (2016)

the brainstem motor-related areas and motor cortex, and sensory feedback from proprioceptive neurons. Coherent movements are achieved by the spatiotemporally appropriate recruitment of MNs through the activity of these diverse presynaptic neurons (Arber 2012; Goulding 2009). Understanding their connectivity rules will provide mechanistic insights into the coordinated activation of muscles. We discuss in this section how RV transsynaptic tracing has begun to reveal complex yet specific connection patterns in the spinal cord and brainstem underlying motor controls.

4.5.1 Methods to Generate Starter Cells in Motor Neurons

As introduced in Sect. 4.2.4, starter cells can be generated in the MNs belonging to a specific motor pool by coinjecting RV*dRG*+RG and a helper AAV expressing RG into a target skeletal muscle during the neonatal period in mice (Stepien et al. 2010) (hereafter we will call this scheme 'muscle coinjection method' for simplicity, Fig. 4.13a). There are three major advantages in the muscle coinjection method. First, just by changing a target muscle, researchers can easily generate starter cells in distinct MN pools. This enabled systematic investigation of the structure of premotor landscapes that regulate antagonistic movements of a joint (Tripodi et al. 2011), and coordinated movements of different limbs (Esposito et al. 2014; Goetz et al. 2015), as we will discuss in Sect. 4.5.2. Second, because the muscle coinjection method does not utilize mouse genetics for generating starter cells, it can be combined with various Cre-mediated viral and genetic tools, such as cell-type specific labeling, lineage tracing, axon projection mapping, and gene conditional knockout. Third, in principle, it can be applied to many mammalian species beyond mice in which Cre transgenic animals are currently unavailable. However, there is also a caveat in this method, with regards to the specificity of RV retrograde tracing. In the central nervous system, RV exhibits remarkable retrograde direction specificity (Ugolini 2011; Miyamichi et al. 2011; Zampieri et al. 2014). An important exception to this general rule was found in the peripheral sensory neurons. RV particles can directly infect proprioceptive sensory neurons in the dorsal root ganglion (DRG) (Tsiang et al. 1989) and ORNs in the olfactory epithelium (Astic et al. 1993), and can then *anterogradely* and transsynaptically spread to the post-synaptic neurons in the spinal cord and in the OB (Zampieri et al. 2014). Although the cellular mechanisms underlying this sensory neuron-specific anterograde RV spread remain elusive, this property of RV should be carefully considered when using the muscle coinjection method, as it can also generate starter cells in the proprioceptive sensory neurons in the DRG that connect to a target muscle.

At the neuromuscular junction, MNs release acetylcholine (ACh) as a neurotransmitter. To genetically restrict the starter cells to MNs, researchers utilized a transgenic mouse line expressing Cre under the promoter of choline acetyltransferase (*ChAT*) gene, which encodes the specific enzyme that synthesizes ACh. To avoid potential 'contamination' of *ChAT* positive non-MN local neurons in the spinal cord (Stepien et al. 2010), researchers injected Cre-dependent AAVs expressing RG and TVA into a target muscle, and RV*dRG*+EnvA into the spinal cord of *ChAT-Cre* mice (Fig. 4.13b; Reardon et al. 2016). In this configuration, proprioceptive sensory neurons in the DRG are genetically excluded from the starter cells because they do not express *ChAT*. With this 'clean' condition, synapse specificity of RV retrograde spread was elegantly demonstrated in known sensory-motor circuits controlling the ankle joint. Two distinct muscles in the ankle, the gastrocnemius (GS) and tibialis anterior (TA) muscles, work antagonistically: GS muscles act as extensors (whose contraction increases the angle of the ankle joint) and TA muscles act as flexors (whose contraction decreases the angle)

(Fig. 4.13b). Proprioceptive sensory neurons monitoring GS muscles connect to the MNs that innervate the same GS muscles (forming a monosynaptic loop), but not to the MNs inverting antagonistic TA muscles, despite the fact that dendrites of these two types of MNs are intermingled in the spinal cord (Frank and Westerfield 1983). If RV can nonsynaptically spread to nearby neurons, retrograde tracing from the MNs innervating GS muscles will mislabel proprioceptive sensory neurons connecting to TA muscles. It turned out that RV-positive sensory endings in the GS muscles were clearly observed, but not in the TA muscles (Reardon et al. 2016), demonstrating a tightly controlled synaptic spread of RV particles in vivo (Fig. 4.13b).

Although utilizing *ChAT-Cre* mice and Cre-dependent AAVs injected into a muscle provides better restriction of starter cells to a defined MN pool, this method compromises the utility of Cre-mediated tools for other labeling and genetic manipulations beyond generating starter cells. The muscle coinjection method may be beneficial if RV anterograde tracing from the proprioceptive starter cells are limited in the experimental condition. Researchers, therefore, compared patterns of the spinal premotor neurons labeled from three defined MN pools by two methods: the muscle coinjection method and *ChAT-Cre*-mediated genetic control of starter MNs (Goetz et al. 2015). Both methods showed very similar spatial distributions of labeled premotor neurons in the spinal cord that were highly characteristic to the starter MN pools. Thus, muscle coinjection method can be reliably used for the MNs tested in this study, and presumably for other MNs. However, careful control experiments should be conducted when the experimental conditions are changed. In the following sections, we will discuss biological findings obtained mainly by using the muscle coinjection method.

4.5.2 Organization of Presynaptic Neurons of a Defined Motor Neuron Pool

The muscle coinjection method allows researchers to assess global 3D distributions of presynaptic neurons that connect to a defined MN pool. Labeled premotor neurons are bilaterally distributed (Fig. 4.13a) across many spinal cord segments, and contain known premotor populations (Stepien et al. 2010; Levine et al. 2014). These premotor neurons consist of multiple subtypes that use different neurotransmitters including glutamate, GABA, glycine, and ACh. Overall distributions of spinal premotor neuron connectivity to an individual MN pool exhibit a high degree of reproducibility across animals. In contrast, analysis of premotor neurons connecting to MN pools with distinct function in motor behavior reveals striking differences in distribution (Tripodi et al. 2011; Goetz et al. 2015). As a remarkable example, Fig. 4.13c shows the spatially segregated distributions of premotor neurons that connect to MNs innervating the antagonistic ankle muscles, extensor GS and flexor TA, along the medial-lateral axis of the spinal cord. Analysis of developmental history revealed that these premotor interneuron populations were derived from common progenitor domains, but segregated by birthdate of neurons. Further, proprioceptive sensory feedback via the DRG was preferentially targeted to medial premotor populations connecting to the extensor MN pools. Together, RV tracing reveals the structural basis for controlling functionally distinct muscles at the premotor circuit level.

The complexity of spinal premotor neurons, including their diverse developmental origins, neurotransmitter types, and connection specificity, poses a major challenge for researchers to functionally dissect them in specific regulations of motor coordination. Here we focus on one example of motor control, the sequential stepping of left and right limbs during walking. This relatively simple locomotor behavior is generated by the rhythmic activity of MNs under the control of spinal neuronal networks known as central pattern generators (CPGs) that comprise of multiple interneuron cell types (Goulding 2009). A gene knockout study demonstrated that EphA4, a tyrosine kinase axon guidance receptor expressed by many excitatory spinal neurons, played an important role in normal walking in mice. Wild-type mice alternate their left and right limbs during walking, while the mutant mice synchronize their limbs (rabbit-like hopping rather than walking). What would be the underlying circuit alterations responsible for walking abnormality in the mutant mice? Researchers have begun to shed light on this issue by using the muscle coinjection method. Labeling of TA flexor premotor neurons in *EphA4* mutant mice led to the identification of abnormal synaptic input to the MNs from contralateral dorsal interneurons, a population defined by a transient expression of a transcription factor Lbx1 during development (Satoh et al. 2016). In an *EphA4* conditional knockout (cKO) mouse where *EphA4* was removed specifically from Lbx1+ neurons during development, approximately seven-times more TA premotor interneurons were labeled in the dorsal contralateral quadrant compared with WT control (Fig. 4.13d). As these Lbx1-lineage spinal interneurons directly receive proprioceptive sensory information, abnormal connectivity in *EphA4* mutant mice can transmit the sensory feedback signals to both sides of the spinal cord, while in wild type mice they are dominantly transmitted to the ipsilateral side. As a result, the cKO mice displayed different locomotor patterns depending on the strength of sensory feedback. While they showed normal alternating gait during walking (with strong sensory feedback), the same mice exhibited synchronous gait strokes during swimming (without strong sensory feedback), akin to a butterfly-like swimming style, in sharp contrast to a crawling-like swimming style with limb alternation observed in wild-type mice. Thus, Lbx1-lineage specific cKO of *EphA4* induces specific connectivity changes in sensory-relay interneurons that influence the robustness of gait choice during locomotion. This work also demonstrates the power of combining the muscle coinjection method with mouse genetics for the analysis of neuronal circuits in specific gene loss-of-function models.

The Lbx1-lineage spinal interneurons are only a part of the CPG network for walking. A homeobox transcription factor Dbx1, which defines the fate of spinal interneurons called V0 interneurons during development, also impacts the left–right alternations during walking (Lanuza et al. 2004). Genetic and cell-type specific ablation of V0 interneurons caused a similar abnormal hopping phenotype at all locomotion speeds, while more selective ablation of inhibitory or excitatory V0 interneurons showed abnormal hopping at a specific range of speed (Talpalar et al. 2013). More comprehensive connection studies in the future should reveal a full picture of spinal pattern generators and their modulators that underlie locomotion in mice.

4.5.3 The Brainstem Nuclei for Motor Controls

In addition to the local spinal premotor interneurons, RV tracing also labels long-distance brainstem nuclei that are directly presynaptic to a defined MN pool in the spinal cord. These nuclei may bridge between the higher brain areas that command motor controls and spinal networks for motor output. A salient example is the identification of premotor neurons in a brainstem nucleus called medullary reticular formation ventral part (MdV), which preferentially innervate MNs controlling forelimb over those regulating hindlimb (Fig. 4.14a; Esposito et al. 2014). MdV premotor neurons are glutamatergic (*vGluT2+*) and innervate a specific subset of forelimb MNs as well as spinal interneurons. To decipher their function in motor control, two lines of experiments were performed. First, Cre-dependent RV transsynaptic tracing using *vGluT2-Cre* mice demonstrated that glutamatergic neurons in the MdV integrated various inputs from upper motor centers including motor cortex, deep cerebellar nuclei (which convey the output signals of the cerebellum), superior colliculus, and other brainstem nuclei. Second, cell-type specific ablation or pharmacogenetic silencing of the MdV glutamatergic neurons markedly impaired skilled motor behaviors, such as accelerating rotarod task, without affecting a simple motor behavior such as running around the home cage. Thus, while simple locomotion seems to be regulated in the spinal pattern generators, skilled behavior requires higher brain centers.

Another important brainstem area that is presynaptic to MNs is the lateral vestibular nucleus (LVe). Neurons in this region convey information from the peripheral vestibular system in the inner ear about movement and orientation of the head. The vestibular feedback signals can be used for continuous postural adjustments during movement, which is necessary for reliable motor behaviors. Researchers applied the muscle coinjection method to limb muscles and found direct premotor neurons in the LVe (Fig. 4.14b; Basaldella et al. 2015). To analyze the type of MNs that LVe neurons provide vestibular feedback signals to, axon projection mapping was used in the LVe, with putative postsynaptic MNs labeled retrogradely from the target muscles. Two layers of specificity for LVe connectivity to MNs were highlighted. LVe axons preferentially contacted extensor (GS) over flexor (TA) motor pools. In addition, LVe axons targeted slow MN subtypes over fast-fatigable subtypes within the extensor pools. What is the importance of this target specificity? Individual MNs vary in size and small MNs innervate relatively few muscle fibers to form motor units that generate small forces (called slow motor

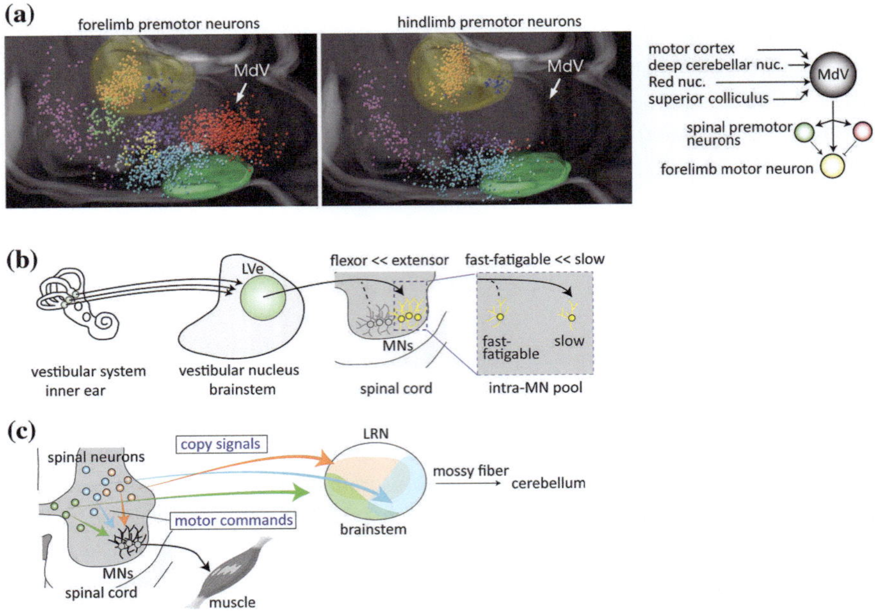

Fig. 4.14 Neuronal circuits in the brainstem motor control areas by RV transsynaptic tracing. **a** *Left images* distribution of brainstem premotor neurons that innervate the forelimb (*left*) or hindlimb (*right*) muscles, revealed by the muscle coinjection method. *Dots* represent premotor neurons, which are differentially colored according to their locations in specific brainstem nuclei. *Yellow area* vestibular nuclei; *green area* inferior olive. Distributions of forelimb and hindlimb premotor neurons are comparable in most brainstem nuclei, except in MdV, which contains almost exclusively forelimb premotor neurons. Right schematic summarizes the connection diagram of MdV, which collects broad motor-related inputs from higher brain centers and innervates forelimb MNs (*yellow circle*), as well as excitatory (*green circle*) and inhibitory (*red circle*) spinal premotor neurons. Adapted with permission from Esposito et al. (2014). **b** Schematic summary of vestibular feedback signals to MNs via lateral vestibular nucleus (*LVe*) in the brainstem. LVe neurons preferentially innervate extensor over flexor MN pools, and within the extensor MN pools, slow over fast-fatigable MNs. Adapted with permission from Basaldella et al. (2015). **c** Schematic summary of spinal premotor projection neurons that send copy signals to the lateral reticular nucleus (*LRN*) in the brainstem, which relays the copy signals to the cerebellum via mossy fibers. Spinal premotor projection neurons, colored according to different progenitor origins, send projections to distinct spatial domains in the LRN. Adapted with permission from Pivetta et al. (2014)

units). This is especially important for sustained muscle contraction, such as the posture maintenance. In contrast, the larger MNs innervate larger muscle fibers that generate more forces, but are easily fatigued (called fast-fatigable motor units), which play important roles in brief exertions of large forces such as running or jumping (Kanning et al. 2010). Thus, LVe neurons are specifically programmed such that the vestibular feedback signals are preferentially conveyed to the subtype of MNs that regulate muscle movements for posture maintenance.

So far we have discussed brainstem to spinal cord connections, but it is known that the opposite flow, from spinal cord to brainstem, also exists. For example, the lateral reticular nucleus (LRN) in the brainstem receives axonal projections from spinal premotor neurons and relays their activities to the cerebellum via mossy fibers. This signal strongly correlates with ongoing spinal intrinsic information ('copy signals') and therefore enables the cerebellum to compare planned actions and ongoing performance of motor control in the spinal cord. To analyze this circuit in detail, researchers applied the muscle coinjection method and detected axonal projections of spinal premotor neurons in specific areas of the LRN (Pivetta et al. 2014). To directly link the spinal input to the cerebellum, TRIO strategy (Sect. 4.2.4) was used and revealed that LRN monosynaptically relayed a diverse set of premotor input associated with functionally distinct MN pools to the cerebellum (Fig. 4.14c). Then, axon projection mapping was used to analyze the topography, the neurotransmitter types, and developmental origin of spinal premotor neurons that project to the LRN. This analysis revealed that both excitatory and inhibitory spinal inputs were highly co-localized within the LRN, but neurons in the different spinal cord domains (cervical or lumbar) innervated distinct LRN domains, showing topographic organization. Also, spinal neurons of different progenitor origins (marked by a transient expression of a specific transcription factor during development) established distinct axonal terminations in the LRN. Collectively, these data provide the anatomical basis upon which functional dissection studies of individual copy signals (Azim et al. 2014; Fink et al. 2014) can be integrated to decipher the principles of accurate motor task execution.

The brainstem also contains MNs regulating cranial muscles. For example, whisking (bilaterally coordinated rhythmic sweeping movements of whiskers to detect the texture, shape, and location of objects) is mediated by a specific type of MNs called vibrissal facial motor neurons (vFMNs) in the lateral facial nucleus in the brainstem. Studies of vFMN premotor pools provide an interesting example of utilizing RV tracing to analyze the developmental timecourse of circuit assembly. Researchers injected RV*dRG-GFP*+RG into the whisker-controlling muscles of a transgenic mouse where all MNs were expressing RG (Takatoh et al. 2013). This strategy permits a side-by-side comparison of labeled premotor neurons before (postnatal day 8) and after (postnatal day 15) the developmental onset of coordinated whisking behaviors. It turned out that the emergence of whisking was associated with the addition of new sets of bilateral excitatory inputs to vFMNs from neurons located in a brainstem nucleus called LPGi (lateral paragigantocellularis). Axon projection mapping indicated that neurons in the LPGi relayed motor commands from the motor cortex to the vFMNs. Therefore, this study shed light on how neuronal circuit assembly during development facilitates the bilateral coordination and cortical control of whisking.

Together, these studies of premotor neurons highlight the rapid advances that RV tracing have provided for uncovering complex yet elegant connection patterns in the spinal cord and brainstem underlying motor controls. These studies also nicely combined RV tracing with other viral and genetic tools, such as lineage tracing, cell-type specific axon projection mapping, gene conditional knockout, and manipulation of neuronal activities. Many techniques and ideas developed in the motor control system should facilitate anatomical, developmental, and functional studies of other neuronal circuits based on RV tracing.

4.6 Future Perspectives

As we have discussed in the above sections, RV-mediated transsynaptic tracing has vastly expanded the scope of neuroanatomy by providing precise information of presynaptic partners to defined neuron populations throughout the brain and spinal cord. We will now end by briefly considering several future directions of transsynaptic tracing techniques.

First, the range of species and developmental stages to which RV tracing can be applied should be extended. Currently, RV tracing has been most successfully used in adult mice. Just like comparative genomics has greatly expanded our understanding of the evolution of gene families, comparative connectomics are essential to illuminate the evolution of neuronal circuits. For example, neuronal circuits regulating social behaviors may drastically change as animals develop in more sophisticated communities. Likewise, circuits regulating feeding and energy expenditure might be altered when animals evolve a specific feeding habit. To achieve these studies, it is necessary to establish RV tracing methods in various mammalian species. To further expand tracing studies beyond mammalian species, new viral tools would be required. For example, a recent study showed that VSV (see Table 4.2) can be a better tracer than RV in some bird and fish species (Mundell et al. 2015). RV tracing also has a potential to reveal developmental processes (Takatoh et al. 2013), but current AAV-based methods require a $\sim$2 week incubation period before RV tracing is initiated. Future technical advances are needed to fully utilize RV tracing in embryonic and early postnatal animals, to decipher the developmental assembly of neuronal circuits in the central brain. In addition to extending retrograde tracing techniques, developing a more efficient and less toxic anterograde transsynaptic tracing system will further facilitate the tracing studies, as currently available anterograde viral vectors are either limited to specific cell types [e.g., sensory neurons (Zampieri et al. 2014; Astic et al. 1993)] or highly toxic to the infected neurons (see Table 4.2).

Second, a standard data acquisition and analysis platform is necessary to compare tracing data from various brain regions and cell types obtained from multiple laboratories. Typically, the acquisition of tracing data has been achieved by serially sectioning the brain, imaging individual sections, manually defining regional boundaries in each section, and counting the number of cells in each region (Weissbourd et al. 2014; Schwarz et al. 2015; Watabe-Uchida et al. 2012; Beier et al. 2015). These processes are not only labor-intensive and require significant expertise of brain anatomy, but also are prone to human-based errors and biases. Ideally, tracing data should be acquired with fully automatic brain imaging methods, such as serial sectioning 2-photon microscopy (Velez-Fort et al. 2014) or optical clearing methods followed by light-sheet microscopy (Lerner et al. 2015; Menegas et al. 2015; Niedworok et al. 2012). An automated data analysis platform for the classification of labeled cells into distinct brain regions should also be established. Often, fine regional boundaries between brain sub-nuclei are not apparent based on the gross morphology of the brain, and remain a subject of dispute among anatomy researchers. Rapid advances in the field of machine learning, as well as improvements in automatic brain imaging methods, may provide powerful solutions to this issue. Once precise methods to image brains, to extract information of labeled cells, and to register them into a common reference brain are established, it will become possible to quantitatively compare input neurons to various starter cell types throughout the brain. A powerful algorithm to compare the 'similarity' of labeled patterns among samples, similar to BLAST in the field of genomics, will be highly useful. This will permit systematic comparisons of connection diagrams in a variety of contexts; for example, male versus female, young versus aged, healthy versus diseased, and before and after animals acquire new tasks. The impacts of genetic modifications (such as conditional gene knockout as pioneered by DeNardo et al. 2015, Satoh et al. 2016) on neuronal circuit assembly and integrity can also be systematically investigated.

Lastly, RV tracing should directly link connectivity to circuit function. One important advantage of RV over other neurotrophic viruses is that neurons survive much longer following viral infection. This enables functional imaging and perturbation of neurons that have been identified based on their connectivity. Indeed, researchers have developed various RV vectors for expressing molecules that manipulate and measure neuronal activity, or express a recombinase (Cre or Flp) to induce expression of a second transgene (Osakada et al. 2011; Reardon et al. 2016). For example, RV-mediated ChR2-expressing cortical neurons elicited light-induced action potentials a month after infection, and GCaMP6f-expressing hippocampal projection neurons showed neuronal burst activities in behaving mice at least two weeks after infection (Reardon et al. 2016). Many of these, or analogous constructs, have begun to dissect the complex neuronal circuits in the brain (see Sect. 4.2.6; Wertz et al. 2015, Yamawaki and Shepherd 2015) and in retina (Yonehara et al. 2013). Future studies will greatly expand this direction to dissect how circuit function is orchestrated by various input neurons in the brain and spinal cord.

References

Adelson JD, Sapp RW, Brott BK, Lee H, Miyamichi K, Luo L, Cheng S, Djurisic M, Shatz CJ (2016) Developmental sculpting of intracortical circuits by MHC Class I H2-Db and H2-Kb. Cereb Cortex 26(4):1453–1463. doi:10.1093/cercor/bhu243

Aghajanian GK, Wang RY (1977) Habenular and other midbrain raphe afferents demonstrated by a modified retrograde tracing technique. Brain Res 122(2):229–242

Aponte Y, Atasoy D, Sternson SM (2011) AGRP neurons are sufficient to orchestrate feeding behavior rapidly and without training. Nat Neurosci 14(3):351–355. doi:10.1038/nn.2739

Arber S (2012) Motor circuits in action: specification, connectivity, and function. Neuron 74(6): 975–989. doi:10.1016/j.neuron.2012.05.011

Arenkiel BR, Hasegawa H, Yi JJ, Larsen RS, Wallace ML, Philpot BD, Wang F, Ehlers MD (2011) Activity-induced remodeling of olfactory bulb microcircuits revealed by monosynaptic tracing. PLoS ONE 6(12):e29423. doi:10.1371/journal.pone.0029423

Astic L, Saucier D, Coulon P, Lafay F, Flamand A (1993) The CVS strain of rabies virus as transneuronal tracer in the olfactory system of mice. Brain Res 619(1–2):146–156

Aston-Jones G, Ennis M, Pieribone VA, Nickell WT, Shipley MT (1986) The brain nucleus locus coeruleus: restricted afferent control of a broad efferent network. Science 234(4777):734–737

Atasoy D, Betley JN, Su HH, Sternson SM (2012) Deconstruction of a neural circuit for hunger. Nature 488(7410):172–177. doi:10.1038/nature11270

Azim E, Jiang J, Alstermark B, Jessell TM (2014) Skilled reaching relies on a V2a propriospinal internal copy circuit. Nature 508(7496):357–363. doi:10.1038/nature13021

Bang SJ, Commons KG (2012) Forebrain GABAergic projections from the dorsal raphe nucleus identified by using GAD67-GFP knock-in mice. J Comp Neurol 520(18):4157–4167. doi:10.1002/cne.23146

Basaldella E, Takeoka A, Sigrist M, Arber S (2015) Multisensory signaling shapes vestibulo-motor circuit specificity. Cell 163(2):301–312. doi:10.1016/j.cell.2015.09.023

Bassareo V, Tanda G, Petromilli P, Giua C, Di Chiara G (1996) Non-psychostimulant drugs of abuse and anxiogenic drugs activate with differential selectivity dopamine transmission in the nucleus accumbens and in the medial prefrontal cortex of the rat. Psychopharmacology 124(4): 293–299

Bates P, Young JAT, Varmus HE (1993) A receptor for subgroup-a rous-sarcoma virus is related to the low-density-lipoprotein receptor. Cell 74(6):1043–1051. doi:10.1016/0092-8674(93)90726-7

Beckstead RM, Domesick VB, Nauta WJ (1979) Efferent connections of the substantia nigra and ventral tegmental area in the rat. Brain Res 175(2):191–217

Beier KT, Saunders A, Oldenburg IA, Miyamichi K, Akhtar N, Luo L, Whelan SP, Sabatini B, Cepko CL (2011) Anterograde or retrograde transsynaptic labeling of CNS neurons with vesicular stomatitis virus vectors. Proc Natl Acad Sci USA 108(37):15414–15419. doi:10.1073/pnas.1110854108

Beier KT, Steinberg EE, DeLoach KE, Xie S, Miyamichi K, Schwarz L, Gao XJ, Kremer EJ, Malenka RC, Luo L (2015) Circuit architecture of VTA dopamine Neurons revealed by systematic input-output mapping. Cell 162(3):622–634. doi:10.1016/j.cell.2015.07.015

Berridge CW, Waterhouse BD (2003) The locus coeruleus-noradrenergic system: modulation of behavioral state and state-dependent cognitive processes. Brain Res Brain Res Rev 42(1): 33–84

Betley JN, Cao ZF, Ritola KD, Sternson SM (2013) Parallel, redundant circuit organization for homeostatic control of feeding behavior. Cell 155(6):1337–1350. doi:10.1016/j.cell.2013.11.002

Bromberg-Martin ES, Matsumoto M, Hikosaka O (2010) Dopamine in motivational control: rewarding, aversive, and alerting. Neuron 68(5):815–834. doi:10.1016/j.neuron.2010.11.022

Callaway EM, Luo L (2015) Monosynaptic circuit tracing with glycoprotein-deleted rabies viruses. J Neurosci Off J Soc Neurosci 35(24):8979–8985. doi:10.1523/JNEUROSCI.0409-15.2015

Card JP, Whealy ME, Robbins AK, Enquist LW (1992) Pseudorabies virus envelope glycoprotein gI influences both neurotropism and virulence during infection of the rat visual system. J Virol 66(5):3032–3041

Cedarbaum JM, Aghajanian GK (1978) Afferent projections to the rat locus coeruleus as determined by a retrograde tracing technique. J Comp Neurol 178(1):1–16. doi:10.1002/cne.901780102

Chuhma N, Tanaka KF, Hen R, Rayport S (2011) Functional connectome of the striatal medium spiny neuron. J Neurosci Off J Soc Neurosci 31(4):1183–1192. doi:10.1523/JNEUROSCI.3833-10.2011

Cowan WM, Gottlieb DI, Hendrickson AE, Price JL, Woolsey TA (1972) The autoradiographic demonstration of axonal connections in the central nervous system. Brain Res 37(1):21–51

Cowley MA, Smart JL, Rubinstein M, Cerdan MG, Diano S, Horvath TL, Cone RD, Low MJ (2001) Leptin activates anorexigenic POMC neurons through a neural network in the arcuate nucleus. Nature 411(6836):480–484. doi:10.1038/35078085

Darvas M, Palmiter RD (2009) Restriction of dopamine signaling to the dorsolateral striatum is sufficient for many cognitive behaviors. Proc Natl Acad Sci USA 106(34):14664–14669. doi:10.1073/pnas.0907299106

Darvas M, Palmiter RD (2010) Restricting dopaminergic signaling to either dorsolateral or medial striatum facilitates cognition. J Neurosci Off J Soc Neurosci 30(3):1158–1165. doi:10.1523/JNEUROSCI.4576-09.2010

DeFalco J, Tomishima M, Liu H, Zhao C, Cai X, Marth JD, Enquist L, Friedman JM (2001) Virus-assisted mapping of neural inputs to a feeding center in the hypothalamus. Science 291 (5513):2608–2613. doi:10.1126/science.1056602

Deisseroth K (2015) Optogenetics: 10 years of microbial opsins in neuroscience. Nat Neurosci 18(9):1213–1225. doi:10.1038/nn.4091

DeNardo LA, Berns DS, DeLoach K, Luo L (2015) Connectivity of mouse somatosensory and prefrontal cortex examined with trans-synaptic tracing. Nat Neurosci 18(11):1687–1697. doi:10.1038/nn.4131

Deshpande A, Bergami M, Ghanem A, Conzelmann KK, Lepier A, Gotz M, Berninger B (2013) Retrograde monosynaptic tracing reveals the temporal evolution of inputs onto new neurons in the adult dentate gyrus and olfactory bulb. Proc Natl Acad Sci USA 110(12):E1152–E1161. doi:10.1073/pnas.1218991110

Esposito MS, Capelli P, Arber S (2014) Brainstem nucleus MdV mediates skilled forelimb motor tasks. Nature 508(7496):351–356. doi:10.1038/nature13023

Etessami R, Conzelmann KK, Fadai-Ghotbi B, Natelson B, Tsiang H, Ceccaldi PE (2000) Spread and pathogenic characteristics of a G-deficient rabies virus recombinant: an in vitro and in vivo study. J Gen Virol 81:2147–2153

Fink AJ, Croce KR, Huang ZJ, Abbott LF, Jessell TM, Azim E (2014) Presynaptic inhibition of spinal sensory feedback ensures smooth movement. Nature 509(7498):43–48. doi:10.1038/nature13276

Frank E, Westerfield M (1983) Development of sensory-motor synapses in the spinal cord of the frog. J Physiol 343:593–610

Franks KM, Russo MJ, Sosulski DL, Mulligan AA, Siegelbaum SA, Axel R (2011) Recurrent circuitry dynamically shapes the activation of piriform cortex. Neuron 72(1):49–56. doi:10.1016/j.neuron.2011.08.020

Fu Y, Tucciarone JM, Espinosa JS, Sheng N, Darcy DP, Nicoll RA, Huang ZJ, Stryker MP (2014) A cortical circuit for gain control by behavioral state. Cell 156(6):1139–1152. doi:10.1016/j.cell.2014.01.050

Garcia I, Quast KB, Huang L, Herman AM, Selever J, Deussing JM, Justice NJ, Arenkiel BR (2014) Local CRH signaling promotes synaptogenesis and circuit integration of adult-born neurons. Dev Cell 30(6):645–659. doi:10.1016/j.devcel.2014.07.001

Gerfen CR, Paletzki R, Heintz N (2013) GENSAT BAC cre-recombinase driver lines to study the functional organization of cerebral cortical and basal ganglia circuits. Neuron 80(6):1368–1383. doi:10.1016/j.neuron.2013.10.016

Goetz C, Pivetta C, Arber S (2015) Distinct limb and trunk premotor circuits establish laterality in the spinal cord. Neuron 85(1):131–144. doi:10.1016/j.neuron.2014.11.024

Goodpasture EW, Teague O (1923) Transmission of the virus of herpes febrilis along nerves in experimentally infected rabbits. J Med Res 44(2):139–184 (137)

Goulding M (2009) Circuits controlling vertebrate locomotion: moving in a new direction. Nat Rev Neurosci 10(7):507–518. doi:10.1038/nrn2608

Greig LC, Woodworth MB, Galazo MJ, Padmanabhan H, Macklis JD (2013) Molecular logic of neocortical projection neuron specification, development and diversity. Nat Rev Neurosci 14(11):755–769. doi:10.1038/nrn3586

Haubensak W, Kunwar PS, Cai H, Ciocchi S, Wall NR, Ponnusamy R, Biag J, Dong HW, Deisseroth K, Callaway EM, Fanselow MS, Luthi A, Anderson DJ (2010) Genetic dissection of an amygdala microcircuit that gates conditioned fear. Nature 468(7321):270–276. doi:10.1038/nature09553

Helmstaedter M, Briggman KL, Turaga SC, Jain V, Seung HS, Denk W (2013) Connectomic reconstruction of the inner plexiform layer in the mouse retina. Nature 500(7461):168–174. doi:10.1038/nature12346

Hioki H, Nakamura H, Ma YF, Konno M, Hayakawa T, Nakamura KC, Fujiyama F, Kaneko T (2010) Vesicular glutamate transporter 3-expressing nonserotonergic projection neurons constitute a subregion in the rat midbrain raphe nuclei. J Comp Neurol 518(5):668–686. doi:10.1002/cne.22237

Hnasko TS, Perez FA, Scouras AD, Stoll EA, Gale SD, Luquet S, Phillips PE, Kremer EJ, Palmiter RD (2006) Cre recombinase-mediated restoration of nigrostriatal dopamine in dopamine-deficient mice reverses hypophagia and bradykinesia. Proc Natl Acad Sci USA 103(23):8858–8863. doi:10.1073/pnas.0603081103

Hnasko TS, Chuhma N, Zhang H, Goh GY, Sulzer D, Palmiter RD, Rayport S, Edwards RH (2010) Vesicular glutamate transport promotes dopamine storage and glutamate corelease in vivo. Neuron 65(5):643–656. doi:10.1016/j.neuron.2010.02.012

Hnasko TS, Hjelmstad GO, Fields HL, Edwards RH (2012) Ventral tegmental area glutamate neurons: electrophysiological properties and projections. J Neurosci Off J Soc Neurosci 32(43):15076–15085. doi:10.1523/JNEUROSCI.3128-12.2012

Igarashi KM, Ieki N, An M, Yamaguchi Y, Nagayama S, Kobayakawa K, Kobayakawa R, Tanifuji M, Sakano H, Chen WR, Mori K (2012) Parallel mitral and tufted cell pathways route distinct odor information to different targets in the olfactory cortex. J Neurosci Off J Soc Neurosci 32(23):7970–7985. doi:10.1523/JNEUROSCI.0154-12.2012

Kanning KC, Kaplan A, Henderson CE (2010) Motor neuron diversity in development and disease. Annu Rev Neurosci 33:409–440. doi:10.1146/annurev.neuro.051508.135722

Kasthuri N, Hayworth KJ, Berger DR, Schalek RL, Conchello JA, Knowles-Barley S, Lee D, Vazquez-Reina A, Kaynig V, Jones TR, Roberts M, Morgan JL, Tapia JC, Seung HS, Roncal WG, Vogelstein JT, Burns R, Sussman DL, Priebe CE, Pfister H, Lichtman JW (2015) Saturated reconstruction of a volume of neocortex. Cell 162(3):648–661. doi:10.1016/j.cell.2015.06.054

Kato HK, Chu MW, Isaacson JS, Komiyama T (2012) Dynamic sensory representations in the olfactory bulb: modulation by wakefulness and experience. Neuron 76(5):962–975. doi:10.1016/j.neuron.2012.09.037

Kato HK, Gillet SN, Peters AJ, Isaacson JS, Komiyama T (2013) Parvalbumin-expressing interneurons linearly control olfactory bulb output. Neuron 80(5):1218–1231. doi:10.1016/j.neuron.2013.08.036

Kim EJ, Juavinett AL, Kyubwa EM, Jacobs MW, Callaway EM (2015) Three types of cortical layer 5 neurons that differ in brain-wide connectivity and function. Neuron 88(6):1253–1267. doi:10.1016/j.neuron.2015.11.002

Kiritani T, Wickersham IR, Seung HS, Shepherd GM (2012) Hierarchical connectivity and connection-specific dynamics in the corticospinal-corticostriatal microcircuit in mouse motor cortex. J Neurosci Off J Soc Neurosci 32(14):4992–5001. doi:10.1523/JNEUROSCI.4759-11.2012

Kobayakawa K, Kobayakawa R, Matsumoto H, Oka Y, Imai T, Ikawa M, Okabe M, Ikeda T, Itohara S, Kikusui T, Mori K, Sakano H (2007) Innate versus learned odour processing in the mouse olfactory bulb. Nature 450(7169):503–508. doi:10.1038/nature06281

Kohara K, Pignatelli M, Rivest AJ, Jung HY, Kitamura T, Suh J, Frank D, Kajikawa K, Mise N, Obata Y, Wickersham IR, Tonegawa S (2014) Cell type-specific genetic and optogenetic tools reveal hippocampal CA2 circuits. Nat Neurosci 17(2):269–279. doi:10.1038/nn.3614

Krashes MJ, Koda S, Ye C, Rogan SC, Adams AC, Cusher DS, Maratos-Flier E, Roth BL, Lowell BB (2011) Rapid, reversible activation of AgRP neurons drives feeding behavior in mice. J Clin Investig 121(4):1424–1428. doi:10.1172/JCI46229

Krashes MJ, Shah BP, Madara JC, Olson DP, Strochlic DE, Garfield AS, Vong L, Pei H, Watabe-Uchida M, Uchida N, Liberles SD, Lowell BB (2014) An excitatory paraventricular nucleus to AgRP neuron circuit that drives hunger. Nature 507(7491):238–242. doi:10.1038/nature12956

Lafon M (2005) Rabies virus receptors. J Neurovirol 11(1):82–87. doi:10.1080/13550280590900427

Lammel S, Hetzel A, Hackel O, Jones I, Liss B, Roeper J (2008) Unique properties of mesoprefrontal neurons within a dual mesocorticolimbic dopamine system. Neuron 57(5):760–773. doi:10.1016/j.neuron.2008.01.022

Lammel S, Ion DI, Roeper J, Malenka RC (2011) Projection-specific modulation of dopamine neuron synapses by aversive and rewarding stimuli. Neuron 70(5):855–862. doi:10.1016/j.neuron.2011.03.025

Lammel S, Lim BK, Ran C, Huang KW, Betley MJ, Tye KM, Deisseroth K, Malenka RC (2012) Input-specific control of reward and aversion in the ventral tegmental area. Nature 491(7423):212–217. doi:10.1038/nature11527

Lanuza GM, Gosgnach S, Pierani A, Jessell TM, Goulding M (2004) Genetic identification of spinal interneurons that coordinate left-right locomotor activity necessary for walking movements. Neuron 42(3):375–386

Lerner TN, Shilyansky C, Davidson TJ, Evans KE, Beier KT, Zalocusky KA, Crow AK, Malenka RC, Luo L, Tomer R, Deisseroth K (2015) Intact-brain analyses reveal distinct information carried by SNc dopamine subcircuits. Cell 162(3):635–647. doi:10.1016/j.cell.2015.07.014

Levine AJ, Hinckley CA, Hilde KL, Driscoll SP, Poon TH, Montgomery JM, Pfaff SL (2014) Identification of a cellular node for motor control pathways. Nat Neurosci 17(4):586–593. doi:10.1038/nn.3675

Li Y, Stam FJ, Aimone JB, Goulding M, Callaway EM, Gage FH (2013) Molecular layer perforant path-associated cells contribute to feed-forward inhibition in the adult dentate gyrus. Proc Natl Acad Sci USA 110(22):9106–9111. doi:10.1073/pnas.1306912110

Liu S, Plachez C, Shao Z, Puche A, Shipley MT (2013) Olfactory bulb short axon cell release of GABA and dopamine produces a temporally biphasic inhibition-excitation response in external tufted cells. J Neurosci Off J Soc Neurosci 33(7):2916–2926. doi:10.1523/JNEUROSCI.3607-12.2013

Lo L, Anderson DJ (2011) A cre-dependent, anterograde transsynaptic viral tracer for mapping output pathways of genetically marked neurons. Neuron 72(6):938–950. doi:10.1016/j.neuron.2011.12.002

Magnus CJ, Lee PH, Atasoy D, Su HH, Looger LL, Sternson SM (2011) Chemical and genetic engineering of selective ion channel-ligand interactions. Science 333(6047):1292–1296. doi:10.1126/science.1206606

Marder E (2012) Neuromodulation of neuronal circuits: back to the future. Neuron 76(1):1–11. doi:10.1016/j.neuron.2012.09.010

Margolis EB, Mitchell JM, Ishikawa J, Hjelmstad GO, Fields HL (2008) Midbrain dopamine neurons: projection target determines action potential duration and dopamine D(2) receptor inhibition. J Neurosci Off J Soc Neurosci 28(36):8908–8913. doi:10.1523/JNEUROSCI.1526-08.2008

Margolis EB, Toy B, Himmels P, Morales M, Fields HL (2012) Identification of rat ventral tegmental area GABAergic neurons. PLoS ONE 7(7):e42365. doi:10.1371/journal.pone.0042365

Mebatsion T, Schnell MJ, Cox JH, Finke S, Conzelmann KK (1996a) Highly stable expression of a foreign gene from rabies virus vectors. Proc Natl Acad Sci USA 93(14):7310–7314. doi:10.1073/pnas.93.14.7310

Mebatsion T, Konig M, Conzelmann KK (1996b) Budding of rabies virus particles in the absence of the spike glycoprotein. Cell 84(6):941–951. doi:10.1016/S0092-8674(00)81072-7

Menegas W, Bergan JF, Ogawa SK, Isogai Y, Umadevi Venkataraju K, Osten P, Uchida N, Watabe-Uchida M (2015) Dopamine neurons projecting to the posterior striatum form an anatomically distinct subclass. eLife 4:e10032. doi:10.7554/eLife.10032

Mirenowicz J, Schultz W (1996) Preferential activation of midbrain dopamine neurons by appetitive rather than aversive stimuli. Nature 379(6564):449–451. doi:10.1038/379449a0

Miyamichi K, Amat F, Moussavi F, Wang C, Wickersham I, Wall NR, Taniguchi H, Tasic B, Huang ZJ, He Z, Callaway EM, Horowitz MA, Luo L (2011) Cortical representations of olfactory input by trans-synaptic tracing. Nature 472(7342):191–196. doi:10.1038/nature09714

Miyamichi K, Shlomai-Fuchs Y, Shu M, Weissbourd BC, Luo L, Mizrahi A (2013) Dissecting local circuits: parvalbumin interneurons underlie broad feedback control of olfactory bulb output. Neuron 80(5):1232–1245. doi:10.1016/j.neuron.2013.08.027

Mori K, Sakano H (2011) How is the olfactory map formed and interpreted in the mammalian brain? Annu Rev Neurosci 34:467–499. doi:10.1146/annurev-neuro-112210-112917

Müller CP, Jacobs BL (2010) Handbook of the behavioral neurobiology of serotonin. Elsevier, Amsterdam

Mundell NA, Beier KT, Pan YA, Lapan SW, Goz Ayturk D, Berezovskii VK, Wark AR, Drokhlyansky E, Bielecki J, Born RT, Schier AF, Cepko CL (2015) Vesicular stomatitis virus enables gene transfer and transsynaptic tracing in a wide range of organisms. J Comp Neurol 523(11):1639–1663. doi:10.1002/cne.23761

Nagai T, Satoh K, Imamoto K, Maeda T (1981) Divergent projections of catecholamine neurons of the locus coeruleus as revealed by fluorescent retrograde double labeling technique. Neurosci Lett 23(2):117–123

Nakashiba T, Cushman JD, Pelkey KA, Renaudineau S, Buhl DL, McHugh TJ, Barrera VR, Chittajallu R, Iwamoto KS, McBain CJ, Fanselow MS, Tonegawa S (2012) Young dentate granule cells mediate pattern separation, whereas old granule cells facilitate pattern completion. Cell 149(1):188–201. doi:10.1016/j.cell.2012.01.046

Namburi P, Beyeler A, Yorozu S, Calhoon GG, Halbert SA, Wichmann R, Holden SS, Mertens KL, Anahtar M, Felix-Ortiz AC, Wickersham IR, Gray JM, Tye KM (2015) A circuit mechanism for differentiating positive and negative associations. Nature 520(7549):675–678. doi:10.1038/nature14366

Nassi JJ, Cepko CL, Born RT, Beier KT (2015) Neuroanatomy goes viral! Front Neuroanat 9:80. doi:10.3389/fnana.2015.00080

Ni Y, Nawabi H, Liu X, Yang L, Miyamichi K, Tedeschi A, Xu B, Wall NR, Callaway EM, He Z (2014) Characterization of long descending premotor propriospinal neurons in the spinal cord. J Neurosci Off J Soc Neurosci 34(28):9404–9417. doi:10.1523/JNEUROSCI.1771-14.2014

Niedworok CJ, Schwarz I, Ledderose J, Giese G, Conzelmann KK, Schwarz MK (2012) Charting monosynaptic connectivity maps by two-color light-sheet fluorescence microscopy. Cell reports 2(5):1375–1386. doi:10.1016/j.celrep.2012.10.008

Ogawa SK, Cohen JY, Hwang D, Uchida N, Watabe-Uchida M (2014) Organization of monosynaptic inputs to the serotonin and dopamine neuromodulatory systems. Cell Rep 8 (4):1105–1118. doi:10.1016/j.celrep.2014.06.042

Oh SW, Harris JA, Ng L, Winslow B, Cain N, Mihalas S, Wang Q, Lau C, Kuan L, Henry AM, Mortrud MT, Ouellette B, Nguyen TN, Sorensen SA, Slaughterbeck CR, Wakeman W, Li Y, Feng D, Ho A, Nicholas E, Hirokawa KE, Bohn P, Joines KM, Peng H, Hawrylycz MJ, Phillips JW, Hohmann JG, Wohnoutka P, Gerfen CR, Koch C, Bernard A, Dang C, Jones AR, Zeng H (2014) A mesoscale connectome of the mouse brain. Nature 508(7495):207–214. doi:10.1038/nature13186

Osakada F, Mori T, Cetin AH, Marshel JH, Virgen B, Callaway EM (2011) New rabies virus variants for monitoring and manipulating activity and gene expression in defined neural circuits. Neuron 71(4):617–631. doi:10.1016/j.neuron.2011.07.005

Oyibo HK, Znamenskiy P, Oviedo HV, Enquist LW, Zador AM (2014) Long-term Cre-mediated retrograde tagging of neurons using a novel recombinant pseudorabies virus. Front Neuroanat 8:86. doi:10.3389/fnana.2014.00086

Petreanu L, Huber D, Sobczyk A, Svoboda K (2007) Channelrhodopsin-2-assisted circuit mapping of long-range callosal projections. Nat Neurosci 10(5):663–668. doi:10.1038/nn1891

Pivetta C, Esposito MS, Sigrist M, Arber S (2014) Motor-circuit communication matrix from spinal cord to brainstem neurons revealed by developmental origin. Cell 156(3):537–548. doi:10.1016/j.cell.2013.12.014

Pollak Dorocic I, Furth D, Xuan Y, Johansson Y, Pozzi L, Silberberg G, Carlen M, Meletis K (2014) A whole-brain atlas of inputs to serotonergic neurons of the dorsal and median raphe nuclei. Neuron 83(3):663–678. doi:10.1016/j.neuron.2014.07.002

Poo C, Isaacson JS (2009) Odor representations in olfactory cortex: "sparse" coding, global inhibition, and oscillations. Neuron 62(6):850–861. doi:10.1016/j.neuron.2009.05.022

Rancz EA, Franks KM, Schwarz MK, Pichler B, Schaefer AT, Margrie TW (2011) Transfection via whole-cell recording in vivo: bridging single-cell physiology, genetics and connectomics. Nat Neurosci 14(4):527–532. doi:10.1038/nn.2765

Reardon TR, Murray AJ, Turi GF, Wirblich C, Croce KR, Schnell MJ, Jessell TM, Losonczy A (2016) Rabies virus CVS-N2c strain enhances retrograde synaptic transfer and neuronal viability. Neuron 89:1–14. doi:10.1016/j.neuron.2016.01.004

Rong L, Gendron K, Strohl B, Shenoy R, Wool-Lewis RJ, Bates P (1998) Characterization of determinants for envelope binding and infection in Tva, the subgroup A avian sarcoma and leukosis virus receptor. J Virol 72(6):4552–4559

Room P, Postema F, Korf J (1981) Divergent axon collaterals of rat locus coeruleus neurons: demonstration by a fluorescent double labeling technique. Brain Res 221(2):219–230

Roth BL (2016) DREADDs for neuroscientists. Neuron 89(4):683–694. doi:10.1016/j.neuron.2016.01.040

Satoh D, Pudenz C, Arber S (2016) Context-dependent gait choice elicited by EphA4 Mutation in Lbx1 spinal interneurons. Neuron 89:1–13

Saunders A, Granger AJ, Sabatini BL (2015) Corelease of acetylcholine and GABA from cholinergic forebrain neurons. eLife 4. doi:10.7554/eLife.06412

Schiemann J, Schlaudraff F, Klose V, Bingmer M, Seino S, Magill PJ, Zaghloul KA, Schneider G, Liss B, Roeper J (2012) K-ATP channels in dopamine substantia nigra neurons control bursting and novelty-induced exploration. Nat Neurosci 15(9):1272–1280. doi:10.1038/nn.3185

Schmued LC, Fallon JH (1986) Fluoro-Gold: a new fluorescent retrograde axonal tracer with numerous unique properties. Brain Res 377(1):147–154

Schnell MJ, Mebatsion T, Conzelmann KK (1994) Infectious rabies viruses from cloned cDNA. EMBO J 13(18):4195–4203

Schnutgen F, Doerflinger N, Calleja C, Wendling O, Chambon P, Ghyselinck NB (2003) A directional strategy for monitoring Cre-mediated recombination at the cellular level in the mouse. Nat Biotechnol 21(5):562–565. doi:10.1038/nbt811

Schwab ME, Javoy-Agid F, Agid Y (1978) Labeled wheat germ agglutinin (WGA) as a new, highly sensitive retrograde tracer in the rat brain hippocampal system. Brain Res 152(1): 145–150

Schwarz LA, Miyamichi K, Gao XJ, Beier KT, Weissbourd B, DeLoach KE, Ren J, Ibanes S, Malenka RC, Kremer EJ, Luo L (2015) Viral-genetic tracing of the input-output organization of a central noradrenaline circuit. Nature 524(7563):88–92. doi:10.1038/nature14600

Shepherd GM (2004) The synaptic organization of the brain, 5th edition, 5th edn. Oxford University Press, Oxford

Sosulski DL, Bloom ML, Cutforth T, Axel R, Datta SR (2011) Distinct representations of olfactory information in different cortical centres. Nature 472(7342):213–216. doi:10.1038/nature09868

Soudais C, Laplace-Builhe C, Kissa K, Kremer EJ (2001) Preferential transduction of neurons by canine adenovirus vectors and their efficient retrograde transport in vivo. FASEB J Off Publ Federation Am Soc Exp Biol 15(12):2283–2285. doi:10.1096/fj.01-0321fje

Spaete RR, Frenkel N (1982) The herpes simplex virus amplicon: a new eucaryotic defective-virus cloning-amplifying vector. Cell 30(1):295–304

Steinbusch HW, van der Kooy D, Verhofstad AA, Pellegrino A (1980) Serotonergic and non-serotonergic projections from the nucleus raphe dorsalis to the caudate-putamen complex in the rat, studied by a combined immunofluorescence and fluorescent retrograde axonal labeling technique. Neurosci Lett 19(2):137–142

Stepien AE, Tripodi M, Arber S (2010) Monosynaptic rabies virus reveals premotor network organization and synaptic specificity of cholinergic partition cells. Neuron 68(3):456–472. doi:10.1016/j.neuron.2010.10.019

Stettler DD, Axel R (2009) Representations of odor in the piriform cortex. Neuron 63(6):854–864. doi:10.1016/j.neuron.2009.09.005

Stokes CC, Isaacson JS (2010) From dendrite to soma: dynamic routing of inhibition by complementary interneuron microcircuits in olfactory cortex. Neuron 67(3):452–465. doi:10.1016/j.neuron.2010.06.029

Stuber GD, Hnasko TS, Britt JP, Edwards RH, Bonci A (2010) Dopaminergic terminals in the nucleus accumbens but not the dorsal striatum corelease glutamate. J Neurosci Off J Soc Neurosci 30(24):8229–8233. doi:10.1523/JNEUROSCI.1754-10.2010

Sun N, Cassell MD, Perlman S (1996) Anterograde, transneuronal transport of herpes simplex virus type 1 strain H129 in the murine visual system. J Virol 70(8):5405–5413

Sun Y, Nguyen AQ, Nguyen JP, Le L, Saur D, Choi J, Callaway EM, Xu X (2014) Cell-type-specific circuit connectivity of hippocampal CA1 revealed through Cre-dependent rabies tracing. Cell reports 7(1):269–280. doi:10.1016/j.celrep.2014.02.030

Swanson LW, Hartman BK (1975) The central adrenergic system. An immunofluorescence study of the location of cell bodies and their efferent connections in the rat utilizing dopamine-beta-hydroxylase as a marker. J Comp Neurol 163(4):467–505. doi:10.1002/cne.901630406

Takatoh J, Nelson A, Zhou X, Bolton MM, Ehlers MD, Arenkiel BR, Mooney R, Wang F (2013) New modules are added to vibrissal premotor circuitry with the emergence of exploratory whisking. Neuron 77(2):346–360. doi:10.1016/j.neuron.2012.11.010

Talpalar AE, Bouvier J, Borgius L, Fortin G, Pierani A, Kiehn O (2013) Dual-mode operation of neuronal networks involved in left-right alternation. Nature 500(7460):85–88. doi:10.1038/nature12286

Tamamaki N, Yanagawa Y, Tomioka R, Miyazaki J, Obata K, Kaneko T (2003) Green fluorescent protein expression and colocalization with calretinin, parvalbumin, and somatostatin in the GAD67-GFP knock-in mouse. J Comp Neurol 467(1):60–79. doi:10.1002/cne.10905

Tripodi M, Stepien AE, Arber S (2011) Motor antagonism exposed by spatial segregation and timing of neurogenesis. Nature 479(7371):61–66. doi:10.1038/nature10538

Tsiang H, Lycke E, Ceccaldi PE, Ermine A, Hirardot X (1989) The anterograde transport of rabies virus in rat sensory dorsal root ganglia neurons. J Gen Virol 70(Pt 8):2075–2085. doi:10.1099/0022-1317-70-8-2075

Ugolini G (1995) Specificity of rabies virus as a transneuronal tracer of motor networks: transfer from hypoglossal motoneurons to connected second-order and higher order central nervous system cell groups. J Comp Neurol 356(3):457–480. doi:10.1002/cne.903560312

Ugolini G (2011) Rabies virus as a transneuronal tracer of neuronal connections. Adv Virus Res 79:165–202. doi:10.1016/B978-0-12-387040-7.00010-X

Ugolini G, Kuypers HG, Simmons A (1987) Retrograde transneuronal transfer of herpes simplex virus type 1 (HSV 1) from motoneurones. Brain Res 422(2):242–256

Veenman CL, Reiner A, Honig MG (1992) Biotinylated dextran amine as an anterograde tracer for single- and double-labeling studies. J Neurosci Methods 41(3):239–254

Velez-Fort M, Rousseau CV, Niedworok CJ, Wickersham IR, Rancz EA, Brown AP, Strom M, Margrie TW (2014) The stimulus selectivity and connectivity of layer six principal cells reveals cortical microcircuits underlying visual processing. Neuron 83(6):1431–1443. doi:10.1016/j.neuron.2014.08.001

Vercelli A, Repici M, Garbossa D, Grimaldi A (2000) Recent techniques for tracing pathways in the central nervous system of developing and adult mammals. Brain Res Bull 51(1):11–28

Vivar C, Potter MC, Choi J, Lee JY, Stringer TP, Callaway EM, Gage FH, Suh H, van Praag H (2012) Monosynaptic inputs to new neurons in the dentate gyrus. Nat Commun 3:1107. doi:10.1038/ncomms2101

Wall NR, Wickersham IR, Cetin A, De La Parra M, Callaway EM (2010) Monosynaptic circuit tracing in vivo through Cre-dependent targeting and complementation of modified rabies virus. Proc Natl Acad Sci USA 107(50):21848–21853. doi:10.1073/pnas.1011756107

Wall NR, De La Parra M, Callaway EM, Kreitzer AC (2013) Differential innervation of direct- and indirect-pathway striatal projection neurons. Neuron 79(2):347–360. doi:10.1016/j.neuron.2013.05.014

Waller A (1850) Experiments on the section of the glossopharyngeal and hypoglossal nerves of the frog, and observations of the alterations produced thereby in the structure of their primitive fibres. Phil Trans R Soc Lond 140:423–429

Wang D, He X, Zhao Z, Feng Q, Lin R, Sun Y, Ding T, Xu F, Luo M, Zhan C (2015) Whole-brain mapping of the direct inputs and axonal projections of POMC and AgRP neurons. Front Neuroanat 9:40. doi:10.3389/fnana.2015.00040

Waselus M, Galvez JP, Valentino RJ, Van Bockstaele EJ (2006) Differential projections of dorsal raphe nucleus neurons to the lateral septum and striatum. J Chem Neuroanat 31(4):233–242. doi:10.1016/j.jchemneu.2006.01.007

Watabe-Uchida M, Zhu L, Ogawa SK, Vamanrao A, Uchida N (2012) Whole-brain mapping of direct inputs to midbrain dopamine neurons. Neuron 74(5):858–873. doi:10.1016/j.neuron.2012.03.017

Weissbourd B, Ren J, DeLoach KE, Guenthner CJ, Miyamichi K, Luo L (2014) Presynaptic partners of dorsal raphe serotonergic and GABAergic neurons. Neuron 83(3):645–662. doi:10.1016/j.neuron.2014.06.024

Wertz A, Trenholm S, Yonehara K, Hillier D, Raics Z, Leinweber M, Szalay G, Ghanem A, Keller G, Rozsa B, Conzelmann KK, Roska B (2015) PRESYNAPTIC NETWORKS. Single-cell-initiated monosynaptic tracing reveals layer-specific cortical network modules. Science 349(6243):70–74. doi:10.1126/science.aab1687

White JG, Southgate E, Thomson JN, Brenner S (1986) The structure of the nervous system of the nematode *Caenorhabditis elegans*. Philos Trans R Soc Lond B Biol Sci 314(1165):1–340

Wickersham IR, Lyon DC, Barnard RJ, Mori T, Finke S, Conzelmann KK, Young JA, Callaway EM (2007a) Monosynaptic restriction of transsynaptic tracing from single, genetically targeted neurons. Neuron 53(5):639–647. doi:10.1016/j.neuron.2007.01.033

Wickersham IR, Finke S, Conzelmann KK, Callaway EM (2007b) Retrograde neuronal tracing with a deletion-mutant rabies virus. Nat Methods 4(1):47–49. doi:10.1038/nmeth999

Xia Y, Driscoll JR, Wilbrecht L, Margolis EB, Fields HL, Hjelmstad GO (2011) Nucleus accumbens medium spiny neurons target non-dopaminergic neurons in the ventral tegmental area. J Neurosci Off J Soc Neurosci 31(21):7811–7816. doi:10.1523/JNEUROSCI.1504-11.2011

Yamawaki N, Shepherd GM (2015) Synaptic circuit organization of motor corticothalamic neurons. J Neurosci Off J Soc Neurosci 35(5):2293–2307. doi:10.1523/JNEUROSCI.4023-14.2015

Yonehara K, Balint K, Noda M, Nagel G, Bamberg E, Roska B (2011) Spatially asymmetric reorganization of inhibition establishes a motion-sensitive circuit. Nature 469(7330):407–410. doi:10.1038/nature09711

Yonehara K, Farrow K, Ghanem A, Hillier D, Balint K, Teixeira M, Juttner J, Noda M, Neve RL, Conzelmann KK, Roska B (2013) The first stage of cardinal direction selectivity is localized to the dendrites of retinal ganglion cells. Neuron 79(6):1078–1085. doi:10.1016/j.neuron.2013.08.005

Yoshihara Y, Mizuno T, Nakahira M, Kawasaki M, Watanabe Y, Kagamiyama H, Jishage K, Ueda O, Suzuki H, Tabuchi K, Sawamoto K, Okano H, Noda T, Mori K (1999) A genetic approach to visualization of multisynaptic neural pathways using plant lectin transgene. Neuron 22(1):33–41

Young JA, Bates P, Varmus HE (1993) Isolation of a chicken gene that confers susceptibility to infection by subgroup A avian leukosis and sarcoma viruses. J Virol 67(4):1811–1816

Zampieri N, Jessell TM, Murray AJ (2014) Mapping sensory circuits by anterograde transsynaptic transfer of recombinant rabies virus. Neuron 81(4):766–778. doi:10.1016/j.neuron.2013.12.033

Zemanick MC, Strick PL, Dix RD (1991) Direction of transneuronal transport of herpes simplex virus 1 in the primate motor system is strain-dependent. Proc Natl Acad Sci USA 88(18): 8048–8051

Zhan C, Zhou J, Feng Q, Zhang JE, Lin S, Bao J, Wu P, Luo M (2013) Acute and long-term suppression of feeding behavior by POMC neurons in the brainstem and hypothalamus, respectively. J Neurosci Off J Soc Neurosci 33(8):3624–3632. doi:10.1523/JNEUROSCI.2742-12.2013

Zhang S, Xu M, Kamigaki T, Hoang Do JP, Chang WC, Jenvay S, Miyamichi K, Luo L, Dan Y (2014) Selective attention. Long-range and local circuits for top-down modulation of visual cortex processing. Science 345(6197):660–665. doi:10.1126/science.1254126

Zhao C, Deng W, Gage FH (2008) Mechanisms and functional implications of adult neurogenesis. Cell 132(4):645–660. doi:10.1016/j.cell.2008.01.033

Chapter 5
Live Imaging of Connectivity in Developing Neural Circuits in *Drosophila*

Mehmet Neset Özel and Peter Robin Hiesinger

Abstract How neural circuits assemble during development influences functional adult circuit architecture, specificity, and variability. Live observation of brain development reveals stochastic and dynamic processes that help to understand functional constraints in the adult circuitry. In the first part of this chapter, we will explore what live imaging tells us about how dynamic processes create and constrain circuit specificity. In the second part of this chapter, we provide a current view of how live observation can be achieved in intact *Drosophila* brains in comparison to developmental imaging in other species. The goal of this chapter is to provide both the context and tools to understand neural circuits as a function of their developmental context.

5.1 The Developmental Context: From Dynamics to Synaptic Specificity

Key questions in modern neuroscience include "*how do neural circuits work?*" and "*how do neural circuits form?*". Approaches to answer these questions require overlapping information concerning neuronal morphology and connectivity. When Roger Sperry referred to an "unadaptable rigidity" of mechanisms that drive the development of visuomotor connectivity (Sperry 1943), he provided a theory that was to define developmental neuroscience for decades to come. The chemoaffinity theory has evolved to include the idea of wiring codes that neurons use to make connections with their targets based on specific molecular markers. In this section, we will discuss a conceptual framework that builds on and expands the chemoaffinity theory to include dynamic and stochastic processes, which are best studied through developmental live imaging.

The search for guidance molecules and the elucidation of their molecular mechanisms has been hugely successful, especially during the last 20 years

M.N. Özel · P.R. Hiesinger (✉)
Freie Universitaet Berlin, Königin-Luise-Strasse 1-3, 14195 Berlin, Germany
e-mail: robin.hiesinger@fu-berlin.de

A. Çelik and M.F. Wernet (eds.), *Decoding Neural Circuit Structure and Function*, DOI 10.1007/978-3-319-57363-2_5

(for review see Kolodkin and Tessier-Lavigne 2011; Raper and Mason 2010; Yogev and Shen 2014). As more molecules and mechanisms were discovered, the ideas of how guidance cues and codes function became more nuanced. Numerous molecules have been shown to function repeatedly at different places and times, in combinations and in gradients (Sanes and Zipursky 2010; Yogev and Shen 2014). Importantly, not all molecules that exhibit attractive or repulsive binding function as guidance cues; for example, thousands of isoforms of the repulsive, homophilic cell adhesion molecule Dscam ensure self-avoidance of dendritic branches from the same neuron, but provide no specific directional cue for any neuronal extension where to grow or make synapses. Instead of functioning as cues, Dscam isoforms (similar to vertebrate Protocadherins) execute a simple pattern formation process based on self-avoidance (Kise and Schmucker 2013). In another example, the widely expressed cell adhesion molecule N-Cadherin is not required as a targeting cue in *Drosophila* photoreceptors, but can function in the stabilization of growth cones (Ozel et al. 2015). Hence, attractive and repulsive molecules can play important roles in neural circuit assembly without specifying target areas or cells.

The two examples above highlight another important extension of the original chemoaffinity theory: both Dscam-mediated self-avoidance and N-Cadherin-mediated stabilization contribute to wiring specificity using dynamic and stochastic processes. Apart from the random Dscam isoform choice of individual neurons, self-avoidance leads to spreading of dendritic branches only if individual branches non-deterministically grow such that self-avoidance can act on them. As a consequence, every neuron, like every snowflake and every apple tree, has a uniquely different branching pattern. Similarly, every growth cone has a unique branching pattern of dynamically extending and retracting filopodia. In the case of *Drosophila* R7 photoreceptor axons, this random pattern of extensions and retractions seems to be required for N-Cadherin-mediated growth cone stabilization (Ozel et al. 2015). These cases exemplify how dynamic and stochastic growth is in fact necessary for the molecules to execute their function. This observation significantly extends on the early versions of the chemoaffinity theory based on molecular matchmaking cues. The idea of matchmaking is more deterministic: from the perspective of a specific cue a stochastic process is more likely to represent noise that the system would try to minimize, rather than a necessary part of the molecular mechanism. Neural circuit assembly is likely to employ both mechanisms: pattern formation based on stochastic growth as well as molecular specification through matchmaking; examples for both have been firmly established (Hassan and Hiesinger 2015; Kolodkin and Tessier-Lavigne 2011; Yogev and Shen 2014).

To understand the role of dynamic and stochastic processes as part of neural circuit assembly, we have recently proposed a rules-based framework to help incorporate the mechanisms and roles of molecules like Dscam and N-Cadherin as part of developmental algorithms underlying brain wiring (Hassan and Hiesinger 2015). In the example of Dscam, in this framework, the focus is on the level of the rule 'self-avoidance' as part of a larger developmental program, rather than the molecular mechanism of homophilic, repulsive binding that executes that rule (or the molecule itself for that matter: the same rule can be carried out by different molecules, as exemplified by Protocadherins in vertebrates). In this case, the rule can be formulated as: (1) grow stochastic filopodial extensions and branches, (2) stop individual filopodia from growing further when contacting other filopodia from the same neuron. This rule is not sufficient without additional constraints, e.g., probability of branching, inter-branch spacing, etc., but it captures the essence of the developmental process executed by the homophilic repulsion function of Dscam.

To understand the interplay of stochastic growth and molecular mechanisms in executing a rule like self-avoidance, it is beneficial to think of neurons and neuronal processes as individual entities that explore the environment, advance to targets and compete with each other. Lichtman and Fraser (2001) proposed the analogy between such a setting and a football (or similar sports) game. Each player has its abilities, restrictions as well as his or her own agenda. If the players each follow the rules, these games will create an ordered structure without any external supervision; however, the exact outcome is not scripted. Referees and game plans can help structure the game, but are not strictly required: even a backyard football game with variable numbers of players and imprecise playing field can work out wonderfully.

How do we figure out the rules of the game called 'brain wiring', which is arguably much more complicated than any sports we have ever invented? An established approach is to disrupt specific genes, in particular those encoding presumptive guidance cues, and study the end results of the effects on the circuit. *Drosophila* has been a particularly useful system using this approach, in part due to the development of a technique to render individual neurons mutant in an otherwise wild-type animal (Lee and Luo 1999). If the disrupted gene is indeed a molecular matchmaking cue, then loss or gain in individual neurons predictively rewire connectivity, as has been shown in several examples, e.g., the teneurins in Drosophila (Hong et al. 2012) or type II Cadherins in the vertebrate retina (Duan et al. 2014). In many cases, however, loss and gain of function studies for single genes led to surprising, less instructive outcomes. Following up with the soccer analogy, a perturbation may, for example, restrict the usage of hands for each player one by one, revealing that only in the case of the goalie this causes the eventual score to change. Here, the analogy highlights what gene perturbations may indirectly provide insight into the rules of the game. Many rules have been discovered through carefully conducted gene perturbation experiments, including the discovery of self-avoidance through mutant analyses of Dscams. If the goal is the characterization of the rule, however, it is not a priori clear that a gene perturbation experiment is the shortest path to uncover the rule. For example, the rule on usage of hands in a soccer game may also be deduced from observation of the unperturbed game. Comparing static pictures from different points of different games would not easily reveal this rule. Static pictures can provide important information on the game but will fail to capture stochastic and dynamic actions that do not stereotypically happen at an exact time point for every single member of a 'player' type. Since stochastic and dynamic actions are key to pattern formation rules like self-avoidance, live observation is of particular importance to the discovery of rules underlying brain wiring. Ultimately, a combination of gene and cell perturbation experiments with live observation yields the highest likelihood of uncovering principles underlying the development of connectivity.

As we will discuss in the second section of this chapter, during the past 15 years our ability to image live neurons forming circuits in their natural environments has significantly improved, especially in the model organisms worm, fly, zebrafish, and mouse. Yet, we have only begun to characterize the dynamic properties of developing neural circuits. In the early parts of brain development, live imaging has already been very useful for the study of neural stem cell migration (Lerit et al. 2014; Ortega and Costa 2016). When it comes to studying the development of circuit connectivity, we need to look at neurons that develop two types of dynamic structures: axonal growth cones and dendritic extensions. In both cases stochastic extension and retraction dynamics of filopodia underlie what appear to be robust choices in the adult connectivity. The next subsections will address our current state of knowledge about the role these subcellular filopodial structures have in vivo and how they might relate to the establishment of synaptic specificity.

5.1.1 Growth Cone Guidance and Early Filopodia

When Cajal studied chick embryos to show that axons grow out of neurons, he discovered what he called a "cone-like lump with a peripheral base" with thorny processes at the tips of commissural axons (Ramón y Cajal 1890). This description came to define the features of the 'textbook growth cone': a widened terminal with filopodia and lamellipodia at its tip. The growth cone has received plenty of attention over the decades as the presumptive structure that detects guidance cues and actively advances to the target (for review see Raper and Mason 2010; Vitriol and Zheng 2012). Filopodia have been suggested as agents for the detection of guidance cues (Rajnicek et al. 2006; Zheng et al. 1996) and in vitro evidence was provided in favor of filopodia forming a clutch mechanism for growth cone movement by acting as 'sticky fingers' (Chan and Odde 2008; Heidemann et al. 1990). It has largely remained unclear, why and how stochastic extension and retraction dynamics execute these processes.

Most of our understanding of the dynamics of growth cones is based on in vitro systems. However, some in vivo live imaging data has already provided glimpses into the functions of filopodia that do not easily fall into the categories 'searching agents' or 'sticky fingers'. Some of the first live imaging studies in intact tissue revealed that growth cones tend to adapt simple, streamlined forms while extending and more complex forms (like the classical growth cone) when they pause or reach their targets, e.g., in mammalian retina preparations (Godement et al. 1994) and intact zebrafish embryos (Jontes et al. 2000; Kaethner and Stuermer 1994). These observations have been interpreted as growth cones adapting to complex, high filopodial activity forms at 'decision regions', where there are multiple cues and the axon needs to make a decision on which direction to go; then it quickly advances towards that direction without many filopodia (Mason and Erskine 2000). On the other hand, it has been shown that filopodia are at least partially dispensable for axon navigation but essential for terminal arborization of retinal axons (Dwivedy et al. 2007). Recent findings on developing R7 photoreceptor growth cones in intact *Drosophila* eye–brain complexes revealed a similar pattern (Ozel et al. 2015). These axons exhibit streamlined structures while extending, but expand into more complicated filopodial structures once they stabilize at the target layer. Interestingly, when their attachment to that layer was genetically impaired, these growth cones went, at unpredictable time points, through a gradual filopodial collapse followed by regaining of motility by the axon tip. The close temporal link between filopodia formation and axon stabilization suggests that in this case filopodia might function as an adhesion surface for stabilization, rather than being important for guidance or extension. However, a direct causal link between the two processes is yet to be established.

Finally, rather than directing the growth cone to the target, filopodia can also function in guidance by extending to a target, expanding and becoming the new axon terminal/growth cone. This was observed in vitro (O'Connor et al. 1990) and in intact *Drosophila* embryos (Murray et al. 1998). A similar behavior is observed for developing R8 photoreceptor growth cones in *Drosophila* as they relocate from their temporary position to the medulla layer M3 at mid-pupal development (Ozel et al. 2015). These growth cones extend a single filopodium to deeper layers, which is initially very dynamic with almost complete retractions and re-extensions but eventually stabilizes its tip in the correct target layer, expands, and ultimately forms the adult R8 axonal terminal prior to synaptogenesis.

The few selected examples discussed here highlight the origin of an important aspect of adult circuit connectivity: The precise axon positions, dendritic branch points, and their contacts in the adult may not only be slightly imprecise due to biological noise, but be the result of necessary stochastic processes based on filopodial dynamics during growth cone guidance, stabilization, and synaptic partner identification. Especially on this latter aspect, important insight comes from filopodia on axons and dendrites at later developmental stages, as discussed in the following section.

5.1.2 Synapse Specification and Late Axonal and Dendritic Filopodia

After a growth cone reaches its target area, the main body of the axon no longer advances; therefore, it may be more accurate to classify the filopodia at this stage as 'axonal filopodia'. These are not limited to the tip of the axon and have been linked to axonal branching (reviewed in Gallo 2011). Dendrites also form filopodia, and these perhaps constitute the class of filopodia that has so far been most closely linked to the establishment of synaptic connectivity.

Work by Stephen Smith and colleagues have provided pioneering insights into the roles of axonal and dendritic filopodia during brain wiring. Live imaging of tectal neuron dendritic arbors in intact developing zebrafish revealed that young arbors create many transient filopodia, some of which become the sites of de novo synapse formation. In turn, formation of these synapses directly stabilizes the respective filopodia, turning it into a stable branch (Niell et al. 2004). Live observation helped to establish a link between the processes of synapse formation and filopodial stabilization without perturbation experiments. Filopodia without a synapse never persisted longer than an hour and the stabilized filopodia only retracted if its synapses were eliminated. Similarly, at the presynaptic partners of these cells (retinal axons), new branches extend preferentially from newly formed synaptic sites and no branch is stabilized over an hour without a synaptic site present (Meyer and Smith 2006). Together, these observations support the synaptotropic model (Vaughn 1989), whereby stabilization through synapse formation guides axonal and dendritic extension. Importantly, this process can only work if axonal and dendritic arbors provide initially stochastically extended filopodial processes to select from. A process based on the rule of selection and stabilization precludes precise positioning of pre- and postsynaptic partners, but not synaptic specificity, in adult circuit connectivity.

These data showcase links between synapse formation and filopodial dynamics and how the former can direct the latter to bias axonal and dendritic arbors towards stabilized connection sites. However, they are only our first glimpses into the roles of filopodial dynamics in brain wiring in intact, developing brains—and they have largely been limited to wild type. Important questions remain, for example: How are some synapses selected to stabilize while others are lost? What are the rules and mechanisms underlying synapse-mediated branch stabilization? Watching the dynamics of axonal and dendritic dynamics in wild-type and mutant neurons are necessary approaches to answer these questions and understand the rules that establish adult connectivity.

5.2 The Imaging Approach: Watching Circuit Assembly Live

Live imaging protocols for *Drosophila* are available for embryos (Evans et al. 2010; Reed et al. 2009), larval neuromuscular junctions (Schmid and Sigrist 2008) and brains (Cabernard and Doe 2013). Investigating the dynamics of how adult neural circuits form requires long-term imaging of developing pupal brains at high spatiotemporal resolution. Until recently, fast and high-resolution imaging in developing *Drosophila* brains was largely restricted to short (about 1 h) imaging periods (Williamson and Hiesinger 2010). In this section, we will describe two techniques that we recently described for long-term, high-resolution imaging of intact, developing pupal brains: unobtrusive intravital imaging in intact pupae and ex vivo imaging in developing eye–brain complexes. We will also discuss available microscopy options along with technical concerns related to the imaging. Finally, we will provide a brief overview of techniques in other model systems that allow long-term imaging of neural circuit formation.

5.2.1 *Intravital Imaging of the Developing Drosophila Brain*

We refer to intravital imaging as a completely non-invasive technique that does not interfere measurably with development or function. We have recently performed intravital imaging of developing brains through the eyes of pupae (Langen et al. 2015). This method is, to our knowledge, the only known example of completely non-invasive imaging of pupal to adult brain development in flies, and may allow imaging of deeper brain structures as well.

During the first half of pupal development the *Drosophila* eye is mostly transparent and the center of the eye is largely not obstructed by lipid droplets or light-scattering tissues. In addition, during the pupal stages there are no muscle contractions, eliminating the need for anesthesia or physical immobilization. This allows monitoring of the developing optic lobe in a completely non-invasive manner for a limited time window of brain development (Fig. 5.1). We exploited these facts to build an imaging chamber (Fig. 5.2), which allowed us to perform long-term, high-resolution imaging of photoreceptor terminals in the lamina neuropil (Langen et al. 2015). The method described here allowed imaging of these growth cones at the resolution of filopodial dynamics for up to 24 h, and the pupae continued normal development thereafter to adulthood.

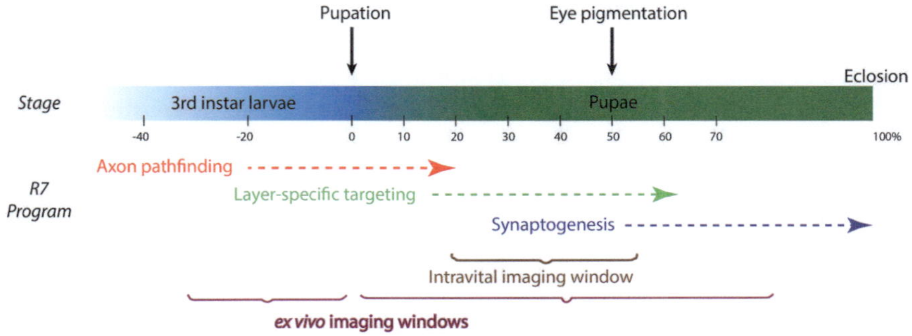

Fig. 5.1 Timeline of *Drosophila* development from late larval stage to adulthood. As a reference, different developmental steps of R7 photoreceptor axons are overlaid. Intravital imaging can be performed between $P + 15\%$ and 55% of the pupal development, while *ex vivo* imaging can be performed throughout the pupal development as well as during 3rd instar larval stage

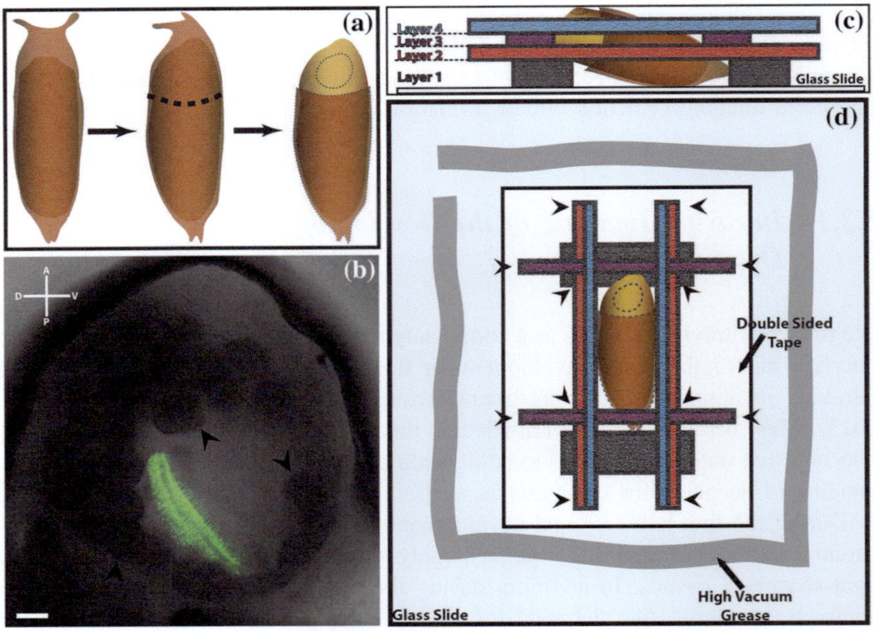

Fig. 5.2 **a** Pupa is aligned such that the right eye is imaged after the cuticle around the head is dissected. **b** Transmitted light image of the right eye of a pupa. Lamina neurons are genetically labeled with GFP, revealing the position of the lamina layer of the optic lobe. *Arrowheads* mark the lipid bodies. **c** *Side* view of the imaging chamber. Four layers of filter paper are shown in different *colors* for clarity. Pupa is kept in position as the whole body axis makes an angle with the plane of the glass slide. **d** *Top* view of the imaging chamber. Layers are shown in the same *colors* as in **c**. *Arrowheads* show the positions where 4% agarose is applied. Images courtesy of E. Agi

A brief protocol summary for intravital imaging through the pupal eye is provided here: First, a part of the pupal case is removed from the head section (Fig. 5.2a); the pupa is mounted on its side at an angle that exposes the eye as the highest point (Fig. 5.2c). Around the pupa an elevated barrier is constructed with filter paper to keep the chamber moist; agarose is used as cement at junctions of the layers (Fig. 5.2d). Vacuum grease holds the cover slip in place. A drop of HL3 solution is put on a cover slip before it is placed on the eye. It is important to make sure that the eye makes firm contact with the cover slip by applying gentle pressure on the cover slip with the forceps while avoiding bursting the eye. Additional technical details have been published (Langen et al. 2015, Supplementary Experimental Procedures).

The chief advantage of this method is the ability to remove the pupae from the chamber after imaging and rear them to be healthy adults, which can then be analyzed to test whether normal development was affected by the imaging conditions. This level of non-invasiveness, however, comes with certain limitations.

First, after $P + 50\%$ of pupal development the fly eye starts to accumulate pigment, which effectively prevents brain imaging. Using white eyed flies (which can be generated even in the presence of *white+* transgenes through a combination of *brown* with either *cinnabar, scarlet,* or *vermillion*) (Kim et al. 2013), can extend the imaging window up to $P + 60\%$. However, the thickening of the eyes increases light scattering, complicating deep brain imaging after this point even in white-eyed flies. Early in development, the intravital imaging time window through the pupal eye is limited to around $P + 15\%$. Until this time point, the pupal case is attached to the cuticle; as a result, removal of the pupal case to expose the eyes before that stage is very difficult.

Second, having to image through a defined window, i.e., the eye, spatially limits the accessible brain areas. Imaging deep in the optic lobe and beyond significantly reduces the signal-to-noise ratio. Finally, both sample movement as well as the intrinsic morphological rearrangements inside the developing pupae often results in significant drift, complicating the experiment and requiring time consuming post-processing.

In summary, the currently available method for intravital imaging of the developing *Drosophila* brain is perfectly suited to study developmental processes of the eye, lamina or distal medulla between $P + 15\%$ and $P + 55\%$.

5.2.2 *Ex Vivo Imaging of Developing Eye–Brain Complexes*

It has long been known that pupal eye–brain complexes of *Drosophila* can recapitulate certain events of metamorphosis when removed from the body and placed under well-controlled culture conditions (Gibbs and Truman 1998). This has been exploited to maintain brains in culture over long periods (Ayaz et al. 2008). An adaptation of this method to perform high-resolution time-lapse imaging required particular consideration of sample drift and phototoxicity as chief problems over long periods at subcellular resolution. To tackle these issues, we recently developed an imaging chamber that can be built with minimal effort (Ozel et al. 2015).

A **brief protocol summary for ex vivo brain imaging is provided here**: We use a combination of embedding in low-melting agarose and a closed system that virtually eliminates sample drift. In addition, placement of the brain right under a coverslip allows access to high-powered objectives with minimal working distance to maximize signal and therefore reduce the required laser power, which is crucial for keeping the tissue healthy during extended periods of imaging (Fig. 5.3b). A detailed step-by-step protocol for building the culture imaging chamber has been published (Ozel et al. 2015). Briefly, pupal eye–brain complexes are dissected in chilled Schneider's *Drosophila* Medium and placed within a drop of 0.4% (in culture medium) low-melting agarose on a Sylgard layer in a petri dish lid. Next, a circular coverslip (4 cm diameter) is placed on the drop; 200–250 μm-thick (depending on the brain's age) spacers are used to prevent the coverslip from crushing the brain. After the polymerization of agarose, the remaining space under the

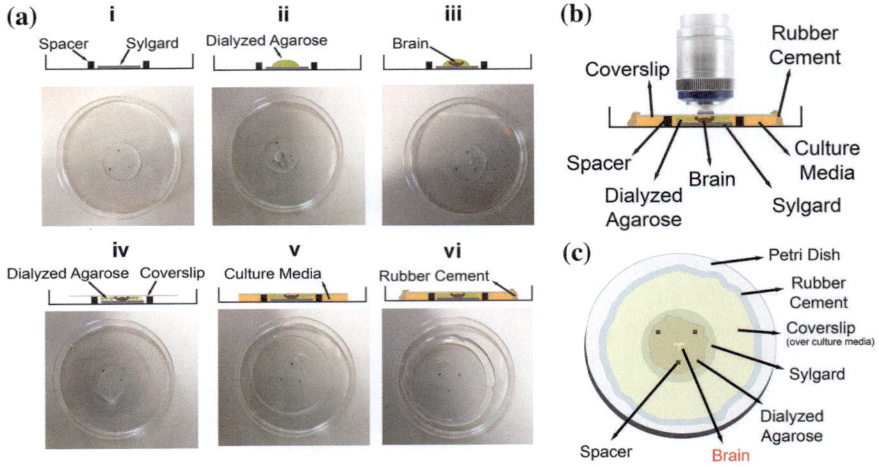

Fig. 5.3 **a** Step-by-step construction of the imaging chamber. (*i*) Spacers are placed on the Sylgard layer in a *triangle* formation. (*ii*) A drop of diluted dialyzed agarose is pipetted onto the Sylgard. (*iii*) Dissected eye-brain complex is placed into the agarose drop. (*iv*) The mix is covered with a coverslip. (*v*) After the agarose polymerization, remaining space under the coverslip is filled with the culture media; (*vi*) and sealed completely with rubber cement. Schematic of the final chamber **b** from the *side* and **c** the *top*. Adapted from Ozel et al. (2015). DOI:10.7554/eLife.10721.004

coverslip is filled with oxygenized culture medium before the chamber is sealed using rubber cement (Fig. 5.3a). We use a culture medium that is based on Schneider's medium and includes 10% fetal bovine serum (FBS), 10 μg/ml insulin, 1% penicillin/streptomycin mix and 1 μg/ml 20-Hydroxyecdysone (20-HE). Usage of a closed system limits the amount of available oxygen and nutrients to the amount of culture media that fits under the coverslip (approximately 500 μl). We have found this is sufficient to support normal development for 24 h. A similar approach has recently been published using an open chamber and culture medium that included ascorbic acid and higher concentrations of ecdysone (Rabinovich et al. 2015). If longer imaging times are necessary, both systems can be adapted to a perfusion chamber (Williamson and Hiesinger 2010).

The ex vivo system addresses two main limitations inherent to intravital imaging. First, every part of the brain is accessible to high-resolution imaging. Second, brains at all stages of pupal development as well as larval brains can be imaged using this technique. However, we found that cultures that start after $P + 60\%$ are often less healthy, possibly because it becomes increasingly difficult to dissect intact eye–brain complexes as the eyes start to fuse with the head cuticle. Furthermore, it does not seem possible to recapitulate the events at the onset of metamorphosis in culture and cultures that start earlier than $P + 10\%$ require a higher concentration of 20-HE (Rabinovich et al. 2015). Finally, previous measurements showed that ecdysteroid levels during metamorphosis peak around $P + 40\%$ and then start to drop (Paul Bainbridge and Bownes 1988). Consistent with this data, we have found that it is necessary to exclude 20-HE from cultures that start after $P + 50\%$ (Ozel et al. 2015). Failure to do so appears to slow down eye pigmentation and induces aberrant filopodial formations on R7 photoreceptor axonal terminals.

Despite its advantages, ex vivo imaging inherently retains one main drawback: not being in vivo. Certain developmental events depend on surrounding tissues, e.g., eye disc expansion or lamina rotation. In addition, some developmental processes occur slower or faster compared to in vivo controls, requiring careful calibration and controls (Ozel et al. 2015).

5.2.3 Microscopy Systems for Developmental Live Imaging

Despite the recent advances in newer technologies like super-resolution or light-sheet microscopy, imaging of small regions deep in living tissue currently still remains the domain of multiphoton microscopy. However, our ex vivo imaging chamber is also compatible with other light microscopy approaches, including standard confocal systems (Zschatzsch et al. 2014), particularly for short-term experiments and superficial brain areas (close to the coverslip). Resonant scanning systems, which provide 10–20 times higher scan speeds compared to conventional systems (Art and Goodman 1993), reduce phototoxicity and are preferable for live imaging. Increased noise that emanates from high laser scan speeds typically need to be reduced by averaging; importantly, even if the total scan time may not be improved significantly when using high averaging, photobleaching is significantly lower at the same time-lapse speed. It is better to scan the same point 20 times for x amount of time rather than scanning it once for $20x$ amount of time because the former allows fluorescent molecules to relax back to the ground state, preventing them from potentially being hit by another photon in their excited, unstable state and rendering them dysfunctional (Borlinghaus 2006).

Confocal microscopes use a pinhole to exclude out-of-focus light. As a result, it excites fluorophores all over the sample, but detects only a fraction of them (Webb 1996). Multiphoton microscopes, on the other hand, utilize simultaneous absorption of two lower energy (higher wavelength) photons. Because excitation rate in this case depends on the second power of light intensity, excitation outside of the focal plane becomes extremely unlikely (Denk et al. 1990), i.e., all of the excited fluorophores can now be detected. The elimination of the need to spatially restrict detected light is the key advantage of two-photon microscopy. First, sending the fluorescence light back through the scanning pathway (descanning) is no longer necessary. By using non-descanned detectors (NDD) close to the objective, valuable photons that would otherwise be lost at every mirror, lens, and filter can now be detected. Second, scattered light, which is very common in thick, living samples and would be excluded by a confocal pinhole, can now be detected. Detection efficiency of a two-photon microscope does not decrease with depth as dramatically as with confocal microscopy, because scattered photons eventually leave the sample and can be counted. Finally, infrared light used for two-photon excitation scatters less than visible light, further increasing depth penetration.

Together, these advantages currently allow higher spatiotemporal resolution for small regions of interest in deep tissue than light-sheet microscopy. In contrast, light-sheet microscopy offers much faster image acquisition of large fields of view at high and even super-resolution (Heddleston and Chew 2016; Hu et al. 2014; Lemon and Keller 2015). However, multi-channel imaging with a two-photon microscope can be complicated. Traditionally, most commercial NDD units have a maximum of two channels, but quad-channel NDD units have recently become commercially available. The maxima for most red fluorophores fall outside the range where standard Ti::Sapphire lasers provide sufficient power (700–1000 nm). Far-red excitation can be achieved with a dedicated second laser or solutions such as OPO (optical parametric oscillator), which extend the excitation range to 1300 nm and beyond.

5.2.4 Fluorescent Markers for Developmental Live Imaging

Along with the advancements in microscopy, the so-called 'GFP revolution' (Chalfie et al. 1994; Heim et al. 1994; Prasher et al. 1992) is responsible for the fact that live imaging as presented in this chapter is possible. Here, we will briefly discuss genetically encoded fluorescent proteins best suited to image neural circuit development using multiphoton microscopy in *Drosophila* brains.

5.2.4.1 Blue/Green Fluorescent Proteins

Despite the development of a plethora of new variants of GFP, the early original EGFP (Heim et al. 1995) remains one of the best fluorophores for multiphoton microscopy. It is a monomer, has low phototoxicity and high two-photon brightness. In vitro measurements place its two-photon excitation maxima at 927 nm (Drobizhev et al. 2011), but in practice we observed only little variation between 900–940 nm in *Drosophila* tissue. A more recent CFP variant, mTFP1 (Ai et al. 2006), has twice the measured two-photon brightness of GFP at the excitation maxima of 875 nm (Drobizhev et al. 2011), but we currently have no experimental data on its use as a protein tag. For imaging of thin membrane processes, we also require a good membrane tag along with a bright fluorophore. Popular tags include myristoylation (myr), CD4 and CD8. In our experience, CD4 performs best in labeling thin structures among the three (Han et al. 2011).

5.2.4.2 Red/Far-Red Fluorescent Proteins

Most available red fluorescent proteins are variants of DsRed, which unfortunately is a multimer and therefore not suitable as a protein tag. However, several newer variants (e.g., mCherry) behave like (or close to) monomers. tdTomato has the highest two-photon brightness of available red fluorophores. Its measured two-photon excitation maxima is 1050 nm (Drobizhev et al. 2011), but we have observed a stronger signal around 1100 nm in fly tissue. This could be related to lower tissue attenuation at these wavelengths. Fortunately, tdTomato can also be excited at 950 nm at decent (but not optimal) levels. This is close to the optimum excitation range for GFP and permits two-color imaging even with a single laser. However, tdTomato is a dimer-like DsRed and is therefore not recommended as a protein tag. In contrast, mKate2 is a monomeric far-red fluorescent molecule that is suitable for protein labeling. It is an improved, threefold brighter version of TagFP635 (mKate) (Shcherbo et al. 2009). It has a measured two-photon excitation maxima at 1140 nm and higher two-photon brightness than any other monomeric red protein (Drobizhev et al. 2011). It also has very high photo- and pH stability. On the other hand, highly red-shifted excitation properties make its usage impractical without a high wavelength laser.

5.3 Developmental Brain Imaging in Other Model Systems

In the final section of this chapter, we will provide a brief, non-comprehensive comparison to similar imaging approaches in other model systems where long-term live imaging of neural circuit development is feasible.

5.3.1 Caenorhabditis Elegans *Embryo Mount*

The nematode *C. elegans* is an excellent genetic model system and the entire embryonic development takes only 14 h. The simplicity of the system (even the adult animals have an invariant set of 302 neurons), small size, fast life cycle, and powerful genetic tools make *C. elegans* an excellent model to study molecular mechanisms of fundamental events (e.g., synaptogenesis). Due to their transparency, live imaging of these embryos is very feasible and the methods have been available for almost 20 years (Mohler and White 1998). As opposed to the *Drosophila* techniques presented in this chapter, live imaging in worms is scalable to very high sample sizes to perform screens. However, to our knowledge, the available methods are limited to the embryonic stage. The particularly small size of *C. elegans* is an advantage when imaging the entire animal, well suited for light-sheet microscopy, and does not require multiphoton microscopy. On the other hand, *C. elegans* provides only a limited model for synaptic specification processes in dense brain regions.

5.3.2 Zebrafish Embryos and Larvae

Zebrafish (*D. rerio*) is an excellent model system for developmental live imaging and comes with the chief advantage (over *Drosophila*) of being a vertebrate. Both its embryos (Dynes and Ngai 1998) and larvae (Niell et al. 2004) are transparent and amenable to long-term live imaging. As discussed in the first part of this chapter, the key studies that provided insight into the relationship between filopodial dynamics and synaptogenesis in vivo came from this system (Meyer and Smith 2006; Niell et al. 2004). Zebrafish embryos develop rapidly; axon growth and synapse formation can already be observed by 24 hpf. Larvae are free swimming and need to be immobilized by embedding in 1% agarose. It may be necessary to include PTU (phenylthiocarbamide) to inhibit pigmentation but live imaging can be performed at least up to 10 dpf for 24 h sessions after which the larvae remain healthy and continue to develop normally (Niell et al. 2004). The challenges of the zebrafish system include fewer genetic tools compared to *C. elegans* or *Drosophila*. For example, sparse labeling of individual, genetically manipulated neurons is not easy. Despite an adaptation of the Gal4/UAS system over 15 years ago (Koster and Fraser 2001), most imaging protocols require injection of plasmid DNA encoding the fluorescent markers to embryos and then screening potentially large numbers of them to identify those suitable.

5.3.3 Frog Tadpoles

Xenopus tadpoles, like zebrafish larvae, are mostly transparent (though their embryos are not), making it possible to perform imaging on the intact animal (Keller 1978). This model has been extensively used to study axonal, dendritic, and synaptic dynamics, particularly during retinotectal circuit formation (Alsina et al. 2001; Elul et al. 2003; Li et al. 2011; Manitt et al. 2009); but also in other circuits, for review see (Erdogan et al. 2016). Traditionally, it has been limited to the early stages of development where tadpoles do not have significant locomotion because the requirement for anesthetics at the later stages is detrimental to long-term imaging. However, a technique that involves constant flow of the anesthetic MS-222 and thereby allowing imaging up to 48 h has recently been described (Hamilton and Henry 2014). Historically, one of the main reasons for imaging *Xenopus* development has been the significantly larger sizes of their cells (as well as their growth cones) than virtually any other model system; making it very suitable for studying subcellular events and protein localization. Its drawbacks are similar to the zebrafish system, i.e., limited selection of genetic tools.

5.3.4 *In Ovo Imaging of Chick Embryogenesis*

It is possible to image the development of chick embryos by inserting an imaging window into the egg, whose details are discussed elsewhere (Kulesa and Fraser 2000). Chick embryos have a great tradition for developmental biology, thanks to their evolutionary vicinity to mammals and easy accessibility of the embryos. Nevertheless, aside from not having a particularly robust genetic toolbox, in ovo imaging is limited to a time window from stage-8 embryos to 5-day old embryos. This window allows the imaging of early nervous system development such as neural crest patterning and somite formation, but not the formation of neural circuits. Extending the imaging window beyond 5 days requires tackling the challenges associated with natural movements of the embryo due to development, causing drastic drifts, as well as the movements due to the heartbeat (Kulesa et al. 2010).

5.3.5 *Imaging in Mammalian Systems*

Mammalian systems such as mice or rats have the highest biological relevance to human physiology and development. Live imaging protocols for nervous system development are available in particular for mice. In utero development may obstruct high-resolution imaging of intact developing mammalian brains; however mouse embryos can be cultured ex vivo until the E10 stage, for up to 24 h periods (Jones et al. 2002). With this technique, processes like cell migration and differentiation can be observed and it has been used to study neurulation (Pyrgaki et al. 2010); but more advanced stages of brain development, i.e., the time when neurons start to establish connections, are not easily accessible.

An important exception is the retina, which is particularly accessible in retinal whole-mount cultures of the mouse eye (Williams et al. 2013). The mammalian retina has an extensive neural circuitry comparable in complexity to the *Drosophila* optic lobe lamina and medulla combined (Sanes and Zipursky 2010). Live imaging can be performed while these circuits develop. Retinal whole-mounts preserve the intrinsic circuitry and can be maintained for several days, but it is recommended to stay within the first 36 h of initial dissection (Williams et al. 2013). Several dynamic processes have been studied with great success in this system, including (1) cell migration and the establishment of horizontal cell territories during early neonatal development (Huckfeldt et al. 2009), (2) dynamics of bipolar cell layer-specific axonal targeting and stabilization at P5 stage (Morgan et al. 2006), and (3) activity-dependent synapse formation between bipolar and retinal ganglion cells at the P9 stage, in combination with electrophysiology (Kerschensteiner et al. 2009).

References

Ai HW, Henderson JN, Remington SJ, Campbell RE (2006) Directed evolution of a monomeric, bright and photostable version of Clavularia cyan fluorescent protein: structural characterization and applications in fluorescence imaging. Biochem J 400:531–540

Alsina B, Vu T, Cohen-Cory S (2001) Visualizing synapse formation in arborizing optic axons in vivo: dynamics and modulation by BDNF. Nat Neurosci 4:1093–1101

Art JJ, Goodman MB (1993) Rapid scanning confocal microscopy. Methods Cell Biol 38:47–77

Ayaz D, Leyssen M, Koch M, Yan J, Srahna M, Sheeba V, Fogle KJ, Holmes TC, Hassan BA (2008) Axonal injury and regeneration in the adult brain of Drosophila. J Neurosci Off J Soc Neurosci 28:6010–6021

Borlinghaus RT (2006) MRT letter: high speed scanning has the potential to increase fluorescence yield and to reduce photobleaching. Microsc Res Tech 69:689–692

Cabernard C, Doe CQ (2013) Live imaging of neuroblast lineages within intact larval brains in Drosophila. Cold Spring Harb Protoc 2013:970–977

Chalfie M, Tu Y, Euskirchen G, Ward WW, Prasher DC (1994) Green fluorescent protein as a marker for gene expression. Science 263:802–805

Chan CE, Odde DJ (2008) Traction dynamics of filopodia on compliant substrates. Science 322:1687–1691

Denk W, Strickler JH, Webb WW (1990) Two-photon laser scanning fluorescence microscopy. Science 248:73–76

Drobizhev M, Makarov NS, Tillo SE, Hughes TE, Rebane A (2011) Two-photon absorption properties of fluorescent proteins. Nat Methods 8:393–399

Duan X, Krishnaswamy A, De la Huerta I, Sanes JR (2014) Type II cadherins guide assembly of a direction-selective retinal circuit. Cell 158:793–807

Dwivedy A, Gertler FB, Miller J, Holt CE, Lebrand C (2007) Ena/VASP function in retinal axons is required for terminal arborization but not pathway navigation. Development 134:2137–2146

Dynes JL, Ngai J (1998) Pathfinding of olfactory neuron axons to stereotyped glomerular targets revealed by dynamic imaging in living zebrafish embryos. Neuron 20:1081–1091

Elul TM, Kimes NE, Kohwi M, Reichardt LF (2003) N- and C-terminal domains of beta-catenin, respectively, are required to initiate and shape axon arbors of retinal ganglion cells in vivo. J Neurosci Off J Soc Neurosci 23:6567–6575

Erdogan B, Ebbert PT, Lowery LA (2016) Using *Xenopus laevis* retinal and spinal neurons to study mechanisms of axon guidance in vivo and in vitro. Semin Cell Dev Biol 51:64–72

Evans IR, Zanet J, Wood W, Stramer BM (2010) Live imaging of Drosophila melanogaster embryonic hemocyte migrations. J Vis Exp 36

Gallo G (2011) The cytoskeletal and signaling mechanisms of axon collateral branching. Dev Neurobiol 71:201–220

Gibbs SM, Truman JW (1998) Nitric oxide and cyclic GMP regulate retinal patterning in the optic lobe of Drosophila. Neuron 20:83–93

Godement P, Wang LC, Mason CA (1994) Retinal axon divergence in the optic chiasm: dynamics of growth cone behavior at the midline. J Neurosci Off J Soc Neurosci 14:7024–7039

Hamilton PW, Henry JJ (2014) Prolonged in vivo imaging of *Xenopus laevis*. Dev Dyn 243:1011–1019

Han C, Jan LY, Jan YN (2011) Enhancer-driven membrane markers for analysis of nonautonomous mechanisms reveal neuron-glia interactions in Drosophila. Proc Natl Acad Sci USA 108:9673–9678

Hassan BA, Hiesinger PR (2015) Beyond molecular codes: simple rules to wire complex brains. Cell 163:285–291

Heddleston JM, Chew TL (2016) Light sheet microscopes: novel imaging toolbox for visualizing life's processes. Int J Biochem Cell Biol 80:119–123

Heidemann SR, Lamoureux P, Buxbaum RE (1990) Growth cone behavior and production of traction force. J Cell Biol 111:1949–1957

Heim R, Prasher DC, Tsien RY (1994) Wavelength mutations and posttranslational autoxidation of green fluorescent protein. Proc Natl Acad Sci U S A 91:12501–12504

Heim R, Cubitt AB, Tsien RY (1995) Improved green fluorescence. Nature 373:663–664

Hong W, Mosca TJ, Luo L (2012) Teneurins instruct synaptic partner matching in an olfactory map. Nature 484:201–207

Hu YS, Zimmerley M, Li Y, Watters R, Cang H (2014) Single-molecule super-resolution light-sheet microscopy. ChemPhysChem 15:577–586

Huckfeldt RM, Schubert T, Morgan JL, Godinho L, Di Cristo G, Huang ZJ, Wong RO (2009) Transient neurites of retinal horizontal cells exhibit columnar tiling via homotypic interactions. Nat Neurosci 12:35–43

Jones EA, Crotty D, Kulesa PM, Waters CW, Baron MH, Fraser SE, Dickinson ME (2002) Dynamic in vivo imaging of postimplantation mammalian embryos using whole embryo culture. Genesis 34:228–235

Jontes JD, Buchanan J, Smith SJ (2000) Growth cone and dendrite dynamics in zebrafish embryos: early events in synaptogenesis imaged in vivo. Nat Neurosci 3:231–237

Kaethner RJ, Stuermer CA (1994) Growth behavior of retinotectal axons in live zebrafish embryos under TTX-induced neural impulse blockade. J Neurobiol 25:781–796

Keller RE (1978) Time-lapse cinemicrographic analysis of superficial cell behavior during and prior to gastrulation in Xenopus laevis. J Morphol 157:223–247

Kerschensteiner D, Morgan JL, Parker ED, Lewis RM, Wong RO (2009) Neurotransmission selectively regulates synapse formation in parallel circuits in vivo. Nature 460:1016–1020

Kim H, Kim K, Yim J (2013) Biosynthesis of drosopterins, the red eye pigments of *Drosophila melanogaster*. IUBMB Life 65:334–340

Kise Y, Schmucker D (2013) Role of self-avoidance in neuronal wiring. Curr Opin Neurobiol 23:983–989

Kolodkin AL, Tessier-Lavigne M (2011) Mechanisms and molecules of neuronal wiring: a primer. Cold Spring Harb Perspect Biol 3

Koster RW, Fraser SE (2001) Tracing transgene expression in living zebrafish embryos. Dev Biol 233:329–346

Kulesa PM, Fraser SE (2000) In ovo time-lapse analysis of chick hindbrain neural crest cell migration shows cell interactions during migration to the branchial arches. Development 127:1161–1172

Kulesa PM, Bailey CM, Cooper C, Fraser SE (2010) In ovo live imaging of avian embryos. Cold Spring Harb Protoc 2010:pdb–prot5446

Langen M, Agi E, Altschuler DJ, Wu LF, Altschuler SJ, Hiesinger PR (2015) The developmental rules of neural superposition in *Drosophila*. Cell 162:120–133

Lee T, Luo L (1999) Mosaic analysis with a repressible cell marker for studies of gene function in neuronal morphogenesis. Neuron 22:451–461

Lemon WC, Keller PJ (2015) Live imaging of nervous system development and function using light-sheet microscopy. Mol Reprod Dev 82:605–618

Lerit DA, Plevock KM, Rusan NM (2014) Live imaging of *Drosophila* larval neuroblasts. J Vis Exp 89

Li J, Erisir A, Cline H (2011) In vivo time-lapse imaging and serial section electron microscopy reveal developmental synaptic rearrangements. Neuron 69:273–286

Lichtman JW, Fraser SE (2001) The neuronal naturalist: watching neurons in their native habitat. Nat Neurosci 4(Suppl):1215–1220

Manitt C, Nikolakopoulou AM, Almario DR, Nguyen SA, Cohen-Cory S (2009) Netrin participates in the development of retinotectal synaptic connectivity by modulating axon arborization and synapse formation in the developing brain. J Neurosci Off J Soc Neurosci 29:11065–11077

Mason C, Erskine L (2000) Growth cone form, behavior, and interactions in vivo: retinal axon pathfinding as a model. J Neurobiol 44:260–270
Meyer MP, Smith SJ (2006) Evidence from in vivo imaging that synaptogenesis guides the growth and branching of axonal arbors by two distinct mechanisms. J Neurosci Off J Soc Neurosci 26:3604–3614
Mohler W, White J (1998) Multiphoton laser scanning microscopy for four-dimensional analysis of *Caenorhabditis elegans* embryonic development. Opt Express 3:325–331
Morgan JL, Dhingra A, Vardi N, Wong RO (2006) Axons and dendrites originate from neuroepithelial-like processes of retinal bipolar cells. Nat Neurosci 9:85–92
Murray MJ, Merritt DJ, Brand AH, Whitington PM (1998) In vivo dynamics of axon pathfinding in the *Drosophilia* CNS: a time-lapse study of an identified motorneuron. J Neurobiol 37:607–621
Niell CM, Meyer MP, Smith SJ (2004) In vivo imaging of synapse formation on a growing dendritic arbor. Nat Neurosci 7:254–260
O'Connor TP, Duerr JS, Bentley D (1990) Pioneer growth cone steering decisions mediated by single filopodial contacts in situ. J Neurosci Off J Soc Neurosci 10:3935–3946
Ortega F, Costa MR (2016) Live imaging of adult neural stem cells in rodents. Front Neurosci 10:78
Ozel MN, Langen M, Hassan BA, Hiesinger PR (2015) Filopodial dynamics and growth cone stabilization in Drosophila visual circuit development. eLIFE 4
Paul Bainbridge S, Bownes M (1988) Ecdysteroid titers during Drosophila metamorphosis. Insect Biochem 18:185–197
Prasher DC, Eckenrode VK, Ward WW, Prendergast FG, Cormier MJ (1992) Primary structure of the *Aequorea victoria* green-fluorescent protein. Gene 111:229–233
Pyrgaki C, Trainor P, Hadjantonakis AK, Niswander L (2010) Dynamic imaging of mammalian neural tube closure. Dev biol 344:941–947
Rabinovich D, Mayseless O, Schuldiner O (2015) Long term ex vivo culturing of Drosophila brain as a method to live image pupal brains: insights into the cellular mechanisms of neuronal remodeling. Front Cell Neurosci 9:327
Rajnicek AM, Foubister LE, McCaig CD (2006) Growth cone steering by a physiological electric field requires dynamic microtubules, microfilaments and Rac-mediated filopodial asymmetry. J Cell Sci 119:1736–1745
Ramón y Cajal S (1890) A quelle epoque apparaissent les expansions des cellule nerveuses de la moelle epinere du poulet. Anat Anzerger 5:609–613
Raper J, Mason C (2010) Cellular strategies of axonal pathfinding. Cold Spring Harb Perspect Biol 2:a001933
Reed BH, McMillan SC, Chaudhary R (2009) The preparation of Drosophila embryos for live-imaging using the hanging drop protocol. J Vis Exp 25
Sanes JR, Zipursky SL (2010) Design principles of insect and vertebrate visual systems. Neuron 66:15–36
Schmid A, Sigrist SJ (2008) Analysis of neuromuscular junctions: histology and in vivo imaging. Methods Mol Biol 420:239–251
Shcherbo D, Murphy CS, Ermakova GV, Solovieva EA, Chepurnykh TV, Shcheglov AS, Verkhusha VV, Pletnev VZ, Hazelwood KL, Roche PM et al (2009) Far-red fluorescent tags for protein imaging in living tissues. Biochem J 418:567–574
Sperry RW (1943) Effect of 180 degree rotation of the retinal field on visuomotor coordination. J Exp Zool 92:263–279
Vaughn JE (1989) Fine structure of synaptogenesis in the vertebrate central nervous system. Synapse 3:255–285
Vitriol EA, Zheng JQ (2012) Growth cone travel in space and time: the cellular ensemble of cytoskeleton, adhesion, and membrane. Neuron 73:1068–1081
Webb RH (1996) Confocal optical microscopy. Rep Prog Phys 59:427

Williams PR, Morgan JL, Kerschensteiner D, Wong RO (2013) In vitro imaging of retinal whole mounts. Cold Spring Harb Protoc 2013
Williamson WR, Hiesinger PR (2010) Preparation of developing and adult Drosophila brains and retinae for live imaging. J Vis Exp
Yogev S, Shen K (2014) Cellular and molecular mechanisms of synaptic specificity. Annu Rev Cell Dev Biol 30:417–437
Zheng JQ, Wan JJ, Poo MM (1996) Essential role of filopodia in chemotropic turning of nerve growth cone induced by a glutamate gradient. J Neurosci Off J Soc Neurosci 16:1140–1149
Zschatzsch M, Oliva C, Langen M, De Geest N, Ozel MN, Williamson WR, Lemon WC, Soldano A, Munck S, Hiesinger PR et al (2014) Regulation of branching dynamics by axon-intrinsic asymmetries in tyrosine kinase receptor signaling. eLIFE 3:e01699

Part II
Behavior: The Behavioral Contributions of Identified Cell Types

Chapter 6
Manipulation of Neural Circuits in *Drosophila* Larvae

Ibrahim Tastekin and Matthieu Louis

Abstract *Drosophila* has proven to be an extraordinarily prolific model organism to study the integrated function of neural circuits. This success largely stems from the development of powerful genetic tools to monitor and to manipulate the activity of identified neurons in the fly nervous system. However, establishing causal relationships between the activity of a given neuron and the expression of a behavior remains challenging both at a technical and at a conceptual level. First, the characterization of behavioral phenotypes still lacks standardization in the field. Here, we illustrate the importance of quantitative analysis of behaviors as complex as sensory navigation (chemotaxis). Second, experimenters are often confronted with the absence of suitable reagents to exclusively label their neurons of interest. A driver line associated with an interesting loss- or gain-of-function phenotype often covers a heterogeneous group of neurons. In the present chapter, we describe how reagents freely available to the fly community can be combined to nail down the relationships between phenotypic traits and the activity of single neurons.

6.1 Introduction

The *Drosophila* larva is a premier model organism to delineate computational principles underlying how neural circuits transform sensory inputs into stereotyped behaviors. Traditionally, neuroscientists study circuit computation by breaking

I. Tastekin · M. Louis (✉)
EMBL/CRG Systems Biology Research Unit, Centre for Genomic Regulation (CRG), The Barcelona Institute of Science and Technology, Dr. Aiguader 88, 08003 Barcelona, Spain
e-mail: mlouis@lifesci.ucsb.edu

I. Tastekin · M. Louis
Universitat Pompeu Fabra (UPF), 08002 Barcelona, Spain

M. Louis
Neuroscience Research Institute and Department of Molecular, Cellular and Developmental Biology, University of California Santa Barbara, 93106 Santa Barbara, California, USA

A. Çelik and M.F. Wernet (eds.), *Decoding Neural Circuit Structure and Function*, DOI 10.1007/978-3-319-57363-2_6

neural circuits down into their core components—individual neurons (invertebrates) or neuronal cell types (vertebrates)—and by testing the necessity and sufficiency of individual neurons to execute a behavior. The combination of community-based reconstruction of the whole larval brain connectivity based on light and electron microscopy (Li et al. 2014; Schneider-Mizell et al. 2016) and the presence of sophisticated genetic tools makes the larva particularly suited to progress from "circuit mapping" to a holistic "circuit cracking" (Olsen and Wilson 2008). The larva combines other advantages for circuit cracking: it has a small heat capacity facilitating thermogenetic manipulations. It is mostly transparent, which is convenient for optogenetic gain-of-function experiments and live imaging. The larva displays stereotyped behaviors on a timescale considerably slower than adult flies (Green et al. 1983). In addition, foraging in the larva can be studied on two-dimensional substrates as basic as an agarose slab instead of complex tri-dimensional environments. As a result, tracking naturalistic behaviors is technically simpler in the larva than in the adult fly.

While the numerical complexity of the nervous system of the larva is reduced by one order of a magnitude compared to its the adult fly counterpart (10,000 vs. 100,000 neurons), the *Drosophila* larva exhibits sensory-driven reorientation maneuvers in chemical, light, and temperature gradients as well as robust escape behaviors in response to threatening stimuli (Hwang et al. 2007; Luo et al. 2010; Kane et al. 2013; Zhang et al. 2013; Ebrahim et al. 2015). The larva is also capable of forming and retrieving associative memory (Gerber and Stocker 2007). The control of reorientation behavior is plastic: it can be modulated by memory traces (Schleyer et al. 2015). Genetic tools provide access to visualize and manipulate the function of small groups or even individual neurons. These tools can be efficiently combined with electron microscopy (EM) reconstruction of the entire larval brain to build circuit-level connectivity diagrams (Ohyama et al. 2015; Schneider-Mizell et al. 2016; Zwart et al. 2016). One can perform "circuit epistasis" by hierarchically manipulating different cell types revealed by EM connectivity diagrams (Ohyama et al. 2015). Altogether, recent advances in the field of larval neurobiology have created unprecedented opportunities to unravel the operation of neural circuits and to test mechanistic hypotheses with a spatiotemporal resolution that will soon match the same standards as in *C. elegans*.

The main objective of this book chapter is to review current genetic tools to manipulate neural functions in the *Drosophila* larva. First, we will draw the attention of the reader on the promises and the limitations of existing tools to study the function of individual neurons. Second, we will discuss the importance of quantifying behavior to search for the neural correlates of sensorimotor functions (Egnor and Branson 2016). Third, we will describe clonal gain-of-function strategies to dissect the contribution of distinct groups of neurons labeled by a driver line associated with a phenotype of interest. This method is intended to make the most out of driver lines with expression patterns that cover more than a couple of neurons—a problem "*Drosophilists*" frequently face when they analyze the neural mechanisms underlying the organization of behavior.

6.2 Genetic Targeting of Neurons in the *Drosophila* Larva

In *Drosophila,* high stereotypy of morphology and connectivity of individual cell types allow the analysis of neural function at a population level. Transgenic expression of reporters and/or effectors in subsets of neurons via binary expression systems has been widely used to visualize and to functionally manipulate specific neurons (Venken et al. 2011). Recently, two large collections of Gal4 driver lines (Pfeiffer et al. 2008; Bidaye et al. 2014) have been created and made accessible to the fly community to label reproducible subsets of neurons in the *Drosophila* brain. Despite the fact that these driver lines label relatively small numbers of neurons compared to their predecessors (e.g., the so-called Kyoto collection), anatomical and behavioral experiments often necessitate targeting even smaller subsets of neurons—ideally single neurons. Stochastic labeling methods such as flip-out, MARCM (Venken et al. 2011) and multicolor flip-out (MCFO) (Nern et al. 2015) have been used to characterize the morphology of single neurons using Gal4 driver lines. For behavioral studies, intersectional expression combining Gal4, Gal80, LexA expression systems, and the Split-Gal4 technique (Luan et al. 2006; Pfeiffer et al. 2010) are now routinely used to restrict expression to predefined subsets of neurons (Aso et al. 2014; Hampel et al. 2015; Ohyama et al. 2015). However, these intersectional techniques are limited by the existence of driver lines with overlapping expression patterns. In the following sections, we will describe how driver lines with expression patterns including more than one neuron can be exploited to draw hypotheses about the link between connectivity and function in specific neurons.

6.3 Inferring Function by Manipulating the Activity of Genetically Labeled Neurons

Traditionally, behavioral experiments are conducted to test the necessity or sufficiency of neurons to execute a certain type of behavior in *Drosophila* (Vogelstein et al. 2014; Tastekin et al. 2015). In many cases, the necessity of a neuron to control a given function is probed by (i) hyperpolarizing the neuron upon overexpression of the inward-rectifier potassium ion channel Kir2.1 (Baines et al. 2001) or by (ii) blocking synaptic transmission with tetanus toxin light chain (TNT) (Sweeney et al. 1995) or the temperature-sensitive dynamin mutant *shibire* (Thum et al. 2006). Caution must be taken while interpreting the results that follow the expression of an effector that is supposed to inhibit neural function. One should keep in mind that TNT impairs the release of neurotransmitter by cleaving neuronal synaptobrevin, a protein necessary for calcium-dependent vesicle fusion (Sweeney et al. 1995; Baines et al. 2001). As a result, TNT does not affect synaptic transmission mediated by pathways independent of synaptobrevin (Thum et al. 2006). In addition, it has been argued that blockage of synaptic transmission affects the electrical development of neurons (Baines et al. 2001). Accordingly, prolonged expression of TNT might lead to compensatory effects at the neuronal and/or circuit level. While UAS constructs inserted in different genomic sites can produce different expression patterns (Aso et al. 2014), expression pattern of a given driver line can vary depending on the reporter it is coupled with (Fig. 6.1). In light of this, co-expressing TNT and fluorescent indicators by using different UAS transgenes does not guarantee a perfect correlation in the resulting expression patterns. One should therefore remember that the expression of a fluorescent indicator might not faithfully reproduce that of TNT. The fact that a tagged version of TNT does not exist makes it difficult to determine whether TNT is expressed in the targeted neurons. Fortunately, a GFP-tagged version of Kir2.1 exists. Although constant hyperpolarization might lead to compensatory effects at the circuit level, it has been shown that expression of Kir2.1 does not lead to a change in the electrical properties of at least two types of motor neurons in the larva (aCC and RP2), suggesting that Kir2.1 expression does not change the electrical properties of a neuron (Baines et al. 2001).

In comparison with TNT and Kir2.1, the dominant-negative allele *shibire*^{ts} offers temporal control, which permits to overcome the compensatory and developmental effects of chronic inhibition. Using *shibire*^{ts}, synaptic release can be reversibly blocked under restrictive temperatures (29–34 °C). One has to remain careful while interpreting the effects of manipulations involving *shibire*^{ts} since the expression of this reagent can induce morphological changes (Gonzalez-Bellido et al. 2009). Another caveat with the use of temperature changes is the interference with innate temperature-driven behaviors (thermotaxis). The outcome of temperature increases is therefore composite: it results from the effects of synaptic transmission block and the innate response to thermal stimulation. Moreover, heat convection induced by temperature changes in the assay can perturb the geometry of any odor gradient it might enclose. For this reason, it is preferable to avoid using effectors requiring temperature changes while testing the necessity of specific sets of neurons to direct orientation behaviors such as chemotaxis (Fig. 6.2a). Toxins (e.g., diphtheria toxin) and proapoptotic genes (e.g., *Reaper* and *Head involution defective*) are more rarely used to block neural function by inducing cell death. Their lack of popularity is mainly due to the detrimental effects that cell death can have on the development of the rest of the brain. For a more detailed discussion of the reagents commonly used to dissect neural function, we refer the reader to two thorough reviews (Simpson 2009; Venken et al. 2011). Upon applications of effectors inducing a loss of function, the effects of impairing the function of a given neuron or a neuronal subset should be always interpreted at the circuit level. In addition, the nonlinear dynamics generated by networks of interconnected cells imply that neural circuits must produce complex behaviors that cannot be inferred from the effects of blocking parts of the circuits.

Sufficiency is usually defined by whether activation of a given neuron triggers a certain type of behavior or the response of a putative downstream partner. Acute activation of neurons in the larva has been successfully accomplished by using thermogenetic and optogenetic tools (Pulver et al. 2009). Targeted expression of

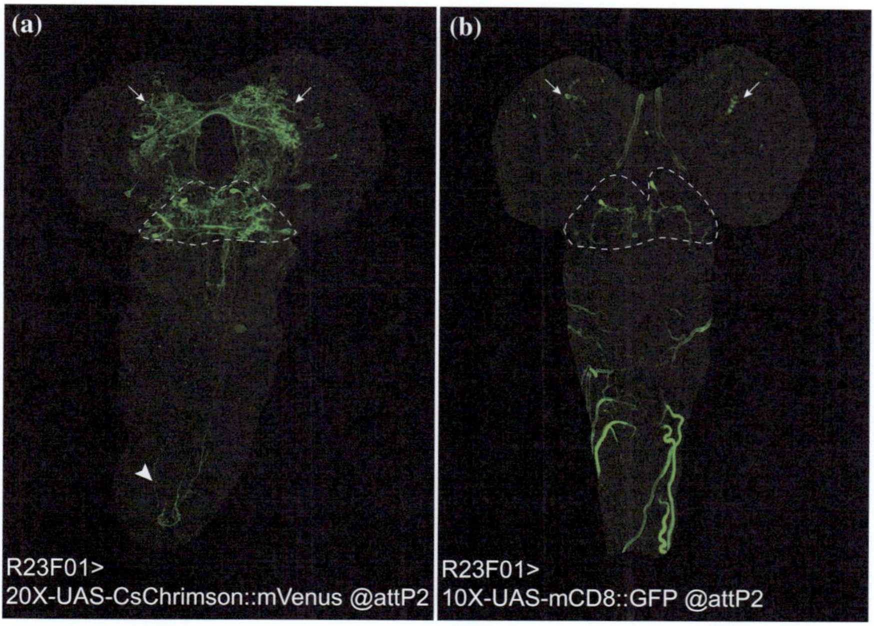

Fig. 6.1 Variability in the expression pattern of the same GAL4 driver line reported by two different UAS transgenes inserted in the same landing site. **a** Expression pattern of R23F01>20X-UAS-IVS-CsChrimson::mVenus. Both R23F01 and 20X-UAS-IVS-CsChrimson:: mVenus transgenes are inserted at the attP2 landing site on the 3rd chromosome. *Dashed line* encloses the subesophageal zone (*SEZ*). Note the high level of expression of the reporter in the SEZ. *Arrows* highlight expression in the brain lobes. The large *arrowhead* indicates the axon of a descending neuron from the SEZ. **b** Expression pattern of R23F01>10X-UAS-IVS-mCD8::GFP (retrieved from http://flweb.janelia.org/cgi-bin/flew.cgi). Both R23F01 and 10X-UAS-IVS-mCD8::GFP transgenes are inserted at the attP2 landing site on the 3rd chromosome. *Dashed line* encloses the subesophageal zone (*SEZ*). In contrast with panel **a**, only two neurons are labeled in the SEZ and very few Kenyon cells are labeled in each brain lobe. The picture shown in panel **b** is courtesy of the Truman lab (Li et al. 2014). It is reproduced with the permission of the author

Drosophila TrpA1 (dTrpA1) channel (Rosenzweig et al. 2008) has been widely used to activate neurons upon temperature increases. Although this tool has proved to be useful to induce stereotypic behavioral sequences in adult flies (von Philipsborn et al. 2011; Marella et al. 2012), it lacks both temporal resolution and control over the intensity ranges of the neural activity. This is particularly important as the level and timing of a gain in neural activity might trigger distinct behavioral output due to complex circuit interactions. It has to be noticed that continuous activation of dTrpA1 might lead to a depolarization block in some neurons via rapid depolarization (Inagaki et al. 2014). Furthermore, temperature manipulations necessary to activate neurons might create behavioral interferences induced by innate responses to temperature changes, as indicated above. In recent work on the sensorimotor control of larval chemotaxis (Tastekin et al. 2015), we were unable to

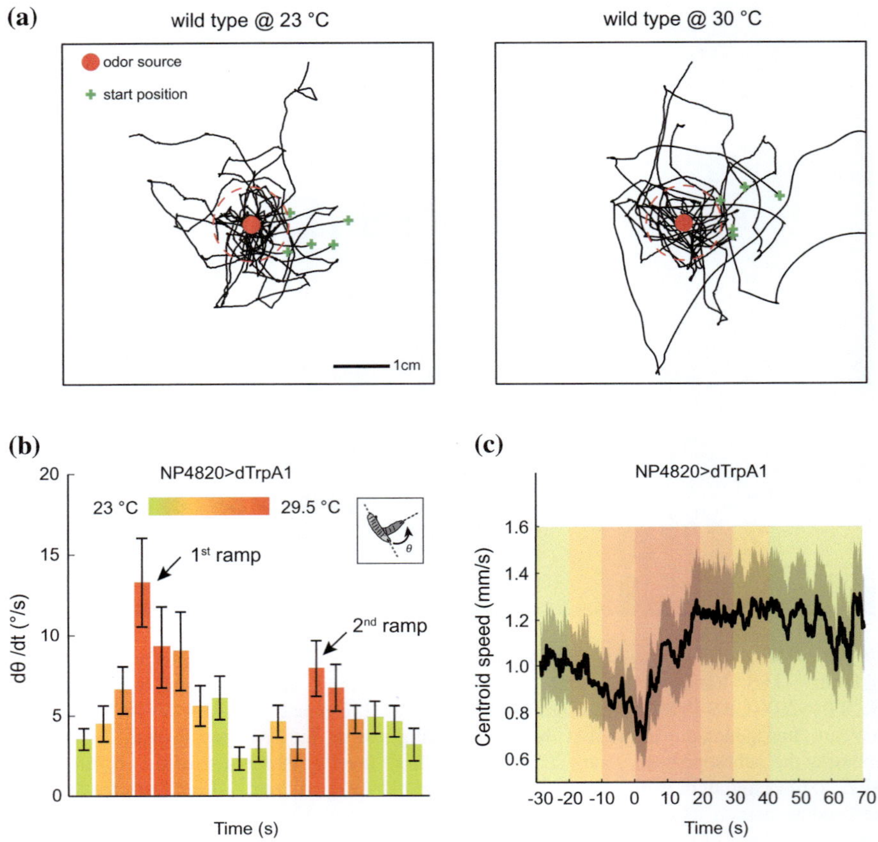

Fig. 6.2 Thermogenetic gain-of-function manipulations in the *Drosophila* larva. **a** Effect of temperature on larval chemotaxis. An odor gradient is formed by using a point odor source (*red dots*, 10 μL of a 100-μM solution of ethyl butyrate). Trajectories from 5 representative larvae were plotted for 23 °C (*left panel*) and 30 °C (*right panel*). Note that wild type larvae tend to stay closer to the odor source when they are allowed to chemotax at 23 °C. **b** Thermogenetic activation of NP4820-labeled neurons by expressing dTrpA1 reagent. Temperature is raised slowly from 23 to 29.5 °C in a period of 30 s and subsequently decreased back to 23 °C. The temperature ramp is repeated twice. Activation of NP4820-labeled neurons led to a transient increase in average head angular speed (head sweeps) during the first temperature increase phase of the temperature (*arrow*). However robust head sweeps cannot be elicited during the second increase in temperature (*arrow* labeled as 2nd ramp). Each *bar* indicates average head angular speed binned in 10-s windows. *Error bars* indicate standard errors of the mean. **c** Thermogenetic activation of NP4820-labeled neurons leads to a transient increase in head sweeps followed by fast crawling. Increase in centroid speed is observed shortly after thermogenetic activation. The *shaded boxes* of different *colors* (see *horizontal* heat map *bar* in **b** for corresponding temperature scale) represent the windows of time during which the temperature was brought from 23 to 29.5 °C

trigger a reproducible gain-of-function phenotype using thermogenetics (Fig. 6.2b) while optogenetic activation led to a strong and reliable phenotype in single larvae. In the next paragraphs, we will argue that the use of optogenetics has multiple advantages compared to thermogenetics.

Due to its superior temporal precision, optogenetic activation has become increasingly adopted for gain-of-function manipulations aiming to test sufficiency (Fenno et al. 2011). Until recently, the performances of Channelrhodopsin 2 (ChR2), a blue light gated ion channel, were limited in *Drosophila* for several reasons including the low penetrance of blue light through the cuticle of adult flies and the innate responses of the adults and the larvae. In spite of this limitation, ChR2 has been successfully applied to study proboscis extension, escape responses, learning, locomotor activity (Schroll et al. 2006; Gordon and Scott 2009; Zimmermann et al. 2009; Matsunaga et al. 2013), and orientation behaviors (Zhang et al. 2007; Gepner et al. 2015; Hernandez-Nunez et al. 2015; Schulze et al. 2015). Since the function of ChR2 necessitates its coupling with the chromophore all-trans retinal that is not endogenously produced by *Drosophila*, larvae must be grown in food complemented with all-trans retinal. Note however that a small amount of retinal is present in regular fly food (Claire McKellar, personal communication). Recent development of red-shifted optogenetic tools (ReaChr, CsChrimson and ChrimsonR) (Inagaki et al. 2014; Klapoetke et al. 2014) enabled deeper penetration of light as well as minimal innate response to visual stimulation, which opened new avenues in *Drosophila* optogenetics. It has been shown that ChrimsonR has relatively higher off kinetics compared to CsChrimson and it can produce sustained trains of spikes when activated at moderately high frequencies (20 Hz) (Klapoetke et al. 2014).

Drosophila larvae are averse to blue light during most of their development (Kane et al. 2013). Abrupt changes in blue light intensity lead to increased turning. On the other hand, we observed that wild type *Drosophila* larvae show minimal to no response to changes in light intensity at 625 nm when they are fed on food with all-trans retinal (Fig. 6.3a, b) while 0.3–3 W/m^2 is sufficient to induce paralysis when CsChrimson is expressed in most of the motor neurons using OK6-Gal4 (Fig. 6.3f). In our hands, much higher light intensities had to be applied to activate brain interneurons (10–18 W/m^2, Fig. 6.3c, d). We observed that reproducible behavioral responses could be elicited over several trials using CsChrimson. However, we noted occasional time-dependent decreases in behavioral response upon the application of prolonged light stimulations (data not shown). This dampening of the gain of function is probably due to the off kinetics of CsChrimson and its slower recovery. It might also be related to the dynamics of the host neuron(s) independently of the effector. Therefore, the kinetics of the effector—whether it is CsChrimson, ChrimsonR or ChR2—should be carefully considered when choosing the duration and frequency of optogenetic stimulations. In case stimulation at high frequencies is required, ChrimsonR should be favored over CsChrimson.

In the *Drosophila* larva, large-scale screens testing loss of functions (necessity) and gain of functions (sufficiency) have been performed to identify neural correlates

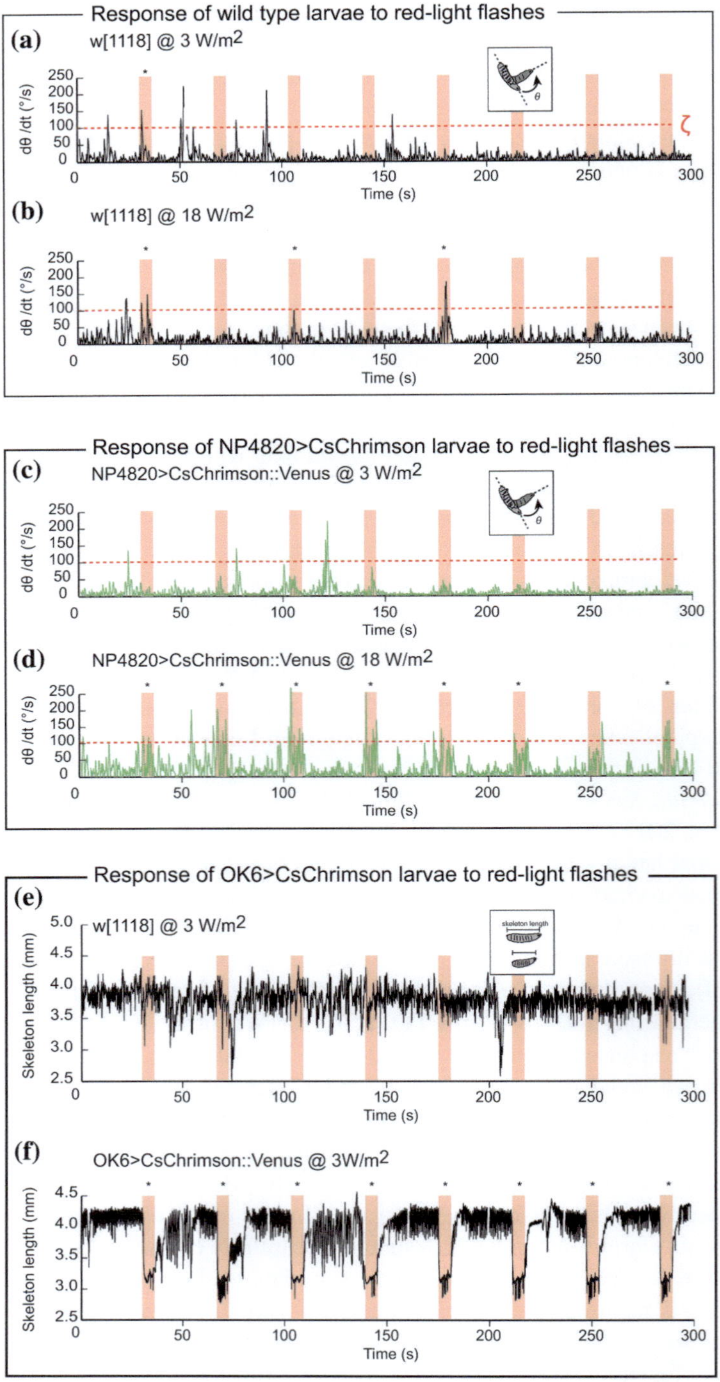
Response of wild type larvae to red-light flashes
(a)
w[1118] @ 3 W/m2
dθ /dt (°/s)
ζ
Time (s)
(b)
w[1118] @ 18 W/m2
dθ /dt (°/s)
Time (s)
Response of NP4820>CsChrimson larvae to red-light flashes
(c)
NP4820>CsChrimson::Venus @ 3 W/m2
dθ /dt (°/s)
Time (s)
(d)
NP4820>CsChrimson::Venus @ 18 W/m2
dθ /dt (°/s)
Time (s)
Response of OK6>CsChrimson larvae to red-light flashes
(e)
w[1118] @ 3 W/m2
Skeleton length (mm)
skeleton length
Time (s)
(f)
OK6>CsChrimson::Venus @ 3W/m2
Skeleton length (mm)
Time (s)

◂**Fig. 6.3** Innate sensitivity of larvae to *red light* and acute optogenetic activation of neural activity using CsChrimson. **a** We probe the response of the control larvae subjected to 6-s flashes of *red light* (625 nm) at an intensity of 3 W/m^2. The behavioral response is defined by quantifying head sweeps as a function of the angular speed. A head sweep is considered to be a "cast" when the absolute value of the angular speed exceeded a threshold ζ of 100°/s. For more information about the method used to determine the value of threshold on the head angular speed, see Fig. 6.5. The genotype used is w[1118], which corresponds to one of the most common genetic background for transgenics. At low light intensity, w[1118] only occasionally performs head sweeps that qualify as head casts (*stars* 1 out of 8 flashes). **b** Same as panel **a** with a higher intensity (18 W/m^2) of *red light*. Head sweeps more frequently qualify as head casts than at a lower intensity of 3 W/m^2 (*stars* 3 out of 8 flashes). **c** Optogenetic activation of the NP4820-labeled neurons with 3 W/m^2 of *red light* (625 nm) upon expression of CsChrimson. Same pattern of light flashes as shown in panel **a**. The larva does not respond to *red light* at this intensity. **d** Same as **c** with 18 W/m^2 intensity. Robust head casts are observed as a function of absolute head angular speed (*stars* 6 out of 8 flashes). In panels **c** and **d**, larvae were raised on regular fly food with a concentration of 0.5 mM all-trans retinal. **e** Response of control larvae w[1118] subjected to 6-s flashes of *red light* at an intensity of 3 W/m^2. Quantification of the behavioral response by the length of larva's skeleton. *Red light* flashes of 3 W/m^2 intensity do produce a significant decrease in body length. **f** Optogenetic activation of OK6 neurons with 3 W/m^2 of *red light*. OK6 covers most of the motor neurons in the VNC. As a result of the global activation of motor neurons, muscles across the body length contract simultaneously leading to significant decrease in the skeleton length

of behavioral control (Vogelstein et al. 2014; Tastekin et al. 2015; Clark et al. 2016; Yoshikawa et al. 2016). The number of neurons typically covered by a driver line that led to a phenotype ranged from one to a few dozens. Instead of treating each labeled neuron as a separate unit, it is convenient to group neurons by lineages. Lineages from circuit elements can be viewed as the anatomical building blocks of the brain (Hartenstein et al. 2015). For some of the hits identified in screens, an interesting behavioral phenotype could not be mapped on a single lineage due to the existence of multiple lineages covered by the driver line yielding the phenotype of interest. In these cases, alternative strategies have been deployed to restrict the phenotype to the activation/silencing of a single cell type. For example, Ohyama and colleagues successfully utilized a combination of two binary expression systems (Gal4 and LexA) together with the Split Gal4 technique to narrow down the mapping of a behavioral phenotype onto one or a small set of lineages (Ohyama et al. 2015). This approach relies on the existence or the generation of combinations of Split Gal4 lines, which is not always possible.

In recent work, we adopted a different strategy to uncover circuit elements participating in the sensory control of the timing of turning maneuvers (Tastekin et al. 2015). In this study, we used a densely expressed Gal4 driver line (multiple cell types with more than 50 neurons in the brain lobes and the subesophageal zone). Our attempts to confine the expression of the driver line to a few neurons using traditional Gal80 and lexA intersections could only lead to the conclusion that one or more cells out of a group of ~15 located in the subesophageal zone (SEZ) are responsible for the gain of function phenotype (triggering of a turning maneuver). To enhance the precision of the circuit-function mapping, we applied an acute gain-of-function strategy combined with random labeling of neurons. We

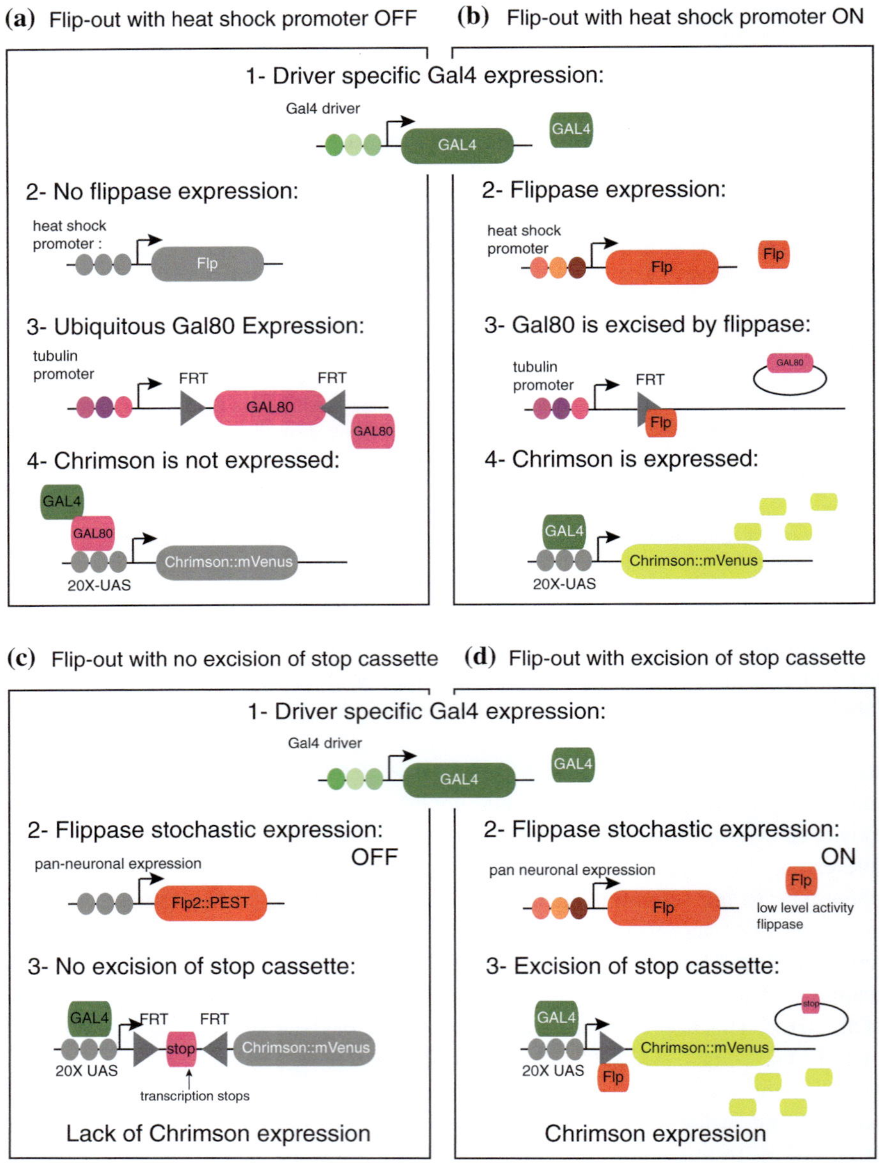

induced stochastic expression of Chrimson::mVenus in small subsets of neurons by combining the original densely expressed Gal4 driver line with a Gal80 driver whose expression was conditioned by a probabilistic flip-out recombination under the control of a heat shock promoter (Fig. 6.4a, b). After performing acute activation of each clone, we visualized the expression of Chrimson protein in individual clones using standard immunostaining against mVenus protein. In this way,

◂**Fig. 6.4** Two different flip-out intersectional strategies to stochastically express CsChrimson:: mVenus in clones. **a, b** "Flip-out" strategy mediated by heat shock (*hs*) promoter. In panel **a**, the hs promoter is OFF. As a consequence, Gal80 flanked by FRT is ubiquitously expressed under the control of tubulin promoter and inhibits Gal4-UAS dependent expression of CsChrimson:: mVenus. In panel **b**, the hs promoter is ON, which drives expression of the flippase protein. Flippase excises the Gal80 sequence, thereby abolishing ubiquitous Gal80 expression. Gal4 can bind to the UAS sequences and drive CsChrimson::mVenus expression in a cell-specific manner. **c, d** "Flip-out" strategy using pan-neuronal expression of low-level activity version of flippase (Flp2::PEST, for details see Nern et al. 2015). A transcriptional stop cassette flanked with FRT was placed between UAS and CsChrimson::mVenus sequences preventing Gal4-dependent expression of CsChrimson::mVenus in the absence of sufficient flippase activity (panel **c**). In panel **d**, the higher level of activity of flippase in some cells is sufficient to excise the stop cassette upstream from the coding sequence of CsChrimson::mVenus. As a result, CsChrimson::mVenus is expressed in a subset of cells of the original pattern labeled by the Gal4 driver

we could directly monitor the expression of the effector (CsChrimson). This approach is more reliable than indirectly assessing the expression of an effector (e.g., TNT) through an additional reporter (e.g., UAS-GFP). As described in the next section, we devised a statistical method to correlate gain-of-function behaviors with expression patterns of CsChrimson. Below, we will detail this approach as it represents a useful alternative to infer circuit function relationships associated with Gal4 lines expressed in multiple lineages when sparse driver lines do not exist to reduce the expression pattern of the original driver line.

6.4 Stochastic Labeling of Neurons Using Flip-Out Approach

The flip-out method has been widely used to stochastically visualize subsets of neurons covered by Gal4 drivers (Venken et al. 2011). We employed a similar strategy based on the "FLP-FRT" recombination system (Fig. 6.4). Following a first variant of this approach, the expression of Gal80 flanked by FRT sequences is induced ubiquitously by a tubulin promoter (FLP-out Gal80) (Gordon and Scott 2009). When the flippase (FLP) recombinase is stochastically expressed under the control of a heat shock promoter, it stochastically induces excision of the FRT-Gal80-FRT cassette downstream of the tubulin promoter. As a result, Gal80 is not expressed in the subset of neurons where the recombination took place, thereby allowing full activity of Gal4 and expression of the effector (e.g., Chrimson:: mVenus). This method was initially applied to stochastically silence/activate neurons involved in proboscis extension in adult flies (Gordon and Scott 2009). It enables lineage-independent expression of effectors in different combinations of neurons and it is possible to optimize the probability of flip-out events by changing the strength and duration of the heat shock. Thus, one can roughly control the number of cells in which the effector expression is allowed by the heat shock induced loss of Gal80.

Fig. 6.5 Stochastic labeling of subsets of neurons in gain-of-function clones of the NP4820-Gal4 ▶ driver line. **a, b** Two clones showing expression in subsets of neurons upon heat-shock-dependent stochastic expression (see method described in Fig. 6.4a, b). *Arrows* indicate neurons that are functional in larvae at the third developmental stage. Stars indicate immature secondary lineage neurons that are unlikely to be functional at the third instar stage. Panel **a** features a clone without a behavioral phenotype while the brain displayed in panel **b** demonstrated a strong gain-of-function phenotype (see text for the explanation of positive and negative phenotypes). **a'**, **b'** Quantification of the behavioral phenotype observed upon acute optogenetic gain of function of the clones shown in panels **a** and **b**, respectively. The trace of panel **a'** is associated with a negative phenotype since the larva does not respond to any of the light flashes. Panel **b'** is associated with a positive phenotype since the larva demonstrated a strong increase in head angular speed ($d\theta/dt$) that exceeded the threshold ζ for 7 out of the 8 flashes. **c** Receiver operating characteristic *curve* used to define a binary classifier for efficient detection of behaviorally positive clones. "Responsivity" is defined as the number of flashes during which head angular speed exceeds a certain threshold value (ζ). Responsivity ranges between 1 and 8. In panel **c**, the ROC analysis corresponds to a responsivity of 6. The ROC is plotted for different ζ values ranging from 10 to 180°/s (*gray circles*). Optimal classification (true positive TP rate is as high as possible and false positive FP rate is as low as possible) was obtained at responsivity = 6 and for a ζ value of 100°/s (*red circle*). Using these criteria, the number of false positives expected for a batch of 70 tested larvae is $0.04 \times 70 = 3$ individuals. **d** Comparison of the ROC corresponding to a responsivity of 5 flashes (*red curve*) and 6 flashes (*blue curve*). For both responsivity, the ROC is shown for values of ζ ranging from 10 to 180°/s

One disadvantage of this flip-out approach is that hs-flp transgene and the tubulin promoter-FRT-Gal80-FRT transgene cannot be combined in the same fly stock since excision of Gal80 might occur in the germ line and lead to an irreversible loss of Gal80 in the offspring. For this reason, a new transgene must be generated for each Gal4 driver line. In order to activate and visualize the neurons that are stochastically labeled, we used red-shifted opsin CsChrimson (Klapoetke et al. 2014) fused to fluorescent protein mVenus (a collection of CsChrimson:: mVenus effector inserted in different landing sites have been generated by Vivek Jayaraman; these reagents are available from the Bloomington stock center). For thermogenetic activation and subsequent visualization of the effector, dTrpA1:: myc-tag fusion can be used ($dTRPA1^{myc}$) (von Philipsborn et al. 2011).

An alternative flip-out strategy relies on the expression of a weakened version of FLP recombinase under the control of a pan-neuronal driver (Nern et al. 2015). In this method, expression of the FLP recombinase is restricted to the differentiated neurons by driving the expression of FLP with the promoter of N-synaptobrevin gene (R57C10) (Jenett et al. 2012). Instead of flanking ubiquitously expressed Gal80, a transcriptional stop cassette flanked by FRT (Wong et al. 2002) was introduced between UAS and CsChrimson::mVenus ($dTRPA1^{myc}$ or $dTRPA1^{mcherry}$ in case of thermogenetics) (von Philipsborn et al. 2011; Asahina et al. 2014). Low-level pan-neuronal expression of weakened FLP recombinase is expected to yield stochastic expression of CsChrimson::mVenus protein in a small subset of neurons (Fig. 6.4c, d). Unlike the Gal80-based method, R57C10-FLP and UAS-FRT-stop-FRT-CsChrimson::mVenus transgenes can be combined in a single

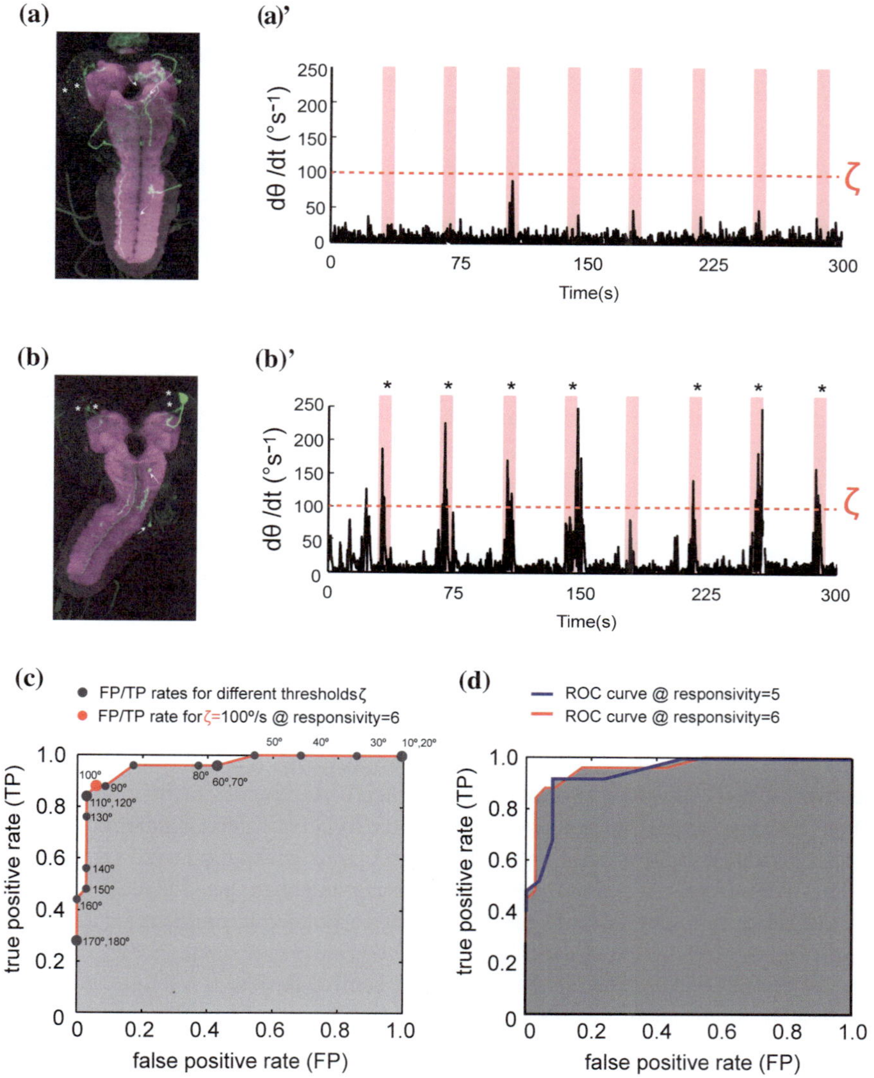

fly stock since FLP expression is restricted to differentiated neurons. Thus, one can easily combine this fly stock with any Gal4 line to perform stochastic gain-of-function experiments, which makes this approach more suited for screening purposes. For both flip-out methods, we were able to reliably visualize the morphology of the labeled neurons by performing traditional immunostaining against mVenus protein with anti-GFP antibodies (Fig. 6.5a, b).

6.5 Acute Activation of Stochastically Labeled Neurons: Stochastic Gain of Function of Neurons in the NP4820 Driver Line

Drosophila larval chemotaxis mainly involves alternation of runs (forward movements by waves of peristaltic contractions) and lateral head sweeps (head casts) followed by directed turns (Gomez-Marin et al. 2011). In a loss-of-function behavioral screen, we identified a Gal4 driver line (NP4820) with a reasonably sparse expression pattern. Activating these neurons by thermogenetics induced transient increases in head sweeps suggesting that NP4820-positive neurons are involved in run-to-turn transitions. Unfortunately, NP4820 labels multiple cell types in the brain lobes, the SEZ and the ventral nerve cord (VNC). Therefore, we applied a stochastic activation method to define which neurons covered by the NP4820 line are responsible for triggering turning maneuvers.

We opted for optogenetic activation for several reasons. First, we observed that larvae with neural activation induced by thermogenetics (dTrpA1) failed to maintain the gain-of-function behavior—an increase in head sweeps—over several seconds. Upon thermogenetic activation of NP4820 neurons, larvae engaged in fast crawling after a short bout of increase in head angular speed (Fig. 6.2c). It is possible that fast crawling results from innate avoidance triggered by high temperatures. It is equally plausible that strong activation of the neurons expressing dTrpA1 leads to a depolarization block in the neurons inducing head sweeps. Second, we were not able to reliably induce head-sweep behavior over several trials (Fig. 6.2b). To limit the identification of false positives, the reproducibility of the behavioral response over several trials is crucial. A separate technical constraint came from the fact that we could not use the blue light activated Channelrhodopsin 2 as the excitation light evoked strong head sweeps in wild type larvae (Kane et al. 2013). Therefore, we expressed CsChrimson::mVenus in NP4820 neurons. NP4820>CsChrimson::mVenus larvae robustly responded to multiple flashes of red light (Fig. 6.3d). In contrast with NP4820>CsChrimson::mVenus larvae, wild type larvae only rarely responded to a series of consecutive flashes of red light at low and moderately high intensities (3 and 18 W/m^2, Fig. 6.3a, b). We reasoned that optogenetic activation using CsChrimson would fulfill the conditions to perform stochastic gain of functions.

We took advantage of the basal leakiness of the heat shock promoter at 23 °C to induce low levels of flippase expression. With this reagent, we restricted CsChrimson expression to 1–5 neurons in individual larvae (clones, Fig. 6.5a, b). Each clone was tested with a stimulation protocol of 8 flashes of 6 s at an intensity of 18 W/m^2 and a wavelength of 625 nm (Fig. 6.3d). Individual flashes were separated by 30 s. Unlike with dTrpA1, we did not observe a decrease in the average head angular speed during consecutive gains of function (data not shown). Larval brains were dissected, fixed and immunostained for anatomical assessment with confocal imaging immediately after the behavioral experiments. To determine the behavioral phenotype of a clone, we implemented a statistical framework based on receiver operating characteristic (ROC) to optimize a binary classifier (Duda et al. 2001) that could discriminate individual larvae that showed the gain-of-function behavior (true positive, TP clones, Fig. 6.5b') from the negative clones (true negative, TN, Fig. 6.5a'). We combined two conditions to define TP and FP (false positive), and calculate the rate of each class of events. First, we tested different threshold values (ζ) on the head angular speed—the head angular speed reveals the increase in head activity associated with turning events—and computed the TP and FP rates for each threshold value (Fig. 6.5c). Second, we evaluated how the TP and FP rates changed for different criteria on the number of flashes leading to an increase in the head angular speed that exceeded the threshold (ζ) (Fig. 6.5d).

Using this approach, we were able to draw eight ROC curves each representing different criteria on the minimum number of expected responses to the 8 light flashes (two of them are shown in Fig. 6.5d). The best performance was observed at a threshold ζ of 100°/s of the head angular speed and with a minimum response to 6 out of the 8 light flashes (responsivity = 6, Fig. 6.5c). We made use of this classifier to define the phenotype of each clone. This classifier was also used to define the expected rate of TP and TN by testing positive and negative controls (original Gal4 line driving expression of CsChrimson::mVenus and parental control devoid of Gal4 driver, respectively). Upon behavioral tests, larvae of the positive and negative controls as well as individual clones were immunostained against the mVenus protein tagging CsChrimson using a commercial antibody against GFP (product number: A-11120, Invitrogen). We tested a total of 70 gain-of-function clones out of which we identified 10 positive hits. This ratio of 10/70 was well above the expected FP rate (3–4 larvae out of 70, for calculation see Fig. 6.5c). The expression pattern of a light-responsive positive clone and a light-indifferent negative clone is illustrated in Fig. 6.5a, b. Finally, we determined the groups of neurons that were labeled more frequently than expected from the FP rate. Those neurons were assumed to be responsible for the gain-of-function phenotype. This stochastic gain-of-function strategy allowed us to narrow down the neurons responsible for the control of run-to-turn transitions to three neurons in the SEZ that were not present in any of the negative clones that had been imagined.

6.6 Closing Remarks

Neural circuits form the computational units of brains. The pace at which neural circuits are identified and functionally studied has largely accelerated in *Drosophila* after the creation of large collections of driver lines that cover sparse subsets of neurons (Pfeiffer et al. 2008; Bidaye et al. 2014). The expression patterns of a large fraction of these two collections have been reported in the adult fly (Jenett et al. 2012) as well as in the larva (Li et al. 2014). Given the relatively small number of neurons that form the larval nervous system (approximately 10,000 neurons), hopes are high that a driver line labeling each neuron can be identified. With the ability to monitor genetically labeled neurons and to reproducibly interfere with their function, "*Drosophilists*" have now at their disposal an extraordinary toolkit to ask how neural circuits contribute to the organization of stereotyped behaviors in the larva. However, experience has shown that this toolkit is imperfect in multiple ways: most Gal4 lines label neurons belonging to more than one lineage. Expression patterns are far from being deterministic: significant variability can be observed across individuals. While these limitations should not undermine the success of massive efforts to characterize the function of neurons of the larval brain in an unbiased way (Vogelstein et al. 2014; Tastekin et al. 2015), they call for caution in the interpretation of functional manipulations.

Variability in the expression pattern of driver lines should be viewed as the rule rather than the exception. Consequently, the action of an effector might differ substantially across individual larvae. It should be common practice to characterize the expression pattern of a given driver with different reporters (Fig. 6.1). The consistency of expression patterns should be compared across different samples as well. Inter-individual variability in the expression of an effector can produce phenotypic diversity at the level of a population of larvae undergoing the same loss-of-function or gain-of-function manipulations. In the case of thermogenetics and optogenetics, the gain-of-function manipulations might also be affected by the signal that gates neural activity—a change in temperature or light intensity (Figs. 6.2a, b and 6.3a, b). The contribution of innate responses should be accounted for and, if significant, it should be subtracted from the behavior induced by the effector. In our experience, this type of analysis necessitates to be grounded in rigorous computational analysis of behavior (Egnor and Branson 2016). In light of the variability inherent to behavioral control, searching for the neural correlate of a particular phenotype must start with the definition of metrics that robustly characterize the manifestation of a certain behavior. In the absence of such quantitative metrics, a screen or more refined manipulations are unlikely to yield conclusive results. The reader should also bear in mind that behaviors tend to form a continuum that cannot always be approximated by discrete states or actions (Szigeti et al. 2015). In larvae, forward runs can be easily told apart from stops and backward runs. By contrast, the difference between head casts and turns is more arbitrary.

The typical absence of driver lines labeling a single neuron has also led the field to develop strategies to pin down the expression pattern of a line with broad coverage. The most elegant approach consists in intersecting two different driver lines with Spit-Gal4 to restrict Gal4 activity to a single neuron (Aso et al. 2014; Hampel et al. 2015; Ohyama et al. 2015). This method, however, relies on the generation of complementary lines, which is often not feasible. In their absence, we argue that the expression pattern of the original driver line can still be reduced through clonal strategies. We reviewed two variants of the flip-out method and illustrated its application to conduct clonal gain-of-function manipulations. Through this approach, we were able to map a phenotype—the sensorimotor control of turning maneuver—onto three neurons located in the SEZ whereas the original Gal4 lines labeled over 50 neurons in different regions spanning the mushroom bodies and the VNC (Tastekin et al. 2015). Interestingly the three remaining neurons included one descending neuron that projects to the VNC.

Although the flip-out method did not allow us to refine the mapping beyond this resolution, nothing guarantees that the phenotype arises from a single neuron. As stated at the beginning of this section, brains are organized by a network of neural circuits rather than isolated cells that carry each a different function. Extrapolating the function of a neural circuit through the manipulation of single cells might be limited since the function of individual neurons is often multiplex. The challenge that lies ahead of the reconstruction of neural circuits is to monitor the integrated function of specific circuits to explain the properties that emerge from their interactions.

References

Asahina K, Watanabe K, Duistermars BJ, Hoopfer E, Gonzalez CR, Eyjolfsdottir EA, Perona P, Anderson DJ (2014) Tachykinin-expressing neurons control male-specific aggressive arousal in Drosophila. Cell 156(1–2):221–235

Aso Y, Hattori D, Yu Y, Johnston RM, Iyer NA, Ngo TT, Dionne H, Abbott LF, Axel R, Tanimoto H, Rubin GM (2014) The neuronal architecture of the mushroom body provides a logic for associative learning. eLIFE 3:e04577

Baines RA, Uhler JP, Thompson A, Sweeney ST, Bate M (2001) Altered electrical properties in Drosophila neurons developing without synaptic transmission. J Neurosci 21(5):1523–1531

Bidaye SS, Machacek C, Wu Y, Dickson BJ (2014) Neuronal control of Drosophila walking direction. Science 344(6179):97–101

Clark MQ, McCumsey SJ, Lopez-Darwin S, Heckscher ES, Doe CQ (2016) Functional genetic screen to identify interneurons governing behaviorally distinct aspects of Drosophila larval motor programs. G3 (Bethesda) 6(7):2023–2031

Duda RO, Hart PE, Stork DG (2001) Pattern classification. Wiley, New York

Ebrahim SA, Dweck HK, Stokl J, Hofferberth JE, Trona F, Weniger K, Rybak J, Seki Y, Stensmyr MC, Sachse S, Hansson BS, Knaden M (2015) Drosophila avoids parasitoids by sensing their semiochemicals via a dedicated olfactory circuit. PLoS Biol 13(12):e1002318

Egnor SE, Branson K (2016) Computational analysis of behavior. Annu Rev Neurosci 39:217–236

Fenno L, Yizhar O, Deisseroth K (2011) The development and application of optogenetics. Annu Rev Neurosci 34:389–412

Gepner RM, Skanata M, Bernat NM, Kaplow M, Gershow M (2015) Computations underlying Drosophila photo-taxis, odor-taxis, and multi-sensory integration. eLIFE 4

Gerber B, Stocker RF (2007) The Drosophila larva as a model for studying chemosensation and chemosensory learning: a review. Chem Senses 32(1):65–89

Gomez-Marin A, Stephens GJ, Louis M (2011) Active sampling and decision making in Drosophila chemotaxis. Nat Commun 2:441

Gonzalez-Bellido PT, Wardill TJ, Kostyleva R, Meinertzhagen IA, Juusola M (2009) Overexpressing temperature-sensitive dynamin decelerates phototransduction and bundles microtubules in Drosophila photoreceptors. J Neurosci 29(45):14199–14210

Gordon MD, Scott K (2009) Motor control in a Drosophila taste circuit. Neuron 61(3):373–384

Green CH, Burnet B, Connolly KJ (1983) Organization and patterns of inter- and intraspecific variation in the behaviour of Drosophila larvae. Anim Behav 31(1):282–291

Hampel S, Franconville R, Simpson JH, Seeds AM (2015) A neural command circuit for grooming movement control. eLIFE 4:e08758

Hartenstein V, Younossi-Hartenstein A, Lovick JK, Kong A, Omoto JJ, Ngo KT, Viktorin G (2015) Lineage-associated tracts defining the anatomy of the Drosophila first instar larval brain. Dev Biol 406(1):14–39

Hernandez-Nunez L, Belina J, Klein M, Si G, Claus L, Carlson JR, Samuel ADT (2015) Reverse-correlation analysis of navigation dynamics in Drosophila larva using optogenetics. eLIFE 4
Hwang RY, Zhong L, Xu Y, Johnson T, Zhang F, Deisseroth K, Tracey WD (2007) Nociceptive neurons protect Drosophila larvae from parasitoid wasps. Curr Biol 17(24):2105–2116
Inagaki HK, Jung Y, Hoopfer ED, Wong AM, Mishra N, Lin JY, Tsien RY, Anderson DJ (2014) Optogenetic control of Drosophila using a red-shifted Channelrhodopsin reveals experience-dependent influences on courtship. Nat Methods 11(3):325–332
Jenett A, Rubin GM, Ngo TT, Shepherd D, Murphy C, Dionne H, Pfeiffer BD, Cavallaro A, Hall D, Jeter J, Iyer N, Fetter D, Hausenfluck JH, Peng H, Trautman ET, Svirskas RR, Myers EW, Iwinski ZR, Aso Y, DePasquale GM, Enos A, Hulamm P, Lam SC, Li HH, Laverty TR, Long F, Qu L, Murphy SD, Rokicki K, Safford T, Shaw K, Simpson JH, Sowell A, Tae S, Yu Y, Zugates CT (2012) A GAL4-driver line resource for Drosophila neurobiology. Cell Rep 2(4):991–1001
Kane EA, Gershow M, Afonso B, Larderet I, Klein M, Carter AR, de Bivort BL, Sprecher SG, Samuel AD (2013) Sensorimotor structure of Drosophila larva phototaxis. Proc Natl Acad Sci U S A 110(40):E3868–E3877
Klapoetke NC, Murata Y, Kim SS, Pulver SR, Birdsey-Benson A, Cho YK, Morimoto TK, Chuong AS, Carpenter EJ, Tian Z, Wang J, Xie Y, Yan Z, Zhang Y, Chow BY, Surek B, Melkonian M, Jayaraman V, Constantine-Paton M, Wong GK, Boyden ES (2014) Independent optical excitation of distinct neural populations. Nat Methods 11(3):338–346
Li HH, Kroll JR, Lennox SM, Ogundeyi O, Jeter J, Depasquale G, Truman JW (2014) A GAL4 driver resource for developmental and behavioral studies on the larval CNS of Drosophila. Cell Rep 8(3):897–908
Luan H, Peabody NC, Vinson CR, White BH (2006) Refined spatial manipulation of neuronal function by combinatorial restriction of transgene expression. Neuron 52(3):425–436
Luo L, Gershow M, Rosenzweig M, Kang K, Fang-Yen C, Garrity PA, Samuel AD (2010) Navigational decision making in Drosophila thermotaxis. J Neurosci 30(12):4261–4272
Marella S, Mann K, Scott K (2012) Dopaminergic modulation of sucrose acceptance behavior in Drosophila. Neuron 73(5):941–950
Matsunaga T, Fushiki A, Nose A, Kohsaka H (2013) Optogenetic perturbation of neural activity with laser illumination in semi-intact Drosophila larvae in motion. J Vis Exp 77:e50513
Nern A, Pfeiffer BD, Rubin GM (2015) Optimized tools for multicolor stochastic labeling reveal diverse stereotyped cell arrangements in the fly visual system. Proc Natl Acad Sci U S A 112(22):E2967–E2976
Ohyama T, Schneider-Mizell CM, Fetter RD, Aleman JV, Franconville R, Rivera-Alba M, Mensh BD, Branson KM, Simpson JH, Truman JW, Cardona A, Zlatic M (2015) A multilevel multimodal circuit enhances action selection in Drosophila. Nature 520(7549):633–639
Olsen SR, Wilson RI (2008) Cracking neural circuits in a tiny brain: new approaches for understanding the neural circuitry of Drosophila. Trends Neurosci 31(10):512–520
Pfeiffer BD, Jenett A, Hammonds AS, Ngo TT, Misra S, Murphy C, Scully A, Carlson JW, Wan KH, Laverty TR, Mungall C, Svirskas R, Kadonaga JT, Doe CQ, Eisen MB, Celniker SE, Rubin GM (2008) Tools for neuroanatomy and neurogenetics in Drosophila. Proc Natl Acad Sci U S A 105(28):9715–9720
Pfeiffer BD, Ngo TT, Hibbard KL, Murphy C, Jenett A, Truman JW, Rubin GM (2010) Refinement of tools for targeted gene expression in Drosophila. Genetics 186(2):735–755
Pulver SR, Pashkovski SL, Hornstein NJ, Garrity PA, Griffith LC (2009) Temporal dynamics of neuronal activation by Channelrhodopsin-2 and TRPA1 determine behavioral output in Drosophila larvae. J Neurophysiol 101(6):3075–3088
Rosenzweig M, Kang K, Garrity PA (2008) Distinct TRP channels are required for warm and cool avoidance in Drosophila melanogaster. Proc Natl Acad Sci U S A 105(38):14668–14673
Schleyer M, Reid SF, Pamir E, Saumweber T, Paisios E, Davies A, Gerber B, Louis M (2015) The impact of odor-reward memory on chemotaxis in larval Drosophila. Learn Mem 22(5):267–277

Schneider-Mizell CM, Gerhard S, Longair M, Kazimiers T, Li F, Zwart MF, Champion A, Midgley FM, Fetter RD, Saalfeld S, Cardona A (2016) Quantitative neuroanatomy for connectomics in Drosophila. eLIFE 5

Schroll C, Riemensperger T, Bucher D, Ehmer J, Voller T, Erbguth K, Gerber B, Hendel T, Nagel G, Buchner E, Fiala A (2006) Light-induced activation of distinct modulatory neurons triggers appetitive or aversive learning in Drosophila larvae. Curr Biol 16(17):1741–1747

Schulze A, Gomez-Marin A, Rajendran VG, Lott G, Musy M, Ahammad P, Deogade A, Sharpe J, Riedl J, Jarriault D, Trautman ET, Werner C, Venkadesan M, Druckmann S, Jayaraman V, Louis M (2015). Dynamical feature extraction at the sensory periphery guides chemotaxis. eLIFE 4

Simpson JH (2009) Mapping and manipulating neural circuits in the fly brain. Adv Genet 65:79–143

Sweeney ST, Broadie K, Keane J, Niemann H, O'Kane CJ (1995) Targeted expression of tetanus toxin light chain in Drosophila specifically eliminates synaptic transmission and causes behavioral defects. Neuron 14(2):341–351

Szigeti B, Deogade A, Webb B (2015) Searching for motifs in the behaviour of larval Drosophila melanogaster and *Caenorhabditis elegans* reveals continuity between behavioural states. J R Soc Interface 12(113):20150899

Tastekin I, Riedl J, Schilling-Kurz V, Gomez-Marin A, Truman JW, Louis M (2015) Role of the subesophageal zone in sensorimotor control of orientation in Drosophila larva. Curr Biol 25(11):1448–1460

Thum AS, Knapek S, Rister J, Dierichs-Schmitt E, Heisenberg M, Tanimoto H (2006) Differential potencies of effector genes in adult Drosophila. J Comp Neurol 498(2):194–203

Venken KJ, Simpson JH, Bellen HJ (2011) Genetic manipulation of genes and cells in the nervous system of the fruit fly. Neuron 72(2):202–230

Vogelstein JT, Park Y, Ohyama T, Kerr RA, Truman JW, Priebe CE, Zlatic M (2014) Discovery of brainwide neural-behavioral maps via multiscale unsupervised structure learning. Science 344(6182):386–392

von Philipsborn AC, Liu T, Yu JY, Masser C, Bidaye SS, Dickson BJ (2011) Neuronal control of Drosophila courtship song. Neuron 69(3):509–522

Wong AM, Wang JW, Axel R (2002) Spatial representation of the glomerular map in the Drosophila protocerebrum. Cell 109(2):229–241

Yoshikawa S, Long H, Thomas JB (2016) A subset of interneurons required for Drosophila larval locomotion. Mol Cell Neurosci 70:22–29

Zhang W, Ge W, Wang Z (2007) A toolbox for light control of Drosophila behaviors through Channelrhodopsin 2-mediated photo activation of targeted neurons. Eur J Neurosci 26(9):2405–2416

Zhang W, Yan Z, Jan LY, Jan YN (2013) Sound response mediated by the TRP channels NOMPC, NANCHUNG, and INACTIVE in chordotonal organs of Drosophila larvae. Proc Natl Acad Sci U S A 110(33):13612–13617

Zimmermann G, Wang LP, Vaughan AG, Manoli DS, Zhang F, Deisseroth K, Baker BS, Scott MP (2009) Manipulation of an innate escape response in Drosophila: photoexcitation of acj6 neurons induces the escape response. PLoS One 4(4):e5100

Zwart MF, Pulver SR, Truman JW, Fushiki A, Fetter RD, Cardona A, Landgraf M (2016) Selective inhibition mediates the sequential recruitment of motor pools. Neuron 91(3):615–628

Chapter 7
Targeted Manipulation of Neuronal Activity in Behaving Adult Flies

Stefanie Hampel and Andrew M. Seeds

Abstract The ability to control the activity of specific neurons in freely behaving animals provides an effective way to probe the contributions of neural circuits to behavior. Wide interest in studying principles of neural circuit function using the fruit fly *Drosophila melanogaster* has fueled the construction of an extensive transgenic toolkit for performing such neural manipulations. Here we describe approaches for using these tools to manipulate the activity of specific neurons and assess how those manipulations impact the behavior of flies. We also describe methods for examining connectivity among multiple neurons that together form a neural circuit controlling a specific behavior. This chapter provides a resource for researchers interested in examining how neurons and neural circuits contribute to the rich repertoire of behaviors performed by flies.

7.1 Introduction

The study of behavior often requires watching an animal move its body to accomplish different tasks. This is because the performance of any behavior requires movement of some part of the body. For example, a fruit fly moves a wing to sing a courtship song, a fish moves its tail fin to swim away from a predator, or a human moves his or her fingers to type the letters that make up a book chapter. To study behavior, one must ultimately confront the problem of how to observe the movements being performed and how to quantify them (Anderson and Perona

---

S. Hampel (✉) · A.M. Seeds (✉)
Janelia Research Campus, Howard Hughes Medical Institute, 19700 Helix Dr., Ashburn, VA 20147, USA
e-mail: stef.hampel@gmail.com

A.M. Seeds
e-mail: seeds.andrew@gmail.com

S. Hampel · A.M. Seeds
Institute of Neurobiology, University of Puerto Rico-Medical Sciences Campus, San Juan, PR 00901, USA

A. Çelik and M.F. Wernet (eds.), *Decoding Neural Circuit Structure and Function*, DOI 10.1007/978-3-319-57363-2_7

2014; Egnor and Branson 2016). Additionally, these very movements may cause technical challenges when researchers wish to manipulate the activity of specific neurons while simultaneously assessing the effect of the manipulation on a behavior.

The field of neuroscience has recently seen a number of innovations that better equip researchers to study neural circuit function in freely moving animals. The first are tools for expressing any protein coding sequence of interest in behaviorally relevant neurons (Venken et al. 2011b; Huang and Zeng 2013). Many of these expression systems are integrated into the genome or, in the case of some vertebrates, introduced into specific brain regions using viruses. The second are tools and techniques developed for the "remote control" of neuronal activity. For example, light-gated ion channels enable the manipulation of the activity of specific neurons using light (Boyden et al. 2005; Bernstein et al. 2012). Third, methods have been developed for the recording, classification, and quantification of behavior. For example, machine vision-based tracking of animal movement has greatly improved the consistency and throughput of behavioral analyses (Anderson and Perona 2014; Egnor and Branson 2016). Finally, additional tools have been developed to examine the functional relationships among different neurons that each contribute to a given behavior. This includes assessment of the functional connectivity between neurons using genetically encoded indicators of neural activity (Broussard et al. 2014). This suite of innovations now empowers researchers to make substantial progress in probing the functions of neural circuits across a range of different species (White 2016).

These tools can be combined in the fruit fly (*Drosophila melanogaster*) to greatly simplify dissection of the behavioral contributions of specific neurons. This is in part because of the ease with which flies containing multiple transgenes can be generated to enable the manipulation of specific neurons, often at single cell resolution. With such exquisite specificity, the range of different tools for visualizing and manipulating behaviorally relevant neurons can be brought to bear on questions of how neural circuits control behavior. Another advantage of flies is their amenability to large-scale screens for identifying previously unknown behaviorally relevant neurons. Such screens offer the prospect of uncovering different types of neurons that together constitute the neural circuit mediating a particular behavior. Collectively, these different tools and approaches can be used to study the rich set of innate behaviors performed by flies such as walking, flight, grooming, feeding, mating, fighting, and escape. Moreover, behavioral and circuit-based studies in flies have provided new insights into basic topics in neuroscience such as learning and memory, sensory-motor integration, neuromodulation, sleep, behavioral choice, behavioral sequencing, motor control, and sensory systems (Huston and Jayaraman 2011; Yoshihara and Ito 2012; Kaun et al. 2012; Perry and Barron 2013; Pavlou and Goodwin 2013; Wilson 2013; Tataroglu and Emery 2014; Borst 2014; Borgmann and Büschges 2015; Owald et al. 2015; Hoopfer 2016; Masek and Keene 2016; McKellar 2016). Given that many of the tools used to study neural circuits have only recently become available, we anticipate that the coming years will experience a rapid growth in our understanding of how neural circuits within the fruit fly nervous system are organized to produce particular behaviors.

The aim of this chapter is to provide a theoretical and practical resource for both beginning and experienced researchers who are interested in studying the roles of neurons and neural circuits in fruit flies. We describe how to identify and manipulate the activity of behaviorally relevant neurons in freely behaving flies. This includes information about expression systems, reagents for manipulating neural activity, behavioral hardware, rearing flies for neural manipulation experiments, and assessing the behavioral impact of neural manipulations. Further, we describe methods for examining how different identified neurons are organized into neural circuits to collectively control behavior. We include discussions about the advantages and disadvantages of different reagents and approaches so that the reader can make informed decisions about the experimental approaches that best suit their needs. To illustrate different techniques, we refer to experiments where researchers study neurons whose activation elicits specific behaviors. However, the approaches discussed here can be applied to the study of many other aspects of nervous system function.

7.2 Binary Expression Systems

Critical for probing the neural basis of behavior in flies are binary expression systems, such as GAL4/UAS, that allow for visualization and manipulation of behaviorally relevant neurons (Venken et al. 2011b; del Valle Rodríguez et al. 2012). GAL4 is a yeast-derived transcription factor that binds to its upstream activating sequence (UAS) to drive transcription of any coding sequence of interest placed under the control of UAS. Binary expression systems are designed to take advantage of the enhancer activity of *Drosophila* genomic regulatory elements that control when and where genes will be expressed in the body (Brand and Perrimon 1993; Duffy 2002). Genomic enhancers are used to direct expression of GAL4 in different subsets of neurons in enhancer trap or enhancer fusion transgenic flies. Enhancer traps arise when a transposable element (for example, P element or Minos) containing a minimal transcriptional promoter upstream of the GAL4 coding sequence is randomly inserted into different locations in the genome. The expression pattern of GAL4 is then directed by the minimal promoter in conjunction with genomic regulatory elements that are local to the transposable element's insertion site (O'Kane and Gehring 1987; Brand and Perrimon 1993). Enhancer traps are historically the most common method for driving expression in neural subsets.

Enhancer fusions contain genomic regulatory elements that are fused with a minimal transcriptional promoter and GAL4, and then inserted into the genome. Most enhancer fusion transgenic fly lines were made with the enhancer fusion inserted into a defined genomic location that contains the *attP* target sequence for *phiC31* mediated site-specific integration (Groth et al. 2004; Pfeiffer et al. 2008; Jenett et al. 2012) (Barry Dickson, personal communication). Many transgenic lines have been generated, each containing a stably inserted *attP* sequence at a different genomic location to provide different possible sites for integration (Groth et al. 2004; Venken et al. 2006; Bischof et al. 2007; Markstein et al. 2008; Ni et al. 2009; Knapp et al. 2015). Ideally the *attP* site is transcriptionally neutral such that only the specific genomic regulatory elements in the enhancer fusion direct the expression pattern of GAL4. Insertion of the enhancer fusion into an *attP* site offers the advantage that it avoids unwanted behavioral consequences of random insertions into different genomic locations. This is in contrast to the transposable element method used for enhancer traps that can introduce mutations (Spradling et al. 1999). Importantly, enhancer fusion transgenic fly lines can easily be remade because the sequences of the genomic fragments and *attP* insertion sites are known (Pfeiffer et al. 2008). In contrast, enhancer trap-based lines cannot be easily remade in the case the fly stock is lost.

Thousands of enhancer trap and enhancer fusion GAL4 transgenic fly lines have been generated, whose collective neural expression patterns cover most, if not all of the neurons in the nervous system (Yoshihara and Ito 2000; Gohl et al. 2011; Jenett et al. 2012) (Barry Dickson, personal communication). Some express GAL4 in specific neuronal types such as dopaminergic neurons, whereas others express in an assortment of neurons. The expression patterns of many GAL4 lines are publicly accessible (e.g., http://flweb.janelia.org/cgi-bin/flew.cgi, http://stockcenter.vdrc.at/control/main) and stocks for several large collections are available at the Bloomington *Drosophila* Stock Center and Vienna *Drosophila* Resource Center (http://flystocks.bio.indiana.edu/Browse/gal4/gal4_main.htm, http://stockcenter.vdrc.at/control/main). Additional binary expression systems have been developed for use in *Drosophila* such as LexA/LexAop and QF/QUAS (Lai and Lee 2006; Potter et al. 2010; Pfeiffer et al. 2010; Riabinina et al. 2015). The advantage of having multiple binary expression systems is that they can be used for independent expression of different coding sequences in the same fly.

GAL4 and other binary expression systems can be used to express any coding sequence of interest in subsets of neurons in the *Drosophila* CNS (central nervous system) (Fig. 7.1a). This is accomplished by crossing GAL4 lines to flies containing a transgene with UAS-fused upstream of the coding sequence. For example, the expression pattern of a GAL4 line is visualized using a UAS-fused gene for a fluorescent protein such as *green fluorescent protein* (*UAS-GFP*, Fig. 7.1b). A variety of fluorescent proteins of different colors have been identified and developed for visualizing neurons. Fluorescent proteins can also be fused to proteins that are targeted to different parts of a neuron such as the membrane, nucleus, or synapses. Membrane targeting is useful for visualizing the morphology of an entire neuron, whereas synaptic targeting marks its inputs and outputs. Reagents for visualizing neurons have been previously reviewed in (Venken et al. 2011b; Sivanantharajah and Zhang 2015). In addition to visualizing subsets of neurons,

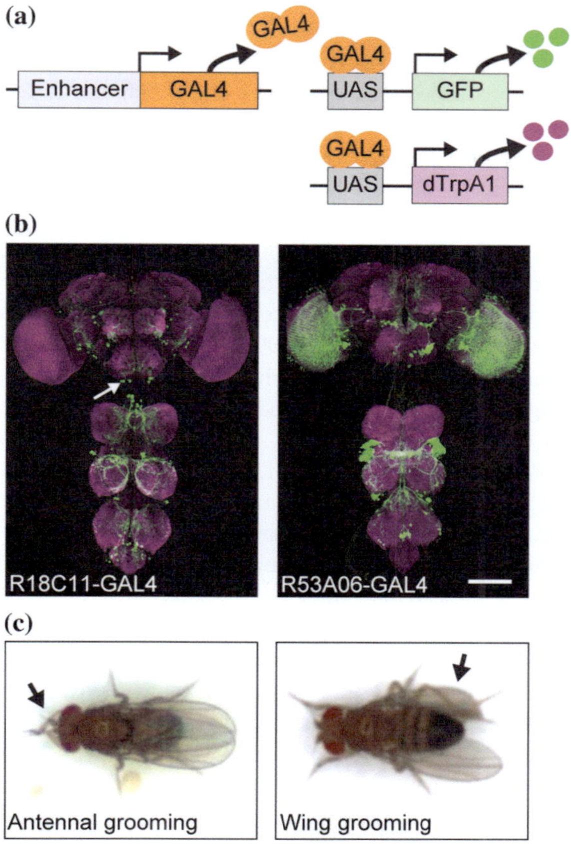

Fig. 7.1 GAL4-mediated expression for visualizing neurons and manipulating their activity. **a** The GAL4/UAS system: GAL4 transcription factor binds to the UAS sequence to direct transcription of a gene of interest (*UAS-GFP* and *UAS-dTrpA1* shown). Genomic regulatory elements that function as transcriptional enhancers direct the expression of GAL4 to particular cells in the nervous system. **b** Two different example enhancer GAL4 fusion expression patterns visualized by expression of GFP in the brain and ventral nervous system. Samples were co-stained with anti-GFP (*green*) and an antibody that marks synapses, anti-Bruchpilot (*magenta*), for visualizing the neuropile. *Scale bar* 100 µm. Images are published in the following references (Seeds et al. 2014; Hampel et al. 2015). **c** Thermogenetic activation of neurons within each pattern using dTrpA1 can elicit grooming of the head (*left* R18C11-GAL4) or wing (*right* R53A06-GAL4). Flies were imaged while on a temperature-controlled peltier plate. *Arrows* point to the legs performing the grooming movements

GAL4 can be used to express ion channels for manipulating their activity. This enables experiments to probe the role(s) of specific neurons in behavior (Fig. 7.1c, discussed in Sect. 7.3). Moreover, the activity of neurons can be visualized during behavior or in response to sensory stimuli by expressing genetically encoded indicators of neural activity (discussed in Sect. 7.8).

7.3 Thermo- and Optogenetic Neural Activation

The activity of neurons can be manipulated using a range of different neural activators or inhibitors (Venken et al. 2011b; Inagaki et al. 2014; Klapoetke et al. 2014). We focus our discussion on neural activators that can be induced acutely using temperature or light. These cation channels have the advantage that they can be kept inactive throughout development of the nervous system and then acutely induced in the adult. *Thermogenetic* activation relies on expression cation channels of the *transient receptor potential* (TRP) family whose conductances change significantly in the presence of warmth or cold (dTrpA1 and TRPM8 respectively) (Hamada et al. 2008; Peabody et al. 2009). The type of *optogenetic* activation described in this chapter relies on channelrhodopsins that are induced by particular wavelengths of light, such as blue light-induced Channelrhodopsin-2 (ChR2) and red light-induced CsChrimson or ReaChR (Boyden et al. 2005; Lin et al. 2013b; Inagaki et al. 2014; Klapoetke et al. 2014; Dawydow et al. 2014). The conductance changes of these channels in response to temperature or light allow for the "remote" activation of neurons with tight temporal control in intact and freely behaving flies.

A critical consideration in designing neural activation experiments is the temporal dynamics of the induced channel activities and how these activities impact the firing behavior of neurons. Some experiments call for a long duration of neural activation over the course of many seconds or minutes. Thermogenetic activation via dTrpA1 can drive neuron spiking and affect behavior over long time courses when exposed to constant warmth (Parisky et al. 2008; Pulver et al. 2009; Bernstein et al. 2012; Seeds et al. 2014; Hoopfer et al. 2015). However, some dTrpA1-activated neurons may show spike frequency adaptation. For example, the continual activation of Gr5a gustatory receptor neurons causes a decay in their spiking within a few seconds, and they are no longer able to elicit a proboscis extension reflex (Inagaki et al. 2014). In contrast, brief thermogenetic activation of these same neurons using an infrared laser can elicit this reflex (Keene and Masek 2012). This suggests that the spike frequency adaptation of Gr5a neurons can be avoided using pulsed rather than constant thermogenetic activation. In conclusion, different thermogenetic experiments indicate that longer duration dTrpA1 activation may cause adaptation in some types of neurons but not in others. Tonic exposure of channelrhodopsins to light has also been shown to cause spike frequency adaptation, which can be circumvented using pulsed rather than constant photostimulation (Pulver et al. 2009; Inagaki et al. 2014). However, one form of ChR2 called ChR2-XXL has been developed that has a slower time course of channel closure, thus enabling longer lasting neural activation (Dawydow et al. 2014). In the case where experiments call for more precise and shorter lasting neural activation, channelrhodopsins offer temporal precision of neural activation in the millisecond time scale, as opposed to hundreds of milliseconds for thermogenetic channels (Boyden et al. 2005; Bath et al. 2014).

It is important to keep in mind that the temperature or light changes required for thermo- and optogenetic activation can cause unwanted secondary effects on behavior. For example, higher temperatures may increase grooming behavior or reduce mating (Seeds et al. 2014; Vaughan et al. 2014). Similarly, optogenetic activation can introduce confounding behavioral artifacts. For example, flies can see the light pulses, which can elicit startle responses, potentially disrupting ongoing behaviors (Klapoetke et al. 2014). In the case where vision is not required for the behavior, these light-induced behavioral artifacts have been avoided by using blind flies with a homozygous allele of the *norpA* gene (de Vries and Clandinin 2013). Exposing flies to the smallest necessary temperature change or lowest light power necessary for neural activation can minimize these secondary behavioral effects.

Red light-activated channelrhodopsins offer significant advantages over those that are blue light activated. As mentioned above, flies can see the blue light used to activate ChR2 (Yamaguchi et al. 2010), which may cause confounding behavioral responses to the light. Blue light also penetrates poorly through the fly cuticle, making it difficult to deliver enough light to activate ChR2-expressing neurons in the CNS (Inagaki et al. 2014). Of note, ChR2-XXL is more sensitive to lower blue light levels than ChR2, which enables blue light activation of neurons in the CNS (Dawydow et al. 2014). In contrast, CsChrimson and ReaChR are activated by red light that readily penetrates the cuticle and effectively activates neurons in the CNS (Inagaki et al. 2014; Klapoetke et al. 2014). Red light is also less visible to flies, and therefore causes fewer behavioral effects than blue light (Inagaki et al. 2014; Klapoetke et al. 2014). In this respect, CsChrimson has an advantage over ReaChR, in that its peak wavelength sensitivity is further red shifted by about 45 nm (Klapoetke et al. 2014). This longer wavelength is less visible to flies, further reducing the behavioral artifacts associated with optogenetic activation.

7.4 Rearing Flies for Thermo- and Optogenetic Behavioral Experiments

Specific conditions should be met when rearing flies for use in thermo- or optogenetic experiments. For thermogenetic experiments, flies need to be reared at temperatures that do not activate dTrpA1 or TRPM8 to avoid ectopic neural activation during development (<25 °C for dTrpA1, >18 °C for TRPM8) (Hamada et al. 2008; Peabody et al. 2009). Channelrhodopsin-expressing flies should be kept in dark containers and/or their vials wrapped in aluminum foil to prevent neurons from being activated by ambient room light. Flies also need to be fed the channelrhodopsin cofactor, all-*trans*-retinal, because they do not produce appreciable endogenous levels. Only flies expressing the ChR2 mutant ChR2-XXL do not require all-*trans*-retinal food supplementation (Dawydow et al. 2014), possibly because its high affinity for all-*trans*-retinal enables it to access low endogenous concentrations. Flies expressing other channelrhodopsins are typically reared on food supplemented with all-*trans*-retinal at concentrations ranging from 0.1 to 0.5 mM (de Vries and Clandinin 2013; Inagaki et al. 2014; Klapoetke et al. 2014;

von Reyn et al. 2014; Hoopfer et al. 2015; Ohyama et al. 2015). Flies have been reared on regular food and transferred to all-*trans*-retinal-supplemented food a few days before they were used for experiments. However, we find that flies expressing channelrhodopsin in some neural types that were reared without all-*trans*-retinal-supplemented food can show motor defects or lethality (Hampel, Seeds, and Hibbard, unpublished observations). This indicates that feeding flies all-*trans*-retinal throughout development protects some neurons from potentially detrimental effects of channelrhodopsin overexpression.

While temperature and light exposure are specific to the inducers of neural activity, more general factors must be considered when rearing flies for behavioral experiments. First, the housing conditions in which flies are reared can affect their behavior. For example, flies reared in isolation are more aggressive than if they are group housed (Hoffmann 1990; Ueda and Kidokoro 2002; Wang et al. 2008). Housing also determines whether flies have mated, which affects many aspects of social behavior. For example, males with previous mating experience modify their courtship behavior to increase their chance of future mating success (Saleem et al. 2014). Mated females show increased rejection toward males that court them (Connolly and Cook 1973). Therefore, it is important to consider whether flies are reared in isolation or in groups when designing experiments. Second, the time of day can affect behaviors such as locomotion, eclosion, feeding, and mating (Allada and Chung 2010). To ensure consistency of behavioral experiments, many groups use circadian-entrained flies that are all tested at the same time of day (Vinayak et al. 2013). Third, hunger influences fly behaviors such as food searching (Root et al. 2011; Gruber et al. 2013), innate avoidance (Bräcker et al. 2013), learning (Krashes et al. 2009), and locomotion (Knoppien et al. 2000; Lee and Park 2004; Albin et al. 2015). Furthermore, the food recipe and number of flies reared on a particular volume of food can affect behavior (Guo et al. 1996). Fourth, flies experience age-dependent changes in behavior such as their propensity to mate or the degradation of their locomotor activity (Grotewiel et al. 2005). Therefore, many experiments use flies that are all the same age. Fifth, flies are often anesthetized using CO_2; however, its use in preparing flies for behavioral experiments can have dramatic and long-lasting effects on many different behaviors (Nicolas and Sillans 1989; Seiger and Kink 1993). Long recovery times (i.e., 24 h) have been recommended to mitigate the effects of CO_2 on behavior (Greenspan 2004); however, other experiments indicate that this may not be long enough (Barron 2000). Cold anesthesia can be used in place of CO_2, as it is reported to cause less severe behavioral side effects (Barron 2000). One way to circumvent the behavioral effects of anesthesia is to transfer flies using an aspirator (Zaninovich et al. 2013). Sixth, genetic background can affect behavior. For example, different sub-strains of Canton Special (CS), a wild-type strain that is frequently used as a control for behavioral experiments, show remarkably different behavior in the same experimental paradigm (Colomb and Brembs 2014). This shows the importance of controlling for genetic background in behavioral experiments. Backcrosses into a common genetic background will ensure consistency between control and experimental flies.

7.5 Hardware for Thermo- and Optogenetic Behavioral Experiments

Different systems have been designed for thermo- or optogenetic manipulation of neural activity in freely behaving flies. These systems have several common features: (1) hardware for delivering thermo- or optogenetic activation to the flies, (2) chambers that are permissive to the behavior being studied, (3) hardware for recording the behavior, (4) and a design that enables practical experimental implementation (e.g., getting flies in and out, cleaning, etc.). We discuss features of behavioral systems that have been designed to address these different issues below.

An advantage of thermogenetic systems is that they can be relatively simple to build at low cost. The different systems used for warming or cooling flies include temperature-controlled rooms or chambers, water baths, heating blocks or pads, or peltier plates (Marella et al. 2012; Mann et al. 2013; Flood et al. 2013; Seeds et al. 2014; Sun et al. 2014; Hampel et al. 2015). Chambers that house the flies have been designed with features that facilitate rapid warming such as mesh floors, or heat-conducting floors that are in direct contact with a heating element (Seeds et al. 2014; Harris et al. 2015). Alternatively, the flies and/or chambers can be pre-warmed prior to an experiment (Keleman et al. 2012; Burke et al. 2012; Seeds et al. 2014; Asahina et al. 2014). These simple systems expose the fly's entire body to temperature induction that typically occurs over a time frame of seconds to minutes. More spatially specific and rapid induction can be achieved by using infrared lasers that produce heat when focused on a particular body part and thereby activate neurons within the targeted area. For example, proboscis extension can be elicited when an infrared laser focused on the mouthparts of an immobilized fly induces dTrpA1 in sugar-sensing gustatory receptor neurons (Keene and Masek 2012). Taking this approach one step further, a computer vision-targeted infrared laser has been devised to precisely warm specific body parts on a freely walking fly, such as the antennae, head, or thorax, to locally activate neurons in different parts of the nervous system (e.g., antennal sensory neurons, brain, or ventral nervous system) (Bath et al. 2014).

Optogenetic systems illuminate flies with particular wavelengths of light. This requires a light source (laser- or LED-based) to illuminate flies from above or below and a means of controlling the intensity and timing of light exposure. Different examples and designs are available for building relatively simple and inexpensive optogenetic systems for whole-fly illumination (Pulver et al. 2011; de Vries and Clandinin 2013; Inagaki et al. 2014; Klapoetke et al. 2014). Figure 7.2 shows an example optogenetics system with components commonly found in other systems. Systems have also been developed that direct laser pulses to specific body parts of a moving fly (Wu et al. 2014). Other systems restrict light to particular regions of a chamber allowing flies to "choose" whether the neurons are activated (or not) by moving into (or out of) the illuminated region, or coupling neural activation with particular localized features of a larger chamber, such as odors (Suh et al. 2007; Lin et al. 2013a; Aso et al. 2014b; Klapoetke et al. 2014; Lin et al. 2015).

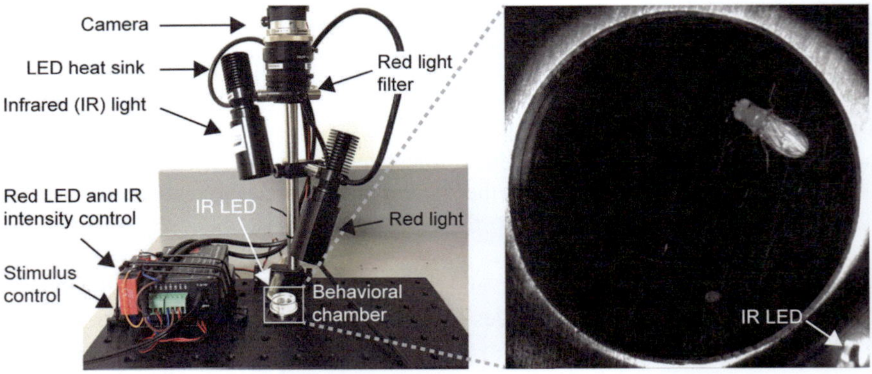

Fig. 7.2 Basic optogenetic apparatus (*left*). Freely behaving fly in an IR-illuminated chamber (*right*). The *IR LED* indicates that red light illumination is in progress for optogenetic activation

Flies must be contained for observation in a behavioral chamber for experiments where they need to be freely behaving. It is important to consider that behavior can be greatly affected by different chamber features such as the size, accessibility of the walls and ceiling to the flies, and environmental conditions (e.g., presence of food, other flies, etc.). For example, a male fly will spend less time courting a female in a large chamber than in a small chamber wherein the male is always in proximity to the female (Ewing and Ewing 1984; Zawistowski and Richmond 1987; Griffith and Ejima 2009). Chambers with low ceilings restrict flies to the floor for studying non-flight behaviors such as locomotion or grooming (Seeds et al. 2014; Triphan et al. 2016). Of note, chamber heights that are too low restrict the performance of behaviors that involve a raised posture, such as copulation or aggression (Hotta and Benzer 1976; Simon and Dickinson 2010). A systematic study of chamber conditions has revealed that a 3.5 mm chamber height is permissive to most non-flight behaviors performed by *Drosophila melanogaster* (Simon and Dickinson 2010). A chamber with a low ceiling, narrow walls, and a "dead end" has been designed that prevents flies from turning around, thus forcing them to walk backwards (Bidaye et al. 2014). In contrast, chambers can be designed with high ceilings to permit flight (Reynolds and Frye 2007; Straw et al. 2011; Ardekani et al. 2013; van Breugel and Dickinson 2014).

Environmental conditions within chambers also affect the behaviors of flies. For example, a fly may vary its behavior if there are other flies in the chamber, and this may be influenced by whether the flies are male or female. Chambers containing male–female combinations promote courtship, whereas male–male combinations can promote aggression. Furthermore, flies are more likely to show aggressive behavior when a female or a food source is present, and the amount of food within the chamber can influence the probability and nature of aggressive behavior (Hoffmann 1987; Chen et al. 2002; Lim et al. 2014). Flies can also leave pheromones or other chemicals behind in chambers, which can affect the social behaviors of new inhabitants (Suh et al. 2004; Wang and Anderson 2010; Lin et al. 2015).

Thus, cleaning chambers between experiments will reduce residual pheromones that could influence behavioral outcomes in future experiments (Zawistowski and Richmond 1987). We have also found that static electricity within chambers can cause increased grooming, which may disrupt the performance of a particular behavior of interest. For example, a static electric field might pull the antennae out of place, a condition that elicits antennal grooming (Hampel et al. 2015). Treatment of behavioral chambers with antistatic agents (e.g., UltraSpray, United SCP) resolves this problem. These examples demonstrate the importance of considering environmental conditions in designing chambers for behavioral experiments.

Although the behaviors of flies can be measured without directly observing their movements (Mendes et al. 2013; Itskov et al. 2014; Seeds et al. 2014; Albin et al. 2015; Egnor and Branson 2016), here we focus on video recording and annotating the movements of flies. Chambers should not only allow flies to perform the behavior of interest, but their design must also facilitate the recording of fly movements such that specific features of the movements can be annotated. In particular, the height and width of chambers are important factors to consider with respect to the camera set up. Chamber height determines the vertical space in which flies can move, and different heights require different methods to ensure that flies are visible at all times. For chambers with high ceilings, methods have been developed for tracking flies in three-dimensional space through the use of multiple cameras and computational reconstruction of flight trajectories (Straw et al. 2011; Ardekani et al. 2013; van Breugel and Dickinson 2014). For non-flight behaviors, one camera can be used in conjunction with a low ceiling to restrict flies to the focal length of the camera lens. The chamber width and camera resolution should be empirically determined to ensure that the recorded video captures features of the fly that are important for observing the behavior being studied. For example, measurements of walking trajectory do not require as high of an image resolution or frame rate as observation of the legs performing grooming movements. Thus, chamber size should be large enough for viewing the behavior, but also small enough to enable recording of a high-quality video. In designing the behavioral chambers, one has to ensure that flies can be well viewed for behavioral annotation. For example, flies tend to cluster at the periphery of chambers with vertical walls, and will often climb onto the walls and ceilings (Simon and Dickinson 2010). This means that fly bodies will be recorded in multiple orientations (e.g., dorsal and ventral), which can complicate behavioral annotation. Flies can be effectively restricted to the floor of the chamber by treating the walls and ceilings with slippery transparent coatings [e.g., Sigmacote (Sigma-Aldrich), Rain X (SOPUS products), or SurfaSil (Thermo Fisher Scientific)] and designing chambers to have walls that slope at an angle to the floor (Simon and Dickinson 2010; Goda et al. 2014; Hoopfer et al. 2015).

Proper illumination of flies within the chamber is critical for obtaining high-quality video recordings. Special consideration should be given to the type of illumination used for optogenetic experiments because visible light may activate neurons expressing channelrhodopsins. Therefore, many optogenetic recording setups visualize the flies using infrared illumination (Inagaki et al. 2014; Hoopfer et al. 2015; Hampel et al. 2015). Bear in mind that flies are unable to see infrared light and are therefore effectively blind under these conditions, which is a problem for studying any behavior that relies on vision. Furthermore, filters on the camera can prevent the bouts of photostimulation from interfering with visualization of the fly in the video (Fig. 7.2). In contrast, recordings of thermogenetic experiments can be made using visible or infrared light. The illumination source can be positioned above or below the behavior chamber. For set ups in which the camera is mounted above the behavior chamber, illumination from above allows for clear visualization of the fly body (Seeds et al. 2014; Hampel et al. 2015), whereas illumination from below produces a high-contrast silhouette (Branson et al. 2009; Simon and Dickinson 2010).

7.6 Annotating Behaviors Elicited by Neural Manipulations

One effective strategy for identifying behaviorally relevant neurons and determining how they control behavior is to use thermo- or optogenetics to manipulate the activity of neurons within the expression patterns of different GAL4 lines (e.g., enhancer-GAL4/*UAS-dTrpA1*). Such experiments require that behavioral performance is readily recognizable so that it can be effectively annotated and quantified. To illustrate techniques and approaches for quantifying the behaviors of freely moving flies in response to neural manipulations, we focus on experiments with neurons whose activation elicits specific behaviors. Although we focus on neural activation, similar experimental approaches that block neural activity have also revealed how specific movements are elicited (Gordon and Scott 2009; Mann et al. 2013). Furthermore, the approaches described here can be applied to the study of other neural circuit functions and behaviors.

Behavior can be quantified through manual annotation of video recordings or by machine vision-based behavioral tracking (Egnor and Branson 2016). Manual annotation involves watching videos of an experiment and annotating behavior “by eye” according to criteria that the researcher uses to define what a particular behavior is. For example, walking behavior could be defined as when a fly moves more than one body length with no pause in leg movements (Seeds et al. 2014). This manual approach can be upgraded with the aid of software that enables marking of behavioral events within the time course of the video such as VCode (Hagedorn et al. 2008), JWatcher (Blumstein and Daniel 2007), or BORIS (Friard and Gamba 2016). Although manual annotation is an effective way to study behavior, it has the drawbacks of being labor-intensive and highly dependent on the judgment of the researcher. This subjective aspect has the potential for inconsistency in how the same video might be annotated by two different researchers, or even by the same researcher at different times. Therefore, machine vision-based behavioral tracking algorithms have been developed to improve annotation consistency, throughput rate, and quantitative analysis of behavior (Anderson and

Perona 2014; Egnor and Branson 2016). Such algorithms track multiple statistics of the trajectories of flies and/or their body parts through time (e.g., translational speed, angular speed, or distance to another fly), which can be used to define classifiers for particular behaviors (Dankert et al. 2009; Branson et al. 2009; Robie et al. 2010; Straw et al. 2011; Tsai and Huang 2012; Donelson et al. 2012; Schusterreiter and Grossmann 2013; Ardekani et al. 2013; Bidaye et al. 2014; Dell et al. 2014; Berman et al. 2014). These statistics can also be fed into supervised machine-learning algorithms, such as the *Janelia Automatic Animal Behavior Annotator* (JAABA), where the researcher can train new behavior classifiers by manually annotating a small amount of video based on their own intuition about the behavior (Branson et al. 2009; Kabra et al. 2013). Other methods have been developed that do not presuppose human-assigned behaviors but instead assign behavioral events based on statistically defined structure of the tracked movements (Berman et al. 2014). These different approaches can provide detailed descriptions of behavior through time, allowing for the extraction of quantitative statistics such as the behavioral duration, frequency, and probability of transitioning between different behaviors.

One major advantage of flies is their amenability to high-throughput screens of hundreds or thousands of GAL4 lines to identify those that direct expression in behaviorally relevant neurons. Because neural activation can elicit strong behavioral phenotypes that are easy to distinguish from controls, visual observation has been an effective way to screen many GAL4 lines quickly. For example, all flies of a particular GAL4 line will perform the same grooming behavior when the targeted neurons are activated (Flood et al. 2013; Seeds et al. 2014; Hampel et al. 2015). Flood *et al.* carried out a screen of 835 GAL4 lines to catalog different behaviors that can be independently elicited with thermogenetic neural activation, including grooming, flight, feeding, and egg laying (Flood et al. 2013). This work provides a great example of the different behaviors that can be easily observed with neural activation. Other screens have focused on identifying GAL4 lines that elicit specific behaviors such as grooming, courtship song, feeding, or walking (von Philipsborn et al. 2011; Seeds et al. 2014; Bidaye et al. 2014; Hampel et al. 2015; Albin et al. 2015). Tracking algorithms have been used to screen for GAL4 lines that express in neurons involved in aggression and climbing (Asahina et al. 2014; Hoopfer et al. 2015; Triphan et al. 2016), and are anticipated to be increasingly used for identifying lines involved in other behaviors. These different screens have proven to be effective for identifying behaviorally relevant neurons. Furthermore, the unbiased screening of different GAL4 lines has led to the identification of neurons that would not have been anticipated *a priori* to underlie particular behaviors.

Once GAL4 lines are identified, different experimental approaches can be employed for more detailed studies of how activation of neurons within the GAL4 pattern elicits behavior. Below are examples of approaches that have been taken. Experiments have been designed that take advantage of the precise temporal control of neural activators to assess the dynamics of neural activation on behavior. This has revealed how neurons can elicit persistent behaviors or cause alterations in behavioral states (Inagaki et al. 2014; Bath et al. 2014; Hoopfer et al. 2015;

Hampel et al. 2015). Different odorant or food conditions have been used to study their contributions to feeding, attraction, avoidance, social behaviors, and learning (Gao et al. 2013; Aso et al. 2014b; Lim et al. 2014; Ramdya et al. 2014; Albin et al. 2015). Wild type, mutant, and flies with neural manipulations have been used to test different cues of conspecific flies that can affect social behaviors (Wang et al. 2011; Hoopfer et al. 2015). "Fly robots" have been used to decipher how different visual and tactile cues affect social behaviors (Pan et al. 2012; Agrawal et al. 2014; Ramdya et al. 2014; Clowney et al. 2015). Flies have been coated in dust to elicit competing grooming behaviors and study the mechanisms of behavioral choice (Seeds et al. 2014). Notably, many of these different behavioral experiments were done after using approaches described below to refine GAL4 expression patterns to only the behaviorally relevant neurons.

7.7 Intersectional Techniques for Identifying Behaviorally Relevant Neurons

In the previous section, we described how GAL4 lines have been identified that express in neurons whose activation is sufficient to elicit specific behaviors. However, since GAL4 lines often express in a large population of neurons that includes both the neurons that are able to elicit a behavior of interest and other neurons that are not involved in the behavior, it becomes a challenge to attribute the behavioral effect of the neural manipulation to any particular neurons. But this is exactly what is necessary to understand how the nervous system is organized to produce behavior. How can one effectively subdivide a population of neurons in a GAL4 pattern to isolate just those that elicit the behavior? The wealth of transgenic tools in *Drosophila* offers different solutions to this challenge.

There are numerous examples of two different enhancer trap or enhancer fusion lines that drive expression in a shared neuron that is involved in a behavior of interest and in unshared neurons that are not involved in the behavior (example shown in Fig. 7.3a). Positive intersectional techniques can be employed to reproducibly target expression only in these shared neurons between two enhancer patterns (Fig. 7.3b). In contrast, negative intersectional techniques involve suppressing expression at this intersection, effectively reducing the number of neurons within a particular GAL4 expression pattern (Suster et al. 2004; Sivanantharajah and Zhang 2015). We focus on positive intersectional techniques here, as they are the most widely used for reproducibly targeting and manipulating specific neurons. We also discuss the advantages and disadvantages one should consider when deciding on a particular method.

The two main strategies for producing positive intersections are *split GAL4* (spGAL4) and recombinase-based. spGAL4 takes advantage of the fact that GAL4 has two modular domains that are both necessary to drive transcription, a *DNA binding domain* (DBD) and a transcriptional *activation domain* (AD). These two

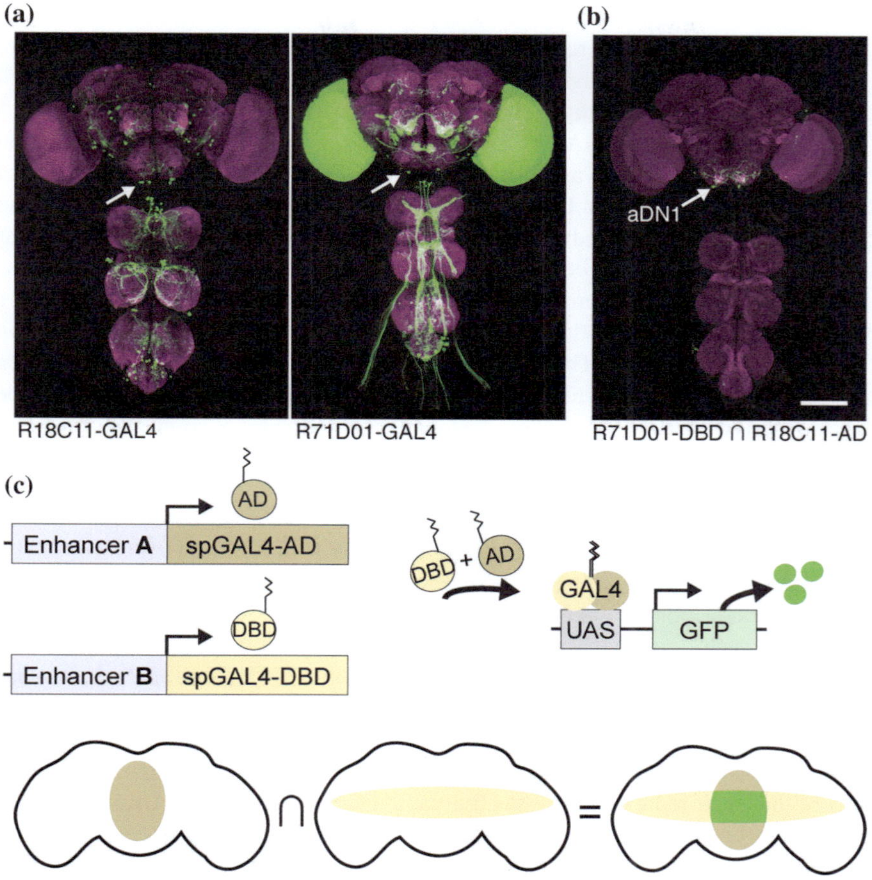

Fig. 7.3 spGAL4 intersectional strategy for identifying behaviorally relevant neurons from broader expression patterns. **a** CNS expression patterns of two different GAL4 lines that elicit antennal grooming with neural activation expressing GFP (*left* and *middle*). *White arrows* point to the only common neurons between the two patterns that elicit antennal grooming (Hampel et al. 2015). *Scale bar* 100 μm. **b** GFP expression pattern of a positive intersection between the two enhancer-driven patterns shown in (**a**) (spGAL4 intersectional method used). The intersection targets a single neuron (aDN1, *white arrows*) whose activation elicits antennal grooming. **c** spGAL4 system: enhancer elements drive expression of the GAL4 activation domain (spGAL4-AD) and the GAL4 DNA binding domain (spGAL4-DBD) in different patterns. The neurons labeled in both expression patterns will express both domains, which together reconstitute GAL4 through leucine zippers. Reconstituted GAL4 binds to the UAS sequence and drives expression of a downstream coding sequence, such as GFP as shown in (**a**)

domains can be co-expressed as separate proteins fused to leucine zipper motifs, and the zippers mediate heterodimer formation to reconstitute the transcriptional activity of GAL4 (Luan et al. 2006; Pfeiffer et al. 2010). The trick is that each spGAL4 domain can be expressed under the control of a different enhancer, and only when these two enhancers express in the same neurons will the two domains heterodimerize to reconstitute active GAL4 (Fig. 7.3c). One advantage of spGAL4 is that it drives highly reproducible expression of UAS-controlled transgenes in the intersected neurons, which contrasts with recombinase-based methods (described below). spGAL4 requires only three transgenes: (1) an enhancer-spGAL4-AD, (2) an enhancer-spGAL4-DBD, and (3) a coding sequence of interest expressed under UAS control. A *split LexA* (spLexA) system has also been developed that takes advantage of the modularity of LexA in the same way as the spGAL4 system (Ting et al. 2011).

A second strategy for producing a positive intersection uses *flippase* (FLP), a recombinase that catalyzes the removal of sequences between two *FLP recognition target sequences* (FRTs). There are several ways in which this activity has been harnessed to enable intersected neurons to express a UAS-controlled transgene. One method is to remove a FRT-flanked stop sequence from a UAS-controlled transgene so that a protein of interest is expressed in cells that are positive for both GAL4 and FLP (e.g., *UAS-FRT-Stop-FRT-GFP*, Fig. 7.4a) (Stockinger et al. 2005; von Philipsborn et al. 2011; Rezával et al. 2012; Alekseyenko et al. 2013). This method requires only three transgenes: (1) an enhancer-FLP line, (2) an enhancer-GAL4 line, and (3) a coding sequence of interest expressed under UAS-FRT-Stop-FRT control. A collection of enhancer trap FLP lines have been produced that can be used for such positive intersections (Bohm et al. 2010). Alternatively, enhancer-LexA lines can be used to drive expression of FLP (*LexAop-FLP*) in place of an enhancer-FLP line (Shang et al. 2008). Another method for performing FLP-based intersections relies on the strong repression of GAL4 activity by its natural regulator in yeast, GAL80. When GAL80 is ubiquitously expressed in all cells of the fly, it represses GAL4 activity and blocks activation of UAS-controlled transgenes (Lee and Luo 1999). GAL4 activity can be restored in cells in which GAL80 is removed, and this can be achieved when the GAL80 coding sequence is flanked by FRTs and the cells express FLP (Fig. 7.4b) (Gordon and Scott 2009; Bohm et al. 2010). Thus, expressing GAL4 and FLP in two partially overlapping enhancer patterns removes GAL80 and allows GAL4 to activate a UAS-controlled transgene in the intersectional cells (Shang et al. 2008; Bohm et al. 2010; Pool et al. 2014). Notably, this method requires four transgenes: (1) a ubiquitously expressed, FRT-flanked GAL80 cassette, (2) an enhancer-FLP line, (3) an enhancer-GAL4 line, and (4) a coding sequence of interest expressed under UAS control.

The FLP-based methods described above require the use of a FLP line with a known pattern of expression (or a collection of enhancer-FLP lines that can be screened), but FLP can also be used in the absence of such lines. An early approach that is labor-intensive but can be essential in some circumstances is to express FLP under heat-shock control in the context of a ubiquitously expressed FRT-flanked GAL80 transgene, to again allow GAL4 activity in cells in which the GAL80 is removed. The key to this approach is to use a regimen of FLP heat-shock induction that results in a low frequency of GAL80 removal from cell to cell so that specific neurons with GAL4 activity are stochastic from one fly to the next (Gordon and Scott 2009; Marella et al. 2012; Mann et al. 2013). These flies must also have

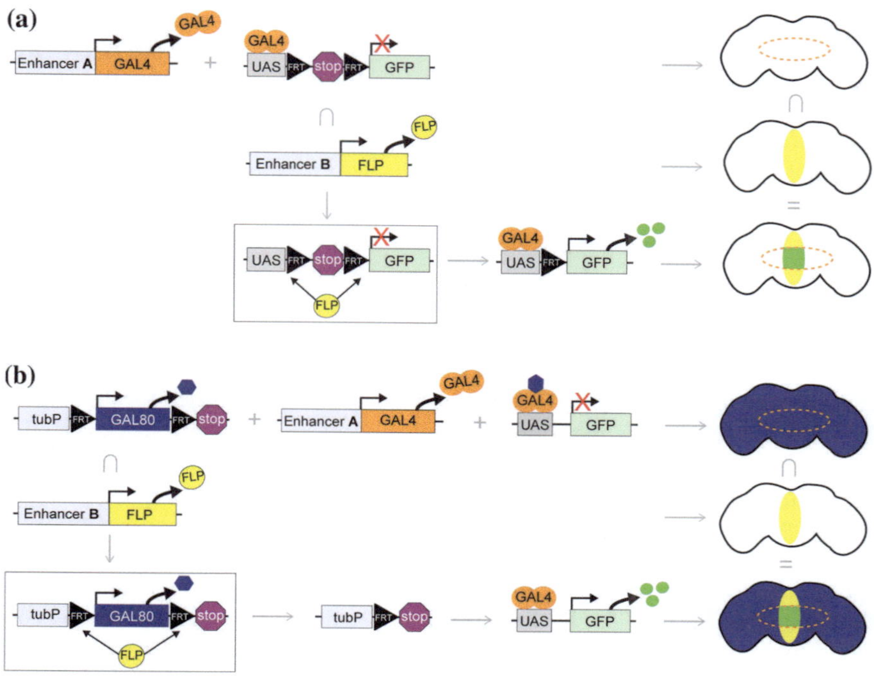

Fig. 7.4 FLP-based intersectional strategies for refining broader expression patterns. **a** A flip-in strategy requires three transgenes. Regulatory enhancer elements drive the expression of GAL4 and FLP in partially overlapping cells. The UAS-fused gene of interest (GFP shown) is expressed in the GAL4 pattern, but only translated in FLP expressing cells that can remove a FRT-flanked stop codon. **b** The GAL80 flip-out method requires four transgenes. Regulatory enhancer elements drive the expression of GAL4 and FLP in partially overlapping patterns. Expression of the repressor GAL80, which inhibits GAL4 activity by binding to its transcriptional activation domain, is driven pan-neurally by the tubulin regulatory sequences (tubP). Only cells that express FLP eliminate GAL80 repression to GAL4 through recombination of FRT sites that flank the GAL80 gene. Therefore, only cells that co-express GAL4 and FLP will transcribe the gene of interest (GFP shown). Figure inspired by Sivanantharajah and Zhang (2015)

transgenes to manipulate neuronal activity (e.g., *UAS-dTrpA1*) and to report the activity of GAL4 in the manipulated neurons (e.g., *UAS-GFP*). Each fly is then tested for whether a neural manipulation affects the behavior of interest, and is subsequently dissected so that its CNS can be imaged to identify which neurons had GAL4 activity. By correlating the behavioral output of a large number of flies with the anatomical location of the neurons that were manipulated, behaviorally relevant neurons from a broader GAL4 expression pattern are identified based on whether they show positive expression and affect the behavior of interest.

The intersectional method employed may depend on the type of GAL4 line that was used to manipulate the neurons of interest (e.g., enhancer trap or enhancer fusion). FLP-based methods have the advantage of being highly versatile in this respect because they can be implemented with any GAL4 line (Shang et al. 2008;

Sivanantharajah and Zhang 2015). An alternate way to implement positive intersections using any GAL4 line of interest is to drive expression of the LexA-DBD under UAS control (*UAS-LexA-DBD*) (Ting et al. 2011). Because spLexA was constructed with leucine zippers and works with the same activation domains that are used for spGAL4, it can be used in conjunction with available AD fly lines that have been constructed for use with spGAL4 (Luan et al. 2006; Ting et al. 2011).

In contrast to the FLP- and spLexA-based methods described above, spGAL4 requires the construction of new fly lines that express spGAL4 halves in the same neurons as the original GAL4 lines of interest. In the case where the GAL4 line is an enhancer fusion, the genomic enhancer is fused to either spGAL4-AD or spGAL4-DBD (Pfeiffer et al. 2008, 2010). The enhancer-spGAL4 fusion construct is then inserted into an *attP* site in the genome using *phiC31* integrase. Such constructs should generally be inserted into the same *attP* site used for the original GAL4 line, because this increases the likelihood of recapitulating that expression pattern (Pfeiffer et al. 2010). Unfortunately, inserting both spGAL4 halves into the same *attP* site can lead to transvection across paired chromosomes, causing each of the spGAL4 halves to be partially or fully expressed in the enhancer pattern of the opposite half (Mellert and Truman 2012). This can undermine the intersectional approach by causing the reconstituted spGAL4 to have a broader expression pattern than anticipated (Fig. 7.5a, b). Because of this and other potential problems conferred by transvection, it is advisable to only use one transgene per *attP* site in a particular fly genotype (Mellert and Truman 2012). One *attP* combination that has been successfully used for spGAL4-AD and DBD insertions is *attP2* and *attP40* (Aso et al. 2014a; Hampel et al. 2015) (Fig. 7.5c); however, the quality of this site pair can vary depending on the enhancer fusion used. Importantly, enhancer fusions inserted at *attP* locations other than where they were originally characterized can yield different expression patterns because of genomic positional effects (Pfeiffer et al. 2010). Therefore, if insertion of a spGAL4-half at one *attP* site does not recapitulate the expected expression pattern, the spGAL4-half may need to be inserted into a different site. A number of different *attP* integration sites have been generated to facilitate this (Groth et al. 2004; Venken et al. 2006; Bischof et al. 2007; Markstein et al. 2008; Ni et al. 2009; Knapp et al. 2015).

spGAL4 halves can be swapped with GAL4 in enhancer trap lines as well. One particular enhancer trap collection called the *integrase swappable in vivo targeting element* (InSITE) system has been designed specifically for such a purpose. InSITE lines are constructed such that GAL4 can be replaced with other genes such as spGAL4 halves via recombinase-mediated cassette exchange, thus enabling its expression in the same pattern (Gohl et al. 2011). This system is advantageous because spGAL4 flies can be generated genetically through simple fly crosses, in contrast to the enhancer fusions that require the generation of new transgenic flies through embryo injections. The disadvantage of InSITE is that spGAL4 replacements can only be done using InSITE enhancer trap lines. Another approach has been developed for use with a collection of *Minos-Mediated Integration Cassette* (MiMIC) transposons, where GAL4, spGAL4, or any other coding sequence can be expressed in the pattern of a native gene of interest (Venken et al. 2011a;

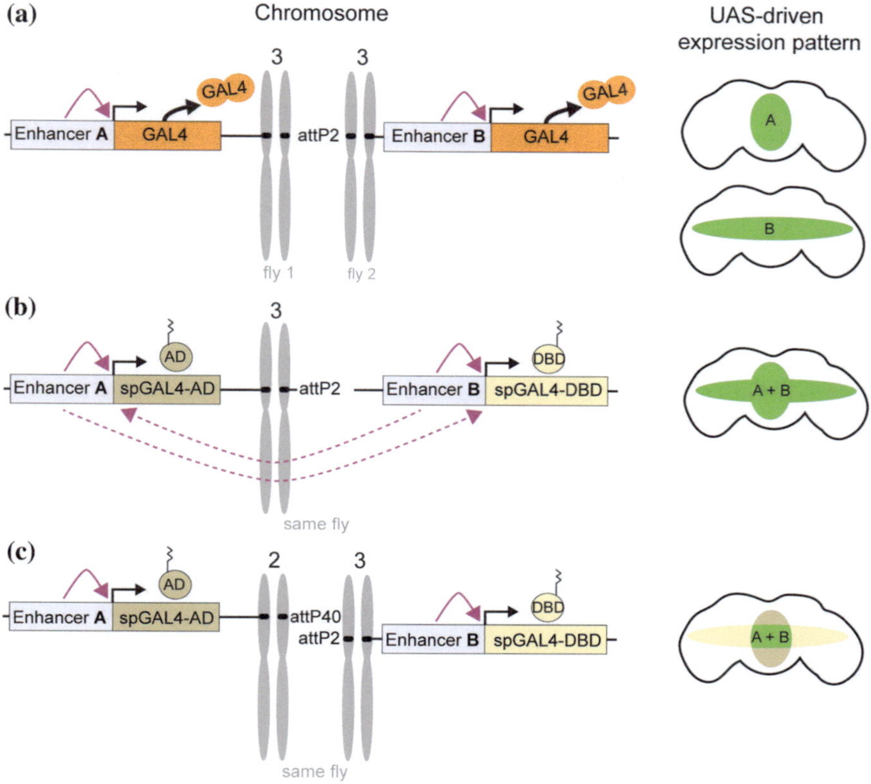

Fig. 7.5 The importance of using appropriate genomic landing sites to avoid transvection. **a** Two different enhancer-GAL4 transgenes (Enhancer A-GAL4; Enhancer B-GAL4) inserted in the same genomic landing site (*attP2*) in two different flies drive expression of a gene of interest in independent patterns (*green*). **b** Two different enhancer-transgenes (Enhancer A-spGAL4-AD; Enhancer B-spGAL4-DBD) inserted in the same genomic landing site (*attP2*) within the same fly can each influence the activation of either transgene's minimal promoter in either *cis* (*magenta arrows*), or *trans* (*magenta dashed arrows*), or in *cis* and *trans* combined. One possible outcome of the latter scenario is shown: Enhancers A and B activate transcription of both spGAL4-AD and -DBD to reconstitute GAL4 in both enhancer-driven expression patterns (*magenta arrows* and *dashed arrows*). See the following reference for more detailed discussion of possible outcomes of transvection (Mellert and Truman 2012). **c** Two different enhancer-transgenes (Enhancer A-spGAL4-AD; Enhancer B-spGAL4-DBD) inserted in distinct genomic landing sites (*attP40* and *attP2*) in the same fly drive transcription of the spGAL4-AD and spGAL4-DBD in the corresponding pattern of each respective enhancer. This leads to reconstitution of GAL4 only in cells that are shared between both enhancer expression patterns

Diao et al. 2015). Finally, an exciting new technique called *homology assisted CRISPR Knock-in* (HACK) offers the possibility of replacing GAL4 from any existing enhancer trap or enhancer fusion line with a coding sequence of interest by performing two simple crosses (Lin and Potter 2016). Although we are not aware of a case where HACK has been used to replace GAL4 with spGAL4 halves, we anticipate that this technique will become a method of choice for performing such swaps.

Positive intersections can be implemented in a couple of different ways depending on the circumstance of the particular experiment. In the case where two identified enhancer patterns are suspected to share a common behaviorally relevant neuron, an intersection can be performed with these two patterns (e.g., enhancer-1-spGAL4-DBD and enhancer-2-spGAL4-AD). In a different scenario, a single enhancer pattern is identified that contains a neuron of interest, but there is no known second enhancer pattern that could be used to produce a positive intersection. In this case, the enhancer-spGAL4 of interest (e.g., enhancer-1-spGAL4-DBD) can be screened against a library of spGAL4-ADs to identify combinations that target the neuron(s) of interest (Luan et al. 2006, 2012). In the case where a FLP-based method is being used, a GAL4 line that expresses in a neuron of interest is screened against a FLP enhancer trap library (Bohm et al. 2010; Rezával et al. 2012; Pool et al. 2014). Although it is feasible that intersectional approaches can refine expression patterns to a single neuron (Fig. 7.3a, b), in many cases the neurons of interest are not the only ones targeted by a particular spGAL4 or FLP/FRT pattern. This complicates the interpretation as to which neuron in the pattern is responsible for the phenotype. Therefore, it is optimal to identify multiple intersectional combinations that target the same neuron of interest but differ in their "contaminating" neurons. This strengthens an argument that the behavioral phenotypes caused by manipulating neural activity are due to the particular shared neuron (Tuthill et al. 2013; Hampel et al. 2015).

7.8 Assessing Functional Connectivity Among Neurons

The approaches described above can reveal individual neurons that elicit stereotyped behaviors when activated. In some cases, different neurons have been identified that elicit the same behavior. For example, at least four different neuronal types can elicit grooming of the antennae with neural activation (Hampel et al. 2015). This raises the question of whether they may be part of the same functionally connected circuit. Below, we describe methods for piecing together how such neurons are organized and function as circuits. This involves analyzing the proximity between neurons and determining their functional connectivity. In other cases, individual neurons have been isolated, but the identities of other connected neurons remain unclear. We discuss techniques that can be employed to identify such missing components of behavioral circuits. This section is intended to provide a brief overview of different circuit-mapping techniques with references for further reading on particular topics of interest.

One way to assess whether different neurons interact is to use light microscopy data to determine whether their neurites are in close proximity. Ideally this is achieved by examining the neurons in the same brain. Such neuroanatomical mapping is difficult because it is technically challenging to independently visualize more than two neurons in the same fly (discussed more below). Computational methods can be employed to align confocal Z-stack images of different neurons to visualize their gross anatomical relationships. That is, different neurons are registered to a "standard" brain and then displayed within the same 3-dimensional space. One widely used free software for registering brain confocal stacks is the *computational morphometry toolKit* (CMTK) (https://www.nitrc.org/projects/cmtk/), and others are available such as BrainAligner (Peng et al. 2011). One problem that arises is that the 3-dimensional renderings of neurons become increasingly messy as more are added, especially when each confocal stack contains multiple neurons. Therefore, additional tools such as the free software FluoRender enable the display of selected neurons from these alignments for the clearest possible visualization of the putative circuit (Wan et al. 2012). Another such free software package is Volocity (by PerkinElmer), and commercially available programs include Amira (by FEI) and Imaris (by Bitplane). These different tools can be used to examine the spatial relationships between neurons to gain a first approximation about whether their neurites are in close enough proximity to form synaptic connections (Fig. 7.6a). Additionally, computational alignment provides a means to display many neurons in the same brain for figures or movies to visualize multi-neuron structures (Aso et al. 2014a; Hampel et al. 2015). Furthermore, the computational alignment of individual neurons to a standard brain affords the possibility to use a program called NBLAST to perform a number of different useful searches (Costa et al. 2016). For example, NBLAST can identify neurons having the same or overlapping morphology as a particular neuron of interest, or to identify GAL4 lines that express in a particular neuron.

The proximity between two neurons can be examined in the same brain using different binary expression systems. For example, one neuron can express GFP using GAL4/UAS (GAL4 or a spGAL4 combination), whereas another neuron can express a different colored fluorescent protein using another binary system such as the LexA/LexAop or QF/QUAS (Fig. 7.6b). It is possible that three binary expression systems could be employed to independently express in three different neurons, however it would be challenging to get the minimum of six required transgenes into the same fly. Generating fly lines that have multiple transgenes inserted as an array into the same genomic location would be one way to do such an experiment (Knapp et al. 2015). Another technique that provides information about the proximity between two neurons is called *GFP reconstitution across synaptic partners* (GRASP) (Feinberg et al. 2008; Gordon and Scott 2009). This technique can be used to determine if two neurons had membrane contacts during development and/or in the developed brain.

The ability to independently target expression in two neurons whose neurites are in close proximity enables testing whether activation of the putative upstream neuron excites or suppresses activity of the downstream neuron. The activity of neurons can be assessed through the use of different genetically encoded indicators of neural activity (Grienberger and Konnerth 2012; Masuyama et al. 2012; Broussard et al. 2014; Fosque et al. 2015; Gao et al. 2015; Dana et al. 2016). For example, *genetically encoded calcium indicators* (GECIs) increase their

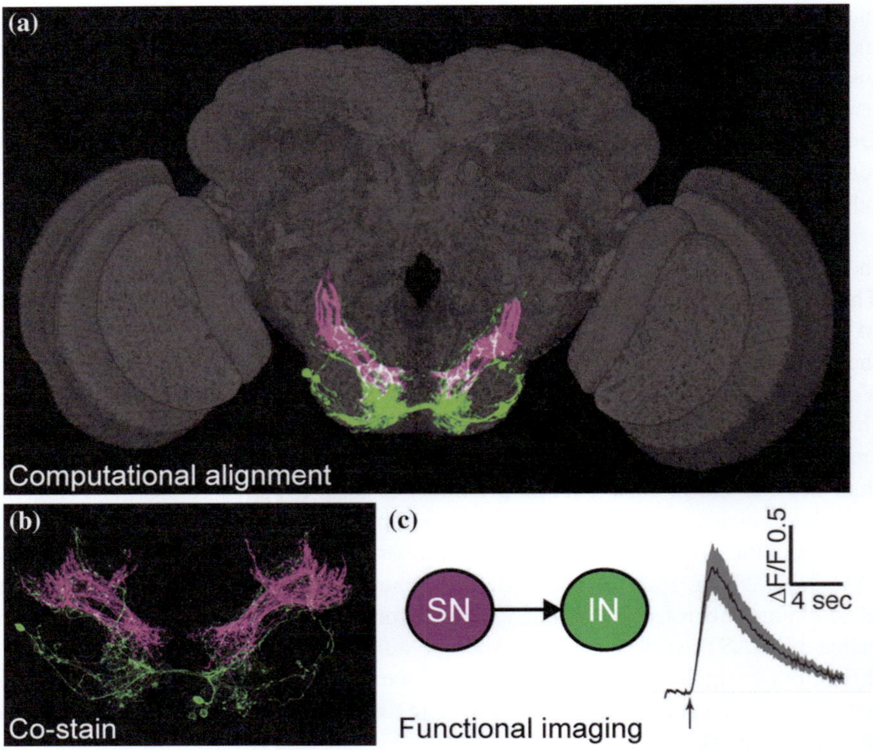

Fig. 7.6 Approaches to assess neural proximity. **a** Computationally aligned neurons (*green* and *magenta*) from two individual brains to a standard brain (CTMK and FluoRender software). **b** Antibody co-stain using GFP and RFP (*green* and *magenta* respectively) that expressed in two neurons within the same brain (same neurons as shown in **a**). **c** Schematic for testing connectivity between neurons. Antennal mechanosensory neurons (SN) expressing CsChrimson are activated with *red light* and simultaneously changes of fluorescence are measured in putative downstream interneurons (IN) expressing GCaMP using two-photon calcium imaging (response trace on *right*). Some of this data is published in the following reference Hampel et al. (2015)

fluorescence as calcium levels rise in active neurons. One example of how to assess functional connectivity is to use LexA/LexAop to express the neural activator CsChrimson in a putative upstream neuron while GAL4/UAS is used to express the GECI GCaMP6 in the putative downstream neuron (Chen et al. 2013; Klapoetke et al. 2014). The upstream neuron is then optogenetically activated while the downstream neuron expressing GCaMP6 is monitored for changes in fluorescence (Fig. 7.6c). These experiments can be done with the nervous system kept in the body or using a dissected brain (Kallman et al. 2015; Zhou et al. 2015; Clowney et al. 2015; Hampel et al. 2015; Cohn et al. 2015; Shirangi et al. 2016). The functional connectivity between neurons that is demonstrated by measuring GECI responses cannot indicate whether the connections are direct, and it is always possible that intermediate neurons are involved. To distinguish between the possibilities of monosynaptic versus polysynaptic connectivity, as well as to characterize the physiological properties of the functional connection, electrophysiology experiments are required (Gruntman and Turner 2013; Kohl et al. 2013; Fişek and Wilson 2014; Tuthill and Wilson 2016).

Whereas activation of specific neurons enables testing whether they are sufficient to elicit a particular behavior, blocking the activity of these neurons enables testing whether they are necessary for the performance of the behavior. Different reagents that are available for inhibiting neurons are discussed in the following references (Venken et al. 2011b; Sivanantharajah and Zhang 2015). Taking this approach one step further, two binary expression systems can be used to independently manipulate different neurons and test necessity and sufficiency for behavior with respect to their functional connectivity. For example, sensory neurons that elicit antennal grooming can be activated while simultaneously blocking synaptic transmission of putative downstream interneurons that express tetanus toxin light chain (Hampel et al. 2015). The inhibition of some interneurons reduced or abolished antennal grooming, showing that those neurons are normally necessary to elicit the full antennal grooming response to sensory neuron activation. Assessing the relative roles of different neurons in this way can reveal how they are organized into a neural circuit that underlies a particular behavior.

Individual neurons involved in a particular behavior are often identified without knowing the identities of their postsynaptic partners. One strategy to identify such downstream circuitry uses enhanced versions of photoactivatable GFP (PA-GFP), which photoconvert from a low-fluorescence to a high-fluorescence form under two-photon illumination (Patterson and Lippincott-Schwartz 2002; Ruta et al. 2010). A small region of axonal arborizations of the identified neuron is subjected to two-photon illumination with the expectation that the locally associated dendrites of its postsynaptic partner, which expresses PA-GFP, is labeled by photoconverted PA-GFP that also diffuses throughout the neuronal processes and cell body (Datta et al. 2008; Ruta et al. 2010; Fişek and Wilson 2014; Clowney et al. 2015). This approach has been used to identify second, third, and fourth order neurons in a pheromone circuit (Ruta et al. 2010), demonstrating the impressive utility of PA-GFP-based approaches for circuit mapping. Furthermore, functional connectivity among the neurons can be assessed as the putative upstream neurons are activated while testing for responses in the PA-GFP-identified downstream neurons (Ruta et al. 2010; Fişek and Wilson 2014; Clowney et al. 2015).

7.9 Complementary Approaches

We close by highlighting approaches for interrogating neural circuit function that complement those discussed above. First, are large-scale efforts to identify and target individual neurons in specific brain regions rather than identifying neurons using the above described behavioral screening methods. For example, spGAL4 combinations have been identified that express in the different neurons innervating the fly mushroom body (Aso et al. 2014a) and visual system (Tuthill et al. 2013; Nern et al. 2015). The activity of each neuron type was then systematically manipulated to assess their contributions to different behaviors (Tuthill et al. 2013; Aso et al. 2014b). Such approaches facilitate the systematic dissection of neurons that make up specific brain regions, and the assessment of their contributions to behavior. Second, serial section *electron microscopy* (EM) images of brain volumes are increasingly being reconstructed, which can reveal neural circuit connectivity to near completion (Saalfeld et al. 2009; Takemura et al. 2013; Ohyama et al. 2015; Berck et al. 2016; Schneider-Mizell et al. 2016). Importantly, this structural information can be used to pose hypotheses about circuit function that can then be tested in conjunction with genetic approaches described above for targeting and manipulating different circuit components (Ohyama et al. 2015). The complementary use of genetic tools and EM microscopy has great potential for studies of circuit connectivity and function. Finally, different approaches have been developed for recording the activity of specific neurons in behaving flies. This includes the use of GECIs or whole-cell patch clamp recordings as flies are either tethered or freely moving in a chamber (Maimon et al. 2010; Seelig et al. 2010; Haberkern and Jayaraman 2016; Grover et al. 2016). In this chapter we have described a set of tools and techniques that collectively enable the interrogation of neural circuit function at different levels of inquiry, from individual neurons, to neural circuits, to behavior.

Acknowledgements The Howard Hughes Medical Institute supported this work. We thank Claire McKellar, Eric Hoopfer, David Mellert, and Carmen Robinett for comments on the manuscript. Karen Hibbard provided input on all-*trans*-retinal feeding of flies. Steve Sawtelle provided advice on optogenetic behavioral systems. We thank Gerry Rubin for supporting our work at Janelia Research Campus.

References

Agrawal S, Safarik S, Dickinson M (2014) The relative roles of vision and chemosensation in mate recognition of *Drosophila melanogaster*. J Exp Biol 217:2796–2805. doi:10.1242/jeb.105817

Albin SD, Kaun KR, Knapp J-M et al (2015) A subset of serotonergic neurons evokes hunger in adult *Drosophila*. Curr Biol 25:2435–2440. doi:10.1016/j.cub.2015.08.005

Alekseyenko OV, Chan Y-B, Li R, Kravitz EA (2013) Single dopaminergic neurons that modulate aggression in *Drosophila*. Proc Natl Acad Sci 110:6151–6156. doi:10.1073/pnas.1303446110

Allada R, Chung BY (2010) Circadian organization of behavior and physiology in *Drosophila*. Annu Rev Physiol 72:605–624. doi:10.1146/annurev-physiol-021909-135815

Anderson DJ, Perona P (2014) Toward a science of computational ethology. Neuron 84:18–31. doi:10.1016/j.neuron.2014.09.005

Ardekani R, Biyani A, Dalton JE et al (2013) Three-dimensional tracking and behaviour monitoring of multiple fruit flies. J R Soc Interface 10:20120547–20120547. doi:10.1098/rsif.2012.0547

Asahina K, Watanabe K, Duistermars BJ et al (2014) Tachykinin-expressing neurons control male-specific aggressive arousal in *Drosophila*. Cell 156:221–235. doi:10.1016/j.cell.2013.11.045

Aso Y, Hattori D, Yu Y et al (2014a) The neuronal architecture of the mushroom body provides a logic for associative learning. Elife 3:e04577–e04577. doi:10.7554/eLife.04577

Aso Y, Sitaraman D, Ichinose T et al (2014b) Mushroom body output neurons encode valence and guide memory-based action selection in *Drosophila*. Elife 3:e04580. doi:10.7554/eLife.04580

Barron AB (2000) Anaesthetising *Drosophila* for behavioural studies. J Insect Physiol 46:439–442

Bath DE, Stowers JR, Hörmann D et al (2014) FlyMAD: rapid thermogenetic control of neuronal activity in freely walking *Drosophila*. Nat Methods 11:756–762. doi:10.1038/nmeth.2973

Berck ME, Khandelwal A, Claus L et al (2016) The wiring diagram of a glomerular olfactory system. Elife. doi:10.7554/eLife.14859

Berman GJ, Choi DM, Bialek W, Shaevitz JW (2014) Mapping the stereotyped behaviour of freely moving fruit flies. J R Soc Interface. doi:10.1098/rsif.2014.0672

Bernstein JG, Garrity PA, Boyden ES (2012) Optogenetics and thermogenetics: technologies for controlling the activity of targeted cells within intact neural circuits. Curr Opin Neurobiol 22:61–71. doi:10.1016/j.conb.2011.10.023

Bidaye SS, Machacek C, Wu Y, Dickson BJ (2014) Neuronal control of *Drosophila* walking direction. Science 344:97–101. doi:10.1126/science.1249964

Bischof JJ, Maeda RKR, Hediger MM et al (2007) An optimized transgenesis system for *Drosophila* using germ-line-specific phiC31 integrases. Proc Natl Acad Sci 104:3312–3317. doi:10.1073/pnas.0611511104

Blumstein DT, Daniel JC (2007) Quantifying behavior the JWatcher way. Sinauer Associates Incorporated, Sunderland, MA

Bohm RA, Welch WP, Goodnight LK et al (2010) A genetic mosaic approach for neural circuit mapping in *Drosophila*. Proc Natl Acad Sci 107:16378–16383. doi:10.1073/pnas.1004669107

Borgmann A, Büschges A (2015) Insect motor control: methodological advances, descending control and inter-leg coordination on the move. Curr Opin Neurobiol 33:8–15. doi:10.1016/j.conb.2014.12.010

Borst A (2014) Fly visual course control: behaviour, algorithms and circuits. Nat Rev Neurosci 15:590–599. doi:10.1038/nrn3799

Boyden ES, Zhang F, Bamberg E et al (2005) Millisecond-timescale, genetically targeted optical control of neural activity. Nat Neurosci 8:1263–1268. doi:10.1038/nn1525

Brand AH, Perrimon N (1993) Targeted gene expression as a means of altering cell fates and generating dominant phenotypes. Development 118:401–415

Branson K, Robie AA, Bender J et al (2009) High-throughput ethomics in large groups of *Drosophila*. Nat Methods 6:451–457. doi:10.1038/nmeth.1328

Bräcker LB, Siju KP, Varela N et al (2013) Essential role of the mushroom body in context-dependent CO_2 avoidance in *Drosophila*. Curr Biol 23:1228–1234. doi:10.1016/j.cub.2013.05.029

Broussard GJ, Liang R, Tian L (2014) Monitoring activity in neural circuits with genetically encoded indicators. Front Mol Neurosci 7:97. doi:10.3389/fnmol.2014.00097

Burke CJ, Huetteroth W, Owald D et al (2012) Layered reward signalling through octopamine and dopamine in *Drosophila*. Nature 492:433–437. doi:10.1038/nature11614

Chen S, Lee AY, Bowens NM et al (2002) Fighting fruit flies: a model system for the study of aggression. Proc Natl Acad Sci U S A 99:5664–5668. doi:10.1073/pnas.082102599

Chen T-W, Wardill TJ, Sun Y et al (2013) Ultrasensitive fluorescent proteins for imaging neuronal activity. Nature 499:295–300. doi:10.1038/nature12354

Clowney EJ, Iguchi S, Bussell JJ et al (2015) Multimodal chemosensory circuits controlling male courtship in *Drosophila*. Neuron 87:1036–1049. doi:10.1016/j.neuron.2015.07.025

Cohn R, Morantte I, Ruta V (2015) Coordinated and compartmentalized neuromodulation shapes sensory processing in *Drosophila*. Cell 163:1742–1755. doi:10.1016/j.cell.2015.11.019

Colomb J, Brembs B (2014) Sub-strains of *Drosophila* Canton-S differ markedly in their locomotor behavior. F1000Res 3:176–176. doi:10.12688/f1000research.4263.1

Connolly K, Cook R (1973) Rejection responses by female *Drosophila melanogaster*: their ontogeny, causality and effects upon the behaviour of the courting male. Behaviour 44:142–166

Costa M, Manton JD, Ostrovsky AD et al (2016) NBLAST: rapid, sensitive comparison of neuronal structure and construction of neuron family databases. Neuron. doi:10.1016/j.neuron.2016.06.012

Dana H, Mohar B, Sun Y et al (2016) Sensitive red protein calcium indicators for imaging neural activity. Elife. doi:10.7554/eLife.12727

Dankert H, Wang L, Hoopfer ED et al (2009) Automated monitoring and analysis of social behavior in *Drosophila*. Nat Methods 6:297–303. doi:10.1038/nmeth.1310

Datta SR, Vasconcelos ML, Ruta V et al (2008) The *Drosophila* pheromone cVA activates a sexually dimorphic neural circuit. Nature 452:473–477. doi:10.1038/nature06808

Dawydow A, Gueta R, Ljaschenko D et al (2014) Channelrhodopsin-2-XXL, a powerful optogenetic tool for low-light applications. Proc Natl Acad Sci 111:13972–13977. doi:10.1073/pnas.1408269111

de Vries SEJ, Clandinin T (2013) Optogenetic stimulation of escape behavior in *Drosophila melanogaster*. J Visualized Exp. doi:10.3791/50192

del Valle Rodríguez A, Didiano D, Desplan C (2012) Power tools for gene expression and clonal analysis in *Drosophila*. Nat Methods 9:47–55. doi:10.1038/nmeth.1800

Dell AI, Bender JA, Branson K et al (2014) Automated image-based tracking and its application in ecology. Trends Ecol Evol (Amst) 29:417–428. doi:10.1016/j.tree.2014.05.004

Diao F, Ironfield H, Luan H et al (2015) Plug-and-play genetic access to *Drosophila* cell types using exchangeable exon cassettes. Cell Rep 10:1410–1421. doi:10.1016/j.celrep.2015.01.059

Donelson NC, Donelson N, Kim EZ et al (2012) High-resolution positional tracking for long-term analysis of *Drosophila* sleep and locomotion using the "tracker" program. PLoS ONE 7: e37250–e37250. doi:10.1371/journal.pone.0037250

Duffy JB (2002) GAL4 system in *Drosophila*: a fly geneticist's swiss army knife. genesis 34:1–15. doi:10.1002/gene.10150

Egnor SER, Branson K (2016) Computational analysis of behavior. Annu Rev Neurosci 39:217–236. doi:10.1146/annurev-neuro-070815-013845

Ewing LS, Ewing AW (1984) Courtship in *Drosophila* melanogaster: behaviour of mixed-sex groups in large observation chambers. Behaviour 90:184–202

Feinberg EH, VanHoven MK, Bendesky A et al (2008) GFP reconstitution across synaptic partners (GRASP) defines cell contacts and synapses in living nervous systems. Neuron 57:353–363. doi:10.1016/j.neuron.2007.11.030

Fişek M, Wilson RI (2014) Stereotyped connectivity and computations in higher-order olfactory neurons. Nat Neurosci 17:280–288. doi:10.1038/nn.3613

Flood TF, Gorczyca M, White BH et al (2013) A large-scale behavioral screen to identify neurons controlling motor programs in the *Drosophila* brain. G3 3:1629–1637. doi:10.1534/g3.113.006205

Fosque BF, Sun Y, Dana H et al (2015) Labeling of active neural circuits in vivo with designed calcium integrators. Science 347:755–760. doi:10.1126/science.1260922

Friard O, Gamba M (2016) BORIS: a free, versatile open-source event-logging software for video/audio coding and live observations. Methods Ecol Evol. doi:10.1111/2041-210X.12584

Gao XJ, Potter CJ, Gohl DM et al (2013) Specific kinematics and motor-related neurons for aversive chemotaxis in *Drosophila*. Curr Biol 23:1163–1172. doi:10.1016/j.cub.2013.05.008

Gao XJ, Riabinina O, Li J et al (2015) A transcriptional reporter of intracellular Ca^{2+} in *Drosophila*. Nat Neurosci 18:917–925. doi:10.1038/nn.4016

Goda T, Leslie JR, Hamada FN (2014) Design and analysis of temperature preference behavior and its circadian rhythm in *Drosophila*. J Visualized Exp e51097–e51097. doi:10.3791/51097

Gohl DM, Silies MA, Gao XJ et al (2011) A versatile in vivo system for directed dissection of gene expression patterns. Nat Methods 8:231–237

Gordon MD, Scott K (2009) Motor control in a *Drosophila* taste circuit. Neuron 61:373–384. doi:10.1016/j.neuron.2008.12.033

Greenspan RJ (2004) Fly pushing, 2nd edn. Cold Spring Harbor Laboratory Press, Cold Spring Harbor

Grienberger C, Konnerth A (2012) Imaging calcium in neurons. Neuron 73:862–885. doi:10.1016/j.neuron.2012.02.011

Griffith LC, Ejima A (2009) Courtship learning in *Drosophila melanogaster*: diverse plasticity of a reproductive behavior. Learn Memory 16:743–750. doi:10.1101/lm.956309

Grotewiel MS, Martin I, Bhandari P, Cook-Wiens E (2005) Functional senescence in *Drosophila melanogaster*. Ageing Res Rev 4:372–397

Groth AC, Fish M, Nusse R, Calos MP (2004) Construction of transgenic *Drosophila* by using the site-specific integrase from phage φC31. Genetics 166:1775–1782

Grover D, Katsuki T, Greenspan RJ (2016) Flyception: imaging brain activity in freely walking fruit flies. Nat Methods 13:569–572. doi:10.1038/nmeth.3866

Gruber F, Knapek S, Fujita M et al (2013) Suppression of conditioned odor approach by feeding is independent of taste and nutritional value in *Drosophila*. Curr Biol 23:507–514. doi:10.1016/j.cub.2013.02.010

Gruntman E, Turner GC (2013) Integration of the olfactory code across dendritic claws of single mushroom body neurons. Nat Neurosci 16:1821–1829. doi:10.1038/nn.3547

Guo A, Li L, Shou-zhen X et al (1996) Conditioned visual flight orientation in *Drosophila*: dependence on age, practice, and diet. Learn Memory 3:49–59

Haberkern H, Jayaraman V (2016) Studying small brains to understand the building blocks of cognition. Curr Opin Neurobiol 37:59–65. doi:10.1016/j.conb.2016.01.007

Hagedorn J, Hailpern JM, Karahalios K (2008) VCode and VData: illustrating a new framework for supporting the video annotation workflow. AVI 317–321. doi:10.1145/1385569.1385622

Hamada FN, Rosenzweig M, Kang K et al (2008) An internal thermal sensor controlling temperature preference in *Drosophila*. Nature 454:217–220. doi:10.1038/nature07001

Hampel S, Franconville R, Simpson JH, Seeds AM (2015) A neural command circuit for grooming movement control. Elife. doi:10.7554/eLife.08758

Harris RM, Pfeiffer BD, Rubin GM, Truman JW (2015) Neuron hemilineages provide the functional ground plan for the *Drosophila* ventral nervous system. Elife. doi:10.7554/eLife.04493

Hoffmann AA (1987) A laboratory study of male territoriality in the sibling species *Drosophila melanogaster* and *D. simulans*

Hoffmann AA (1990) The influence of age and experience with conspecifics on territorial behavior in *Drosophila melanogaster*

Hoopfer ED (2016) Neural control of aggression in Drosophila. Curr Opin Neurobiol 38:109–118. doi:10.1016/j.conb.2016.04.007

Hoopfer ED, Jung Y, Inagaki HK et al (2015) P1 interneurons promote a persistent internal state that enhances inter-male aggression in *Drosophila*. Elife. doi:10.7554/eLife.11346

Hotta Y, Benzer S (1976) Courtship in *Drosophila* mosaics: sex-specific foci for sequential action patterns. Proc Natl Acad Sci 73:4154–4158

Huang ZJ, Zeng H (2013) Genetic approaches to neural circuits in the mouse. Annu Rev Neurosci 36:183–215. doi:10.1146/annurev-neuro-062012-170307

Huston SJ, Jayaraman V (2011) Studying sensorimotor integration in insects. Curr Opin Neurobiol 21:527–534. doi:10.1016/j.conb.2011.05.030

Inagaki HK, Jung Y, Hoopfer ED et al (2014) Optogenetic control of *Drosophila* using a red-shifted channelrhodopsin reveals experience-dependent influences on courtship. Nat Methods 11:325–332. doi:10.1038/nmeth.2765

Itskov PM, Moreira J-M, Vinnik E et al (2014) Automated monitoring and quantitative analysis of feeding behaviour in *Drosophila*. Nat Commun 5:4560. doi:10.1038/ncomms5560

Jenett AA, Rubin GM, Ngo T-TB et al (2012) A GAL4-driver line resource for *Drosophila* neurobiology. Cell Rep 2:991–1001. doi:10.1016/j.celrep.2012.09.011

Kabra M, Robie AA, Rivera-Alba M et al (2013) JAABA: interactive machine learning for automatic annotation of animal behavior. Nat Methods 10:64–67. doi:10.1038/nmeth.2281

Kallman BR, Kim H, Scott K (2015) Excitation and inhibition onto central courtship neurons biases *Drosophila* mate choice. Elife 4:e11188. doi:10.7554/eLife.11188

Kaun KR, Devineni AV, Heberlein U (2012) *Drosophila melanogaster* as a model to study drug addiction. Hum Genet 131:959–975. doi:10.1007/s00439-012-1146-6
Keene AC, Masek P (2012) Optogenetic induction of aversive taste memory. Neuroscience 222:173–180. doi:10.1016/j.neuroscience.2012.07.028
Keleman K, Vrontou E, Krüttner S et al (2012) Dopamine neurons modulate pheromone responses in *Drosophila* courtship learning. Nature 489:145–149. doi:10.1038/nature11345
Klapoetke NC, Murata Y, Kim SS et al (2014) Independent optical excitation of distinct neural populations. Nat Methods 11:338–346. doi:10.1038/nmeth.2836
Knapp J-M, Chung P, Simpson JH (2015) Generating customized transgene landing sites and multi-transgene arrays in *Drosophila* using phiC31 integrase. Genetics. doi:10.1534/genetics.114.173187
Knoppien P, van der Pers JNC, van Delden W (2000) Quantification of locomotion and the effect of food deprivation on locomotor activity in *Drosophila*. J Insect Behav 13:27–43. doi:10.1023/A:1007759424777
Kohl J, Ostrovsky AD, Frechter S, Jefferis GS (2013) A bidirectional circuit switch reroutes pheromone signals in male and female brains. Cell 155:1610–1623. doi:10.1016/j.cell.2013.11.025
Krashes MJ, DasGupta S, Vreede A et al (2009) A neural circuit mechanism integrating motivational state with memory expression in *Drosophila*. Cell 139:416–427. doi:10.1016/j.cell.2009.08.035
Lai S-L, Lee T (2006) Genetic mosaic with dual binary transcriptional systems in *Drosophila*. Nat Neurosci 9:703–709. doi:10.1038/nn1681
Lee G, Park JH (2004) Hemolymph sugar homeostasis and starvation-induced hyperactivity affected by genetic manipulations of the adipokinetic hormone-encoding gene in *Drosophila melanogaster*. Genetics 167:311–323
Lee T, Luo L (1999) Mosaic analysis with a repressible cell marker for studies of gene function in neuronal morphogenesis. Neuron 22:451–461
Lim RS, Eyjólfsdóttir E, Shin E et al (2014) How food controls aggression in *Drosophila*. PLoS ONE 9:e105626. doi:10.1371/journal.pone.0105626
Lin C-C, Potter CJ (2016) Editing transgenic DNA components by inducible gene replacement in *Drosophila melanogaster*. Genetics. doi:10.1534/genetics.116.191783
Lin C-C, Prokop-Prigge KA, Preti G, Potter CJ (2015) Food odors trigger *Drosophila* males to deposit a pheromone that guides aggregation and female oviposition decisions. Elife. doi:10.7554/eLife.08688
Lin H-H, Chu L-A, Fu T-F et al (2013a) Parallel neural pathways mediate CO_2 avoidance responses in *Drosophila*. Science 340:1338–1341. doi:10.1126/science.1236693
Lin JY, Knutsen PM, Muller A et al (2013b) ReaChR: a red-shifted variant of channelrhodopsin enables deep transcranial optogenetic excitation. Nat Neurosci 16:1499–1508. doi:10.1038/nn.3502
Luan H, Peabody NC, Vinson CR, White BH (2006) Refined spatial manipulation of neuronal function by combinatorial restriction of transgene expression. Neuron 52:425–436. doi:10.1016/j.neuron.2006.08.028
Luan H, Diao F, Peabody NC, White BH (2012) Command and compensation in a neuromodulatory decision network. J Neurosci 32:880–889. doi:10.1523/JNEUROSCI.3707-11.2012
Maimon G, Straw AD, Dickinson MH (2010) Active flight increases the gain of visual motion processing in *Drosophila*. Nat Neurosci 13:393–399. doi:10.1038/nn.2492
Mann K, Gordon MD, Scott K (2013) A pair of interneurons influences the choice between feeding and locomotion in *Drosophila*. Neuron 79:754–765. doi:10.1016/j.neuron.2013.06.018
Marella S, Mann K, Scott K (2012) Dopaminergic modulation of sucrose acceptance behavior in *Drosophila*. Neuron 73:941–950. doi:10.1016/j.neuron.2011.12.032
Markstein M, Pitsouli C, Villalta C et al (2008) Exploiting position effects and the gypsy retrovirus insulator to engineer precisely expressed transgenes. Nat Genet 40:476–483. doi:10.1038/ng.101

Masek P, Keene AC (2016) Gustatory processing and taste memory in *Drosophila*. J Neurogenet 30:112–121. doi:10.1080/01677063.2016.1185104

Masuyama KK, Zhang YY, Rao YY, Wang JWJ (2012) Mapping neural circuits with activity-dependent nuclear import of a transcription factor. J Neurogenet 26:89–102. doi:10.3109/01677063.2011.642910

McKellar CE (2016) Motor control of fly feeding. J Neurogenet 30:101–111. doi:10.1080/01677063.2016.1177047

Mellert DJ, Truman JW (2012) Transvection is common throughout the *Drosophila* genome. Genetics 191:1129–1141. doi:10.1534/genetics.112.140475

Mendes CS, Bartos I, Akay T et al (2013) Quantification of gait parameters in freely walking wild type and sensory deprived *Drosophila melanogaster*. Elife 2:e00231. doi:10.7554/eLife.00231

Nern A, Pfeiffer BD, Rubin GM (2015) Optimized tools for multicolor stochastic labeling reveal diverse stereotyped cell arrangements in the fly visual system. Proc Natl Acad Sci 112:E2967–E2976. doi:10.1073/pnas.1506763112

Ni J-Q, Liu L-P, Binari R et al (2009) A *Drosophila* resource of transgenic RNAi lines for neurogenetics. Genetics 182:1089–1100. doi:10.1534/genetics.109.103630

Nicolas G, Sillans D (1989) Immediate and latent effects of carbon dioxide on insects. Annu Rev Entomol 34:97–116

O'Kane CJ, Gehring WJ (1987) Detection in situ of genomic regulatory elements in *Drosophila*. Proc Natl Acad Sci 84:9123

Ohyama T, Schneider-Mizell CM, Fetter RD et al (2015) A multilevel multimodal circuit enhances action selection in *Drosophila*. Nature 520:633–639. doi:10.1038/nature14297

Owald D, Lin S, Waddell S (2015) Light, heat, action: neural control of fruit fly behaviour. Philos Trans R Soc B. doi:10.1098/rstb.2014.0211

Pan Y, Meissner GW, Baker BS (2012) Joint control of *Drosophila* male courtship behavior by motion cues and activation of male-specific P1 neurons. Proc Natl Acad Sci U S A 109:10065–10070. doi:10.1073/pnas.1207107109

Parisky KM, Agosto J, Pulver SR et al (2008) PDF cells are a GABA-responsive wake-promoting component of the *Drosophila* sleep circuit. Neuron 60:672–682. doi:10.1016/j.neuron.2008.10.042

Patterson GH, Lippincott-Schwartz J (2002) A photoactivatable GFP for selective photolabeling of proteins and cells. Science 297:1873–1877. doi:10.1126/science.1074952

Pavlou HJ, Goodwin SF (2013) Courtship behavior in Drosophila melanogaster: towards a "courtship connectome". Curr Opin Neurobiol 23:76–83. doi:10.1016/j.conb.2012.09.002

Peabody NC, Pohl JB, Diao F et al (2009) Characterization of the decision network for wing expansion in *Drosophila* using targeted expression of the TRPM8 channel. J Neurosci 29:3343–3353. doi:10.1523/JNEUROSCI.4241-08.2009

Peng H, Chung P, Long F et al (2011) BrainAligner: 3D registration atlases of *Drosophila* brains. Nat Methods 8:493–498. doi:10.1038/nmeth.1602

Perry CJ, Barron AB (2013) Neural mechanisms of reward in insects. Annu Rev Entomol 58:543–562. doi:10.1146/annurev-ento-120811-153631

Pfeiffer BD, Jenett A, Hammonds AS et al (2008) Tools for neuroanatomy and neurogenetics in *Drosophila*. Proc Natl Acad Sci 105:9715–9720. doi:10.1073/pnas.0803697105

Pfeiffer BD, Ngo T-TB, Hibbard KL et al (2010) Refinement of tools for targeted gene expression in *Drosophila*. Genetics 186:735–755. doi:10.1534/genetics.110.119917

Pool A-H, Kvello P, Mann K et al (2014) Four GABAergic interneurons impose feeding restraint in *Drosophila*. Neuron 83:164–177. doi:10.1016/j.neuron.2014.05.006

Potter CJ, Tasic B, Russler EV et al (2010) The Q system: a repressible binary system for transgene expression, lineage tracing, and mosaic analysis. Cell 141:536–548. doi:10.1016/j.cell.2010.02.025

Pulver SR, Pashkovski SL, Hornstein NJ et al (2009) Temporal dynamics of neuronal activation by Channelrhodopsin-2 and TRPA1 determine behavioral output in *Drosophila* larvae. J Neurophysiol 101:3075–3088. doi:10.1152/jn.00071.2009

Pulver SR, Hornstein NJ, Land BL, Johnson BR (2011) Optogenetics in the teaching laboratory: using channelrhodopsin-2 to study the neural basis of behavior and synaptic physiology in *Drosophila*. Adv Physiol Educ 35:82–91. doi:10.1152/advan.00125.2010

Ramdya P, Lichocki P, Cruchet S et al (2014) Mechanosensory interactions drive collective behaviour in *Drosophila*. Nature. doi:10.1038/nature14024

Reynolds AM, Frye MA (2007) Free-flight odor tracking in *Drosophila* is consistent with an optimal intermittent scale-free search. PLoS ONE 2:e354. doi:10.1371/journal.pone.0000354

Rezával C, Pavlou HJ, Dornan AJ et al (2012) Neural circuitry underlying *Drosophila* female postmating behavioral responses. Curr Biol 22:1155–1165. doi:10.1016/j.cub.2012.04.062

Riabinina O, Luginbuhl D, Marr E et al (2015) Improved and expanded Q-system reagents for genetic manipulations. Nat Methods 12:219–222, 5 p following 222. doi:10.1038/nmeth.3250

Robie AA, Straw AD, Dickinson MH (2010) Object preference by walking fruit flies, *Drosophila melanogaster*, is mediated by vision and graviperception. J Exp Biol 213:2494–2506. doi:10.1242/jeb.041749

Root CM, Ko KI, Jafari A, Wang JW (2011) Presynaptic facilitation by neuropeptide signaling mediates odor-driven food search. Cell 145:133–144. doi:10.1016/j.cell.2011.02.008

Ruta V, Datta SR, Vasconcelos ML et al (2010) A dimorphic pheromone circuit in *Drosophila* from sensory input to descending output. Nature 468:686–690. doi:10.1038/nature09554

Saalfeld S, Cardona A, Hartenstein V, Tomancak P (2009) CATMAID: collaborative annotation toolkit for massive amounts of image data. Bioinformatics 25:1984–1986. doi:10.1093/bioinformatics/btp266

Saleem S, Ruggles PH, Abbott WK, Carney GE (2014) Sexual experience enhances *Drosophila melanogaster* male mating behavior and success. PLoS ONE 9:e96639. doi:10.1371/journal.pone.0096639

Schneider-Mizell CM, Gerhard S, Longair M et al (2016) Quantitative neuroanatomy for connectomics in *Drosophila*. Elife. doi:10.7554/eLife.12059

Schusterreiter C, Grossmann W (2013) A two-fly tracker that solves occlusions by dynamic programming: computational analysis of *Drosophila* courtship behaviour. EURASIP J Image Video Process 2013:64. doi:10.1214/aoms/1177698950

Seeds AM, Ravbar P, Chung P et al (2014) A suppression hierarchy among competing motor programs drives sequential grooming in *Drosophila*. Elife 3:e02951

Seelig JD, Chiappe ME, Lott GK et al (2010) Two-photon calcium imaging from head-fixed *Drosophila* during optomotor walking behavior. Nat Methods 7:535–540. doi:10.1038/nmeth.1468

Seiger MB, Kink JF (1993) The effect of anesthesia on the photoresponses of four sympatric species of *Drosophila*. Behav Genet 23:99–104

Shang Y, Griffith LC, Rosbash M (2008) Light-arousal and circadian photoreception circuits intersect at the large PDF cells of the *Drosophila* brain. Proc Natl Acad Sci 105:19587–19594. doi:10.1073/pnas.0809577105

Shirangi TR, Wong AM, Truman JW, Stern DL (2016) Doublesex regulates the connectivity of a neural circuit controlling *Drosophila* male courtship song. Dev Cell 37:533–544. doi:10.1016/j.devcel.2016.05.012

Simon JC, Dickinson MH (2010) A new chamber for studying the behavior of *Drosophila*. PLoS ONE 5:e8793. doi:10.1371/journal.pone.0008793

Sivanantharajah L, Zhang B (2015) Current techniques for high-resolution mapping of behavioral circuits in *Drosophila*. J Comp Physiol A 201:895–909. doi:10.1007/s00359-015-1010-y

Spradling AC, Stern D, Beaton A et al (1999) The Berkeley *Drosophila* Genome Project gene disruption project: single P-element insertions mutating 25% of vital *Drosophila* genes. Genetics 153:135–177

Stockinger P, Kvitsiani D, Rotkopf S et al (2005) Neural circuitry that governs *Drosophila* Male Courtship Behavior. Cell 121:795–807. doi:10.1016/j.cell.2005.04.026

Straw AD, Branson K, Neumann TR, Dickinson MH (2011) Multi-camera real-time three-dimensional tracking of multiple flying animals. J R Soc Interface 8:395–409. doi:10.1098/rsif.2010.0230

Suh GSB, Wong AM, Hergarden AC et al (2004) A single population of olfactory sensory neurons mediates an innate avoidance behaviour in *Drosophila*. Nature 431:854–859. doi:10.1038/nature02980

Suh GSB, Ben-Tabou de Leon S, Tanimoto H et al (2007) Light activation of an innate olfactory avoidance response in *Drosophila*. Curr Biol 17:905–908. doi:10.1016/j.cub.2007.04.046

Sun F, Wang Y, Zhou Y et al (2014) Identification of neurons responsible for feeding behavior in the *Drosophila* brain. Sci China Life Sci 57:391–402. doi:10.1007/s11427-014-4641-2

Suster ML, Seugnet L, Bate M, Sokolowski MB (2004) Refining GAL4-driven transgene expression in *Drosophila* with a GAL80 enhancer-trap. genesis 39:240–245. doi:10.1002/gene.20051

Takemura S-Y, Bharioke A, Lu Z et al (2013) A visual motion detection circuit suggested by *Drosophila* connectomics. Nature 500:175–181. doi:10.1038/nature12450

Tataroglu O, Emery P (2014) Studying circadian rhythms in *Drosophila melanogaster*. Methods 68:140–150. doi:10.1016/j.ymeth.2014.01.001

Ting C-Y, Gu S, Guttikonda S et al (2011) Focusing transgene expression in *Drosophila* by coupling Gal4 with a novel split-LexA expression system. Genetics 188:229–233. doi:10.1534/genetics.110.126193

Triphan T, Nern A, Roberts SF et al (2016) A screen for constituents of motor control and decision making in *Drosophila* reveals visual distance-estimation neurons. Sci Rep 6:27000. doi:10.1038/srep27000

Tsai H-Y, Huang Y-W (2012) Image tracking study on courtship behavior of *Drosophila*. PLoS ONE 7:e34784. doi:10.1371/journal.pone.0034784

Tuthill JC, Wilson RI (2016) Parallel transformation of tactile signals in central circuits of *Drosophila*. Cell 164:1046–1059. doi:10.1016/j.cell.2016.01.014

Tuthill JC, Nern A, Holtz SL et al (2013) Contributions of the 12 neuron classes in the fly lamina to motion vision. Neuron 79:128–140. doi:10.1016/j.neuron.2013.05.024

Ueda A, Kidokoro Y (2002) Aggressive behaviours of female *Drosophila melanogaster* are influenced by their social experience and food resources. Physiol Entomol 27:21–28. doi:10.1046/j.1365-3032.2002.00262.x

van Breugel F, Dickinson MH (2014) Plume-tracking behavior of flying *Drosophila* emerges from a set of distinct sensory-motor reflexes. Curr Biol 24:274–286. doi:10.1016/j.cub.2013.12.023

Vaughan AG, Zhou C, Manoli DS, Baker BS (2014) Neural pathways for the detection and discrimination of conspecific song in *D. melanogaster*. Curr Biol 24:1039–1049. doi:10.1016/j.cub.2014.03.048

Venken KJT, He Y, Hoskins RA, Bellen HJ (2006) P[acman]: a BAC transgenic platform for targeted insertion of large DNA fragments in *D. melanogaster*. Science 314:1747–1751. doi:10.1126/science.1134426

Venken KJT, Schulze KL, Haelterman NA et al (2011a) MiMIC: a highly versatile transposon insertion resource for engineering *Drosophila melanogaster* genes. Nat Methods 8:737–743

Venken KJT, Simpson JH, Bellen HJ (2011b) Genetic manipulation of genes and cells in the nervous system of the fruit fly. Neuron 72:202–230. doi:10.1016/j.neuron.2011.09.021

Vinayak P, Coupar J, Hughes SE et al (2013) Exquisite light sensitivity of *Drosophila melanogaster* cryptochrome. PLoS Genet 9:e1003615. doi:10.1371/journal.pgen.1003615

von Philipsborn AC, Liu T, Yu JY et al (2011) Neuronal control of *Drosophila* courtship song. Neuron 69:509–522. doi:10.1016/j.neuron.2011.01.011

von Reyn CR, Breads P, Peek MY et al (2014) A spike-timing mechanism for action selection. Nat Neurosci 17:962–970. doi:10.1038/nn.3741

Wan Y, Otsuna H, Chien C-B, Hansen C (2012) FluoRender: an application of 2D image space methods for 3D and 4D confocal microscopy data visualization in neurobiology research. IEEE Pacific visualization symposium, pp 201–208

Wang L, Anderson DJ (2010) Identification of an aggression-promoting pheromone and its receptor neurons in *Drosophila*. Nature 463:227–231. doi:10.1038/nature08678

Wang L, Dankert H, Perona P, Anderson DJ (2008) A common genetic target for environmental and heritable influences on aggressiveness in *Drosophila*. Proc Natl Acad Sci 105:5657–5663. doi:10.1073/pnas.0801327105

Wang L, Han X, Mehren J et al (2011) Hierarchical chemosensory regulation of male-male social interactions in *Drosophila*. Nat Neurosci 14:757–762. doi:10.1038/nn.2800

White BH (2016) What genetic model organisms offer the study of behavior and neural circuits. J Neurogenet 30:54–61. doi:10.1080/01677063.2016.1177049

Wilson RI (2013) Early olfactory processing in *Drosophila*: mechanisms and principles. Annu Rev Neurosci 36:217–241. doi:10.1146/annurev-neuro-062111-150533

Wu M-C, Chu L-A, Hsiao P-Y et al (2014) Optogenetic control of selective neural activity in multiple freely moving *Drosophila* adults. Proc Natl Acad Sci 111:5367–5372. doi:10.1073/pnas.1400997111

Yamaguchi S, Desplan C, Heisenberg M (2010) Contribution of photoreceptor subtypes to spectral wavelength preference in *Drosophila*. Proc Natl Acad Sci 107:5634–5639. doi:10.1073/pnas.0809398107

Yoshihara M, Ito K (2000) Improved Gal4 screening kit for large-scale generation of enhancer-trap strains. Drosoph Inf Serv 83:199–202

Yoshihara M, Ito K (2012) Acute genetic manipulation of neuronal activity for the functional dissection of neural circuits-a dream come true for the pioneers of behavioral genetics. J Neurogenet 26:43–52. doi:10.3109/01677063.2012.663429

Zaninovich OA, Kim SM, Root CR et al (2013) A single-fly assay for foraging behavior in *Drosophila*. J Visualized Exp e50801. doi:10.3791/50801

Zawistowski S, Richmond RC (1987) Experience mediated reduction in courtship of *Drosophila melanogaster* in large and small chambers. J Comp Psychol 101:90–93

Zhou C, Franconville R, Vaughan AG et al (2015) Central neural circuitry mediating courtship song perception in male *Drosophila*. Elife. doi:10.7554/eLife.08477

Chapter 8
Paradigms for the Quantification of Behavioral Responses in Zebrafish

Chiara Cianciolo Cosentino and Stephan C.F. Neuhauss

Abstract The increasing popularity of the zebrafish (*Danio rerio*) as a vertebrate model organism has made it the most genetically studied vertebrate, only surpassed by the mouse. Zebrafish popularity stems from its favorable biological properties such as its high fecundity, rapid development, and (as larva) optical transparency. Recent years have seen the development of an impressive genetic toolbox for the zebrafish. While earlier geneticists had to rely on mutant strains generated by random chemical mutagenesis, zebrafish researchers have now the full complement of modern genetic tools at their fingertips. This includes efficient transposon-mediated transgenesis and CRISPR/Cas9-mediated genome editing. These recent genetic advances in combination with the optical properties of the larva enable sophisticated neural circuit analyses, unsurpassed in any other vertebrate organisms.

8.1 Introduction

Already early in development zebrafish larvae display a wide range of behavioral responses. The underlying neural circuits are composed of relatively few cells, which make their analysis much simpler than in larger animals. Anatomical structures are easy to visualize, due to both the transparency of the larvae and the ability to selectively label specific cells or cell types by transgenic technology. Such labeling techniques are also suited to explore functional anatomy using activity sensing dyes (currently mainly GCAMP calcium sensors) and use the power of optogenetics to selectively stimulate neural circuits. Thus the zebrafish model system holds great potential to directly relate neural circuit activity with behavioral outputs.

A prerequisite for such comprehensive analyses is the quantitative assessment of behaviors. Most of behavioral analysis in zebrafish was initially motivated by the

C. Cianciolo Cosentino · S.C.F. Neuhauss (✉)
Institute of Molecular Life Sciences, University of Zurich,
Winterthurerstrasse 190, 8057 Zurich, Switzerland
e-mail: stephan.neuhauss@imls.uzh.ch

A. Çelik and M.F. Wernet (eds.), *Decoding Neural Circuit Structure and Function*, DOI 10.1007/978-3-319-57363-2_8

desire to use simple behaviors as screening assays for genetic or pharmacological screens. Indeed the small size and the possibility to screen thousands of larvae in parallel inspired large-scale chemical screens. The field has now matured to study components of behavior at high spatial and temporal resolution and the use of virtual behavior paradigms. In this chapter we will discuss behavioral approaches currently used in zebrafish with an emphasis on those that lend themselves to a quantitative analysis.

8.2 Larval and Adult Tracking

The analysis of zebrafish behavior owes much to the development of tracking technologies. It is getting increasingly easier to use commercial cameras to track animals in real time using a variety of software options (many of them open source). The availability of this technology has spurred a large number of studies of various zebrafish behaviors with the common denominator of tracking the position of the animal over time.

8.2.1 Spontaneous Movements

Zebrafish embryos develop rapidly and a number of spontaneous locomotor behaviors, or at least behaviors not triggered by externally provided stimulation, are apparent early on in development and are summarized below.

8.2.1.1 Spontaneous Coiling

The first overt motor behavior in zebrafish is the spontaneous periodic flexions of the trunk observed at around 17 hours post fertilization (hpf), when the embryo still resides inside the chorion. This coiling behavior is sensory input independent and controlled by a neural network within the spinal cord. The contractions are initiated by pacemaker ipsilateral caudal (IC) neurons in the rostral spinal cord generating a periodic gap-junction synchronized depolarization in the spinal neurons, leading to ipsilateral contractions of the trunk musculature (Saint-Amant and Drapeau 2000; Drapeau et al. 2002; Brustein et al. 2003). This simple locomotor pattern can easily be analyzed and quantified in multi-well microplates with automated image acquisition procedures, and has been used in both small-scale (Selderslaghs et al. 2010, 2013) and high-content (Raftery et al. 2014) screening assays for the identification of developmental neurotoxins. The main advantages of this assay, compared to other more complex zebrafish-based locomotor behaviors, are the shorter duration of the test, the low costs, and the possibility to efficiently analyze large numbers of animals (Raftery et al. 2014; Saint-Amant and Drapeau 2000, 2001).

8.2.1.2 Swimming Behavior

By the end of the first week of development, larval zebrafish already possess a repertoire of various stereotyped locomotor behaviors with specific kinematic properties that have been characterized in detail with the use of high-speed video cameras at high temporal resolution of 1000 frames per second (Wolman and Granato 2012). The swimming behavior of larval zebrafish is characterized by burst and slow swims. Slow swims (scoot) consist of rostro-caudal propagation of mild bends with minimal side-to-side movement of the head and minimal bending of the front half of the body, whereas the burst swimming pattern is characterized by higher speed and larger bend angles, with ample side-to-side movement of the head and a more rostral bend location (Budick and O'Malley 2000). Turning behaviors can also be distinguished into two types, routine and escape turns. Routine turns are spontaneous slow-speed turns, in which the fish makes a bend of approximately 60°, with a slow angular velocity, distinct from high-velocity turns associated with escape responses described below (Burgess et al. 2010; Budick and O'Malley 2000; McElligott and O'Malley 2005b).

8.2.2 Sensory Induced Movements

At the early stages of nervous system development the zebrafish embryo shows only spontaneous movements, but already starting at 16 days post fertilization (dpf), the embryo starts to interact with the environment and responds to different sensory inputs. Initially these inputs are exclusively tactile, since the primary sensory system with Rohon–Beard sensory neurons is the first to develop. Responses to other stimuli, such as sound and light, can only occur when the necessary sensory organs become functional.

8.2.2.1 Startle Response

The startle response is an evolutionarily highly conserved reaction of an organism to unexpected or aversive stimuli. In zebrafish the tactile startle responses can first be elicited at around 27 hpf (Saint-Amant and Drapeau 1998). Two different forms of startle response, namely the C- and S-start (named after the specific bending of the body axis), can be observed. The C-start can be divided into two stages, where initially the body is bend into a C-shape, orienting the fish away from the stimulation site, followed by a propulsive tail stroke that propels the fish away from the stimulus, often referred to as a burst swim (Fetcho 1991). The S-start is characterized by a rostral bend away from the stimulus and a caudal bend on the opposite side of the body. In larval zebrafish, it is thought that head stimulation generally elicits C-starts, while tail stimulation can evoke both C- and S-starts (Liu et al. 2012). Startle responses are mediated by the Mauthner (M) cells, a pair of large, fast conducting reticulospinal interneurons with somata located on either side of the brainstem. The Mauthner neurons and their serial homologs, MiD2 cm and MiD3 cm cells, are all activated in response to head stimulation (Kimmel et al. 1981; Kohashi and Oda 2008; Maaswinkel and Li 2003; Liu et al. 2012), whereas only M-cells are necessary to drive a fast startle in response to tail stimulation (O'Malley et al. 1996). Being a well-studied stereotyped and robust behavior, the startle response has proven to be ideally suited for screening embryos for motility defects. Indeed this behavior has been exploited in a seminal large-scale genetic screen for motility defects, isolating more than 100 mutant strains with defects affecting sensory and motor systems (Granato et al. 1996). Besides tactile stimulation, there are a number of other stimuli that can trigger escape and avoidance responses in zebrafish larvae, including acoustic and vibrational stimuli, water flow sensed by the air cells of the later line system (Kohashi and Oda 2008), abrupt changes in light intensities, gradient of light, moving objects, and various chemicals (Saint-Amant and Drapeau 1998; Colwill and Creton 2011).

The acoustic startle response has been used by Burgess and Granato in an elegant screen aimed at identifying genes regulating the process of sensorimotor gating by measuring prepulse inhibition (PPI) of the startle response in zebrafish (Burgess and Granato 2007b).

Another good example of the use of this behavior comes from another study (Nicolson et al. 1998), which lead to the isolation of mutant lines affecting the function of the sensory cells of the inner ear. In the fish, the escape reflex triggered by an acoustic stimulus has also been used in a high-throughput screen isolating fish with hearing defects (Bang et al. 2002).

8.2.2.2 High Temporal Resolution Analysis

All the stereotyped movements described above are characterized by a large range of tailbeat frequencies, which require the use of high-speed video cameras at 1000 frames per second to capture them accurately. Such analyses will be of particular interest for complex behaviors that may then be divided into singular components, akin to a fine-grained ethogram (e.g., Budick and O'Malley 2000; McElligott and O'Malley 2005a; Borla et al. 2002). Since the underlying neural circuits of these singular components are different, a detailed analysis holds the potential to unravel modular neural networks that can be combined to specify behavioral responses.

8.2.2.3 Light- and Vision-Induced Responses

Zebrafish need to respond to sensory stimuli with an appropriate behavioral response. Since at larval stages vision is the primary sense, it comes as no surprise that there are a number of visually evoked quantifiable behaviors. Due to the human emphasis on vision, our culture has developed sophisticated tools for visual stimulation, unrivaled by any other sensory modality. Hence, we can manipulate visual properties, such as brightness and hue with great precision, enabling us to probe different properties of vision in the zebrafish.

Photomotor response (PMR)
The earliest response of zebrafish to light is the photomotor response (PMR), where dark-adapted embryos starting at 30 hpf display light-induced coiling and swimming behavior, followed by a refractory period in which embryos do not respond to a second light stimulus. Fascinatingly, this response occurs at a developmental stage where no visual organs, such as eyes and pineal gland, are formed. Subsequent studies demonstrated that this response resides in hindbrain neurons using an opsin-based light-sensing mechanism (Kokel et al. 2013). This simple behavior is very well suited for large-scale screening of neuroactive compounds, since a surprising variety of changes in the behavioral profile, such as length of the activation and refractory phase, can be evaluated (Kokel et al. 2010).

Phototaxis
Phototaxis, an evolutionary ancient behavior already present in prokaryotes, describes the locomotor movement of the whole animal toward or away from a source of light. Larval zebrafish can show both positive and negative phototactic behaviors, depending on the adaptation state and the intensity of the light source (Brockerhoff et al. 1995; Orger and Baier 2005; Burgess et al. 2010). The rate and direction of phototaxis is determined by the relative intensity of the light stimulus (Burgess and Granato 2007a; Burgess et al. 2010). Low-intensity light stimuli will induce positive phototaxis, where the larvae will first turn toward and then rapidly approach the light stimulus. Larvae will respond to high light intensity stimuli with negative phototaxis, by turning and moving slowly away from the illuminated area (Burgess and Granato 2007a). Burgess and coworkers deciphered the neural logic by showing that the retinal ON pathway mediates the increase in swimming behavior, while the OFF pathway is involved in the directed turning response (Burgess et al. 2010; reviewed in Mueller et al. 2010a).

Visual motor response (VMR)
The visual motor response (VMR) is a visual startle response that can be recorded in free-swimming larvae (older than 4 dpf). It consists of a robust increase in locomotor activity in response to a sharp increase or decrease in illumination (Emran et al. 2007, 2008). This behavior is driven by retina-mediated vision. Studies have shown that the ON channel mediates the VMR (Emran et al. 2007), while this behavior is independent of motion vision (Maurer et al. 2010). In the most commonly used configuration, the VMR assay tracks the spikes of locomotion of the zebrafish larvae in response to multiple alternating 30 min light ON and light OFF periods. Upon receiving a light stimulus, fish exhibit a brief increase of motor activity, the startle response, followed by a return to a lower than baseline activity. This assay ideally lends itself to large-scale tracking and is commonly used for screens related to both vision and locomotion.

Optomotor response (OMR)
Zebrafish have a tendency to follow broad movements in the surround, likely as an adaption to keep a stable position in moving water. This intrinsic behavior can be utilized for the optomotor response (OMR), which is robustly displayed in fish older than 6 dpf. In the laboratory, this behavior is commonly evoked either by placing the free-swimming animals in a small tank surrounded by a moving drum fitted with stripes (Clark 1981; Bilotta 2000) or by displaying apparently moving stripes on a flipped monitor fitted with a transparent container on top (Orger and Baier 2005; Neuhauss et al. 1999). The later assay is well suited for group testing, since larvae do not yet display shoaling behavior, while adults have to be tested individually (Fig. 8.1). While this assay has been widely used to probe different

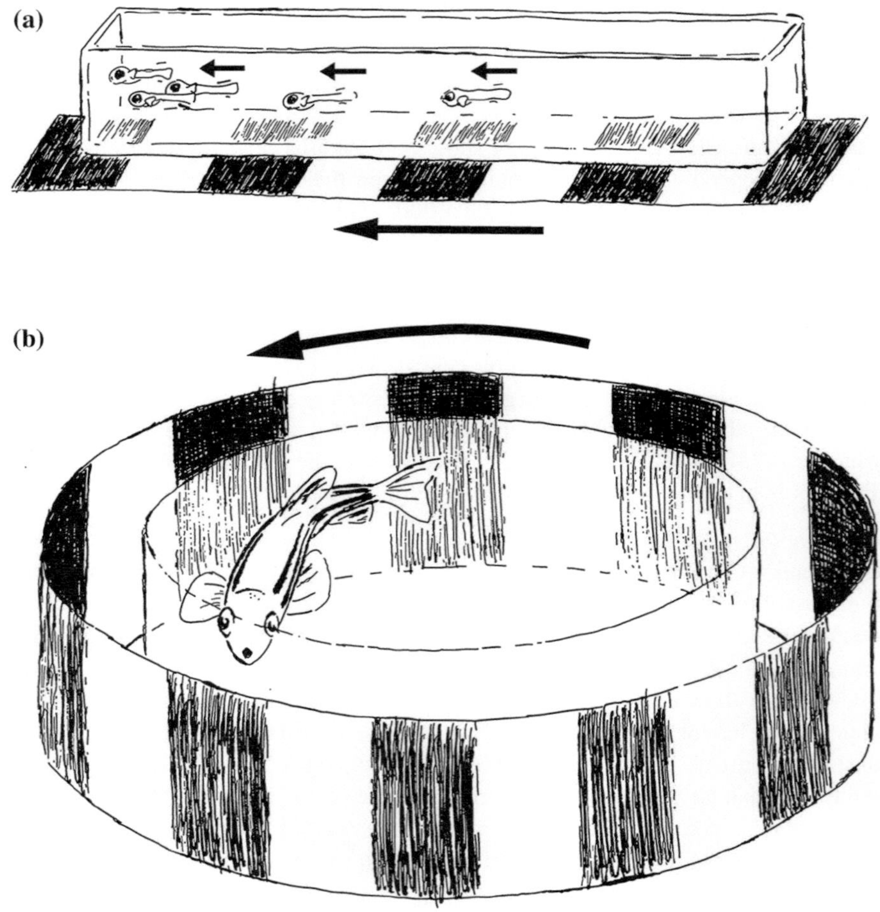

Fig. 8.1 Optomotor response assay (OMR) in zebrafish larvae (**a**) and adults (**b**). **a** Population screening for larval OMR, **b** adult OMR. Actual size of larvae and adult fish is proportionally much smaller than depicted in the diagram (adapted from Roeser and Baier 2003; Neuhauss 2003)

aspects of motion processing (reviewed in Fleisch and Neuhauss 2006), it is less useful for chemical screens, since it depends on locomotion in general, which is more readily disturbed by exogenous chemical compounds than vision.

8.2.2.4 Chemically Induced Locomotion

Apart from the above-mentioned sensory stimuli, basically all sensory stimuli can, under the appropriate conditions, induce changes in zebrafish locomotor behavior. These can include noxious chemical stimuli (Prober et al. 2008) and olfactory cues (Kermen et al. 2013). A long known example is the anxiety-related response of fish to an olfactory perceived Schreckstoff ("fear substance"), which will be discussed in more details below (Frisch von 1938). Such studies represent promising tools for the development of high-throughput behavioral paradigms for drugs and mutation screening aimed at the analysis of the biological mechanisms of fear (avoidance behavior) in vertebrates.

8.2.2.5 High-Throughput Parallel Tracking

In the preceding section we often referred to the possibility to adapt the various behavioral assays for automatic high-throughput screens. This has been realized for a number of assays for larvae, with the obvious advantage that many larvae can be scored in parallel over a long period of time without the experimenter being present. Moreover, thanks to data-sifting algorithms, such analyses offer the opportunity to detect patterns that are invisible to the human observer.

All these assays depend on computerized video-tracking tools that record the animal's position over time. Most published systems use tracking in two dimensions, but 3D tracking for adults is now already frequently used (Maaswinkel et al. 2013, 2015; Moravec et al. 2015). For larval tracking, recording in multi-well plates makes it possible to examine a large number of motor events and facilitates quantitative analysis of behavior at medium and high-throughput (Burgess and Granato 2007a; Burgess et al. 2009). The user can choose between a number of commercial and open-source tracking systems.

The possibility of high-throughput tracking has drawn considerable interest for behavioral screens aimed at the discovery and characterization of new neuroactive molecules, also owing to the advantage of testing a behaving vertebrate, while still doing (from a regulatory point of view) an in vitro experiment.

The first two high-throughput screens for small molecules able to modulate behavior already demonstrated the enormous potential of such screens (Rihel et al. 2010, Kokel et al. 2010). Rihel and colleagues examined the effects of nearly 4000 small molecules on larval zebrafish sleep/wake behavior. It has been shown that zebrafish exhibit locomotor pattern characteristic of sleep-like state, resting predominantly at night and showing an increased arousal threshold during the rest bouts (reviewed in Rihel and Schier 2012). For the screen, single 4 dpf larvae were placed in the wells of a 96-well plate, and exposed to different small molecules dissolved directly in the water for 3 days. Automated tracking allowed the analysis of different parameters of sleep/wake behaviors, including the average locomotor activity during day and night, the length and the number of sleep bouts, and the length of sleep latency. The quantitative behavioral data for each compound was broken into multiple parameters and converted into a behavioral profile used for hierarchical clustering. The clustering organized the compounds based on phenotype. Of the 4000 compounds screened with this method, more than 450 were found to significantly alter sleep/wake behavior.

In the second high-throughput study, Kokel and collaborators used the larval zebrafish photomotor response (PMR) to screen the effect of 14,000 small molecules (Kokel et al. 2010). The group performed an automated screening, which led to the identification of hundreds of unique chemicals able to alter the PMR, causing increased activity, increased sedation, response latency, or refractory period. These behavioral patterns were used to identify new psychotropic chemicals and to predict their molecular targets. Together with the high-throughput assay performed by Rihel et al., these results suggest that behavioral profiling can be a useful tool to understand neuronal signaling, rapidly identify novel neuroactive compounds and predict their mechanisms of action (Rihel and Schier 2012). Following these seminal studies, a number of recent studies have expanded such chemical screens to antipsychotic compounds (Bruni et al. 2016), and autism-related neurochemistry (Hoffman et al. 2016).

8.2.3 Complex Behavior

So far, we have mostly dealt with simple and reflexive behaviors. Such behaviors are ideally suited for large-scale screenings and will doubtlessly be of tremendous value for drug discovery. Their advantage is that they are largely independent of the internal state of the animal and lack motivational components. Complex behaviors that combine many of the simpler behaviors described are harder to study but highly relevant for human behavior, including disorders such as psychiatric diseases. Studies of complex behaviors in zebrafish have only recently been initiated, frequently taking inspiration from well-established rodent behavioral paradigms. In the following part we will focus on three zebrafish behaviors relevant for neural circuit analysis: prey capture, epilepsy and anxiety.

8.2.3.1 Prey Capture

Prey capture is an innate behavior that can be observed in larval zebrafish starting at 4 dpf. This behavior is visually induced and relays on a more complex processing of visual information, including cognitive components. Before initiating prey capture, zebrafish faced with a moving stimulus in their visual field need to assess their own feeding state and the size of the object, in order to decide if to prey on the object or initiate an escape response (Filosa et al. 2016).

Prey capture is highly dependent on vision at larval stages and depends on activation of neurons in the optic tectum (Muto and Kawakami 2013; Roeser and Baier 2003). The coordinated motor sequence is unique to hunting behavior and consists of unilateral tail bends into a J shape. This orients larvae toward their prey (typically paramecia), followed by convergent saccades of the eyes and a final capture swim and a bite (Bianco and Engert 2015; Borla et al. 2002; McElligott and O'Malley 2005a; Patterson et al. 2013).

This sequence of locomotor behaviors is distinct from navigational swimming and escape behavior. Therefore, J-bends and convergent eye movements have been taken as indication of hunting behavior in behavioral assays.

8.2.3.2 Epilepsy

Within the last few years, the zebrafish has emerged as a relevant model vertebrate for epilepsy research (Grone and Baraban 2015; Cunliffe 2016). Epilepsy is a group of neurological diseases that is caused by hyperactivity of the brain or parts thereof, and characterized by epileptic seizures. The disease is surprisingly common in human populations with a sizable fraction of patients being unresponsive to common anti-epileptic drugs. Hence, there is a large medical motivation to both understand the neurological basis of the disease and to find novel drugs to medicate unresponsive patients.

Baraban et al. (2005) were the first to demonstrate that seizures can be generated in larval zebrafish by simply exposing the animals to Pentylenetetrazole (PTZ), a common convulsant agent. Treatment with PTZ reliably elicited stereotyped and concentration-dependent seizure-like behaviors in zebrafish larvae at 7 dpf than could be quantified by video-tracking and image analysis (Baraban et al. 2007). The authors divided the behavior into three stages. Following exposure to PTZ, the larvae initially dramatically increased their swimming activity (Stage I); this was followed by a rapid "whirlpool-like" circling swim behavior (Stage II) and culminated in a loss of posture and immobility for 1–3 s (Stage III). The latencies to the beginning of stages I, II, or III were concentration dependent. PTZ is now widely used to induce acute seizures in zebrafish, but it is only one of a wide range of chemicals that have been shown to induce seizures and convulsions in zebrafish (reviewed in Stewart et al. 2012). The stereotyped behaviors induced by PTZ and other convulsing agents closely resemble those induced in mammals (Jirsa et al. 2014). Such behaviors have been successfully used in large-scale mutagenesis screens aimed at identifying gene mutations that confer seizure resistance (Baraban et al. 2007), in assays aimed at evaluating the seizure liability of early-stage drug development (Winter et al. 2008), or for the screening of new potential anti epileptic drugs (AEDs) (Baraban et al. 2013; Siebel et al. 2015). In larval models, animals can be placed in multi-well plates and, thanks to the embryo permeability, chemicals can be added directly to the media. The larvae can be recorded simultaneously using video-tracking software and a top view camera (Baraban et al. 2005), allowing the concurrent analysis of several conditions in a short period of time. The brain electrical activity during epilepsy can also be recorded in agarose-immobilized larvae, allowing the generation of electro-encephalograms (EEG) (Hortopan et al. 2010; Stewart et al. 2012), or by direct in vivo imaging using genetically encoded calcium indicator dies (Baxendale et al. 2012). In adult fish, chemicals can either be added to the medium or injected intraperitoneally. Observations take place in transparent tanks, where seizures are measured using a special scoring system either manually or with video-recording for automated analyses (Wong et al. 2010).

8.2.3.3 Anxiety

Anxiety-related behavior clearly confers selective advantage for an animal to avoid a potentially harmful situation. This adaptive behavior also lies at the core of a number of human mental disorders. Zebrafish display anxiety-related behaviors to a number of stimuli, which can be used to search for neuroactive substances modifying the behavior or eventually to uncover the underlying neuronal basis of these behaviors.

Historically, the first anxiety-linked behaviors in ostariophysi fish has been described in 1938 by Karl von Frisch (1938). Fish from this superorder (where the zebrafish and other cyprinids belong to) display a characteristic anxiety-like behavior to a Schreckstoff (fear substance) found in club cells of the skin of conspecifics. Recently, the chemical nature of this alarm signal has been elucidated (Mathuru et al. 2012). In this behavior, which is mimicked by a number of other alarm substances, adult fish display erratic movements with rapid changes in direction and acceleration and deceleration episodes (Speedie and Gerlai 2008).

Another evolutionary conserved anxiety-associated behavior often described in zebrafish is thigmotaxis, the movement toward a solid object. In zebrafish this is displayed as the preference to stay in close proximity to the edge of the tank, also termed "wall hugging". Across species this is taken as an index of anxiety and widely used for drug screening. Thigmotaxis is readily measurable by tracking larval zebrafish and by recording the time spend in a defined zone close to the wall of the circular well. This behavior has been validated using known anxiogenic (e.g., caffeine) or anxiolytic (e.g., diazepam) drugs and was utilized to screen for such compounds in medium-throughput studies (Richendrfer et al. 2012; Schnörr et al. 2012).

A modification used in adult zebrafish is the "novel tank diving test", where fish tend to stay closer to the bottom of an unfamiliar tank before venturing into the upper water column after acclimatization. This behavior is described as the zebrafish equivalent of the popular open field test in rodents. Again this behavior can be drastically modified by anxiogenic and anxiolytic drugs and will be an important assay to screen for mutant strains with potential effects on fear-related behavior.

8.3 Eye Tracking

In all the behaviors discussed so far, the position of the whole animal is recorded. Already at larval stages the eye of the zebrafish is large enough that its movements can reliably be recorded. This movement has a more dominant sensory component, since only the eye muscles are needed for its execution. The reflexive optokinetic response can reliably be recorded starting at 4 dpf and has been used to uncover a number of mutant strains affected in vision (Brockerhoff et al. 1995, 1998; Neuhauss et al. 1999; Gross et al. 2005). Interestingly, in all these genetic screens not a single mutation leading to selective eye muscle dysfunction has been isolated. This behavioral paradigm was one of the first assays where eye position was automatically tracked over time, allowing accurate measurement of acuity and other visual performance properties (Rinner et al. 2005; Beck et al. 2004; Huang and Neuhauss 2008). Since the larva is embedded in a viscous medium preventing whole body movements, this paradigm is ideally suited to measure neuronal activity while the larvae is performing the behavior (Portugues et al. 2014). Some efforts have been reported to adapt the assay for high-throughput applications (Mueller et al. 2011), but embedding the larvae is a time-consuming step that still has to be performed manually.

Since eye movements are suppressed by body movements, it became more difficult to measure the OKR in adult fish, but a number of holders to impair movements without affecting eye movements have been reported (Tappeiner et al. 2012; Beck et al. 2004; Mueller and Neuhauss 2010b).

8.4 Virtual Behavior

With the exception of eye tracking, all of the discussed zebrafish behaviors are manifested in the locomotion of the whole body. This poses a hurdle for imaging neural networks during execution of the behavior, since the exposure times needed with currently available calcium-sensitive fluorescent probes will only result in blurred images. Therefore, imaging in behaving fish is either done at a time where the fish stops moving (Muto and Kawakami 2013) or various embedding techniques are used. For instance, hindbrain activity during swimming can be assayed be immobilizing the head of the larva, while the tail is free to move (Fetcho and O'Malley 1997). Other preparations immobilize the whole larvae in agarose and measure neuronal responses to stimulation.

An innovative approach is immobilizing the larvae while changing the stimulation in accordance to its own (fictive) movement by reafference (Portugues and Engert 2011; Engert 2012). This paradigm was applied to image neuronal activity on immobilized larvae during navigation in virtual environments. Information from electric recordings from motor nerves were used to update the visual stimulation in

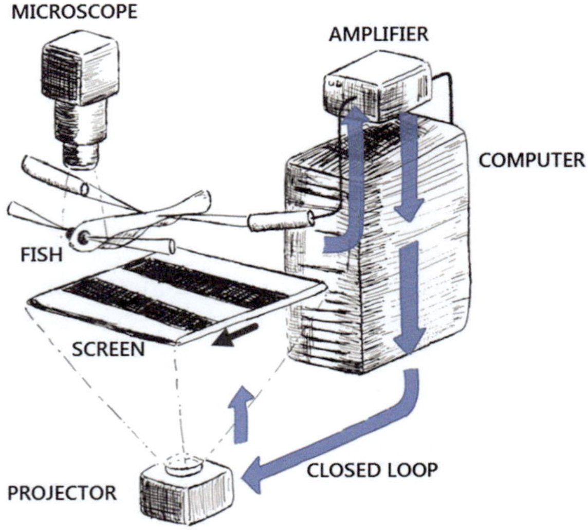

Fig. 8.2 Schematic of the experimental setup of the virtual motor adaptation assay in a immobilized larval zebrafish supported by suction pipettes and with two recording pipettes at both sides of the body. A moving grating is shown *below* the larva and its behavior is monitored with a high-speed camera (adapted from Hughes 2013)

a closed-loop configuration, effectively creating a virtual environment through which the immobilized larvae fictively swims (Ahrens et al. 2012, 2013) (Fig. 8.2).

Prey capture behavior is also uniquely suited to use virtual behavior techniques to record brain activity (Trivedi and Bollmann 2013). Two recent studies exemplify the power of this approach (Semmelhack et al. 2014; Bianco and Engert 2015). Both groups evoked hunting behavior in tethered larvae with various visual stimuli, while simultaneously recoding neuronal activity with two-photon imaging. Bianco and Engert use convergent saccades as readout for hunting behavior. They identified populations of neurons in the optic tectum that are likely involved in visual prey recognition. Interestingly, the neurons involved in the perceptions of prey and those involved in initiating hunting behavior are distinct populations, and activation of the prey perceiving neural network only occasionally activate the network involved in hunting behavior. This preparation can now be used to tease out which networks are involved in decision-making. The modulatory role of signals from the hypothalamic–pituitary–interrenal axis and the serotonergic system on such decisions have been recently reported (Filosa et al. 2016). In a second study Semmelhack and colleagues used the triggering of J-turns as proxy for hunting behavior to implicate two types of ganglion cells that with projections to a pretectal area (AF7) and collateral to the optic tectum are involved in prey capture behavior. Neurons with arbors in AF7 project to multiple sensory and premotor areas of the brain. Hence, selectivity for prey-like stimuli is decoded already in the retina (Semmelhack et al. 2014). Both complementary studies identified neural circuits involved in prey detection highlighting the tremendous potential of virtual behavioral paradigms for neural circuit analysis.

8.5 Outlook

The study of zebrafish behavior has come a long way. We now have a catalogue (Kalueff et al. 2013) of observable behaviors in zebrafish that can be used to understand the underlying neuronal networks responsible for its execution. Genetic screens have revealed some of the underlying genes needed to construct these networks and drug screens resulted in important insights into the neuropharmacology of these networks. The rapid advances in microscopic and software techniques will give us the opportunity to not only screen large amounts of larvae for behavioral modifications, but also to link these observations to the activity of single neurons in the behaving animal. In the near future we will witness even more instances where the precise connectivity of behavior mediating neural networks will be elucidated. The zebrafish is unrivaled among vertebrates in its potential to precisely dissect neuronal circuits underlying behavior.

Acknowledgements We apologize to all the authors whose work we could not cite due to space limitations. We are grateful to Selin Özgut and Matthias Gesemann for comments on the manuscript. CCC was supported by the RiMED foundation; work in the authors' laboratory is supported by the Swiss National Science Foundation (31003A_173083). Special thanks to Irene Ojeda Naharros for artwork on both figures.

References

Ahrens MB, Li JM, Orger MB, Robson DN, Schier AF, Engert F, Portugues R (2012) Brain-wide neuronal dynamics during motor adaptation in zebrafish. Nature 485(7399):471–477. doi:10.1038/nature11057

Ahrens MB, Huang KH, Narayan S, Mensh BD, Engert F (2013) Two-photon calcium imaging during fictive navigation in virtual environments. Front Neural Circuits 7 104. doi:10.3389/fncir.2013.00104

Bang PI, Yelick PC, Malicki JJ, Sewell WF (2002) High-throughput behavioral screening method for detecting auditory response defects in zebrafish. J Neurosci Methods 118(2):177–187

Baraban SC, Taylor MR, Castro PA, Baier H (2005) Pentylenetetrazole induced changes in zebrafish behavior, neural activity and c-fos expression. Neuroscience 131(3):759–768. doi:10.1016/j.neuroscience.2004.11.031

Baraban SC, Dinday MT, Castro PA, Chege S, Guyenet S, Taylor MR (2007) A large-scale mutagenesis screen to identify seizure-resistant zebrafish. Epilepsia 48(6):1151–1157. doi:10.1111/j.1528-1167.2007.01075.x

Baraban SC, Dinday MT, Hortopan GA (2013) Drug screening in Scn1a zebrafish mutant identifies clemizole as a potential Dravet syndrome treatment. Nat Commun 4:2410. doi:10.1038/ncomms3410

Baxendale S, Holdsworth CJ, Santoscoy M, Paola L, Harrison Michael RM, Fox J, Parkin CA et al (2012) Identification of compounds with anti-convulsant properties in a zebrafish model of epileptic seizures. Dis Models Mech 5(6):773–784. doi:10.1242/dmm.010090

Beck JC, Gilland E, Tank DW, Baker R (2004) Quantifying the ontogeny of optokinetic and vestibuloocular behaviors in zebrafish, medaka, and goldfish. J Neurophysiol 92(6):3546–3561. doi:10.1152/jn.00311.2004

Bianco IH, Engert F (2015) Visuomotor transformations underlying hunting behavior in zebrafish. Curr Biol (CB) 25(7):831–846. doi:10.1016/j.cub.2015.01.042

Bilotta J (2000) Effects of abnormal lighting on the development of zebrafish visual behavior. Behav Brain Res 116(1):81–87

Borla MA, Palecek B, Budick S, O'Malley DM (2002) Prey capture by larval zebrafish: evidence for fine axial motor control. Brain Behav Evol 60(4):207–229

Brockerhoff SE, Hurley JB, Janssen-Bienhold U, Neuhauss SC, Driever W, Dowling JE (1995) A behavioral screen for isolating zebrafish mutants with visual system defects. Proc Natl Acad Sci U S A 92(23):10545–10549

Brockerhoff SE, Dowling JE, Hurley JB (1998) Zebrafish retinal mutants. Vision Res 38(10): 1335–1339

Bruni G, Rennekamp AJ, Velenich A, McCarroll M, Gendelev L, Fertsch E et al (2016) Zebrafish behavioral profiling identifies multitarget antipsychotic-like compounds. Nat Chem Biol 12(7):559–566. doi:10.1038/nchembio.2097

Brustein E, Saint-Amant L, Buss RR, Chong M, McDearmid JR, Drapeau P (2003) Steps during the development of the zebrafish locomotor network. J Physiol Paris 97(1):77–86. doi:10.1016/j.jphysparis.2003.10.009

Budick SA, O'Malley DM (2000) Locomotor repertoire of the larval zebrafish: swimming, turning and prey capture. J Exp Biol 203(Pt 17):2565–2579

Burgess HA, Granato M (2007a) Modulation of locomotor activity in larval zebrafish during light adaptation. J Exp Biol 210(Pt 14):2526–2539. doi:10.1242/jeb.003939

Burgess HA, Granato M (2007b) Sensorimotor gating in larval zebrafish. J Neurosci (Official J Soc Neurosci) 27(18):4984–4994. doi:10.1523/JNEUROSCI.0615-07.2007

Burgess HA, Johnson SL, Granato M (2009) Unidirectional startle responses and disrupted left-right co-ordination of motor behaviors in robo3 mutant zebrafish. Genes Brain Behav 8(5):500–511. doi:10.1111/j.1601-183X.2009.00499.x

Burgess HA, Schoch H, Granato M (2010) Distinct retinal pathways drive spatial orientation behaviors in zebrafish navigation. Curr biol (CB) 20(4):381–386. doi:10.1016/j.cub.2010.01.022

Clark DT (1981) Visual responses in developing zebrafish (*Brachydanio rerio*). Ph.D. dissertation, University of Oregon Press, Eugene, OR

Colwill RM, Creton R (2011) Imaging escape and avoidance behavior in zebrafish larvae. Rev Neurosci 22(1):63–73. doi:10.1515/RNS.2011.008

Cunliffe VT (2016) Building a zebrafish toolkit for investigating the pathobiology of epilepsy and identifying new treatments for epileptic seizures. J Neurosci Methods 260:91–95. doi:10.1016/j.jneumeth.2015.07.015

Drapeau P, Saint-Amant L, Buss RR, Chong M, McDearmid JR, Brustein E (2002) Development of the locomotor network in zebrafish. Prog Neurobiol 68(2):85–111

Emran F, Rihel J, Dowling JE (2008) A behavioral assay to measure responsiveness of zebrafish to changes in light intensities. J Visualized Exp(JoVE) (20). doi:10.3791/923

Emran F, Rihel J, Adolph AR, Wong KY, Kraves S, Dowling JE (2007) OFF ganglion cells cannot drive the optokinetic reflex in zebrafish. Proc Natl Acad Sci U S A 104(48): 19126–19131. doi:10.1073/pnas.0709337104

Engert F (2012) Fish in the matrix: motor learning in a virtual world. Front Neural Circuits 6:125. doi:10.3389/fncir.2012.00125

Fetcho JR (1991) Spinal network of the Mauthner cell. Brain Behav Evol 37(5):298–316

Fetcho JR, O'Malley DM (1997) Imaging neuronal networks in behaving animals. Curr Opin Neurobiol 7(6):832–838

Filosa A, Barker AJ, Dal Maschio M, Baier H (2016) Feeding state modulates behavioral choice and processing of prey stimuli in the zebrafish tectum. Neuron 90(3):596–608. doi:10.1016/j.neuron.2016.03.014

Fleisch VC, Neuhauss Stephan CF (2006) Visual behavior in zebrafish. Zebrafish 3(2):191–201. doi:10.1089/zeb.2006.3.191

Granato M, Van Eeden FJ, Schach U, Trowe T, Brand M, Furutani-Seiki M et al (1996) Genes controlling and mediating locomotion behavior of the zebrafish embryo and larva. Development (Cambridge, England) 123:399–413

Grone BP, Baraban SC (2015) Animal models in epilepsy research: legacies and new directions. Nat Neurosci 18(3):339–343. doi:10.1038/nn.3934

Gross JM, Perkins BD, Amsterdam A, Egana A, Darland T, Matsui JI et al (2005) Identification of zebrafish insertional mutants with defects in visual system development and function. Genetics 170(1):245–261. doi:10.1534/genetics.104.039727

Hoffman EJ, Turner KJ, Fernandez JM, Cifuentes D, Ghosh M, Ijaz S et al (2016) Estrogens suppress a behavioral phenotype in zebrafish mutants of the autism risk gene, CNTNAP2. Neuron 89(4):725–733. doi:10.1016/j.neuron.2015.12.039

Hortopan GA, Dinday MT, Baraban SC (2010) Zebrafish as a model for studying genetic aspects of epilepsy. Dis Models Mech 3(3–4):144–148. doi:10.1242/dmm.002139

Huang Y-Y, Neuhauss Stephan CF (2008) The optokinetic response in zebrafish and its applications. Front Biosci J Virtual Libr 13:1899–1916

Hughes, Virginia (2013) Mapping brain networks: fish-bowl neuroscience. Nature 493 (7433): 466–468.DOI: 10.1038/493466a

Jirsa VK, Stacey WC, Quilichini PP, Ivanov AI, Bernard C (2014) On the nature of seizure dynamics. Brain J Neurol 137(Pt 8):2210–2230. doi:10.1093/brain/awu133

Kalueff AV, Gebhardt M, Stewart AM, Cachat JM, Brimmer M, Chawla JS et al (2013) Towards a comprehensive catalog of zebrafish behavior 1.0 and beyond. Zebrafish 10(1):70–86. doi:10.1089/zeb.2012.0861

Kermen F, Franco LM, Wyatt C, Yaksi E (2013) Neural circuits mediating olfactory-driven behavior in fish. Front Neural Circuits 7:62. doi:10.3389/fncir.2013.00062

Kimmel CB, Sessions SK, Kimmel RJ (1981) Morphogenesis and synaptogenesis of the zebrafish Mauthner neuron. J Comp Neurol 198(1):101–120. doi:10.1002/cne.901980110

Kohashi Tsunehiko, Oda Yoichi (2008) Initiation of Mauthner- or non-Mauthner-mediated fast escape evoked by different modes of sensory input. J Neurosci (Official J Soc Neurosci) 28(42):10641–10653. doi:10.1523/JNEUROSCI.1435-08.2008

Kokel D, Bryan J, Laggner C, White R, Cheung Chung YJ, Mateus R et al (2010) Rapid behavior-based identification of neuroactive small molecules in the zebrafish. Nat Chem Biol 6(3):231–237. doi:10.1038/nchembio.307

Kokel D, Dunn TW, Ahrens MB, Alshut R, Cheung Chung YJ, Saint-Amant L et al (2013) Identification of nonvisual photomotor response cells in the vertebrate hindbrain. J Neurosci (Official J Soc Neurosci) 33(9):3834–3843. doi:10.1523/JNEUROSCI.3689-12.2013

Liu YC, Bailey I, Hale ME (2012) Alternative startle motor patterns and behaviors in the larval zebrafish (*Danio rerio*). J Comp Physiol (A Neuroethology Sens Neural Behav Physiol) 198(1):11–24. doi:10.1007/s00359-011-0682-1

Maaswinkel H, Li L (2003) Spatio-temporal frequency characteristics of the optomotor response in zebrafish. Vision Res 43(1):21–30

Maaswinkel H, Zhu L, Weng W (2013) Using an automated 3D-tracking system to record individual and shoals of adult zebrafish. J Visualized Exp (JoVE) 82:50681. doi:10.3791/50681

Maaswinkel H, Zhu L, Weng W (2015) A small-fish model for behavioral-toxicological screening of new antimalarial drugs: a comparison between erythro- and threo-mefloquine. BMC Res Notes 8:122. doi:10.1186/s13104-015-1088-x

Mathuru AS, Kibat C, Cheong WF, Shui G, Wenk MR, Friedrich RW, Jesuthasan S (2012) Chondroitin fragments are odorants that trigger fear behavior in fish. Curr biol (CB) 22(6): 538–544. doi:10.1016/j.cub.2012.01.061

Maurer CM, Schonthaler HB, Mueller KP, Neuhauss Stephan CF (2010) Distinct retinal deficits in a zebrafish pyruvate dehydrogenase-deficient mutant. J Neurosci (Official J Soc Neurosci) 30(36):11962–11972. doi:10.1523/JNEUROSCI.2848-10.2010

McElligott MB, O'Malley DM (2005a) Prey tracking by larval zebrafish: axial kinematics and visual control. Brain Behav Evol 66(3):177–196. doi:10.1159/000087158

McElligott MB, O'Malley DM (2005b) Prey tracking by larval zebrafish: axial kinematics and visual control. Brain Behav Evol 66(3):177–196. doi:10.1159/000087158

Moravec CE, Li E, Maaswinkel H, Kritzer MF, Weng W, Sirotkin HI (2015) Rest mutant zebrafish swim erratically and display atypical spatial preferences. Behav Brain Res 284:238–248. doi:10.1016/j.bbr.2015.02.026

Mueller KP, Neuhauss Stephan CF (2010a) Behavioral neurobiology: how larval fish orient towards the light. Curr Biol (CB) 20(4):159–161. doi:10.1016/j.cub.2009.12.028

Mueller KP, Neuhauss Stephan CF (2010b) Quantitative measurements of the optokinetic response in adult fish. J Neurosci Methods 186(1):29–34. doi:10.1016/j.jneumeth.2009.10.020

Mueller KP, Schnaedelbach Oliver DR, Russig HD, Neuhauss Stephan CF (2011) Visiotracker, an innovative automated approach to oculomotor analysis. J Visualized Exp (JoVE) 56. doi:10.3791/3556

Muto A, Kawakami K (2013) Prey capture in zebrafish larvae serves as a model to study cognitive functions. Front Neural Circuits 7:110. doi:10.3389/fncir.2013.00110

Neuhauss SC, Biehlmaier O, Seeliger MW, Das T, Kohler K, Harris WA, Baier H (1999) Genetic disorders of vision revealed by a behavioral screen of 400 essential loci in zebrafish. J Neurosci (Official J Soc Neurosci) 19(19):8603–8615

Neuhauss, Stephan CF (2003) Behavioral genetic approaches to visual system development and function inzebrafish. J Neurobiol 54(1):148–160. DOI: 10.1002/neu.10165

Nicolson T, Rusch A, Friedrich RW, Granato M, Ruppersberg JP, Nusslein-Volhard C (1998) Genetic analysis of vertebrate sensory hair cell mechanosensation: the zebrafish circler mutants. Neuron 20(2):271–283

O'Malley DM, Kao YH, Fetcho JR (1996) Imaging the functional organization of zebrafish hindbrain segments during escape behaviors. Neuron 17(6):1145–1155

Orger MB, Baier H (2005) Channeling of red and green cone inputs to the zebrafish optomotor response. Vis Neurosci 22(3):275–281. doi:10.1017/S0952523805223039

Patterson BW, Abraham AO, MacIver MA, McLean DL (2013) Visually guided gradation of prey capture movements in larval zebrafish. J Exp Biol 216(16):3071–3083. doi:10.1242/jeb.087742

Portugues R, Engert F (2011) Adaptive locomotor behavior in larval zebrafish. Front Syst Neurosci 5:72. doi:10.3389/fnsys.2011.00072

Portugues R, Feierstein CE, Engert F, Orger MB (2014) Whole-brain activity maps reveal stereotyped, distributed networks for visuomotor behavior. Neuron 81(6):1328–1343. doi:10.1016/j.neuron.2014.01.019

Prober DA, Zimmerman S, Myers BR, McDermott BM, Rihel J, Kim SH, Caron S et al (2008) Zebrafish TRPA1 channels are required for chemosensation but not for thermosensation or mechanosensory hair cell function. J Neurosci (Official J Soc Neurosci) 28(40):10102–10110. doi:10.1523/JNEUROSCI.2740-08.2008

Raftery TD, Isales GM, Yozzo KL, Volz DC (2014) High-content screening assay for identification of chemicals impacting spontaneous activity in zebrafish embryos. Environ Sci Technol 48(1):804–810. doi:10.1021/es404322p

Richendrfer H, Pelkowski SD, Colwill RM, Creton R (2012) On the edge: pharmacological evidence for anxiety-related behavior in zebrafish larvae. Behav Brain Res 228(1):99–106. doi:10.1016/j.bbr.2011.11.041

Rihel J, Schier AF (2012) Behavioral screening for neuroactive drugs in zebrafish. Dev Neurobiol 72(3):373–385. doi:10.1002/dneu.20910

Rihel J, Prober DA, Arvanites A, Lam K, Zimmerman S, Jang S et al (2010) Zebrafish behavioral profiling links drugs to biological targets and rest/wake regulation. Science (New York, N.Y.) 327(5963):348–351. doi:10.1126/science.1183090

Rinner O, Rick JM, Neuhauss Stephan CF (2005) Contrast sensitivity, spatial and temporal tuning of the larval zebrafish optokinetic response. Invest Ophthalmol Vis Sci 46(1):137–142. doi:10.1167/iovs.04-0682

Roeser T, Baier H (2003) Visuomotor behaviors in larval zebrafish after GFP-guided laser ablation of the optic tectum. J Neurosci (The Official Journal of the Society for Neuroscience) 23(9): 3726–3734

Saint-Amant L, Drapeau P (1998) Time course of the development of motor behaviors in the zebrafish embryo. J Neurobiol 37(4):622–632

Saint-Amant L, Drapeau P (2000) Motoneuron activity patterns related to the earliest behavior of the zebrafish embryo. J Neurosci (The Official Journal of the Society for Neuroscience) 20(11): 3964–3972

Saint-Amant L, Drapeau P (2001) Synchronization of an embryonic network of identified spinal interneurons solely by electrical coupling. Neuron 31(6):1035–1046

Schnörr SJ, Steenbergen PJ, Richardson MK, Champagne DL (2012) Measuring thigmotaxis in larval zebrafish. Behav Brain Res 228(2):367–374. doi:10.1016/j.bbr.2011.12.016

Selderslaghs Ingrid WT, Hooyberghs J, De Coen W, Witters HE (2010) Locomotor activity in zebrafish embryos: a new method to assess developmental neurotoxicity. Neurotoxicol Teratol 32(4):460–471. doi:10.1016/j.ntt.2010.03.002

Selderslaghs Ingrid WT, Hooyberghs J, Blust R, Witters HE (2013) Assessment of the developmental neurotoxicity of compounds by measuring locomotor activity in zebrafish embryos and larvae. Neurotoxicol Teratol 37:44–56. doi:10.1016/j.ntt.2013.01.003

Semmelhack JL, Donovan JC, Thiele TR, Kuehn E, Laurell E, Baier H (2014) A dedicated visual pathway for prey detection in larval zebrafish. eLife 3. doi:10.7554/eLife.04878

Siebel AM, Menezes FP, Da Costa Schaefer I, Petersen BD, Bonan CD (2015) Rapamycin suppresses PTZ-induced seizures at different developmental stages of zebrafish. Pharmacol Biochem Behav 139(Pt B):163–168. doi:10.1016/j.pbb.2015.05.022

Speedie N, Gerlai R (2008) Alarm substance induced behavioral responses in zebrafish (Danio rerio). Behav Brain Res 188(1):168–177. doi:10.1016/j.bbr.2007.10.031

Stewart AM, Desmond D, Kyzar E, Gaikwad S, Roth A, Riehl R, et al. (2012) Perspectives of zebrafish models of epilepsy: what, how and where next? Brain Res Bull 87(2–3):135–143. doi:10.1016/j.brainresbull.2011.11.020

Tappeiner C, Gerber S, Enzmann V, Balmer J, Jazwinska A, Tschopp M (2012) Visual acuity and contrast sensitivity of adult zebrafish. Front Zool 9(1):10. doi:10.1186/1742-9994-9-10

Trivedi CA, Bollmann JH (2013) Visually driven chaining of elementary swim patterns into a goal-directed motor sequence: a virtual reality study of zebrafish prey capture. Front Neural Circuits 7:86. doi:10.3389/fncir.2013.00086

Von Frisch Karl (1938) On the psychology of schooling fish. vol 26, 1938th edn. Naturwissenschaft, pp 601–606

Winter M, Redfern WS, Hayfield AJ, Owen SF, Valentin J-P, Hutchinson TH (2008) Validation of a larval zebrafish locomotor assay for assessing the seizure liability of early-stage development drugs. J Pharmacol Toxicol Methods 57(3):176–187. doi:10.1016/j.vascn.2008.01.004

Wolman M, Granato M (2012) Behavioral genetics in larval zebrafish: learning from the young. Dev Neurobiol 72(3):366–372. doi:10.1002/dneu.20872

Wong K, Stewart A, Gilder T, Wu N, Frank K, Gaikwad S et al (2010) Modeling seizure-related behavioral and endocrine phenotypes in adult zebrafish. Brain Res 1348:209–215. doi:10.1016/j.brainres.2010.06.012

Chapter 9
The Use of Computational Modeling to Link Sensory Processing with Behavior in *Drosophila*

Jan Clemens and Mala Murthy

Abstract Understanding both how the brain represents information and how these representations drive behaviour are major goals of systems neuroscience. Even though genetic model organisms like *Drosophila* grant unprecedented experimental access to the nervous system for manipulating and recording neural activity, the complexity of natural stimuli and natural behaviours still poses significant challenges for solving the connections between neural activity and behaviour. Here, we advocate for the use of computational modelling to complement (and enhance) the *Drosophila* toolkit. We first lay out a modelling framework for making sense of the relation between natural sensory stimuli, neuronal responses, and natural behaviour. We then highlight how this framework can be used to reveal how neural circuits drive behaviour, using selected case studies.

9.1 The Challenge

A major goal of systems neuroscience is to understand how the brain represents information, and how those representations are used to drive behavior. Animal brains have evolved to solve particular problems, such as detecting the movements of prey or the features of a suitable mate and changing patterns of locomotion accordingly. Studying these natural behaviors allows systems neuroscientists access to the (potentially conserved) computations and neural mechanisms underlying sensory processing, decision-making, and motor control in these animal model systems.

The genetic model organism *Drosophila melanogaster* exhibits a range of robust and complex behaviors such as acoustic communication during courtship (Coen et al. 2014, 2016; Clemens et al. 2015a), detection and integration of multisensory cues to locate food sources (van Breugel and Dickinson 2014; Bell and Wilson 2016), visually guided flight control (Clark et al. 2011; Aptekar et al. 2012; de Vries and Clandinin 2012; Silies et al. 2014), and avoidance of threats (Reyn et al. 2014).

J. Clemens · M. Murthy (✉)
Princeton Neuroscience Institute, Princeton University, Princeton, NJ, USA
e-mail: mmurthy@princeton.edu

A. Çelik and M.F. Wernet (eds.), *Decoding Neural Circuit Structure and Function*, DOI 10.1007/978-3-319-57363-2_9

Recording *Drosophila* behaviors can be done in high-throughput, and its numerically compact and largely hardwired nervous system facilitates identifying the underlying neurons. In combination with an unparalleled genetic toolkit for manipulating and monitoring neuronal activity during behavior, flies represent an ideal model system for solving neural mechanisms that link sensory processing with behavior.

Here, we advocate for the use of computational modeling to complement (and enhance) the *Drosophila* toolkit. Modeling, in particular, allows one to make sense of the highly complex relation between natural sensory stimuli, neuronal responses, and natural behavior. For instance, for natural behaviors, both sensory stimuli and motor outputs can vary over multiple temporal and spatial scales and the underlying transformations can be highly nonlinear—this presents a challenge for determining what aspects of the sensory world drive a particular response. This is especially acute for social interactions, in which the behavioral output of one animal constitutes the most relevant sensory stimulus for the other animal.

In addition to illuminating the relation between sensory stimuli and behavior, computational modeling can also reveal how neurons represent behaviorally relevant stimuli and how these neural representations are read out to produce appropriate motor patterns—in short, how sensory information is encoded and decoded by the brain. Ideally, to solve these relationships, neural activity is recorded in behaving animals. Correlated neuronal and behavioral variability can then be exploited for inferring relationships between the two (Britten et al. 1996; Parker and Newsome 1998), and neuronal *decoding* models can identify the dynamical and nonlinear relationship between neuronal codes and behavior (Haefner et al. 2013). Recent advances have now facilitated combining neural and behavioral recordings in head-fixed *Drosophila* (Seelig et al. 2010; Kim et al. 2015), but head-fixing can limit the behavioral repertoire of the animal, particularly for social behaviors. In these cases, modeling is particularly useful since it permits linking data sets from separate experiments—neural recordings from fixed animals with behavioral recordings from freely moving animals. Neural *encoding* models can be derived from recordings in fixed, non-behaving animals—these models can then be used for predicting neural responses to the sensory stimuli from behavioral data sets. Thus, computational models can serve as stand-ins for recording neural activity during behavior and thus facilitate overcoming experimental hurdles when linking neural codes and natural behaviors (Parnas et al. 2013; Schulze et al. 2015; Clemens et al. 2015a; Badel et al. 2016). Here, we detail this approach using data from *Drosophila* (both adults and larvae), and we discuss selected studies that highlight both the challenges and advantages associated with computational modeling in this model system.

9.2 The Approach

Understanding how the brain generates behavior in response to sensory stimuli involves bridging three different levels of description and can conceptually be separated into three steps (Fig. 9.1a):

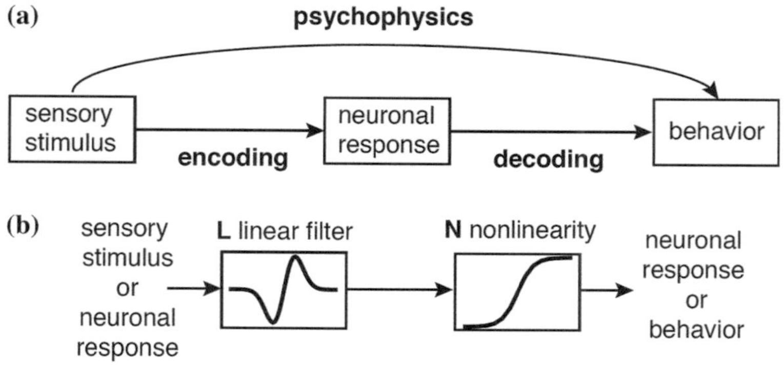

Fig. 9.1 **a** Linking sensory stimuli, neuronal responses and behavior can conceptually be separated into three steps: First, linking sensory stimuli and behavior to identify the stimulus features driving behavior (psychophysics). Second, linking sensory stimuli and neuronal responses to describe how behaviorally relevant stimulus features are represented in the brain (encoding). Third, linking neuronal responses with behavior to understand how neuronal representations are read out and transformed into behavior (decoding). **b** Linear-nonlinear (*LN*) models are useful in all three steps, since they describe the input–output transformation with two computational steps: First, a linear filter (*L*) acts as a template that is compared to the input. In a second step the filtered input is transformed to the output using a fixed nonlinearity (*N*), which can mimic a threshold or saturation in the input–output transformation

1. Linking stimuli and behavior ("psychophysics") to identify the stimulus features driving behavior.
2. Linking stimuli and neural codes ("encoding") to determine how the behaviorally relevant stimulus features are represented in the brain.
3. Linking neural codes and behavior ("decoding") to reveal how neuronal activity is read out and transformed into behavior.

Interestingly, a single modeling framework has been successfully employed in all three steps, so-called linear-nonlinear (LN) models (Fig. 9.1b). This class of models has been largely developed in the context of linking stimuli and neural codes (step 2 above) (e.g., Schwartz et al. 2006; Pillow et al. 2008; Sharpee 2013; Aljadeff et al. 2016), but has also recently found applications in predicting behavior from both sensory stimuli and neuronal responses (e.g., Kato et al. 2014; Coen et al. 2014; Schulze et al. 2015; Clemens et al. 2015a). LN models treat the brain as a black box, i.e., they reduce the complex action of neural networks, individual neurons, and ion channels to a sequence of relatively simple computations. In this light, it is surprising this framework works as often as it does for characterizing neural mechanisms.

In general, LN models describe the transformation from one or more temporally or spatially varying inputs to an output in two computational steps: a linear filter and a nonlinearity (Fig. 9.1b). In the first step, a linear filter h acts as a template that is compared to the input s and thus constitutes the stimulus pattern most effective in driving the output s'. Mathematically, the comparison between the filter and the input is implemented through "convolution", where the input at each point in time (τ) and/or space (x, y) is multiplied with the template and then summed. By making time explicit in the filter, the LN model can thus capture complex dynamical relationships.

$$s'(t) = \sum_{x,s,\tau} s(x, y, t - \tau)\,\mathrm{h}(x, y, \tau)$$

In the second step of the LN model a static nonlinearity (NL) maps the filtered stimulus to the output—the neuronal or behavioral response (Chichilnisky 2001; Schwartz et al. 2006; Aljadeff et al. 2016). This part mimics properties of the input–output relationship such as thresholds (describing the minimally effective stimulus required to produce a response) or response saturation.

This separation of an input–output transformation into a dynamical linear and a static nonlinear part is a powerful feature of LN models. It facilitates model identification from often limited and noisy experimental data as well as the interpretation of model structure. For example, the linear filter helps identify spatial and dynamical properties of an input driving the response (Fig. 9.2a–c) (Suh and Baccus 2014). A unilobed linear filter indicates that integration or smoothing

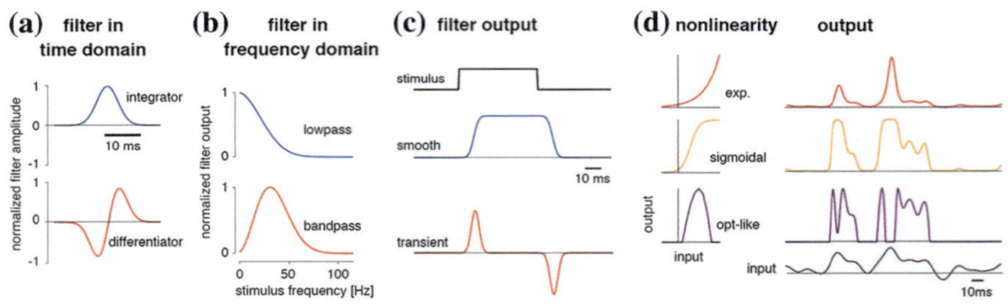

Fig. 9.2 **a** The action of linear filters can be best understood as lying between two extremes: a unilobed filter with integrating properties (*blue*) and a bi-lobed filter with differentiating properties (*red*). Typically, experimentally derived filters constitute a mixture between these purely integrating and purely differentiating filters. **b** The frequency domain representation, obtained by applying a Fourier transform to the filters in (**a**), reveals that the two filters have fundamentally different frequency transfer properties: integrating filters (*blue*) mainly let pass low frequencies, while differentiators (*red*) are selective for a limited range of frequencies, rejecting both very low and very high frequencies. **c** The response properties of integrating and differentiating filters become obvious when looking at responses to a step-like input (*black*): integrators sum up the input over their duration and reject high frequencies associated with sharp edges in the stimulus, effectively smoothing the stimulus. By contrast, differentiators only encode stimulus changes but not sustained stimulus epochs and hence only respond transiently during the step's on- and offset. **d** The nonlinearity (NL) in an LN model transforms the filtered input (*bottom*) to the output. An accelerating NL, e.g., an exponential function (*red*), amplifies strong inputs and accentuates stimulus differences. By contrast, a saturating nonlinearity, e.g., a sigmoidal function (*orange*), compresses large values and thereby reduces differences in the input. An optimum-like, bell-shaped NL (*purple*) produces outputs only for intermediate input values

underlies the response. By contrast, a biphasic filter—with both a positive and a negative lobe—has differentiating properties, indicating that the response relies on transients in the stimulus. More complex filters like Gabor filters, which correspond to short sinusoidals whose amplitude is modulated by a Gaussian, have bandpass properties—confining responses only to a narrow range of the input spectrum (Clemens and Hennig 2013).

The nonlinearity in the LN model corresponds to a tuning curve for the filter—or the selectivity of the response for filter outputs. A monotonically increasing NL indicates that response magnitude is determined by stimulus magnitude, while a unimodal NL indicates that only a limited range of filter outputs drives the response (Fig. 9.2d). Separating input–output transformations thus helps to discriminate changes in overall response gain from changes in stimulus selectivity. For example, motivation, context, or attention can change the overall gain of responses—and therefore affect the nonlinearity—without changing the feature selectivity of the response represented by the filter (Baccus and Meister 2002; Rabinowitz et al. 2011; Clemens et al. 2015b).

How can one identify the two components of an LN model from experimental data? Usually, linear filters and nonlinearities are estimated in separate steps. The standard approach for estimating the first stage of the model—the filter—is by averaging all stimuli leading to a response (Fig. 9.3a–b). This is most straightforward for LN models of neuronal responses: the so-called spike-triggered average (STA) is computed by aligning the stimuli preceding all spikes and averaging them.

$$c_{STA}(\tau) = \sum_{\{t_{\text{event}}\}} r(t_{\text{event}}) s(t_{\text{event}} - \tau)$$

where r is a binary vector that is 1 if $t = t_{\text{event}}$ and 0 otherwise, s is the stimulus, τ is the temporal delay and c_{STA} is the STA filter. This can also be applied to behavioral events (in lieu of spikes) like the onset of an escape behavior (Reyn et al. 2014) or the individual pulses of a fly song (Coen et al. 2016). A simple generalization enables applying the STA approach to continuous outputs like neuronal calcium levels or the amplitude of the animal's communication signal (Kato et al. 2014; Coen et al. 2016). In these cases, the filter c is estimated by weighting the stimulus by an analog response value r.

$$c(\tau) = \sum_{t} r(t) s(t - \tau)$$

The STA is the special case in which the response r is either 0 (no event occurred) or 1 (event occurred). The advantage of this approach is that the filter can be directly computed from stimulus response pairs, without the need for computationally expensive optimization procedures. While the STA has intuitive appeal and is useful for visualizing the relation between input and output, it provides an accurate estimate of the stimulus feature driving the response only for inputs with uncorrelated structure (such as white noise) (Paninski 2003; Sharpee 2013).

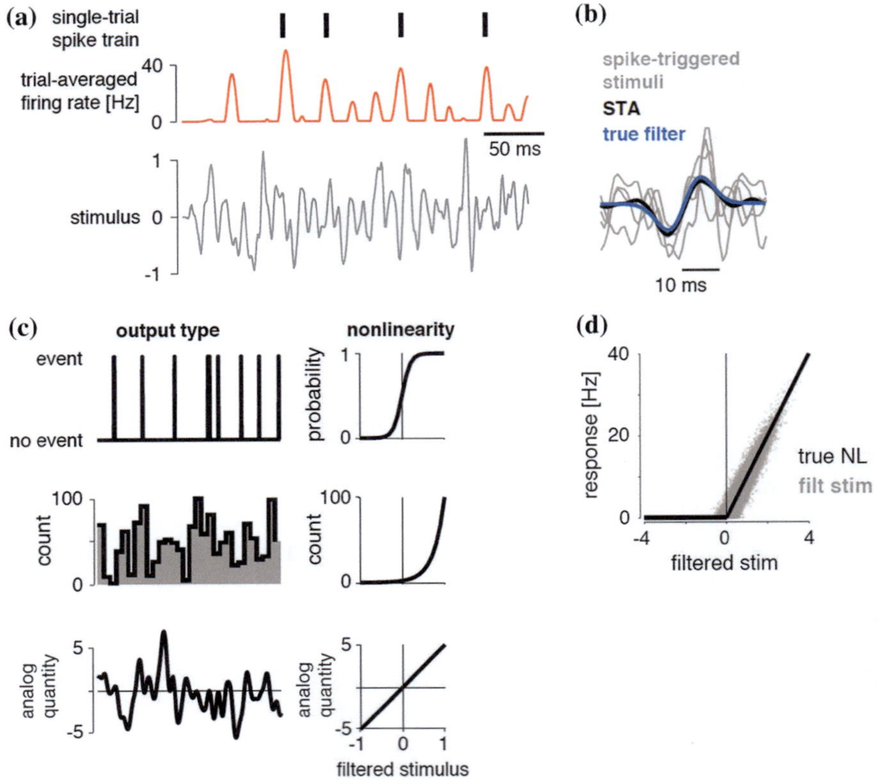

Fig. 9.3 **a** To illustrate the process of identifying an LN model from experimental data, we generated a trial-averaged firing rate (*red*) and a sparse spike train (*black lines, top*) from a white noise stimulus (*gray*) using an LN model with a bandpass filter and a simple, threshold-linear nonlinearity. **b** Extracting all stimuli preceding a spike yields the so-called spike-triggered stimulus ensemble (*STE gray*, shown for the 4 spikes in A, spike is to right—at the end—of the stimulus snippet). The average of this STE is the spike-triggered average (*STA black*, average over 200 spikes), which resembles the true filter (*blue*) used for generating the spikes. **c** The nonlinearity (NL) in the LN model can be obtained using two approaches. In the GLM approach, the nonlinearity is fixed and only the filter can vary. The specific form of the NL is usually dictated by the type of output to be predicted. For binary events (e.g., spike/no spike), the NL is a logistic function. For event rates (e.g., firing rate), the NL is an exponential function. For continuous outputs with positive and negative values the standard NL is usually a linear function. **d** LN models can also be estimated using a free-form NL. One way of identifying the NL in these cases is by filtering the stimulus with the STA and plotting the filtered stimulus versus the actual response (*gray*). The relationship can then be binned and averaged or fitted using a suitable parametric function. For the example shown here, the resulting plot recovers the thresholding NL (*black*) used for generating the response in **a**

However, natural stimuli are highly structured and characterized by higher-order correlations associated with edges in visual scenes or transients in acoustic signals (Simoncelli and Olshausen 2001; Machens and Zador 2003).

Alternatives to the STA method have been developed that permit identifying the filter without the strong requirement of uncorrelatedness of input statistics (Truccolo et al. 2005; Pillow et al. 2008; Sharpee 2013; Williamson et al. 2015; Aljadeff et al. 2016). These methods find the filter by directly minimizing the mismatch between the prediction of the model and the experimentally measured response using numerical optimization techniques in which the filter shape is modified gradually over many iterations to improve model fit. Efficient implementations of the optimization procedure are standard in all common statistical analysis packages and often include a procedure termed *regularization*, which involves adding a penalty term to the error function that is minimized to find the optimal filter. The penalty biases the filter to have desired or biologically "reasonable" properties. For instance, stimulus correlations lead to filter structure that does not contribute to model fit, but rather impede interpretation. By penalizing filters with many non-zero or large values, the impact of stimulus correlations can be minimized, greatly facilitating the interpretation of filter structure and improving model performance for stimuli with different correlation structure (Mineault et al. 2009; Park and Pillow 2011).

Generalized linear models (GLMs) are a powerful implementation of this approach, in which filter shape is free to vary, but the nonlinearity is constrained to a specific parametric form (Aljadeff et al. 2016)—most commonly the NLs used are logistic, exponential or linear (Fig. 9.3c). Pre-defining the type of nonlinearity in the second stage of the LN model reduces the number of free model parameters and hence the amount of data needed for fitting the model, making GLMs especially suitable when data is scarce or noisy. However, in some cases it may be necessary to fit a model with an unconstrained nonlinearity. In these cases, the filtered stimulus is usually plotted against the actual responses (Fig. 9.3d) and the relationship between the two is estimated either by binning and averaging or by fitting a suitable parametric function (Chichilnisky 2001).

9.2.1 Model Evaluation and Selection

Model performance is usually quantified as the match between actual data and model predictions, using a quantity such as the mean-square error or the correlation coefficient. For GLMs that predict probabilities, models should be evaluated based on the model likelihood (see Aljadeff et al. 2016 for details). Alternatively, when the output is the probability of a binary event—e.g., whether a spike will occur or not—one can predict events from the model output by setting a decision threshold (e.g., if $p > 0.5$ generate the event, otherwise generate no event) and use methods from signal detection theory (e.g., the percentage of correct predictions) to assess model performance (Coen et al. 2014). However, care needs to be taken when the event to be predicted is extremely rare—e.g., when only 1% of the inputs induce an event. A model that always predicts non-events would be correct in 99% of the cases. To improve model fit, such data sets should be "balanced", by selecting random subsets of data for fitting and evaluation such that events and non-events occur with equal frequency (Coen et al. 2014).

If a model has many free parameters it can reflect properties of the data set used for fitting, not the general input–output relation of the system under study. The model performance calculated from the same data used to train the model can thus greatly overestimate the model quality. The impact of this so-called "overfitting" is usually assessed via cross-validation—computing model performance with data not used for fitting (e.g., Schönfelder and Wichmann 2012; Aljadeff et al. 2016). Practically, this is done by splitting the data into a training set for estimating model parameters (usually 80% of data) and a test set for quantifying model performance (usually 20%). This split can be randomized and repeated—parameters and performance values from different runs can be averaged and their variance estimated to statistically compare the parameters or performances of models obtained for different experimental conditions.

The ability to statistically compare models is essential for model selection—a method of selecting the simplest model that can explain the response. Generally, there exists a trade-off between model complexity and performance—a model with many parameters usually performs better than a simpler one. Several measures of model fit—such as the Akaike or the Bayesian information criterion (Zucchini 2000)—approach this problem by penalizing models with many parameters. In simple cases, one can select models based on a simple cutoff criterion—e.g., that a more complex model needs to outperform the simple one by at least 10% (Coen et al. 2014).

9.2.2 Extensions of the Standard Linear-Nonlinear Model

The relatively simple, standard LN model with a single linear filter and a static nonlinearity can be extended to account for more complex input–output transformations. Several extensions explicitly take into account possible nonlinear interactions between multiple filters operating on a single stimulus, like spike-triggered covariance analysis (Schwartz et al. 2006) or generalized quadratic models (Rajan et al. 2013). However, with greater model power comes the need for more data to identify model parameters, rendering these methods impractical when data are limited. Often, neuronal or behavioral responses are non-stationary or dependent on internal factors (e.g., behavioral states). While the standard GLM framework can still accommodate these cases (Pillow et al. 2008; Fründ et al. 2014), more complex types of non-stationarities affecting the stimulus response mapping can be modeled by combining LNs with hidden Markov models (HMM), in which the HMM models the switching between states and one LN model per state captures state-specific input–output transformations (Escola et al. (2011); see Wiltschko et al. (2015) for a related approach).

Another feature of neural computations is that they often bridge several time scales. For instance, the decision to engage in a behavior is often based on the detection of a stimulus on short time scales (milliseconds) while the resulting motor output evolves on much longer time scales (seconds or minutes). Models that exploit this separation of time scales usually contain an explicit integration stage. This approach has been applied successfully to study the behavioral evaluation of courtship songs (Clemens and Hennig 2013; Clemens et al. 2015a). When extended with a behavioral threshold, such models resemble so-called "drift diffusion models", which integrate sensory information—in this case the output of the LN model—to a decision boundary, the crossing of which corresponds to decision-making (Brunton et al. 2013; DasGupta et al. 2014; Clemens et al. 2014).

9.3 Linking Sensory Stimuli with Behavior

Fitting a model to link sensory cues and behaviors requires measuring (1) the putatively relevant sensory stimuli and (2) the behavioral output to be predicted. A number of studies using *Drosophila* have examined the impact of visual, chemosensory, or thermal cues on navigational decisions (Gomez-Marin et al. 2011; Clark et al. 2011; Censi et al. 2013; Klein et al. 2014; Gepner et al. 2015; Schulze et al. 2015). In these studies, the detailed recording of fly movement parameters with a quantitative, model-supported analysis of the behavior revealed the navigational strategies with which *Drosophila* negotiates its environment to find food and avoid noxious chemical, heat, or light stimuli. For example, Censi et al. (2013) examined the visual features driving navigation in adult *Drosophila* during free-flight in a circular arena. They tracked the position and angle of flies in 3D and estimated the stimulus features—given by the distance and angle from the arena's wall—that triggered rapid changes in direction (or saccades). They found that flies turn away from the wall when approaching it, e.g., they tended to turn right when the wall was to their left. Given that their model could predict more than 90% of the saccades, the authors concluded that the majority of flight decisions in their arena were driven by visual cues. Gomez-Marin et al. (2011) used behavioral manipulation and computational modeling to show how *Drosophila* larvae find the source of an attractive odorant. The first step in understanding the computations underlying this behavior was to reconstruct the larva's sensory environment—in this case the concentration of odorants at the chemosensory organ in the head. This was done, by tracking the larval head position and determining—either through measurement or through physical modeling—the local concentration of odorants the larva receives at each moment in time. This highlights the importance of reconstructing the stimulus relative to the animal's frame of reference. An animal's sensory experience depends on its position relative to the stimulus source, on how the stimulus propagates through the medium, and on the properties of their sense organs (e.g., their directionality). *Drosophila* larval behaviors can easily be classified into three relatively stereotyped motor programs during chemotaxis: phases of straight locomotion (termed "runs"), intense side-to-side movement of the head ("head casting"), and fast reorientation events ("turns"). Models that explain the relation between chemosensory cues and navigation behaviors usually use the rate or probability of performing one of these behaviors as a function of the stimulus (Gomez-Marin et al. 2011; Gepner et al. 2015; Schulze et al. 2015; Hernandez-Nunez et al. 2015). Based on an accurate description of the larva's sensory experience as well as of its complete body posture during navigation, the authors determined the features of the odor concentration gradient that triggered turning and head casting using STA-like analyses. This revealed the behavioral program controlling navigation: decreases in concentration stop a run and induce head casting. Upon realizing that it has moved away from the source, the larva actively samples the concentration gradient by casting its head, which translates weak spatial differences in concentration into strong temporal differences. When perceiving an increase in concentration during the head cast, the larva initiates the turn, therefore reorienting it towards higher concentration. This study exemplifies the power of modeling for revealing how different behavioral modules—running, head casting, and turning—are engaged in a stimulus-dependent manner to support successful navigation behavior.

Modeling is also useful to characterize sensorimotor relationships during more complex behaviors, like social interactions. During courtship, *Drosophila* males chase females and produce a song by vibrating their wings. The song's pattern is evaluated by the female and informs her mating decision (Coen et al. 2014; Clemens et al. 2015a). Coen et al. (2014) investigated the origin of the male's highly variable song pattern. The song consists of two, relatively stereotyped modes—"pulse" and "sine"—which are sequenced into variable bouts, stemming from the fact that the duration of each song mode is not fixed and that males rapidly transition between the two modes. Coen et al. (2014) asked whether the variable bout structure is purely stochastic—e.g., due to noise in the song-producing circuits—or whether it is the outcome of the male dynamically and adaptively shaping song structure in response to sensory cues. The authors tracked the movement of the male and female, along with recording the song produced by the male—the movements and interactions of the pair served as inputs to a GLM with regularization. The regularization during model fitting (Mineault et al. 2009) helped the authors reduce the impact of strong correlations among different inputs, in addition to minimizing temporal correlations within a single input. This model-based analysis of the behavior revealed that almost every aspect of song patterning—song starts, the choice of song mode, or song ends—are predicted by the movements and interactions of the flies. For instance, the distance between male and female is most predictive of song starts: the filter that predicts whether or not a song start will occur reveals that the male only begins a song in sine mode when near the female. In a follow-up study, using the same GLM method, the authors showed that not only do males bias toward pulse mode (the louder song mode) when far away from the female, but they also modulate the amplitude of each pulse within this mode relative to distance to the female (Coen et al. 2016). Overall, the model-based analysis of courtship interactions revealed that the diversity present in song structure results from the male dynamically and adaptively controlling when and what to sing in response to changing feedback from the female.

These three case studies showcase that modeling can facilitate extracting information from large and complex behavioral data sets. In providing a detailed and quantitative description of the mapping between the animal's sensory experience and its behavior, these studies (1) reveal the computations that govern behavior and (2) lay the groundwork for identifying the neural circuits and neural codes underlying the behavior.

9.4 Linking Sensory Stimuli with Neural Responses—Neuronal Encoding Models

How are behaviorally relevant stimulus features represented by the brain? Modeling—especially the use of LN models—has a long tradition in characterizing neural responses to sensory stimuli for many model systems and sensory modalities (see e.g. Schwartz et al. 2006; Aljadeff et al. 2016). In the context of *Drosophila* systems neuroscience, LN models have provided insight into the dynamical stimulus features driving neuronal responses in the olfactory system (Nagel and Wilson 2011; Martelli et al. 2013), visual system (de Vries and Clandinin 2012; Seelig and Jayaraman 2013; Behnia et al. 2014), and auditory system (Clemens et al. 2015a). These LN models provide information about what sensory information is available to the brain for driving behaviors and in which format it is represented. Moreover, the LN model of a neuron—obtained from responses recorded in fixed animals, for example—can serve as a surrogate for neural recordings when it is not feasible to record neural activity during behavior. This is particularly useful in organisms like *Drosophila* with genetically and morphologically identifiable cell types, whose response properties are stereotyped in most, but not all (e.g., Murthy et al. 2008), cases.

As an example, (Clemens et al. 2015a) used an LN model extension to characterize the encoding of sound in the fly brain. The authors recorded intracellularly from a set of 10, morphologically diverse neurons from early stages of the fly's auditory pathway: the antennal mechanosensory and motor center (AMMC), which is the projection area of the auditory receptor neurons, and the wedge and the ventrolateral protocerebrum, which receives its main auditory inputs from the AMMC. Sound is largely encoded via graded membrane voltage (Vm) responses, not spikes. The authors hence fit LN models to the neuronal membrane voltage changes. A standard LN model was not sufficient to reproduce the Vm traces of these neurons due to prominent adaptation, and hence the authors used an extension—a so-called "adaptive LN model"—in which the stimulus is first processed by an adaptation stage and the adapted stimulus is then fed into a standard LN model with a filter and a nonlinearity. Crucially, the model, which was fitted to neuronal responses for artificial song pulse trains, generalized well to predict Vm responses to natural courtship song, suggesting that these computational steps are sufficient to describe the neural code for song processing in these auditory neurons. While adaptation parameters were relatively diverse, reflecting the diversity of adaptation time scales and strengths for the neurons in the data set, the shapes of the linear filters and nonlinearities were similar across the models of all 10 cell types, which is surprising given the large morphological diversity of the neurons in the data set. The linear filters were biphasic—with a dominant positive lobe followed by a shallow but long negative lobe. The positive lobe conveys low-pass filter properties—smoothing the stimulus envelope—while the negative lobe has a weakly differentiating effect—it accentuates transients like the onsets or offsets of song bouts (Fig. 9.2a–c). Analysis of female behavioral responses to male song during natural courtship revealed that female slowing—a proxy of her willingness to mate—was strongly correlated with the duration of song bouts over timescales of tens of seconds. The temporal properties of the neuronal code for song in the auditory system are well-suited for extracting information about bout structure. Bout starts are encoded in positive Vm transients due to adaptation, negative Vm transients encode bout ends due to the negative filter lobe, and the amount of song is represented in sustained responses during bouts due to the positive filter lobe.

Schulze et al. (2015) used models to show how responses of larval olfactory sensory neurons (OSN) encode the temporal pattern of odorant concentration as encountered during chemotaxis. Instead of using a linear-nonlinear model to represent the dynamical codes underlying concentration coding in these OSN, they employed a model of the signal transduction cascade involved in the transformation of odorant concentration to OSN firing rate. Similar models have been used to model neural codes in other chemosensory systems (Kato et al. 2014) or in the vertebrate retina (Ozuysal and Baccus 2012). This encoding model revealed the temporal stimulus features represented in the time-varying OSN firing rate: for positive concentration gradients, OSNs encode the slope; for negative gradients, OSNs act as offset detectors. These temporal features of olfactory stimuli are known to guide chemotaxis behaviors (see above, Gomez-Marin et al. 2011): the duration of runs as well as the initiation and direction of turns are guided by positive slopes in concentration, while stopping and head casting occur upon abrupt decreases of concentration. Again, the properties of the dynamical codes in OSNs directly support navigation behavior. While these two studies highlight the power of careful quantification of behavior for interpreting neural codes, these interpretations must ultimately be tested by linking neural codes directly with behavior via decoding models.

9.5 Linking Neural Responses with Behavior—Neuronal Decoding Models

Having identified both the sensory cues driving behavior and how these cues are encoded by the brain, the next task is to determine how neuronal codes are transformed into behavior. Modeling helps (1) to infer the brain's readout strategies and (2) to identify the properties of neural codes most relevant for driving a particular behavior. As laid out in the Introduction, the neural responses for fitting decoding models can be directly recorded during the behavior, or, if not feasible, obtained either by replay of the animal's sensory experience while recording from a fixed preparation or by generating surrogate neuronal responses through virtual recordings from neural encoding models.

Several studies of *Drosophila* have employed decoding models to investigate the relationship between chemosensory codes and behavior (Parnas et al. 2013; Gepner et al. 2015; Hernandez-Nunez et al. 2015; Badel et al. 2016; Bell and Wilson 2016). For example, (Hernandez-Nunez et al. 2015) asked how the activity of gustatory receptor neurons (GRN) affects chemotaxis behavior. Optogenetic activation with temporal white noise patterns and STA analyses revealed the filters underlying the transition from run to turn behavior (and from turn to run behavior). Comparing the STA filters obtained from single-GRN activation experiments to those for multi-GRN activation experiments revealed that the magnitude as well as the shape of the filters for multi-GRN activation could not be predicted from a simple, linear combination of single-GRN STA filters. This demonstrated that gustatory information is integrated nonlinearly between individual GRNs in the context of behavior.

Modeling was also used to infer algorithms underlying the integration of sensory cues across modalities. Gepner et al. (2015) studied how visual and chemosensory cues are integrated during larval orientation behavior. Visual stimuli were provided using blue light and chemosensory neurons were activated optogenetically (via csChrimson, Klapoetke et al. 2014) using red light, which is outside the sensitivity range of the photoreceptors. Stimulation with temporal white noise patterns allowed the authors to infer the features driving behavioral responses (turn rates) for light and fictive chemosensory stimulation via STA analyses. This revealed that visual cues increased turn rates, while attractive chemosensory cues decreased turn rates. Using a decoding approach, they then determined how these two opposing cues that feed into the same motor output (turn rate) are combined. Specifically, they tested whether the two modalities were combined using "early integration" versus "independent pathways". That is, whether the filtered signals from each modality were linearly combined to yield a single "turn rate" signal, or whether each modality generated its own, independent turn rate, which was then summed. Their modeling approach lent strong support to the early integration model.

Other studies have examined odortaxis in adult *Drosophila* and used decoding models to determine how the neural population code in the fly's antennal lobe (AL) is read out by downstream circuits. Each glomerulus in the AL receives exclusive input from OSNs expressing a single odorant receptor. Second-order projection neurons (PN) typically innervate single glomeruli and link the AL to higher-order olfactory centers in the lateral horn (for innate behaviors) and the mushroom body (for learned behaviors) (Jefferis et al. 2007; Aso et al. 2014). Similar to Hernandez-Nunez et al. (2015), Bell and Wilson (2016) asked how glomerular activation is translated into walking behavior to odors (odortaxis). Specifically, they tested whether the behavioral effects of multiple OSN activation sum linearly or nonlinearly. Interestingly, responses to pairwise activation frequently did not match the sum of the responses to the activation of the individual OSN in a pair. Instead, the responses to pairwise OSN activation often equaled the responses to the stronger OSN in the pair, suggesting a "max" pooling of OSN activity by downstream circuits. In a similar study, Badel et al. (2016) used odorant stimulation—not optogenetic activation of individual OSNs—to directly link naturalistic glomerular codes for odors to odortaxis in flying *Drosophila*. They showed that the glomerular code is relatively linear, since the neuronal responses to mixtures can be predicted from the responses to their constituents. This was consistent with the observation of linearity in the behavioral responses to mixtures. However, by examining behavioral responses to odors presented in sequences, they found that odortaxis over longer timescales depends on the olfactory "context"—the other odors in the sequence. This context effect can be strong and even switch the valence of an individual odor from attractive to aversive and is inconsistent with a linear glomerular code. The authors then used an extension of their linear decoding model that includes canonical computations—mean subtraction and divisive normalization—to reproduce the context effect. These computations are likely implemented in higher-order olfactory areas downstream of the AL glomeruli.

The models mentioned above infer behavioral readout strategies from a *static* description of neuronal population activity—but modeling can also facilitate reading out *dynamical* neuronal responses. For example, Schulze et al. (2015) used a nonlinear dynamic encoding model of *Drosophila* larval OSNs (see above) to relate neural responses in OSNs to behavioral dynamics (time-varying turn rates). The decoding model used was a GLM and consisted of a single linear weight followed by a sigmoidal transform. To determine whether OSN dynamics were essential for driving behavioral dynamics, the authors predicted behavior either directly from the stimulus or from model OSN responses. Taking into account OSN dynamics greatly improved behavioral predictions, suggesting the neuronal dynamics in the sensory periphery strongly shape behavioral dynamics.

In the Clemens et al. (2015a) study (see above), the authors discovered that female slowing during courtship, a proxy of her willingness to mate, was strongly associated with long timescale features of the male's courtship song—in particular, the average duration of song bouts over timescales of tens of seconds. The authors then used a decoding model to link the neuronal responses to song in the auditory system with the female slowing response. The highly interactive nature of the courtship chase prevented recording neural activity in the auditory system during behavior and the authors instead relied on an encoding model to reconstruct the neural representation of song recorded during courtship. The encoding model (see above) consisted of two computations: an adaptation stage, which produced positive transients at the start of each song bout, and a linear filter with a positive and a negative lobe, which produced sustained neural responses during a bout and negative neural responses at the end of each bout. Using a decoder, the authors asked: (i) How is this neuronal representation of bout structure read out? and (ii) What encoder computations are crucial for producing the behavior? They started with a decoder that transformed the reconstructed neuronal response to female speed in two computational steps. First, a sigmoidal nonlinearity transformed the neuronal responses and mimicked common nonlinearities in putative downstream circuits like thresholds or saturation. Second, an integration stage linked the timescales of neuronal computations in the auditory system (a few hundred milliseconds) to the long behavioral timescales (tens of seconds). By manipulating the encoding model, the authors determined the computations in the auditory system essential for reproducing the behavior. Although both adaptation and biphasic filtering were necessary to reproduce the neuronal responses to song in the auditory system, the behavior mainly relied on aspects of the neuronal response associated with biphasic filtering—sustained neural responses during a bout and negative neural responses at the end of each bout. While the integral of the sustained response corresponded to the amount of song, the integral of the negative offset response encoded the number of bouts (since each bout in the integration time window produced the same, stereotypical offset response). Thus, the decoder predicted behavior by combining song amount and bout number. Because bout duration, which the behavioral analysis had identified as being most strongly associated with the female's slowing behavior, is given by the ratio of these two quantities, the authors modified their decoder to directly compute bout duration from the neuronal responses. This decoder almost perfectly reproduced the behavioral relation between bout duration and female speed during natural courtship. Overall, the decoding analysis revealed (1) the aspects of the neural representation of song in the auditory system likely to be essential for generating behavior (in this case, biphasic filtering) and (2) probable computations (such as integration and division) that transform neural codes in the auditory system into behavior. The approach highlights how computational modeling can both help to overcome the experimental difficulties associated with recording neuronal activity during interactive, social behaviors and to generate hypotheses regarding the computations underlying sensorimotor transformations.

9.6 Experimental Tests of Computational Models

The above examples reveal how modeling can provide insight into the neural computations that transform natural stimuli into behavior. However, to establish causality, models should be tested experimentally. Experimental tests can involve manipulations of sensory stimuli, of known sensory pathways, or of the activity of subsets of neurons. This is greatly facilitated in *Drosophila* via genetic tools for activating or inactivating genes or neuronal activity during behavior. For instance, based on behavioral analyses, Ramdya et al. (2015) posited that collective behavior in *Drosophila* relies on mechanosensory cues. While individual flies only poorly avoid aversive odorants, groups of flies, by bumping into each other, "push" each other out of the aversive odor zone. The authors genetically activated and inactivated leg mechanosensory neurons to demonstrate the sufficiency and necessity of mechanosensory cues for this collective behavior. That is, inactivation abolished the behavior while activation was sufficient to induce flies to move. Similarly, the decoding model by Badel et al. (2016) (see above) linked the olfactory population code in the fly's antennal lobe to odortaxis and predicted that individual glomeruli have only a small impact on navigational decisions. Consistent with this notion, they found that inactivating single glomeruli negligibly changed behavioral responses.

Another strategy for testing models fitted to natural behavior is to use artificial, controlled stimuli. This approach has a long tradition for testing models of visual motion-processing in *Drosophila*, where models are tested by monitoring the behavioral responses of tethered flies to artificial visual stimuli (Eichner et al. 2011; Fitzgerald et al. 2015; Leonhardt et al. 2016). This strategy was also employed by Coen et al. (2016), who used GLMs to identify the distance between males and females during courtship as the most significant predictor of the male's song amplitude. The authors tested their model using a virtual reality setup in which the visual cues available to the male could be precisely manipulated. The fly was tethered and allowed to walk on an air-suspended ball in front of a computer screen. Artificial visual stimuli—a black square moving on a white background with movement statistics similar to those encountered during courtship by the male—elicited song amplitude modulation as a function of the size of the square, a visual feature strongly correlated with distance. Similarly, van Breugel and Dickinson (2014) used stimulus manipulations to test the behavioral observation that olfactory and visual stimuli interact during odortaxis in free flight. Specifically, they placed high-contrast visual cues within a flight tunnel to show that attractive olfactory stimuli increase the visual saliency of objects.

In the light of correlations between sensory cues, models derived from natural behaviors often have difficulties discriminating between internally and externally generated cues (Censi et al. 2013). In Coen et al. (2014), GLMs predicted an association between male forward velocity and song mode—faster males were more likely to produce the so-called "pulse song" versus "sine song". However, the nature of the sensory cue associated with male forward velocity was unclear. Was the cue the optic flow generated from the male's own motion? Was it a signal internally generated, like an efference copy of the motor drive or proprioceptive feedback from the muscles or joints? Using flies carrying mutations that ablate all photoreceptor cells, the authors found that models built on blind male data performed similarly to models built on wild type male data—in both cases, the male's own motion effectively predicted song mode choice. To more explicitly test for a link between male locomotor circuits and song patterning, the authors fixed the male flies in place and induced singing by optogenetically activating song command neurons. Comparing the optogenetically induced song of fixed versus freely moving males, they found that males that cannot move produced more "pulse song", demonstrating that interfering with locomotion alters singing. These examples highlight how experimental tests reduce ambiguity regarding which sensory cues and neural circuits drive behavior.

9.7 Prospects

The case studies cited above highlight how the use of computational modeling powerfully complements the *Drosophila* genetic toolkit for solving open questions in systems neuroscience. We expect that methodological innovations—both experimental and theoretical—will further increase the utility of computational modeling in this model system. Imaging technology now permits recording from larger numbers of *Drosophila* neurons simultaneously (Bouchard et al. 2015; Harris et al. 2015; Lu et al. 2016; Aimon et al. 2016) or from subsets of neurons in freely behaving flies (Grover et al. 2016). Computational models applied to such data will facilitate both interpreting population neural activity and connecting neural activity with behavior. In parallel, the use of unsupervised classification methods has revealed stereotyped structure in animal behavior—an animal's movements over time can be described as sequences of discrete behavioral modules (Vogelstein et al. 2014; Berman et al. 2014; Berman et al. 2016; Wiltschko et al. 2015). The task of computational modeling will now be to determine how sensory cues and internal states affect behavioral sequencing, and how neural codes underlie the choice of behavioral modules. In conclusion, combining the wealth of genetic tools to dissect the neural circuits underlying behavior in *Drosophila* with advances in machine learning and computational modeling now makes it more feasible than ever to link sensory processing, neural representations, and behavior in this system.

References

Aimon S, Katsuki T, Grosenick L, Broxton M, Deisseroth K, Sejnowski TJ, Greenspan RJ (2016) Linking stimuli and behavior with fast near-whole brain recordings in adult Drosophila. bioRxiv 033803. doi:10.1101/033803

Aljadeff J, Lansdell BJ, Fairhall AL, Kleinfeld D (2016) Analysis of neuronal spike trains, deconstructed. Neuron 91:221–259. doi:10.1016/j.neuron.2016.05.039

Aptekar JW, Shoemaker PA, Frye MA (2012) Figure tracking by flies is supported by parallel visual streams. Curr Biol 22:482–487. doi:10.1016/j.cub.2012.01.044

Aso Y, Hattori D, Yu Y, Johnston RM, Iyer NA, Ngo T-T, Dionne H, Abbott L, Axel R, Tanimoto H, Rubin GM (2014) The neuronal architecture of the mushroom body provides a logic for associative learning. eLife. doi:10.7554/eLife.04577

Baccus SA, Meister M (2002) Fast and slow contrast adaptation in retinal circuitry. Neuron 36:909–919. doi:10.1016/S0896-6273(02)01050-4

Badel L, Ohta K, Tsuchimoto Y, Kazama H (2016) Decoding of context-dependent olfactory behavior in Drosophila. Neuron. doi:10.1016/j.neuron.2016.05.022

Behnia R, Clark DA, Carter AG, Clandinin TR, Desplan C (2014) Processing properties of ON and OFF pathways for Drosophila motion detection. Nature 512:427–430. doi:10.1038/nature13427

Bell JS, Wilson RI (2016) Behavior reveals selective summation and max pooling among olfactory processing channels. Neuron. doi:10.1016/j.neuron.2016.06.011

Berman GJ, Choi DM, Bialek W, Shaevitz JW (2014) Mapping the stereotyped behaviour of freely moving fruit flies. J R Soc Interface 11. doi:10.1098/rsif.2014.0672

Berman GJ, Bialek W, Shaevitz JW (2016) Predictability and hierarchy in Drosophila behavior

Bouchard MB, Voleti V, Mendes CS, Lacefield C, Grueber WB, Mann RS, Bruno RM, Hillman EMC (2015) Swept confocally-aligned planar excitation (SCAPE) microscopy for high-speed volumetric imaging of behaving organisms. Nat Photonics 9:113–119. doi:10.1038/nphoton.2014.323

Britten KH, Newsome WT, Shadlen MN, Celebrini S, Movshon JA (1996) A relationship between behavioral choice and the visual responses of neurons in macaque MT. Vis Neurosci 13: 87–100

Brunton BW, Botvinick MM, Brody CD (2013) Rats and humans can optimally accumulate evidence for decision-making. Science 340:95–98. doi:10.1126/science.1233912

Censi A, Straw AD, Sayaman RW, Murray RM, Dickinson MH (2013) Discriminating external and internal causes for heading changes in freely flying Drosophila. PLoS Comput Biol 9: e1002891. doi:10.1371/journal.pcbi.1002891.s002

Chichilnisky EJ (2001) A simple white noise analysis of neuronal light responses. Netw Comput Neural Syst 12:199–213

Clark DA, Bursztyn L, Horowitz MA, Schnitzer MJ, Clandinin TR (2011) Defining the computational structure of the motion detector in Drosophila. Neuron 70:1165–1177. doi:10.1016/j.neuron.2011.05.023

Clemens J, Hennig RM (2013) Computational principles underlying the recognition of acoustic signals in insects. J Comput Neurosci 35:75–85. doi:10.1007/s10827-013-0441-0

Clemens J, Krämer S, Ronacher B (2014) Asymmetrical integration of sensory information during mating decisions in grasshoppers. Proc Natl Acad Sci U S A 111:16562–16567. doi:10.1073/pnas.1412741111

Clemens J, Girardin CC, Coen P, Guan X-J, Dickson BJ, Murthy M (2015a) Connecting neural codes with behavior in the auditory system of Drosophila. Neuron 87:1332–1343. doi:10.1016/j.neuron.2015.08.014

Clemens J, Rau F, Hennig RM, Hildebrandt KJ (2015b) Context-dependent coding and gain control in the auditory system of crickets. Eur J Neurosci 42:2390–2406. doi:10.1111/ejn.13019

Coen P, Clemens J, Weinstein AJ, Pacheco DA, Deng Y, Murthy M (2014) Dynamic sensory cues shape song structure in Drosophila. Nature 507:233–237. doi:10.1038/nature13131

Coen P, Xie M, Clemens J, Murthy M (2016) Sensorimotor transformations underlying variability in song intensity during Drosophila courtship. Neuron 89:629–644. doi:10.1016/j.neuron.2015.12.035

DasGupta S, Ferreira CH, Miesenböck G (2014) FoxP influences the speed and accuracy of a perceptual decision in Drosophila. Science 344:901–904. doi:10.1126/science.1252114

de Vries SEJ, Clandinin TR (2012) Loom-sensitive neurons link computation to action in the Drosophila visual system. Curr Biol 22:353–362. doi:10.1016/j.cub.2012.01.007

Eichner H, Joesch M, Schnell B, Reiff DF, Borst A (2011) Internal structure of the fly elementary motion detector. Neuron 70:1155–1164. doi:10.1016/j.neuron.2011.03.028

Escola S, Fontanini A, Katz D, Paninski L (2011) Hidden Markov models for the stimulus–response relationships of multistate neural systems. Neural Comput 23:1071–1132. doi:10.1162/neco_a_00118

Fitzgerald JE, Clark DA, Carandini M (2015) Nonlinear circuits for naturalistic visual motion estimation. eLife 4:e09123. doi:10.7554/eLife.09123

Fründ I, Wichmann FA, Macke JH (2014) Quantifying the effect of intertrial dependence on perceptual decisions. J Vis 14:9. doi:10.1167/14.7.9

Gepner R, Mihovilovic Skanata M, Bernat NM, Kaplow M, Gershow M (2015) Computations underlying Drosophila photo-taxis, odor-taxis, and multi-sensory integration. eLife 4:599. doi:10.7554/eLife.06229

Gomez-Marin A, Stephens GJ, Louis M (2011) Active sampling and decision making in Drosophila chemotaxis. Nat Commun 2:441. doi:10.1038/ncomms1455

Grover D, Katsuki T, Greenspan RJ (2016) Flyception: imaging brain activity in freely walking fruit flies. Nat Methods. doi:10.1038/nmeth.3866

Haefner RM, Gerwinn S, Macke JH, Bethge M (2013) Inferring decoding strategies from choice probabilities in the presence of correlated variability. Nat Neurosci 16:235–242. doi:10.1038/nn.3309

Harris DT, Kallman BR, Mullaney BC, Scott K (2015) Representations of taste modality in the Drosophila brain. Neuron 86:1449–1460. doi:10.1016/j.neuron.2015.05.026

Hernandez-Nunez L, Belina J, Klein M, Si G, Claus L, Carlson JR, Samuel AD, Calabrese RL (2015) Reverse-correlation analysis of navigation dynamics in Drosophila larva using optogenetics. eLife 4:e06225. doi:10.7554/eLife.06225

Jefferis GSXE, Potter CJ, Chan AM, Marin EC, Rohlfing T, Maurer CR, Luo L (2007) Comprehensive maps of Drosophila higher olfactory centers: spatially segregated fruit and pheromone representation. Cell 128:1187–1203. doi:10.1016/j.cell.2007.01.040

Kato S, Xu Y, Cho CE, Abbott LF, Bargmann CI (2014) Temporal responses of *C. elegans* chemosensory neurons are preserved in behavioral dynamics. Neuron 81:616–628. doi:10.1016/j.neuron.2013.11.020

Kim AJ, Fitzgerald JK, Maimon G (2015) Cellular evidence for efference copy in Drosophila visuomotor processing. Nat Neurosci 18:1247–1255. doi:10.1038/nn.4083

Klapoetke NC, Murata Y, Kim SS, Pulver SR, Birdsey-Benson A, Cho YK, Morimoto TK, Chuong AS, Carpenter EJ, Tian Z, Wang J, Xie Y, Yan Z, Zhang Y, Chow BY, Surek B, Melkonian M, Jayaraman V, Constantine-Paton M, Wong GK-S, Boyden ES (2014) Independent optical excitation of distinct neural populations. Nat Methods 11:338–346. doi:10.1038/nmeth.2836

Klein M, Afonso B, Vonner AJ, Hernandez-Nunez L, Berck M, Tabone CJ, Kane EA, Pieribone VA, Nitabach MN, Cardona A, Zlatic M, Sprecher SG, Gershow M, Garrity PA, Samuel ADT (2014) Sensory determinants of behavioral dynamics in Drosophila thermotaxis. Proc Natl Acad Sci U S A 112:E220–E229. doi:10.1073/pnas.1416212112

Leonhardt A, Ammer G, Meier M, Serbe E, Bahl A, Borst A (2016) Asymmetry of Drosophila ON and OFF motion detectors enhances real-world velocity estimation. Nat Neurosci 19:706–715. doi:10.1038/nn.4262

Lu R, Sun W, Liang Y, Kerlin A, Bierfeld J, Seelig J, Wilson DE, Scholl B, Mohar B, Tanimoto M, Koyama M, Fitzpatrick D, Orger MB, Ji N (2016) Video-rate volumetric functional imaging of the brain at synaptic resolution. bioRxiv. doi:10.1101/058495

Machens CK, Zador AM (2003) Auditory modeling gets an edge. J Neurophysiol 90:3581–3582. doi:10.1152/jn.00832.2003

Martelli C, Carlson JR, Emonet T (2013) Intensity invariant dynamics and odor-specific latencies in olfactory receptor neuron response. J Neurosci 33:6285–6297. doi:10.1523/JNEUROSCI.0426-12.2013

Mineault PJ, Barthelmé S, Pack CC (2009) Improved classification images with sparse priors in a smooth basis. J Vis 9:1–24

Murthy M, Fiete I, Laurent G (2008) Testing odor response stereotypy in the Drosophila mushroom body. Neuron 59:1009–1023. doi:10.1016/j.neuron.2008.07.040

Nagel KI, Wilson RI (2011) Biophysical mechanisms underlying olfactory receptor neuron dynamics. Nat Neurosci 14:208–216. doi:10.1038/nn.2725

Ozuysal Y, Baccus SA (2012) Linking the computational structure of variance adaptation to biophysical mechanisms. Neuron 73:1002–1015. doi:10.1016/j.neuron.2011.12.029

Paninski L (2003) Convergence properties of three spike-triggered analysis techniques. Netw Comput Neural Syst 14:437–464

Park M, Pillow JW (2011) Receptive field inference with localized priors. PLoS Comput Biol 7:e1002219. doi:10.1371/journal.pcbi.1002219

Parker AJ, Newsome WT (1998) Sense and the single neuron: probing the physiology of perception. Annu Rev Neurosci 21:227–277. doi:10.1146/annurev.neuro.21.1.227

Parnas M, Lin AC, Huetteroth W, Miesenböck G (2013) Odor discrimination in Drosophila: from neural population codes to behavior. Neuron 79:932–944. doi:10.1016/j.neuron.2013.08.006

Pillow JW, Shlens J, Paninski L, Sher A, Litke AM, Chichilnisky EJ, Simoncelli EP (2008) Spatio-temporal correlations and visual signalling in a complete neuronal population. Nature 454:995–999. doi:10.1038/nature07140

Rabinowitz NC, Willmore BDB, Schnupp JW, King AJ (2011) Contrast gain control in auditory cortex. Neuron 70:1178–1191. doi:10.1016/j.neuron.2011.04.030

Rajan K, Marre O, Tkačik G (2013) Learning quadratic receptive fields from neural responses to natural stimuli. Neural Comput. doi:10.1162/NECO_a_00463

Ramdya P, Lichocki P, Cruchet S, Frisch L, Tse W, Floreano D, Benton R (2015) Mechanosensory interactions drive collective behaviour in Drosophila. Nature 519:233–236. doi:10.1038/nature14024

Schönfelder VH, Wichmann FA (2012) Sparse regularized regression identifies behaviorally-relevant stimulus features from psychophysical data. J Acoust Soc Am 131:3953–3969. doi:10.1121/1.3701832

Schulze A, Gomez-Marin A, Rajendran VG, Lott G, Musy M, Ahammad P, Deogade A, Sharpe J, Riedl J, Jarriault D, Trautman ET, Werner C, Venkadesan M, Druckmann S, Jayaraman V, Louis M (2015) Dynamical feature extraction at the sensory periphery guides chemotaxis. eLife. doi:10.7554/eLife.06694

Schwartz O, Pillow JW, Rust NC, Simoncelli EP (2006) Spike-triggered neural characterization. J Vis 6:484–507. doi:10.1167/6.4.13

Seelig JD, Jayaraman V (2013) Feature detection and orientation tuning in the Drosophila central complex. Nature. doi:10.1038/nature12601

Seelig JD, Chiappe ME, Lott GK, Dutta A, Osborne JE, Reiser MB, Jayaraman V (2010) Two-photon calcium imaging from head-fixed Drosophila during optomotor walking behavior. Nat Methods 7:535–540. doi:10.1038/nmeth.1468

Sharpee TO (2013) Computational identification of receptive fields. Annu Rev Neurosci 36:103–120. doi:10.1146/annurev-neuro-062012-170253

Silies M, Gohl DM, Clandinin TR (2014) Motion-detecting circuits in flies: coming into view. Annu Rev Neurosci 37:307–327. doi:10.1146/annurev-neuro-071013-013931

Simoncelli EP, Olshausen BA (2001) Natural image statistics and neural representation. Annu Rev Neurosci 24:1193–1216. doi:10.1146/annurev.neuro.24.1.1193

Suh B, Baccus SA (2014) Building blocks of temporal filters in retinal synapses. PLoS Biol 12: e1001973. doi:10.1371/journal.pbio.1001973

Truccolo W, Eden UT, Fellows MR, Donoghue JP, Brown EN (2005) A point process framework for relating neural spiking activity to spiking history, neural ensemble, and extrinsic covariate effects. J Neurophysiol 93:1074–1089. doi:10.1152/jn.00697.2004

van Breugel F, Dickinson MH (2014) Plume-tracking behavior of flying Drosophila emerges from a set of distinct sensory-motor reflexes. Curr Biol 24:274–286. doi:10.1016/j.cub.2013.12.023

Vogelstein JT, Park Y, Ohyama T, Kerr R, Truman JW, Priebe CE, Zlatic M (2014) Discovery of brainwide neural-behavioral maps via multiscale unsupervised structure learning. Science. doi:10.1126/science.1250298

von Reyn CR, Breads P, Peek MY, Zheng GZ, Williamson WR, Yee AL, Leonardo A, Card GM (2014) A spike-timing mechanism for action selection. Nat Neurosci 17:962–970. doi:10.1038/nn.3741

Williamson RS, Sahani M, Pillow JW (2015) The equivalence of information-theoretic and likelihood-based methods for neural dimensionality reduction. PLoS Comput Biol 11: e1004141. doi:10.1371/journal.pcbi.1004141

Wiltschko AB, Johnson MJ, Iurilli G, Peterson RE, Katon JM, Pashkovski SL, Abraira VE, Adams RP, Datta SR (2015) Mapping sub-second structure in mouse behavior. Neuron 88:1121–1135. doi:10.1016/j.neuron.2015.11.031

Zucchini W (2000) An introduction to model selection. J Math Psychol 44:41–61. doi:10.1006/jmps.1999.1276

Chapter 10
Motor-Driven Modulation in Visual Neural Circuits

Terufumi Fujiwara and Eugenia Chiappe

Abstract Experiments in anesthetized, immobile animals have contributed to the classical view that sensory and motor functions in the brain are separated processes. However, under natural conditions, the nervous system and the body of a moving animal interact continuously, and it is from this interaction that neural circuits in the brain form an internal representation of the sensory world. We move our head to detect and localize the source of a sound, we move our eyes to scan a visual scene; likewise, tactile sensation is based on our body's movement, and olfaction occurs in the context of sniffing. Sensory and motor components of a sensory modality are intimately connected to each other during an active process. How this relation is implemented across sensorimotor circuits, and how motor–sensory coordination improves sensation, are questions that still remain unclear. Recent technological advances have made possible to record neural activity from sensory areas while animals walk or fly—behavioral conditions in which sensation most frequently happens. From these studies, performed both in insects and mammals, it has become apparent that the neural dynamics of primary sensory areas are readily influenced by ongoing locomotion. In this chapter, we discuss work dissecting different components of the locomotion-dependent modulations, focusing on visual circuits in flies and mice. The presence of these locomotive-related signals in early visual centers strongly suggests that motor–sensory coordination is dynamic, diverse, and adaptable to the behavioral situation of the animal.

10.1 Introduction

Sensation typically happens while animals move around and exploit natural habitats. As a consequence, movement affects the temporal and spatial pattern of sensory excitation—the sensory sampling. Motor-driven sensory sampling may

T. Fujiwara · E. Chiappe (✉)
Sensorimotor Integration Laboratory, Champalimaud Neuroscience Programme,
Av Brasilia s/n (Doca de Pedrouços), Lisbon 1400-038, Portugal
e-mail: eugenia.chiappe@neuro.fchampalimaud.org

A. Çelik and M.F. Wernet (eds.), *Decoding Neural Circuit Structure and Function*, DOI 10.1007/978-3-319-57363-2_10

jeopardize the detection of events in the world; yet, animals do move their body to uncover the source of the external stimuli. For example, we move our head to localize a sound or an odor source, or move our eyes and body to perceive the 3D structure of a visual landscape, and locate a target within it. The sensory and motor components in every sensory modality are intimately related to each other, and these relations form the bases of an internal representation of the sensory world (Ringach 2009; Wurtz et al. 2011). Therefore, motor–sensory coordination is not detrimental but fundamental to guide an animal's interaction with its environment. How motor–sensory coordination is implemented within neural circuits in the brain remains unclear.

In order to understand how neural circuits construct an internal representation of the external sensory world, it is important first to consider what challenges the brain faces when sensation occurs in the context of movement. During motor-driven sensation the brain must differentiate sensory signals occurring in the environment, known as *exafference* (von Holst and Mittelstaedt 1950), from those that inevitably result from an animal's own movements, so-called *reafference* (von Holst and Mittelstaedt 1950). To differentiate exafference from reafference, nervous systems across the animal kingdom monitor self-movement. Many sensory systems signal self-movement: the vestibular and proprioceptive systems inform the brain about body and head movements, motion vision conveys information about eye, head or body movements (see below). In addition, premotor circuits route copies of motor commands to different areas of the brain, including sensory pathways. These internally generated signals, known as corollary discharge (CD, Sperry 1950) or efference copies (von Holst and Mittelstaedt 1950) can specifically inhibit the incoming reafference stream (for a review on this topic, we refer the reader to Crapse and Sommer 2008). Such cancelation may happen at early stages of sensory processing (Poulet and Hedwig 2002; Schneider et al. 2014), and as a result, downstream neurons within the circuit only respond to exafference. This type of inhibitory effect by CD is often observed in reflexive systems that must be prevented from self-activation (Davis et al. 1973; Sillar and Roberts 1988; Roy and Cullen 2004; Kim et al. 2015), or sensory systems that must be prevented from self-induced desensitization (Voss et al. 2006; Chagnaud et al. 2015).

Under certain circumstances sensory reafference should not be canceled out (Sommer and Wurtz 2002; Hendricks et al. 2012). In the context of navigation and other oriented behaviors, the brain relies on an accurate internal estimate of self-movement that is based on multisensory reafferent information (Sommer and Wurtz 2002; Whitlock et al. 2008; Franklin and Wolpert 2011; Requarth et al. 2014; Coen et al. 2016). For example, when a stationary prey is detected, a mantis moves its upper body from side to side while keeping the gaze constant. This self-induced motion parallax is combined with proprioceptive information from neck sensors to accurately estimate the distance to the prey, and plan a strike jump (Poteser et al. 1998; Kral 2012). Multimodal reafferent information calculates physical parameters of the environment (in this example, the absolute distance to an object) that can be estimated neither with a single sensory pathway, nor in the absence of movement. In addition to multisensory reafference, CD signals of intended body movements can be also integrated with multimodal information to update internal models of ongoing movement (Franklin and Wolpert 2011; Requarth et al. 2014), and of space (Stackman et al. 2003; Sommer and Wurtz 2006; Whitlock et al. 2008). In summary, the brain monitors ongoing movement from CD signals and/or reafference information; in so doing, the brain maintains the dynamic range of sensory systems despite self-excitation, and constructs internal representations of space and self-movement that command oriented behaviors.

One fundamental and complex oriented behavior is locomotion. Different limbs, the body and the head need all be coordinated together for postural control, and for the control of direction and speed of the movement of the body. Here self-movement estimation during locomotion is not only critical for navigation, but also for planning the next locomotive steps to approach an attractive target, or to runaway from a dangerous source. Vision can provide critical information about the translational and angular movements of the body: the retina of an animal moving about is stimulated by visual flow. The self-generated pattern of visual flow, called optic-flow field, contains information related to the trajectory of the animal (Koenderink 1986). Therefore, the brain could use optic-flow processing networks to estimate the ongoing course: the wide distribution of optic-flow sensitive circuits in both invertebrate and vertebrate species would imply this might be the case (Duffy and Wurtz 1991a, b; Grasse and Cynader 1991; Wall and Smith 2008; Masseck and Hoffmann 2009; Kubo et al. 2014). But vision alone provides ambiguous body-movement information, because optic-flow fields are confounded by changes in gaze and/or by the movement of objects in the world. In primates, optic-flow processing networks also receive vestibular and eye-movement related information (Newsome et al. 1988; Bradley et al. 1996). The convergence of visual and nonvisual information is thought to be critical for the perception of heading (Britten and van Wezel 1998; Angelaki et al. 2011). At present, it is unclear whether other nonvisual signals related to body movement also arrive to these brain areas when the animal is engaged in locomotion. The presence of body-movement related signals in this network could improve the internal estimate of heading, and could also allow for an internal estimate of the velocity of the moving body (Koenderink 1986).

Recent technological advances have made possible to record from sensory areas in walking and flying flies, or running mice. Studies in walking and flying flies have demonstrated that optic-flow processing neurons are modulated by locomotion (Figs. 10.1b and 10.2b, e) (Chiappe et al. 2010; Maimon et al. 2010; Jung et al. 2011; Suver et al. 2012). In rodents, optic-flow processing neurons are found at the accessory optic system, a set of brainstem areas implicated in gaze stabilization (Simpson 1984; Yonehara et al. 2009; Distler and Hoffmann 2011; Dhande et al. 2013). While it is still unclear whether locomotion modulates any of these networks, recent work has shown that visual neurons in the thalamus and the primary visual cortex of the mouse are modulated by movement of the animal (Fig. 10.1a) (Niell and Stryker 2010; Keller et al. 2012; Ayaz et al. 2013; Bennett et al. 2013; Polack et al. 2013; Saleem et al. 2013; Erisken et al. 2014; Fu et al. 2014). Locomotion is always accompanied by an increase of arousal, which broadly

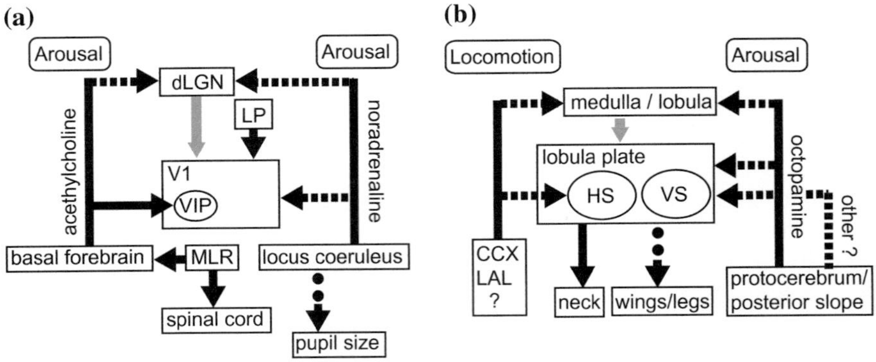

Fig. 10.1 Proposed pathways for locomotion-induced modulations in visual circuits in mice (**a**), and in flies (**b**). **a** Changes in arousal levels during locomotion modulate primary visual cortex (V1) activity through cholinergic (Fu et al. 2014; Lee et al. 2014) and noradrenergic (Polack et al. 2013) pathways. Cholinergic modulations preferentially act on VIP positive cells (Fu et al. 2014). The mesencephalic locomotor related area (*MLR*) activates the basal forebrain, and nonspecific thalamic nuclei (Lee et al. 2014); therefore, it is thought to mediate the cholinergic-induced activity modulations. The locus coeruleus, housing noradrenergic neurons, controls pupil dilations (Joshi et al. 2016) known to correlate with arousal levels (Vinck et al. 2015). Thalamic projections to V1 such as the dorsal lateral geniculus nucleus (*dLGN*) and the pulvinar area (LP), also receive motor-related signals during locomotion (Erisken et al. 2014; Roth et al. 2016). **b** The arousal-induced modulations in optic-flow processing neurons of the fly lobula plate (LoP), the horizontal and vertical system cells (HS- and VS-cells), are directed largely through octopaminergic pathways (Suver et al. 2012), but other pathways may also play a role. In addition, HS-cells selectively receive saccade-related potentials (Kim et al. 2015). These signals precede saccades, suggesting they may be centrally generated (i.e., CD signals). The fly's central complex (*CCX*) or lateral accessory lobe (*LAL*)—areas thought to control turning behavior (Pfeiffer and Homberg 2014; Strauss and Berg 2010), may direct this modulation. HS-cells and VS-cells are known to have direct connections with neck motor neurons in blowflies (Strausfeld et al. 1987; Strausfeld and Sayen 1985). The activity of HS-cells can also drive wing movements (Haikala et al. 2013) and may also drive leg movements (Heisenberg et al. 1978). There is evidence that arousal-like modulations are already present at upstream visual circuits (medulla and lobula circuits, Tuthill et al. 2014). For both schematics, gray arrows indicate visual pathways; solid and the dashed arrows indicate identified and proposed/potential pathways, respectively

changes the state of the brain (Harris and Thiele 2011; McGinley et al. 2015; Nelson and Mooney 2016). Importantly, brain states control the gain of sensory processing, learning, and memory (Lee and Dan 2012); therefore, the modulations associated with locomotion in visual interneurons could result from an arousal-induced change of the state of the network. Indeed, many aspects of the modulations described in V1 during locomotion correlate to changes in arousal levels (McGinley et al. 2015; Vinck et al. 2015). Similarly, locomotion could exert its modulation through changes in the arousal levels of the fly during flight (Suver et al. 2012; Tuthill et al. 2014). On the other hand, locomotion itself, in the form of motor feedback, could send specific signals to visual interneurons about the movement of the body through space (Keller et al. 2012; Saleem et al. 2013; Erisken et al. 2014; Fu et al. 2014; Roth et al. 2016).

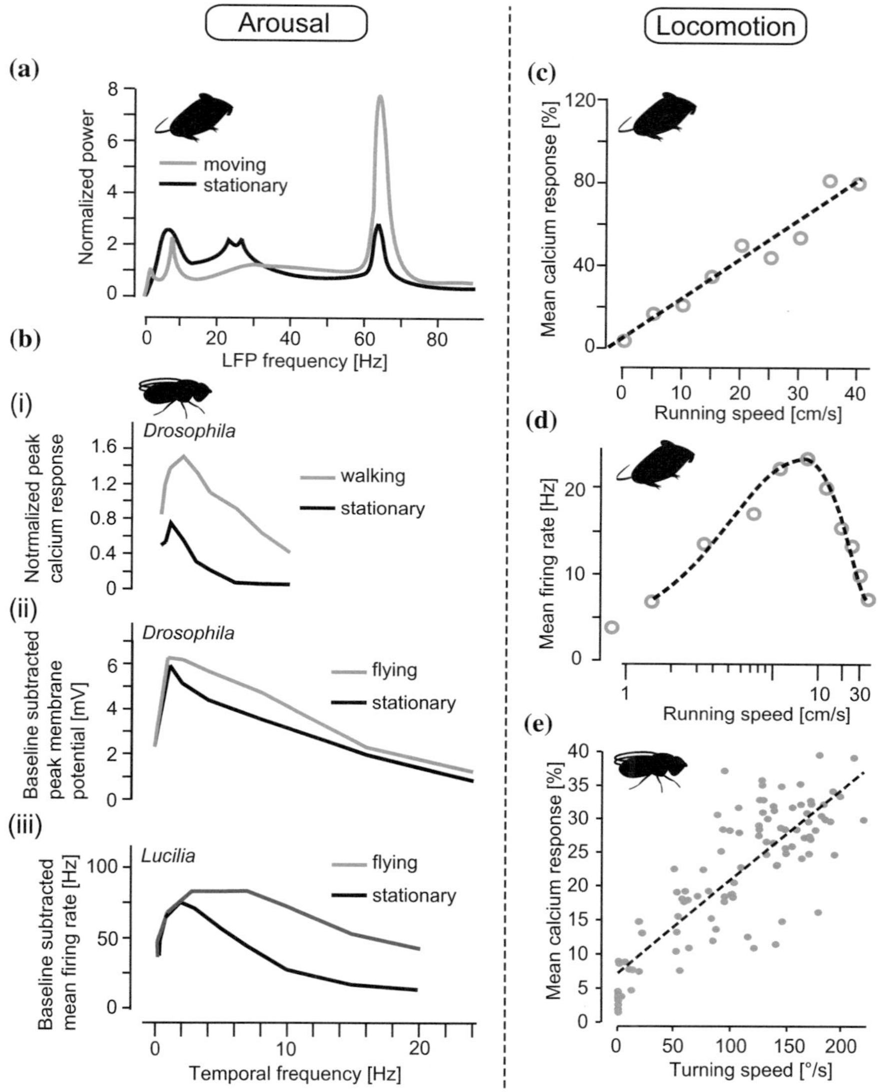

Fig. 10.2 Examples of arousal- and motor feedback-induced phenomena in visual circuits. **a**–**b** Arousal- and **c**–**d** locomotion-induced phenomena. **a** Mean local field potential (*LFP*) power spectrum of mouse V1 during movement (*gray*) and immobility (*black*). **b** (*i*) Mean temporal frequency tuning curves in *Drosophila* HS-cells to wide-field motion stimuli as measured with calcium imaging during walking (*gray*) and stationary (*black*) trials. (*ii*) Similar as (*i*) but from whole-cell electrophysiological recordings from VS-cells during flight. (*iii*) Similar as in (*ii*) but from extracellular recordings of H1-cell activity, an optic-flow sensitive cell in blowflies (*Lucilia*). **c** Running speed tuning of V1 neurons, measured with calcium imaging in layer 2/3 neurons that are sensitive to a mismatched between behavior and the expected visual feedback. **d** Example of running speed tuning of neurons from deep layers of V1 obtained with extracellular recordings in dark conditions. **e** Turning speed tuning of a *Drosophila* HS-cell to wide-field motion stimulus measured with calcium imaging. Data are adapted with permission from: Niell and Stryker (2010) (**a**), Chiappe et al. (2010) (**b**(*i*) and **e**); Suver et al. (2012) (**b**(*ii*)), Jung et al. (2011) (**b**(*iii*)), Keller et al. (2012) (**c**), Saleem et al. (2013) (**d**)

What the functional role of these locomotive-related signals is in early visual processing areas in flies and mice remains unclear. Answering this question requires dissecting the mechanisms by which locomotion-related signals affect sensory processing. For this, genetic model organisms such as the fruitfly (*Drosophila melanogaster*), the mouse (*Mus musculus*), and the zebrafish (*Danio rerio*) present a unique opportunity because of the experimental possibility to apply genetic tools in specific, genetically identified population of cells simultaneous with physiology and behavior, either in head-fixed or freely moving conditions (Dombeck et al. 2007; Maimon et al. 2010; Seelig et al. 2010; Niell and Stryker 2010; Schneider et al. 2014; Grover et al. 2016). Here, we review studies looking at locomotion-dependent modulations on visual processing in flies and rodents, where the application of genetic-based approaches has been developing rapidly. With the current experimental evidence, and to clarify the functional implication of these prominent modulations, we first discuss the phenomena caused by changes in the arousal state of an animal, and then we look into the evidence for locomotive-specific phenomena.

10.2 Arousal State Modulations During Locomotion

10.2.1 Arousal State Modulations in Rodents

The population dynamics of neurons in the lateral geniculate nucleus of the thalamus (LGN) and the primary visual cortex (V1) are modulated by locomotion (Fig. 10.1a) (Niell and Stryker 2010; McGinley et al. 2015). Concomitant with locomotion is the change in the arousal state of the animal. Therefore the origin of the locomotive-induced modulations may be simply ascribed to changes in the arousal state of the animal rather than by locomotive-related motor feedback (see below). In rodents and other mammals, the animal's arousal correlates positively with pupil size (Erisken et al. 2014; Reimer et al. 2014; Vinck et al. 2015; McGinley et al. 2015). Pupil dilation precedes the onset of locomotion, and it is sustained after locomotion termination. Furthermore, changes in pupil size can also be triggered by artificial mechanical stimuli, such as a puff of air, which is presumably associated with a startled response of the animal. Using the difference in time between the onset/offset of the pupil dilation versus that of locomotion, and the artificial induction of pupil dilation with no overt movement, Vinck and colleagues identified the aspects of locomotive modulations that were specifically linked to changes in arousal levels (Vinck et al. 2015).

Changes in arousal levels induce a global switch in cortical neural activity associated with changes in the network coding capacity (Harris and Thiele 2011; McGinley et al. 2015). In V1, low-frequency (2–10 Hz) local field potentials (LFPs) prevail during quiet awake states, but are decreased with increases in arousal levels preceding the onset of locomotion (Fig. 10.2a), or when arousal changes are artificially induced (Vinck et al. 2015). This reduction is accompanied by weakened pairwise correlations in firing activity among V1 neurons (Erisken et al. 2014; Vinck et al. 2015). The decorrelation of the firing activity of V1 neurons during high levels of arousal is presumably associated with the development of gamma band oscillations (30–80 Hz) (Niell and Stryker 2010; Bennett et al. 2013; Reimer et al. 2014; McGinley et al. 2015; Vinck et al. 2015), a frequency band thought to be related to activate sensory processing (Poulet and Petersen 2008). Furthermore, spontaneous firing rates of regular (presumably excitatory) and fast (likely inhibitory) spiking neurons also decline with the increment of arousal levels (Vinck et al. 2015). The decline of spontaneous activity may be related to both a decrease in variance of the membrane potential of the cells (Bennett et al. 2013; Polack et al. 2013), as well as an overall depolarization of the baseline membrane potential level preceding the onset of locomotion (Polack et al. 2013). Importantly, a decrease in spontaneous activity may be advantageous to improve the signal-to-noise ratio (the ratio between evoked/spontaneous cell activity) of the neuron's responses, which can impact cortical coding of sensory stimuli, as well as the performance of an animal during a sensory discrimination task (Cohen and Maunsell 2009; Goard and Dan 2009; Pinto et al. 2013). Altogether, these studies show that changes in arousal levels modify the activity of visual cortical networks in mice in a manner that resembles the physiological correlates of attention in primates: augmenting the sensory processing capacity of the network (Cohen and Maunsell 2009; Goard and Dan 2009; Harris and Thiele 2011).

The mesencephalic reticular formation (MRF) in the brainstem has long been implicated in the control of arousal, REM sleep and initiation of locomotion (Moruzzi and Magoun 1949; Shik et al. 1966; Steriade et al. 1990; Kobayashi and Isa 2002; Goetz et al. 2016). Artificial activation of this brainstem area in anesthetized animals induces desynchronization of low-frequency oscillations of the electroencephalogram (EEG) signal (Moruzzi and Magoun 1949), a phenomenon that resembles the physiological correlates of alertness (Steriade et al. 1993). Within the MRF is the mesencephalic locomotor region (MLR). MLR sends projections to descending motor pathways (Garcia-Rill and Skinner 1987a, b). Accordingly, electrical activation of MLR induces locomotion (Shik et al. 1966, Mori et al. 1978). In addition, MLR sends projections to the thalamus and basal forebrain (BF) (Nauta and Kuypers 1958), and these ascending projections may partially elicit the physiological correlates of alertness mediated by MRF (Fig. 10.1a) (Moruzzi and Magoun 1949; Goard and Dan 2009). In a recent study, Lee and colleagues used optogenetic tools to stimulate glutamatergic neurons in MLR (Lee et al. 2014). As expected, they showed that the activation of these neurons induces running (Fig. 10.1a) (Lee et al. 2014; Roseberry et al. 2016). In addition, optogenetic activation of MLR increases the gain of visual responses and enhances gamma oscillations in V1 (Lee et al. 2014). Importantly, under certain experimental conditions, stimulation of MLR induces similar effects in V1 without overt movement. Strikingly, this result parallels the modulation of V1 neurons by increased arousal levels (Vinck et al. 2015). Furthermore, optogenetic activation of axon terminals of BF–projecting MLR neurons, or of cholinergic neurons within

BF mimics cortical enhancement in V1 (Pinto et al. 2013; Lee et al. 2014), suggesting that a pathway including MLR and BF may mediate the arousal effect in V1 neurons during locomotion (Fig. 10.1a). In support of this idea, it was recently shown that electric stimulation of BF activates vasoactive intestinal peptide (VIP) positive cells in V1 via nicotinic receptors (Fig. 10.1a) (Alitto and Dan 2013). Indeed, the activity of VIP neurons in V1 correlate with running, and this correlation declines after application of nicotinic receptor antagonists (Fu et al. 2014). VIP neurons in auditory and somatosensory cortices are similarly activated upon locomotion (Eggermann et al. 2014; Nelson and Mooney 2016). Therefore, the modulation observed in the visual cortex by increases in arousal state seems to share common mechanisms across different sensory cortices. However, cholinergic activity does not account for all the phenomena associated with changes in arousal levels. For example, the baseline membrane potential boost, and the decrease in membrane potential variance upon running onset depend on noradrenergic inputs to V1 (Fig. 10.1a) (Polack et al. 2013).

Cholinergic activity in sensory cortices has been linked to arousal, attention, memory encoding (Hasselmo and Sarter 2011), and performance in visual discrimination tasks (Cohen and Maunsell 2009; Goard and Dan 2009; Pinto et al. 2013). Locomotion also improves the performance of a discrimination task in conditions of low contrast (Bennett et al. 2013). This improvement in behavioral performance is likely originating from the increase in the signal-to-noise ratio of the activity of V1 neurons (Vinck et al. 2015). Alternatively, changes in the spatial integration properties of the receptive fields of V1 neurons induced by locomotion could also improve behavioral performance, and broadening spatial receptive fields might be a way to adaptively coordinate spatial attention during active movement (Ayaz et al. 2013). However, it remains unclear whether the changes in the spatial properties of V1 neurons receptive fields can be ascribed to arousal-related modulations or to phenomena associated with motor feedback.

10.2.2 Arousal State Modulations in Visual Circuits in Flies

Octopamine (OA) in insects increases the general level of arousal of the animal (Roeder 2005). Many of the modulations of neural activity observed during the transition from quiescence into movement can be reproduced with OA agonists in non-behaving preparations. This may imply that such modulatory activity is associated with changes in arousal levels. However, in insects, it has not been possible to decouple arousal-related modulations from modulations induced by overt movement yet. Therefore, it is still possible that some aspect of OA-based modulations in visual circuits may be related to locomotion-induced motor feedback.

Motion vision is critical for the detection of moving objects in the world, and for monitoring self-movement (Lappe et al. 1999). During flight or walking, the retina is excited with visual flow, known as optic-flow. Optic-flow processing neurons are thought to be important for monitoring self-movement to correct deviations from the intended course. In flies, optic-flow processing neurons are located in the Lobula Plate (LoP) of the fly brain. Because of their large dendritic arbors, expanding across the retinotopical organization of the LoP, these neurons are called Lobula Plate Tangential cells (LPTCs) (Borst 2014; Silies et al. 2014). In each hemisphere of the *Drosophila* brain, there are three horizontal system cells (HS-cells) (Scott et al. 2002; Schnell et al. 2010), and six vertical system cells (VS-cells) processing optic-flow along the yaw, or the roll (and pitch) axis, respectively (Hausen 1982a, b; Joesch et al. 2008). These LPTCs show graded membrane potential changes upon wide-field visual motion stimuli, they are direction selective, and they are tuned to a preferred velocity of the moving visual stimuli. For example, HS-cells depolarize with front to back motion, and hyperpolarize with back to front motion, and the direction-selective response (the difference between the magnitude of the preferred- and the nulled-direction response) peaks at a best stimulus velocity (Joesch et al. 2008; Schnell et al. 2010). Likewise, VS-cells depolarize with downward motion and hyperpolarize with upward motion, and they also display temporal tuning properties. Recently, it has become possible to record from these neurons while the fly is walking (Seelig et al. 2010) or flying (Maimon et al. 2010). Experiments performed under these conditions have shown that locomotion modulates the activity of HS- and VS-cell. Flight induces a boost in the membrane potential of VS-cells in the absence of visual motion stimuli (Maimon et al. 2010; Suver et al. 2012). Walking and flight also induce an increase in the magnitude of the response of HS- and VS-cells to visual motion stimuli (Fig. 10.2b). Furthermore, during walking, this response increase scales monotonically with walking speed (Fig. 10.2e) (Chiappe et al. 2010; Longden et al. 2014). Last but not least, the temporal tuning properties of LPTCs are modified during locomotion. Walking and flight induce a shift in the sensitivity of HS- and VS-cells towards larger stimulus velocities (Chiappe et al. 2010; Suver et al. 2012), a phenomenon also observed in LPTCs of other fly species (Fig. 10.2b) (Jung et al. 2011; Longden et al. 2014), which may be related to the role of these networks during locomotion.

The prime candidate for these behavioral-state dependent changes in the activity of LPTCs has been OA, a neuromodulator released during flight in insects (Fig. 10.1b) (Orchard et al. 1993). Consistent with this idea, electrophysiological recordings from LPTCs and other early visual neurons in the presence of OA or OA agonists (chlordimeform, CDM) produce increased responses to visual stimuli resembling the modulation induced by locomotion in these neurons (Longden and Krapp 2009, 2010; Rien et al. 2012; Tuthill et al. 2014). This phenomenon could be partly explained by a decrease in the adaptation to motion stimuli in motion-sensitive circuits upstream from LPTCs (Longden and Krapp 2010; de Haan et al. 2012; Rien et al. 2012; Lüders and Kurtz 2015). Indeed, T4 and T5 cells, cognate presynaptic neurons to LPTCs, show flight-induced increased response to visual stimuli (Schnell et al. 2014).

OA neurons project to LoP and other areas of the visual system of the fly (Busch et al. 2009). Therefore, the activity of these OA neurons could mediate the modulations observed during locomotion. Indeed, *Drosophila* OA neurons show an increase in activity during flight (Suver et al. 2012). Using *Drosophila*'s genetic tools, Suver and colleagues artificially activated OA neurons, while they recorded the activity of VS-cells in flying and non-flying flies (Suver et al. 2012). Activation of OA neurons reproduced some of the flight effects; it induced an increased visual response in VS-cells (Suver et al. 2012). Conversely, silencing OA neurons by overexpressing an inward-rectified potassium channel (Kir2.1) abolishes the flight-induced increased visual response. Under this experimental condition, the flight-induced baseline membrane potential depolarization remains intact. This result, together with the observation that the time course of the baseline shift, and of the increased response amplitude are different, supports the idea that these locomotion-induced phenomena originate from different mechanisms. What the source of the locomotion-induced fast increase in baseline membrane potential is remains unclear (Fig. 10.1b).

The observed locomotion-induced modulations in LPTCs, i.e., the depolarization of the baseline membrane potential, the increased response to visual stimuli, and the change in the temporal sensitivity of the neuron, are in line with the idea that locomotion—perhaps through a change in arousal state, controls the gain and the temporal properties of visual processing. LPTCs have been thought to be responsible for a stabilization reflex known as the optomotor response. Imagine that a sudden gust of wind pushes a flying fly into an unintended course. Optic-flow resulting from this passive body drift will excite HS- or VS-cells depending on the direction of movement. Activity in these neurons are thought to compensate for the detour, bringing the fly back to her intended course. Although it is unclear how exactly the activity of these neurons controls compensatory body movements, there is some evidence indicating that HS-cells may control ipsilateral head and body yaw movements (Fig. 10.1b) (Blondeau and Heisenberg 1982; Haikala et al. 2013). Consistent with this proposed function, *Drosophila* mutants that lack LPTCs, or flies with ablated LPTCs show an attenuated optomotor reflex (Heisenberg et al. 1978; Geiger and Nässel 1981; Hausen and Wehrhahn 1983). Therefore, the locomotion-induced modulations on the activity of LPTCs are likely controlling the gain of the transformation of visual motion stimuli into a body compensatory movement. The observation that the gain of the head optomotor response is increased during flight supports this idea (Haag et al. 2010). Moreover, it has been recently shown that the activity of OA neurons may control flight acceleration upon visual motion stimulation (van Breugel et al. 2014).

The temporal properties of visual motion processing in LPTCs are modulated by the behavioral state of the fly (Fig. 10.2b). This locomotion-induced modulation may be beneficial for the function of LPTCs in the control of locomotion. The statistics of motion visual stimuli during locomotion versus quiescence are different. During quiescence, motion stimuli will be elicited by movement of objects in the world, or by movements of the surface where the fly is landed, usually elicited by a wind breeze. At the onset of locomotion, motion visual stimuli will be largely generated by the animal's own movement. Electrophysiological recordings in non-behaving preparations from LPTCs have shown that the temporal sensitivity of these neurons to stimuli that elicits robust optomotor response could not account for the temporal properties of the behavior: the optimal stimulus velocity for the behavior is at least 4X faster than the optimal stimulus velocity for the neuron in quiescence (Götz and Wenking 1973; Schnell et al. 2010). Therefore, it has been difficult to reconcile these disparate results until the recent discovery of the locomotion-induced temporal modulation of visual processing in LTPCs and/or in upstream neurons (Seelig et al. 2010; Chiappe et al. 2010; Tuthill et al. 2014).

The modulations in the activity of LPTCs by the transition from quiescence into locomotion could be related to changes in the arousal state of the animal, or by specific locomotion-related motor feedback. While it has not been possible to decouple these two phenomena in insects, for example, there is still no "pupil dilation"—like measurements monitoring changes in arousal independent of overt movement, several lines of evidence indicate that the described modulations could be related to changes in fly alertness. Increases in OA levels have been associated with increases in arousal levels (Roeder 2005). The activity of OA neurons increases during flight (Suver et al. 2012). Artificial activation of OA reproduces many of the effects triggered by flight, and silencing these neurons reduces the response amplitude of LPTCs to visual stimuli without affecting flight per se. What makes OA neurons active during flight, and whether these neurons are also involved in the walking-induced modulations of HS-cells are questions that remain poorly understood.

10.3 Locomotion-Specific Modulations

10.3.1 Locomotion-Specific Modulations in Rodents

Recent evidence indicates that early visual circuits in mice receive locomotion-related motor feedback. However, because the majority of the work has been performed in conditions that did not decouple a change in arousal from overt movement, the interpretation of the results from these studies cannot exclude the possibility that some of the phenomena noted below could still relate to a change in the arousal level of the animal.

Locomotion induces increased evoked responses in layer 2/3 V1 neurons (Niell and Stryker 2010), and increased spontaneous activity in both regular and fast-spiking neurons across layers in V1 (Vinck et al. 2015). The latter effect was decoupled from changes in the arousal state of the mouse that accompanies the onset of locomotion (Vinck et al. 2015). Other studies confirmed these initial observations, and importantly, revealed a high level of heterogeneity in the effect of locomotion on V1 activity (Andermann et al. 2011; Keller et al. 2012; Bennett et al. 2013; Polack et al. 2013; Saleem et al. 2013; Erisken et al. 2014; Fu et al. 2014). For example, some neurons are excited while others are suppressed by locomotion (Polack et al. 2013; Saleem et al. 2013; Erisken et al. 2014; Fu et al. 2014). This variability could be partially accounted by differences in the effect of modulation across different cortical layers (Andermann et al. 2013; Erisken et al. 2014), or among different cell types (Polack et al. 2013; Fu et al. 2014). Nevertheless, such heterogeneity would not be expected if the source of the modulation were related to a global change in arousal levels. Interestingly, V1 neurons respond to running even in darkness (Fig. 10.2c, d) (Keller et al. 2012; Andermann et al. 2013; Saleem et al. 2013; Erisken et al. 2014), with firing rates that can be modulated monotonically with the increase or decrease of running speed, or can display a bell-shaped running-speed tuning (Saleem et al. 2013; Erisken et al. 2014). Such nonlinear speed tuning property indicates the presence of a motor-related specific feedback signal rather than a global arousal-related modulation. Other evidence for locomotion-related motor feedback comes from virtual reality experiments in mice. Keller and colleagues adopted a closed-loop virtual reality system where the visual-motor coupling is interrupted by a brief halt of the visual feedback (Keller et al. 2012). In this environment, a subset of V1 neurons selectively responds to a mismatch between running and the expected visual-flow feedback, revealing potential predictive motor-related signals in the visual cortex.

It is still unclear what the source of these motor-related feedback signals is. Both lower-order thalamic projections to V1, through the dorsal lateral geniculus (dLGN) (Erisken et al. 2014; Roth et al. 2016), or higher-order thalamic projections to V1, through the lateral posterior nucleus (LP, the pulvinar equivalent in mice), contain neurons sensitive to the running speed of the animal (Roth et al. 2016). These observations are not surprising given that both dLGN and LP receive inputs from motor-related areas (Grieve et al. 2000; Baldwin et al. 2011; Wurtz et al. 2011). Considering that MLR can initiate locomotion, and can activate thalamic and BF circuits (Nauta and Kuypers 1958), it is tempting to speculate that MRL could also deliver locomotion-related motor feedback to V1 (Fig. 10.1a). Another possibility could be feedback from other cortical areas to V1. These feedback signals have been characterized in the somatosensory system (Lee et al. 2008; Petreanu et al. 2009, 2012; Lee et al. 2013; Zagha et al. 2013) in the context of active whisking, and in primary auditory cortex (Nelson et al. 2013; Nelson and Mooney 2016). Therefore, it could be the case that V1 receives locomotor signals from cortico-thalamic feedback (Sillito et al. 2006). In addition, the motor-related signals could arrive from the superior colliculus, through its projection to dLGN (Erisken et al. 2014; Liang et al. 2015; Roth et al. 2016). These potential different sources of modulations might explain the heterogeneous sensitivity of V1 neurons to the locomotion-induced modulations.

The effect of locomotion-related modulations is diverse among V1 neurons even within the same cortical layers (Keller et al. 2012; Saleem et al. 2013; Erisken et al. 2014). In some cases, locomotion motor feedback-related signals are not simply gating or modulating visual processing, these signals induce well-defined running speed tuning curves in V1 neurons in darkness (Fig. 10.2d). In addition, some neurons in V1 selectively respond to the mismatch between running and the expected visual feedback (Keller et al. 2012; Roth et al. 2016), suggesting that self-generated visual inputs might be actively suppressed in V1. This might be a specific property of locomotion-related feedback; however, Saleem and collegues, found that many V1 neurons code the sum of running and visual-flow speeds instead of their mismatch (Saleem et al. 2013; see also Roth et al. 2016). The integration of running speed and visual feedback may be a useful mechanism for an internal estimate of the animal's own speed during navigation (Whitlock et al. 2008; Kropff et al. 2015). It remains to be seen how these apparent contradictory interactions between visual and motor-related signals, a positive integration versus a sensitivity to a mismatch, interrelate within networks in V1, and what the specific function of these interactions is for visuomotor processing and visually guided behaviors.

10.3.2 Locomotion-Specific Modulations in Flies

Much less is known about locomotion-related motor feedback in visual circuits in flies. During walking, the response amplitude of HS-cells to visual stimuli increases monotonically with walking speed (Fig. 10.2e) (Chiappe et al. 2010; in blowflies: Longden et al. 2014), suggesting that these neurons may receive quantitative walking-related information.

More recently, an intriguing locomotion-specific modulation was found in a subclass of HS-cells, the HSN cell, while a fly was flying. Explorative flight is characterized by segments of straight course interrupted by abrupt changes in heading, known as saccades (Schilstra and Hateren 1999; Muijres et al. 2014). In the absence of visual stimuli, recordings from HSN showed that when the fly made a saccade the membrane potential of the neuron was depolarized or hyperpolarized. The direction of the modulation depended on the direction of the saccade, suggesting the presence of extra-retinal, direction-selective signals (Kim et al. 2015). These direction-selective signals were opposite to the visual direction-selective responses of the cell, such that, in the presence of visual motion stimuli, saccades provoked a cancelation of the response of the HSN to visual stimuli (Kim, et al. 2015). Because the saccade-related potentials precede the saccades, these signals are thought to represent CD signals with an apparent cancelation function.

LPTCs are thought to trigger the optomotor reflex to stabilize the course of the fly during segments of straight, forward movement. A long-standing question has been how the flies can escape from the optomotor reflex during voluntary turns, for example, during saccades (von Holst and Mittelstaedt 1950). Here, a CD signal predicting the sensory consequence of the abrupt saccade may be used to silence optomotor inducing networks (von Holst and Mittelstaedt 1950). The motor-related signals in HSNs could serve such a cancelation function, blocking the stabilization reflex specifically when the fly desires to change course. It is still not clear what the source of these signals is; however, reasonable candidates are the central complex and the lateral accessory lobe of the fly brain, premotor areas shown to be associated to turning behavior (Strauss and Berg 2010; Pfeiffer and Homberg 2014). Indeed, visual responses in central complex neurons are also modulated during locomotion (Weir et al. 2014; Seelig and Jayaraman 2015; Weir and Dickinson 2015). It will be interesting to investigate if there is any causal relation between the motor-related modulations observed in HS-cells and the ones observed in such higher brain centers.

10.4 Conclusion

The dynamics of visual neurons is influenced by the behavioral context of an animal (Fig. 10.1). Movement of the body through space generates modulatory signals such that the activity of sensory circuits change in correspondence with the arousal level of the animal. In rodents, many aspects of this arousal-induced modulation resemble the attention-related phenomena described in homolog circuits in primates. Both processes involve a similar control of activity oscillations and correlations among cortical neurons, which are mediated by common receptors and neuromodulatory systems (Fig. 10.1a) (Harris and Thiele 2011). This may seem at first surprising because arousal is a global state while spatial attention is a local process, highlighting one location of the visual field over others. However, broadly speaking, locomotion and attention can be both considered as active internal states. As such, these behavioral contexts may share basic cognitive factors, for example, the commonly observed changes in cortical state. More striking is the fact that some commonalities in state-dependent modulations on visual activity are beginning to emerge from studies in flies (Fig. 10.1b). Like attention in primates, and locomotion in rodents, walking and flight in flies induce changes in the response gain and stimulus selectivity of visual interneurons (Chiappe et al. 2010; Maimon et al. 2010; Jung et al. 2011), and these phenomena are also under the control of neuromodulation (Suver et al. 2012). Thus, modulation by behavioral context is a common feature of visual processing.

One interesting aspect of the state-dependent modulations is how multiple coexisting active internal states may interact to control the dynamics of visual or other sensory circuits. Studies of attention in monkeys have been exclusively focused on eye movements. Because more ethological relevant behaviors, such as locomotion, are difficult to control in a laboratory setting, it is still not clear whether and how locomotion modulates visual activity in monkeys, and if so, how attention and locomotion interact to control the dynamics of visual circuits, and the performance of visual guided behaviors. On the other hand, it is possible that in mice (Gill et al. 2000), and even perhaps in insects (Hoy 1989; van Swinderen 2007; Wiederman and O'Carroll 2013), one could test the interaction of attention-like and locomotion-induced phenomena in visual circuits with the development of suitable behavioral paradigms. Using modern techniques, such as specific labeling, and recording and perturbing neural activity in a genetically identified population of neurons, these experiments would allow for a detailed understanding of what aspects of the modulations are shared between alertness and attention, and which ones are not. Moreover, these experiments could also test what the function of either the arousal- or the attention-induced modulations is for visuomotor processing, and for animal behavior: two important elements that are currently unsatisfactorily understood.

Equally prominent in rodents are the locomotion-related motor feedback signals arriving to visual circuits from bottom-up and top-down inputs (Fig. 10.2c–e). These signals were uncovered while mice run in darkness, and showed a variety of modulations on visual activity. It is still unclear what the origin of these signals is. Both proprioceptive information as well as CD signals could deliver such modulations to V1 neurons. During running, some V1 neurons are excited, others are inhibited, and yet others are "band-pass" modulated by the speed of the mouse. Moreover, some neurons display a cooperative interaction between visual stimuli and running speed, while others display mismatching sensitivity between the two signals, suggesting a predictive component of the modulation. Although it is not known what the function of these motor feedback signals is for V1 computation, the variety in motor feedback modulations suggest that in V1 both multimodal self-movement estimation, and error-detection functions may be coexisting. Importantly, these functions may reflect neural processes related to an internal monitoring of ongoing movement, and/or an update of internal models of the ongoing movement. Indeed, these two computations are critical to drive visually guided spatial oriented behaviors. In *Drosophila*, there is some evidence supporting the idea that visual neurons in this species also receive motor feedback signals in the context of walking (Fig. 10.2e). Future work performing electrophysiology in flies walking in darkness should determine if this is indeed the case. Interestingly, in the context of a saccade during flight, optic-flow processing neurons receive CD-like signals that cancel the visual responses of the neurons. This cancelation function may be another example of how the brain in a moving animal controls reflexes that can be self-induced by voluntary behavior. It is of high priority is to determine what the source of these modulations is. Circuits thought to be related to control voluntary turning behavior, such as the central complex and the lateral accessory lobe, would be the prime candidates for testing their role in directing motor-related signals to visual circuits. Using the ever-expanding genetic tools of *Drosophila*, and taking advantage of her relatively numerically simpler brain than that of mice or primates, one could apply a systematic strategy to perturb the activity of different populations of neurons and examine their role in generating the motor-related signals in visual neurons.

Modern techniques have uncovered that locomotion induces changes in the state of sensory circuits, and have determined that the state of the brain follows a continuing from sleep, to quiescence, to behavior. To understand the function of this state-dependent modulation, it is crucial to decouple arousal components from those generated by motor feedback during locomotion. Because neural circuits may be sensitive to the relation between self-movement and self-generated sensory stimuli, fine control of the properties of the sensory stimuli, as well as a more detailed characterization of the behavior is required to understand the mechanisms and function of locomotion-induced modulation. In addition, recording from more naturalistic situations (i.e., in closed-loop conditions) would be also critical to understand how neural circuits process self-generated versus externally generated stimuli while animals interact with an environment. All these preconditions are addressable by combining head-fixed preparations with virtual reality technology (Dombeck and Reiser 2012; see also: Minderer et al. 2016).

In summary, recording from, and perturbing the activity of genetically identified populations of neurons while animals behave in controlled environments will be essential to test how downstream circuits in the brain read out motor–sensory coordination to ultimately construct an internal representation of the sensory world that guides animal behavior. In this manner, a multidisciplinary effort combining realistic models of sensory circuit function, and experiments in genetic model organisms, will pave the way to solve mechanistic questions, and to generate hypothesis about the principles of motor–sensory coordination that can be tested in other animals, and in other scenarios.

References

Alitto HJ, Dan Y (2013) Cell type specific modulation of neocortical activity by basal forebrain input. Front Syst Neurosci 6:79

Andermann ML, Kerlin AM, Roumis DK, Glickfeld LL, Reid RC (2011) Functional specialization of mouse higher visual cortical areas. Neuron 72:1025–1039

Andermann ML, Gilfoy NB, Goldey GJ, Sachdev RN, Wolfel M, McCormick DA, Reid RC, Levene MJ (2013) Chronic cellular imaging of entire cortical columns in awake mice using microprisms. Neuron 80:900–913

Angelaki DE, Gu Y, Deangelis GC (2011) Visual and vestibular cue integration for heading perception in extrastriate visual cortex. J Physiol 589:825–833

Ayaz A, Saleem AB, Scholvinck ML, Carandini M (2013) Locomotion controls spatial integration in mouse visual cortex. Curr Biol 23:890–894

Baldwin MK, Wong P, Reed JL, Kaas JH (2011) Superior colliculus connections with visual thalamus in gray squirrels *Sciurus carolinensis*: evidence for four subdivisions within the pulvinar complex. J Comp Neurol 519:1071–1094

Bennett C, Arroyo S, Hestrin S (2013) Subthreshold mechanisms underlying state dependent modulation of visual responses. Neuron 80:350–357

Blondeau J, Heisenberg M (1982) The three-dimensional optomotor torque system of *Drosophila melanogaster*. J Comp Physiol 145:321–329

Borst A (2014) Fly visual course control behaviour algorithms and circuits. Nat Rev Neurosci 15:590–599

Bradley DC, Maxwell M, Andersen RA, Banks MS, Shenoy KV (1996) Mechanisms of heading perception in primate visual cortex. Science 273:1544–1547

Britten KH, van Wezel RJ (1998) Electrical microstimulation of cortical area MST biases heading perception in monkeys. Nat Neurosci 1:59–63

Busch S, Selcho M, Ito K, Tanimoto H (2009) A map of octopaminergic neurons in the Drosophila brain. J Comp Neurol 513:643–667

Chagnaud BP, Banchi R, Simmers J, Straka H (2015) Spinal corollary discharge modulates motion sensing during vertebrate locomotion. Nat Commun 6:7982

Chiappe ME, Seelig JD, Reiser MB, Jayaraman V (2010) Walking modulates speed sensitivity in Drosophila motion vision. Curr Biol 20:1470–1475

Coen P, Xie M, Clemens J, Murthy M (2016) Sensorimotor transformations underlying variability in song intensity during Drosophila courtship. Neuron 89:629–644

Cohen MR, Maunsell JH (2009) Attention improves performance primarily by reducing interneuronal correlations. Nat Neurosci 12:1594–1600
Crapse TB, Sommer MA (2008) Corollary discharge across the animal kingdom. Nat Rev Neurosci 9:587–600
Davis WJ, Siegler MVS, Mpitos GJ (1973) Distributed neuronal oscillators and efference copy in the feeding system of Pleurobranchaca. J Neurophysiol 36:258–274
de Haan R, Lee YJ, Nordstrom K (2012) Octopaminergic modulation of contrast sensitivity. Front Integr Neurosci 6:55
Dhande OS, Estevez ME, Quattrochi LE, El-Danaf RN, Nguyen PL, Berson DM, Huberman AD (2013) Genetic dissection of retinal inputs to brainstem nuclei controlling image stabilization. J Neurosci 33:17797–17813
Distler C, Hoffmann KP (2011) Visual pathway for the optokinetic reflex in infant macaque monkeys. J Neurosci 31:17659–17668
Dombeck DA, Khabbaz AN, Collman F, Adelman TL, Tank DW (2007) Imaging large scale neural activity with cellular resolution in awake mobile mice. Neuron 56:43–57
Dombeck DA, Reiser MB (2012) Real neuroscience in virtual worlds. Curr Opin Neurobiol 22:3–10
Duffy CJ, Wurtz RH (1991a) Sensitivity of MST neurons to optic flow stimuli. I. A continuum of response selectivity to large field stimuli. J Neurophysiol 65:1329–1345
Duffy CJ, Wurtz RH (1991b) Sensitivity of MST neurons to optic flow stimuli II Mechanisms of response selectivity revealed by small field stimuli. J Neurophysiol 65:1346–1359
Eggermann E, Kremer Y, Crochet S, Petersen CC (2014) Cholinergic signals in mouse barrel cortex during active whisker sensing. Cell Rep 9:1654–1660
Erisken S, Vaiceliunaite A, Jurjut O, Fiorini M, Katzner S, Busse L (2014) Effects of locomotion extend throughout the mouse early visual system. Curr Biol 24:2899–2907
Franklin DW, Wolpert DM (2011) Computational mechanisms of sensorimotor control. Neuron 72:425–442
Fu Y, Tucciarone JM, Espinosa JS, Sheng N, Darcy DP, Nicoll RA, Huang ZJ, Stryker MP (2014) A cortical circuit for gain control by behavioral state. Cell 156:1139–1152
Garcia-Rill E, Skinner RD (1987a) The mesencephalic locomotor region. II. Projections to reticulospinal neurons. Brain Res 411:13–20
Garcia-Rill E, Skinner RD (1987b) The mesencephalic locomotor region. I. Activation of a medullary projection site. Brain Res 411:1–12
Geiger G, Nässel DR (1981) Visual orientation behaviour of flies after selective laser beam ablation of interneurones. Nature 293:398–399
Gill TM, Sarter M, Givens B (2000) Sustained visual attention performance-associated prefrontal neural activity: evidence for cholinergic activity. J Neurosci 20:4745–4757
Goard M, Dan Y (2009) Basal forebrain activation enhances cortical coding of natural scenes. Nat Neurosci 12:1444–1449
Goetz L, Piallat B, Bhattacharjee M, Mathieu H, David O, Chabardès S (2016) On the role of the pedunculopontine nucleus and mesencephalic reticular formation in locomotion in nonhuman primates. J Neurosci 36:4917–4929
Götz KG, Wenking H (1973) Visual control of locomotion in the walking fruit fly Drosophila. J Comp Physiol 85:235–266
Grasse KL, Cynader MS (1991) The accessory optic system in frontal-eyed animals. Macmillan, New York, pp 111–139
Grieve KL, Acuna C, Cudeiro J (2000) The primate pulvinar nuclei vision and action. Trends Neurosci 23:35–39
Grover D, Katsuki T, Greenspan RJ (2016) Flyception: imaging brain activity in freely walking fruit flies. Nat Methods 13:569–572
Haag J, Wertz A, Borst A (2010) Central gating of fly optomotor response. Proc Nat Acad Sci U S A 107:20104–20109
Haikala V, Joesch M, Borst A, Mauss AS (2013) Optogenetic control of fly optomotor responses. J Neurosci 33:13927–13934
Harris KD, Thiele A (2011) Cortical state and attention. Nat Rev Neurosci 12:509–523

Hasselmo ME, Sarter M (2011) Modes and models of forebrain cholinergic neuromodulation of cognition. Neuropsychopharmacology 36:52–73

Hausen K (1982a) Motion sensitive interneurons in the optomotor system of the fly. I. The horizontal cells: structure and signals. Biol Cybern 45:143–156

Hausen K (1982b) Motion sensitive interneurons in the optomotor system of the fly. II. The horizontal cells: receptive field organization and response characteristics. Biol Cybern 46:67–79

Hausen K, Wehrhahn C (1983) Microsurgical lesion of horizontal cells changes optomotor yaw responses in the blowfly *Calliphora erythrocephala*. Proc R Soc Lond B 219:211–216

Heisenberg M, Wonneberger R, Wolf R (1978) Optomotor-blindh31—a Drosophila mutant of the lobula plate giant neurons. J Comp Physiol 124:287–296

Hendricks M, Ha H, Maffey N, Zhang Y (2012) Compartmentalized calcium dynamics in a C elegans interneuron encode head movement. Nature 487:99–103

Hoy RR (1989) Startle, categorical response, and attention in acoustic behavior of insects. Annu Rev Neurosci 12:355–375

Joesch M, Plett J, Borst A, Reiff DF (2008) Response properties of motion sensitive visual interneurons in the lobula plate of *Drosophila melanogaster*. Curr Biol 18:368–374

Jung SN, Borst A, Haag J (2011) Flight activity alters velocity tuning of fly motion sensitive neurons. J Neurosci 31:9231–9237

Keller GB, Bonhoeffer T, Hubener M (2012) Sensorimotor mismatch signals in primary visual cortex of the behaving mouse. Neuron 74:809–815

Kim AJ, Fitzgerald JK, Maimon G (2015) Cellular evidence for efference copy in Drosophila visuomotor processing. Nat Neurosci 18:1247–1255

Kobayashi Y, Isa T (2002) Sensory motor gating and cognitive control by the brainstem cholinergic system. Neural Netw 15:731–741

Koenderink JJ (1986) Optic flow. Vision Res 26:161–179

Kral K (2012) The functional significance of mantis peering behaviour. Eur J Entomol 109:295–301

Kropff E, Carmichael JE, Moser MB, Moser EI (2015) Speed cells in the medial entorhinal cortex. Nature 523:419–424

Kubo F, Hablitzel B, Maschio MD, Driever W, Baier H, Arrenberg AB (2014) Functional architecture of an optic flow responsive area that drives horizontal eye movements in zebrafish. Neuron 81:1344–1359

Lappe M, Bremmer F, van den Berg AV (1999) Perception of self motion from visual flow. Trends Cogn Sci 3:329–336

Lee SH, Dan Y (2012) Neuromodulation of brain states. Neuron 76:209–222

Lee S, Carvell GE, Simons DJ (2008) Motor modulation of afferent somatosensory circuits. Nat Neurosci 11:1430–1438

Lee S, Kruglikov I, Huang ZJ, Fishell G, Rudy B (2013) A disinhibitory circuit mediates motor integration in the somatosensory cortex. Nat Neurosci 16:1662–1670

Lee AM, Hoy JL, Bonci A, Wilbrecht L, Stryker MP, Niell CM (2014) Identification of a brainstem circuit regulating visual cortical state in parallel with locomotion. Neuron 83:455–466

Liang F, Xiong XR, Zingg B, Ji XY, Zhang LI, Tao HW (2015) Sensory cortical control of a visually induced arrest behavior via corticotectal projections. Neuron 86:755–767

Longden KD, Krapp HG (2009) State dependent performance of optic flow processing interneurons. J Neurophysiol 102:3606–3618

Longden KD, Krapp HG (2010) Octopaminergic modulation of temporal frequency coding in an identified optic flow processing interneuron. Front Syst Neurosci 4:153

Longden KD, Muzzu T, Cook DJ, Schultz SR, Krapp HG (2014) Nutritional state modulates the neural processing of visual motion. Curr Biol 24:890–895

Lüders J, Kurtz R (2015) Octopaminergic modulation of temporal frequency tuning of a fly visual motion sensitive neuron depends on adaptation level. Front Integr Neurosci 9:36

Maimon G, Straw AD, Dickinson MH (2010) Active flight increases the gain of visual motion processing in Drosophila. Nat Neurosci 13:393–399

Masseck OA, Hoffmann KP (2009) Comparative neurobiology of the optokinetic reflex. Ann N Y Acad Sci 1164:430–439

McGinley MJ, Vinck M, Reimer J, Batista-Brito R, Zagha E, Cadwell CR, Tolias AS, Cardin JA, McCormick DA (2015) Waking state: rapid variations modulate neural and behavioral responses. Neuron 87:1143–1161

Minderer M, Harvey CD, Donato F, Moser E (2016) Neuroscience: virtual reality explored. Nature 533:324–325

Mori S, Nishimura H, Kurakami C, Yamamura T, Aoki M (1978) Controlled locomotion in the mesencephalic cat distribution of facilitatory and inhibitory regions within pontine tegmentum. J Neurophysiol 41:1580–1591

Moruzzi G, Magoun HW (1949) Brain stem reticular formation and activation of the EEG. Electroencephalogr Clin Neurophysiol 1:455–473

Muijres FT, Elzinga MJ, Melis JM, Dickinson MH (2014) Flies evade looming targets by executing rapid visually directed banked turns. Science 344:172–177

Nauta WJH, Kuypers HGJM (1958) Some ascending pathways in the brain stem reticular formation. In: Jasper HH, Proctor LD, Knighton RS, Noshay WC, Costello RT (eds) Reticular formation of the brain (pp. 3-30), Little, Brown: Oxford, England, 766 pp

Nelson A, Mooney R (2016) The basal forebrain and motor cortex provide convergent yet distinct movement related inputs to the auditory cortex. Neuron 90:635–648

Nelson A, Schneider DM, Taktoh J, Sakurai K, Wang F, Mooney R (2013) A circuit for motor cortical modulation of auditory cortical activity. J Neurosci 33:14342–14353

Newsome WT, Wurtz RH, Komatsu H (1988) Relation of cortical areas MT and MST to pursuit eye movements II Differentiation of retinal from extraretinal inputs. J Neurophysiol 60:604–620

Niell CM, Stryker MP (2010) Modulation of visual responses by behavioral state in mouse visual cortex. Neuron 65:472–479

Orchard I, Ramirez JM, Lange AB (1993) A multifunctional role for octopamine in locust flight. Annu Rev Entomol 38:227–249

Petreanu L, Mao T, Sternson SM, Svoboda K (2009) The subcellular organization of neocortical excitatory connections. Nature 457:1142–1145

Petreanu L, Gutnisky DA, Huber D, Xu NL, O'Connor DH, Tian L, Looger L, Svoboda K (2012) Activity in motor sensory projections reveals distributed coding in somatosensation. Nature 489:299–303

Pfeiffer K, Homberg U (2014) Organization and functional roles of the central complex in the insect brain. Annu Rev Entomol 59:165–184

Pinto L, Goard MJ, Estandian D, Xu M, Kwan AC, Lee SH, Harrison TC, Feng G, Dan Y (2013) Fast modulation of visual perception by basal forebrain cholinergic neurons. Nat Neurosci 16:1857–1863

Polack PO, Friedman J, Golshani P (2013) Cellular mechanisms of brain state dependent gain modulation in visual cortex. Nat Neurosci 16:1331–1339

Poteser M, Pabst MA, Kral K (1998) Proprioceptive contribution to distance estimation by motion parallax in praying mantid. J Exp Biol 201:1483–1491

Poulet JF, Hedwig B (2002) A corollary discharge maintains auditory sensitivity during sound production. Nature 418:872–876

Poulet JF, Petersen CC (2008) Internal brain state regulates membrane potential synchrony in barrel cortex of behaving mice. Nature 454:881–885

Reimer J, Froudarakis E, Cadwell CR, Yatsenko D, Denfield GH, Tolias AS (2014) Pupil fluctuations track fast switching of cortical states during quiet wakefulness. Neuron 84:355–362

Requarth T, Kaifosh P, Sawtell NB (2014) A role for mixed corollary discharge and proprioceptive signals in predicting the sensory consequences of movements. J Neurosci 34:16103–16116

Rien D, Kern R, Kurtz R (2012) Octopaminergic modulation of contrast gain adaptation in fly visual motion sensitive neurons. Eur J Neurosci 36:3030–3039

Ringach DL (2009) Spontaneous and driven cortical activity: implications for computation. Curr Opin Neurobiol 19:439–444

Roeder T (2005) Tyramine and octopamine: ruling behavior and metabolism. Annu Rev Entomol 50:447–477

Roseberry TK, Lee AM, Lalive AL, Wilbrecht L, Bonci A, Kreitzer AC (2016) Cell type specific control of brainstem locomotor circuits by basal ganglia. Cell 164:526–537

Roth MM, Dahmen JC, Muir DR, Imhof F, Martini FJ, Hofer SB (2016) Thalamic nuclei convey diverse contextual information to layer of visual cortex. Nat Neurosci 19:299–307

Roy JE, Cullen KE (2004) Dissociating self generated from passively applied head motion neural mechanisms in the vestibular nuclei. J Neurosci 24:2102–2111

Saleem AB, Ayaz A, Jeffery KJ, Harris KD, Carandini M (2013) Integration of visual motion and locomotion in mouse visual cortex. Nat Neurosci 16:1864–1869

Schilstra C, Hateren JH (1999) Blowfly flight and optic flow. I. Thorax kinematics and flight dynamics. J Exp Biol 202:1481–1490

Schneider DM, Nelson A, Mooney R (2014) A synaptic and circuit basis for corollary discharge in the auditory cortex. Nature 513:189–194

Schnell B, Joesch M, Forstner F, Raghu SV, Otsuna H, Ito K, Borst A, Reiff DF (2010) Processing of horizontal optic flow in three visual interneurons of the Drosophila brain. J Neurophysiol 103:1646–1657

Schnell B, Weir PT, Roth E, Fairhall AL, Dickinson MH (2014) Cellular mechanisms for integral feedback in visually guided behavior. Proc Nat Acad Sci U S A 111:5700–5705

Scott EK, Raabe T, Luo L (2002) Structure of the vertical and horizontal system neurons of the lobula plate in Drosophila. J Comp Neurol 454:470–481

Seelig JD, Jayaraman V (2015) Neural dynamics for landmark orientation and angular path integration. Nature 521:186–191

Seelig JD, Chiappe ME, Lott GK, Dutta A, Osborne JE, Reiser MB, Jayaraman V (2010) Two photon calcium imaging from head fixed Drosophila during optomotor walking behavior. Nat Methods 7:535–540

Shik ML, Severin FV, Orlovsky GN (1966) Control of walking and running by means of electrical stimulation of the mid-brain. Biophysics 11:756–765

Joshi S, Yin Li, Rishi M. Kalwani, Joshua I. Gold (2016) Relationships between pupil diameter and neuronal activity in the locus coeruleus, colliculi, and cingulate cortex. Neuron 89:221–234

Silies M, Gohl DM, Clandinin TR (2014) Motion detecting circuits in flies coming into view. Annu Rev Neurosci 37:307–327

Sillito AM, Cudeiro J, Jones HE (2006) Always returning feedback and sensory processing in visual cortex and thalamus. Trends Neurosci 29:307–316

Sillar KT, Roberts A (1988) A neuronal mechanism for sensory gating during locomotion in a vertebrate. Nature 331:262–265

Simpson JI (1984) The accessory optic system. Annu Rev Neurosci 7:13–41

Sommer MA, Wurtz RH (2002) A pathway in primate brain for internal monitoring of movements. Science 296:1480–1482

Sommer MA, Wurtz RH (2006) Influence of the thalamus on spatial visual processing in frontal cortex. Nature 444:374–377

Sperry RW (1950) Neural basis of the spontaneous optokinetic response produced by visual inversion. J Comp Physiol Psychol 43:482–489

Stackman RW, Golob EJ, Bassett JP, Taube JS (2003) Passive transport disrupts directional path integration by rat head direction cells. J Neurophysiol 90:2862–2874

Steriade M, Datta S, Pare D, Oakson G, Dossi RC (1990) Neuronal activities in brain–stem cholinergic nuclei related to tonic activation processes in thalamocortical systems. J Neurosci 10:2541–2559

Steriade M, McCormick DA, Sejnowski TJ (1993) Thalamocortical oscillations in the sleeping and aroused brain. Science 262:679–685

Strauss R, Berg C (2010) The central control of oriented locomotion in insects—towards a neurobiological model. IEEE world congress on computational intelligence, pp 3919–3926

Suver MP, Mamiya A, Dickinson MH (2012) Octopamine neurons mediate flight induced modulation of visual processing in Drosophila. Curr Biol 22:2294–2302

Tuthill JC, Nern A, Rubin GM, Reiser MB (2014) Wide field feedback neurons dynamically tune early visual processing. Neuron 82:887–895
van Breugel F, Suver MP, Dickinson MH (2014) Octopaminergic modulation of the visual flight speed regulator of Drosophila. J Exp Biol 217:1737–1744
van Swinderen B (2007) Attention-like processes in Drosophila require short-term memory genes. Science 315:1590–1593
Vinck M, Batista-Brito R, Knoblich U, Cardin JA (2015) Arousal and locomotion make distinct contributions to cortical activity patterns and visual encoding. Neuron 86:740–754
von Holst E, Mittelstaedt H (1950) Das Reafferenzprinzip. Wechselwirkungen zwischen Zentralnervensystem und Peripherie. Naturwissenschaften 37:464–476
Voss M, Ingram JN, Haggard P, Wolpert DM (2006) Sensorimotor attenuation by central motor command signals in the absence of movement. Nat Neurosci 9:26–27
Wall MB, Smith AT (2008) The representation of egomotion in the human brain. Curr Biol 18:191–194
Weir PT, Dickinson MH (2015) Functional divisions for visual processing in the central brain of flying Drosophila. Proc Nat Acad Sci U S A 112:E5523–E5532
Weir PT, Schnell B, Dickinson MH (2014) Central complex neurons exhibit behaviorally gated responses to visual motion in Drosophila. J Neurophysiol 111:62–71
Whitlock JR, Sutherland RJ, Witter MP, Moser MB, Moser EI (2008) Navigating from hippocampus to parietal cortex. Proc Nat Acad Sci U S A 105:14755–14762
Wiederman SD, O'Carroll DC (2013) Selective attention in an insect neuron. Curr Biol 23:156–161
Wurtz RH, McAlonan K, Cavanaugh J, Berman RA (2011) Thalamic pathways for active vision. Trends Cogn Sci 15:177–184
Yonehara K, Ishikane H, Sakuta H, Shintani T, Nakamura-Yonehara K, Kamiji NL, Usui S, Noda M (2009) Identification of retinal ganglion cells and their projections involved in central transmission of information about upward and downward image motion. PLoS ONE 4:e4320
Zagha E, Casale AE, Sachdev RN, McGinley MJ, McCormick DA (2013) Motor cortex feedback influences sensory processing by modulating network state. Neuron 79:567–578

Chapter 11
Causal Circuit Explanations of Behavior: Are Necessity and Sufficiency Necessary and Sufficient?

Alex Gomez-Marin

There are empirical methods and conceptual confusions.
Our training and core practices concern research methods;
The discipline is and always has been deeply skeptical of philosophy.
We emphasize methods for the verification of hypotheses and
Minimize the analysis of the concepts entailed by the hypotheses. (...)
All the empiricism in the world can't salvage a bad idea.
(Hogan 2001).

Abstract In the current advent of technological innovation allowing for precise neural manipulations and copious data collection, it is hardly questioned that the explanation of behavioral processes is to be chiefly found in neural circuits. Such belief, rooted in the exhausted dualism of cause and effect, is enacted by a methodology that promotes "necessity and sufficiency" claims as the goal-standard in neuroscience, thus instructing young students on what shall reckon as explanation. Here I wish to deconstruct and explicate the difference between what is done, what is said, and what is meant by such causal circuit explanations of behavior. Well-known to most philosophers, yet ignored or at least hardly ever made explicit by neuroscientists, the original grand claim of "understanding the brain" is imperceptibly substituted by the methodologically sophisticated task of empirically establishing counterfactual dependencies. But for the twenty-first century neuroscientist, after so much pride, this is really an excess of humility. I argue that to upgrade intervention to explanation is prone to logical fallacies, interpretational leaps and carries a weak explanatory force, thus settling and maintaining low standards for intelligibility in neuroscience. To claim that behavior is explained by a "necessary and sufficient" neural circuit is, at best, misleading. In that, my critique (rather than criticism) is indeed mainly negative. Positively, I briefly suggest some

A. Gomez-Marin (✉)
Behavior of Organisms Laboratory, Instituto de Neurociencias
CSIC-UMH, Alicante, Spain
e-mail: a.gomezmarin@umh.es

A. Çelik and M.F. Wernet (eds.), *Decoding Neural Circuit Structure and Function*, DOI 10.1007/978-3-319-57363-2_11

available alternatives for conceptual progress, such as adopting circular causality (rather than lineal causality in the flavor of top-down reductionism), searching for principles of behavior (rather than taking an arbitrary definition of behavior and rushing to dissect its "underlying" neural mechanisms), and embracing process philosophy (rather than substance-mechanistic ontologies). Overall, if the goal of neuroscience is to understand the relation between brain and behavior then, in addition to excruciating neural studies (one pillar), we will need a strong theory of behavior (the other pillar) and a solid foundation to establish their relation (the bridge).

11.1 Explaining Explanation (Prelude)

A cat is chasing a mouse all over the house. When we observe this, or any other phenomenon, we can be in the presence of it for its own sake—we can know it in its immediacy. If we are curious; however, we are soon compelled to ask what is going on. Our quest for understanding begins. The rational mind seeks an explanation.

But, what does it mean to explain a phenomenon? The canonical approach says that to explain is to find the cause.[1] It is known since Aristotle that the notion of causality has a quadruple structure: understanding why something is what it is requires to identify its material cause (what it is made of, namely, the tangible substrate from which something can take place; i.e., nerve cells, muscle, bone), formal cause (what it is to be, namely, what something can become without contradicting itself; i.e., being a mouse and a cat, a predator and a prey), efficient cause (what produces it, namely, the source of change; i.e., hunger, scent, sight), and final cause (what it is for, namely, that for which it becomes and perfects itself; i.e., the need to escape the mouse, sustain life, the good). Science, in its reaction against the teleology of the nineteenth century, eschewed the final cause.[2] Since the formal cause appears uninteresting or incomprehensible, and the material cause is taken for granted, this left the twentieth-century scientist with the main pre-occupation of dissecting efficient causes. In the case of the cat, to find out what brought about chasing behavior.

---

[1]More generally, to explain is to substitute fact by abstraction. The notion of causality is a vast topic, and this is not the place to try to give a complete account. It is of interest to briefly note that it has been claimed that «in advanced sciences (…) the word 'cause' never occurs» (Russell 1913). Indeed, modern physics has succeeded by finding laws (quantitative invariant relations of variables) rather than causes (chains of antecedent–consequent events).

[2]«Teleology has been discredited chiefly because it was defined to imply a cause subsequent in time to a given effect. When this aspect of teleology was dismissed, however, the associated recognition of the importance of purpose was also unfortunately discarded. Since we consider purposefulness a concept necessary for the understanding of certain modes of behavior we suggest that a teleological study is useful if it avoids problems of causality and concerns itself merely with an investigation of purpose.» (Rosenblueth et al. 1943).

When understanding becomes the exercise of determining efficient causality, and efficient causality only, one is then compelled to ask: what *produced* the behavior of the cat[3]? A child would easily reply that the mouse is the cause; another would say that it is because the cat is hungry that it is going after the mouse. For almost all neuroscientists, it is obvious that it is the cat's brain what caused the cat's body to chase the mouse's body with its little mouse brain inside. Our task here is to expose what is meant when one claims that "neural circuit X causes behavior Y," and then determine to what extent this constitutes a satisfactory endpoint to the explanation of behavior itself.

Interestingly, from an evolutionary perspective, one may dare to claim that the converse is true: isn't behavior what produced the brain? (Not to mention that bacteria behave, plants behave and robots behave, none with brain cells). In other words, causality in biology goes both ways. It is only because we prefer to concentrate on the short timescales required by our laboratory experiments that we stress proximate physiological causes at the expense of overlooking ultimate ones, including development and evolution.[4] Tinbergen's four questions remind us that to understand behavior one needs to consider more than what is just going on here-and-now,[5] but to seriously take into account context and history as nested timescales.[6]

If to explain is to determine efficient causality, let's examine its common definition in more detail: the cause of an effect is the set of factors that produce,

[3]Causality need not be equated with necessary connection: «It can grant that there are situations in which, given the initial conditions and no interference, only one result will accord with the laws of nature; but it will not see general reason, in advance of discovery, to suppose that any given course of things has been so determined. So it may grant that in many cases difference of issue can rightly convince us of a relevant difference of circumstances; but it will deny that, quite generally, this *must* be so.» (Anscombe 1971). Put plainly: «not being determined does not imply not being caused.» (ibid).

[4]«We suggest that in many cases in biology, the causal link might be bidirectional: A causes B through a fast-acting physiological process, while B causes A through a slowly accumulating evolutionary process. Furthermore, many trained biologists tend to consistently focus at first on the fast-acting direction, and overlook the slower process in the opposite direction. (…) While A is a proximate cause of B, B may have prevailed even before A, and may have ultimately affected A. So why is the reasoning of many biologists seemingly more prone to focus at first on the effect acting on the short-term, physiological time scale explanation and not on the processes that take millions of years to manifest themselves? Is it because of the biologists' training?» (Karmon and Pilpel 2016).

[5]«Huxley likes to speak of "the three major problems of Biology": that of causation, that of survival value, and that of evolution—to which I should like to add a fourth, that of ontogeny» (Tinbergen 1963) or «behavior is part and parcel of the adaptive equipment of animals; that, as such, its short-term causation can be studied in fundamentally the same way as that of other life processes; that its survival value can be studied just as systematically as its causation; that the study of its ontogeny is similar to that of the ontogeny of structure; and that the study of its evolution likewise follows the same lines as that of evolution of form.» (ibid).

[6]«It is now time to ask about its phylogenetic origins. Not because an historical explanation could replace the efficient causes of dynamics, but in order to see how these dynamics came to be actualised.» (Kortmulder 1998, p. 123).

bring about, or make the effect happen. The average neuroscientist, then, reckoning the mouse as a stimulus, chasing as a response, and hunger as an internal state, would devote the quest to try to figure out their neural substrate (as we will argue, he or she is not to blame—but still accountable—for the use of a poor behavioral conceptualization and a weak neuro-behavioral nexus). Then, in the effort to spatially localize efficient causality—a textbook example of the so-called mereological fallacy (to ascribe to a part what only applies to the whole)—it is inside the skull of the cat where the real deal is taken to be. And therefore it is there where it ought to be sought.

We glimpse a sophisticated offshoot of behaviorism, which could be called "neuralism." Behaviorism, in its obsession against the mental, insisted that only what was purely observable and measureable as external acts performed by the organism should be worthy of scientific study. When it became possible to start looking at the inside, neurophysiology added a window to study behavior by making the inside nearly comparably as observable as the outside. Because, no doubt, the inside is uncharted fascinating territory, twenty-first century neuralism surpassed twentieth-century behaviorism. At the same time, the former carried and extended the most characteristic bias of the latter: the idea that only what is directly observable—now the behavior of the nervous system—is what relevantly the matter is. Thus, neuralism, in its indifference for (if not disdain toward) animal behavior—the very same phenomenon it itself had set at the very core of its agenda (the cat chasing the mouse)—inverted the imbalance: to the degree that one makes sense of what is going on inside, what is going on outside can be rendered as mere contingency. Content ("things held") was once more divorced from context ("things woven"). Differing in what scientists *can* do (now to manipulate and measure the activity inside), both neuralism and behaviorism coincide in what scientists *want* to do (to establish input–output relations, within a "black box," or not). For the most part, reflexology still dominated their thoughts; stimulus and response as the only realities.[7] Inheriting the successful "response function" theory of physics, scientists in the life and social sciences tried it as an ansatz to the study of the behavior of organisms: vary external (or internal) conditions, and relate them to observed changes in output. Rather than stressing one side of the inside–outside dichotomy, the bias of behaviorism is actually better defined as the idea that behavior must be conceived as a lineal input–output process.[8] For neuralism it is the same. Behavior is not regarded as a process of its own right, but as a by-product of neural activity. Then, facts about behavior are deemed as "just phenomena" (and one often hears: "we are interested in mechanisms!"), which amounts to treating behavior as an

[7]«precisely on the condition of limiting oneself strictly to the identity or difference of responses in the presence of such and such given stimuli» (Merleau-Ponty 1942, p. 183).

[8]«Are we not brought back to the classical problems which behaviorism tried to eliminate by leveling behavior to the unique plane of physical causality?» (ibid, p. 131).

"epi-phenomenon" of neural mechanisms. Dawkins' reflection on the neurophysiologist's nirvana is worthy of repetition forty years later.[9]

Indeed, changes in the brain "go together with" changes in behavior. Despite lacking two-photon microscopy and ignorant about the existence of neurons, a millennial papyrus reports what could perhaps be considered the first observed neural-motor lesion correlation in history: the realization that what is inside the head "has something to do" with behavior.[10] But, why do we take for granted that behavior is understood to the extent that some part of the brain is shown to "cause" it? And, perhaps most importantly, how does that idea determine our scientific agenda?

A great deal of neuroscience has become "circuit cracking." Since its inception, our training as neuroscientists reinforces the idea that given "a behavior," our job essentially consists in pinning down the neurons responsible for it. This is, of course, reasonable. Yet, note that "a behavior" usually implies whatever one wishes to call "a behavior," namely, an unexamined arbitrariness most recently sanctioned by the relative ease at quantifying "what animals do;" any choice of ours is justified as long as we "put a number on behavior" (a critical point we would not have space to fully address here).

Indeed, if neuroscience is understood literally as "neural science" (the science primarily concerned with the properties of neural tissue), one may do without behavior, at least for a while. Yet, if neuroscience's ultimate goal is to explain how nervous activity *enables* or *supports* behavioral processes, then we should be quite preoccupied with how much "filling in the (neural) gaps" *per se* entails understanding of behavior all.

---

[9]«If we look far into the future of our science, what will it mean to say we 'understand' the mechanism of behavior? The obvious answer is what may be called the neurophysiologist's nirvana: the complete wiring diagram of the nervous system of a species, every synapse labeled as excitatory or inhibitory; presumably, also a graph, for each axon, of nerve impulses as a function of time during the course of each behavior pattern. This ideal is the logical end point of much contemporary neuroanatomical and neurophysiological endeavor, and because we are still in the early stages, the ultimate conclusion does not worry us. But it would not constitute understanding of how behavior works in any real sense at all. No man could hold such a mass of detail in his head. Real understanding will only come from distillation of general principles at a higher level, to parallel for example the great principles of genetics— particulate inheritance, continuity of germ-line and non-inheritance of acquired characteristics, dominance, linkage, mutation, and so on. Of course neurophysiology has been discovering principles for a long time, the all-or-none nerve impulse, temporal and spatial summation and other synaptic properties, y-efferent servo-control and so on. But it seems possible that at higher levels some important principles may be anticipated from behavioral evidence alone. The major principles of genetics were all inferred from external evidence long before the internal molecular structure of the gene was even seriously thought about.» (Dawkins 1976).

[10]«If thou examinest a man having a smash of his skull, under the skin of his head, while there is nothing at all upon it, thou shouldst palpate his wound. Shouldst thou find that there is a swelling protruding on the out side of that smash which is in his skull, while his eye is askew because of it, on the side of him having that injury which is in his skull; (and) he walks shuffling with his sole, on the side of him having that injury which is in his skull.» (The Edwin Smith Surgical Papyrus 1930).

You may have noticed how the beginning of a standard neuroscience presentation must include "the neural mechanisms of behavior" rhetoric no-matter-what in the first slide (rhetoric, and thus concerned with techniques and skills on how to succeed in the public sphere and advance one's career rather than with the subject matter; persuasion before precision and allure before clarity). Invariably, the speaker must announce a "circuits-of-behavior reduction." The behavioral phenomenon is not only taken for granted, but also deemed as trivial. Dissecting its neural basis is what guides curiosity and drives research (and attracts funding). In fact, by conflating phenomenon with appearance, mechanism appears real while phenomenon becomes epiphenomenal. One should feel fortunate if behavior survives one more slide. If it does, it is usually in the form of an awkward hybrid of anthropomorphic psychological constructs and ad hoc quantitative indices, more or less automated and refined. From then on, one is free to abandon the phenomenon (aka, the cat chasing the mouse) in order to move into to the real deal as soon as possible: the neural circuits.

If the "how" is postulated to be found in neural circuits as the explanatory cause of behavior—and note that it could be sought elsewhere still as a "how," for instance in biomechanics, metabolism, etc.—then one is thinking about a particular notion of efficient cause: «the necessary and sufficient condition for the appearance of something», that «at the presence of which the effect follows, and at whose removal the effect disappears.»[11]

11.2 Counterfactual Dependence (Fugue)

Indeed, upon "certain interventions" at the neural level, "certain changes" take place at the behavioral level (or the following pervasive hypothesis: "If I did stuff, things would happen."). This theoretical paradigm can be methodologically pursued with great efficacy by means of necessity and sufficiency (N&S) tests. The speed, precision, and selectivity of current neural interventions have established such decomposition method as the dominant *explicit* experimental procedure and *implicit* conceptual framework to address brain-behavior relations. The N&S approach to explanation is empowered by today's "manipulate and measure" (M&M) doctrine. Both operate under the presupposition that, if one controls the input as well as possible and then records the output as well as possible, any problem is in principle solvable. From this perspective, improved instrumentation and data collection are the essence of progress.

In other words, the explanation of behavior seems to be contained in claims such as "circuit activity X is *necessary* for behavioral process Y" (or, "if X had not happened, then Y would not have happened") in combination, when possible, with claims such as "circuit activity X is *sufficient* for behavioral process Y" (or, "if X were to happen, then Y would also happen").

[11]See Galileo's definition of cause in Bunge (2009, p. 33).

The N&S approach is legitimate in principle, popular in practice, procedurally simple, methodologically powerful, and conceptually straightforward. While its virtues are often celebrated (sometimes hyped), its problematic points are hardly acknowledged, at least in neuroscience forums. Here I try to explicate the difference between what is *done*, what is *said*, and what is *meant* when "circuit X explains behavior Y."

Failure of behavioral function upon inhibition of neural function is interpreted as the latter being a necessary condition for the former—the circuit is thus claimed to be indispensable, or necessary. Respectively, emulation of behavioral function upon activation of neural function is interpreted as the latter being a sufficient condition for the former—the circuit is thus claimed to be enough, or sufficient. If both N&S hold, the circuit is claimed to be *the cause* of the behavior, which is then regarded as pretty much *explained*. But «all the empiricism in the world can not salvage a bad idea» (Hogan 2001).

We must ask to what extent N&S reflect facts or their interpretation[12] (note that different interpretations may generate different experiments). In conflict with the above prescription, in reality, co-occurrence of behavioral activity upon neural activation is one thing, and circuit sufficiency is another. Similar concerns must be raised when making necessity claims. Let's see why, and where these leaps of interpretation lie.

The problem with this simplistic model of causality is that it defines necessary and sufficient in order to create a specific effect. It asserts its own formal cause into the process, isolating that phenomenon from any real process. So it defines conditions for its self-validation. When setting up conditions that demonstrate cause and effect relations that we create, we have imitated the principle, but it is still to be seen whether this reveals the actual conditions for the existence of the natural phenomenon. To put it plainly, our causal manipulations do not produce the behavioral effect: they reproduce it in a given context.

Necessary expresses "what is needed," while sufficient expresses "what meets the need" for something to occur. Necessary is what is required, compulsory, indispensable, not susceptive of being waved. Sufficient is what is enough,[13] adequate, unwilling to tolerate any more of something. Remaining ambiguous about whether those needs are primarily of the scientist or of the animal whose circuit is to be claimed necessary and sufficient, N&S conditions are harder to understand than to believe.

[12]«statements of necessity and sufficiency are not fundamental truths about neural mechanisms, but rather are interpretations of experimental outcomes» (DiDomenico and Eaton 1988).

[13]«Now "sufficient condition" is a term of art whose users may therefore lay down its meaning as they please. So they are in their rights to rule out the query: "May not the sufficient conditions of an event be present, and the event yet not take place?" For "sufficient condition" is so used that if the sufficient conditions for X are there, X occurs. But at the same time, the phrase cozens the understanding into not noticing an assumption. For "sufficient condition" sounds like: "enough." One can ask: "May there not be enough to have made something happen—and yet it not have happened?"» (Anscombe 1971).

Our insistence for objectivity can be an excess of pretense for disinterestedness after so much anthropocentrism. In accounting for the behavior of the cat, it is not the cat that is placed at the center of the explanatory effort, but our own activity as humans, with our biases, interests, and habits. We say "circuit X is sufficient" for the cat to behave but what we really mean—and tragically omit—is that "it is sufficient for us to activate circuit X" in order to observe the cat's natural behavior.[14] In other words, we are not concerned with the myriad of processes that nature puts in confluence, but only with the narrow element that we need to insert in nature in order to emulate its principle.[15] Thus, the sufficient condition belongs more to us than to the cat. Remaining mostly ignorant about how or why the phenomenon occurs, what is brought to the foreground (thus, all that counts in practice) is what is "enough for me to do" so that the cat chases the mouse once more. Similarly, when we say "circuit X is necessary" for the cat to behave, again we imply that it is imperative for nature to have that circuit at work so as to be able to produce the cat's behavior, while all we showed—and all we can really say—is that "it is necessary for us" to remove circuit X so as to be able to block the natural phenomenon. Still remaining in ignorance (now tamed by the great feeling of control that intervention brings), what is brought to the foreground is what is "indispensable for me to do" so that the cat doesn't chase the mouse this time. Causal explanations of this flavor, then, turn out to be more necessary (and sufficient) for the neuroscientist than for neuroscience; circuit activation and inhibition are more the means for the neuroscientist to try to explain the cat's behavior than for the cat to try to catch the mouse.[16]

[14]Causal accounts reflect the notion of liability in court: "the judge decides that an individual is liable to a certain amount for an action he has *caused.*"

[15]«the lights went on when Mrs. Smith turned the switch (…). It is evident that the statements are all singular, rather than general or lawlike. Moreover, in none of them is the occurrence tacitly assumed to be "the cause" a sufficient condition for the event alleged to be its effect. For example, turning a switch does not suffice to produce illumination, since many other conditions must be satisfied for this to happen. In making such causal statements, it is, of course, possible that we know what these further conditions are, and take them for granted without mentioning them explicitly; but this is rarely the case, and we are usually able to cite only a few of these conditions, without knowing all of them. In either case, what we are doing is designating as the cause of an event just one item, selected from what is tacitly supposed to be its full complement of necessary and sufficient conditions, because the item is deemed important for various reasons.» (Nagel 1965).

[16]An interesting caveat—revealing an arbitrary preference for a particular level of organization and working in conjunction with a reduced notion of causality—is the following: if circuit activation is said to produce the behavior of the animal, it could also be said that the behavior of the scientist has produced the animal's circuit activation in the first place, or «[a]t most it would show that experience could be produced by means of interaction between a probing scientist and a healthy animal. We haven't yet imagined a case in which experience emerges from the causal effects of neural activity alone» (Noë 2004, p. 211).

The sufficient condition *is by itself not sufficient* because the effect it is shown to produce depends on the conditions in which the cause takes place.[17] This is, each of the elements of the sufficient condition may be necessary to prove emulation, while none of them alone is strictly sufficient. That condition, being all that may be required for the scientist to elicit behavior, is not all what is required for behavior to occur. The sufficient condition is then "relatively sufficient." Moreover, since necessity in biology is nearly always relative, the necessary condition is "contingent necessary."[18] It would then be more accurate to qualify N&S as *relative-contingent* N&S.

Circuit N&S do not occur in the void, but in (and "because of") the particular conditions and relations to other neural and nonneural elements and processes (genetics, biomechanics). One would wish but in fact cannot leave out the contingency of the experimental design (the task that is chosen by the experimenter), nor the temporal constraints of the test (whether manipulations allow time for the organism to adapt, learn or recover). The immediacy of the laboratory conditions in which we test reality often involves shrinking the spatiotemporal spread of behavior.[19] We are often ignorant of (and, unfortunately, sometimes indifferent to) the actual conditions for the occurrence of the behavioral phenomenon we are trying to explain.[20] We can not specify either all the circumstances for which the so-called cause will not cause its effect anymore.[21] "X sufficient for Y" suggests but

[17]«other factors, without which the event would not occur, are all assumed to be constant and given.» (Eaton and DiDomenico 1985).

[18]«although in the sense of "cause" under discussion the cause of an event is a necessary (or indispensable) condition for its occurrence, the necessity may be only relative. (…) there may be more than one set of sufficient conditions for an event's occurrence (…). Accordingly, in the sense of "cause" under discussion, the cause of an event is in general neither a sufficient nor an absolutely necessary condition for the event's occurrence. The cause may be called a "contingently necessary" condition» (Nagel 1965).

[19]«There is something in the context of the experiment which goes beyond the stimuli and responses directly found within it. There is, for example, the problem which the experimenter has set and his deliberate arrangement of apparatus and selection of conditions with a view to disclosure of facts that bear upon it. There is also an intent on the part of the subject. Now I am not making this reference to "problem," "selective arrangement" and "intent" or purpose in order to drag in by the heels something mental over and beyond the behavior. The object is rather to call attention to a definite characteristic of behavior, namely, that it is not exhausted in the immediate stimuli-response features of the experimentation.» (Dewey 1930).

[20]«The point I want to stress is that we seldom have enough information to state explicitly the full set of sufficient conditions for the occurrence of concrete events. The most we can hope to accomplish in such situations is to state what are the best only "important" indispensable conditions, such that if they are realized the occurrence of the designated events is made "probable;" and we thereby take for granted that the remaining conditions essential for the occurrence of the events are also realized, even when we do not know what those remaining conditions are.» (Nagel 1965).

[21]«They realize that if you take a case of cause and effect, and relevantly describe the cause A and the effect B, and then construct a universal proposition, "Always, given A, a B follows," you usually would not get anything true. You have got to describe the absence of circumstances in which A would not cause a B. But the task of excluding all such circumstances can not be carried out.» (Anscombe 1971).

does not mean "sufficient in a vacuum," yet it implies "everything else being equal." Not knowing the context, what we find in one may not apply in another. Disregard contingency, still this view of causation is dependent on it.

Behavioral absence upon circuit inhibition, and its presence upon activation, can generate false positives and false negatives, and produce spurious conclusions (see Eaton and DiDomenico 1985). Results can have alternative interpretations due to the system's plasticity and redundancy. A neuron or circuit may be N&S for the chain of processes to continue, but not for the end result, due to shared responsibility and multiple realizability. Activation and inactivation can have secondary side effects. Moreover, perturbations can be opposed and compensated by the organism.

Another limitation of the N&S approach is that most of its claims are not univocal, but statistical. They are valid only upon averaging of trials and pooling of individuals. Being true for everybody, they are not true for everyone. They are the «confused result of many instances» (Whitehead 1933, p. 112). Even accepting the power of M&M under the framework of N&S tests, most of the time we can not "bring about Y by doing X". Variability is then conflated with noise and deemed as undesirable both for the scientist and for the animal. Yet, when control is studied from the perspective of the animal, behavioral variability may be in service of constancy.[22]

The popularity of the N&S approach in neuroscience is in part fed back by the amounts of neural and behavioral data one can now produce at ease. However, the availability of unlimited data, even within putative infinitely precise M&Ms, will not suffice to understand the brain, behavior and their relation. The humbling effort of applying that logic to the functioning of a simple microprocessor cast huge concerns about the validity of such expectation,[23] calling into scrutiny the habit of collecting facts *for the sake of it*, in case we may need them in the future.[24] Given the scarcity of brain-behavior theories, we don't know yet what the right levels of granularity and abstraction are. Without a conceptual thrust to explain behavior, it is unlikely we shall be able to "understand the brain." Call it behavioral chauvinism if you wish; the fact is that brains are for behavior.

[22]«Instead of asking how a particular experimental manipulation alters the subsequent behavior of an organism, one might instead ask how an experimental manipulation alters the parameters of the system. This is a subtly different question, but the difference is important, and requires that the parameters of the system be understood to begin with. Understanding what variables organisms may be controlling necessitates that organisms be understood on their own terms before they are used as model systems to answer larger questions.» (Bell 2014).

[23]«We argue that the analysis of this simple system implies that we should be far more humble at interpreting results from neural data analysis. It also suggests that the availability of unlimited data, as we have for the processor, is in no way sufficient to allow a real understanding of the brain.» (Jonas and Kording 2016).

[24]«the rule "collect truth for truth's sake" may be justified when the truth is unchanging; but when the system is not completely isolated from its surroundings, and is undergoing secular changes, the collection of truth is futile, for it will not keep.» (Ashby 1958).

Notwithstanding the value in manipulation, when our interest is not so much to understand the world but to control it, we end up learning what we do to things rather than what things do. Overall, we don't understand what or why behavior is, but a particular way of inducing its occurrence in hopefully similar experimental conditions. Things get much more delicate when such "things" are animals, because animals are living organisms and their behavior is, as we will see later, self-caused.[25] Actually, what organisms do is to control their perceptions by means of opposing (our and other) disturbances (see below and Powers 1973). In the lab, our concern with what *we* can make animals do easily precludes us from knowing what *they* do and why they do it.

But for many, to explain is to successfully intervene: "A causes B if control of A renders B controllable." If I can outsource the happenings of B to my intervention on A, I will then expect a change in A to be followed by a change in B. Similarly, for many others, to understand is taken as to be able to fix what one tries to understand,[26] which is an inversion of Stuart Mill's famous prescription: "to find out how something works, look to see how it fails." Building, though, is more challenging than intervening or fixing: to build a model *that behaves* is much more insightful than to make a model *of behavior*. In other words, to "abstract" a behavioral process (via classification) or to "extrapolate" from one case to many (via statistics) is less powerful than to have a generative model.[27] But from the interventionist point of view, explanation is that practical information relevant to manipulation, which does not necessarily imply generalization for prediction, and which is quite different than grasping the concrescence of behavior. Be that as it may, to intervene *is not* to explain.[28]

The natural urge to upgrade correlation to causation (a still surprisingly common flaw) has gone further: it has replaced, when possible, neural correlates (NCs) with neural counterfactuals (NCFs), with the concomitant urgency to upgrade intervention to explanation. This makes us believe that what caused the comportment of

---

[25]«Behavior, as a relation between a living system operating as a whole and the medium operating as an independent entity, does not take place in the anatomy/physiological domain of the organism, but depends on it. (…) the behavior that a living system exhibits is neither determined by it nor by the medium alone, even when a particular structural change in a living system may specifically interfere with its ability to generate a particular behavior» (Maturana 1995).

[26]«To understand what this flaw is, I decided to follow the advice of my high school mathematics teacher, who recommended testing an approach by applying it to a problem that has a known solution. (…) I started to contemplate how biologists would determine why my radio does not work and how they would attempt to repair it.» (Lazebnik 2002).

[27]«An explanation is the proposition of a generative mechanism or process which, if allowed to operate, gives rise, as a result of its operation, to the experience to be explained.» (Maturana 1995).

[28]«If a biologist or another scientists is interested in control or manipulation, and many of them are, then this is an epistemic goal for that biologist doing science. It is an important goal, and it is important to learn how to intervene and manipulate in experimental settings. Much scientific training and subsequent practice involves pursuing that goal. Having these as goals for one's science and scientific practice is not the same as finding a mechanism nor is it the same as explaining things by mechanisms. In fact, controlling is not explaining at all.» (Machamer 2004).

the organism is our action upon it, by means of altering certain conditions that trigger this or that output. We fail to understand the behavior is control *of* the organism *by* the organism. Such abuse and misuse of causal thinking is susceptible in and typical of a kind of industrialized science, where the balance between cutting-edge instrumentation and intellectual depth is tilted. When convenience is king, we may be tempted to sacrifice understanding.[29]

The M&M approach, under the N&S paradigm, has more important and far-reaching provisos. It conflates experimental results with the particular interpretation it makes of them.[30] Running with virtually no theory (something to be covered in more detail in another occasion), it appears to exempt one from making explicit any pre-suppositions for the explanation of the phenomenon; no abstract idea beyond the minimal conjecture that "a change" between two conditions might be observed (and if it is not observed, one can always vary the conditions until it is). Such a "nought of question" easily dissimulates itself under the answers provided by statistical hypothesis testing: testing for rejection of the null hypothesis is transformed into the rejection of testing a null hypothesis (a camouflaging that is not the statistician's fault). Lack of conceptual insight often collapses into a "Shakespearean two-alternative forced choice:" to be or not to be … statistically significant.

Now it is clear why the interventionist method at the circuit level doesn't teach us much about the structure of behavior, but mainly whether it occurs or not (regardless of its biological significance), which confronts us with a grave problem: anything can be called behavior.[31] In other words, when searching for NCFs, "anything goes" for behavior. Erecting the neural circuit causation approach as the conceptual framework to understand behavior reduces the exploration of behavioral theories to the exploitation of handling protocols. The search for principles of behavior is substituted by the development of molecular and cellular techniques. Empirical work takes completely over theoretical work. Concepts are reduced to methods and experimentation becomes data collection. Then, explaining behavior consists in coarsely describing it, saying "because," and then describing in detail some of the molecular and cellular substrates that "produce" it. If we agree to call

[29]«Its first characteristic is that its ultimate aim is not understanding but the purely practical one of control. If a system is too complex to be understood, it may nevertheless still be controllable. For to achieve this, all that the controller wants to find is some action that gives an acceptable result; he is concerned only with what happens, not with why it happens. Often, no matter how complex the system, what the controller wants is comparatively simple: has the patient recovered? —have the profits gone up or down? —has the number of strikes gone up or down?» (Ashby 1958).

[30]«This operational approach excludes alternative methodologies because the N&S causal definition is the same as the methodology for its own demonstration.» (DiDomenico and Eaton 1988).

[31]«the methodological connection to behavior is undefined since the Command Neuron Experiment allows virtually any phenomenon to be used as "behavior"» (ibid).

behavior "anything I can put a number on," it is then likely that we will discover many neural substrates of foggy constructs.[32]

A deeper source of confusion and misunderstanding is the belief that the N&S approach belongs to the practice of a kind of "immaculate perception," devoid of and proudly untainted by any bias of interpretation. Being nowadays most popular and least consciously practiced "epsilon of theory"—self-limited to what the eye can literally see, not what the mind can think—it preaches the M&M gospel of our times: "when we map it all, we shall explain it all." But, such lack of premise is nothing but a premise of lack.[33]

Pay attention to the Q&A after a seminar, and you may observe the ease with which the speaker happily turns down speculation when asked "what would you expect?" by uttering "we don't know" and then comfortably completing the statement by saying "we will see what happens when we do this and that." As if not having any idea of what could be going on, and leaving it all to future M&Ms, was a virtue. Their null model is then merely "a difference" in its minimal conceptual sense, namely, that upon changes in one variable we expect something different in another. When, upon scratching, the *null* hypothesis is hardly more than "to check the effects of X on Y," we are in the presence of the *dull* hypothesis: put a thousand cats in a thousand boxes with a thousand mice, and randomly turn a thousand circuits off and on, until behavior disappears and appears again. Lack of insight, both before and after the experiment, is compensated by figures that seek to impress (i.e., dozens of automated assays in parallel, hundreds of neurons recorded simultaneously, thousands of animals tested, millions of frames generated, billions of dollars spent). This is a naïve understanding of the idea that "more is better."[34]

Of course one cannot in principle be opposed to data, more data, and much more data. In practice, though, this can lead to "chaos in the brickyard" (Forscher 1963). Scientists can be considered builders of explanations by means of assembling facts. Such "bricks," difficult and expensive to make, were essential for the edifice not to collapse. As more emphasis was put into the brick-making system, data became easier and faster to produce. Very good bricks were certainly made. And many more were produced ahead of demand, pending the decision of what type of

[32]A biologist friend of mine once told me:"what I care about is to pin down the genes of a behavior, behavior being anything I can put a number so that it will allow me to get to my genes." Replace genes (or gene networks) with neurons (or neural circuits) and the punch-line is the same: behavior as a pretext for genetic and neural manipulations in the age of technocratic science.

[33]«The scientist cannot make the rejoinder here that he thinks without ontological background. To believe that one is not doing metaphysics or to want to abstain from doing it is always to imply an ontology, but an unexamined one —just as governments run by "technicians" do not make political policy, but never fail to have one— and often the worst of all.» (De Waelhens 1942).

[34]«this is no doubt due to practical preoccupations and notably to a representation of productivity and labor in terms of scale. The more masons work, the higher the building rises. The more copyists copy, the longer their copy becomes. Fabricating labor is the only labor in which the amplification of the product is quantitatively, spatially proportional to the progress of action. (…) A technician's and artisan's thought willingly concentrates on this demiurgic elaboration of the indeterminate.» (Jankélévitch 2015 p. 168).

building one would be supposed to erect with them. Ultimately, there were so many bricks everywhere that they started to defeat their purpose. «It became difficult to complete a useful edifice because, as soon as the foundations were discernible, they were buried under an avalanche of random bricks. And, saddest of all, sometimes no effort was made even to maintain the distinction between a pile of bricks and a true edifice» (Forscher 1963). Data, for data's sake, overrules theory and, thus, nullifies itself.

This radical belief in induction,[35] intertwined with a conscious eagerness of not pursuing hypothesis,[36] misses the fact that any description starts with comparison which, in turn, assumes a particular perspective.[37] Indifference is impossible. There is always something in the data which we turn our attention to and away from. Accordingly, bringing forth the contributions of our subjectivity is an honest attitude, prone to less confusion than insisting on unbiased approaches (often confusing unsupervised with unbiased), and hoping to be able to produce results that

[35]The belief that «[s]cientific discovery, or the formulation of scientific theory, starts in with the untarnished and unembroidered evidence of the senses. It starts with simple observation —simple, unbiased, unprejudiced, naive, or innocent observation—and out of this sensory evidence, embodied in the form of simple propositions or declarations of fact, generalizations will grow up and take shape, almost as if some process of crystalization or condensation were taking place. Out of a disorderly array of facts, an orderly theory, an orderly general statement, will somehow emerge» (Medawar 1978).

[36]«The belief underlying Mass Observation was apparently this: that if one could only record and set down the actual raw facts about what people do and what people say in pubs, in trains, when they make love to each other, when they are playing games, and so on, then somehow, from this wealth of information, a great generalization would inevitably emerge. Well, in point of fact, nothing important emerged from this approach. (…) [T]he starting point of induction is philosophic fiction. There is no such thing as unprejudiced observation. Every act of observation we make is biased. What we see or otherwise sense is a function of what we have seen or sensed in the past. The second point is this: Scientific discovery or the formulation of the scientific idea on the one hand, and demonstration or proof on the other hand, are two entirely different notions.» (ibid).

[37]«In the use of language, for instance, we depend on the fact that names have been given to objects, qualities, and relations, which fix certain similarities and differences in the flow of experience as boundaries containing it, dividing it, directing it. Whenever we describe, we class things or properties or events together or apart on the basis of the similarities and differences marked by the words we choose. Consequently, to the extent that science begins with description, it begins with comparison. But no two things, no two qualities, no two events are alike in all respects, or alike in none. Any description singles out some similarities and differences to the exclusion of others, which could be the basis of alternative descriptions. Consequently, a demand for a complete description of anything amounts to a contradiction in terms. A demand for a pure description would be equally incoherent, for, of necessity, the similarities and differences that we pick out when we describe anything will depend on what we intend the description for, our expectations about the matter in question, considerations of relevance to some focus of interest, and other prior assumptions. Comparison necessarily assumes perspective.» (Beer 1980).

equate with their interpretation. Pretense of absolute is an absolute pretense. And dismissing that as "just philosophy" is just a philosophy of dismissal.[38]

If what I am saying is accurate, it will also be unpopular, because most biological sciences, and in particular neuroscience, have great investments ("promissory notes" à la Feyerabend) in counterfactual dependence as a proxy for explanation. The step-by-step interventionist approach is comforting to the intellect because it provides a monotonic succession of measurable chains. The M&M *modus operandi* and its recipe are this: try to keep everything fixed, change one thing at a time,[39] and see what varies. But, as suggested above and shown below, exploiting the relationship between an independent variable and a dependent variable presupposes a lineal notion of causation, which is inadequate in the study of animal behavior.[40] In biology (the study of living organisms!) the method of investigation must respect the system being studied. Organisms cannot be properly studied as closed physical systems. William James example opposing magnets and lovers[41] emphasizes that one of the essential properties of living organisms is that they can produce the same ends *with* (and *by*) different means. Inert systems, in contrast, tend to produce different ends even with very similar means. The former are characterized by convergence to a goal, the latter by sensitivity to initial conditions.[42] This leads us into a serious appraisal of what *the behavior of organisms* is, and what it is not; the profound difference between a cat chasing a mouse and a leaf blown by the wind.

---

[38]«While often explicitly denying the relevance of philosophy to its operations, psychology has implicitly used the philosophical assumptions of a seventeenth-century ontological dualism, a nineteenth-century epistemological empiricism, and an early twentieth-century neopositivism, to build a standard orthodox approach to the resolution of the antinomies. (…) the product of the acceptance of some basic ontological and epistemological —hence philosophical—assumptions. These assumptions begin with the idea of splitting reason from observation, and follow with the epistemological notion that knowledge and, indeed, reason itself originates in observation and only observation. These assumptions then lead to a particular definition of scientific method as entailing observation, causation, and induction–deduction, and only observation, causation, and induction–deduction. Sometimes, the split is found in explicit and implicit attacks on theory, as in a particular rhetoric that states that all theories must be induced directly from observations (i.e., must be "data based" or "data driven"). It is also found in a dogmatic retort given to any reflective critique —"that's just philosophy."» (Overton 2006).

[39]«until about 1925, the rule "vary only one factor at a time" was regarded as the very touchstone of the scientific method.» (Ashby 1958).

[40]«What we have is a circuit, not an arc or broken segment of a circle. This circuit is more truly termed organic than reflex, because the motor response determines the stimulus, just as truly as sensory stimulus determines the movement.» (Dewey 1896, p. 363).

[41]«If now we pass from such actions as these to those of living things, we notice a striking difference. Romeo wants Juliet as filings want a magnet; and if no obstacles intervene he moves toward her by as straight a line as they. But Romeo and Juliet, if a wall be built between them, do not remain idiotically pressing their faces against its opposite sides like the magnet and the filings with the [obstructing] card. Romeo soon finds a circuitous way, by scaling the wall or otherwise, of touching Juliet's lips directly. With the filings the path is fixed; whether it reaches the end depends on accidents. With the lover it is the end which is fixed; the path may be modified indefinitely.» (James 1890, p. 6).

[42]«A profound difference between most inanimate and living systems can be expressed by the concept of equifinality» (Von Bertalanffy 1950).

11.3 Beyond Lineal Causality (Allegro)

In our era of non-risky science due to scientific careers at risk, little can change if what is rewarded (and thus selected for) is confined to the conceptual template: "I study the role of brain region X in the behavioral construct Y"—even if we add some glam bio-tech and really good selling skills ... Such a research program provides "many results" which, upon close inspection, are often *questionless answers*. A lot is done but nothing is really tested, and so one can never be wrong. Note how many titles and abstracts in neuroscience journals and seminars reflect (at the same time that conceal) this by means of "filler verbs" that project a sense of understanding and explanation.[43] This lack of (ability or interest for?) a solid theory of behavior and of brain-behavior relations is turning the "explanatory gap" between circuits and behavior into an "explanatory jump." Since neuronal circuits are nominated to *explain behavior*, we were compelled to ask two apparently unnecessary questions: What does it mean to *explain*? And, what is *behavior*? All the effort made in this essay went in the direction of not taking for granted that turning neurons on-and-off and subsequently observing behaviors on-and-off is a satisfying schema for explanation in current neuroscience. Let me conclude by sketching, very briefly, some ideas and alternatives[44] on N&S and M&Ms, and their associated ideologies.

An organism is cause and effect of itself—one can say it louder but not clearer. Behavior is control. Behavior is, therefore, a circular-causal process. Then, the study of living beings requires going beyond lineal causality.[45]

---

[43]Amongst the most used verbs we find: "involves, reflects, reveals, mediates, is associated with, contributes to, shapes, modulates, alters, regulates, drives, determines, generates, produces, encodes, underlies, induces, enables, ensures, supports, promotes, suppresses, inhibits, prevents, disrupts, controls, and causes." The pet expression is perhaps "X plays a role in Y," or "The role of X in Y." Certainly, lack of coffee "plays a role" in my writing of this piece, and gravity "mediates" the mouse escape from the paws of the cat.

[44]«But it is not sufficient to oppose a description to reductive explanations since the latter could always challenge these descriptive characteristics of human action as being only apparent. It would be necessary to bring to light the abuse of causal thinking in explanatory theories and at the same time to show positively how the physiological and sociological dependencies which they rightly take into account ought to be conceived.» (Merleau-Ponty 1942, p. 176).

[45]«In describing the physical or organic individual and its milieu, we have been led to accept the fact that their relations were not mechanical, but dialectical. A mechanical action, whether the word is taken in a restricted or looser sense, is one in which the cause and the effect are decomposable into real elements which have a one-to-one correspondence. In elementary actions, the dependence is unidirectional; the cause is the necessary and sufficient condition of the effect considered in its existence and its nature; and, even when one speaks of reciprocal action between two terms, it can be reduced to a series of unidirectional determinations. On the contrary, as we have seen, physical stimuli act upon the organism only by eliciting a global response which will vary qualitatively when the stimuli vary quantitatively; which respect to the organism they play the role of occasions rather than of cause; the reaction depends on their vital significance rather than on the material properties of the stimuli. Hence, between the variables upon which conduct actually depends and this conduct itself there appears a relation of meaning, an intrinsic relation. Once cannot assign a moment in which the world acts on the organisms, since the very effect of this "action" expresses the internal law of the organism.» (Merleau-Ponty 1942, p. 160).

The foundations to understand feedback control can actually be found way back in the history of Egyptian inventions.[46] Explanations of behavior need to abandon the "push-pull" idea of lineal causality and embrace the "go-round" notion of circular causality (ironically, the etymology of circuit is to "go-round").

If we must cease to look for cause and effect in behavior, what shall we look for then? When it comes to studying the behavior of organisms, lineal causality is to circular causality what arithmetics is to algebra. Namely, rather than being engaged in finding all possible combinations between stimulus and response, the real quest is to find stable relationships, whatever the value of the variables[47]; to find relations, rather than list all connections. Data collection then becomes experimentation as well as conceptual testing.

Accordingly, to treat behavior as a dependent variable, and stimulation or neural manipulation as an independent variable, ignores feedback, which is the essential ingredient in the process that constitutes behavior. The M&Ms approach (precise control manipulates one variable in the system, and precise monitoring captures whatever the change), which by itself cannot suffice, is still valuable if N&S tests are upgraded to the Test for Controlled Variables (Marken 2001). The ability of the organism to oppose disturbances, granted by circular causality via negative feedback loops, allows revisiting the taboo of teleology in the science of behaving systems.[48] Rather than concentrating on models of behavior, let us conceive models that behave, and then set up our experiments so as to allow control, not of the animal, but by the animal. In a word, to see behavior from the animal's perspective.[49]

---

[46]«Ktesibios's water clock required a steady, unvarying flow of water to measure accurately the steady, unvarying flow of time. But because water flows more quickly from a full container and more slowly when it is less full, Ktesibios had to devise a way to keep the vessel at a constant level while water was flowing from it into the clock mechanism. As he did this in a manner not unlike that of the modern flush toilet to which it is assumed the reader has handy access, I will use this more modern invention instead of the water clock as our first example of a feedback-control device.» (Cziko 2000).

[47]«When you learn to see behavior in terms of relationships among variables instead of causal connections between one event and another, you can see invariance where formerly you could see only specific causal connections. You can see that when someone builds a fire in the fireplace, two people may show "the same behavior," even though one of them takes off a sweater while another opens a window.» (Powers 1998).

[48]«I would say that a system is teleological if it has a mechanism which enables it to maintain a specific property despite environmental changes. (…) It must have certain types of compensating mechanisms — what we call essentially a negative feedback. (…) I would not say that a simple pendulum which moves in such a way that it strives to achieve the lowest potential energy is a teleological system. There are no compensating effects in the pendulum. Hence, as a system, it does not have an internal structure that enables it to compensate for environmental changes.» (Nagel 1965).

[49]«rather than concentrating on what an animal is doing, what may be more relevant is what the animal is trying to perceive. This double inversion (from the experimenter's point of view to the animal's and from action to perception) has three potentially critical implications for the study of

Another way to phrase this substantially different alternative is the following: adapt the method of study to the object of study (rather than the current Procrustean opposite). In other words, and following good-old-fashioned Uexkullian zoology: treat the animal as a subject. This requires that the means of studying an organism respect the possibility to understand it. Otherwise, neuroscience becomes neural science: the study of the properties of neural tissue and the side effects on changes at the macroscopic level under the moniker of "behavior"—still fascinating yet uncoupled from its significance within the whole, which is the living organism.[50] Deliberate zoomorphism is preferable to unconscious anthropomorphism, which is often a kind of technomorphism—our tools do not tell us what things are or how they work; they enable us to answer properly posed questions about that.

More generally, circular causality exemplifies the essential role of top-down causation[51] in biology (a huge and central topic which we cannot cover here). However, the behavior-to-circuits reduction reinforces the idea that causality's arrow can only be upwards, at the same time that it encourages the belief that the real thing is happening inside, not outside (as we mentioned at the beginning of our discussion, neuroscience is in danger of replacing twentieth-century behaviorism with twenty-first century technology-reinforced neuralism).

Circular causality implies that the behavior of animal is, to a great extent, self-caused. An important source of confusion may lie in conflating input-and-output with cause-and-effect. Note that when there is no distinction between cause and effect (circular causality), there can still be a difference between input and output (something comes in, something comes out—and in again, and out again). Contrary to "chain-like" thinking, effects are simultaneous with causes, not instantaneous.

One can turn this whole contradiction into complementary opposition: lineal causality is a particular case of circular causality upon breaking the loop.[52] But note that even if we artificially break the loop in the lab, the animal is still a creature in closed-loop. There is no escape from this: the challenge is to understand and apply

(Footnote 49 continued)

cognition: first, motor output is a side effect of perceptual control and therefore quantitative data collection data is insufficient by itself, second, averaging across individuals may smear out control variables, and third, restrained experimental setups may not let animals control the relevant inputs; freedom in the requisite dimensions is essential. In short, control (circular causality) and subjectivity (animal centrism) may be essential ingredients in behavioral and cognitive neuroscience.» (Gomez-Marin and Mainen 2016).

[50]«Could not the application of physico-chemical methods possibly mean, in principle, such a destruction of the organism… Can it really teach us something about the functioning of the organism?"» (Goldstein 1934, p. 109).

[51]«five essentially different classes of top-down influence can be identified, and their existence demonstrated by many real-world examples. They are: algorithmic top-down causation; top-down causation via nonadaptive information control, top-down causation via adaptive selection, top-down causation via adaptive information control and intelligent top-down causation.» (Ellis 2011).

[52]«We have a general manipulative technique for making anything hot: we put it on a fire.» (Von Wright 1971).

the notion of causation in closed-loop, and then to realize its asymmetry: that the animal controls its perception of the environment more effectively than the world controls its behavior. It is remarkable how, during the twentieth century, the scientific study of behavior and its neural underpinnings exploited the idea that animal behavior could and should be studied precisely by *not* letting animals behave…

Already 120 years ago, Dewey, in a stroke of genius, brought "the effect of output on input" to its ultimate consequences. He was able to articulate the notion of "circular act" by realizing that «the so-called response is not merely to the stimulus; it is *into* it» (Dewey 1896). In other words, he saw that «the expression of every impulse stimulates other experiences and these react into the original impulse and modify it» (Dewey 1894, p. 14). Foreseeing the pioneering work of the cybernetics movement (which, paradoxically, missed its own point about "control in the animal" by concentrating on "information in the machine"), he asked «What shall we term that which is not sensation-followed-by-idea-followed-by-movement» (Dewey 1896), thus anticipating the idea of «back-reference of an experience to the impulse which induces it» (Dewey 1894, p. 15). Dewey restored the conceptual equilibrium between movement and sensation.[53] In the same year (and in the same book!) we find a very similar pioneering account of perception and action as alternating, rather than alternative, views.[54]

A notable exception to the usual conception of causation is Whitehead's philosophy of organism which, by abandoning «the notion of an actual entity as the unchanging subject of change» (Whitehead 1929, p. 29), goes beyond the rational

[53]«The real beginning is with the act of seeing; it is looking and not a sensation of light. (…) In other words, we now have an enlarged and transformed coordination; the act is seeing no less than before, but it is now seeing-for-reaching purposes. There is still a sensori-motor circuit, one with more content or value, not a substitution of a motor response for a sensory stimulus. (…) it is a seeing-of-a-light-that-means-pain-when-contact-occurs.» (Dewey 1896).

[54]«In relation to each other inside the act of attention, most discussions of the subject appear to make the ear process merely a stimulus to which the hand adjustment is merely a response. But the question arises, What holds the ear to its work? Why does the reagent maintain his listening attitude? It may be replied that it is 'because he is told to. "But he is not told to listen any more than he is told to move his hand. If the telling suffices in one case it should in the other. Moreover, he is not merely to listen, or even to listen just for the click, but to listen for the click as a pressure signal. It is this character of the click as a signal for pressure that keeps up the interest in it and the attention to it. (…) The hand therefore is stimuli as well as response to the ear, and the latter is response as well as stimulus to the hand. Each is both stimulus and response to the other. The distinction of stimulus and response is therefore not one of content, the stimulus being identified with the ear, the response with the hand, but one of function, and both offices belong equally to each organ. (…) it must be kept in mind that this latter is a distinction falling inside the act, not between the hand movement considered as the act, and the sound considered as its external stimulus or "cause." In a word, the reagent reacts as much with his ear as he does with his hand." (Angell and Moore 1896, p. 252).

exercise of determination of successors by antecedents.[55] The implications of Whitehead's process ontology in the study of brain-behavior relationships would require a whole chapter by itself, if not a whole book. We cannot underscore enough here its great importance as a solid foundation to erect an organism-centric scientific study of topics that concern neuroscience in particular, and biology in general. Much more to be said in the future.

Causal-manipulative approaches are prone to reductionist and mechanistic frames. However, a re-elaboration of the notion of mechanism, with a tamed flavor of reductionism, has been proposed: it does without a decomposition of the system under study into "parts" and "interactions" and, instead, defines a mechanism in terms of "entities" together with their "activities."[56] Such a neo-mechanistic approach is based on (and at the same time distinguished from) a blend of substance and process ontologies.[57] One may regard it as a smooth transition from the static notion of mechanism to something more akin to process. Yet, just because one can draw an arrow from A to B, it does follow that the arrow is a process-explanation (quite the contrary). Regrettably, in order to avoid the full implications of process philosophy, such a mingle becomes dualistic at best and self-contradictory at worst.[58]

Let me briefly mention quantum causality, where causes can be nonlocal to their effects. «The inference to (efficient) causation is always based on an *interpretation* of observed correlations (…), correlations for which both types of classical efficient causation can be rule out, even experimentally» (Atmanspacher 2014). Entanglement (both truly quantum and also classical, by means of epistemic inaccessibility) remains an alternative as intriguing as interesting: «this does not mean that the correlations have no reason at all or are just 'causeless' (…). But their cause is not of the efficient variety—in Aristotle's terminology, it comes closest to the notion of formal causation.» (ibid)

[55]«The concrescence of each individual actual entity is internally determined and is externally free». More explicitly: «The doctrine of the philosophy of organism is that, however far the sphere of efficient causation be pushed in the determination of components of a concrescence (…) beyond the determination of these components there always remains the final reaction of the self-creative unity of the universe. This final reaction completes the self-creative act by putting the decisive stamp of creative emphasis upon the determinations of efficient cause.» (Whitehead 1929, p. 47).

[56]«Mechanisms are entities and activities organized such that they are productive of regular changes from start or set up to finish or termination conditions» (Machamer, Darden and Craver 2000).

[57]«We meant thereby to distinguish our position, or way of thinking, from a substance ontology and from process ontology, and we chose "entity" and "activity" because these terms seemed to carry fewer historical and philosophical presuppositions than "substance" and "process."» (Machamer 2004).

[58]«Process philosophers would have us redefine all entities in terms of combining processes, but this seems a bit too strange. Therefore, we (MDC 2000) decided to be dualist.» (ibid).

To end, let us insist on the following essential truism: the study of the behavior of organisms is crucial in order to understand the behavior of organisms.[59]

11.4 Know What You Mean and Say It (Finale)

I took the liberty and challenge of not speaking to the choir nor spitting to the noir.[60] This critique is certainly a "minority report," and it may be deemed too abstract and philosophical for the mainstream neuroscientist. But science is not only about data and tools, but primarily about the concepts that such data and tools allow to probe.[61]

Here it was deemed necessary to halt and ask, in the midst of such great efforts in manipulating neural circuits, to what extent they *explain* of behavior. Deconstructing the most common phraseology found in neuroscience, we identified, traced, and articulated the primacy of some basic assumptions about what explanation is and what behavior is. We refrained from opposing neural chauvinism with behavioral chauvinism.

We followed the original call for understanding and realized that the strength in the initial notion of explanation runs down a gradient of concessions and approximations until we are left with hardly more than *counterfactual dependence*. Overall, we concluded that neural circuit necessity and sufficiency may not be necessary or sufficient to explain behavior. Paraphrasing Woese, there is nothing wrong with N&S per se. Wrong is when N&S comes to define explanation neuroscience.

---

[59]And this shall include, together with theory, the appreciation for descriptive science:«Simply describing what we see is not considered very scientific nowadays and 'descriptive science' has become a derogatory term. We must have a hypothesis to test, or better, a controlled experiment that can be performed to identify ecological rules and laws. However, if "ecological rules" were followed by all systems, unexpected things would not happen, which is evidently not the case. Deviations from rules are the main determinants of history, but we cannot test something that is unexpected. As such, our quest to identify rules and regularities could be preventing us from understanding the history of these systems. Paradoxically, we aim at understanding historical systems while using ahistorical approaches! We need a means of reporting these contingencies so that we can better understand the historical trajectory of ecological systems.» (Boero 2013).

[60]«Both the criteria of plausibility and of scientific value tend to enforce conformity, while the value attached to originality encourages dissent. This internal tension is essential in guiding and motivating scientific work. The professional standards of science must impose a framework of discipline and at the same time encourage rebellion against it. They must demand that, in order to be taken seriously, an investigation should largely conform to the currently predominant beliefs about the nature of things, while allowing that in order to be original it may to some extent go against these.» (Polanyi 1962).

[61]«Science is impelled by two main factors, technological advance and a guiding vision. A properly balanced relationship between the two is key to the successful development of a science: without the proper technological advances the road ahead is blocked. Without a guiding vision there is no road ahead; the science becomes an engineering discipline, concerned with temporal practical problems.» (Woese 2004).

Neuroscience is a vibrant endeavor, yet it is hard to see where we are going with what we are doing, what we mean by understanding, or what counts as a true insight. From cellular neurobiology to cognitive neuroscience, no matter how disparate their interests and different their communities, we all ultimately subscribe, one way or another, to the "neurons explain behavior" mantra. Trying to bring clarity and honesty about what we *can not* currently explain—and most likely *would not* be able to solve—with our methodological procedures and conceptual frameworks is one more reminder for humility, after so much pride. One thing is clear: the amount of understanding we can gain about the neural *implementation* of behavioral processes depends on the quality of prior investigations of such behavioral processes, and also on the conceptual grounding that bridges the relation between neural and behavioral levels of organization.

It is surprising how little effort is made (by scientists) to explain what is actually meant by (scientific) explanation. Interwoven attitudes reflect ingrained habits of the intellect: the idea that pristine observation, when combined with powerful manipulation, shall disclose the truth about things—even when the pursuit of truth is actually deemed as impossible in principle and irrelevant in practice. Following Bacon's dictum (and in contrast with Aristotle's) one should "torture" *nature* until she spits out her answers. Paradoxically, the force required in such an interrogatory (engineering new tools, analyzing big data, trading necessity–sufficiency tests for explanation, selling our findings, getting grants, etc.) drains the capacity to consciously recall what we wanted to ask *her* in the first place. This sort of scientific amnesia can only perpetuate itself unless we are willing to equip ourselves with the kind of literacy that shall make precise the difference between what we do versus what we claim we do. In sum, *know what you mean and say it.*

Acknowledgements I thank Björn Brembs, André Brown, Adam Calhoun, Udi Fonio, Asif Ghazanfar, José Gomes Pinto, Gordon Globus, Eyal Gruntman, Rod Hemsell, Johannes Jaeger, Konrad Kording, Gonçalo Lopes, Adam Matic, Laura Navío, Joana Rigato, Troy Shirangi, and Ibrahim Tastekin, for feedback. My views need not coincide with theirs.

References

Angell JR, Moore AW (1896) Studies from the psychological laboratory of the University of Chicago: I. reaction-time: a study in attention and habit. Psychol Rev 3(3):245

Anscombe GEM (1971) Causality and determination. In: Sosa E, Tooley M (eds) Causation, 1993 Oxford, pp 88–104

Ashby WR (1958) Requisite variety and its implications for the control of complex systems. Cybernetica 1(2):83–99

Atmanspacher H (2014) Roles of causation and meaning for interpreting correlations. J Anal Psychol 59(3):429–434

Beer CG (1980) Perspectives on animal behavior comparisons. In: Comparative methods in psychology. Erlbaum Hillsdale, New Jersey, pp 17–64

Bell HC (2014) Behavioral variability in the service of constancy. Int J Comp Psychol 27(2)

Boero F (2013) Observational articles: a tool to reconstruct ecological history based on chronicling unusual events. F1000Research 2:168
Bunge M (2009) Causality and modern science, 4th revised edn. Transaction Publishers, New Jersey
Cziko GA (2000) The things we do: using the lessons of bernard and darwin to understand the what, how, and why of our behavior. The MIT Press
Dawkins R (1976) Hierarchical organisation: a candidate principle for ethology. In: Bateson PPG, Hinde RA (eds) Growing points in ethology. Cambridge University Press
De Waelhens A (1942) A philosophy of the ambiguous, foreword to the second French edition of the structure of behavior by Merleau-Ponty. Beacon Press, Boston
Dewey J (1894) The study of ethics: a syllabus. Register Publishing Company
Dewey J (1896) The reflex arc concept in psychology. Psychol Rev 3(4):357
Dewey J (1930) Conduct and experience. In: Murchison C (ed) Psychologies of 1930. The International University Series in Psychology. Clark University Press, Worcester, Massachusetts
DiDomenico R, Eaton RC (1988) Seven principles for command and the neural causation of behavior. Brain Behav Evol 31(3):125–140
Eaton RC, DiDomenico R (1985) Command and the neural causation of behavior: a theoretical analysis of the necessity and sufficiency paradigm. Brain Behav Evol 27(2–4):149–164
Ellis GF (2011) Top-down causation and emergence: some comments on mechanisms. Interface focus, rsfs.2011.0062
Forscher BK (1963) Chaos in the brickyard. Science 142(3590):339
Goldstein K (1934) The Organism. Zone Books, New York
Gomez-Marin A, Mainen ZF (2016) Expanding perspectives on cognition in humans, animals, and machines. Curr Opin Neurobiol 37:85–91
Hogan R (2001) Wittgenstein was right. Psychol Inq 12(1):27
James W (1890) The principles of psychology(1950), Dover, New York
Jankélévitch V (2015) Henri Bergson. Duke University Press
Jonas E, Kording K (2017) Could a neuroscientist understand a microprocessor? PLOS Comput Biol 13(1):e1005268. doi:10.1371/journal.pcbi.1005268
Karmon A, Pilpel Y (2016) Biological causal links on physiological and evolutionary time scales. Elife 5:e14424
Kortmulder K (1998) Play and evolution: second thoughts on the behavior of animals. Int Books, The Netherlands
Lazebnik Y (2002) Can a biologist fix a radio?—Or, what I learned while studying apoptosis. Cancer Cell 2(3):179–182
Machamer P (2004) Activities and causation: the metaphysics and epistemology of mechanisms. Int Stud Philos Sci 18(1):27–39
Machamer P, Darden L, Craver CF (2000) Thinking about mechanisms. Philos Sci 67(1):1–25
Marken RS (2001) Controlled variables: psychology as the center fielder views it. Am J Psychol 114(2):259
Maturana HR (1995) Biology of self-consciousness. In: Consciousness: distinction and reflection, Bibliopolis, pp 145–175
Medawar PB (1978) Is the scientific paper fraudulent? yes; it misrepresents scientific thought. Sci Books SR/August 1:42–43
Merleau-Ponty M (1942) The Structure of Behavior (trans: Fischer A)(1963). Boston, Beacon Press
Nagel E (1965) Types of causal explanation in science. In: Daniel L (ed) Cause and Effect. The Free Press, New York
Noë A (2004) Action in perception. MIT press
Overton WF (2006) Developmental psychology: philosophy, concepts, methodology. In: Richard ML (ed) Handbook of child psychology, vol 1, Wiley, New Jersey
Polanyi M (1962) The republic of science: its political and economic theory. Minerva 1:54–74
Powers WT (1973) Behavior: the control of perception. Aldine, Chicago

Powers WT (1998) About stimulus response theory and perceptual control theory. Post to the control systems group network
Rosenblueth A, Wiener N, Bigelow J (1943) Behavior, purpose and teleology. Philos Sci 10(1): 18–24
Russell B (1913) On the notion of cause. Proc Aristotelian Soc New Ser 13:1–26
The Edwin Smith Surgical Papyrus (1930) In: Breasted JH (ed) University of Chicago Press, Chicago
Tinbergen N (1963) On aims and methods of ethology. Z für Tierpsychologie 20(4):410–433
Von Bertalanffy L (1950) The theory of open systems in physics and biology. Science 111(2872): 23–29
Von Wright GH (1971) Explanation and understanding. Cornell University Press, New York
Whitehead AN (1929) Process and reality (1978). In: David RG, Donald S (ed) New York
Whitehead AN (1933) Adventures of ideas (1967). The Free Press. New York
Woese CR (2004) A new biology for a new century. Microbiol Mol Biol Rev 68(2):173–186

Part III
Physiology: Visualizing the Activity of Identified Circuit Elements

Chapter 12
Visualization of Synapses and Synaptic Plasticity in the *Drosophila* Brain

Thomas Riemensperger, Florian Bilz and André Fiala

Abstract *Drosophila* represents a favorable model organism to analyze neuronal circuits underlying behavior. This is mainly due to the versatile genetic tools by which transgene expression can be targeted to virtually any neuronal population in the brain. Fluorescent sensor proteins enable one to monitor the physiological parameters correlating with the function of neurons, and a number of actuator proteins exist that can be used to selectively manipulate distinct neurons. However, the mode of operation of neuronal circuits for determining behavior is not only based on a static connectivity between neurons, but also on the physiological properties of synaptic transmission and their plasticity. Techniques to detect many synapses at high resolution across many neurons in vivo and to access their physiology and plasticity are required. Here, we summarize recent genetic approaches to visualize synapses and to analyze synaptic plasticity in the *Drosophila* brain.

12.1 Introduction

A key task in current neuroscience is to dissect the often enormously complex brain circuits in terms of structure and connectivity between individual neurons in order to understand how information is encoded and processed (Luo et al. 2008). One ultimate goal is to uncover how complex behavior is organized by nervous systems and their neuronal networks. For this endeavor "model animals" are advantageous that are genetically accessible, i.e., whose germ line cells can be genetically transformed and DNA-encoded probes be stably expressed in defined populations

T. Riemensperger (✉) · F. Bilz · A. Fiala (✉)
Department of Molecular Neurobiology of Behavior, Johann-Friedrich-Blumenbach-Institute for Zoology and Anthropology, Georg-August-University Göttingen, Julia-Lermontowa-Weg 3, 37077 Göttingen, Germany
e-mail: triemen@gwdg.de

A. Fiala
e-mail: afiala@gwdg.de

A. Çelik and M.F. Wernet (eds.), *Decoding Neural Circuit Structure and Function*, DOI 10.1007/978-3-319-57363-2_12

of neurons. Within the range of well-established model organisms the fruit fly *Drosophila melanogaster* is, in terms of neuronal complexity, situated between the nematode *Caenorhabditis elegans* with its limited, stereotypic number of 302 neurons (White et al. 1986) and a limited behavioral repertoire, and genetically manipulatable mammals like mice, with their millions of neurons and high numerical redundancy (Luo et al. 2008). Despite the relatively low number of ~100,000 brain neurons (Chiang et al. 2011) in comparison with mice, the behavioral repertoire and neuronal complexity of *Drosophila* is rich enough to allow for addressing interesting questions and for conceptual comparisons with mammals. Neurons and their synaptic connections in the *Drosophila* central nervous system are very often stereotypically identifiable between individual animals, and their gross synaptic connections are often genetically determined. This situation has motivated the generation of many complementary genetic expression systems that can be used to target transgenes with high precision to almost all types of neurons (Venken et al. 2011), and indeed a large proportion of neurons can be unambiguously identified and specifically manipulated or monitored experimentally.

Experimenters analyzing if and how distinct neurons contribute to the organization of a particular behavior can typically follow two strategies. First, one can manipulate neurons selectively, e.g., through optogenetic or thermogenetic techniques, and observe the effects on behavior (Fiala et al. 2010; Owald et al. 2015; Riemensperger et al. 2016). The second approach relies on monitoring physiological parameters associated with neuronal function, e.g., intracellular Ca^{2+} dynamics (Riemensperger et al. 2012). These recordings can be correlated with a sensory input or a behavioral action, or an input–output analysis of distinct neuronal circuits can shed light on their information processing properties. In addition, detailed ultrastructural characterizations of the synaptic connectivity between large numbers of neurons is beginning to emerge, not least because of technical progress in electron microscopy that allows one to visualize extended parts of the nervous system and the exact synaptic connectivity in three dimensions (Briggman and Denk 2006). However, one complication relies on the fact that in most cases behavior is dynamic. For example, many types of behavior are controlled, or at least influenced by, the circadian rhythm (Frenkel and Ceriani 2011). Motivational factors, e.g., the animal's feeding status, in many cases determine behavioral output (Su and Wang 2014). Sensory perception and the behavioral responses elicited by sensory stimuli are subject to experience such as adaptation or learning (e.g., Twick et al. 2014; Fiala 2007). Aging also influences an animal's behavior, and *Drosophila* serves as a model organism for analyzing the neuronal and behavioral effects of it (Saitoe et al. 2005). All of these phenomena rely on the intrinsic plasticity of the nervous system, e.g., modifications of the excitability of neurons or structural and functional changes in synaptic transmission at very different time scales. Synapses represent computational units that to a large degree determine the physiological and computational properties of the neuronal circuits they are part of (Abbott and Regehr 2004). Therefore, the analysis of the dynamic, plastic properties of synapses is a prerequisite for understanding the mode of operation of complex neuronal circuits function. Here, we highlight three common approaches to visualize synaptic connections and synaptic plasticity in the *Drosophila* central nervous system.

12.2 Visualizing Synapses and Their Plasticity Through the Expression of Tagged Synaptic Proteins

For many years electron microscopy has been and still is the ultimate method for confirming synaptic contacts, i.e., by demonstrating at an ultrastructural level whether presynaptic active zones, synaptic vesicles and postsynaptic densities are present at the connection between two neurons (Meinertzhagen and Lee 2012). Whereas electron microscopy offers unmatched spatial precision, it is restricted to fixed tissue, and a comparison between many animals is problematic. And electrophysiological recordings with their exceptional temporal accuracy and physiological informative value are usually limited to small numbers of cells. The ectopic expression of pre- or postsynaptic proteins tagged with fluorescent markers has been established as a very useful approach to analyze synaptic connectivity across larger populations of neurons using light microscopy, and often in an in vivo situation. This approach utilizes the large number of proteins that are very specifically targeted to synaptic compartments. The first fluorescence-tagged synaptic proteins were generated on the basis of integral vesicle membrane proteins, i.e., GFP-tagged Synaptobrevin (Ito et al. 1998; Estes et al. 2000; Zhang et al. 2002) and GFP-tagged Synaptotagmin (Zhang et al. 2002). These markers opened a route to determine the directionality of synaptic connections in the *Drosophila* brain (e.g., Ito et al. 1998), to observe vesicle dynamics, and to compare their localization or quantity between wildtype animals and mutants (e.g., Estes et al. 2000). The active zone protein Bruchpilot (BRP) represents an example of a presynaptic scaffold protein that is a component of T-shaped structures at presynaptic active zones, i.e., T-bars (Wagh et al. 2006; Kittel et al. 2006). This structural presynaptic protein could be tagged with fluorescent markers without affecting synaptic function, enabling one to localize presynapses and to observe structural long-term changes in the density of presynaptic structures (Kremer et al. 2010). Fluorescence-tagged receptor proteins offer the possibility to monitor the structure and modification of postsynapses. For example, a GFP-tagged subunit of the *Drosophila* nicotinic acetylcholine receptor (Dα7) (Leiss et al. 2009) has been used to analyze structural plasticity in Kenyon cells of the mushroom body (Kremer et al. 2010). Subunits of a glutamate receptor (GluR-IIA and GluR-IIB) have also been tagged with fluorescence proteins to observe their dynamics at the larval neuromuscular junction (Rasse et al. 2005). Of course, receptor proteins as postsynaptic markers are restricted to specific synapses functioning on the basis of the particular transmitter and receptor. A more general postsynaptic marker is the postsynaptic density protein Discs large (Dlg), the *Drosophila* homologue of the vertebrate PSD-95, the GFP-tagging of which has provided a tool to visualize postsynapses (Zhang et al. 2007). For instance, Donlea et al. (2014) used Dlg-GFP to correlate visual experience and aging with plastic changes in postsynapses of clock neurons controlling the circadian rhythm of *Drosophila*.

A general limitation of these techniques relies on the ectopic expression of a labeled synaptic protein in addition to its endogenous counterpart. If the expression of tagged proteins is driven by promoters other than the endogenous ones for the respective synaptic proteins, expression levels and localization might differ between tagged and endogenous proteins. To ensure that this is not the case, tagged synaptic proteins can be expressed under control of the endogenous promoter and in the null mutant background for the endogenous gene (e.g., Rasse et al. 2005). To facilitate this, Chen et al. (2014) have conceived a technique named synaptic tagging with recombination (STaR). The authors have inserted modified gene loci into the fly's genome such that a tag is inserted downstream of the locus of the respective protein, but behind a stop codon flanked by recombinase recognition sites. A recombinase can be expressed in a defined population of cells, thereby removing the stop codon and inducing the expression of the tagged gene. This has one advantage: The tagged proteins, in the case of this study, which was focused on the fly's visual system, the presynaptic BRP and a postsynaptic histamine-gated Cl^{-}-channel (ort) are expressed under the same regulatory genetic machinery as the endogenous gene (Chen et al. 2014). The analysis of synaptic structures using fluorescence-tagged protein and conventional light microscopy is of course limited in spatial resolution. Synaptic substructures, e.g., active zones, vesicle clusters, or postsynaptic densities, are typically smaller in size than the diffraction barrier of light microscopy allows resolving. Super-resolution microscopy such as STED (Klar et al. 2000) or Localization Microscopy (Rust et al. 2006; Betzig et al. 2006; Heilemann et al. 2008) is therefore an important advancement to detail subcellular synaptic structures and their plastic modifications. However, fusing synaptic proteins with a fluorescence protein might alter their functional properties, which always must be controlled for.

12.3 Visualizing Cell–Cell Contacts and Synapses Through Split GFP Reconstitution

The expression of genetically encoded fluorescent proteins targeted to synaptic compartments can help to localize synaptic structures in a given subset of cells. However, the density of synapses in the central nervous system of *Drosophila* is often too high to be resolved by conventional light microscopy. This is greatly facilitated by GFP reconstitution across synaptic partners (GRASP), a technique that has been invented and described by Feinberg et al. (2008) in *C. elegans*. A fast folding variant of green fluorescent protein (GFP) (Pedelacq et al. 2006) is split into two complementary fragments, both of which are non-fluorescent on their own.

One fragment (spGFP1-10) includes the first 10 of 11 strands of the GFP β-barrel, the second fragment (spGFP11) the remaining 11th strand. If both fragments complement each other, which occurs if they are in very close proximity, the complete barrel-shaped GFP is reconstituted and regains its fluorescent property (Cabantous et al. 2005). If the two fragments are directed to the extracellular site of neuronal membranes, the fluorescence can serve as a detector for close proximity between neurons. Feinberg et al. (2008) have used the transmembrane domain of the CD4 from human T cells protein to demonstrate the functionality of this approach in muscle cells of *C. elegans*. Moreover, they managed to generate more specific markers for synaptic contacts by fusing one or two split GFP fragments to synapse-specific transmembrane proteins, e.g., Neuroligin-1 (NLG-1). The approach of using broadly localized membrane-tethered, CD4-fused split GFP reconstitution was adopted for *Drosophila* first by Gordon and Scott (2009). Here, the functionality to mark close proximity between cells was demonstrated by expressing one fragment of split GFP in olfactory receptor neurons and the other fragment in second-order olfactory projection neurons. In this study the authors have used this technique to exclude a direct connection between motor neurons involved in a taste-induced motor response and sugar-sensitive gustatory sensory neurons (Gordon and Scott 2009). A most impressive application of this technique has been reported by Gorostiza et al. (2014) who analyzed synaptic contacts that rhythmically assemble or disassemble in the course of the day. Based on the finding that clock neurons controlling the flies' circadian rhythm undergo structural remodeling on a daily basis (e.g., Fernández et al. 2008), they analyzed direct cell–cell contacts. The authors found that contacts between a distinct group of clock neurons (sLNvs) and their target cells visualized using split GFP reconstitution show changes in the course of the day (Gorostiza et al. 2014). From a technical point of view, this work demonstrates, first, that structural synaptic plasticity occurring in the range of hours can be detected by this method. Second, this finding provides also a good argument against one serious objection against this technique, i.e., the risk that GFP reconstitution across two neurons might artificially stabilize or even artificially establish cell–cell contacts. Another example for an application of GRASP in *Drosophila* is the analysis of contacts between extrinsic and intrinsic neurons of the mushroom body. The mushroom body is a central brain structure critically involved in adaptive behavior such as associative learning (Heisenberg 2003; Fiala 2007), but also in the control of sleep (Bushey and Cirelli 2011) and overall locomotor activity (Martin et al. 1998). All of these behaviors involve the action of aminergic neurons that modulate activity and/or synaptic output of intrinsic mushroom body neurons (Kenyon cells). Pech et al. (2013) used the technique of split GFP reconstitution to detail which aminergic neurons connect to Kenyon cells in close physical proximity, and where exactly it is on the entire mushroom body structure that they do so. Figure 12.1 exemplifies that cell–cell contacts between aminergic neurons and Kenyon cells can be visualized using split GFP reconstitution. Aminergic neurons innervating the mushroom body have also provided a cellular model for the analysis of neurodegeneration and aging. Visualizing morphological changes of physical contacts between mushroom body

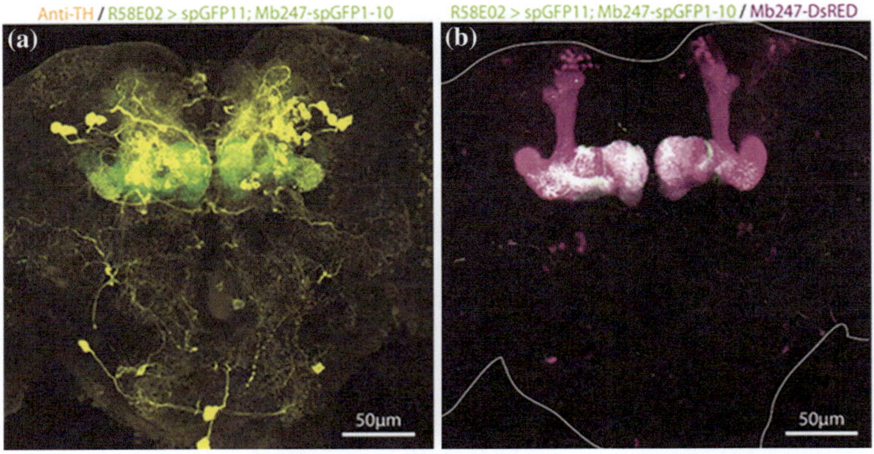

Fig. 12.1 Split GFP reconstitution between a specific population of dopaminergic neurons and intrinsic neurons of the *Drosophila* mushroom body (Kenyon cells). **a** spGFP11 is expressed under control of the driver line E58E02-Gal4 that drives gene expression in a cluster of dopaminergic neurons. sp1-10GFP is expressed in Kenyon cells under control of the promoter Mb247 (Pech et al. 2013). In addition, dopaminergic neurons are stained with an antibody against tyrosine hydroxylase (anti-TH). **b** In the same preparation the red fluorescent marker DsRED is expressed in Kenyon cells under control of the promoter mb247 and depicted here in *magenta*. The reconstituted split GFP fluorescence is superimposed in *green color*. The overlap appears in *white color*

Kenyon cells and their dopaminergic interaction partners using split GFP reconstitution revealed that in aging flies or under conditions mimicking Parkinson's disease-like cellular effects, locomotion defects are accompanied by a progressive loss of dopamine neuron connectivity to the mushroom bodies (Riemensperger et al. 2013). Similarly, Tonoki and Davis (2015) used the GRASP technique to show that age-dependent learning deficits are accompanied by gradual reduction of connectivity within the circuit between dorsal paired medial serotoninergic neurons and mushroom body Kenyon cells.

These examples demonstrate that the GRASP technique is a very valid tool to analyze neuron-to-neuron connections and their plastic modifications. One obvious limitation of targeting the split GFP fragments to the extracellular site of a membrane protein ubiquitously present along the entire membranes of neurons (e.g., CD4), is that only close proximity between neurons is detected. It remains unclear whether this proximity indeed represents synaptic contacts between neurons. Complementary tools, e.g., the expression of tagged synaptic proteins, are helpful to analyze whether or not the GRASP fluorescence indeed represents functional chemical synapses. But conversely, the lack of a reconstituted GFP fluorescence is a valid argument that synaptic contacts are absent between two given neurons. However, two important technical advancements have been made recently. First, split GFP variants of various fluorescent colors have been described (Macpherson et al. 2015; Li et al. 2016), which offers the possibility to label more than one type of connection in the same preparation if different expression systems are simultaneously used. Second, a chimeric protein between one split GFP and the presynaptic membrane protein neuronal Synaptobrevin (n-Syb) has been generated, which acts as an unambiguous marker for chemical synapses, and not only for physical proximity (Macpherson et al. 2015). Since n-Syb is a protein of the vesicle membrane, and the split GFP fragment is fused with the intravesicular C-terminus of n-Syb, it is exposed to the synaptic cleft only after vesicle exocytosis. Therefore, this construct provides also a measure of activity-dependent vesicle fusion, as demonstrated by Macpherson et al. (2015) for the fly's olfactory, visual and thermosensory system.

12.4 Visualizing Activity and Plasticity Using Targeted Fluorescence Sensors

Synaptic plasticity can occur at many different time scales, ranging from short-term modulation of the efficiency in synaptic transmission within milliseconds over intracellular structural rearrangements in the range of hours and days to gross morphological alterations occurring over hours to days or the animal's entire lifetime. Analyzing the dynamics in the efficiency of synaptic transmission is, therefore, a prerequisite for understanding how synaptic connections shape the mode of function of a neuronal circuit. The first genetically encoded fluorescence sensor for synaptic transmission, SynaptopHluorin, was invented by Miesenböck et al. (1998). A pH-sensitive variant of GFP is targeted to the lumen of synaptic vesicles by fusing it to the intravesicular terminus of Synaptobrevin. The very low intravesicular pH value increases through vesicle exocytosis, which enhances the fluorescence emission of SynaptopHluorin. This sensor has been widely used in *Drosophila*, e.g., to visualize odor-evoked synaptic transmission (Ng et al. 2002) or changes of odor representations in the antennal lobe in the course of associative olfactory learning (Yu et al. 2004). A red variant of pHluorin (pHTomato) targeted to the lumen of synaptic vesicles by fusing it with rat Synaptophysin (Li and Tsien 2012) has been applied in *Drosophila* by Pech et al. (2015). This red variant can be used simultaneously with green fluorescent sensors or markers in the same or different neurons (Pech et al. 2015). Limitations of pHluorins are the relatively low signal-to-noise ratio and poor temporal resolution, which is a result of the stochastic nature of exo- and endocytosis events and the small percentage of releasable vesicles. The term "synaptic activity" encompasses a number of physiological processes, ranging from changes in membrane potential at the presynaptic site, Ca^{2+} influx, vesicle exo-/endocytosis, transmitter release to transmitter–receptor interaction and postsynaptic potential changes at the postsynaptic site. An appropriate sensor for synaptic activity would ideally detect a parameter that changes transiently and to a large degree. Ca^{2+} imaging using cytosolic fluorescent Ca^{2+} sensors represents a very widely used technique in *Drosophila* neuroscience

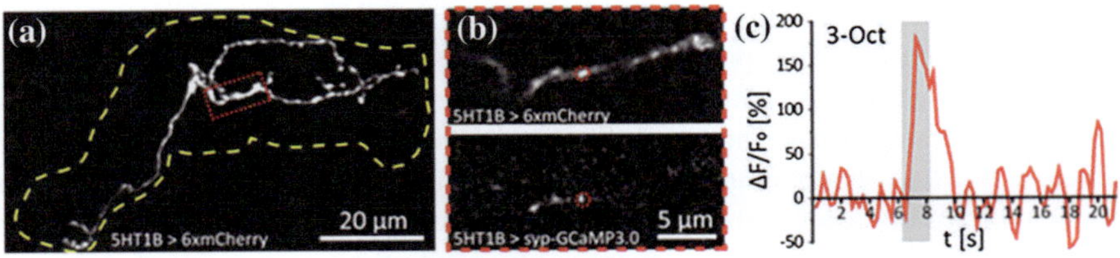

Fig. 12.2 Odor-evoked presynaptic Ca^{2+} dynamics in one bouton of a single Kenyon cell of the *Drosophila* mushroom body. The MARCM technique (Lee and Luo 1999) was used to express the red fluorescence protein mCherry and the presynaptically targeted Ca^{2+} sensor Synaptophysin-GCaMP3 (Pech et al. 2015) in a single γ-lobe Kenyon cell under control of the driver line 5HT1B-Gal4. **a** Immunohistochemical staining of a single γ-lobe Kenyon cell expressing mCherry visualized using confocal microscopy. The *yellow dashed line* depicts the outline of the entire γ-lobe). The *red box* shows an axonal branch shown in **b**. **b** Two-photon Ca^{2+}-imaging (method according to Dipt et al. 2014). The mCherry fluorescence (*top*) was used to trace the axonal branch, Synaptophysin-GCaMP3 signals (*bottom*) were monitored simultaneously. **c** Ca^{2+} dynamics of one bouton (*red circle* in **b**) showing an increase in presynaptic Ca^{2+} in response to stimulation with the odorant 3-octanol (*gray area*)

(Riemensperger et al. 2012). The green fluorescent Ca^{2+} sensor GCaMP3 (Tian et al. 2009) has been targeted to presynapses by fusing it with rat Synaptophysin (Dreosti et al. 2009; Li et al. 2011). This construct is readily functional in *Drosophila* and has been used to analyze experience-dependent changes in synapses of second-order projection neurons of the olfactory pathway (Pech et al. 2015). An example of the visualization of presynaptic Ca^{2+} dynamics in one Kenyon cell of the mushroom body is shown in Fig. 12.2. Likewise, Cohn et al. (2015) used the modular concept of tagging synaptic proteins with a fluorescence sensor, and fused the presynaptic protein Synaptotagmin with the Ca^{2+} sensor GCaMP6 (Cohn et al. 2015). This sensor was used in an impressive study to monitor the spatially confined dopamine-dependent modulation of synaptic activity in mushroom body Kenyon cells (Cohn et al. 2015). In order to monitor postsynaptic Ca^{2+} dynamics, Pech et al. (2015) have created a sensor in which GCaMP3 was fused with the postsynaptic density protein dHomer (Diagana et al. 2002). This sensor provides a means to accurately and selectively visualize postsynaptic Ca^{2+} influx and observe experience-dependent plasticity induced by long-term stimulation with an odor in olfactory projection neurons (Pech et al. 2015).

In conclusion, the development of DNA-encoded fluorescence sensors to not only localize pre- and postsynaptic structures, but also to determine the functionality and strength of synaptic contacts, e.g., by visualizing pre- and postsynaptic Ca^{2+} dynamics in vivo and across many synapses, provide an important step to advance the functional dissection of neuronal circuits underlying behavior.

Acknowledgements This work was supported by the Deutsche Forschungsgemeinschaft (SFB 889/B04) and the German Ministry of Research and Education via the Bernstein Center for Computational Neuroscience Göttingen (01GQ1005A). We are grateful to Robert J. Kittel for helpful comments on the manuscript.

References

Abbott LF, Regehr WG (2004) Synaptic computation. Nature 431:796–803

Betzig E, Patterson GH, Sougrat R, Lindwasser OW, Olenych S, Bonifacino JS, Davidson MW, Lippincott-Schwartz J, Hess HF (2006) Imaging intracellular fluorescent proteins at nanometer resolution. Science 313:1642–1645

Briggman KL, Denk W (2006) Towrads neural circuit reconstruction with volume electron microscopy techniques. Curr Opin Neurobiol 16:562–570

Bushey D, Cirelli C (2011) From genetics to structure to function: exploring sleep in Drosophila. Int Rev Neurobiol 99:213–244

Cabantous S, Terwilliger TC, Waldo GS (2005) Protein tagging and detection with engineered self-assembling fragments of green fluorescent protein. Nat Biotechnol 23:102–107

Chen Y, Akin O, Nern A, Tsui CYK, Pecot MY, Zipursky SL (2014) Cell-type-specific labeling of synapses in vivo through synaptic tagging with recombination. Neuron 81:280–293

Chiang AS, Lin CY, Chuang CC, Chang HM, Hsieh CH, Yeh CW, Shih CT, Wu JJ, Wang GT, Chen YC, Wu CC, Chen GY, Ching YT, Lee PC, Lin CY, Lin HH, Wu CC, Hsu HW, Huang YA, Chen JY, Chiang HJ, Lu CF, Ni RF, Yeh CY, Hwang JK (2011) Three-dimensional reconstruction of brain-wide wiring networks in Drosophila at single-cell resolution. Curr Biol 21:1–11

Cohn R, Morantte I, Ruta V (2015) Coordinated and compartmentalized neuromodulation shapes sensory processing in Drosophila. Cell 163:1742–1755

Diagana TT, Thomas U, Prokopenko SN, Xiao B, Worley PF, Thomas JB (2002) Mutation of Drosophila homer disrupts control of locomotor activity and behavioral plasticity. J Neurosci 22:428–436

Dipt S, Riemensperger T, Fiala A (2014) Optical calcium imaging using DNA-encoded fluorescence sensors in transgenic fruit flies, *Drosophila melanogaster*. Methods Mol Biol 1071:195–206

Donlea JM, Ramanan N, Silverman N, Shaw PJ (2014) Genetic rescue of functional senescence in synaptic and behavioral plasticity. Sleep 37:1427–1437

Dreosti E, Odermatt B, Dorostkar MM, Lagnado L (2009) A genetically encoded reporter of synaptic activity in vivo. Nat Methods 6:883–889

Estes PS, Ho GL, Narayanan R, Ramaswami M (2000) Synaptic localization and restricted diffusion of a Drosophila neuronal synaptobrevin—green fluorescent protein chimera in vivo. J Neurogenet 13:233–255

Feinberg EH, Vanhoven MK, Bendesky A, Wang G, Fetter RD, Shen K, Bargmann CI (2008) GFP Reconstitution Across Synaptic Partners (GRASP) defines cell contacts and synapses in living nervous systems. Neuron 57:353–363

Fernández MP, Berni J, Ceriani MF (2008) Circadian remodeling of neuronal circuits involved in rhythmic behavior. PLoS Biol 6:e69

Fiala A (2007) Olfaction and olfactory learning in Drosophila: recent progress. Curr Opin Neurobiol 17:720–726

Fiala A, Suska A, Schlüter OM (2010) Optogenetic approaches in neuroscience. Curr Biol 20: R897–R903

Frenkel L, Ceriani MF (2011) Circadian plasticity: from structure to behavior. Int Rev Neurobiol 99:107–138

Gordon MD, Scott K (2009) Motor control in a Drosophila taste circuit. Neuron 61:373–384

Gorostiza EA, Depetris-Chauvin A, Frenkel L, Pírez N, Ceriani MF (2014) Circadian pacemaker neurons change synaptic contacts across the day. Curr Biol 24:2161–2167

Heilemann M, van de Linde S, Schüttpelz M, Kasper R, Seefeldt B, Mukherjee A, Tinnefeld P, Sauer M (2008) Subdiffraction-resolution fluorescence imaging with conventional fluorescent probes. Angew Chem Int Ed Engl 47:6172–6176

Heisenberg M (2003) Mushroom body memoir: from maps to models. Nat Rev Neurosci 4:266–275

Ito K, Suzuki K, Estes P, Ramaswami M, Yamamoto D, Strausfeld NJ (1998) The organization of extrinsic neurons and their implications in the functional roles of the mushroom bodies in *Drosophila melanogaster* Meigen. Learn Mem 5:52–77

Klar TA, Jakobs S, Dyba M, Egner A, Hell SW (2000) Fluorescence microscopy with diffraction resolution barrier broken by stimulated emission. Proc Natl Acad Sci U S A 97:8206–8210

Kittel RJ, Wichmann C, Rasse TM, Fouquet W, Schmidt M, Schmid A, Wagh DA, Pawlu C, Kellner RR, Willig KI, Hell SW, Buchner E, Heckmann M, Sigrist SJ (2006) Bruchpilot promotes active zone assembly, Ca^{2+} channel clustering, and vesicle release. Science 312:1051–1054

Kremer MC, Christiansen F, Leiss F, Paehler M, Knapek S, Andlauer TF, Förstner F, Kloppenburg P, Sigrist SJ, Tavosanis G (2010) Structural long-term changes at mushroom body input synapses. Curr Biol 20:1938–1944

Lee T, Luo L (1999) Mosaic analysis with a repressible cell marker for studies of gene function in neuronal morphogenesis. Neuron 22:451–461

Leiss F, Koper E, Hein I, Fouquet W, Lindner J, Sigrist S, Tavosanis G (2009) Characterization of dendritic spines in the Drosophila central nervous system. Dev Neurobiol 69:221–234

Li Y, Tsien RW (2012) pHTomato, a red, genetically encoded indicator that enables multiplex interrogation of synaptic activity. Nat Neurosci 15:1047–1053

Li H, Foss SM, Dobryy YL, Park CK, Hires SA, Shaner NC, Tsien RY, Osborne LC, Voglmaier SM (2011) Concurrent imaging of synaptic vesicle recycling and calcium dynamics. Front Mol Neurosci 4:34

Li Y, Guo A, Li H (2016) CRASP: CFP reconstitution across synaptic partners. Biochem Biophys Res Commun 469:352–356

Luo L, Callaway EM, Svoboda K (2008) Genetic dissection of neural circuits. Neuron 57:634–660

MacPherson LJ, Zaharieva EE, Kearney PJ, Alpert MH, Lin TY, Turan Z, Lee CH, Gallio M (2015) Dynamic labelling of neural connections in multiple colours by trans-synaptic fluorescence complementation. Nat Comm 6:e10024

Martin JR, Ernst R, Heisenberg M (1998) Mushroom bodies suppress locomotor activity in *Drosophila melanogaster*. Learn Mem 5:179–191

Meinertzhagen IA, Lee CH (2012) The genetic analysis of functional connectomics in Drosophila. Adv Genet 80:99–151

Miesenböck G, De Angelis DA, Rothman JE (1998) Visualizing secretion and synaptic transmission with pH-sensitive green fluorescent proteins. Nature 394:192–195

Ng M, Roorda RD, Lima SQ, Zemelman BV, Morcillo P, Miesenböck G (2002) Transmission of olfactory information between three populations of neurons in the antennal lobe of the fly. Neuron 36:463–474

Owald D, Lin S, Waddell S (2015) Light, heat, action: neural control of fruit fly behaviour. Philos Trans R Soc Lond B Biol Sci 370:20140211

Pech U, Pooryasin A, Birman S, Fiala A (2013) Localization of the contacts between Kenyon cells and aminergic neurons in the *Drosophila melanogaster* brain using SplitGFP reconstitution. J Comp Neurol 521:3992–4026

Pech U, Revelo NH, Seitz KJ, Rizzoli SO, Fiala A (2015) Optical dissection of experience-dependent pre- and postsynaptic plasticity in the Drosophila brain. Cell Rep 10:2083–2095

Pedelacq JD, Cabantous S, Tran T, Terwilliger TC, Waldo GS (2006) Engineering and characterization of a superfolder green fluorescent protein. Nat Biotechnol 24:79–88

Rasse TM, Fouquet W, Schmid A, Kittel RJ, Mertel S, Sigrist CB, Schmidt M, Guzman A, Merino C, Qin G, Quentin C, Madeo FF, Heckmann M, Sigrist SJ (2005) Glutamate receptor dynamics organizing synapse formation in vivo. Nat Neurosci 8:898–905

Riemensperger T, Pech U, Dipt S, Fiala A (2012) Optical calcium imaging in the nervous system of *Drosophila melanogaster*. Biochim Biophys Acta 1820:1169–1178

Riemensperger T, Issa AR, Pech U, Coulom H, Nguyễn MV, Cassar M, Jacquet M, Fiala A, Birman S (2013) A single dopamine pathway underlies progressive locomotor deficits in a Drosophila model of Parkinson disease. Cell Rep 5:952–960

Riemensperger T, Kittel RJ, Fiala A (2016) Optogenetics in Drosophila neuroscience. Methods Mol Biol 1408:167–175

Rust MJ, Bates M, Zhuang X (2006) Sub-diffraction-limit imaging by stochastic optical reconstruction microscopy (STORM). Nat Methods 3:793–795

Saitoe M, Horiuchi J, Tamura T, Ito N (2005) *Drosophila* as a novel animal model for studying the genetics of age-related memory impairment. Rev Neurosci 16:137–149

Su CY, Wang JW (2014) Modulation of neural circuits: how stimulus context shapes innate behavior in Drosophila. Curr Opin Neurobiol 29:9–16

Tian L, Hires SA, Mao T, Huber D, Chiappe ME, Chalasani SH, Petreanu L, Akerboom J, McKinney SA, Schreiter ER, Bargmann CI, Jayaraman V, Svoboda K, Looger LL (2009) Imaging neural activity in worms, flies and mice with improved GCaMP calcium indicators. Nat Methods 6:875–881

Tonoki A, Davis RL (2015) Aging impairs protein-synthesis-dependent long-term memory in Drosophila. J Neurosci 35:1173–1180

Twick I, Lee JA, Ramaswami M (2014) Olfactory habituation in Drosophila—odor encoding and its plasticity in the antennal lobe. Prog Brain Res 208:3–38

Venken KJ, Simpson JH, Bellen HJ (2011) Genetic manipulation of genes and cells in the nervous system of the fruit fly. Neuron 72:202–230

Wagh DA, Rasse TM, Asan E, Hofbauer A, Schwenkert I, Dürrbeck H, Buchner S, Dabauvalle MC, Schmidt M, Qin G, Wichmann C, Kittel R, Sigrist SJ, Buchner E (2006) Bruchpilot, a protein with homology to ELKS/CAST, is required for structural integrity and function of synaptic active zones in Drosophila. Neuron 49:833–844

White JG, Southgate E, Thomson JN, Brenner S (1986) The structure of the nervous system of the nematode *Caenorhabditis elegans*. Philos Trans R Soc Lond B Biol Sci 314:1–340

Yu D, Ponomarev A, Davis RL (2004) Altered representation of the spatial code for odors after olfactory classical conditioning; memory trace formation by synaptic recruitment. Neuron 42:437–449

Zhang YQ, Rodesch CK, Broadie K (2002) Living synaptic vesicle marker: Synaptotagmin-GFP. Genesis 34:134–142

Zhang Y, Guo H, Kwan H, Wang JW, Kosek J, Lu B (2007) PAR-1 kinase phosphorylates Dlg and regulates its postsynaptic targeting at the Drosophila neuromuscular junction. Neuron 53:201–215

Chapter 13
Whole-Brain Imaging Using Genetically Encoded Activity Sensors in Vertebrates

Andreas M. Kist, Laura D. Knogler, Daniil A. Markov, Tugce Yildizoglu and Ruben Portugues

Abstract In the mid-twentieth century, the development of electrophysiology revolutionized the way that the brain could be studied, allowing scientists to advance beyond anatomy and neuroethology and address questions involving brain function. These recordings offered a temporally and spatially high-resolution readout of the activity of single cells and enabled a detailed understanding of the input–output function of individual neurons. Nevertheless, understanding the brain one neuron at a time seems like a daunting task. Over the last two decades, a considerable amount of research has focused on understanding the brain at the mesoscale of brain circuits and networks, trying to bridge the gap from single neurons to the function of the whole brain in generating behavior. This is a large, open and exciting field that encompasses theory, computational models, behavioral studies, genetic manipulations and many more approaches. Importantly, the current interest in brain circuits is fueled by the development of new techniques that allow us to acquire data relevant to addressing network function and the activity of large populations of neurons. In this chapter, we present an introduction to whole-brain, single-cell resolution imaging in a behaving vertebrate model organism, the larval zebrafish. We describe the fundamental concepts developed during the last five years that are important for understanding large-scale imaging techniques in vertebrates from experimental design to data acquisition and analysis.

Andreas M. Kist and Laura D. Knogler contributed equally.

A.M. Kist · L.D. Knogler · D.A. Markov · T. Yildizoglu · R. Portugues (✉)
Research Group of Sensorimotor Control, Max Planck Institute of Neurobiology, Martinsried 82152, Germany
e-mail: rportugues@neuro.mpg.de

A. Çelik and M.F. Wernet (eds.), *Decoding Neural Circuit Structure and Function*, DOI 10.1007/978-3-319-57363-2_13

13.1 Indicators and Genetics

13.1.1 Genetically Encoded Activity Indicators

Many studies in systems neuroscience involving functional imaging of brain activity use genetically encoded activity indicators that report different correlates of neuronal activity by changes in fluorescence (for a review see Lin and Schnitzer 2016). These indicators allow researchers to monitor neural activity optically under a fluorescence microscope, often in a non-invasive way. While many different sensors exist for functional imaging, genetically encoded calcium indicators (GECIs) are by far the most widely used regardless of the model organism, although indicators for voltage, vesicular release, and different neurotransmitters are also available. Whole-brain imaging studies in zebrafish to date have exclusively used GECIs, in particular, variants of the GFP/calmodulin-derived GCaMP (Chen et al. 2013).

An important consideration of functional calcium imaging is the slow timescales over which neural activity can be resolved. This derives from the fact that changes in calcium concentration and subsequent changes in calcium indicator fluorescence occur more slowly than action potentials. For example, while a single action potential takes ~3–5 ms, the cytosolic calcium concentration that accompanies it peaks ~10 ms after its initiation. This concentration then decreases exponentially with a half decay time of 50–70 ms as calcium is buffered and cleared from the cytosol (Helmchen et al. 1997; Koester and Sakmann 2000). The GECI kinetics further spread the signal out in time due to the slow binding and release of calcium. For example, the response of the calcium indicator GCaMP6s to a single action potential has a half decay time of ~550 ms in cell culture (Chen et al. 2013). This smearing out of the activity over these relatively long time scales gives us a longer time over which we can sample and register changes in fluorescence. This enables us to increase the field of view and to collect signals from many neurons in the brain.

The timing of sparse action potentials and spike rates can be recovered from calcium signals with moderate accuracy by using deconvolution algorithms (Theis et al. 2016), however, the relationship between spikes and calcium signals is nonlinear and varies across neurons (Harris et al. 2016). Typically, the calcium activity traces that emerge from large-scale imaging studies serve as the basis for analyses such as linear regression and clustering (but see Deneux et al. 2016) and there is currently little advantage to making inferences about spike rates.

GECIs are continuously being engineered and are being successfully used in imaging experiments in zebrafish and other vertebrates. These indicators differ in their decay times and signal-to-noise ratios and are constantly improving (see Rose et al. 2014; Lin and Schnitzer 2016 for review). In addition, a new calcium indicator called CaMPARI designed by Fosque et al. (2015) offers the opportunity to image activity across the nervous system in freely behaving animals. Behavioral experiments in freely moving larval zebrafish evoke a richer repertoire of behaviors than in restrained fish, however, calcium imaging in this experimental setting is very demanding. This technique thus enables the identification of neural activity from more natural behaviors. Briefly, this protein undergoes efficient and irreversible green-to-red conversion only when elevated intracellular calcium and experimenter-controlled illumination coincide, providing a temporally precise "snapshot" of brain activity offline. Thus, CaMPARI can be used to image neural activity from the entire brain with a high spatial resolution, but for only single (or few) timepoints.

13.1.2 Pan-neuronal and Targeted Expression of GECIs

Using existing genetic tools, one can generate transgenic animals that express selected genetically encoded activity indicators in the whole brain or in a specific cell population of interest. For whole-brain imaging experiments in larval zebrafish, activity indicators are typically expressed under the pan-neuronal promoter *elavl3* (*HuC*) (Fig. 13.1a; Kim et al. 1996). It should be noted that the transcriptional activity of this promoter decreases after ~7 days post-fertilization (Sato et al. 2006), making experiments more challenging at later stages of development. The expression of GCaMP can also be targeted to a neuronal subpopulation of interest using either cell-specific promoters or one of the many available enhancer trap lines in combination with the Gal4/UAS system (Scott et al. 2007). In mammals, *synapsin I* can be used as a pan-neuronal driver for GECIs (Thiel et al. 1991; Tian et al. 2009) while the *thy1* promoter can be used to target subsets of neurons in different brain areas (Feng et al. 2000; Chen et al. 2012). Other promoters such as *vglut2* can be used to specifically label excitatory (glutamatergic) neurons across the brain (Borgius et al. 2010), simplifying the interpretation of population activity. Genetic tricks can also be used to restrict activity indicators to subcellular regions in order to allow the subsequent segmentation of individual neurons. For example, a histone fusion (typically histone H2B, Freeman et al. 2014; Vladimirov et al. 2014) or nucleus localization sequence (abbreviated as NLS; Schrödel et al. 2013; Kim et al. 2014) in the GECI construct (e.g. NLS-GCaMP) restricts the fluorescent signal to the smaller volume of the nucleus (Fig. 13.1b), which may be easier to automatically segment using machine vision algorithms.

13.2 Whole-Brain Imaging Setup

The larval zebrafish has an archetypal vertebrate brain plan though it has only about 100,000 neurons in total (Butler and Hodos 2005). Its dimensions, 800 × 400 300 μm, make it ideally suited for fitting the whole brain simultaneously under the field of view of high-end objectives, thus allowing the imaging of populations of neurons and circuits distributed throughout the brain. This can be done using a variety of techniques. The two most common ones are two-photon scanning microscopy (Ahrens et al. 2012; Portugues et al. 2014) and light-sheet microscopy (Ahrens et al. 2013; Panier et al. 2013), although alternative volumetric imaging approaches such as

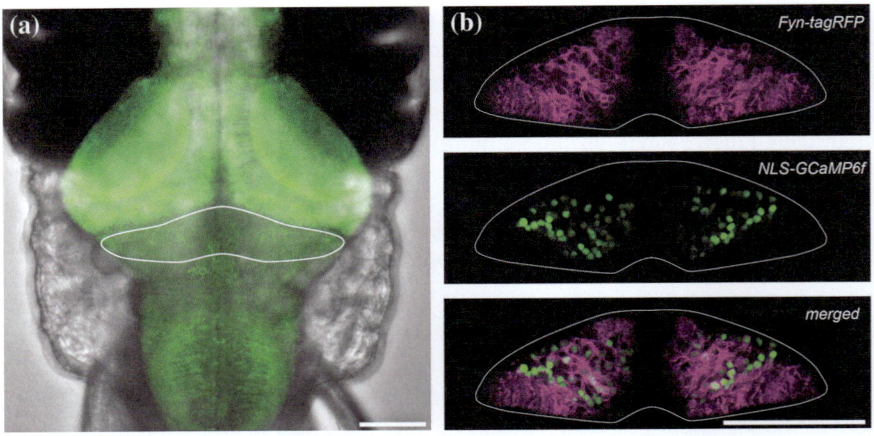

Fig. 13.1 Expression of genetically encoded calcium indicators in the larval zebrafish brain. **a** Confocal *z*-projection of a transgenic larval zebrafish head at 6 days post-fertilization (dpf) expressing GCaMP6f (*green*) under control of the pan-neuronal promoter *elavl3* (*HuC*). **b** Close-up image of a single *z*-plane of the cerebellum of a 7 dpf larval zebrafish expressing membrane-tagged RFP (Fyn-tagRFP, *magenta*) and NLS-GCaMP6f (*green*) under the control of a Purkinje cell-specific promoter. Scale bars = 100 μm

light-field microscopy are available (Prevedel et al. 2014). We will now present an overview of the former two techniques and illustrate them with some sample data from whole-brain imaging experiments in larval zebrafish.

13.2.1 Microscopy Systems

Two-photon microscopy enables optical sectioning by restricting the probability of fluorophore excitation to a small area at the focal plane (Fig. 13.2a, b). The laser is scanned across the focal plane point-by-point to generate an image (for further reading see Denk et al. 1990). A 3D volume spanning the whole brain is then generated by combining these planes. Two-photon imaging has been used extensively over the last decade to study neural population activity in larval zebrafish (reviewed by Renninger and Orger 2013). Light-sheet microscopy, in contrast, is a more recently developed technique that works by shining a thin sheet of excitation light perpendicular to the optical axis that simultaneously illuminates the whole focal plane of the objective (Fig. 13.2c, d). Methods to generate the sheet include fast scanning by galvanometer mounted mirrors (Keller et al. 2008) and the use of cylindrical lenses (Huisken et al. 2004). By underfilling the objective back aperture, the point-spread function is elongated in z in relation to our sample horizontal axis (x). To achieve dynamic volume acquisition in either of these methods, one can move the sample through the focal plane, or manipulate the optics in order to move the focal plane through the stationary sample to achieve fast whole-brain imaging (Huisken and Stainier 2009; Ahrens et al. 2013; Bennett and Ahrens 2016).

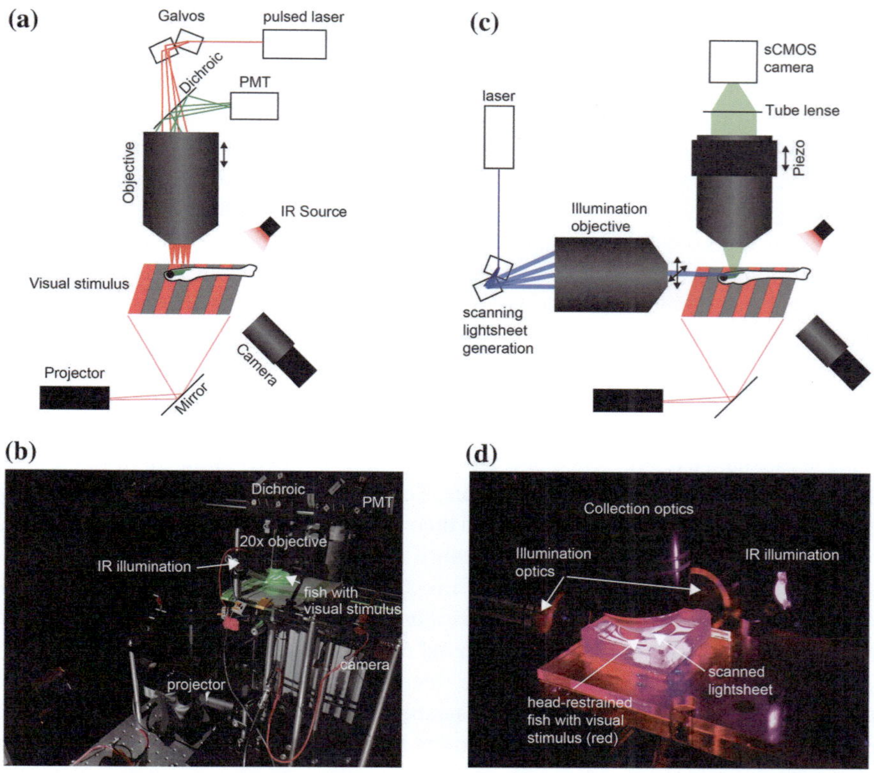

Fig. 13.2 Setups used for whole-brain imaging in larval zebrafish with simultaneous behavioral tracking. **a** Simplified scheme of a two-photon scanning microscope using galvanometers for scanning. A visual stimulus is projected from below and behavior is tracked using a camera and infrared illumination. Given the small size of the zebrafish brain, a 20× water-dipping objective provides a large enough field of view and working distance to image nearly the entire brain in three dimensions, at single-cell resolution. **b** Custom built two-photon microscope. **c** Simplified scheme of a scanned light-sheet microscope with visual stimulation from below and behavior tracking using a camera. **d** Custom built light-sheet microscope during an experiment

Promising new arrangements which remove the mechanical piezo are also being developed (Bouchard et al. 2015). Finally, light-field microscopy allows optical sectioning by using a microlens array that focuses light from different planes within the sample to different areas on the sensor, thus enabling the experimenter to acquire different planes simultaneously (Prevedel et al. 2014).

13.2.2 Simultaneous Behavioral Tracking

To study the brain circuits underlying behavior it is necessary to monitor behavior while recording neuronal activity. In rodents, this can be done using head mounted miniature microscopes (Ghosh et al. 2011; Helmchen et al. 2013). Improved resolution is obtained with larger fixed microscopes, for which it is necessary to immobilize the animal relative to the optical path in order to minimize motion artifacts that reduce data quality. In addition, this immobilization must be as compatible with normal physiology of the animal as possible, allowing it to behave normally. Head-fixed setups for mice that allow stable imaging were introduced a decade ago (Dombeck et al. 2007) and have revolutionized the field.

In an analogous way, it is possible to restrain larval zebrafish by embedding them in low-melting point agarose without the need for surgery. This sufficiently immobilizes the animal even after removing the agarose from around the tail leaving it free to produce swimming behaviors. Larvae are typically mounted in 35 mm Petri dishes for two-photon experiments and in custom-made chambers for light-sheet imaging, which minimize aberration of the excitation laser presented from the side. In this preparation, it is straightforward to record tail motion with a high speed camera placed outside the main excitation and emission light path. The eyes of the larva can also be freed to investigate oculomotor behaviors. These methods allow for online recording and tracking of the motor output that can then be fed back to the visual stimulus in real time, creating a closed-loop virtual reality (Portugues and Engert 2011). This is analogous to virtual environments for mice, where displacement of the animal is measured by its movement on a styrofoam ball and is fed back to update its position and orientation within a computer-rendered maze (Dombeck et al. 2010).

Alternatively, electrophysiological recordings of "fictive" behavior from peripheral nerves from spinal motor neurons in paralyzed animals can provide information about motor output (Masino and Fetcho 2005; Ahrens et al. 2012). These methods also allow for online recording and tracking of motor output and enable closed-loop experiments in which visual feedback can be presented in real time.

13.2.3 Presentation of Sensory Stimuli

The agarose-embedded zebrafish preparation is relatively simple and allows for both the presentation of a variety of sensory stimuli as well as the simultaneous tracking of behavior while imaging. Due to the highly visual nature of zebrafish larvae and the ease of stimulus control, most studies to date involve the presentation of visual stimuli. These stimuli can be projected onto the visual field of the animal from below using a commercial projector (see Fig. 13.2). In order to optically separate the stimulus presentation from the fluorescence, it is common to present red stimuli and image neural activity in the green channel, while using blue or infrared lasers for excitation in light-sheet or two-photon imaging, respectively. Fortunately, zebrafish larvae respond to red stimuli well (Orger and Baier 2005). Nevertheless, it is also possible to present visual stimuli coinciding with pauses in imaging, such as dead times when the camera is spooling a frame or when the fast scanning two-photon mirror is turning or flying-back. This allows for the display of visual stimuli of any color and frees the red channel to be used for imaging purposes. Additional components can easily be implemented in these imaging setups to evoke behavior and neural activity in response to other sensory stimuli such as mechanosensation (Lacoste et al. 2015), water movement (Groneberg et al. 2015), and olfaction (Esposti et al. 2013). In these experiments, TTL triggers and image timestamping can be used to ensure the synchronization of image acquisition with both stimulus delivery and behavioral measurements.

13.3 Whole-Brain Data Acquisition and Handling

Since these datasets are very large and require many processing and analysis steps, it is particularly important to optimize the acquisition to ensure the quality of the data being recorded (Harris et al. 2016). Here, we address some key considerations for data acquisition that can affect data quality in whole-brain imaging experiments.

13.3.1 Light Collection

Two-photon scanning microscopy scans the infrared excitation laser point-by-point within a plane in the sample (Fig. 13.2a) and emitted photons are focused using collection lenses and detected by large and sensitive photomultiplier tubes (PMTs). In zebrafish whole-brain experiments, the speed at which scanning can occur is limited by the number of photons that are emitted by the sensor and detected. The electrical current from the PMT can then be digitized in one of two ways. First, it can be integrated for a given time and the total signal can then be assigned as the fluorescence arising from a given location (x, y, z). The second method, referred to as photon-counting, can be implemented by hardware or software and consists of counting the number of threshold crossings above a given current, assuming each one corresponds to an individual photon and assigning this as the number of photons emitted from this location.

In contrast, in light-sheet microscopy light is collected from the illuminated plane by highly sensitive cameras (sCMOS or EM-CCD). The illuminated plane is imaged by the microscope onto the sensor such that the location information is preserved (x, y). Without having to raster scan in x and y, data acquisition is much faster than in two-photon scanning microscopy.

13.3.2 Sampling

Acquisition speeds need to sample at a temporal frequency that is appropriate to both the kinetics of the activity indicator and the temporal dynamics of neural activity wishing to be observed. The acquisition speeds available with different microscopes should therefore be considered during the experimental design with respect to the Nyquist–Shannon sampling theorem. The spatial resolution in sampling, which is diffraction-limited, is also an important consideration. Ideally, the signal from an individual neuron should span several voxels to ensure coverage and proper SNR. The smallest neurons in larval zebrafish are at least five microns in diameter, so a resolution of about 1 μm per pixel in every dimension will ensure that a neuron will span 50–100 voxels in total. This resolution is important for some of the segmentation approaches we introduce below.

13.3.3 Data Formats

The raw data that results from these experiments is a set of fluorescence values that can be assigned to a three-dimensional location (x, y, z) and time. Recall that these values arose from integrating a current, counting photons or reading out a pixel value from a camera. In order to understand how much data is generated in these experiments, let us consider the following estimates: In a two-photon experiment, scanning 200 planes within a sample, each plane for 1 min, at a resolution of 512×512 pixels at 3 Hz, and saving each fluorescence value as a 16 bit integer will generate about 20 GB of data. In a light-sheet setup, where one is capable of scanning imaging at 40 Hz with a resolution of 1024×1024 pixels, a similar amount of data is generated in about 4 min at the same bit depth.

In order to deal with data streaming, storage and subsequent analysis, it is convenient to choose and appropriate bit depth that balances maximizing dynamic range and takes into account the resolution of the system. Furthermore, lossless compression is an important method to reduce the physical space needed on a hard drive. However, as reading and writing these files takes more time than saving uncompressed files, the process is typically performed after the experiment. File formats for handling large datasets are for example the HDF5 format (also used in the new Matlab mat files) and the Keller lab block size format (KLB, Amat et al. 2015). The advantage of these formats is that they allow the user to dynamically access a subset of frames and channels if the whole file does not fit into memory and offer compression options as well.

13.4 Processing Large-Scale Datasets

Following acquisition, several steps of processing are applied to ensure the data is properly aligned across the volume, registered to a reference anatomy (if desired), and that interesting features can be extracted from regions of interest (ROIs) in an automated way from the large dataset and interpreted in a biologically meaningful context. Here, we describe the basic workflow of image processing for whole-brain activity datasets.

13.4.1 Image Alignment and Registration

To correct for small movements or drift in (*x*, *y*), each frame within a plane is aligned to the time-averaged image in that plane by translation. Often, multiple alignment procedures can be used. This not only improves the quality of the data but also makes segmentation, which is described below, easier. If a motion artifact is too large to be corrected (for example, due to a particularly vigorous behavior), this frame can be removed from the stack or substituted by the average across the stack. Once all frames within a plane are corrected, it may be necessary to align all the planes to each other and, having done this, the within-plane aligned imaging data is shifted with these per-plane displacements. For light-sheet data, the registered views of one time point can be combined into a single volume by averaging or by multiview deconvolution (Preibisch et al. 2014). Once a dataset is properly aligned, it can be registered through a combination of affine and non-affine (e.g., warp) transformations to a reference anatomy. The widely used Computational Morphometry Toolkit (CMTK; http://www.nitrc.org/projects/cmtk/) is freely available and offers fast algorithms to process various types of imaging data (Rohlfing and Maurer 2003). This series of transformations allows the user to morph the voxel/ROI coordinates of each experiment to the reference anatomy coordinates.

Registration across individuals has been demonstrated to work effectively in zebrafish, enabling the comparison of activity maps across individual fish with micrometer resolution (Portugues et al. 2014). This study demonstrated that the stereotypy of activity maps in the larval zebrafish is extremely high for the optokinetic reflex, a well-known visuomotor behavior. As more researchers adopt whole-brain functional imaging approaches, one can envision combining these diverse studies into a comprehensive atlas of larval zebrafish whole-brain activity. These functional data can furthermore be merged with complementary maps such as the Z-Brain atlas, which contains confocal stacks of total brain activity during a behavioral period as reported by endogenous mitogen-activated protein kinase expression (Randlett et al. 2015). Functional activity can also be mapped onto detailed anatomical and gene expression data of the nervous system such as found in the ViBE-Z atlas (Ronneberger et al. 2012). Cross-modal mapping of light-sheet functional imaging data to the Z-Brain atlas via bridging references has already been demonstrated successfully for neural activity maps of spontaneous, fictive locomotion in larval zebrafish (Dunn et al. 2016).

13.4.2 Feature Extraction

The majority of early calcium imaging experiments used manual ROI selection to extract signals from individual neurons that were initially reported as simple changes in fluorescence relative to baseline ($\Delta F/F$). Since then, the field has shifted toward using the z-scored (normalized) fluorescence signal. This measures the signal in relation to the standard deviation, and better reports the signal to noise of the measurement. Furthermore, the datasets acquired with whole-brain volumetric imaging contain tens of thousands of neurons, making manual feature extraction infeasible. We outline below several methods that have been designed to automatically extract spatially bounded regions of activity in as unbiased a way as possible, enabling us to reason about the biological relevance of the obtained signals.

13.4.2.1 Voxel-Wise Correlations

In the first whole-brain experiments in larval zebrafish, Ahrens et al. (2012) developed a method to automatically detect ROI centers as local maxima in a statistic related to the $\Delta F/F$ signal and segmenting out a cell-sized region around it. An alternative approach was developed by Portugues et al. (2014). This is based on the correlation in fluorescence of neighboring voxels that are co-active, for example because they are part of the same neuron. This method is a useful way to find ROIs involved in sensorimotor behavior without any constraints on how the activity should relate to sensory or motor-related stimuli.

Briefly, the activity trace of each individual voxel in a volumetric data set is compared to the summed activity of the neighboring voxels within a cube. The size of the cube is chosen to consist of a typical cell size, for example the 124 voxels which form a $5 \times 5 \times 5$ cube at $0.8 \times 0.8 \times 1$ μm resolution. As discussed earlier, the resolution of the imaging must therefore be sufficiently high such that the dimensions of one neuron spans several voxels. In this way, a three-dimensional map of correlation coefficients can be constructed which is the same size as the original volume. This now provides a basis for building ROIs (see below). When volumetric two-photon imaging is used such that the whole-brain activity is not acquired simultaneously but reconstructed across z-planes, performing correlations with neighboring voxels across planes can be used to identify robust (for example, time-locked sensory) responses. If activity varies across planes, as is the case of spontaneous activity or variable motor responses, then correlation calculations should be restricted to neighboring voxels within the same plane.

Once a three-dimensional stack of correlation values has been obtained, it is used as the starting point to build ROIs. First, the voxel with the global maximum correlation value is identified and set as a seed for the first ROI. Secondly, its fluorescence time-series is then correlated with the fluorescence time-series of its closest 26 neighbors (in three dimensions). If the correlation is high, the neighboring voxel is added to the ROI, otherwise it is not. The algorithm then proceeds iteratively, incorporating more voxels into the ROI when their fluorescence correlates strongly with the fluorescence of the already added ones. The first ROI stops growing when no more voxels have a high correlation or when a given ROI size is reached. Having finished the first ROI one can repeat the process by identifying the second highest global correlation value and growing a second ROI with it as a new seed.

13.4.2.2 Linear Regression

When the goal of an experiment is to relate neural activity to a specific sensory stimulus or behavior, a complementary or alternative approach to unbiased voxel-wise correlations is to build a set of regressors and look for neural activity related to these regressors (Miri et al. 2011a). This approach allows the experimenter to perform targeted analyses on the data, for instance, finding neurons that are activated by a certain sensory stimulus or combination of stimuli. A drawback of this approach is that the analysis is limited by the user-defined regressors—unlike in the previous method, here the number of possible types of neural responses depends on the number of regressors being tested. With this in mind, it is a good idea to think carefully about the potential diversity of responses (e.g., laterality, timing, nonlinearities) and to be thorough while designing regressors.

Briefly, a set of regressors corresponding to stimulus/behavioral variables of interest are created, convolved with the kinetics of the activity indicator, and orthonormalized (Miri et al. 2011a). Intuitively, each regressor is the expected fluorescence that a neuron would show if it was perfectly coding for the particular variable linearly. Linear regression is then performed using a matrix of these vectors to fit each voxel time-series, giving a set of correlation coefficients (weights) for each regressor as well as the residual error. This strategy allows for multiple regressors to contribute to a voxel's overall activity, which can model the time-series more exactly than relying on finding one best-fitting regressor. One drawback is that linear regression alone may still require significant user intervention or knowledge about morphology when trying to segment individual neurons from these results (Miri et al. 2011b).

Regression-based methods can also be applied to already segmented data by the procedure described above, as demonstrated in Portugues et al. (2014) and pictured in Fig. 13.3. Here, ROIs were built using unbiased voxel-wise correlations and subsequently assigned a graded sensory versus motor identity based on linear regression to salient sensory (a drifting visual grating) and motor (swimming) variables.

13.4.2.3 Principal Component Analysis, Independent Component Analysis, and Non-negative Matrix Factorization

Other approaches to finding neuronal activity responses in large-scale functional imaging data rely on signal processing methods that aim to identify meaningful underlying components within a noisy multivariate signal. Full imaging datasets are of a very high dimensionality: an X by Y pixel image monitored at T timepoints will have XYT dimensions. An extremely useful first step is therefore to apply a dimensionality reduction method. Using principal component analysis (PCA), groups of pixels that co-vary can be identified. Each principal component will be an X by Y image, but the whole image will have a single time course. PCA identifies these components, which are orthogonal by construction, in order of decreasing variance, which is natural if we assume that the signal being sought is larger than the noise. The number N of relevant principal components needs to be determined

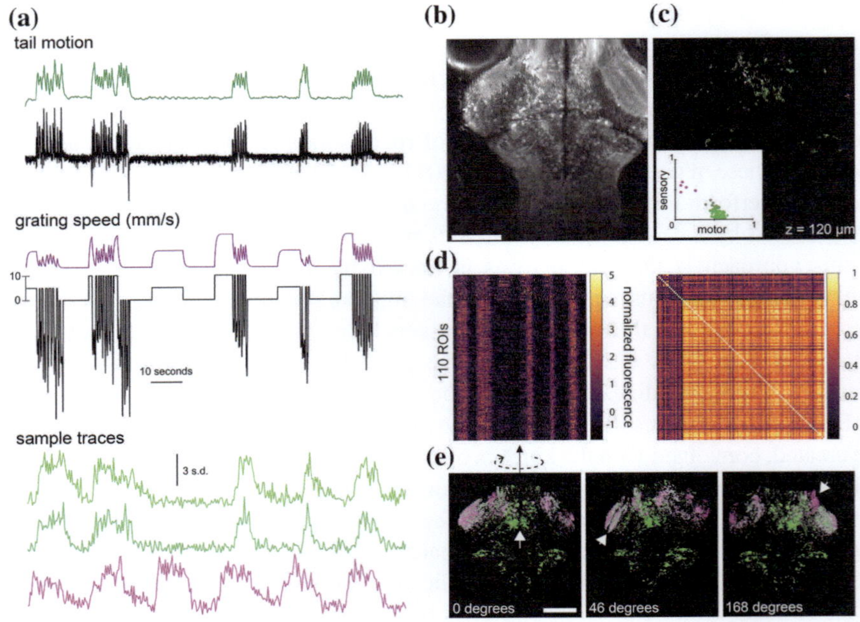

Fig. 13.3 Two-photon scanning microscopy identification of sensorimotor related units. **a** A behaving larval zebrafish (expressing GCaMP6s pan-neuronally) with its head restrained in agarose swam (*top black trace*) in response to a forward drifting visual grating. Each plane was imaged for two minutes with a two-photon scanning microscope (one such plane is shown in (**b**)). Ten second trials of forward movement (at 5 and 10 mm/s) were separated by ten seconds of static grating. This was a closed-loop experimental paradigm, such that the swimming of the larval zebrafish was fed instantaneously into the grating speed to mimic the real life relative velocity between stimulus and swimming larva (*middle black trace*). The behavior (tail motion) and stimulus (grating speed) were convolved with the calcium indicator kernel as described in the literature to create a motor and sensory regressor (*green* and *purple traces* above the *black traces*). An unbiased automated functional correlation algorithm (as described in the text) was used to segment active ROIs, the normalized fluorescence of three of which are shown at the *bottom*, color-coded by their relative correlation with the motor and sensory regressors. **c** 110 ROIs were identified in this one plane and their anatomical location is shown. Each ROI is color-coded for its correlation with sensory and motor regressors. **d** The fluorescence traces for all 110 ROIs in this plane, arranged from high sensory correlation to high motor correlation (*left*). The cross-correlelogram for the 110 units (*right*). **e** All 2300 units identified across all planes in this one fish, color-coded as in (**c**). The field of view is as in (**b**) and (**c**) and is displayed in 3 roll angles to better observe the three-dimensional structure of the units. Salient points, labeled with arrows, show the nucleus of the medial longitudinal fasciculus (*nMLF*) which has been implicated with the control of swimming (*left panel*), layering in the optic tectum neuropil (*middle panel*) and a retinal ganglion cell arborization field (*right panel*). Scale bars = 100 μm

for each particular experiment, but this will typically be in the order of 10–20 and much smaller than T. The dimension of this representation will therefore be N $(XY + T)$. PCA is useful to identify neural activity and separate it from a noisy background, but as a large part of the activity of the neurons may be correlated, it frequently cannot be used to identify them individually.

Independent component analysis (ICA), on the other hand, can be used to extract both spatial and temporal components from the data, as in the work by Mukamel et al. (2009). They designed a two-step approach to identifying ROIs in large data sets. This method first performs dimensionality reduction using PCA to remove background noise from the signal, then runs spatiotemporal ICA on the "cleaned" data to yield components termed spatial filters that ideally each correspond to a cellular signal. The assumption underlying this method is that the statistics of these independent signal sources are distinct (and non-Gaussian) and this allows the method to tease them apart. Image segmentation can also be used at this point to further separate signals that have similar activity patterns but are not spatially contiguous. This approach is relatively fast and has been used successfully to analyze large-scale neural data sets of whole-brain zebrafish activity acquired with light-field (Prevedel et al. 2014) and light-sheet microscopy (Tomer et al. 2015). Figure 13.4 shows ROIs from a sample light-sheet experiment identified through the main, non-noise components of PCA.

In experiments where multiple signals, or neurons, may be present in a single voxel, neither unbiased correlations nor PCA/ICA are likely to yield good results because nonlinear de-mixing is needed to identify these multiple and overlapping signals. Non-negative matrix factorization (NMF) provides an algorithm for better separating overlapping signals from close cellular substructures, though it is more

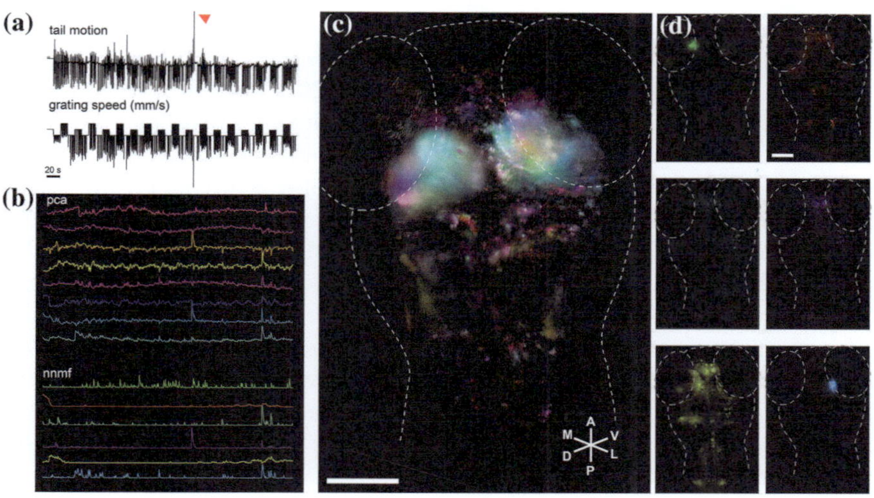

Fig. 13.4 The same fish as in Fig. 13.3 imaged under a light-sheet microscope. **a** The experimental protocol is analogous to Fig. 13.3 and elicits swimming behavior in response to moving gratings as well as a single turn (*red arrowhead*). The total duration of the experiment was 6 min. **b** Above, the time-courses of the eight main principal components for the whole brain are shown color-coded. Below, the data was de-noised using 20 principal components and non-negative matrix factorization was subsequently applied to the de-noised data. The top six components are shown. **c** PCA components are mapped onto the 3D anatomy with an outline of the head and eyes shown for reference. Color code same as in **b**. **c** The six NMF components mapped onto the anatomy of an individual plane. Scale bars = 100 µm

computationally demanding than PCA/ICA processing (Maruyama et al. 2014). This method has the added advantage that the temporal coefficients which make the time-series and describe the activity are always positive and are therefore easy to interpret biologically. Elaborations of this algorithm are evolving to improve the spatial localization of the signals with constrained NMF (CNMF; Pnevmatikakis et al. 2016), while downsampling and parallel computing strategies aim to make processing times more reasonable for the large datasets emerging from simultaneous volumetric imaging (Freeman et al. 2014; Friedrich et al. 2015). A sample NMF analysis is shown in Fig. 13.4 on PCA-cleaned data to show the type of components that are extracted using these different methods.

13.4.3 Identifying Neural Ensembles

With these methods, it should be possible to segment whole-brain imaging data into unique ROIs, whether these are voxels, neuropil, neurons, or neuronal ensembles. Yet one still needs to make biological sense of these disparate activity profiles. The strategy for understanding whole-brain activity maps once again depends on both the experimental setup and the type of data to emerge from the study.

Correlation coefficients corresponding to different regressors may be directly visualized as a color-coded map on the anatomy, allowing responses to be represented as a continuum (Fig. 13.3c, e). This is a logical way of showing neural activity that spans a broad response range, for example with respect to behavioral timing (Portugues et al. 2014), laterality tuning (Dunn et al. 2016), or sensorimotor identity (Fig. 13.3c, e). In contrast, other experiments may elicit activity that falls into discrete response categories. Several studies have used *k*-means clustering to sort neural responses into distinct groups, such as neurons that have a specific involvement in one aspect of motor adaptation (Ahrens et al. 2012), or neurons belonging to different sensorimotor processing streams (Panier et al. 2013). Methods such as PCA and NMF may extract components that show distinct spatial localizations corresponding to functionally-related neural ensembles (Fig. 13.4c, d). Finally, anatomical segmentation of activity signals by brain region can provide insights into local circuit activity and connectivity, such the antiphase oscillations of neurons shown in the zebrafish hindbrain (Ahrens et al. 2013). As the acquisition and analysis of large-scale data sets improves, so too must our efforts to integrate these vast amounts of data into a coherent set of ideas into how a biological system is working.

13.5 Biological Findings Emerging from Large-Scale Functional Imaging

Neural activity arising from distributed networks acts dynamically across large volumes (Ahrens et al. 2012, 2013; Panier et al. 2013; Portugues et al. 2014; Prevedel et al. 2014; Kato et al. 2015). In order to understand the coordinated activity of large assemblies of neurons or capture global population dynamics of the vertebrate brain one needs to sample brain-wide activity. Here we would like to highlight a few whole-brain and large-scale imaging studies in vertebrates that have led to important biological insights and consider future directions for the field.

13.5.1 Sensorimotor Transformations

Despite the practical, reductionist approach often employed in neurobiology, it is important to remember that behavior emerges from the activity of the nervous system as a whole (Bargmann and Marder 2013). Thanks to the compact size of the larval zebrafish, neural responses to sensory inputs can be imaged as these signals are processed across the brain and are transmitted to the motor networks that drive behavior. Portugues et al. (2014) used simple rotating visual stimuli known to induce an optokinetic reflex (OKR; Beck et al. 2004) while measuring brain-wide activity as well as the behavioral output of the animal. This study reveals a sparse and distributed, yet highly organized, pattern of sensory and motor responses. While this activity map is not sufficient to explain the neural underpinning of the OKR, it nonetheless provides a scaffold for the flow of information through these circuits and highlights areas of interest for targeted recordings and manipulations. As our understanding of the zebrafish behavioral repertoire becomes more comprehensive and our behavioral quantification more detailed and precise, so does our ability to correlate brain-wide neural activity with unique phases or units of sensorimotor behaviors.

13.5.2 State-Dependent Behaviors

The activity of neural circuits is dependent not only on the pattern of input but also on the current state of the network to make behavioral choices (Izhikevich 2007). Two recent studies have imaged brain activity in the context of spontaneous behavior in the larval zebrafish in order to look for structured spontaneous neural activity that could reveal something about the animal's internal state. Romano et al. (2015) found that tectal neurons were spontaneously active in assemblies that resembled sensory-evoked responses. They proposed that the brain has preferred network states that aid in visuomotor behaviors important for survival, such as prey tracking. Dunn et al. (2016) recorded the spontaneous activity profile across the entire brain during exploratory behavior. They observed that fish produced a seemingly random turning behavior in the absence of a sensory stimulus, thus influencing foraging strategies in an environment with few external cues. Using whole-brain light-sheet imaging in a paralyzed, head-fixed animal while recording fictive motor output, they showed that stochastic transitions between internal states, driven by small, mutually inhibiting populations of neurons, promote turning to either the left or right. Targeted investigations of correlated changes in local circuit activity and behavior during states such as hunger (Filosa et al. 2016) or motivation related to odor attraction/avoidance (Krishnan et al. 2014) have revealed major shifts in local neural representations that we can now try to understand at the level of global brain patterning in the behaving animal.

13.5.3 Population Coding

Large-scale imaging from all (or most) neurons in a population also allows us to gain insight into how the brain uses population coding during neural processing, which can reveal high-dimensional representations not visible in the activity of single neurons (Averbeck et al. 2006). Critically, there are many long-standing theories of population coding, such as the role of the cerebellar granule cell layer in producing sparse, high-dimensional representations (Marr and Thach 1969; Albus 1971), that are now being experimentally tested for the first time with these imaging methods (L. Knogler, unpublished data). Furthermore, repeated presentations of the same stimulus can elicit highly variable ("noisy") responses that are correlated across neurons, yet it is not well understood how noise correlations affect coding at the population level (Averbeck et al. 2006). Decoding models incorporating shared variability has been shown to recover more information than their independent counterparts, suggesting that small, structured correlations between pairs or small groups of neurons may have a significant impact on coding efficiency at large scales (Pillow et al. 2008).

13.5.4 Large-Scale Imaging in Mammalian Models

Large-scale calcium imaging studies in the behaving mouse, first established by Dombeck et al. (2007), are becoming ever more accessible and popular in systems neuroscience. Thanks to recent advances in optical and computational tools, scientists can now monitor activity in of hundreds of neurons simultaneously in awake and behaving mice, in acute or chronic experiments (Rose et al. 2016), enabling them to study neural activity across ever larger scales in the mammalian brain (see Hamel et al. 2015 for a review).

For example, a recent study by Andermann et al. (2013) implemented microprisms to simultaneously capture neural activity across entire cortical columns in the mouse somatosensory, barrel, or visual cortex during locomotion. This study revealed diverse responses across neurons and cortical layers in correlation with behavior that suggest a basis for the organization of interlaminar assemblies within a cortical column. Large-scale recordings also permit investigations into processes such as learning and memory that are known to involve distributed representations and thus benefit from observations at the population and network level (for review, see Jercog et al. 2016). Work by Peters et al. (2014) for example, showed that a population of neurons ($\sim$200) in the mouse motor cortex has activity that reorganizes substantially during motor learning. Even more importantly, once the behavior was fully learned, the information encoded at the population level remained stable even though individual neurons themselves showed very high variability. These and other findings confirm the intuition that the poorly understood activity of neural ensembles as a population code, rather than the activity of individual neurons themselves, is crucial to understanding how information is encoded and read out by the nervous system.

Concerning the technology, there is constant progress on the issues limiting the scale of calcium imaging in mammalian studies (Hamel et al. 2015). For instance, the use of red fluorescent calcium indicators reduces scattering at longer wavelengths, enabling scientists to record in all six layers of mouse cortex up to 900 μm below the surface of the brain (Tischbirek et al. 2015). Even deeper tissue imaging (up to 1200 μm) of subcortical structures in mice is now possible using three-photon microscopy (Horton et al. 2013). Instrumentation development in addition to new scanning and sampling strategies will also continue to facilitate volumetric imaging by increasing imaging volume and speed (for review, see Ji et al. 2016). Finally, several recent experiments in non-human primates have used GECIs to image activity in cortical neurons (Sadakane et al. 2015; Santisakultarm et al. 2016; Seidemann et al. 2016), laying the groundwork for the future imaging of small neural populations with single-cell resolution in the primate brain.

13.6 Outlook

As shown by the many studies emerging in the field of whole-brain imaging in the larval zebrafish, we are no longer constrained by technology if we wish to record the activity throughout the brain of a behaving vertebrate. As is also becoming clear though, this alone does not tell us how the brain works. Large datasets only provide useful answers when we ask the appropriate questions; therefore, with these techniques it is as critical as ever to design experiments with a clear biological rationale. Furthermore, these data need to be integrated into a theoretical framework that will advance our conceptual understanding of how the brain works (Gao and Ganguli 2015). Exciting studies emerging from whole-brain imaging in the larval zebrafish in particular are making great progress into understanding the relationship between neural activity and simple, innate behaviors. As more laboratories embark on large-scale imaging studies across model organisms, we are optimistic that the wealth of data and theory will lead to insights into the overarching principles of brain activity as they relate to motor and higher order (e.g. cognitive) functions, but there is still plenty of work to be done.

Acknowledgements We thank Vilim Stih, Patricia Cooney and Joel Ryan for helpful comments on the manuscript. LK was funded by the Humboldt Foundation, Carl von Siemens Foundation and the Max-Planck-Gesellschaft. AMK, DM, and TY were funded by IMPRS and the Max-Planck Gesellschaft. RP was funded by the Max-Planck-Gesellschaft.

References

Ahrens MB, Li JM, Orger MB et al (2012) Brain-wide neuronal dynamics during motor adaptation in zebrafish. Nature 485:471–477. doi:10.1038/nature11057

Ahrens MB, Orger MB, Robson DN et al (2013) Whole-brain functional imaging at cellular resolution using light-sheet microscopy. Nat Methods 10:413–420. doi:10.1038/nmeth.2434

Albus JS (1971) A theory of cerebellar function. Math Biosci 10:25–61. doi:10.1016/0025-5564(71)90051-4

Amat F, Höckendorf B, Wan Y et al (2015) Efficient processing and analysis of large-scale light-sheet microscopy data. Nat Protoc 10:1679–1696

Andermann ML, Gilfoy NB, Goldey GJ et al (2013) Chronic cellular imaging of entire cortical columns in awake mice using microprisms. Neuron 80(4). doi:10.1016/j.neuron.2013.07.052

Averbeck BB, Latham PE, Pouget A (2006) Neural correlations, population coding and computation. Nat Rev Neurosci 7:358–366. doi:10.1038/nrn1888

Bargmann CI, Marder E (2013) From the connectome to brain function. Nat Methods 10:483–490. doi:10.1016/j.cub.2012.01.061

Beck JC, Gilland E, Tank DW, Baker R (2004) Quantifying the ontogeny of optokinetic and vestibuloocular behaviors in zebrafish, medaka, and goldfish. J Neurophysiol 92:3546–3561

Bennett DV, Ahrens MB (2016) A practical guide to light sheet microscopy. Methods Mol Biol (Clifton, NJ) 1451:321

Borgius L, Restrepo CE, Leao RN et al (2010) A transgenic mouse line for molecular genetic analysis of excitatory glutamatergic neurons. Mol Cell Neurosci 45:245–257. doi:10.1016/j.mcn.2010.06.016

Bouchard MB, Voleti V, Mendes CS et al (2015) Swept confocally-aligned planar excitation (SCAPE) microscopy for high-speed volumetric imaging of behaving organisms. Nat Photonics 9:113–119

Butler AB, Hodos W (2005) Comparative vertebrate neuroanatomy: evolution and adaptation. John Wiley & Sons

Chen Q et al (2012) Imaging neural activity using Thy1-GCaMP transgenic mice. Neuron 76:297–308

Chen T-W, Wardill TJ, Sun Y et al (2013) Ultrasensitive fluorescent proteins for imaging neuronal activity. Nature 499:295–300. doi:10.1038/nature12354

Denk W, Strickler JH, Webb WW (1990) Two-photon laser scanning fluorescence microscopy. Science 248:73–76

Deneux T, Kaszas A, Szalay G et al (2016) Accurate spike estimation from noisy calcium signals for ultrafast three-dimensional imaging of large neuronal populations in vivo. Nat Commun 7:12190. doi:10.1038/ncomms12190

Dombeck DA, Khabbaz AN, Collman F et al (2007) Imaging large-scale neural activity with cellular resolution in awake, mobile mice. Neuron 56:43–57

Dombeck DA, Harvey CD, Tian L et al (2010) Functional imaging of hippocampal place cells at cellular resolution during virtual navigation. Nat Neurosci 13:1433–1440. doi:10.1038/nn.2648

Dunn TW, Mu Y, Narayan S et al (2016) Brain-wide mapping of neural activity controlling zebrafish exploratory locomotion. Elife 5:1–29. doi:10.7554/eLife.12741

Esposti F, Johnston J, Rosa J et al (2013) Olfactory stimulation selectively modulates the OFF pathway in the Retina of Zebrafish. Neuron 79:97–110. doi:10.1016/j.neuron.2013.05.001

Feng G, Mellor RH, Bernstein M et al (2000) Imaging neuronal subsets in transgenic mice expressing multiple spectral variants of GFP. Neuron 28:41–51. doi:10.1016/S0896-6273(00)00084-2

Filosa A, Barker AJ, Dal Maschio M, Baier H (2016) Feeding state modulates behavioral choice and processing of Prey Stimuli in the Zebrafish Tectum. Neuron 90:596–608

Fosque BF, Sun Y, Dana H et al (2015) Neural circuits. Labeling of active neural circuits in vivo with designed calcium integrators. Science 347:755–760. doi:10.1126/science.1260922

Freeman J, Vladimirov N, Kawashima T et al (2014) Mapping brain activity at scale with cluster computing. Nat Methods 11:941–950. doi:10.1038/nmeth.3041

Friedrich J, Soudry D, Mu Y, et al (2015) Fast constrained non-negative matrix factorization for whole-brain calcium imaging data. Conf Neural Inf Process Syst 1–5

Gao P, Ganguli S (2015) On simplicity and complexity in the brave new world of large-scale neuroscience. Curr Opin Neurobiol 32:148–155. doi:10.1016/j.conb.2015.04.003

Ghosh KK, Burns LD, Cocker ED et al (2011) Miniaturized integration of a fluorescence microscope. Nat Methods 8:871–878

Groneberg AH, Herget U, Ryu S, De Marco RJ (2015) Positive taxis and sustained responsiveness to water motions in larval zebrafish. Front Neural Circuits 9:9. doi:10.3389/fncir.2015.00009

Hamel EJO, Grewe BF, Parker JG, Schnitzer MJ (2015) Cellular level brain imaging in behaving mammals: an engineering approach. Neuron 86:140–159

Harris KD, Quiroga RQ, Freeman J, Smith SL (2016) Improving data quality in neuronal population recordings. Nat Neurosci 19:1165–1174. doi:10.1038/nn.4365

Helmchen F, Borst JG, Sakmann B (1997) Calcium dynamics associated with a single action potential in a CNS presynaptic terminal. Biophys J 72:1458

Helmchen F, Denk W, Kerr JND (2013) Miniaturization of two-photon microscopy for imaging in freely moving animals. Cold Spring Harb Protoc 2013:pdb–top078147

Horton NG, Wang K, Kobat D et al (2013) In vivo three-photon microscopy of subcortical structures within an intact mouse brain. Nat Phot 7:205–209

Huisken J, Swoger J, Bene F Del, et al (2004) Live Embryos by Selective Plane Illumination Microscopy. 13–16. doi:10.1126/science.1100035

Huisken J, Stainier DYR (2009) Selective plane illumination microscopy techniques in developmental biology. Development 136:1963–1975

Izhikevich EM (2007) Dynamical systems in neuroscience: the geometry of excitability and bursting. MIT Press, Cambridge, MA

Jercog P, Rogerson T, Schnitzer MJ (2016) Large-scale fluorescence calcium-imaging methods for studies of long-term memory in behaving mammals. Cold Spring Harb Perspect Biol 8: a021824

Ji N, Freeman J, Smith SL (2016) Technologies for imaging neural activity in large volumes. Nat Neurosci 19:1154–1164

Kato S, Kaplan HS, Schrödel T et al (2015) Global brain dynamics embed the motor command sequence of Caenorhabditis elegans. Cell 163:656–669

Keller PJ, Schmidt AD, Wittbrodt J, Stelzer EHK (2008) Reconstruction of zebrafish early embryonic development by scanned light sheet microscopy. Science 322:1065–1069. doi:10.1126/science.1162493

Kim C-H, Ueshima E, Muraoka O et al (1996) Zebrafish elav/HuC homologue as a very early neuronal marker. Neurosci Lett 216:109–112

Kim CK, Miri A, Leung LC et al (2014) Prolonged, brain-wide expression of nuclear-localized GCaMP3 for functional circuit mapping. Front Neural Circuits 8:1–12. doi:10.3389/fncir.2014.00138

Koester HJ, Sakmann B (2000) Calcium dynamics associated with action potentials in single nerve terminals of pyramidal cells in layer 2/3 of the young rat neocortex. J Physiol 529:625–646

Krishnan S, Mathuru AS, Kibat C et al (2014) The right dorsal habenula limits attraction to an odor in zebrafish. Curr Biol 24:1167–1175. doi:10.1016/j.cub.2014.03.073

Lacoste AMB, Schoppik D, Robson DN, et al (2015) A convergent and essential interneuron pathway for mauthner-cell-mediated escapes. Curr Biol 1–9. doi:10.1016/j.cub.2015.04.025

Lin MZ, Schnitzer MJ (2016) Genetically encoded indicators of neuronal activity. Nat Neurosci 19:1142–1153. doi:10.1038/nn.4359

Marr D, Thach WT (1969) A theory of cerebellar cortex. In: From the Retina to the Neocortex. Springer, pp 11–50

Maruyama R, Maeda K, Moroda H et al (2014) Detecting cells using non-negative matrix factorization on calcium imaging data. Neural Networks 55:11–19. doi:10.1016/j.neunet.2014.03.007

Masino M, Fetcho JR (2005) Fictive swimming motor patterns in wild type and mutant larval zebrafish. J Neurophysiol 93:3177–3188. doi:10.1152/jn.01248.2004
Miri A, Daie K, Burdine RD et al (2011a) Regression-based identification of behavior-encoding neurons during large-scale optical imaging of neural activity at cellular resolution. J Neurophysiol 105:964–980. doi:10.1152/jn.00702.2010
Miri A, Daie K, Arrenberg AB et al (2011b) Spatial gradients and multidimensional dynamics in a neural integrator circuit. Nat Neurosci 14:1150–1159. doi:10.1038/nn.2888
Mukamel EA, Nimmerjahn A, Schnitzer MJ (2009) Automated analysis of cellular signals from large-scale calcium imaging data. Neuron 63:747–760. doi:10.1016/j.neuron.2009.08.009
Orger MB, Baier H (2005) Channeling of red and green cone inputs to the zebrafish optomotor response. Vis Neurosci 22:275–281. doi:10.1007/3-540-35375-5
Panier T, Romano SA, Olive R et al (2013) Fast functional imaging of multiple brain regions in intact zebrafish larvae using selective plane illumination microscopy. Front Neural Circuits 7:1–11. doi:10.3389/fncir.2013.00065
Peters AJ, Chen SX, Komiyama T (2014) Emergence of reproducible spatiotemporal activity during motor learning. Nature 510:263–267
Pillow JW, Shlens J, Paninski L et al (2008) Spatio-temporal correlations and visual signalling in a complete neuronal population. Nature 454:995–999
Pnevmatikakis EA, Soudry D, Gao Y et al (2016) Simultaneous denoising, deconvolution, and demixing of calcium imaging data. Neuron 89:299. doi:10.1016/j.neuron.2015.11.037
Portugues R, Engert F (2011) Adaptive locomotor behavior in larval zebrafish. Front Syst Neurosci 5:72. doi:10.3389/fnsys.2011.00072
Portugues R, Feierstein CE, Engert F, Orger MB (2014) Whole-brain activity maps reveal stereotyped, distributed networks for visuomotor behavior. Neuron 81:1328–1343. doi:10.1016/j.neuron.2014.01.019
Preibisch S, Amat F, Stamataki E et al (2014) Efficient Bayesian-based multiview deconvolution. Nat Methods 11:645–648. doi:10.1038/nmeth.2929
Prevedel R, Yoon Y-G, Hoffmann M et al (2014) Simultaneous whole-animal 3D imaging of neuronal activity using light-field microscopy. Nat Methods 11:727–730. doi:10.1038/nmeth.2964
Randlett O, Wee CL, Naumann EA et al (2015) Whole-brain activity mapping onto a zebrafish brain atlas. Nat Methods 12:1–12. doi:10.1038/nmeth.3581
Renninger SL, Orger MB (2013) Two-photon imaging of neural population activity in zebrafish. Methods 62:255–267
Rohlfing T, Maurer CR (2003) Nonrigid image registration in shared-memory multiprocessor environments with application to brains, breasts, and bees. IEEE Trans Inf Technol Biomed 7:16–25
Romano SA, Pietri T, Pérez-Schuster V, et al (2015) Spontaneous Neuronal Network Dynamics Reveal Circuit's Functional Adaptations for Behavior. Neuron 1–16. doi:10.1016/j.neuron.2015.01.027
Ronneberger O, Liu K, Rath M et al (2012) ViBE-Z: a framework for 3D virtual colocalization analysis in zebrafish larval brains. Nat Methods 9:735–742. doi:10.1038/nmeth.2076
Rose T, Goltstein PM, Portugues R, Griesbeck O (2014) Putting a finishing touch on GECIs. Front Mol Neurosci 7:88
Rose T, Jaepel J, Hübener M, Bonhoeffer T (2016) Cell-specific restoration of stimulus preference after monocular deprivation in the visual cortex. Science 352:1319–1322
Sadakane O, Masamizu Y, Watakabe A et al (2015) Long-term two-photon calcium imaging of neuronal populations with subcellular resolution in adult non-human primates. Cell Rep 13:1989–1999
Santisakultarm TP, Kersbergen CJ, Bandy DK et al (2016) Two-photon imaging of cerebral hemodynamics and neural activity in awake and anesthetized marmosets. J Neurosci Methods 271:55–64
Sato T, Takahoko M, Okamoto H (2006) HuC:Kaede, a useful tool to label neural morphologies in networks in vivo. Genesis 44:136–142. doi:10.1002/gene.20196

Schrodel T, Prevedel R, Aumayr K, Zimmer M, Vaziri A (2013) Brain-wide 3D imaging of neuronal activity in Caenorhabditis elegans with sculpted light. Nat Meth 10:1013–1020
Scott EK et al (2007) Targeting neural circuitry in zebrafish using GAL4 enhancer trapping. Nat Methods 4:323–326
Seidemann E, Chen Y, Bai Y et al (2016) Calcium imaging with genetically encoded indicators in behaving primates. Elife 5:e16178
Theis L, Berens P, Froudarakis E et al (2016) Benchmarking spike rate inference in population calcium imaging. Neuron 90:471–482
Thiel G, Greengard P, Südhof TC (1991) Characterization of tissue-specific transcription by the human synapsin I gene promoter. Proc Natl Acad Sci 88:3431–3435
Tian L, Hires SA, Mao T et al (2009) Imaging neural activity in worms, flies and mice with improved GCaMP calcium indicators. Nat Methods 6:875–881. doi:10.1038/nmeth.1398
Tischbirek C, Birkner A, Jia H et al (2015) Deep two-photon brain imaging with a red-shifted fluorometric Ca^{2+} indicator. Proc Natl Acad Sci 112:11377–11382
Tomer R, Lovett-Barron M, Kauvar I et al (2015) SPED light sheet microscopy: fast mapping of biological system structure and function. Cell 163:1796–1806. doi:10.1016/j.cell.2015.11.061
Vladimirov N, Mu Y, Kawashima T et al (2014) Light-sheet functional imaging in fictively behaving zebrafish. Nat Methods 11:1–2. doi:10.1038/nmeth.3040

Chapter 14
Understanding Mood Disorders Using Electrophysiology and Circuit Breaking

He Liu and Dipesh Chaudhury

Abstract Mood disorders such as major depressive disorders are predicted to increase globally and according to the World Health Organization (WHO), it will become a leading contributor to the global burden of disease over the next few years. Pathophysiology of mood and reward processing leads to mood disorders such as anxiety, depression and addiction. Comorbidity of these disorders in a majority of patients implies that overlapping brain regions most likely regulate these processes. Evidence from the literature described in this chapter suggests that the multiplicity of symptoms related to mood disorders most likely is the result of aberrations in different aspects of normal neural functions ranging from the molecular up to the neural circuit. This review synthesizes findings from rodent studies from which emerges a role for different, yet interconnected, molecular systems and associated neural circuits to the aetiology of depression. Thus, in order to develop more effective and faster acting treatments for mood disorders such as depression, much work is still needed in understanding how exposure to stress lead to the sequence of changes in molecular, genetic/epigenetic processes and eventually neural circuit signalling. Using the combination of animal models of mood disorders together with the development of novel and sophisticated technologies to study molecular, genetic and neural circuit changes, there is a good possibility for the development of newer and better therapeutics for the treatment of mental disorders in the near future.

14.1 Introduction

Pathophysiology of mood and reward processing leads to disorders such as anxiety, depression, and addiction. Furthermore, comorbidity of these disorders in a majority of patients implies that overlapping brain regions most likely regulate these processes. Mood disorders such as major depressive disorders are predicted to increase globally

H. Liu · D. Chaudhury (✉)
Department of Biology, New York University Abu Dhabi, Abu Dhabi, United Arab Emirates
e-mail: dc151@nyu.edu

A. Çelik and M.F. Wernet (eds.), *Decoding Neural Circuit Structure and Function*, DOI 10.1007/978-3-319-57363-2_14

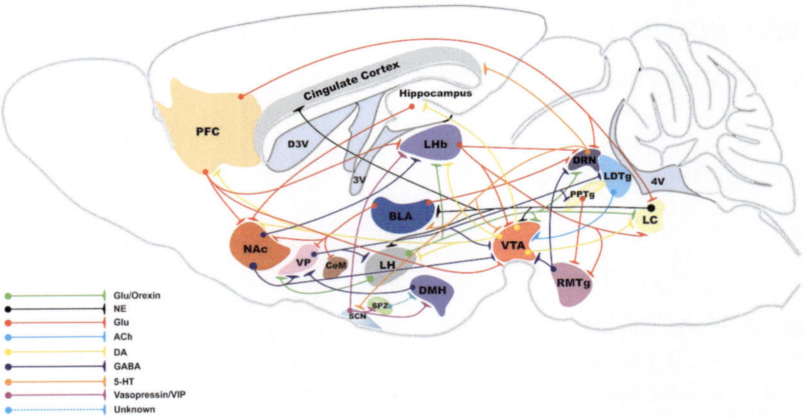

Fig. 14.1 Schematic of the major neural circuit connections involved in encoding for depression-related behaviours. The network displays the complex interplay between numerous neurotransmitters in regulating cellular activity within various brain nuclei. *5-HT* serotonin; *ACh* acetylcholine; *DA* dopamine; *GABA* gamma-aminobutyric acid; *Glu* glutamate; *NE* norepinephrine; *VIP* vasoactive intestinal peptide

and according to the World Health Organization (WHO), it will become a leading contributor to the global burden of disease over the next few years. In light of this, there has been an increase in research into understanding the parts played by various neural circuits in the brain in encoding for social behaviours that may lead to the future development of novel, faster acting therapeutics. Recent groundbreaking studies have shown that deep brain stimulation (DBS), which involves delivery of focal electrical current to specific neural structures within the brain, rapidly alleviates symptoms of depression (Zarate et al. 2013; Holtzheimer and Mayberg 2010). The effectiveness of DBS in alleviating depression in patients when targeted at various brain regions suggests that depression is a neural circuit disorder (Holtzheimer and Mayberg 2010; Sartorius and Henn 2007; Schneider et al. 2013; Berlim et al. 2014; Bogod et al. 2014; Dobrossy et al. 2015). Figure 14.1 shows a representation of some of the regions of the brain and their connections that have been implicated in depression. DBS as a therapeutic option for treatment-resistant patients is based on the novel hypothesis that depression come about as a result of aberrations in communications between specific neural structures of the brain. Thus, focal neuromodulation with DBS is believed to 'reset' neural activity of aberrant circuits associated with depression. At present, the mechanism by which the pathophysiology of neural circuits lead to depression is unclear. However, recent approaches combining animal models of depression together with electrophysiological, optogenetics, chemogenetics and molecular analysis have begun to reveal a complex interplay between various neural circuits and cell types in encoding for anxiety, depression and addiction (Han and Friedman 2012; Chaudhury et al. 2013; Francis et al. 2014, 2015; Russo and Nestler 2013; Lammel et al. 2014; Tye et al. 2011, 2013).

14.2 Heterogeneous Midbrain Dopamine Projections and Functional Regulation

Midbrain dopamine (DA) neurons in the ventral tegmental area (VTA), that sends robust dopaminergic projections to the NAc, play a key role in reward processing (Grace et al. 2007). Furthermore, a wide variety of brain regions such as the ventral hippocampus (is involved in context and focus on task), amygdala (mediates emotional behaviour), and prefrontal cortex (modulates activity throughout the limbic system to regulate decision making and behavioural flexibility) together control motivated behaviours. Each of these systems has overlapping projections to the nucleus accumbens (NAc), part of the midbrain dopamine circuit, where these signals are integrated under the modulatory influence of DA. In vivo, DA neurons have three distinct firing patterns, an inactive hyperpolarized state, a slow (2–10 Hz) single spike, tonic activity, and high frequency, multiple spike, burst or phasic activity (Grace and Bunney 1983). The mesolimbic dopaminergic pathway composed of dopaminergic (DA) neurons in the ventral tegmental area (VTA) and their projections to the nucleus accumbens (NAc) is crucial for the recognition of emotionally salient stimuli such as reward (Koob 2008) and aversion (Wenzel et al. 2015). High frequency, phasic, dopamine neuronal activity and the resulting transients in dopamine release are thought to comprise key learning signals in the brain (Schultz 2007; Robinson et al. 2007) encoding information related to external rewards and cues with resultant appetitive motivational behaviours. Dopamine transients occur spontaneously in several brain regions and are most prominent at the presentation of unexpected stimuli (Rebec et al. 1997; Robinson et al. 2007), rewards (Roitman et al. 2008) and social interaction, considered as a rewarding stimulus for social animals (Robinson et al. 2001, 2002). Social interaction is a complex behaviour essential for stable social dynamics and ultimately survival in many species. Impaired social interactions are a hallmark of several psychiatric disorders including autism, schizophrenia, depression and social anxiety disorders. In addition to congenital causes of impaired social interaction, external social stress in humans and animals is considered a major risk factor in the onset and development of neuropsychiatric disorders such as depression (Charney et al. 2004; El-Sayed et al. 2015; Henriques-Alves and Queiroz 2015). Fast-scan cyclic voltammetry has demonstrated the possible rewarding effect of social interaction, where transient DA release was directly measured in the NAc of rats exposed to novel rats (Robinson et al. 2011). Recent detailed analysis using novel fiberphotometry techniques measured increased VTA-NAc DA cell activity during social interaction (Gunaydin et al. 2014). Furthermore, this signalling was specifically dependent on D1-receptor activation in the NAc (Gunaydin et al. 2014). In light of the importance of DA signalling in modulating motivational behaviours such as social interaction, understanding the effect of stressful social encounters on DA signalling in the brain circuits may be a useful tool in uncovering DA neural circuit disorders associated with depression.

Recent evidence suggests that, DA neurons of the VTA-NAc circuit play a key role in modulating depression-related behaviours (Chaudhury et al. 2013; Tye et al.

2013; Nestler 2002; Berton and Nestler 2006; Berton et al. 2006; Krishnan et al. 2007; Cao et al. 2010; Friedman et al. 2014). Earlier work has shown that the in vitro firing rate and in vivo phasic firing events of VTA DA neurons in the brain reward system are significantly increased in mice exhibiting the susceptible (depressed) phenotype following exposure to the chronic social defeat paradigm (Krishnan et al. 2007; Cao et al. 2010; Anstrom et al. 2009; Razzoli et al. 2011). Conversely, in vivo recordings of rats susceptible to the learned helplessness paradigm exhibited decreased VTA DA neuron activity and that the antidepressant ketamine both rescued those rats previously susceptible to the learned helplessness paradigm and increased VTA DA activity (Belujon and Grace 2014). Two recent optogenetic studies directly demonstrated the role of VTA DA neurons in depression (Chaudhury et al. 2013; Tye et al. 2013). In one study, optogenetic induction of phasic, but not tonic, firing of VTA DA neurons was shown to rapidly induce the susceptible (depressed) phenotype in mice that had previously undergone a weak subthreshold social defeat stress paradigm (Chaudhury et al. 2013). Conversely, the second study showed the opposite effects where phasic activity of VTA DA neurons rescued stress-induced depressive-like behaviour in mice that had undergone chronic mild stress (Tye et al. 2013). The discrepancy between the two studies has been discussed (Lammel et al. 2014; Walsh and Han 2014) and may highlight differential coding processes of VTA DA neurons for strong or weak stressful stimuli (Valenti et al. 2012), which is highly consistent with the fine context-detecting functions of VTA DA neurons (Schnitzer 2002; Walsh et al. 2014). It has recently been shown that rats put through a strong stressful paradigm (restrained stress) exhibited increased firing in the VTA while those mice put through a weaker stress paradigm (mild inescapable stress) exhibited decreased activity (Valenti et al. 2012). Detailed circuit analysis reveals the existence of complex heterogeneity in neural inputs and outputs within brain nuclei such that subregions of a nuclei may receive differential inputs and modulate different downstream targets. These divergences most likely lead to the emergence of complex neuro-computational processes resulting in subtle differential roles of these subcircuits in encoding for aspects of behaviours. Thus, a possible explanation for the differential coding for weak or strong stressors might lie in the existence of functionally distinct subpopulations of VTA DA neurons. Those VTA DA neurons that exhibited decreased firing following weak stress exposure were located primarily in the medial and central portions of the VTA (Valenti et al. 2012). Furthermore, VTA DA neurons located in ventral VTA are excited by noxious foot shock while dorsal VTA DA neural activity is inhibited (O'Donnell and Grace 1994). Differential activity coding has also been demonstrated in mice exposed to the same stressor where for example VTA neurons projecting to the NAc (VTA-NAc) exhibit increased firing while those projecting to mPFC (VTA-mPFC) exhibit decreased firing in mice susceptible to social defeat stress (Chaudhury et al. 2013). Likewise, rapid induction of the depressed phenotype was observed by (1) optical induction of phasic activity in the VTA-NAc circuit and (2) optical inhibition of the VTA-mPFC circuit (Chaudhury et al. 2013). These projection-specific DA neurons exhibit differing physiological properties. DA neurons projecting to NAc exhibit robust I_h currents (hyperpolarization-activated non-selective cation currents) while DA neurons projecting to mPFC lack robust I_h currents (Friedman et al. 2014; Lammel et al. 2011). In addition to the heterogeneity in intrinsic physiological properties, these various subpopulations of VTA DA projection cells receive synaptic inputs from different nuclei within the brain. As described previously this is an example of heterogeneous inputs leading to increased computational power required for complex behaviours. Some major excitatory inputs to VTA come from the laterodorsal tegmentum/pedunculopontine tegmentum (LDTg/PPTg) and the lateral habenula (LHb). The LHb, a nucleus that integrates signalling from the basal frontal cortical areas and midbrain monoaminergic nuclei, has a functional role in different aspects of motivated behaviours (Ji and Shepard 2007; Lecca et al. 2014), where, for example, LHb neurons projecting specifically to the rostra medial tegmental nucleus (RMTg) mediate behavioural avoidance (Stamatakis and Stuber 2012). Furthermore, the LHb is a key neuroanatomical regulator of midbrain reward circuits, where increased activity of LHb projections to VTA is known to encode for depression following exposure to a learned helplessness or chronic social defeat (CSD) stress paradigm (Li et al. 2011; Dipesh Chaudhury 2014). In light of the previous findings that VTA cells projecting to the NAc (VTA-NAc) exhibit increased activity while VTA cells projecting to mPFC (VTA-mPFC) exhibit decreased activity in stress susceptible mice (Chaudhury et al. 2013), together with ongoing work showing increased activity of LHb neurons in stress susceptible mice (Dipesh Chaudhury 2014), raise the intriguing possibility into the existence of heterogeneous regulatory inputs from LHb to the VTA. For example, it could be hypothesized that the increased activity in VTA-NAc circuits in stress susceptible mice may, in part, arise from direct activation from a subset of glutamatergic LHb cells while the decreased activation of VTA-mPFC circuits may arise from feed forward inhibition from another subset of glutamatergic LHb cells first innervating VTA GABA interneurons that go unto inhibit VTA-mPFC circuits. Previous findings had shown evidence of both direct excitatory LHb input to VTA DA neurons and indirect inhibition of VTA DA neurons whereby increased LHb activity first activates GABA neurons in the VTA or RMTg that go unto attenuate VTA DA activity (Ji and Shepard 2007; Lammel et al. 2012). This hypothesis may, in part, explain the heterogeneous nature of VTA DA cell activity in encoding for behaviour, where some reports show increased activity of VTA DA encodes for reward (Grace et al. 2007; Schultz 2016), while others show enhanced activity encode for depression and aversive stimuli (Pezze and Feldon 2004). That aversive stimulation has differential effects on VTA DA neurons activity located at ventral or dorsal portions of the VTA (Brischoux et al. 2009; Ungless and Grace 2012) further supports the notion that neural inputs into regions such as the VTA may have differential anatomical distribution with preferential inputs to VTA DA or VTA GABA neurons. A role for VTA GABA neurons in encoding aversive stimuli was recently shown where footshock increased VTA GABA activity, while simultaneously attenuating VTA DA activity via post-synaptic $GABA_A$ receptor activation (Tan et al. 2012). Furthermore, optogenetic activation VTA GABA neurons induced conditioned place aversion, similar to VTA DA optogenetic inhibition (Tan et al. 2012). These findings for a role of GABA in modulating aversive and depressive behaviours further highlight the importance of GABA signalling in regulating normal neural network activity and that aberration in GABA signalling may lead to the aetiology of mental disorders such as depression as highlighted by the GABAergic deficit hypothesis of major depressive disorders (Luscher et al. 2011).

Neurons from the LDTg synapse primarily on VTA DA neurons projecting to the NAc while neurons from the LHb synapse either unto VTA DA neurons projecting to the mPFC or unto GABAergic neurons in a portion of the VTA also known as the RMTg (Lammel et al. 2012). In addition, these circuits encode opposing behaviours since selective activation of LDTg and LHb inputs to the VTA elicit reward and aversive behaviours respectively (Lammel et al. 2012), while LHb neurons projecting specifically to the RMTg mediate behavioural avoidance (Stamatakis and Stuber 2012). The mesopontine tegmentum consisting of the PPTg and LDTg is a heterogenous brain structure composed of glutamatergic, GABAergic and Cholinergic cells that is involved in a variety of behaviours ranging from locomotion, reward to sleep (Clarke et al. 1996; Maskos 2008; Wang and Morales 2009). Optogenetic activation of cholinergic projections from PPTg and LDTg to VTA was shown to induce conditioned place preference (CPP) while attenuation of these inputs decreased CPP (Xiao et al. 2016). Evidence for multiple neurotransmitters release from the same synapse has been accumulating over the years (Vaaga et al. 2014) and recent findings of basal ganglia (BG) input to the LHb have added to the literature. For example, electrophysiological recordings of post-synaptic LHb cellular responses following optogenetic activation of BG input to the LHb have shown to elicit both excitatory and inhibitory responses and that antidepressant medication changes the balance of the respective inputs to the LHb (Shabel et al. 2014). These findings highlight the importance of maintaining the correct balance in 'normal' conditions and that mental disorders such as depression may arise from the deviation of standard excitatory/inhibitory balance in specific brain regions.

14.3 Reward Processing by Lateral Habenula Circuitry

The LHb and VTA have reciprocal connections such that the VTA sends projections back to the LHb. Recent analysis using a combination of slice electrophysiology and optogenetics found that optogenetic stimulation of VTA terminals in the LHb induced post-synaptic GABA-mediated currents resulting in inhibition of LHb neural activity (Stamatakis et al. 2013). Detailed analysis of VTA-LHb cells highlighted several surprising findings where for example though these cells were TH+, a classical indicator of VTA DA cells; DA was not detected in the LHb following targeted stimulation of the VTA-LHb circuit. Furthermore, electrophysiological analysis found that these cells were typically less excitable than VTA-NAc cells and did not exhibit I_h currents much like VTA-mPFC cells first shown by Lammel and collaborators (Lammel et al. 2011). Detailed circuit analysis further highlighted that activation of the VTA-LHb cells induced increased spontaneous activity in VTA DA cells possibly via a loop involving inhibition of glutamatergic output from LHb leading to subsequent decreased activity of GABAergic cells of the RMTg which in turns removes inhibitory input from RMTg to VTA (Stamatakis et al. 2013). At the behavioural level, activation of this loop, regulating reward-related behaviours, follows the general principal that in specific context, increased VTA DA cell activity encodes for reward-related information (Stamatakis et al. 2013). According to current estimates approximately 65% of VTA neurons are dopaminergic while the remaining 30% are GABAergic and 5% Glutamatergic (Ungless and Grace 2012). As described previously, the participation of the VTA in diverse behaviours most likely reflects differential functional role of heterogenous neural phenotypes. For example, earlier work had shown differential distribution of VTA DA cell in encoding for aversive and rewarding stimuli (Lammel et al. 2012), and now more recent work on VTA glutamatergic neurons has shown that these projections to the LHb encode for aversive conditioning, since optical activation of this VTA glutamatergic neurons induced conditioned place aversion, an effect that was blocked with the infusion of AMPA/NMDA antagonist into the LHb (Root et al. 2014). Overall, it is likely therefore that aberration in the VTA-LHb-VTA loop may lead to depression-like behaviours involving a combination of DA and Glu signalling. In summary, the VTA has robust neural connections with a variety of brain regions associated with complex behaviours such as learning and memory, anxiety, reward and sleep (Fig. 14.2), and recent evidence suggests a role of VTA in the neural circuits related to mood disorders such as depression.

In addition to the role of the LHb in encoding for depression and reward, recent studies had shown that GABAergic projections from the basal forebrain (BF) (including the NAc, Lateral Septum and Diagonal Band Nuclei) to the LHb modulate aggression-like behaviours (Golden et al. 2016). Specifically the authors reported increased activity of BF projections to LHb in aggressive, but not in nonaggressive, mice, as measured by increased c-fos expression. Conversely, electrophysiological analysis showed corresponding decrease in neural activity in the LHb of aggressive, but not in the nonaggressive, mice (Golden et al. 2016). Using optogenetic techniques, the authors were also able to directly correlate the functional role of BF projections to the LHb in encoding for aggression. Here, they found that optical activation of BF projection to LHb decreased LHb neural activity and increased aggression-like behaviour in previously nonaggressive mice while optical inhibition of BF projections to LHb increased LHb neural activity and decreased aggression-like behaviour in previously aggressive mice (Golden et al. 2016). Since regions of the basal forebrain and LHb are implicated in addiction and depression, respectively, this study unravels some of the circuitry that leads to the clinical observations that neuropsychiatric conditions such as substance abuse (Beck et al. 2014) and depression (Martin et al. 2013) are typically co-morbid with aggressive behaviours. The ventral portion of the ventral medial hypothalamus (VMHvl), an area known to be important in attack, was recently shown to play an essential role in aggression seeking (Falkner et al. 2016). Using a combination of self-initiated aggression task (SIA) together with in vivo electrophysiological analysis this group

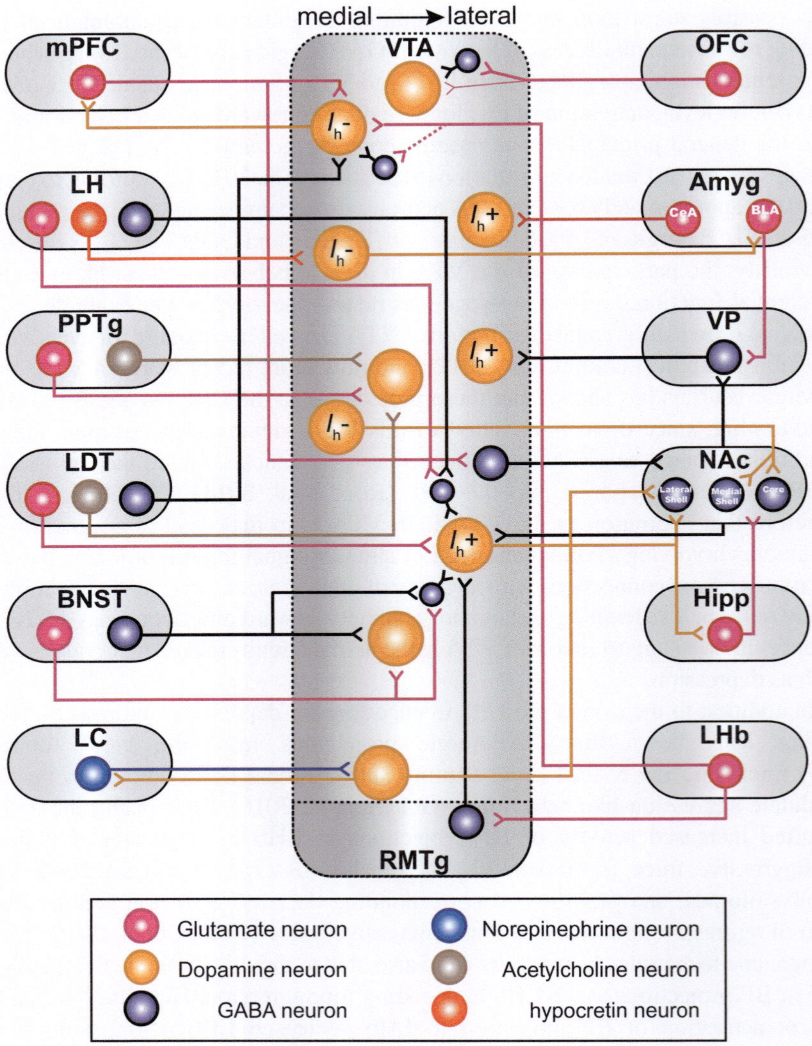

Fig. 14.2 Graphical representation of neural circuits connecting the dopamine reward region, the VTA (Ventral Tegmental Area), and regions associated with mood regulation. These reciprocal connections illustrate how these various regions are able to regulate each other, such that aberrations in signalling may lead to disorders such as depression. *mPFC* medial Prefrontal Cortex, *LH* Lateral Hypothalamus, *DR* Dorsal Raphe, *PPTg* Pedunculopontine Tegmentum, *LDTg* Laterodorsal Tegmentum, *RMTg* Rostromedial Tegmental Nucleus, *LHb* Lateral Habenula, *Hipp* Hippocampus, *NAc* Nucleus Accumbens, *BNST* Bed Nucleus of the Stria Terminalis, *VP* Ventral Pallidum, *LC* Locus Coeruleus, *OFC* Orbitofrontal Cortex, I_h hyperpolarization-activated non-selective cation currents

showed that a subset of VMHvl neurons in aggressive male mice generally exhibited increased activity during aggression seeking. Furthermore, pharmacogenetic inactivation of the VMHvl reduced aggression seeking while optogenetic stimulation accelerated aggression seeking (Falkner et al. 2016). The ability to bidirectionally modulate aggression seeking behaviour further expands the role of specific brain circuits in encoding for negative motivated behaviours and will further help expand our understandings of comorbidity of various symptoms in behavioural disorders.

14.4 Dorsal Raphe Nucleus Connections Implicated in Mood Disorders

Disregulation in signalling of the monoamine neurotransmitter, serotonin, in depression has been known for the last 60 years with the accidently discovery that inhibitors of monoamine breakdown alleviate depression. The dorsal raphe nucleus (DRN), the main source of the brains serotonin (5-HT), is implicated in the pathophysiology and therapeutics of mental disorders such as autism, anxiety and depression (Michelsen et al. 2008; Dolen et al. 2013). The DRN has reciprocal projections with various regions of the brains involved in regulating emotions such as the LHb, hippocampus, hypothalamus, amygdala and cerebral cortex (Michelsen et al. 2008). For example, evidence suggests that serotonergic projections from the DRN may modulate synaptic inputs in the LHb, where it was shown that excitatory BG input to the LHb, that encodes aversive behaviour, is suppressed by serotonin (Shabel et al. 2012). This finding highlights a putative circuit that may be a target for classical monoaminergic uptake blocker antidepressants. Furthermore, DRN neurons are regulated, both phasically and tonically, by excitatory glutamatergic and inhibitory GABAergic neural inputs from various regions implicated in mood disorders (Soiza-Reilly and Commons 2014). In line with the monoamine hypothesis of depression, 5-HT levels were decreased in mice exhibiting the depressed phenotype following chronic mild stress (Yang et al. 2008). Furthermore, evidence suggests a close link between the LHb and DRN circuit in depression since lesion of the LHb alleviated symptoms of depression by increasing 5-HT levels (Yang et al. 2008). The LHb and mPFC are critical regulators of DRN activity, where glutamatergic projections from these regions attenuate DRN activity via a feedforward mechanism involving inhibitory interneurons (Wang and Aghajanian 1977; Ferraro et al. 1996; Challis et al. 2014). Investigations into the micro-circuitry of DRN in encoding for depression showed increased excitability of GABAergic DRN neurons in mice susceptible, but not those resilient, to social defeat stress (Challis et al. 2013). Furthermore, this study found associated decreases in 5-HT neuron activity in the DRN of susceptible, but not resilient mice, implicating that in the DRN, encoding for depression leads to increased GABA activity followed by the subsequent decrease in 5-HT neuron activity (Challis et al. 2013). Follow-up studies showed a direct functional correlation between mPRC-DRN circuit activity and depression-related behaviours, where optogenetic modulation of mPFC neurons projecting to DRN regulated DRN GABAergic interneuron activity leading to changes in depressive-like behaviour in mice that had undergone social defeat stress (Challis et al. 2014). The DRN and VTA are reciprocally interconnected excitatory-inhibitory loop (Guiard et al. 2008). Selective lesion of VTA DA neurons leads to decreases firing of 5-HT neurons in DRN (Guiard et al. 2008), which suggests that DA excites 5-HT cells in the DRN. Furthermore selective lesion of 5-HT neurons in DR leads to increase activity in 36% of VTA DA neurons suggesting that 5-HT neurons inhibit VTA DA neurons. As described for the heterogenous activity of VTA DA neurons to aversive stimuli, DRN 5-HT neurons have been reported to exhibit similar profiles where 5-HT neurons that exhibit baseline rhythmic activity were phasically excited by noxious stimuli while those subsets of 5-HT neurons that exhibited baseline-bursting activity were inhibited (Schweimer and Ungless 2010). Apart from reciprocal modulation between VTA DA and DRN 5-HT cells, recent evidence has highlighted a role for DR DA cells in modulating social behaviours. The DRN is made of heterogenous nuclei of serotonergic neurons, as well as neurons that express GABA, Glutamate and Dopamine (Vasudeva et al. 2011). The functional role of DRN DA neurons is not well known except that it is not involved in rewarding behaviours as observed during intracranial self-stimulation studies (McDevitt et al. 2014). A recent study found evidence that DRN DA cells play a role in encoding for an animal's experience of social isolation (Nieh et al. 2016). Specifically social isolation induced increased AMPA/NMDA ratio, indicative of synaptic potentiation of glutamatergic inputs unto DRN DA cells. Furthermore, optogenetic stimulation of DR DA cells increased motivation for social preference, in previously socially isolated mice. However, surprisingly optogenetic induction of these neurons also induced place avoidance (Nieh et al. 2016). The authors have hypothesized that DRN DA neurons in this pathway do not encode for the rewarding phenomenon of social interaction, as observed in VTA DA cells (Gunaydin et al. 2014; Tsai et al. 2009) but rather modulate a 'loneliness-like' state that regulates social motivation. Furthermore, electrophysiological neural circuit analysis of DRN DA projections has shown that: (a) these neurons innervate the BNST and Central amygdala, two regions associated with anxiety and aversive emotional states (Adhikari 2014) and (b) these neurons also co-release glutamate, as evidenced the presence of short-latency fast AMPA-mediated excitatory post-synaptic currents (EPSC) recorded in cell bodies of Bed Nucleus of the Stria Terminalis (BNST) and central amygdala (CeA) following optical activation of DR DA terminals expressing ChR2. These electrophysiological findings may partially explain the complex and differential effects of DRN DA neurons activity on social preference but not in the rewarding component. Though 'loneliness' is a difficult concept to test in mice, these findings implicate a role of the DR DA neurons in encoding for social behaviour, which is effected in mood disorders. Social isolation, social exclusion or feelings of social disconnection can lead to loneliness which is a strong aversive emotional state in humans and detrimental to physical and mental well-being (Cacioppo et al. 2006; Holt-Lunstad et al. 2010; House et al. 1988). Since serotonergic neurons of the DRN play an important role in the control of sleep-wake states (Espana and Scammell 2011), it would be interesting to determine whether the DRN DA neurons play a role in wake states and how these may affect social preference.

14.5 Locus Coeruleus and Amygdala Involved in Responses to Stress

The locus coeruleus (LC), the major noradrenergic nuclei in the brain, is a vital component of the stress response. In rodents, stress increases LC activity (Curtis et al. 2002) and treatments with antidepressants such as SSRIs reduce LC activity, possibly via pre-synaptic autoreceptor activation (West et al. 2009). Anatomically the LC has reciprocal connections with various regions of the brains also known to be associated with depression (Sara 2009). Significant increases in gene expression levels for NMDA and mGluR subunits were found in the LC of postmortem depressed patients (Chandley et al. 2014) supporting the notion that disrupted glutamatergic-noradrenergic interactions in the LC lead to depression. Electrophysiological measures of basal firing rates of LC neurons in the Wistar Kyoto (WKY) rat, a strain that exhibits depressive and anxiety-like behaviours, were shown to be significantly higher compared to the standard Wistar strain of rats (Bruzos-Cidon et al. 2014). In addition, LC neurons from the WKY were found to be less responsive to the inhibitory effect of Alpha2-adrenoreceptor activation (Bruzos-Cidon et al. 2015). Furthermore, a role for LC dysfunction in depression comes from observations of robust glutamatergic projections from the PFC and LHb, two regions known to exhibit increased activity in depression (Li et al. 2011, 2013; Covington et al. 2010; Thompson et al. 2015). Neural processing of fear learning has recently been shown to pass from the lateral hypothalamus (LH) to the amygdala via the LC (Sears et al. 2013). This study demonstrated that orexin (hypocretin) fibres from the lateral hypothalamus directly depolarize LC neurons via rapid co-release of glutamate and orexin leading to activation of NMDA and orexin-1 receptors, respectively. Furthermore, orexin activation of LC neurons leads to increase noradrenergic signalling, via beta adrenergic receptor in the lateral nucleus of the amygdala leading to enhanced fear memory formation. Disregulation of this LH (orexinergic)-LC (noradrenergic)-amygdala circuit may be another possible mechanism by which depression occurs. Recently two populations of projection-specific neurons were identified in the basolateral amygdala (BLA) that encode for reward and aversion (Namburi et al. 2015). The authors demonstrate that basal lateral amygdala neurons projecting to the NAc exhibit increased synaptic plasticity in mice following reward training while BLA neurons projecting to the CeM exhibited increased synaptic plasticity in mice that had undergone fear conditioning (Namburi et al. 2015). This finding further extends the notions that distinct population of neurons exists within a nuclei encoding for opposing behavioural phenotypes, and that aberration in balance between neural circuits may lead to mental disorders such as depression. Exposure to stress has shown shifts in the balance of network inputs from cortical and thalamic to the BLA. For example, activation of norepinephrine (NE) beta receptors following fear conditioned learning shifts BLA neurons respond more robustly to the faster acting thalamic input over slower cortical inputs (Johnson et al. 2011). The BLA receives inputs from the auditory thalamus and auditory association cortex, where thalamic input encodes sensory information of the conditioned stimulus and the cortical pathway provides a more processed representation of the stimulus (LeDoux 2000). A recent neural circuit study extended the role of DA modulation in the BLA in putatively shifting network balances following exposure to stress. Specifically D2 receptor activation exhibited stronger net excitatory modulation of the auditory pathway and a stronger net inhibitory modulation of the cortical pathway (Chang and Grace 2015). The study of NE Beta and DA receptor modulation (Johnson et al. 2011; Chang and Grace 2015) showing shifts in input towards subcortical pathways is believed to be an evolutionary advantageous fight-or-flight response in animals during acute stress while the slower cortical inputs are involved in evaluating complex environmental stimulus (Johnson et al. 2011; LeDoux 2000). Thus in relation to the two dynamically different processing pathways, it is hypothesized that chronic stress induces maladaptive changes in the brain (McEwen 2007) where for example repeated chronic stress could induce shifts in stronger inputs from cortical inputs where organisms become more ruminative instead of proactive, as is proposed to occur in depression (Belzung et al. 2015). The VTA receives inhibitory inputs from LC and DRN in the form of noradrenergic and serotonergic projections, respectively. Increased LC release of NE onto VTA via chronic optogenetic activation of LC projections of VTA leads to decreased VTA DA activity and makes susceptible mice resilient to CSD (Isingrini et al. 2016). The mechanism of action of NE released from LC occurs via $\alpha 1$ and $\beta 3$ noradrenergic receptors activation on DA cells projecting to NAc (Hongxing Zhang-unpublished observations).

14.6 Lateral Hypothalamus and Ventral Pallidum Encoding Reward Directed Behaviour

The LH sends glutamatergic, GABAergic and/or peptidergic (Orexin/Hypocretin and Neurotensin) inputs to VTA. Previous studies showed that LH GABAergic projection to VTA increases feeding behaviour, while activation of LH glutamatergic projection to VTA regulates inputs to VTA (Nieh et al. 2016). Glutamatergic input from LH to VTA was shown to induce conditioned place avoidance, attenuate social interaction and decrease DA release in NAc as it targets the interneurons in the VTA. Conversely, GABAergic input from LH to VTA was shown to induce conditioned place preference, social interaction and increase DA release in NAc as it leads to the disinhibition of VTA interneurons to encode rewarding behaviour. Therefore, evidence suggests that the net effect of the LH inputs on VTA encodes rewarding behaviour (Nieh et al. 2016).

Moreover, VTA DA neurons receive potent inhibitory inputs from the ventral pallidum (VP) and the BLA, a region associated with stress and fear learning, sends excitatory glutamatergic inputs to the VP (Luscher and Malenka 2012; Radley et al. 2006). As previously described in relation to stress, DA and depression, acute stressors initially activate the DA system, followed 24 h later by potent attenuation of DA activity. Complex afferent polysynaptic inputs involving the BLA-VP circuit may in part be responsible for decreased VTA DA activity during encoding for the different temporal aspects of depression in stress susceptible mice following chronic mild stress (CMS) exposure (Goldwater et al. 2009). Evidence of the BLA-VP-VTA circuit in modulating depression-related behaviours comes from findings showing (a) reversal of decreased VTA DA activity in CMS susceptible rats following blockade of excitatory glutamatergic inputs to the VP (thus removing feed forward inhibitory input from the VP to the VTA, and (b) pharmacological activation of BLA decreases VTA DA activity (Chang and Grace 2014). A thorough understanding of the Dopamine pathways in the brain is therefore crucial for designing selective pharmacological modulation of dopamine release in various diseases such as schizophrenia and mood disorders and reducing the prevalence of side effects of current treatments (Anstrom et al. 2009).

14.7 Decoding Dopamine 1 and Dopamine 2 Subtype Receptors Signalling in Depression

Increasing evidence shows that the nucleus accumbens (NAc), a region typically associated with reward-related behaviours, has a critical role in depression symptomology including reduced motivation and anhedonia (Russo and Nestler 2013; Berton et al. 2006; Krishnan et al. 2007; Christoffel et al. 2011; Vialou et al. 2010; Lim et al. 2015). The transcription factor delta FosB regulates transcription of numerous genes in the NAc (McClung and Nestler 2003; Renthal et al. 2009). Two target genes of delta FosB, AMPA glutamate receptor subunit GluR2 and Sparc-like 1 (SC1) are upregulated in the NAc of mice resilient to CSD stress (Vialou et al. 2010). Furthermore, CHIPsec analysis showed significant binding of delta FosB on the GluR2 promoter and qPCR analysis revealed sustained GluR2 mRNA in NAc of resilient mice (Vialou et al. 2010). GluR2 subunit has profound effects on AMPA receptor function, where GluR2-lacking AMPA receptors are Ca^{2+}-permeable and show greater receptor conductance and strong inward rectification, as compared to GluR2-containing receptors (Bredt and Nicoll 2003). This switch to GluR2-lacking AMPA receptors increases neuronal excitability (Li et al. 2012). Electrophysiological measure showed decrease in both GluR2-mediated currents and increased inward rectification in stress susceptible, but not stress resilient mice, further implicating increased excitability, in a subset of NAc neurons, encodes for depressive-like behaviour (Vialou et al. 2010). Medium spiny neurons (MSNs) of the NAc and dorsal striatum are enriched in D1 or D2 receptors and they send distinct projections to BG and reward structures. NAc D1-MSNs send projections to VP, globus pallidum, VTA and substrantia nigra (SN), while NAc D2-MSNs send projections to VP (Nicola 2007; Smith et al. 2013). These two neural populations work in concert to promote normal behaviour while imbalance in one subtype can promote dysfunctional motivational states (Albin et al. 1989; Maia and Frank 2011; Kravitz and Kreitzer 2012; Lenz and Lobo 2013). The network balance model demonstrates that activation of D1-MSNs leads to positive reward behaviour while activation of D2-MSNs leads to aversive behaviours (Maia and Frank 2011; Carlezon and Thomas 2009; Lobo and Nestler 2011; Freeze et al. 2013). Exposure to CSD was shown to differentially induce expression of the transcription factor delta FosB in the NAc MSNs. Mice susceptible to CSD stress expressed elevation of delta FosB in D2-MSNs while those resilient to CSD stress expressed elevation of delta FosB in D1-MSNs (Lobo et al. 2013). Furthermore, anhedonia following restrained stress is mediated by decreased excitatory synaptic strength of NAc D1-, but not D2-MSNs (Lim et al. 2012). These findings were further extended in a recent study where mice susceptible to CSD exhibited decreased excitatory synaptic inputs into D1, but not D2-MSNs and that chronic chemogenetic attenuation of D1-MSNs, but not D2-MSNs, activity induced depressive-like behaviours in mice previously resilient to CSD stress (Francis et al. 2015). Furthermore repeated optogenetic activation of D2-MSNs induced depressive-like behaviour in mice exposed to a subthreshold social defeat paradigm (Francis et al. 2015). These findings are exciting as it demonstrates: (a) two distinct circuit mechanisms within the NAc encode for depressive-like behaviours; and (b) changes in synaptic signalling in NAc MSNs circuit most likely require long-term molecular changes in order to the expression of depression-like behaviours since chronic, but not acute, optogenetic and chemogenetic manipulations were required to induce susceptibility to stress. Several brain regions that mediate aspects of motivated, goal-directed, behaviours such as the ventral subiculum of the hippocampus (vSub), amygdala and prefrontal cortex send overlapping projections to the NAc, where these inputs are integrated under dopaminergic modulatory control (Grace et al. 2007). These differential inputs onto MSNs together in addition to the differential modulatory role of DA on D1 and D2 direct and indirect pathway highlight the complex interplay of DA with other neurotransmitters in regulating behavioural processes. Furthermore, changes in the balance of DA inputs onto NAc can lead to drastic changes in cognition and emotional state. Furthermore, systems level analysis has shown that DA receptor subtypes differentially regulate inputs to the NAc from the limbic system and PFC, where for example, tonic D2 receptor activation selectively attenuated inputs from the mPFC while phasic DA activity increases NAc neurons responsiveness to limbic inputs via activation of D1 receptors (O'Donnell and Grace 1994; Goto and Grace 2005a, b; West and Grace 2002). Aberration in the balance between these various inputs to the NAc is believed to lead to the pathophysiology of motivated behaviours such as addiction and depression. A recent study using a combination of in vivo electrophysiology recordings in rats that had previously undergone a learned helplessness depression paradigm showed decreased synaptic input from the vSub into the NAc of rats susceptible to the depressive phenotype, and not to the resilient ones (Belujon and Grace 2014). Furthermore, administration of the novel antidepressant ketamine both rescued the depressed phenotype in rats previously susceptible to the learned helplessness paradigm and also increased the vSub synaptic input to the NAc in these rats (Belujon and Grace 2014). At the microcircuit level decreased synaptic activity from vSub projections to the NAc shell but not NAc core specifically induced the depressed phenotype (Belujon and Grace 2014). Analysis of immediate early gene expression as a measure of neural activity found that expression levels of Arc and Egr 1 were both decreased in the ventral hippocampus (vHIP) of mice resilient to CSD stress while Egr 1 expression was increased in mice resilient to CSD (Bagot et al. 2015). Furthermore, physiological measures showed differential circuit specific synaptic adaptations following CSD where, in resilient mice, vHIP afferents projecting to NAc exhibited decreased glutamate release while mPFC afferents projecting to NAc exhibited increased glutamate release (Bagot et al. 2015). Physiologically and anatomically distinct VTA glutamatergic neurons projecting to NAc were shown to have lower expression of D2 receptors as indicated by decreased sensitivity to D2 agonist (Hnasko et al. 2012). These findings further expand the importance of balance in D1 and D2 receptor activation in performing complex neural computational necessary for encoding of mood and reward behaviours. It has been over a decade since it was hypothesized that a subset of VTA DA neurons co-release glutamate (Chuhma et al. 2004; Hnasko et al. 2010; Stuber et al. 2010; Tecuapetla et al. 2010; Hnasko and Edwards 2012). Due to earlier technical limitations, such as problems with selective stimulation of VTA DA cells, this idea was initially greeted with scepticism. However, with the advent of technologies such as optogenetics together with development of transgenic mice, it has been possible to systematically test this hypothesis. Optical activation of VTA DA neurons was shown to induce EPSC's in NAc shell but not in Dorsal Striatum (Stuber et al. 2010). Moreover, that AMPA/NMDA antagonist, but not D1/D2 antagonist, blocked the optically induced EPSC in NAc suggests that glutamate and not DA induced post-synaptic currents (Stuber et al. 2010). Vesicular glutamate transmitters such as VGLUT2 are necessary and sufficient for the exocytototic release of glutamate from neurons (Reimer and Edwards 2004; Takamori 2006). The observation that optogenetic stimulation of VTA DA projection to NAc in VGLU2 conditional knockout mice did not induce EPSC's in NAc shell indicates that glutamate is co-released from VTA DA neurons projecting to NAc shell (Stuber et al. 2010). The BLA plays an important role in emotional learning (Tye et al. 2011; Balleine and Killcross 2006; Maren and Quirk 2004; LeDoux 2003; Tye et al. 2008; Paton et al. 2006; Shabel and Janak 2009). Cue-triggered motivated behaviour is regulated by BLA projections to the NAc (Cador et al. 1989; Ambroggi et al. 2008; Di Ciano and Everitt 2004; Shiflett and Balleine 2010; Setlow et al. 2002). Detailed circuit analysis found that in vivo optical activation of BLA-to-NAc fibres promotes self-stimulation when animals were placed in an operant, nose-poke, task (Stuber et al. 2011). These findings highlight a role of this the BLA-to-NAc circuit, together with previous findings of the VTA-to-NAc circuit in, modulating reward behaviours.

14.8 Dendritic Plasticity Underlying Neuropsychiatric Disorders

Dendritic spine plasticity is a critical element of experience-dependent reorganization of the brain circuits and that maladaptive changes in dendritic spine development are suggested to underlie neuropsychiatric disorders such as depression, anxiety and addiction (Berton and Nestler 2006; Luscher and Malenka 2012; Kauer and Malenka 2007; Holtmaat and Svoboda 2009; Russo et al. 2010). Various stress paradigms lead to significant alterations in neuronal morphology in different cell types (Radley et al. 2006; Goldwater et al. 2009; Shansky et al. 2009). Furthermore, BDNF and its downstream targets such as I_{KB} Kinase (IKK) and the transcription factor nuclear factor kB (NF_{kB}) have been shown to be important regulators of neuronal structure (Chakravarthy et al. 2006; Russo et al. 2009). Mice susceptible to CSD stress exhibit dendritic remodelling (increased stubby spine formation) together with increased excitatory input into MSNs (Christoffel et al. 2011). Furthermore, molecular analysis found increases in both IKK and the associated phosphorylated IkB proteins in the NAc of mice susceptible to social stress (Christoffel et al. 2011). Evidence linking stress regulated changes in spine morphology and activation of IKK activity comes from approaches involving conditional knock down of IKK. Viral-mediated conditional knock down of IKK by expression of IKK dominant negative (IKKdn) protein reversed the formation of stubby spines in mice susceptible to CSD stress (Christoffel et al. 2011). Experience-dependent plasticity and the associated increases in synaptic input are known to lead to changes in spine morphology (Holtmaat and Svoboda 2009). The observations that VTA DA neurons projecting to NAc exhibit increased phasic firing in stress susceptible mice (Chaudhury et al. 2013) and that, the corticortrophin stress hormone gates BDNF signalling in NAc MSNs (Walsh et al. 2014) provides a mechanism by which increased BDNF induces changes in IKK enzyme activity leading to associated changes in dendritic spine morphology in the NAc of stress susceptible mice (Christoffel et al. 2011).

14.9 Comorbidity of Anxiety and Reward Related to Neural Circuits

The comorbidity of anxiety and dysfunctional reward processing such as addiction and reward implicates that common neural circuits contribute to these disparate neuropsychiatric symptoms. The extended amygdala, including the BNST modulates fear and anxiety (Davis et al. 2010; Walker and Davis 2008) and its projections to VTA likely explain the high comorbidity between anxiety, addiction and depression. Studies investigating the functional connectivity between the BNST and VTA have shown that functionally opposing BNST to VTA circuit regulates reward and aversive motivational states (Jennings et al. 2013). The BNST sends a combination of glutamatergic and GABAergic projections to the VTA, whereby aversive stimuli is differentially coded such that exposure to footshock increased activity in BNST glutamatergic projections to the VTA while decreasing activity in BNST GABAergic projections to VTA. Furthermore, optogenetic activation of the glutamatergic projection was shown to induce aversive and axiogenic behaviours while activation of the GABAergic projections induced rewarding and anxiolytic behaviours. Microcircuit analysis of BNST inputs to the VTA showed that glutamatergic neurons innervated VTA glutamate neurons while GABAergic neurons innervated VTA GABA neurons. Future analysis into which sub-populations of VTA neurons these inputs innervate may further elucidate the downstream effects of BNST–VTA coding of the different behavioural responses. Anatomical, behavioural and neuroimaging studies have implicated the BNST in pathological and adaptive forms of anxiety, where for example lesion of the BNST has been reported to lead to decreased anxiety-like behaviour (Walker et al. 2009; Duvarci et al. 2009). More recent studies using a combination of behavioural assays together with in vivo and in vitro electrophysiology and optogenetics have further identified differential roles of BNST subregions in modulating anxiety. For example, a recent study had reported that optogenetic attenuation of a subregion of the BNST, the ovBNST, induced anxiolytic behaviour on an elevated plus maze (EPM) test while similar inhibition of BLA input to another subregion of the BNST, the adBNST, induced anxiogenic behaviours on the EPM test (Kim et al. 2013). Furthermore, functional electrophysiological analysis found that optical activation of BLA input to adBNST induced net excitation in adBNST cell activity, implicating that BLA input to the BNST is glutamatergic in nature while local microcircuit inputs from ovBNST to adBNST were shown to be predominantly GABAergic in nature (Kim et al. 2013). Overall, these studies further highlight the complexity of neural circuits where defined subregions within nuclei can encode for opposing behaviours, where for example increased activity of BLA inputs to the adBNST subregions and decreased activity in the ovBNST subregion encode for anxiolytic behaviour. Plasticity in the projection of VTA DA neurons to the dentate gyrus (DG) in the hippocampus might be involved in addiction since behavioural analysis showed that morphine-induced conditioned place preference (CPP) requires NMDAR activation in VTA and VTA-Hippocampus DG activity (Hu et al. 2014). Since both addiction and depression are comorbid, it would be interesting to further investigate this pathway in the context of depression.

14.10 Homeostatic Synaptic Plasticity in Reward Circuits

Addictive drugs hijack learning processes through remodelling of neural circuits that underlie positive reinforcement such as the mesolimbic DA system. Drug-induced changes in synaptic plasticity have been observed where single exposure to cocaine induces long-term potentiation of AMPA-R, but not NMDA, mediated currents at excitatory synapses onto VTA DA cells (Ungless et al. 2001). Furthermore, cocaine increased AMPA:NMDA ratio in VTA DA cells 24 h after a single injection (Ungless et al. 2001). These studies were further extended where cocaine was shown to decrease NMDA receptor excitatory post-synaptic currents while at the same time increase insertion of GluA2-lacking AMPA receptors in post-synaptic VTA DA neurons (Mameli et al. 2011). These changes in drug-induced synaptic plasticity inverted the rules of activity-dependent plasticity whereby hyperpolarizing currents induced LTP following cocaine treatment (Mameli et al. 2011). Since cocaine-induced plasticity in the VTA is able to trigger changes downstream in the NAc (Conrad et al. 2008; Mameli et al. 2009), it is highly likely that these drug-induced changes at the neural circuit level, leading to abberations in synaptic signalling, may be responsible for behavioural changes such as depression and anxiety associated with drug abuse. Plasticity-related changes have been observed in the form of homeostatic changes within neural circuits related to depression. Recent advances have highlighted possible mechanisms that determine the brain's ability to cope with stress. As described previously, multiple lines of evidence implicate dysregulation in the brain's reward circuit in depression (Chaudhury et al. 2013; Tye et al. 2013; Berton et al. 2006; Krishnan et al. 2007; Nestler and Carlezon 2006). Increased activity of VTA DA neurons has been causally linked to depression-related behaviours (Chaudhury et al. 2013; Krishnan et al. 2007; Cao et al. 2010), where for example increased activity of VTA DA neuron, in stress susceptible mice, is intrinsically induced by upregulation of I_h, an excitatory driving force of VTA DA neurons (Cao et al. 2010; Wanat et al. 2008; Neuhoff et al. 2002),while pharmacological reduction of increased I_h in susceptible mice reverses depression (Cao et al. 2010; Friedman et al. 2014). Furthermore, chronic antidepressant treatment with fluoxetine normalizes this hyperexcitability and decreases I_h in these neurons (Cao et al. 2010). Together, these observations suggest that VTA DA neuron hyperactivity and increased excitatory I_h are both pathophysiological changes in mice susceptible to stress. A recent study confirmed that upregulation of I_h current in VTA DA neurons induced increased activity in these neurons in stress susceptible mice (Friedman et al. 2014). Surprisingly, while in resilient mice activity of these neurons was found to be normal, I_h current was even higher, which was observed in parallel with increased potassium (K^+) current (Friedman et al. 2014). The role of I_h current on mediating the resilient phenotype was determined by overexpression of the hyperpolarization-activated and cyclic nucleotide-gated channel 2 (HCN2), which mediates I_h current, in VTA DA neurons in susceptible mice. The enhancement of I_h current in previously susceptible mice due to overexpression of HCN2 channel is accompanied by compensatory upregulation of inhibitory K^+ currents, leading to normalization of the firing rate of the hyperactive neurons in susceptible mice and inducing the resilient phenotype (Friedman et al. 2014). Moreover, repeated optogenetic activation of VTA DA neurons in susceptible mice that presumably exhibit hyperactivity, was shown to induce the resilient phenotype together with associated decreased spontaneous activity and increased K^+-currents in these cells (Friedman et al. 2014). Furthermore, this was observed specifically in VTA DA neurons projecting to NAc but not mPFC, which correlates with previous projection-specific role in encoding for depression (Chaudhury et al. 2013). Previous molecular analysis had shown that mice resilient to CSD stress exhibited normalized firing activity in VTA DA neurons associated with a corresponding increase in genes coding for subtype of K^+ channels (Krishnan et al. 2007). In primary neuronal cultures, excessive hyperactivity has been shown to induce homeostatic upregulation of inhibitory driving force K^+-mediated currents (Zhang and Shapiro 2012). Therefore, K^+ channels have been considered as new therapeutic targets to naturally mediate active stress-coping or resilience. Both overexpression of KCNQ-type K^+ channel or intra-VTA infusion of KCNQ openers has been shown to lead to normalization of both the depressive-like behaviours and neuronal hyperexcitability (Friedman et al. 2016). In line with these homeostatic plasticity changes related to mood regulation, the previously discussed findings that chronic activation of LC-VTA projection induces the resilient phenotype were shown to be caused by the following homeostatic plasticity related to enhancing the excitatory current I_h and the inhibitory K^+-currents in VTA DA cells. In conclusion, homeostatic plasticity plays a fundamental role in stabilizing neuronal activity in response to excessive perturbations under both physiological (Non et al. 2014; Meaney 2001) and disease conditions (Francis et al. 1999). Observations that VTA DA neurons of resilient mice exhibit upregulation of the excitatory driving force I_h and inhibitory driving force K^+-current suggests homeostatic plasticity play a fundamental role in normalizing the neuronal hyperactivity in promoting natural resilience to stress. Further investigations into homeostatic adaptive mechanisms in DA and other neural circuit pathways leading to natural resilience have potential implication in the development of more naturalistic treatment strategies for mental disorders such as depression.

14.11 Concluding Remarks

Mental illness is a complex disorder such that patients with depression, for example, can exhibit comorbidity to other diseases such as anxiety and addiction. Evidence from the literature described in this chapter suggests that the multiplicity of symptoms related to mood disorders most likely is the result of aberrations in different aspects of normal neural functions ranging from the molecular up to the neural circuit. Furthermore, experimental observations from animal studies that different brain regions exhibit abnormal molecular and neural dynamic activity, together with clinical reports that classical antidepressant medications only work on a subset of patients, indicate that neural pathologies that lead to mood disorders can vary between patients. Thus, in order to develop more effective and faster-acting treatments for mood disorders such as depression, much work is still needed in understanding how exposure to stress lead to the sequence of changes in molecular, genetic/epigenetic processes and eventually neural circuit signaling. Using the combination of animal models of mood disorders together with the development of novel and sophisticated technologies to study molecular, genetic and neural circuit changes, there is a good possibility for the development of newer and better therapeutics for the treatment of mental disorders in the near future.

References

Adhikari A (2014) Distributed circuits underlying anxiety. Front Behav Neurosci 8:112

Albin RL, Young AB, Penney JB (1989) The functional anatomy of basal ganglia disorders. Trends Neurosci 12:366–375

Ambroggi F, Ishikawa A, Fields HL, Nicola SM (2008) Basolateral amygdala neurons facilitate reward-seeking behavior by exciting nucleus accumbens neurons. Neuron 59:648–661

Anstrom KK, Miczek KA, Budygin EA (2009) Increased phasic dopamine signaling in the mesolimbic pathway during social defeat in rats. Neuroscience 161:3–12

Bagot RC, Parise EM, Pena CJ, Zhang HX, Maze I, Chaudhury D, Persaud B, Cachope R, Bolanos-Guzman CA, Cheer JF, Deisseroth K, Han MH, Nestler EJ (2015) Ventral hippocampal afferents to the nucleus accumbens regulate susceptibility to depression. Nat Commun 6:7062

Balleine BW, Killcross S (2006) Parallel incentive processing: an integrated view of amygdala function. Trends Neurosci 29:272–279

Beck A, Heinz AJ, Heinz A (2014) Translational clinical neuroscience perspectives on the cognitive and neurobiological mechanisms underlying alcohol-related aggression. Curr Top Behav Neurosci 17:443–474

Belujon P, Grace AA (2014) Restoring mood balance in depression: ketamine reverses deficit in dopamine-dependent synaptic plasticity. Biol Psychiatry 76:927–936

Belzung C, Willner P, Philippot P (2015) Depression: from psychopathology to pathophysiology. Curr Opin Neurobiol 30:24–30

Berlim MT, McGirr A, Van den Eynde F, Fleck MP, Giacobbe P (2014) Effectiveness and acceptability of deep brain stimulation (DBS) of the subgenual cingulate cortex for treatment-resistant depression: a systematic review and exploratory meta-analysis. J Affect Disord 159:31–38

Berton O, Nestler EJ (2006) New approaches to antidepressant drug discovery: beyond monoamines. Nat Rev Neurosci 7:137–151

Berton O, McClung CA, Dileone RJ, Krishnan V, Renthal W, Russo SJ, Graham D, Tsankova NM, Bolanos CA, Rios M, Monteggia LM, Self DW, Nestler EJ (2006) Essential role of BDNF in the mesolimbic dopamine pathway in social defeat stress. Science 311: 864–868

Bogod NM, Sinden M, Woo C, Defreitas VG, Torres IJ, Howard AK, Ilcewicz-Klimek MI, Honey CR, Yatham LN, Lam RW (2014) Long-term neuropsychological safety of subgenual cingulate gyrus deep brain stimulation for treatment-resistant depression. J Neuropsychiatry Clin Neurosci 26:126–133

Bredt DS, Nicoll RA (2003) AMPA receptor trafficking at excitatory synapses. Neuron 40: 361–379

Brischoux F, Chakraborty S, Brierley DI, Ungless MA (2009) Phasic excitation of dopamine neurons in ventral VTA by noxious stimuli. Proc Natl Acad Sci U S A 106:4894–4899

Bruzos-Cidon C, Miguelez C, Rodriguez JJ, Gutierrez-Lanza R, Ugedo L, Torrecilla M (2014) Altered neuronal activity and differential sensitivity to acute antidepressants of locus coeruleus and dorsal raphe nucleus in Wistar Kyoto rats: a comparative study with Sprague Dawley and Wistar rats. Eur Neuropsychopharmacol 24:1112–1122

Bruzos-Cidon C, Llamosas N, Ugedo L, Torrecilla M (2015) Dysfunctional inhibitory mechanisms in locus coeruleus neurons of the wistar kyoto rat. Int J Neuropsychopharmacol 18:pyu122

Cacioppo JT, Hughes ME, Waite LJ, Hawkley LC, Thisted RA (2006) Loneliness as a specific risk factor for depressive symptoms: cross-sectional and longitudinal analyses. Psychol Aging 21:140–151

Cador M, Robbins TW, Everitt BJ (1989) Involvement of the amygdala in stimulus-reward associations: interaction with the ventral striatum. Neuroscience 30:77–86

Cao JL, Covington HE 3rd, Friedman AK, Wilkinson MB, Walsh JJ, Cooper DC, Nestler EJ, Han MH (2010) Mesolimbic dopamine neurons in the brain reward circuit mediate susceptibility to social defeat and antidepressant action. J Neurosci 30:16453–16458

Carlezon WA Jr, Thomas MJ (2009) Biological substrates of reward and aversion: a nucleus accumbens activity hypothesis. Neuropharmacology 56(Suppl 1):122–132

Chakravarthy S, Saiepour MH, Bence M, Perry S, Hartman R, Couey JJ, Mansvelder HD, Levelt CN (2006) Postsynaptic TrkB signaling has distinct roles in spine maintenance in adult visual cortex and hippocampus. Proc Natl Acad Sci U S A 103:1071–1076

Challis C, Boulden J, Veerakumar A, Espallergues J, Vassoler FM, Pierce RC, Beck SG, Berton O (2013) Raphe GABAergic neurons mediate the acquisition of avoidance after social defeat. J Neurosci 33(13978–13988):13988a

Challis C, Beck SG, Berton O (2014) Optogenetic modulation of descending prefrontocortical inputs to the dorsal raphe bidirectionally bias socioaffective choices after social defeat. Front Behav Neurosci 8:43

Chandley MJ, Szebeni A, Szebeni K, Crawford JD, Stockmeier CA, Turecki G, Kostrzewa RM, Ordway GA (2014) Elevated gene expression of glutamate receptors in noradrenergic neurons from the locus coeruleus in major depression. Int J Neuropsychopharmacol 17:1569–1578

Chang CH, Grace AA (2014) Amygdala-ventral pallidum pathway decreases dopamine activity after chronic mild stress in rats. Biol Psychiatry 76:223–230

Chang CH, Grace AA (2015) Dopaminergic modulation of lateral amygdala neuronal activity: differential D1 and D2 receptor effects on thalamic and cortical afferent inputs. Int J Neuropsychopharmacol 18:pyv015

Charney DS, Dejesus G, Manji HK (2004) Cellular plasticity and resilience and the pathophysiology of severe mood disorders. Dialogues Clin Neurosci 6:217–225

Chaudhury D, Walsh JJ, Friedman AK, Juarez B, Ku SM, Koo JW, Ferguson D, Tsai HC, Pomeranz L, Christoffel DJ, Nectow AR, Ekstrand M, Domingos A, Mazei-Robison MS, Mouzon E, Lobo MK, Neve RL, Friedman JM, Russo SJ, Deisseroth K, Nestler EJ, Han MH (2013) Rapid regulation of depression-related behaviours by control of midbrain dopamine neurons. Nature 493:532–536

Christoffel DJ, Golden SA, Dumitriu D, Robison AJ, Janssen WG, Ahn HF, Krishnan V, Reyes CM, Han MH, Ables JL, Eisch AJ, Dietz DM, Ferguson D, Neve RL, Greengard P, Kim Y, Morrison JH, Russo SJ (2011) IkappaB kinase regulates social defeat stress-induced synaptic and behavioral plasticity. J Neurosci 31:314–321

Chuhma N, Zhang H, Masson J, Zhuang X, Sulzer D, Hen R, Rayport S (2004) Dopamine neurons mediate a fast excitatory signal via their glutamatergic synapses. J Neurosci 24:972–981

Clarke NP, Bolam JP, Bevan MD (1996) Glutamate-enriched inputs from the mesopontine tegmentum to the entopeduncular nucleus in the rat. Eur J Neurosci 8:1363–1376

Conrad KL, Tseng KY, Uejima JL, Reimers JM, Heng LJ, Shaham Y, Marinelli M, Wolf ME (2008) Formation of accumbens GluR2-lacking AMPA receptors mediates incubation of cocaine craving. Nature 454:118–121

Covington HE 3rd, Lobo MK, Maze I, Vialou V, Hyman JM, Zaman S, LaPlant Q, Mouzon E, Ghose S, Tamminga CA, Neve RL, Deisseroth K, Nestler EJ (2010) Antidepressant effect of optogenetic stimulation of the medial prefrontal cortex. J Neurosci 30:16082–16090

Curtis AL, Bello NT, Connolly KR, Valentino RJ (2002) Corticotropin-releasing factor neurones of the central nucleus of the amygdala mediate locus coeruleus activation by cardiovascular stress. J Neuroendocrinol 14:667–682

Davis M, Walker DL, Miles L, Grillon C (2010) Phasic vs sustained fear in rats and humans: role of the extended amygdala in fear vs anxiety. Neuropsychopharmacology 35:105–135

Di Ciano P, Everitt BJ (2004) Direct interactions between the basolateral amygdala and nucleus accumbens core underlie cocaine-seeking behavior by rats. J Neurosci 24:7167–7173

Dipesh Chaudhury HZ, Juarez B, Friedman A, Ku S, Han MH (2014) Lateral habenula projections to a subset of ventral tegmental area neurons rapidly encodes for susceptibility to social defeat stress. Soc Neurosci Abs

Dobrossy MD, Furlanetti LL, Coenen VA (2015) Electrical stimulation of the medial forebrain bundle in pre-clinical studies of psychiatric disorders. Neurosci Biobehav Rev 49:32–42

Dolen G, Darvishzadeh A, Huang KW, Malenka RC (2013) Social reward requires coordinated activity of nucleus accumbens oxytocin and serotonin. Nature 501:179–184

Duvarci S, Bauer EP, Pare D (2009) The bed nucleus of the stria terminalis mediates inter-individual variations in anxiety and fear. J Neurosci 29:10357–10361

El-Sayed AM, Palma A, Freedman LP, Kruk ME (2015) Does health insurance mitigate inequities in non-communicable disease treatment? Evidence from 48 low- and middle-income countries. Health Policy 119:1164–1175
Espana RA, Scammell TE (2011) Sleep neurobiology from a clinical perspective. Sleep 34: 845–858
Falkner AL, Grosenick L, Davidson TJ, Deisseroth K, Lin D (2016) Hypothalamic control of male aggression-seeking behavior. Nat Neurosci 19:596–604
Ferraro G, Montalbano ME, Sardo P, La Grutta V (1996) Lateral habenular influence on dorsal raphe neurons. Brain Res Bull 41:47–52
Francis D, Diorio J, Liu D, Meaney MJ (1999) Nongenomic transmission across generations of maternal behavior and stress responses in the rat. Science 286:1155–1158
Francis TC, Chaudhury D, Lobo MK (2014) Optogenetics: illuminating the neural basis of rodent behavior. Open Access Anim Physiol 6:33–51
Francis TC, Chandra R, Friend DM, Finkel E, Dayrit G, Miranda J, Brooks JM, Iniguez SD, O'Donnell P, Kravitz A, Lobo MK (2015) Nucleus accumbens medium spiny neuron subtypes mediate depression-related outcomes to social defeat stress. Biol Psychiatry 77:212–222
Freeze BS, Kravitz AV, Hammack N, Berke JD, Kreitzer AC (2013) Control of basal ganglia output by direct and indirect pathway projection neurons. J Neurosci 33:18531–18539
Friedman AK, Walsh JJ, Juarez B, Ku SM, Chaudhury D, Wang J, Li X, Dietz DM, Pan N, Vialou VF, Neve RL, Yue Z, Han MH (2014) Enhancing depression mechanisms in midbrain dopamine neurons achieves homeostatic resilience. Science 344:313–319
Friedman AK, Juarez B, Ku SM, Zhang H, Calizo RC, Walsh JJ, Chaudhury D, Zhang S, Hawkins A, Dietz DM, Murrough JW, Ribadeneira M, Wong EH, Neve RL, Han MH (2016) KCNQ channel openers reverse depressive symptoms via an active resilience mechanism. Nat Commun 7:11671
Golden SA, Heshmati M, Flanigan M, Christoffel DJ, Guise K, Pfau ML, Aleyasin H, Menard C, Zhang H, Hodes GE, Bregman D, Khibnik L, Tai J, Rebusi N, Krawitz B, Chaudhury D, Walsh JJ, Han MH, Shapiro ML, Russo SJ (2016) Basal forebrain projections to the lateral habenula modulate aggression reward. Nature 534:688–692
Goldwater DS, Pavlides C, Hunter RG, Bloss EB, Hof PR, McEwen BS, Morrison JH (2009) Structural and functional alterations to rat medial prefrontal cortex following chronic restraint stress and recovery. Neuroscience 164:798–808
Goto Y, Grace AA (2005a) Dopamine-dependent interactions between limbic and prefrontal cortical plasticity in the nucleus accumbens: disruption by cocaine sensitization. Neuron 47:255–266
Goto Y, Grace AA (2005b) Dopaminergic modulation of limbic and cortical drive of nucleus accumbens in goal-directed behavior. Nat Neurosci 8:805–812
Grace AA, Bunney BS (1983) Intracellular and extracellular electrophysiology of nigral dopaminergic neurons–3. Evidence for electrotonic coupling. Neuroscience 10:333–348
Grace AA, Floresco SB, Goto Y, Lodge DJ (2007) Regulation of firing of dopaminergic neurons and control of goal-directed behaviors. Trends Neurosci 30:220–227
Guiard BP, El Mansari M, Merali Z, Blier P (2008) Functional interactions between dopamine, serotonin and norepinephrine neurons: an in-vivo electrophysiological study in rats with monoaminergic lesions. Int J Neuropsychopharmacol 11:625–639
Gunaydin LA, Grosenick L, Finkelstein JC, Kauvar IV, Fenno LE, Adhikari A, Lammel S, Mirzabekov JJ, Airan RD, Zalocusky KA, Tye KM, Anikeeva P, Malenka RC, Deisseroth K (2014) Natural neural projection dynamics underlying social behavior. Cell 157:1535–1551
Han MH, Friedman AK (2012) Virogenetic and optogenetic mechanisms to define potential therapeutic targets in psychiatric disorders. Neuropharmacology 62:89–100
Henriques-Alves AM, Queiroz CM (2015) Ethological evaluation of the effects of social defeat stress in mice: beyond the social interaction ratio. Front Behav Neurosci 9:364
Hnasko TS, Edwards RH (2012) Neurotransmitter corelease: mechanism and physiological role. Annu Rev Physiol 74:225–243

Hnasko TS, Chuhma N, Zhang H, Goh GY, Sulzer D, Palmiter RD, Rayport S, Edwards RH (2010) Vesicular glutamate transport promotes dopamine storage and glutamate corelease in vivo. Neuron 65:643–656
Hnasko TS, Hjelmstad GO, Fields HL, Edwards RH (2012) Ventral tegmental area glutamate neurons: electrophysiological properties and projections. J Neurosci 32:15076–15085
Holt-Lunstad J, Smith TB, Layton JB (2010) Social relationships and mortality risk: a meta-analytic review. PLoS Med 7:e1000316
Holtmaat A, Svoboda K (2009) Experience-dependent structural synaptic plasticity in the mammalian brain. Nat Rev Neurosci 10:647–658
Holtzheimer PE 3rd, Mayberg HS (2010) Deep brain stimulation for treatment-resistant depression. Am J Psychiatry 167:1437–1444
House JS, Landis KR, Umberson D (1988) Social relationships and health. Science 241:540–545
Hu L, Jing XH, Cui CL, Xing GG, Zhu B (2014) NMDA receptors in the midbrain play a critical role in dopamine-mediated hippocampal synaptic potentiation caused by morphine. Addict Biol 19:380–391
Isingrini E, Perret L, Rainer Q, Amilhon B, Guma E, Tanti A, Martin G, Robinson J, Moquin L, Marti F, Mechawar N, Williams S, Gratton A, Giros B (2016) Resilience to chronic stress is mediated by noradrenergic regulation of dopamine neurons. Nat Neurosci 19:560–563
Jennings JH, Sparta DR, Stamatakis AM, Ung RL, Pleil KE, Kash TL, Stuber GD (2013) Distinct extended amygdala circuits for divergent motivational states. Nature 496:224–228
Ji H, Shepard PD (2007) Lateral habenula stimulation inhibits rat midbrain dopamine neurons through a GABA(A) receptor-mediated mechanism. J Neurosci 27:6923–6930
Johnson LR, Hou M, Prager EM, Ledoux JE (2011) Regulation of the fear network by mediators of stress: norepinephrine alters the balance between cortical and subcortical afferent excitation of the lateral amygdala. Front Behav Neurosci 5:23
Kauer JA, Malenka RC (2007) Synaptic plasticity and addiction. Nat Rev Neurosci 8:844–858
Kim SY, Adhikari A, Lee SY, Marshel JH, Kim CK, Mallory CS, Lo M, Pak S, Mattis J, Lim BK, Malenka RC, Warden MR, Neve R, Tye KM, Deisseroth K (2013) Diverging neural pathways assemble a behavioural state from separable features in anxiety. Nature 496:219–223
Koob GF (2008) A role for brain stress systems in addiction. Neuron 59:11–34
Kravitz AV, Kreitzer AC (2012) Striatal mechanisms underlying movement, reinforcement, and punishment. Physiology 27:167–177
Krishnan V, Han MH, Graham DL, Berton O, Renthal W, Russo SJ, Laplant Q, Graham A, Lutter M, Lagace DC, Ghose S, Reister R, Tannous P, Green TA, Neve RL, Chakravarty S, Kumar A, Eisch AJ, Self DW, Lee FS, Tamminga CA, Cooper DC, Gershenfeld HK, Nestler EJ (2007) Molecular adaptations underlying susceptibility and resistance to social defeat in brain reward regions. Cell 131:391–404
Lammel S, Ion DI, Roeper J, Malenka RC (2011) Projection-specific modulation of dopamine neuron synapses by aversive and rewarding stimuli. Neuron 70:855–862
Lammel S, Lim BK, Ran C, Huang KW, Betley MJ, Tye KM, Deisseroth K, Malenka RC (2012) Input-specific control of reward and aversion in the ventral tegmental area. Nature 491: 212–217
Lammel S, Lim BK, Malenka RC (2014) Reward and aversion in a heterogeneous midbrain dopamine system. Neuropharmacology 76 Pt B:351–359
Lecca S, Meye FJ, Mameli M (2014) The lateral habenula in addiction and depression: an anatomical, synaptic and behavioral overview. Eur J Neurosci 39:1170–1178
LeDoux JE (2000) Emotion circuits in the brain. Annu Rev Neurosci 23:155–184
LeDoux J (2003) The emotional brain, fear, and the amygdala. Cell Mol Neurobiol 23:727–738
Lenz JD, Lobo MK (2013) Optogenetic insights into striatal function and behavior. Behav Brain Res 255:44–54
Li B, Piriz J, Mirrione M, Chung C, Proulx CD, Schulz D, Henn F, Malinow R (2011) Synaptic potentiation onto habenula neurons in the learned helplessness model of depression. Nature 470:535–539

Li DP, Byan HS, Pan HL (2012) Switch to glutamate receptor 2-lacking AMPA receptors increases neuronal excitability in hypothalamus and sympathetic drive in hypertension. J Neurosci 32:372–380
Li K, Zhou T, Liao L, Yang Z, Wong C, Henn F, Malinow R, Yates JR 3rd, Hu H (2013) βCaMKII in lateral habenula mediates core symptoms of depression. Science 341:1016–1020
Lim BK, Huang KW, Grueter BA, Rothwell PE, Malenka RC (2012) Anhedonia requires MC4R-mediated synaptic adaptations in nucleus accumbens. Nature 487:183–189
Lim LW, Prickaerts J, Huguet G, Kadar E, Hartung H, Sharp T, Temel Y (2015) Electrical stimulation alleviates depressive-like behaviors of rats: investigation of brain targets and potential mechanisms. Transl Psychiatry 5:e535
Lobo MK, Nestler EJ (2011) The striatal balancing act in drug addiction: distinct roles of direct and indirect pathway medium spiny neurons. Front Neuroanat 5:41
Lobo MK, Zaman S, Damez-Werno DM, Koo JW, Bagot RC, DiNieri JA, Nugent A, Finkel E, Chaudhury D, Chandra R, Riberio E, Rabkin J, Mouzon E, Cachope R, Cheer JF, Han MH, Dietz DM, Self DW, Hurd YL, Vialou V, Nestler EJ (2013) DeltaFosB induction in striatal medium spiny neuron subtypes in response to chronic pharmacological, emotional, and optogenetic stimuli. J Neurosci 33:18381–18395
Lüscher C, Malenka RC (2012) NMDA receptor-dependent long-term potentiation and long-term depression (LTP/LTD). Cold Spring Harb Perspect Biol 4
Luscher B, Shen Q, Sahir N (2011) The GABAergic deficit hypothesis of major depressive disorder. Mol Psychiatry 16:383–406
Maia TV, Frank MJ (2011) From reinforcement learning models to psychiatric and neurological disorders. Nat Neurosci 14:154–162
Mameli M, Halbout B, Creton C, Engblom D, Parkitna JR, Spanagel R, Luscher C (2009) Cocaine-evoked synaptic plasticity: persistence in the VTA triggers adaptations in the NAc. Nat Neurosci 12:1036–1041
Mameli M, Bellone C, Brown MT, Luscher C (2011) Cocaine inverts rules for synaptic plasticity of glutamate transmission in the ventral tegmental area. Nat Neurosci 14:414–416
Maren S, Quirk GJ (2004) Neuronal signalling of fear memory. Nat Rev Neurosci 5:844–852
Martin LA, Neighbors HW, Griffith DM (2013) The experience of symptoms of depression in men vs women: analysis of the national comorbidity survey replication. JAMA Psychiatry 70:1100–1106
Maskos U (2008) The cholinergic mesopontine tegmentum is a relatively neglected nicotinic master modulator of the dopaminergic system: relevance to drugs of abuse and pathology. Br J Pharmacol 153(Suppl 1):S438–S445
McClung CA, Nestler EJ (2003) Regulation of gene expression and cocaine reward by CREB and DeltaFosB. Nat Neurosci 6:1208–1215
McDevitt RA, Tiran-Cappello A, Shen H, Balderas I, Britt JP, Marino RA, Chung SL, Richie CT, Harvey BK, Bonci A (2014) Serotonergic versus nonserotonergic dorsal raphe projection neurons: differential participation in reward circuitry. Cell Rep 8:1857–1869
McEwen BS (2007) Physiology and neurobiology of stress and adaptation: central role of the brain. Physiol Rev 87:873–904
Meaney MJ (2001) Maternal care, gene expression, and the transmission of individual differences in stress reactivity across generations. Annu Rev Neurosci 24:1161–1192
Michelsen KA, Prickaerts J, Steinbusch HW (2008) The dorsal raphe nucleus and serotonin: implications for neuroplasticity linked to major depression and Alzheimer's disease. Prog Brain Res 172:233–264
Namburi P, Beyeler A, Yorozu S, Calhoon GG, Halbert SA, Wichmann R, Holden SS, Mertens KL, Anahtar M, Felix-Ortiz AC, Wickersham IR, Gray JM, Tye KM (2015) A circuit mechanism for differentiating positive and negative associations. Nature 520:675–678
Nestler EJ (2002) From neurobiology to treatment: progress against addiction. Nat Neurosci 5(Suppl):1076–1079

Nestler EJ, Carlezon WA Jr (2006) The mesolimbic dopamine reward circuit in depression. Biol Psychiatry 59:1151–1159

Neuhoff H, Neu A, Liss B, Roeper J (2002) I(h) channels contribute to the different functional properties of identified dopaminergic subpopulations in the midbrain. J Neurosci 22:1290–1302

Nicola SM (2007) The nucleus accumbens as part of a basal ganglia action selection circuit. Psychopharmacology 191:521–550

Nieh EH, Vander Weele CM, Matthews GA, Presbrey KN, Wichmann R, Leppla CA, Izadmehr EM, Tye KM (2016) Inhibitory input from the lateral hypothalamus to the ventral tegmental area disinhibits dopamine neurons and promotes behavioral activation. Neuron 1286–1298

Non AL, Binder AM, Kubzansky LD, Michels KB (2014) Genome-wide DNA methylation in neonates exposed to maternal depression, anxiety, or SSRI medication during pregnancy. Epigenetics Off J DNA Methyl Soc 9:964–972

O'Donnell P, Grace AA (1994) Tonic D2-mediated attenuation of cortical excitation in nucleus accumbens neurons recorded in vitro. Brain Res 634:105–112

Paton JJ, Belova MA, Morrison SE, Salzman CD (2006) The primate amygdala represents the positive and negative value of visual stimuli during learning. Nature 439:865–870

Pezze MA, Feldon J (2004) Mesolimbic dopaminergic pathways in fear conditioning. Prog Neurobiol 74:301–320

Radley JJ, Rocher AB, Miller M, Janssen WG, Liston C, Hof PR, McEwen BS, Morrison JH (2006) Repeated stress induces dendritic spine loss in the rat medial prefrontal cortex. Cereb Cortex 16:313–320

Razzoli M, Andreoli M, Michielin F, Quarta D, Sokal DM (2011) Increased phasic activity of VTA dopamine neurons in mice 3 weeks after repeated social defeat. Behav Brain Res 218:253–257

Rebec GV, Christensen JR, Guerra C, Bardo MT (1997) Regional and temporal differences in real-time dopamine efflux in the nucleus accumbens during free-choice novelty. Brain Res 776:61–67

Reimer RJ, Edwards RH (2004) Organic anion transport is the primary function of the SLC17/type I phosphate transporter family. Pflugers Arch 447:629–635

Renthal W, Kumar A, Xiao G, Wilkinson M, Covington HE 3rd, Maze I, Sikder D, Robison AJ, LaPlant Q, Dietz DM, Russo SJ, Vialou V, Chakravarty S, Kodadek TJ, Stack A, Kabbaj M, Nestler EJ (2009) Genome-wide analysis of chromatin regulation by cocaine reveals a role for sirtuins. Neuron 62:335–348

Robinson DL, Wightman RM (2007) In: Michael AC, Borland LM (eds) Electrochemical methods for neuroscience

Robinson DL, Phillips PE, Budygin EA, Trafton BJ, Garris PA, Wightman RM (2001) Sub-second changes in accumbal dopamine during sexual behavior in male rats. NeuroReport 12:2549–2552

Robinson DL, Heien ML, Wightman RM (2002) Frequency of dopamine concentration transients increases in dorsal and ventral striatum of male rats during introduction of conspecifics. J Neurosci Off J Soc Neurosci 22:10477–10486

Robinson DL, Zitzman DL, Smith KJ, Spear LP (2011) Fast dopamine release events in the nucleus accumbens of early adolescent rats. Neuroscience 176:296–307

Roitman MF, Wheeler RA, Wightman RM, Carelli RM (2008) Real-time chemical responses in the nucleus accumbens differentiate rewarding and aversive stimuli. Nat Neurosci 11:1376–1377

Root DH, Mejias-Aponte CA, Qi J, Morales M (2014) Role of glutamatergic projections from ventral tegmental area to lateral habenula in aversive conditioning. J Neurosci 34:13906–13910

Russo SJ, Nestler EJ (2013) The brain reward circuitry in mood disorders. Nat Rev Neurosci 14:609–625

Russo SJ, Wilkinson MB, Mazei-Robison MS, Dietz DM, Maze I, Krishnan V, Renthal W, Graham A, Birnbaum SG, Green TA, Robison B, Lesselyong A, Perrotti LI, Bolanos CA,

Kumar A, Clark MS, Neumaier JF, Neve RL, Bhakar AL, Barker PA, Nestler EJ (2009) Nuclear factor kappa B signaling regulates neuronal morphology and cocaine reward. J Neurosci 29:3529–3537
Russo SJ, Dietz DM, Dumitriu D, Morrison JH, Malenka RC, Nestler EJ (2010) The addicted synapse: mechanisms of synaptic and structural plasticity in nucleus accumbens. Trends Neurosci 33:267–276
Sara SJ (2009) The locus coeruleus and noradrenergic modulation of cognition. Nat Rev Neurosci 10:211–223
Sartorius A, Henn FA (2007) Deep brain stimulation of the lateral habenula in treatment resistant major depression. Med Hypotheses 69:1305–1308
Schneider TM, Beynon C, Sartorius A, Unterberg AW, Kiening KL (2013) Deep brain stimulation of the lateral habenular complex in treatment-resistant depression: traps and pitfalls of trajectory choice. Neurosurgery 72, ons184–193; discussion ons193
Schnitzer MJ (2002) Biological computation: amazing algorithms. Nature 416:683
Schultz W (2007) Multiple dopamine functions at different time courses. Annu Rev Neurosci 30:259–288
Schultz W (2016) Dopamine reward prediction-error signalling: a two-component response. Nat Rev Neurosci 17:183–195
Schweimer JV, Ungless MA (2010) Phasic responses in dorsal raphe serotonin neurons to noxious stimuli. Neuroscience 171:1209–1215
Sears RM, Fink AE, Wigestrand MB, Farb CR, de Lecea L, Ledoux JE (2013) Orexin/hypocretin system modulates amygdala-dependent threat learning through the locus coeruleus. Proc Natl Acad Sci U S A 110:20260–20265
Setlow B, Holland PC, Gallagher M (2002) Disconnection of the basolateral amygdala complex and nucleus accumbens impairs appetitive pavlovian second-order conditioned responses. Behav Neurosci 116:267–275
Shabel SJ, Janak PH (2009) Substantial similarity in amygdala neuronal activity during conditioned appetitive and aversive emotional arousal. Proc Natl Acad Sci U S A 106:15031–15036
Shabel SJ, Proulx CD, Trias A, Murphy RT, Malinow R (2012) Input to the lateral habenula from the basal ganglia is excitatory, aversive, and suppressed by serotonin. Neuron 74:475–481
Shabel SJ, Proulx CD, Piriz J, Malinow R (2014) Mood regulation. GABA/glutamate co-release controls habenula output and is modified by antidepressant treatment. Science 345:1494–1498
Shansky RM, Hamo C, Hof PR, McEwen BS, Morrison JH (2009) Stress-induced dendritic remodeling in the prefrontal cortex is circuit specific. Cereb Cortex 19:2479–2484
Shiflett MW, Balleine BW (2010) At the limbic-motor interface: disconnection of basolateral amygdala from nucleus accumbens core and shell reveals dissociable components of incentive motivation. Eur J Neurosci 32:1735–1743
Smith RJ, Lobo MK, Spencer S, Kalivas PW (2013) Cocaine-induced adaptations in D1 and D2 accumbens projection neurons (a dichotomy not necessarily synonymous with direct and indirect pathways). Curr Opin Neurobiol 23:546–552
Soiza-Reilly M, Commons KG (2014) Unraveling the architecture of the dorsal raphe synaptic neuropil using high-resolution neuroanatomy. Front Neural Circuits 8:105
Stamatakis AM, Stuber GD (2012) Activation of lateral habenula inputs to the ventral midbrain promotes behavioral avoidance. Nat Neurosci 15:1105–1107
Stamatakis AM, Jennings JH, Ung RL, Blair GA, Weinberg RJ, Neve RL, Boyce F, Mattis J, Ramakrishnan C, Deisseroth K, Stuber GD (2013) A unique population of ventral tegmental area neurons inhibits the lateral habenula to promote reward. Neuron 80:1039–1053
Stuber GD, Hnasko TS, Britt JP, Edwards RH, Bonci A (2010) Dopaminergic terminals in the nucleus accumbens but not the dorsal striatum corelease glutamate. J Neurosci 30:8229–8233
Stuber GD, Sparta DR, Stamatakis AM, van Leeuwen WA, Hardjoprajitno JE, Cho S, Tye KM, Kempadoo KA, Zhang F, Deisseroth K, Bonci A (2011) Excitatory transmission from the amygdala to nucleus accumbens facilitates reward seeking. Nature 475:377–380

Takamori S (2006) VGLUTs: 'exciting' times for glutamatergic research? Neurosci Res 55:343–351

Tan KR, Yvon C, Turiault M, Mirzabekov JJ, Doehner J, Labouebe G, Deisseroth K, Tye KM, Lüscher C (2012) GABA neurons of the VTA drive conditioned place aversion. Neuron 73:1173–1183

Tecuapetla F, Patel JC, Xenias H, English D, Tadros I, Shah F, Berlin J, Deisseroth K, Rice ME, Tepper JM, Koos T (2010) Glutamatergic signaling by mesolimbic dopamine neurons in the nucleus accumbens. J Neurosci 30:7105–7110

Thompson SM, Kallarackal AJ, Kvarta MD, Van Dyke AM, LeGates TA, Cai X (2015) An excitatory synapse hypothesis of depression. Trends Neurosci 38:279–294

Tsai HC, Zhang F, Adamantidis A, Stuber GD, Bonci A, de Lecea L, Deisseroth K (2009) Phasic firing in dopaminergic neurons is sufficient for behavioral conditioning. Science 324:1080–1084

Tye KM, Stuber GD, de Ridder B, Bonci A, Janak PH (2008) Rapid strengthening of thalamo-amygdala synapses mediates cue-reward learning. Nature 453:1253–1257

Tye KM, Prakash R, Kim SY, Fenno LE, Grosenick L, Zarabi H, Thompson KR, Gradinaru V, Ramakrishnan C, Deisseroth K (2011) Amygdala circuitry mediating reversible and bidirectional control of anxiety. Nature 471:358–362

Tye KM, Mirzabekov JJ, Warden MR, Ferenczi EA, Tsai HC, Finkelstein J, Kim SY, Adhikari A, Thompson KR, Andalman AS, Gunaydin LA, Witten IB, Deisseroth K (2013) Dopamine neurons modulate neural encoding and expression of depression-related behaviour. Nature 493:537–541

Ungless MA, Grace AA (2012) Are you or aren't you? Challenges associated with physiologically identifying dopamine neurons. Trends Neurosci 35:422–430

Ungless MA, Whistler JL, Malenka RC, Bonci A (2001) Single cocaine exposure in vivo induces long-term potentiation in dopamine neurons. Nature 411:583–587

Vaaga CE, Borisovska M, Westbrook GL (2014) Dual-transmitter neurons: functional implications of co-release and co-transmission. Curr Opin Neurobiol 29:25–32

Valenti O, Gill KM, Grace AA (2012) Different stressors produce excitation or inhibition of mesolimbic dopamine neuron activity: response alteration by stress pre-exposure. Eur J Neurosci 35:1312–1321

Vasudeva RK, Lin RC, Simpson KL, Waterhouse BD (2011) Functional organization of the dorsal raphe efferent system with special consideration of nitrergic cell groups. J Chem Neuroanat 41:281–293

Vialou V, Robison AJ, Laplant QC, Covington HE 3rd, Dietz DM, Ohnishi YN, Mouzon E, Rush AJ 3rd, Watts EL, Wallace DL, Iniguez SD, Ohnishi YH, Steiner MA, Warren BL, Krishnan V, Bolanos CA, Neve RL, Ghose S, Berton O, Tamminga CA, Nestler EJ (2010) DeltaFosB in brain reward circuits mediates resilience to stress and antidepressant responses. Nat Neurosci 13:745–752

Walker DL, Davis M (2008) Role of the extended amygdala in short-duration versus sustained fear: a tribute to Dr. Lennart Heimer. Brain Struct Funct 213:29–42

Walker DL, Miles LA, Davis M (2009) Selective participation of the bed nucleus of the stria terminalis and CRF in sustained anxiety-like versus phasic fear-like responses. Prog Neuropsychopharmacol Biol Psychiatry 33:1291–1308

Walsh JJ, Han MH (2014) The heterogeneity of ventral tegmental area neurons: projection functions in a mood-related context. Neuroscience 282:101–108

Walsh JJ, Friedman AK, Sun H, Heller EA, Ku SM, Juarez B, Burnham VL, Mazei-Robison MS, Ferguson D, Golden SA, Koo JW, Chaudhury D, Christoffel DJ, Pomeranz L, Friedman JM, Russo SJ, Nestler EJ, Han MH (2014) Stress and CRF gate neural activation of BDNF in the mesolimbic reward pathway. Nat Neurosci 17:27–29

Wanat MJ, Hopf FW, Stuber GD, Phillips PE, Bonci A (2008) Corticotropin-releasing factor increases mouse ventral tegmental area dopamine neuron firing through a protein kinase C-dependent enhancement of Ih. J Physiol 586:2157–2170

Wang RY, Aghajanian GK (1977) Inhibiton of neurons in the amygdala by dorsal raphe stimulation: mediation through a direct serotonergic pathway. Brain Res 120:85–102

Wang HL, Morales M (2009) Pedunculopontine and laterodorsal tegmental nuclei contain distinct populations of cholinergic, glutamatergic and GABAergic neurons in the rat. Eur J Neurosci 29:340–358

Wenzel JM, Rauscher NA, Cheer JF, Oleson EB (2015) A role for phasic dopamine release within the nucleus accumbens in encoding aversion: a review of the neurochemical literature. ACS Chem Neurosci 6:16–26

West AR, Grace AA (2002) Opposite influences of endogenous dopamine D1 and D2 receptor activation on activity states and electrophysiological properties of striatal neurons: studies combining in vivo intracellular recordings and reverse microdialysis. J Neurosci 22:294–304

West CH, Ritchie JC, Boss-Williams KA, Weiss JM (2009) Antidepressant drugs with differing pharmacological actions decrease activity of locus coeruleus neurons. Int J Neuropsychopharmacol 12:627–641

Xiao C, Cho JR, Zhou C, Treweek JB, Chan K, McKinney SL, Yang B, Gradinaru V (2016) Cholinergic mesopontine signals govern locomotion and reward through dissociable midbrain pathways. Neuron 90:333–347

Yang LM, Hu B, Xia YH, Zhang BL, Zhao H (2008) Lateral habenula lesions improve the behavioral response in depressed rats via increasing the serotonin level in dorsal raphe nucleus. Behav Brain Res 188:84–90

Zarate C, Duman RS, Liu G, Sartori S, Quiroz J, Murck H (2013) New paradigms for treatment-resistant depression. Ann N Y Acad Sci 1292:21–31

Zhang J, Shapiro MS (2012) Activity-dependent transcriptional regulation of M-Type (Kv7) K(+) channels by AKAP79/150-mediated NFAT actions. Neuron 76:1133–1146

Chapter 15
Combining Anatomy, Measurements and Manipulation of Neuronal Activity to Interrogate Circuit Function in *Drosophila*

Yvette E. Fisher and Thomas R. Clandinin

Abstract In this chapter we will discuss the application of genetics to the interrogation of neuronal function in fruit flies from a historical and modern perspective. We will review the current state-of-the-art tool kit for circuit dissection including neuronal measurements, manipulation and quantitative behavioral assessment. We will then discuss how these approaches can be productively applied to the interrogation of circuit computation by discussing recent discoveries in visual circuitry. Due to dramatic progress in recent years, these new tools have greatly expanded our understanding of the cell types and algorithmic transformations that underpin the detection of visual motion.

15.1 A Brief History of Neurogenetics in Flies

Modern *Drosophila* neurobiology owes much of its origins to a rich history of neurogenetics that began in the 1960s. This early work sought to use genetics as a toolkit to probe nervous system function. This idea, bold and controversial at the time, posited that complex neuronal function and even behavior could be understood from a genetic perspective (Benzer 1967; Erlenmeyer-Kimling and Hirsch 1961). The scientific dream was that genetic changes could be related to neuronal function and that this would in turn lead to a mechanistic understanding of the underlying processing. Stated in these terms, these original experiments are conceptually very similar to modern studies of *Drosophila* neurobiology, where the goal is to use genetic approaches to alter particular elements within a neuronal system and then perform precise measurements of neuronal function to define the role for that element.

Y.E. Fisher
Department of Neurobiology, Harvard Medical School, Boston, USA

T.R. Clandinin (✉)
Department of Neurobiology, Stanford University, Stanford, USA
e-mail: trc@stanford.edu

A. Çelik and M.F. Wernet (eds.), *Decoding Neural Circuit Structure and Function*, DOI 10.1007/978-3-319-57363-2_15

By starting with clearly defined behaviors such as geotaxis (Erlenmeyer-Kimling and Hirsch 1961), phototaxis (Benzer 1967), optomotor responses (Heisenberg and Götz 1975), circadian rhythmicity (Konopka and Benzer 1971), chemosensory responses (McKenna et al. 1989), and heat-induced paralysis (Siddiqi and Benzer 1976; Suzuki et al. 1971), genetic fly strains could be easily segregated based on their behavioral phenotypes. Early studies performed by Jerry Hirsch and colleagues used a diverse array of wild strain variants to confirm that differences in behavioral tendencies for geotaxis and phototaxis were in fact heritable and likely multigenic (Erlenmeyer-Kimling and Hirsch 1961). This work was followed by an approach that focused on identifying individual mutations, originally spearheaded by Seymour Benzer and William Pak. Benzer, Pak and their contemporaries induced rare mutations within inbred (homogenous) fly strains using ethyl methane sulfonate (EMS) so that changes in behavior could often be attributed to a single induced mutation. This approach was instrumental in identifying many of the genes in the phototransduction pathway (Pak et al. 1970). In addition, a number of critical genes involved in neuronal signaling, nervous system development, and even circadian rhythmicity were also identified using this methodology; including the voltage gated sodium channel *paralytic* (Suzuki et al. 1971), the clock gene *period* (Konopka and Benzer 1971) and the transcription factor *optomotor blind* (*omb*) (Heisenberg et al. 1978). However, due to technical limitations, these studies fell short of the lofty goal of a mechanistic description of the circuits that lead from the processing of a sensory input to the organism's output, behavior. It is now clear that to dissect the circuits underlying even simple behaviors requires the ability to independently manipulate small subsets of the brain or even individual cell types. This is because, beyond a small number of notable exceptions (Dudai and Jan 1976; Hall 1978; Konopka and Benzer 1971), most genes involved in neuronal function play vital roles in many different neurons throughout the brain, making phenotypes obtained from traditional mutations hard to interpret due to their widespread and pleotropic effects. Modern molecular genetics in *Drosophila* has alleviated this dilemma with a wealth of genetic tools that allow tissue- and even neuron-specific manipulation of circuit elements. Thus, the ambitious scientific dream of the neurogenetic pioneers has come into reach during the past decade.

15.2 The Current "Circuit Bashing" Tool Kit

15.2.1 Defining and Targeting Neuronal Elements

15.2.1.1 Anatomy

The fly brain is comprised of approximately 100,000 neurons and within this broad population, neuron types can be grouped based on morphology, functional response properties and molecular expression profile. The exquisite stereotypy of the fly brain makes detailed description of the cellular anatomy particularly useful.

Correspondingly great effort has been spent on defining neuronal anatomy at the level of individual neurons, their axonal and dendritic processes, and their synaptic connections. These identifiable neuronal types can be found again and again with very similar numbers and properties across many individual flies (Chiang et al. 2011; Fischbach and Dittrich 1989; Stocker et al. 1990). The fact that this stereotypy is predominantly maintained down to the level of connectivity and functional signaling properties makes the fly an attractive system in which to study circuit computation as the loci for particular neuronal transformations is likely to be preserved across individuals.

Anatomical descriptions of neuron types have been cataloged for a number of brain regions. The majority of cell types in the fly brain have been characterized using light microscopy combined with stochastic and sparse neuronal labeling that was either achieved, traditionally using Golgi impregnations (Fischbach and Dittrich 1989) or more recently by stochastic labeling using genetic approaches (Chiang et al. 2011; Nern et al. 2015). Recent anatomical efforts have now turned to the use of electron microscopy (EM), with the hope of defining the full wiring diagram between these cataloged neuron types. A columnar region of the early visual areas, the lamina, the medulla, and some of the lobula complex has been extensively imaged, traced and reconstructed (Meinertzhagen and Sorra 2001; Rivera-Alba et al. 2011; Shinomiya et al. 2014; Takemura et al. 2013). Similar EM reconstruction efforts of other brain regions of interest are currently underway. Comfortingly, there is significant agreement between the traditional light microscopy catalog and neurons described by the modern EM reconstruction efforts (Takemura et al. 2013). In the visual system, these connectomics efforts have greatly aided the identification of many of the neurons involved in elementary motion detection (Shinomiya et al. 2014; Takemura et al. 2013). However, while EM methods can make predictions about which connections are functionally important, by the assumption that more synaptic connection sub-serve stronger functional connectivity, this assumption is unlikely to be universally true. Therefore functional validation of the circuit architecture hypothesized from EM approaches is invaluable. Indeed, recent functional studies of motion-detecting circuits suggest significant departures from the circuit hypothesis put forward based on purely anatomical work (Ammer et al. 2015; Takemura et al. 2013). Despite this cautionary note, connectomics data has already proved invaluable by generating and constraining hypotheses for functional studies. Moving forward, an important addition to current data sets of neuron types and their connections will be a full description of the neurotransmitters and neurotransmitter receptors expressed by each of these cell types. Ongoing development and improvements in the methodology for single neuron mRNA sequencing promise these data and much more (Crocker et al. 2016; Tan et al. 2015).

15.2.1.2 Targeting of Genetically Defined Neurons

Modern genetic tools allow for the modular and reproducible targeting of individual cell types using transgenic driver lines. Enhancer trapping technology, originally introduced using the GAL4/UAS system by Brand and Perrimon in the early 1990s, allows the experimenter to essentially highjack the genomes "zip-code" system to target genetically defined neuron types (Brand and Perrimon 1993). By placing a transcription factor sequence, such as Gal4, downstream of a putative genomic enhancer element, the GAL4 protein will be specifically expressed within the tissues targeted by that enhancer sequence. More than ten thousand different driver lines currently exist. These various extensive libraries have taken slightly different approaches to co-opting enhancer elements. Some collections have inserted transcription factor sequences semi-randomly into endogenous genomic sites (Gohl et al. 2011) while parallel approaches achieve cell types specificity by cloning different putative enhancer fragments in front of a transcription factor sequence (Jenett et al. 2012; Kvon et al. 2014; Pfeiffer et al. 2008, 2010). Because of the use of transcription factors, all of these driver line systems are binary, meaning that once captured, a particular neuronal pattern can be targeted by a large number of different genetic effectors that label, measure neuronal responses, or manipulate neuronal activity (i.e., EGFP, GCamp, ChR2) by simply performing a single genetic cross (Fig. 15.1).

While already useful for dissection of the nervous system, these enhancer trap driver lines are rarely fully restricted to a single neuron type. Fortunately, the targeting patterns of these driver lines can often be further refined using an ever-growing number of methods. Expression patterns can be restricted to the shared pattern between two driver lines (AND logic gate) using for example the split Gal4 system (Luan et al. 2006). Alternatively one pattern can be subtracted from another (NOT logic gate), using, for example, the GAL80 repressor (Luan et al. 2006; Suster et al. 2004). Using such refinement, specific labeling of single neuron types can be achieved allowing the study of each neurons role in a particular behavior or circuit computation. In addition to methods that refine patterns in a targeted manner, tools also exist that can make a driver line sparser. Such sparsening is most commonly achieved by combining driver lines with

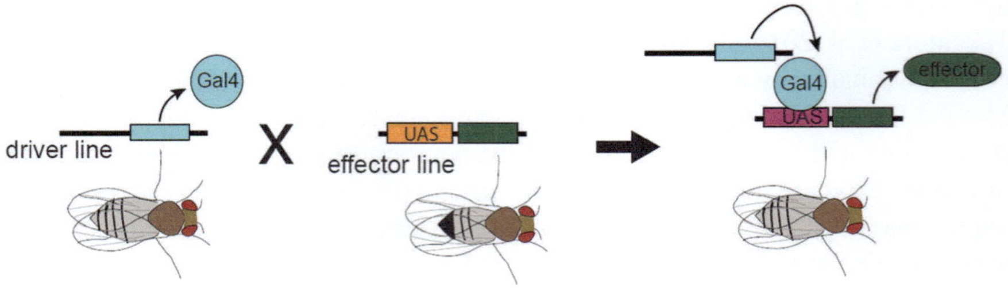

Fig. 15.1 Schematic of an example fly cross to demonstrate the logic of binary expression systems. A nonnative transcription factor e.g., Gal4 (*blue*) is used in combination with its target DNA sequence (e.g., upstream activating sequence (UAS), *orange*) to control the expression of an effector protein. Flies that carry a binary system transcription factor, "driver lines," can be crossed to any appropriate "effector" line to drive expression of the effector in only the specific cells and tissues that have transcription factor expression

recombinase-based logical elements (i.e., Flp/FRT, Cre/LoxP) and can be used to add a stochastic element to the genetic control of a particular driver line (Gordon and Scott 2009; Gruntman and Turner 2013). Thus, targeting effector expression to single cell types, and even single cells, is now largely a routine throughout much of the fly brain.

15.2.2 *Measuring Brain Activity*

The ability to use genetic labeling of particular neuron subclasses opens up the possibility to target the same neurons, reproducibly across individuals, for electrical or optical recordings. In the fly, such recordings are almost invariably performed in vivo. In this experimental setup the head of the fly is secured, a region of the cuticle is dissected away to reveal the neurons of interest, and the fly is mounted for recording under a microscope that allows for imaging or electrophysiology, or both.

15.2.2.1 Electrophysiology

By labeling neurons of interest with a fluorescent protein, cell bodies can be targeted for whole-cell electrophysiological recordings (Joesch et al. 2008; Wilson et al. 2004; Zheng et al. 2006). Using such measurements, membrane voltage or action potentials can be correlated with controlled sensory inputs (Joesch et al. 2008; Wilson et al. 2004), the activity of other neurons (Bhandawat et al. 2007), or even the tethered behavior of the fly (Maimon et al. 2010). Such experiments provide a powerful way to begin to probe the role of a neuron within a particular circuit. Whole-cell patch clamp recording has the added benefit that in addition to providing a precise means of measuring voltage changes within a neuron, it also allows for manipulation of such voltage signals by current injection. Another relative advantage of electrophysiological recording is the high level of temporal precision this approach provides. Unfortunately, recording from neurons with small cell bodies, as is commonly the case for columnar visual system neurons, is extremely difficult and time consuming. As a result, a number of researchers have turned to calcium imaging to interrogate the response properties of columnar visual system neurons.

15.2.2.2 Calcium Imaging

Optical recording of neuronal activity has seen considerable progress in the past decade and has now become a prominent method with which to measure neuronal activity. Genetically encoded indicators of neuronal activity can be easily expressed within the pattern of particular driver lines. The most widely applied of these are genetically encoded calcium indicators (GECI). A wide array of GECIs exist, with the currently most commonly used being the GCaMP family. GCaMPs are circularly permuted GFP molecules that also possess a calcium-binding domain (Looger and Griesbeck 2012). Binding of calcium causes a conformational change that leads to an increase in GFP fluorescence. Thus when GCaMP, or any GECI, is expressed within a neuron of interest it provides a fluorescence signal that is related to changes in intracellular calcium concentration of the labeled neuron. Recent work using directed evolution has led to significant improvements in the utility of GCaMPs by increasing the magnitude of fluorescence signals and improving temporal kinetics (Akerboom et al. 2009, 2012; Chen et al. 2013). Further work has also diversified the color palette of GECIs, leading to the development of two families of red calcium indicators, the RCaMPs and the R-GECOs (Akerboom et al. 2013; Kalko et al. 2011). Application of GECIs for the measurements of neuronal activity in vivo is currently a productive approach for the investigation of neuronal circuits in the fly.

Optical recording provides the ability to measure signals from different subcellular compartments of a neuron. Many molecular mechanisms regulate intracellular calcium concentration within a neuron and these mechanisms are differentially active within the various cellular compartments (Catterall 2011). Thus, when GECIs are applied to the measurement of neuronal activity they provide a complex signal that combines information about changes in membrane potential with information about the underlying cell biological processes that regulate calcium flux. This complexity of calcium is both a challenge and an advantage of recordings using GECIs. Such complexity should be carefully taken into consideration regarding the interpretations of experimental results. Indeed, the impact of calcium signals on the downstream neuronal circuit differs considerably across the diverse compartmental regions of a neuron (Yang et al. 2016). In axon terminals, calcium relates closely to vesicle fusion and thus synaptic output. However, elsewhere in the cell such as the dendritic regions or cell bodies calcium signals can represent postsynaptic currents, action potentials, intracellular calcium regulation, or other conductances. Moving forward, these indicators will be utilized to push for a more nuanced understanding of all of the aspects that contribute to calcium concentrations within diverse cellular compartments. Indeed, recent studies have shed light on the complex contribution of these intertwined processes to neuronal activity and reveal that subcellular neuronal processing can occur in the fly (Yang et al. 2016).

In addition to the ability to record from multiple neuronal sub-compartments, optical approaches also enable simultaneous recordings from large populations of neurons. As commonly used driver lines often label a single cell type within a certain brain region it can be possible to record the neuronal signals from tens to hundreds of these labeled neurons. This data provides an exciting handle on the population activity of a brain region. However, if within these driver lines the processes of labeled neurons are interdigitated it can be difficult to decisively attribute a fluorescence signal to a single neuron or to traces across the sub-compartments of an individual neuron. Stochastic genetic tools that further restrict the labeling of a certain population have been used successfully to circumvent this issue (Gordon and Scott 2009; Gruntman and Turner 2013). Thus, by combining genetic targeting with optical imaging, data can be acquired from a population of labeled neurons or even the sub-compartments of individual neurons.

One ongoing struggle with many applications of calcium imaging concerns the relatively slow kinetics of these signals, relative to electrophysiological approaches. The decreased temporal resolution of calcium imaging experiments has multiple sources. First, measured signals are slower than the underlying changes in calcium concentration because of the kinetics of the indicator. This blurring is produced by both the ON and OFF rates of the dye. Currently, the fastest GCaMP, GCaMP6fast, has a rise time of 11 ms (Helassa et al. 2015). In addition, changes in the underlying signal itself, calcium concentration, occur on slower time scales than changes in voltage, because it depends directly on the diffusion of ions. In some cases this slower calcium signal may actually be the more biologically relevant signal, for example in presynaptic terminals where calcium ions directly trigger neurotransmitter release. However, in other cellular compartments, faster signals such as membrane voltage, synaptic currents, or even neurotransmitter receptor kinetics may be more informative. A better understanding of the subcellular regulation of calcium and voltage signals would be extremely useful in informing the choice of which aspect of neuronal activity to measure for different experimental contexts.

15.2.2.3 Voltage Imaging

Dynamic changes of intracellular membrane potential provide vital information both about a single neuron's activity and about its communication within a broader circuit. For this reason, scientists have been long searching for a way to record this voltage signal optically within a genetically defined set of neurons. In principle, such genetically encoded voltage indicators (GEVIs) have the advantage that they would allow the recording of voltage within small neuronal compartments, which are inaccessible to electrophysiology, as well as allowing the simultaneous recording of voltage dynamics from tens or hundreds of neurons. As a result of considerable effort, such sensors have become usable for in vivo interrogation in the past few years. In flies, the GEVI Arclight was successfully used to measure voltage signals across the dendritic arbors of neurons involved in regulating circadian rhythmicity (Cao et al. 2013) and recently a FRET based optical indicator (Ace2N-2AA-mNeon) was shown to detect individual action potentials within LN neurons of the olfactory system (Gong et al. 2015). Very recently, a new indicator, ASAP2f, has been used to measure signal transformations in the peripheral circuits of the fly visual system using two-photon imaging (Yang et al. 2016). These proof-of-principle experiments, which have successfully recorded voltage signals from genetically defined neurons using GEVIs, now pave the way for further indicator improvements and for the application of these approaches to the interrogation of how voltage signals are regulated across individual neuronal compartments and across a whole circuit.

15.2.3 Quantitative Behavioral Measurements

Another productive way to probe circuit function is by measuring behavior. Behavioral analysis can be broadly grouped into two complementary approaches. Population assays provide a high-throughput way to obtain data about neuronal function, while single fly behavioral assays allow for a nuanced control of the flies sensory experience and a precise read out of an individual's behavior.

A number of fly behaviors are amenable to large population assays that make forward genetic screens possible. For example, screens using visually evoked behaviors have been instrumental in identifying genes involved in phototransduction and in identifying neurons with specializations for particular aspects of visual processing (Katsov and Clandinin 2008; Pak et al. 1970). While classic forms of these assays, such as the "counter-current distribution" or "optomotor Y-maze", identified different phenotypic flies based on sorting (Benzer 1967; Heisenberg and Götz 1975), modern renditions have instead turned to automated tracking and characterization of flies behavior within a population chamber (Branson et al. 2009; Dankert et al. 2009; Katsov and Clandinin 2008; Zhu et al. 2009).

On the other hand, single fly assays provide the ability to precisely control the sensory input and precisely monitor the behavioral output of an individual fly. Many of these assays are performed while the fly is tethered in place but still permitted to either fly or walk, but specialized assays also exist to probe a range of innate fly behaviors. For example, the proboscis extension reflex can be easily evoked when potential food is presented to a semi restrained fly (Gordon and Scott 2009). The walking and turning behavior of a fly can be effectively measured by having a tethered fly walk on an air-cushioned ball. In this design, the balls' movement is read by optical mice and provides a proxy for the rotational and translational movements of the fly (Bahl et al. 2013; Clark et al. 2011; Rister et al. 2007; Silies et al. 2013). Alternatively, the intended course control maneuvers of tethered flight can be measured by monitoring wing beat amplitude, or the torque applied to the tether (Götz 1964; Heisenberg and Götz 1975). For visual experiments, such tethered assays are particularly attractive since they allow controlled stimuli to be presented. Combining these quantitative behavioral assays with the manipulation of genetically defined cell types provides a powerful means to causally assess the contribution of individual circuit elements to a particular aspect of sensory processing or behavioral output. Single fly behavior can also be performed in combination with neuronal recordings of activity, either using calcium imaging or electrophysiology. While extremely technically challenging, these experiments provide the opportunity to correlate circuit function with behavioral output on a trial-by-trial basis. This approach has been successfully used to study how sensory processing is regulated by different behavioral states, and to interrogate the internal representation of self-motion and -navigation (Chiappe et al. 2010; Maimon et al. 2010; Seelig and Jayaraman 2015).

Using tethered or semi-restricted behavior makes the behavioral analysis more approachable because it simplifies the behavioral read out. One considerable concern, however, is that these restricted behaviors are so artificial that interpretations of circuit function obtained in these contexts may not generalize to how the brain computes under more naturalistic conditions. For this reason, there is a desire in the field to study more ethologically relevant behaviors in the fly. Determining what behaviors are ethologically relevant to the long cultivated laboratory fruit fly can be difficult (Dickinson 2014). Yet, recent heroic engineering efforts that now enable real-time automated tracking of freely flying flies in large arenas are likely to aid this goal (Muijres et al. 2015; Straw et al. 2010, 2011).

15.2.4 Manipulating Neuronal Activity

To assess the contribution of a particular element to a biological process, it is useful to break that element and then examine the consequences. As Seymour Benzer first posited, this "approach may be fruitful in tackling the complex structures and events underlying behavior" (Benzer 1967). Much as the classical work performed large population screens to look at the contribution of genetic mutations to neuronal function and behavior. Modern genetic tools allow for a similar approach applied to different circuit elements and there exist an array of genetic tools for this purpose.

15.2.4.1 Silencing Defined Neurons

One straightforward method for silencing a group of neurons is to use a tissue specific driver line to express an effector that disrupts synaptic transmission. In flies, this aim is generally achieve using either Tetanus toxin light chain (TNT) or Shibire (ts) (Kitamoto et al. 2000; Sweeney et al. 1995). Tetanus toxin irreversibly disrupts synaptic transmission by cleaving synaptobrevin (Sweeney et al. 1995). However, expression of Tetanus toxin has been shown to cause vesicle trafficking deficits within the target neuron (Hiesinger et al. 1999). When using TNT, the developmental effects of this prolonged manipulation of synaptic transmission, and potentially cell health, are an important caveat, especially when using driver lines whose expression starts during nervous system development. The temperature sensitive properties of Shibire^{ts}, another synaptic effector, can partially circumvent such concerns by adding temporal control to the manipulation of synaptic function. Shibire encodes dynamin, a GTPase required for synaptic vesicle recycling. When Shibire^{ts}, a temperature inducible and dominant negative form of Shibire, is expressed within target neurons, transmission will remain intact as long as the flies are maintained at the low permissive temperature. Only upon a shift to the high restrictive temperature, will vesicle recycling be blocked and as a consequence synaptic transmission disrupted. Notably, non-specific effects on cell function have also been observed when Shibire^{ts} is expressed at high levels (Gonzalez-Bellido et al. 2009).

Silencing of neuronal activity can also be achieved by altering neuronal excitability. Most notable for this purpose are the inward rectifying channel Kir2.1, which holds neuronal membranes at a more hyperpolarized potential, and optogenetic reagents which allow for light induced hyperpolarization (Inada et al. 2011). In addition, neurons can also be removed from a circuit by selective expression of diphtheria toxin light chain, killing cells of interest (Han et al. 2000). With the refined genetic control afforded by transgenic driver lines it is possible, using these effectors, to silence a single neuron class while leaving the rest of the brain unaffected. In this manner, the contribution of identified neurons for guiding behavior or downstream neuronal activity can be systematically tested. From such studies, much can be inferred about the contribution of each class of neurons to a particular computation. Genetic effectors for silencing of defined neuronal classes have been instrumental in ascribing the causal role of particular neuronal elements in a diverse number of behaviors including, olfactory processing, feeding behavior, optomotor responses, and many more (Gordon and Scott 2009; Root et al. 2008; Silies et al. 2014).

15.2.4.2 Activating Defined Neurons

Another productive avenue for interrogating the nervous system is to artificially activate neurons. A number of genetically encoded reagents allow membrane potential to be controlled by heat, chemical application, or light. In particular, expression of the cation channel TRPA1 is often used to control neuronal excitability using shifts in temperature (Hamada et al. 2008). Optogenetics also allows neuronal activity to be controlled by light. With the many optogenetic variants developed since channelrhodopsin, control can be obtained using a range of different wavelengths creating compatibility with optical imaging. Notably, because light penetrates the fly cuticle, neurons can be manipulated using optogenetics without performing surgery. This is advantageous for the study of complex or group behaviors when it is desirable to have the flies freely moving. For example, the Giant Fiber neurons, which mediate aspects of the startle response, were activated in unrestrained flies using CsChrimson, allowing for the precise characterization of the timing and dynamics of escape behavior (von Reyn et al. 2014). Combining behavioral observation with activation of a particular subclass of neurons can be a compelling way to assess the role of that neuron's activity in producing a given behavior. It is worth noting that the permeability of the cuticle to light does have drawbacks for the investigation of visual processing, as the light intended to activate the exogenous opsin has the potential to activate rhodopsin molecules in the eye. To circumvent this limitation, investigators who study visual processing, or who have concerns that visual activation may alter their phenotype, often work with genetically blinded flies (de Vries and Clandinin 2012). Another useful application of optogenetic reagents is to use them to activate upstream neurons in combination with recording from a downstream circuit element. The precise temporal control provided by optogenetic simulation allows for nuanced investigation of the circuit transformation. For example, this approach has been successfully applied to probe the effect of feedforward network architecture on signal timing and fidelity in olfactory processing (Jeanne and Wilson 2015). Optogenetic activation was also instrumental in the identification of the neurons that mediate null direction inhibition in the motion circuit (Mauss et al. 2014, 2015).

15.2.4.3 Manipulation of Neuronal Signaling

The release and detection of neurotransmitters and neuromodulators tightly shape the quality and sign of neuronal signaling. Tools to manipulate these channels of communication can be instrumental for dissecting circuit logic. A number of pharmacological and genetic tools have been applied toward this goal in *Drosophila*.

Classical pharmacological approaches are a common and productive way to manipulate particular neuronal signaling channels within a circuit, as in many cases the same drugs work in insects as in vertebrate systems. Drugs that can block a number of different neurotransmitter receptors or channels have been described. For example, drugs that block inhibitory receptor signaling (i.e., GABAA, GABAB, GluCl) have been instrumental for uncovering how inhibition shapes olfactory processing (Liu and Wilson 2013; Root et al. 2008; Wilson and Laurent 2005). However, the specificity of the manipulation is a common concern, both about the specificity of the drug itself for the intended receptor target and about the widespread effects that might be caused by global drug application. In some cases it is possible to assuage these concerns by combining pharmacology with other manipulations such as fast injection of neurotransmitter, optogenetic activation of specific circuit elements, or genetic manipulation of the target receptor (Fisher et al. 2015a; Liu and Wilson 2013; Mauss et al. 2014).

The molecules that shape neuronal signaling can also be manipulated genetically. For this purpose, homozygous mutant animals can be used if loss-of-function mutations are viable. For example, a mutation in *Shaking-B*, a gap junction subunit, was used to demonstrate that electrical connections mediate lateral interactions between glomerular odor channels in the olfactory system (Yaksi and Wilson 2010). Unfortunately, mutations in many important neuronal signaling molecules, including voltage gated channels and neurotransmitter receptors, are often lethal. One exception to this rule seems to be in the sensory periphery where mutations in molecules involved in the transduction of sensory signals are often viable. These mutants are extremely useful for the study of sensory systems since they allow the experimenter to manipulate the processing of incoming sensory signals. For example, in the olfactory system such mutations have been used to investigate the contribution of individual odor receptors to neuronal dynamics, using the so-called "empty neuron system" (Dobritsa et al. 2003). In this genetic scheme, a mutation of a certain olfactory receptor is used to remove the ability of a particular olfactory receptor neuron type to respond to odor by mutating its cognate receptor. Then another olfactory receptor is mis-expressed within this same neuron using the Gal4-UAS system. In this manner, the effects of odor receptors on odor response dynamics can be isolated from any heterogeneity between olfactory neuron properties. Using this approach it was shown that odorant receptors determine many aspects of the olfactory neuron response properties, including spontaneous firing rate and odor response dynamics (Hallem et al. 2004). A similar strategy was used in the visual system to artificially shift the spectral properties of a particular subclass of photoreceptors. This work demonstrated that the outer photoreceptors, which were previously thought to be exclusively involved in color vision, in fact provide signals that can also inform motion detection (Wardill et al. 2012).

Neuronal signaling elements can also be manipulated using cell-type specific knockdown based on RNA interference (RNAi). In *Drosophila*, RNAi can be expressed cell-type specifically in combination with transgenic binary expression systems. Large transgenic libraries exist that enable the targeting of almost any gene in the *Drosophila* genome (Dietzl et al. 2007; Ni et al. 2009). Pertaining to neuronal signaling, this method has been utilized to study the contribution of different inhibitory receptors (i.e., GABAAR, GABABR and GluCl) to sensory processing (Freifeld et al. 2013; Liu and Wilson 2013; Root et al. 2008) and to dissect the contribution of sodium spikes to ORN responses (Nagel and Wilson 2011). While transgenic RNAi methods can isolate the contributions of specific neuronal signaling molecules to circuit function in specific cell types, RNAi mediated knockdown is rarely complete. In addition, the extent of RNAi mediated knockdown is highly dependent on the driver line used. For this reason, negative results obtained using RNAi cannot be interpreted. RNAi can also have off-target effects on multiple genes. Thus, while a number of pharmacological and genetic tools exist for the dissection of specific aspects of neuronal signaling, no existing method is perfect. Moving forward, an ideal tool for the study of circuit function would enable gene disruption in a cell-type specific manner with tight temporal control, and would be easily generalized to any gene of interest.

15.3 Using These Tools to Dissect Circuit Function

15.3.1 Introduction

The development of this powerful toolkit has paralleled the functional dissection of a number of different circuit processes in the fly. These studies have greatly informed our understanding of a variety of sensory systems, including vision, olfaction, and audition, learning and memory circuits, courtship behaviors, and many others (Albert and Gopfert 2015; Guven-Ozkan and Davis 2014; Pavlou and Goodwin 2013; Silies et al. 2014; Wilson 2013). Here we focus on the dissection of visual motion detection, a circuit in which satisfying links between behavior, circuit function, anatomy, and computation have emerged over the last few years.

A rich theoretical basis makes visual motion detection an attractive context in which to dissect the logic of how neuronal signals are transformed and processed within biological circuits. At its core, elementary motion detection (EMD) is a relatively simple computation: motion direction can be determined by comparing light signals from at least two different points in visual space over time. Pioneering work by Hassenstein and Reichardt in the 1950s introduced the use of quantitative measurements of insect behavior to interrogate motion-detecting circuits (Hassenstein and Reichardt 1956). Initially using the beetle *Chlorophanus,* this fundamental work formulated the Hassenstein and Reichardt correlator (HRC), an algorithmic description for how neuronal circuits could compute (Hassenstein and Reichardt 1956). This model posits that motion direction can be computed by comparing the light signals detected by two spatially adjacent photoreceptors. In this model, each photoreceptor is specifically sensitive to recent changes in light input, called contrast, at a particular location in space. The contrast signals measured by both photoreceptors are then correlated (multiplied) after one of the two signals is delayed in time (Fig. 15.2a). Because of this temporal delay within the circuit, motion moving in a particular direction, called the preferred direction, will lead to a coincidence between the two signals at the level of the correlation operation. Conversely, when motion is presented in the opposite or non-preferred direction, the structure of the HRC is such that the signals from the two photoreceptors will not interact. Without this coincidence and correlation, motion in the non-preferred direction will produce little to no signal. Finally, to create a strongly directional signal, the outputs of mirror symmetric detectors that are oppositely tuned are subtracted from one another (Fig. 15.2b). Thus, there are three main algorithmic elements in this elegant model: (1) a time delay, (2) a correlation operation (multiplication), and (3) subtraction between the mirror symmetric detectors. Since its formulation, this model has led neuroscientists to ponder first, whether neurons within motion-detecting circuits indeed perform the operations proposed by the HRC, and second, if so, how are such arithmetic operations actually implemented by biological circuits?

It is important to note that two other alternative models of motion detection have also been described. As Barlow and Levick first pointed out, instead of a correlation operation, a Veto operation (AND-NOT) can also be used to produce direction selective signals (Barlow and Levick 1965). Another model, the motion energy model (ME), is favored for describing cortical direction selectivity (Adelson and Bergen 1985). While the ME models make similar predictions to the HRC about the output of intact circuits, the order and operation of various linear and nonlinear transformations differ between the two model types. In particular, ME models become strongly direction selective by squaring the sum of two different spatiotemporal filters that both sample asymmetrically across space and time. These alternative models should be kept in mind as the field continues to interrogate the inner workings of the fly elementary motion-detecting circuitry. That said, all of these models share a number of core elements, including differential temporal processing and a nonlinear interaction between signals from different points in space. Therefore, for this section, the sequential processing steps proposed by the

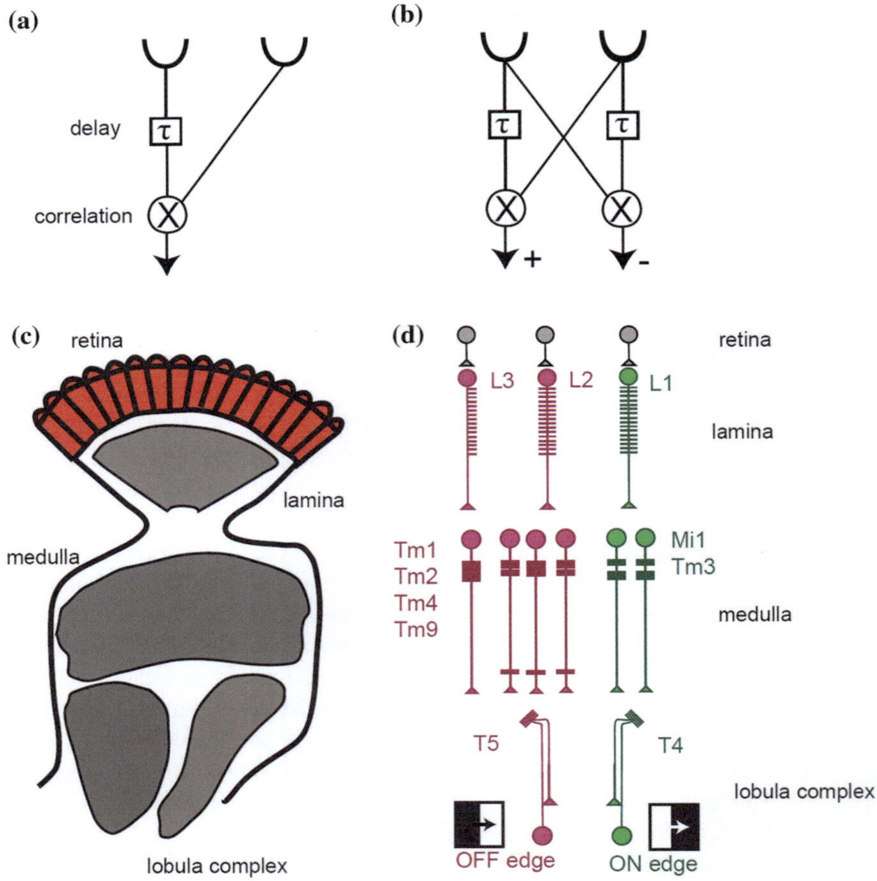

Fig. 15.2 **a** A schematic illustration of a central component of the HRC model. This "half correlator" samples light from two points in space and incorporates a temporal delay and a correlation operation (*multiplication*). This model is sufficient to detect motion and can produce a direction selective output. **b** In the "full correlator" HRC model, two of these mirror symmetric "half correlators" are subtracted from one another to produce a fully opponent signal. One motion direction produces a positive signal while the opposite motion direction produces a negative signal. **c** Schematic of the fruit fly visual system. **d** Schematic of the known neuronal players of elementary motion-detecting circuitry and their proposed connectivity

classic HRC model will serve as an outline to lead a discussion of current understanding of the circuit transformations that underpin motion processing in the fly.

While early work on insect motion vision focused on quantitative behavioral measurements, the discovery of wide-field motion sensitive neurons in the blowfly provided access to the internal workings of insect motion processing (Dvorak et al. 1975; Hausen 1976). These neurons, lobula plate tangential cells (LPTCs), appeared to receive signals from an array of elementary (local) motion-detecting elements. Investigation of LPTCs confirmed a number of core predictions of the original HRC model, including that motion can be evoked by the sequential activation of photoreceptors that sample two adjacent points in visual space (Riehle and Franceschini 1984), and the demonstration that opposing motion signals are subtracted from one another in order to produce strongly directional signals (Egelhaaf et al. 1990; Hausen 1976). However, while LPTCs were studied extensively, the upstream circuits that were hypothesized to actually implement the HRC remained elusive for decades. Indeed, the neurons upstream from the LPTCs are both very small, making them difficult to study using electrophysiology, and both numerous and diverse, requiring a more systematic approach. With the introduction of genetic tools for the measurement and manipulation of neuronal elements in the fruit fly *Drosophila*, and aided by extensive characterization of the anatomy of the visual system, the true structure of elementary motion detection in the fly is now coming into view. A number of the major neuronal players have since been identified and the field is beginning to uncover the algorithmic operations that underpin this quintessential computation.

15.3.2 Defining the Neuronal Elements

The first step toward dissecting the elementary motion-detecting circuitry was to define the neurons involved. The fly visual system is comprised of four neuropils that sequentially process visual information conveyed from the retina (Fig. 15.2c). Following detection of light signals by photoreceptors, visual signals are conveyed first to the lamina. Second order neurons then transmit light information to a large visual neuropil called the medulla. Containing approximately $\sim$60 columnar cell types the medulla was long proposed to be the site of the EMD but, until recent years, it has been the most difficult to systematically interrogate. Transmedulla neurons next send synaptic output to the lobula complex, comprised of the lobula and lobula plate brain regions. The lobula complex is the region in blowflies where wide-field motion sensitive neurons reside, and homologous neurons such as HS, Hx, and VS have now also been identified in fruit flies (Fischbach and Dittrich 1989; Joesch et al. 2008; Scott et al. 2002; Wasserman et al. 2015). While perhaps there is slightly less cellular diversity in fruit flies relative to their larger cousins, much of the underlying response properties and circuit logic at the level of LPTCs appear comparable (Joesch et al. 2008).

15.3.3 Interrogation of Circuit Candidates

A wealth of data about the morphology of the individual neuron types in the visual system of the fruit fly emerged approximately 25 years ago, building on earlier studies in larger flies (Fischbach and Dittrich 1989; Meinertzhagen and O'Neil

1991; Strausfeld 1976; Takemura et al. 2008). An early genetic study used genetic manipulations to determine which lamina neurons were required for motion detection (Rister et al. 2007). This study used genetic silencing of lamina neurons combined with behavior to show that two lamina monopolar cells, L1 and L2, are redundantly required for flies to respond behaviorally to moving stimuli. Further work dissected the specific contributions of these two neurons, either using behavior or downstream physiology. These studies demonstrated that L1 synaptic activity is required for responses to moving ON edges, while blocking L2 activity disrupts responses to moving OFF edges (Clark et al. 2011; Joesch et al. 2010). These data argue that ON and OFF edges are processed within segregated circuits and partially resolved a long-standing dilemma about how biological circuits could implement sign correct multiplication. In particular, since the original HRC model treated ON and OFF edges together, arithmetic multiplication was necessary to predict responses to both types of moving edge. By splitting the system into two separate motion detector circuits, ON and OFF stimuli can be represented by depolarizing neuronal signals, the multiplicative nonlinearity that underpins direction selectivity does not have to be sign correct. Whether measurements were made of optomotor behavior or by electrophysiological recordings from downstream LPTCs, largely similar conclusions were drawn about the contribution of L1 and L2 to the processing of ON and OFF edges. Moreover, genetic tools for neuronal manipulation were essential for defining the inputs into circuitry that detects ON and OFF edges.

An exciting breakthrough regarding the circuitry of elementary motion detection came with the identification of the first small field, columnar and direction selective neurons, T4 and T5. By genetically silencing T4 and T5 neurons using Shibire^{ts} or Kir2.1, it was shown that these neurons are required for motion responses within downstream LPTCs and for optomotor behavior (Bahl et al. 2013; Schnell et al. 2012). Following their identification, calcium imaging demonstrated that T4 and T5 exhibit direction selective responses to ON and OFF edges respectively (Fig. 15.2d) (Maisak et al. 2013). Interestingly, individual T4 or T5 neurons are tuned to one of the four cardinal directions and each of these columnar neurons encodes motion signals for a local region in visual space. In this manner, the full population of over 5000 T4 and T5 neurons, which tile retinotopic space, encode ON and OFF motion in four directions for each location in the flies' field of view (Mauss et al. 2014). Taken together, these findings argue that T4 and T5 and their presynaptic inputs represent the circuit components that define the elementary motion detector.

Which medulla neurons provide functional visual input to T4 and T5? Connectomics data obtained from serial section EM reconstructions produced a detailed dataset describing the neurons that connect L1 to T4 and that link L2 to T5 (Shinomiya et al. 2014; Takemura et al. 2013). Neuronal silencing done in combination with behavior or physiology, have since been used to test these predictions. It has now been confirmed that Mi1 and Tm3 provide a functional link from L1 to T4 (Ammer et al. 2015; Takemura et al. 2013), while the medulla neurons Tm1, Tm2, and Tm4 all provide functional synaptic input to the processing of OFF edges, and likely connect L2 with T5 (Meier et al. 2014; Serbe et al. 2016; Shinomiya et al. 2014). These medulla neuron candidates were identified because they represented the strongest feedforward pathways from L1 and L2 to T4 or T5 (Fig. 15.2d). Moreover, combining connectomic predictions with functional validation based on genetic neuronal silencing has been instrumental in identifying elements within the motion-detecting circuit.

15.3.4 Forward Genetics

A parallel set of studies took an alternative, unbiased approach to discover new circuit components (Katsov and Clandinin 2008; Silies et al. 2013; Tuthill et al. 2013). A set of genetics screens revealed another pathway that originates from the lamina neuron L3 and makes critical contributions to motion processing (Silies et al. 2013; Tuthill et al. 2013). Similar to L2, L3 provides input to circuitry required for the detection of OFF edge motion. This forward genetic screen also identified the neuron Tm9, a downstream synaptic partner of L3 (Fisher et al. 2015b). In parallel, it was shown that Tm9 provides a large number of synaptic inputs onto T5 (Shinomiya et al. 2014). Calcium imaging of T5 performed in combination with synaptic silencing of Tm9 revealed that loss of Tm9 input leads to a significant reduction in the directional responses of T5, demonstrating that it is a major component of dark edge motion detection. Another study has corroborated the central role of Tm9 in OFF edge processing by measuring electrophysiology of downstream LPTCs and behavior (Serbe et al. 2016). Moreover, facile genetics in *Drosophila* enables forward genetic screens, and the identification of the L3-Tm9 pathway for OFF edge motion detection underscores the power of such an approach (Fig. 15.2d).

15.3.5 Defining the Computation at Each Level of the Circuit

Which computations are actually performed at each processing level in the circuit? How does biology implement the mathematical operations that are necessary to compute motion direction? The classical HRC model proposes the combination of three main algorithmic transformations. First, the model requires a temporal "delay" in which signals from neurons that sample light from adjacent points in space must have different temporal filtering properties. Second, these signals need to interact in a nonlinear manner, using correlation (or multiplication). This correlation operation is at the core of the motion computation, and the site of this interaction should reveal directional motion signals. Finally, a subtraction operation between two mirror symmetric motion signals is implemented to produce a strongly direction selective circuit that is excited by one direction of motion and inhibited by motion of the opposite direction. Signatures of each of these transformation steps have now been characterized and ongoing work seeks to put together a complete circuit model of motion detection that includes an understanding at the level of circuit and molecular mechanisms.

15.3.5.1 Temporal Processing (The Delay)

Characterization of the temporal response properties of neurons in the elementary motion circuit has been performed using calcium imaging and electrophysiology combined with controlled visual stimuli designed to interrogate the timing of neuronal signals. Generally two classes of visual stimuli have been utilized. Relatively simple stimuli, such as full field illumination with different temporal intervals can be used to interrogate the onset and offset responses of a neuron to light. Alternatively, stimuli with time-varying contrast signals (most often drawn randomly from a Gaussian distribution) have been used in combination with reverse correlation analysis to extract the linear temporal filtering properties of genetically defined neurons of interest. As most of the early visual system neurons do not spike, reverse correlation is either performed using the signal provided by calcium imaging or the recorded membrane potential.

Using such approaches, many different temporal response properties have been uncovered. Indeed, already at the level of the second order lamina neurons, differential temporal filtering can be observed. In particular, the lamina neurons L1 and L2 are fast, with responses that peak soon after the onset of a luminance change and are quite transient, suggesting that they predominantly encode recent changes in light levels (Clark et al. 2011; Freifeld et al. 2013; Reiff et al. 2010). Meanwhile, the responses of the lamina neuron L3 are much slower and are also more sustained (Silies et al. 2013; Strother et al. 2014). Perhaps consistent with these physiological properties, one behavioral study has suggested that L3 neurons may contribute preferentially to detecting slower motion stimuli (Tuthill et al. 2013). However, while faster and slower signals clearly emerge directly postsynaptic to photoreceptors within the lamina, exactly how these properties map on to circuit models of elementary motion detection remains unclear.

To build a complete picture of the transformations within the early motion circuit it has been vital to characterize the temporal response properties of the columnar neuronal elements of the medulla. T4 and T5 are themselves direction selective while their columnar presynaptic inputs from the medulla are not. These data suggest that direction selectivity arises within T4 and T5 based on a convergence and interaction between slower and faster temporal signals. Thus, both the temporal filtering properties of different medulla interneurons, and how these signals are transmitted onto T4 and T5 is of great interest. Taking a candidate approach based on connectomics data, Rudy Behnia and colleagues interrogated the temporal filtering properties of the main neuronal links from L1 and L2 to T4 and T5 using whole-cell electrophysiological recordings (Behnia et al. 2014). This study used reverse correlation to obtain the linear temporal filters for the neurons Mi1, Tm3,

Tm1, and Tm2. Strikingly, there is an offset in the response latency between proposed feedforward pathways for ON and OFF edge motion detection, Mi1/Tm3 and Tm1/Tm2, respectively. Specifically, the linear filter for Mi1 had a slower time-to-peak compared with Tm3 (18 ms slower) while Tm1 had a slower time to peak versus Tm2 (13 ms slower). Moreover, the entire response waveform of these cells were systematically offset in time, with Mi1 and Tm1 being phase shifted later than Tm3 and Tm2, respectively. These data, taken together with EM work, which revealed a spatially asymmetric wiring between these medulla neurons onto T4 and T5, suggested a simple circuit model: for ON motion detection Mi1 and Tm3 represent the delayed and non-delayed inputs that interact at the level of T4, while for OFF edge motion detection Tm1 and Tm2 would have similar roles as the channels into T5.

Since this circuit proposal, additional complexities have been uncovered. Tm4 and Tm9 are also important for OFF edge motion detection and calcium imaging of these neurons revealed additional diversity in terms of temporal response dynamics (Fisher et al. 2015b; Serbe et al. 2016). Tm9 represents a much slower and sustained signal for the OFF edge pathway than any of the other described circuit elements, while Tm4 has fast kinetics similar to Tm2 (Fisher et al. 2015b; Serbe et al. 2016). In the case of the ON pathway, silencing of neither Mi1 nor Tm3 completely abolishes T4 motion responses suggesting that there must also be substantial contribution from additional columnar elements (Ammer et al. 2015). It would be interesting to uncover the temporal properties of these additional ON pathway neurons.

The original HRC proposed two parallel channels with different temporal filtering that interact to produce a direction selective signal. Conversely, the biological circuit indeed utilizes parallel temporal processing channels, but appears to use more than two channels. This complexity may allow the circuit to be more robust for the detection of visual motion in natural scenes. Or perhaps the contribution of these various input channels allows flexibility so the tuning properties of the circuit can be modulated or subsampled based on the requirements of the organism? Further manipulation studies are needed to probe the exact contribution of these differential temporal filtering properties and individual parallel channels to the encoding of many types of motion stimuli.

15.3.5.2 The Correlation Operation

The HRC model next hypothesizes that signals from two points in space, one delayed and one non-delayed, should interact in a nonlinear manner to produce a directional signal. The identification of the columnar direction selective units, T4 and T5, shed considerable light on the inner workings of the motion circuit (Maisak et al. 2013). Recordings from these neurons further confirmed that the motion circuit processes moving ON and OFF edges within segregated circuits. T4 neurons encode ON edges and T5 respond preferentially to moving OFF edges.

As T4 and T5 neurons likely represent the first direction selective units within the visual system they pinpoint the stage in the circuit where the HRC model proposes a correlation operation should occur. Anatomical work based on EM reconstruction added further support for this notion with the finding that columnar medulla inputs display spatially asymmetric wiring onto T4 and T5. Indeed, in the case of T4 this asymmetric wiring is consistent with direction selective responses. An important next step was to understand the nature of the nonlinear operation that leads to direction selective signals within T4 and T5. Using calcium imaging in combination with apparent motion stimuli, this circuit was shown to perform a nonlinear correlation operation on light signals from adjacent points in space (Fisher et al. 2015a). Therefore, as predicted by the original HRC model, this circuit actually performs a correlation (multiplication-like) operation. However, as is seems a likely outcome for a real biological system, recent studies also reveals additional complexity within these circuit transformations. For example, T4 and T5 have orientation selective receptive fields that are shaped by inhibitory circuitry and the observation that motion signals within T4 and T5 are regulated by the behavioral state of the organism (Fisher et al. 2015a; Leong et al. 2016; Schnell et al. 2014).

15.3.5.3 The Subtraction Operation

Direction selective neurons across a diverse range of organisms are excited by motion in one direction and inhibited by motion in the opposite direction. In the fly, these motion opponent signals emerge within wide-field LPTCs (Egelhaaf et al. 1990; Single et al. 1997). Early characterization of these hyperpolarizing signals suggested that they arise via synaptic inhibition from upstream direction selective units that impinge onto the LPTC dendritic arbor. This inhibition maps elegantly on to the final subtraction step proposed by the HRC. In fruit flies, blocking the synaptic output of T4 and T5 neurons abolished both the excitatory and inhibitory motion evoked responses (Schnell et al. 2012). T4 and T5 neurons are cholinergic and thus likely excitatory, leading to the hypothesis that these cells provide direct excitatory motion signals to LPTCs while also passing directional information on to a relay circuit that is responsible for the inhibitory signal (Mauss et al. 2014). Consistent with this hypothesis, optogenetic activation of T4 and T5 produces a biphasic electrical signal within LPTC neurons that is comprised of a fast monosynaptic excitatory current followed by a slower inhibitory current (Mauss et al. 2014). Ionotropic GABAergic inhibition had long been proposed to mediate this inhibitory current, based on pharmacological studies using the drug picrotoxin (Egelhaaf et al. 1990; Single et al. 1997). Interestingly however, glutamate can also provide inhibitory currents by acting on glutamate-gated chloride channels, and these GluCl channels are also blocked by picrotoxin (Liu and Wilson 2013; Rohrbough et al. 2002). In the lobula plate, direction selectivity is organized in a layer-specific manner, such that layers with opposing motion preference are adjacent to one another, with the anterior most two layers having sensitivity to front to back and back-to-front motion, respectively, while the two posterior layers encode opposing vertical motion. An anatomical screen of Gal4 driver lines identified a class of neurons that span both of the vertical motion layers (Mauss et al. 2014). These glutamatergic LPi neurons are perfectly situated to convey the predicted subtraction between these motion sensitive layers (Mauss et al. 2015). In support of this hypothesis, optogenetic activation of Lpi3-4 neurons evokes inhibitory currents within downward sensitive VS neurons, which is blocked by picrotoxin (Mauss et al. 2015). These LPi neurons exhibit the direction selectivity predicted based on their anatomical pattern of innervation, and blocking their synaptic output abolishes null direction inhibition signals within VS neurons while leaving preferred direction signals intact (Mauss et al. 2015). Thus, LPi neurons are critical circuit element for producing the subtraction step within motion detection pathway. Current evidence supports a model where T4 and T5 neurons provide the DS signal onto LPi, and LPi's in turn, relay this DS signal onto LPTCs neurons of the opposite motion sensitivity by inhibition through glutamate-gated chloride channels.

15.4 Concluding Remarks About Motion Detection

In conclusion, a rich theoretical and experimental history in combination with a wide array of modern anatomical and genetic tools for circuit interrogation has led to rapid dissection of elementary motion processing in recent years. Remarkably many of the neuronal transformations predicted from classical theoretical work can now be observed experimentally within the fruit fly visual system. The next step is to combine all of these disparate experimental results into a complete circuit model of the intact circuit. Ideally such a model would capture important signatures about the underlying biology including cell morphology, signal transformations between synaptic connections, and a description of how voltage and calcium signals are transformed across neuronal compartments. To achieve this goal, more experiments still need to be done to completely define the main neuronal elements within this circuit and to pinpoint the molecular components of the central circuit transformations. Yet these goals seem within reach in the coming years making this a very exciting time for the study of motion detection and for circuits neuroscience in the fly more broadly.

References

Adelson EH, Bergen JR (1985) Spatiotemporal energy models for the perception of motion. J Opt Soc Am 2:284–299

Akerboom J, Rivera JDV, Guilbe MMR, Malavé ECA, Hernandez HH, Tian L, Hires SA, Marvin JS, Looger LL, Schreiter ER (2009) Crystal structures of the GCaMP calcium sensor reveal the mechanism of fluorescence signal change and aid rational design. J Biol Chem 284:6455–6464

Akerboom J, Chen T-W, Wardill TJ, Tian L, Marvin JS, Mutlu S, Calderón NC, Esposti F, Borghuis BG, Sun XR et al (2012) Optimization of a GCaMP calcium indicator for neural activity imaging. J Neurosci 32:13819–13840

Akerboom J, Carreras Calderón N, Tian L, Wabnig S, Prigge M, Tolö J, Gordus A, Orger MB, Severi KE, Macklin JJ et al (2013) Genetically encoded calcium indicators for multi-color neural activity imaging and combination with optogenetics. Front Mol Neurosci 6:2

Albert JT, Gopfert MC (2015) Hearing in *Drosophila*. Curr Opin Neurobiol 34:79–85

Ammer G, Leonhardt A, Bahl A, Dickson BJ, Borst A (2015) Functional specialization of neural input elements to the *Drosophila* ON motion detector. Curr Biol 1–7

Bahl A, Ammer G, Schilling T, Borst A (2013) Object tracking in motion-blind flies. Nat Neurosci 1–11

Barlow HB, Levick WR (1965) The mechanism of directionally selective units in the rabbit retina. J Physiol 178:477–504

Behnia R, Clark DA, Carter AG, Clandinin TR, Desplan C (2014) Processing properties of ON and OFF pathways for *Drosophila* motion detection. Nature 512:427–430

Benzer S (1967) Behavioral mutants of *Drosophila* isolated by countercurrent distribution. Proc Natl Acad Sci 58:1112–1119

Bhandawat V, Olsen SR, Gouwens NW, Schlief ML, Wilson RI (2007) Sensory processing in the *Drosophila* antennal lobe increases reliability and separability of ensemble odor representations. Nat Neurosci 10:1474–1482

Brand AH, Perrimon N (1993) Targeted gene expression as a means of altering cell fates and generating dominant phenotypes. Development 118:401–415

Branson K, Robie AA, Bender J, Perona P, Dickinson MH (2009) High-throughput ethomics in large groups of *Drosophila*. Nat Methods 6:451–457

Cao G, Platisa J, Pieribone VA, Raccuglia D, Kunst M, Nitabach MN (2013) Genetically targeted optical electrophysiology in intact neural circuits. Cell 154:904–913

Catterall WA (2011) Voltage-gated calcium channels. Cold Spring Harb Perspect Biol 1–23

Chen T-W, Wardill TJ, Sun Y, Pulver SR, Renninger SL, Baohan A, Schreiter ER, Kerr RA, Orger MB, Jayaraman V et al (2013) Ultrasensitive fluorescent proteins for imaging neuronal activity. Nature 499:295–300

Chiang A-S, Lin C-Y, Chuang C-C, Chang H-M, Hsieh C-H, Yeh C-W, Shih C-T, Wu J-J, Wang G-T, Chen Y-C et al (2011) Three-dimensional reconstruction of brain-wide wiring networks in *Drosophila* at single-cell resolution. Curr Biol 21:1–11

Chiappe ME, Seelig JD, Reiser MB, Jayaraman V (2010) Walking modulates speed sensitivity in *Drosophila* motion vision. Curr Biol 20:1470–1475

Clark DA, Bursztyn L, Horowitz M, Schnitzer MJ, Clandinin TR (2011) Defining the computational structure of the motion detector in *Drosophila*. Neuron 70:1165–1177

Crocker A, Guan X-J, Murphy CT, Murthy M (2016) Cell-type-specific transcriptome analysis in the *Drosophila* mushroom body reveals memory-related changes in gene expression. Cell Rep 15:1–17

Dankert H, Wang L, Hoopfer ED, Anderson DJ, Perona P (2009) Automated monitoring and analysis of social behavior in *Drosophila*. Nat Methods 6:297–303

de Vries SEJ, Clandinin TR (2012) Loom-sensitive neurons link computation to action in the *Drosophila* visual system. Curr Biol 22:353–362

Dickinson MH (2014) Death valley, *Drosophila*, and the Devonian toolkit. Annu Rev Entomol 59:51–72

Dietzl G, Chen D, Schnorrer F, Su K, Barinova Y, Fellner M, Gasser B, Kinsey K, Oppel S, Scheiblauer S et al (2007) A genome-wide transgenic RNAilibrary for conditional gene inactivation in *Drosophila*. Nature 448

Dobritsa AA, Der Goes Van, Van Naters W, Warr CG, Steinbrecht RA, Carlson JR (2003) Integrating the molecular and cellular basis of odor coding in the *Drosophila* antenna. Neuron 37:827–841

Dudai Y, Jan Y (1976) dunce, a mutant of *Drosophila* deficient in learning. Proc Natl Acad Sci 73:1684–1688

Dvorak DR, Bishop LG, Eckert HE (1975) On the identification of movement detectors in the fly optic lobe. J Comp Physiol A 100:5–23

Egelhaaf M, Borst A, Pilz B (1990) The role of GABA in detecting visual motion. Brain Res 509:156–160

Erlenmeyer-Kimling L, Hirsch J (1961) Measurement of the relations between chromosomes and behavior. Science (80-) 134:1068–1069

Fischbach K-F, Dittrich APM (1989) The optic lobe of *Drosophila melanogaster*. I. A Golgi analysis of wild-type structure. Cell Tissue Res 258:442–475

Fisher YE, Silies M, Clandinin TR (2015a) Orientation selectivity sharpens motion detection in *Drosophila*. Neuron 88:1–14

Fisher YE, Leong JCS, Sporar K, Ketkar MD, Gohl DM, Clandinin TR, Silies M (2015b) A class of visual neurons with wide-field properties is required for local motion detection. Curr Biol 25:3178–3189

Freifeld L, Clark DA, Schnitzer MJ, Horowitz MA, Clandinin TR (2013) GABAergic lateral interactions tune the early stages of visual processing in *Drosophila*. Neuron 78:1075–1089

Gohl DM, Silies MA, Gao XJ, Bhalerao S, Luongo FJ, Lin C, Potter CJ, Clandinin TR (2011) A versatile in vivo system for directed dissection of gene expression patterns. Nat Methods 8:231–237

Gong Y, Huang C, Li JZ, Grewe BF, Zhang Y, Eismann S, Schnitzer MJ (2015) High-speed recording of neural spikes in awake mice and flies with a fluorescent voltage sensor. Science (80-) 350:1361–1366

Gonzalez-Bellido PT, Wardill TJ, Kostyleva R, Meinertzhagen IA, Juusola M (2009) Overexpressing temperature-sensitive dynamin decelerates phototransduction and bundles microtubules in *Drosophila* photoreceptors. J Neurosci 29:14199–14210

Gordon MD, Scott K (2009) Motor control in a *Drosophila* taste circuit. Neuron 61:373–384

Götz K (1964) Optomotorische untersuchung des visuellen systems einiger augenmutanten der fruchtfliege Drosophila. Biol Cybern

Gruntman E, Turner GC (2013) Integration of the olfactory code across dendritic claws of single mushroom body neurons. Nat Neurosci 16:1821–1829

Guven-Ozkan T, Davis RL (2014) Functional neuroanatomy of *Drosophila* olfactory memory formation. Learn Mem 21:519–526

Hall JC (1978) Courtship among males due to a male-sterile mutation in *Drosophila melanogaster*. Behav Genet 8:125–141

Hallem EA, Ho GM, Carlson JR (2004) The molecular basis of odor coding in the *Drosophila* antenna. Cell 117:965–979

Hamada FN, Rosenzweig M, Kang K, Pulver SR, Ghezzi A, Jegla TJ, Garrity PA (2008) An internal thermal sensor controlling temperature preference in *Drosophila*. Nature 454:217–220

Han DD, Stein D, Stevens LM (2000) Investigating the function of follicular subpopulations during *Drosophila* oogenesis through hormone-dependent enhancer-targeted cell ablation. Development 127:573–583

Hassenstein V, Reichardt W (1956) System theoretical analysis of time, sequence and sign analysis of the motion perception of the snout-beetle Chlorophanus. German Z Naturforsch 11:513–524

Hausen K (1976) Functional characterization and anatomical identification of motion sensitive neurons in the lobula plate of the blowfly Calliphora erythrocephala. Zeitschrift Fur Naturforsch Sect C J Biosci 31:629–634

Heisenberg M, Götz KG (1975) The use of mutations for the partial degradation of vision in *Drosophila melanogaster*. J Comp Physiol Series A 98:217–241

Heisenberg M, Wonneberger R, Wolf R (1978) Optomotor-blindH31-a *Drosophila* mutant of the lobula plate giant neurons. J Comp Physiol Series A 124:287–296

Helassa N, Zhang XH, Conte I, Scaringi J, Esposito E, Bradley J, Carter T, Ogden D, Morad M, Torok K (2015) Fast-response calmodulin-based fluorescent indicators reveal rapid intracellular calcium dynamics. Sci Rep 5:15978

Hiesinger PR, Reiter C, Schau H, Fischbach KF (1999) Neuropil pattern formation and regulation of cell adhesion molecules in *Drosophila* optic lobe development depend on synaptobrevin. J Neurosci 19:7548–7556

Inada K, Kohsaka H, Takasu E, Matsunaga T, Nose A (2011) Optical dissection of neural circuits responsible for *Drosophila* larval locomotion with halorhodopsin. PLoS ONE 6:e29019

Jeanne JM, Wilson RI (2015) Convergence, divergence, and reconvergence in a feedforward network improves neural speed and accuracy. Neuron 88:1014–1026

Jenett A, Rubin GM, Ngo TTB, Shepherd D, Murphy C, Dionne H, Pfeiffer BD, Cavallaro A, Hall D, Jeter J et al (2012) A GAL4-driver line resource for *Drosophila* neurobiology. Cell Rep 2:991–1001

Joesch M, Plett J, Borst A, Reiff DF (2008) Response properties of motion-sensitive visual interneurons in the lobula plate of *Drosophila melanogaster*. Curr Biol 18:368–374

Joesch M, Schnell B, Raghu SV, Reiff DF, Borst A (2010) ON and OFF pathways in *Drosophila* motion vision. Nature 468:300–304

Kalko EKV, Dukas R, Ratcliffe JM, Teeling EC, Haven N, Fattu JM, Bates ME, Simmons JA, Riquimaroux H, Surlykke A et al (2011) An expanded palette of genetically encoded Ca^{2+} indicators. Science (80-) 333:1888–1891

Katsov AY, Clandinin TR (2008) Motion processing streams in *Drosophila* are behaviorally specialized. Neuron 59:322–335

Kitamoto T, Xie X, Wu CF, Salvaterra PM (2000) Isolation and characterization of mutants for the vesicular acetylcholine transporter gene in *Drosophila melanogaster*. J Neurobiol

Konopka RJ, Benzer S (1971) Clock mutants of *Drosophila melanogaster*. Proc Natl Acad Sci 68:2112–2116

Kvon EZ, Kazmar T, Stampfel G, Yáñez-Cuna JO, Pagani M, Schernhuber K, Dickson BJ, Stark A (2014) Genome-scale functional characterization of *Drosophila* developmental enhancers in vivo. Nature 512:91–95

Leong JCS, Esch JJ, Poole B, Ganguli S, Clandinin TR (2016) Direction selectivity in *Drosophila* emerges from preferred-direction enhancement and null-direction suppression. J Neurosci 36:8078–8092

Liu WW, Wilson RI (2013) Glutamate is an inhibitory neurotransmitter in the *Drosophila* olfactory system. Proc Natl Acad Sci 110:10294–10299

Looger LL, Griesbeck O (2012) Genetically encoded neural activity indicators. Curr Opin Neurobiol 22:18–23

Luan H, Peabody NC, Vinson CR, White BH (2006) Refined spatial manipulation of neuronal function by combinatorial restriction of transgene expression. Neuron 52:425–436

Maimon G, Straw AD, Dickinson MH (2010) Active flight increases the gain of visual motion processing in *Drosophila*. Nat Neurosci 13:393–399

Maisak MS, Haag J, Ammer G, Serbe E, Meier M, Leonhardt A, Schilling T, Bahl A, Rubin GM, Nern A et al (2013) A directional tuning map of *Drosophila* elementary motion detectors. Nature 500:212–216

Mauss AS, Meier M, Serbe E, Borst A (2014) Optogenetic and pharmacologic dissection of feedforward inhibition in *Drosophila* motion vision. J Neurosci 34:2254–2263

Mauss AS, Pankova K, Arenz A, Nern A, Rubin GM, Borst A (2015) Neural circuit to integrate opposing motions in the visual field. Cell 162:351–362

McKenna M, Monte P, Helfand SL, Woodard C, Carlson J (1989) A simple chemosensory response in *Drosophila* and the isolation of acj mutants in which it is affected. Proc Natl Acad Sci 86:8118–8122

Meier M, Serbe E, Maisak MS, Haag J, Dickson BJ, Borst A (2014) Neural circuit components of the *Drosophila* OFF motion vision pathway. Curr Biol 24:385–392

Meinertzhagen IA, O'Neil SD (1991) Synaptic organization of columnar elements in the lamina of the wild type in *Drosophila melanogaster*. J Comp Neurol 305:232–263

Meinertzhagen IA, Sorra KE (2001) Synaptic organization in the fly's optic lamina: few cells, many synapses and divergent microcircuits. Prog Brain Res 131:53–69

Muijres FT, Elzinga MJ, Iwasaki NA, Dickinson MH (2015) Body saccades of *Drosophila* consist of stereotyped banked turns. J Exp Biol 218:864–875

Nagel KI, Wilson RI (2011) Biophysical mechanisms underlying olfactory receptor neuron dynamics. Nat Neurosci 14:208–216

Nern A, Pfeiffer BD, Rubin GM (2015) Optimized tools for multicolor stochastic labeling reveal diverse stereotyped cell arrangements in the fly visual system. Proc Natl Acad Sci 112:E2967–E2976

Ni J-Q, Liu L-P, Binari R, Hardy R, Shim H-S, Cavallaro A, Booker M, Pfeiffer BD, Markstein M, Wang H et al (2009) A *Drosophila* resource of transgenic RNAi lines for neurogenetics. Genetics 182:1089–1100

Pak WL, Grossfield J, Arnold KS (1970) Mutants in the visual pathway of *Drosophila melanogaster*. Nature 222:351–354

Pavlou HJ, Goodwin SF (2013) Courtship behavior in *Drosophila melanogaster*: towards a "courtship connectome". Curr Opin Neurobiol 23:76–83

Pfeiffer BD, Jenett A, Hammonds AS, Ngo T-TB, Misra S, Murphy C, Scully A, Carlson JW, Wan KH, Laverty TR et al (2008) Tools for neuroanatomy and neurogenetics in *Drosophila*. Proc Natl Acad Sci 105:9715–9720

Pfeiffer BD, Ngo T-TB, Hibbard KL, Murphy C, Jenett A, Truman JW, Rubin GM (2010) Refinement of tools for targeted gene expression in *Drosophila*. Genetics 186:735–755

Reiff DF, Plett J, Mank M, Griesbeck O, Borst A (2010) Visualizing retinotopic half-wave rectified input to the motion detection circuitry of *Drosophila*. Nat Neurosci 13:973–978

Riehle A, Franceschini N (1984) Motion detection in flies: parametric control over ON-OFF pathways. Exp Brain Res 390–394

Rister J, Pauls D, Schnell B, Ting C-Y, Lee C-H, Sinakevitch I, Morante J, Strausfeld NJ, Ito K, Heisenberg M (2007) Dissection of the peripheral motion channel in the visual system of *Drosophila melanogaster*. Neuron 56:155–170

Rivera-Alba M, Vitaladevuni SN, Mischenko Y, Lu Z, Takemura S, Scheffer L, Meinertzhagen IA, Chklovskii DB, de Polavieja GG (2011) Wiring economy and volume exclusion determine neuronal placement in the *Drosophila* Brain. Curr Biol 1–6

Rohrbough J, Rohrbough J, Broadie K, Broadie K (2002) Electrophysiological analysis of synaptic transmission in central neurons of *Drosophila* larvae. J Neurophysiol 88:847–860

Root CM, Masuyama K, Green DS, Enell LE, Nässel DR, Lee C-H, Wang JW (2008) A presynaptic gain control mechanism fine-tunes olfactory behavior. Neuron 59:311–321

Schnell B, Raghu S, Nern A, Borst A (2012) Columnar cells necessary for motion responses of wide-field visual interneurons in *Drosophila*. J Comp Physiol A

Schnell B, Weir PT, Roth E, Fairhall AL, Dickinson MH (2014) Cellular mechanisms for integral feedback in visually guided behavior. Proc Natl Acad Sci 2014:1–8

Scott EK, Raabe T, Luo L (2002) Structure of the vertical and horizontal system neurons of the lobula plate in *Drosophila*. J. Comp. Neurol. 454:470–481

Seelig JD, Jayaraman V (2015) Neural dynamics for landmark orientation and angular path integration. Nature 521:186–191

Serbe E, Meier M, Leonhardt A, Borst A (2016) Comprehensive characterization of the major presynaptic elements to the *Drosophila* OFF motion detector. Neuron 1–13

Shinomiya K, Karuppudurai T, Lin T-Y, Lu Z, Lee C-H, Meinertzhagen IA (2014) Candidate neural substrates for off-edge motion detection in *Drosophila*. Curr Biol 24:1062–1070

Siddiqi O, Benzer S (1976) Neurophysiological defects in temperature-sensitive paralytic mutants of *Drosophila melanogaster*. PNAS 73:3253–3257

Silies M, Gohl DM, Fisher YE, Freifeld L, Clark DA, Clandinin TR (2013) Modular use of peripheral input channels tunes motion-detecting circuitry. Neuron 79:111–127

Silies M, Gohl DM, Clandinin TR (2014) Motion-detecting circuits in flies: coming into view. Annu Rev Neurosci 37:307–327

Single S, Haag J, Borst A (1997) Dendritic computation of direction selectivity and gain control in visual interneurons. J Neurosci 17:6023–6030

Stocker R, Lienhard M, Borst A, Fischbach K-F (1990) Neuronal architecture of the antennal lobe in *Drosophila melanogaster*. Cell Tissue Res 262:9–34

Strausfeld NJ (1976) Atlas of an insect brain. Springer, Berlin

Straw AD, Lee S, Dickinson MH (2010) Visual control of altitude in flying *Drosophila*. Curr Biol 20:1550–1556

Straw AD, Branson K, Neumann TR, Dickinson MH (2011) Multi-camera real-time three-dimensional tracking of multiple flying animals. J R Soc Interface 8:395–409

Strother JA, Nern A, Reiser MB (2014) Direct observation of ON and OFF pathways in the *Drosophila* visual system. Curr Biol 24:976–983

Suster ML, Seugnet L, Bate M, Sokolowski MB (2004) Refining GAL4-driven transgene expression in *Drosophila* with a GAL80 enhancer-trap. Genesis 39:240–245

Suzuki DT, Grigliatti T, Williamson R (1971) A mutation (parats) causing reversible adult paralysis. PNAS 68:890–893

Sweeney ST, Broadie K, Keane J, Niemann H, O'Kane CJ (1995) Targeted expression of tetanus toxin light chain in *Drosophila* specifically eliminates synaptic transmission and causes behavioral defects. Neuron 14:341–351

Takemura S-Y, Lu Z, Meinertzhagen IA (2008) Synaptic circuits of the *Drosophila* optic lobe: the input terminals to the medulla. J Comp Neurol 509:493–513

Takemura S, Bharioke A, Lu Z, Nern A, Vitaladevuni S, Rivlin PK, Katz WT, Olbris DJ, Plaza SM, Winston P et al (2013) A visual motion detection circuit suggested by *Drosophila* connectomics. Nature 500:175–181

Tan L, Zhang KX, Pecot MY, Nagarkar-Jaiswal S, Lee PT, Takemura SY, McEwen JM, Nern A, Xu S, Tadros W et al (2015) Ig superfamily ligand and receptor pairs expressed in synaptic partners in *Drosophila*. Cell 163:1756–1769

Tuthill JC, Nern A, Holtz SL, Rubin GM, Reiser MB (2013) Contributions of the 12 neuron classes in the fly lamina to motion vision. Neuron 79:128–140

von Reyn CR, Breads P, Peek MY, Zheng GZ, Williamson WR, Yee AL, Leonardo A, Card GM (2014) A spike-timing mechanism for action selection. Nat Neurosci 17:1–12

Wardill TJ, List O, Li X, Dongre S, McCulloch M, Ting C-Y, O'Kane CJ, Tang S, Lee C-H, Hardie RC et al (2012) Multiple spectral inputs improve motion discrimination in the *Drosophila* visual system. Science (80-) 336:925–931

Wasserman SM, Aptekar JW, Lu P, Nguyen J, Wang AL, Keles MF, Grygoruk A, Krantz DE, Larsen C, Frye MA (2015) Olfactory neuromodulation of motion vision circuitry in *Drosophila*. Curr Biol 25:467–472

Wilson RI (2013) Early olfactory processing in *Drosophila*: mechanisms and principles. Annu Rev Neurosci 36:217–241

Wilson RI, Laurent G (2005) Role of GABAergic inhibition in shaping odor-evoked spatiotemporal patterns in the *Drosophila* antennal lobe. J Neurosci 25:9069–9079

Wilson RI, Turner GC, Laurent G (2004) Transformation of olfactory representations in the *Drosophila* antennal lobe. Science (80-) 303:366–370

Yaksi E, Wilson RI (2010) Electrical coupling between olfactory glomeruli. Neuron 67:1034–1047

Yang HH, Sun X, Ding X, Lin MZ, Clandinin TR, Yang HH (2016) Subcellular imaging of voltage and calcium signals reveals neural processing in vivo. Cell 1–13

Zheng L, de Polavieja GG, Wolfram V, Asyali MH, Hardie RC, Juusola M (2006) Feedback network controls photoreceptor output at the layer of first visual synapses in *Drosophila*. J Gen Physiol 127:495–510

Zhu Y, Nern A, Zipursky SL, Frye MA (2009) Peripheral visual circuits functionally segregate motion and phototaxis behaviors in the fly. Curr Biol 19:613–619

Part IV
Development: The Molecular Determinants of Cell Type Diversity

Chapter 16
Generation of Neuronal Diversity in the Peripheral Olfactory System in *Drosophila*

Catherine Hueston and Pelin C. Volkan

Abstract It is estimated that there are billions of neurons in the human brain. Each of these neurons is thought to play a distinct role in nervous system function. It is the emergent functional properties of these neurons that allow organisms to detect and process information about the complex sensory environment, and successfully navigate through these environments. To perform this task, many neural circuits representing different modalities of sensory information are organized to optimally encode and discriminate the nature of the stimuli. Diversity of neuronal fates that constitute these circuits is generated during development, as cells with high differentiation potentials, such as neural stem/progenitor/precursor cells, divide, communicate and make fate decisions that gradually diversify and restrict the type of cells they can generate. Sensory systems are a particularly interesting place to understand mechanisms of neural diversity as functional types of sensory neurons are under the constraints brought about by the complexity of the sensory world. In this chapter, we will review our current understanding of mechanisms generating neuronal diversity in the visual and olfactory systems (Van Bortle and Corces in Cell 152(6):1213–1217, 2013; Henikoff in Neuron 86(6):1319–1321, 2015; Mo et al. in Neuron 86(6):1369–1384, 2015).

16.1 Organization of the *Drosophila* Olfactory System

Detection of a broad range of chemosensory signals is essential for the survival of most multicellular organisms. To increase chemosensory sensitivity and specificity, animal sensory systems have evolved a large number of neurons with highly specific functions. The olfactory system of animals, from flies to mammals, is astonishingly diverse and organizationally conserved (Barish and Volkan 2015;

C. Hueston
Department of Neurobiology, Duke University, Durham, NC, USA

P.C. Volkan (✉)
Department of Biology, Duke Institute for Brain Science, Durham, NC, USA
e-mail: pelin.volkan@duke.edu

A. Çelik and M.F. Wernet (eds.), *Decoding Neural Circuit Structure and Function*, DOI 10.1007/978-3-319-57363-2_16

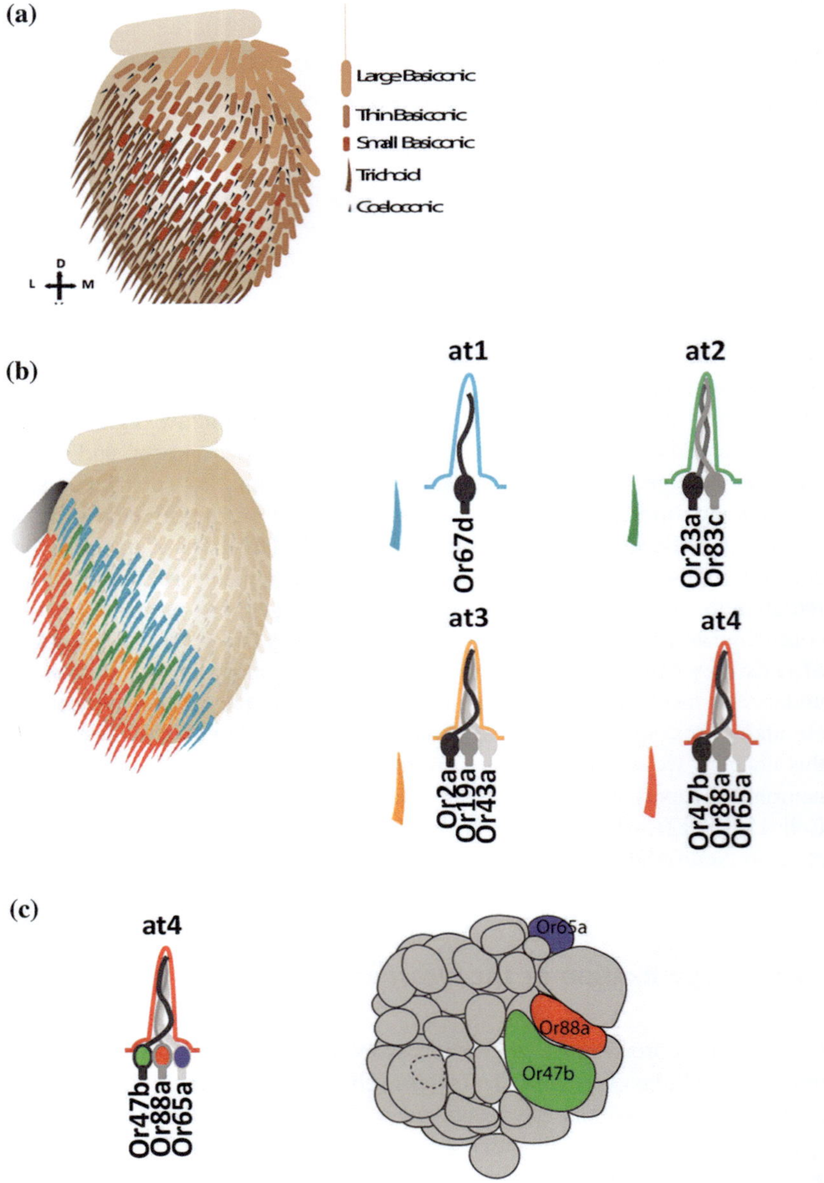
(a)
Large Basiconic
Thin Basiconic
Small Basiconic
Trichoid
Coeloconic
D
L
M
(b)
at1
Or67d
at2
Or23a
Or83c
at3
Or2a
Or19a
Or43a
at4
Or47b
Or88a
Or65a
(c)
at4
Or47b
Or88a
Or65a
Or65a
Or88a
Or47b

◄**Fig. 16.1** Anatomical organization of the *Drosophila* olfactory system. **a** Spatial organization of sensilla types on the antenna. Trichoid, basiconic (including *large, thin, and small basiconics*), and coeloconic sensilla occupy distinct zones. Rare intermediate sensilla are not shown. **b** Sub-segmentation of the trichoid sensilla type into subtypes (at1–at4) is exemplified on the *right*. The ORNs in each subtype express an invariable combination of OR genes (Or19a and Or19b are co-expressed; Or65a, Or65b, and Or65c are co-expressed). Modified from Sakano (2010), Li et al. (2013). **c** Connectivity of at4 ORN axons in the antennal lobe. The neurons and their glomerular targets are *color-coded*

Sakano 2010). However, unlike the mammalian olfactory system, which consists of 1400 olfactory receptor neurons (ORNs), the numerical complexity of neuronal types is much reduced in flies, with 50 adult ORN classes, which makes investigation of the entire system at the resolution of single cells possible. This genetically accessible model system can be exploited to dissect out the dynamics of the process that generates 50 different ORN fates from beginning patterning stages until terminal differentiation in detail.

Each ORN class is defined by the exclusive expression of a single olfactory receptor gene from approximately 80 possibilities in the genome. *Drosophila* ORN cell bodies reside in one of two olfactory organs, the antennae and maxillary palps, clustered within hair-like structures, or sensilla, whose distinct morphologies are distinguishable by electron microscopy as basiconic, trichoid, and coeloconic sensilla (Fig. 16.1) (Sen et al. 2003; Stocker 1994). Basiconic sensilla also have their own morphological subdivisions consisting of large, thin, and small basiconics. Each sensilla type also consists of subtypes, which are defined by the number and the invariant combinations of ORNs they house. For example, trichoid at1 sensilla subtype, houses a single ORN that expresses Or67d, whereas trichoid at4 subtype, contains three ORNs expressing Or47b, Or88a, and Or65a, respectively. The spatial positioning of sensilla subtypes on the antenna and palps is highly conserved (Couto et al. 2005; Fishilevich and Vosshall 2005) (Fig. 16.1).

The axons of all antennal and maxillary palp ORNs enter to the antennal lobe of the brain through the antennal and labial nerve fascicles, respectively (Stocker 1994). Once in the antennal lobe, all ORN axons of the same class converge onto a unique region within the antennal lobe called a "glomerulus", where they form synapses with a small number of projection neurons (Vosshall et al. 2000; Clyne et al. 1999) (Figs. 16.1 and 16.4). High level of neuronal diversity, and the hierarchically invariant organization of both the antenna and the antennal lobes has been a valuable tool for the study of the Drosophila olfactory system. Availability of reporters that mark each of the 50 ORN classes also makes the olfactory system a powerful model for the study of neuronal diversity, connectivity, and circuit identity.

16.1.1 Olfactory Receptors

Odor tuning properties of each class of ORNs are defined by the expression of typically a single olfactory receptor, which is localized to the ciliary dendrites of

ORNs, exposed to the odors that diffuse through pores on the sensillum cuticle. The first class of olfactory receptors were initially cloned using difference cloning, and shortly after their identification, singular expression of olfactory receptors in each ORN was shown using in situ hybridization and Gal4 reporters driven by olfactory receptor promoters (Couto et al. 2005; Fishilevich and Vosshall 2005; Benton et al. 2009; Silbering et al. 2011; Vosshall et al. 1999). *Drosophila* odorant receptors are seven-transmembrane domain proteins similar to the canonical G protein-coupled receptors found in mice (Vosshall et al. 1999; Vosshall 2004). However unlike G protein-coupled receptors, these receptors are inserted into the cell membrane in the reverse orientation with the amino terminus located intracellularly (Benton et al. 2006) and do not function as G protein-coupled receptors. *Drosophila* ORs also require the co-receptor *orco* for localization of ORs to ORN cilia and for successful signal transduction (Larsson et al. 2004). Currently, it is believed that ORs and OR83b physically interact to form a cation channel that is activated by either ionotropic or metabotropic means (Sato et al. 2008; Wicher et al. 2008). Receptor activation leads to signal transduction cascades that electrically activate the ORN.

In addition to ORs, a number of Gustatory receptors are expressed in the antenna, such as *Gr21a* and *Gr63a,* which function as CO_2 receptors, and *Gr10a* expression was also detected in the antenna, yet this receptor so far has not been shown to function in olfaction. More recently, a new class of olfactory receptors, ionotropic receptors (IRs), has been discovered in *Drosophila.* They have a similar structure to ionotropic glutamate receptors, earning them the name ionotropic receptors or IRs (Benton et al. 2009; Silbering et al. 2011; Abuin et al. 2011). These receptors are tuned primarily to acids, ammonia, humidity, and temperature (Silbering et al. 2011; Ai et al. 2010; Enjin et al. 2016). Like canonical ORs, IRs require multiple co-receptors to generate odor-evoked electrophysiological responses and for proper localization (Abuin et al. 2011). IRs and their co-receptors are also thought to form heteromeric ion channels.

16.1.2 Prepatterning of the Olfactory Sensory Tissue During Development

ORN development in *Drosophila* starts with the prepatterning of the larval eye-antennal disc (Figs. 16.2 and 16.3) (Barish and Volkan 2015). The disc initially contains undifferentiated epidermal cells. Developmental patterning genes such as *engrailed*, *wingless* and *hedgehog* spatially pattern the disc in different axes (Royet and Finkelstein 1997; Morata 2001; Chou et al. 2010; Song et al. 2012). Early in development expression of engrailed in the posterior region of the antennal disc, leads to expression of the secreted protein Hh to the anterior regions of the disc in a gradient. Hh activates wingless and Dpp, in a wedge-like expression pattern on opposing sides of the disc. At the center of the disc, gradients of wingless and Dpp meet and this interaction leads to expression of EGF, which diffuses outward in a radial gradient. This gradient initiates the expression of some members of the proximodistal patterning transcription factors, BarH1/2 and Dachshund, at different coordinates within the radial gradient. For example, BarH1/2 is expressed close to the center of the disc, whereas Dachshund gets expressed in the outer regions of the disc (Fig. 16.2b). As the tissue grows with waves of cell divisions, gradients change and new transcription factors, such as Rotund, Bab1, and Apterous get incorporated into the gene regulatory network to further diversify the zones within the disc (Fig. 16.2c). This transcription factor (TF) network consists of Rn, BarH1/H2, along with Bric-à-brac (Bab), Apterous (Ap) and Dachshund (Dac), and patterns the developing olfactory tissue (Fig. 16.2a) (Li et al. 2016). Unique combinations of these five transcription factors pattern the developing olfactory tissue into seven rings with diverse differentiation potentials that give rise to SOPs with diverse fates. Modifications of this network lead to predictable changes in the diversity of sensilla subtypes and ORN pools (Li et al. 2013, 2016). These studies provided a molecular map that defines diverse SOP fates and suggested an early prepatterning gene regulatory network as a modulator of SOP and terminally differentiated ORN diversity.

16.1.3 SOP Selection and Asymmetric Divisions

Once the disc is patterned into zones with unique developmental potentials, sensory organ precursor (SOP) selection occurs through expression of proneural genes (Fig. 16.3). Two proneural genes, Atonal and Amos, both of which encode HLH domain transcription factors have so far been identified. Atonal is required to specify coeloconic sensilla lineages, whereas trichoid and basiconic sensilla identity require Amos function (Gupta and Rodrigues 1997; Jhaveri et al. 2000; zur Lage et al. 2003). Amos mutants lack basiconic and trichoid sensilla, whereas atonal mutants have no coeloconic sensilla. Prepatterning gene *lozenge* (*lz*) activates *amos* expression, yet the regulator of *atonal* in the antennal disc is unknown (Gupta et al. 1998; Goulding et al. 2000). Despite the role for Lz in amos regulation, Lz also has a graded effect on basiconic and trichoid sensilla fates, where trichoid fates require lower levels of lz compared to basiconic fates (Gupta et al. 1998). Even though null alleles of *lz* lack both basiconic and trichoid sensilla, hypomorphic alleles of lz show expansion of some trichoid fates at the expense of basiconic fates.

Selection of SOPs from different fields of a prepatterned disc through proneural gene activity results in approximately 20–25 different types of SOPs, each with a unique differentiation potential to form a distinct sensilla subtype. Thus, each SOP gives rise to both the stereotyped ORN cluster and the non-neuronal support cells of a given sensilla (Fig. 16.3c). This happens as the multipotent SOP divides asymmetrically 2–3 times, to first generate intermediate precursors for neuronal and non-neuronal lineages, followed by additional divisions to form the terminally differentiated, unipotent ORNs. It was shown that Notch signaling functions during these asymmetric divisions in all SOPs, to mediate binary segregation of alternate

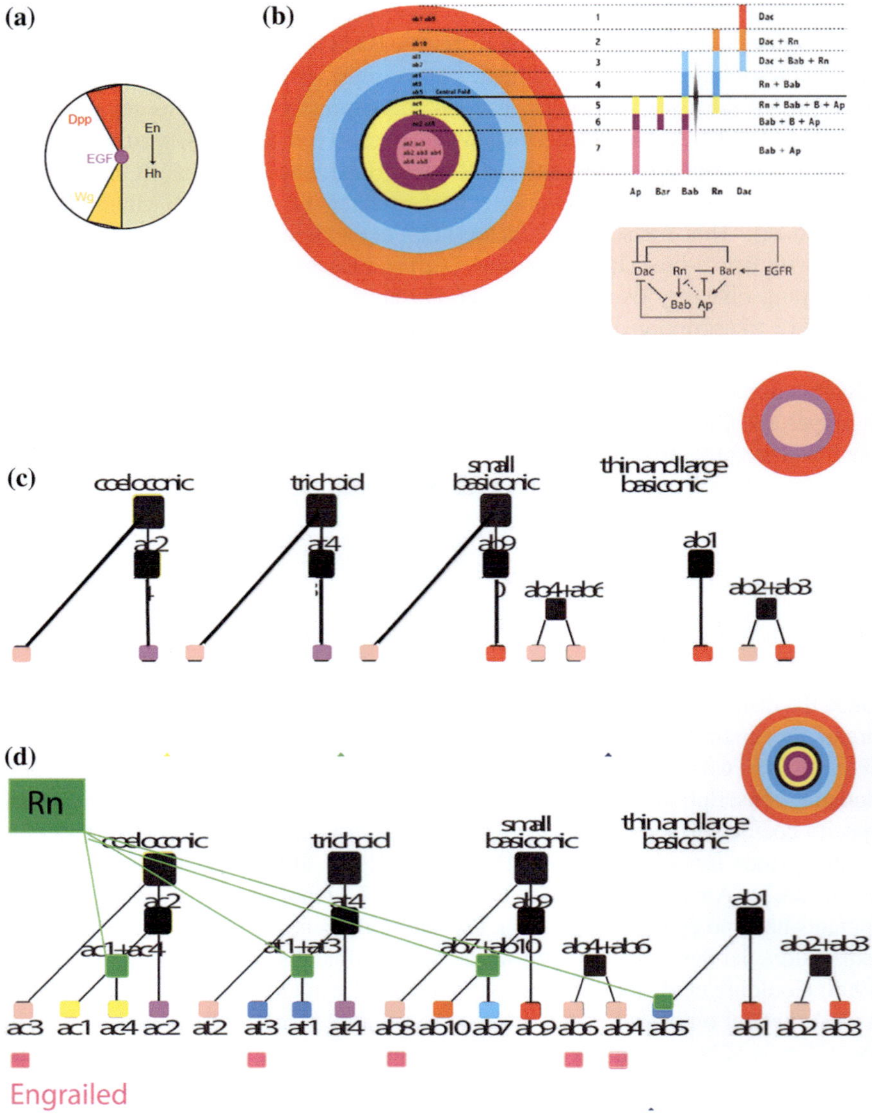
(a)
Dpp
En
EGF
Hh
Wg
(b)
1 Dac
2 Dac + Rn
3 Dac + Bab + Rn
4 Rn + Bab
5 Rn + Bab + B + Ap
6 Bab + B + Ap
7 Bab + Ap
Ap Bar Bab Rn Dac
Dac Rn Bar EGFR
Bab Ap
(c)
coeloconic
trichoid
small basiconic
thin and large basiconic
ac2
at4
ab9
ab4+ab6
ab1
ab2+ab3
(d)
Rn
coeloconic
trichoid
small basiconic
thin and large basiconic
ac2
at4
ab9
ab1
ac1+ac4
at1+at3
ab7+ab10
ab4+ab6
ab2+ab3
ac3 ac1 ac4 ac2 at2 at3 at1 at4 ab8 ab10 ab7 ab9 ab6 ab4 ab5 ab1 ab2 ab3
Engrailed

◂**Fig. 16.2** Antennal disc SOP map and decision tree for ORN diversification. **a** Early patterning of the antennal disc by *engrailed* (*en*), *hedgehog* (*Hh*), *wingless* (*wg*), and *decapentaplagic* (*dpp*) expression. Expression of *en* in the posterior disc activates *Hh* expression, which signals to the anterior disc to activate *wg* and *dpp* in a wedge pattern. *EGF* expression is activated where *dpp* and *wg* gradients meet, and diffuses radially. **b** Third instar antennal disc consists of *7 rings* defined by the combinations and gradients of the gene regulatory network components. The sensilla subtype identities arising from each ring is written inside the disc. The image does not incorporate *lozenge* gradient and *en* expression for simplicity. **c** and **d** Decision tree of sensilla subtype SOPs. In early larval stages prior to expression of *rn*, *Bab1*, and *ap*, listed sensilla subtype SOP fates are determined. **c** *Squares* denote decision points and precursor differentiation potentials based on sensilla type/subtype relationships obtained from our results. **d** Expression of *rn*, in late third instar antennal disc delineate novel ORN fate trajectories from the existing fates shown in **c**. This is followed by *Bab1* and *ap* expression, which yields the *7 rings* within the antennal disc **b**. The sensilla subtype SOP identities in the tree are *color-coded* based on their location on the disc. Engrailed and Lz are not represented for simplicity reasons

fates (Endo et al. 2007). Disruptions to Notch signaling during these divisions result in the duplication of a single ORN fate at the expense of the other in the same sensillum (Fig. 16.3c) (Endo et al. 2007). For example, mutations in *mastermind (mam),* a downstream effector of Notch signaling results in loss of one of two ORN classes in a sensillum (NOTCH ON ORN), and duplication of the other ORN class (NOTCH OFF ORN). On the other hand, mutations in *numb (nb),* an inhibitor of Notch signaling results in duplication of NOTCH ON ORN and loss of NOTCH OFF ORN identity.

The chromatin modifier gene *hamlet* is required to regulate the effects of the Notch signaling pathway by changing histone modifications on downstream effectors of Notch (Endo et al. 2011). *Hamlet* mutants phenocopy loss of Notch signaling, and result in duplication of one ORN fate at the expense of the other in the same sensillum. Hamlet, interacts with the transcriptional repressor C-terminal binding protein (CtBP) to suppress Notch activity by controlling the transcription of the primary target of Notch activation, *Enhancer of split m3* (*E*(*spl*)*m3*) (Endo et al. 2011). The transient expression pattern of Hamlet allows it to interfere with Notch signaling only at the final stages of SOP divisions where OR gene expression is selected by ORNs, without affecting cell fate decisions in previous divisions.

16.2 OR Gene Choice

The organization of the olfactory system and the mutant analyses performed so far suggests that OR choice and connectivity decisions are linked during development of each ORN (Larsson et al. 2004; Kurtovic et al. 2007; Elmore and Smith 2001; Dobritsa et al. 2003). Given that the olfactory receptors do not regulate connectivity in *Drosophila* and the finding that ORN connections are set prior to the onset of olfactory receptor expression suggests that the two programs are separate (Marin et al. 2005). So far, the only gene that differentially regulates olfactory receptor

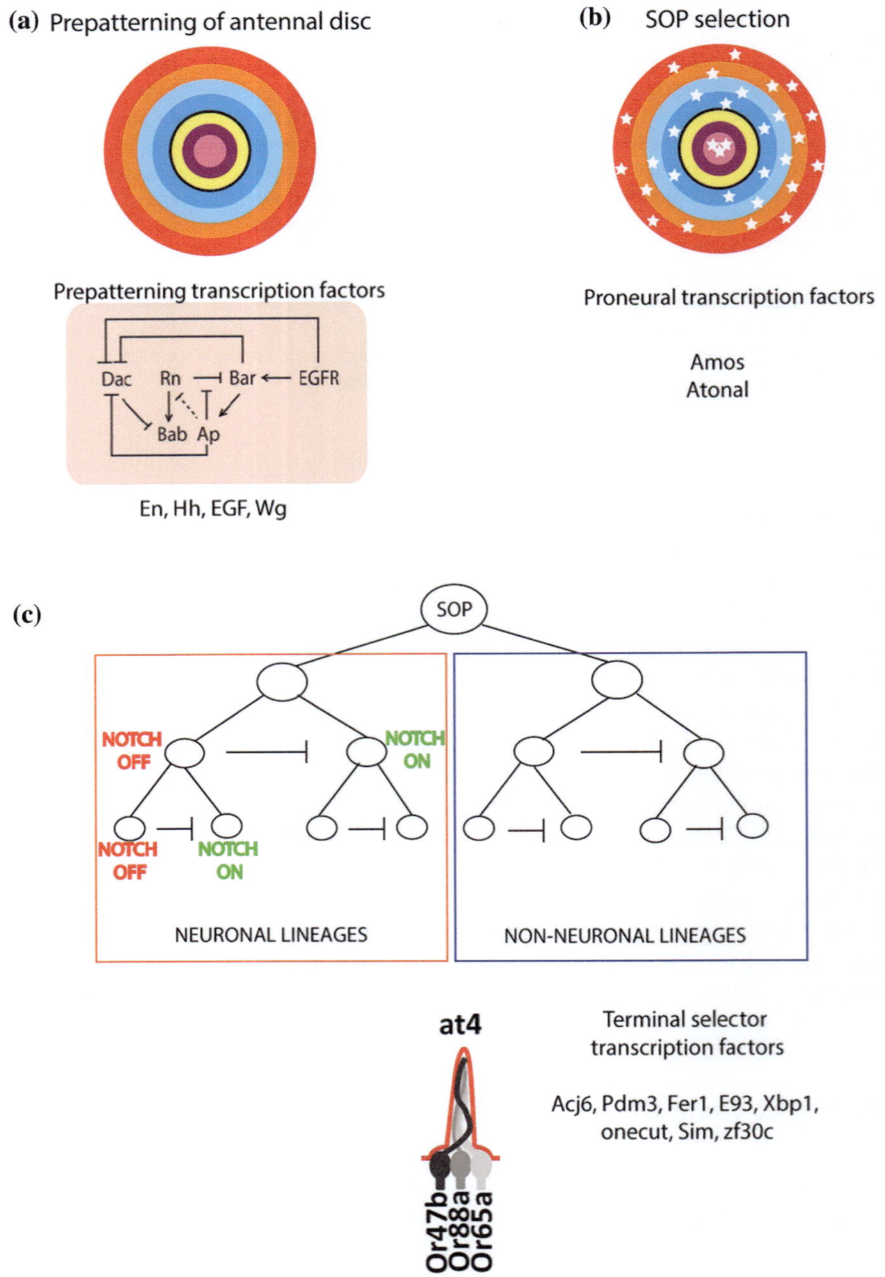
(a) Prepatterning of antennal disc
Prepatterning transcription factors
Dac
Rn
Bar
EGFR
Bab
Ap
En, Hh, EGF, Wg
(b) SOP selection
Proneural transcription factors
Amos
Atonal
(c)
SOP
NOTCH OFF
NOTCH ON
NOTCH OFF
NOTCH ON
NEURONAL LINEAGES
NON-NEURONAL LINEAGES
at4
Or47b
Or88a
Or65a
Terminal selector transcription factors
Acj6, Pdm3, Fer1, E93, Xbp1, onecut, Sim, zf30c

◂**Fig. 16.3** Three stages of ORN development: prepatterning during larval stages using transcription factor networks consisting of listed transcription factors results in distinct rings with diverse differentiation potentials **(a)**. SOPs (*stars*) are selected through expression of proneural genes **(b)**. Precursor cells undergo repeated rounds of asymmetric cell divisions, first to generate precursors for the neuronal and non-neuronal lineages in the sensillum, followed by the generation of different ORNs and the support cells **(c)**. The number of ORNs in a given sensillum relies on sensilla-specific apoptosis (Sen et al. 2004). Notch signaling is utilized at each asymmetric division for binary fate segregation, designating each cell as NOTCH ON or NOTCH OFF based on the requirement of a given fate on Notch signaling

expression among the ORNs in the same sensillum, but not the wiring programs is the chromatin modulatory protein Alhambra. In *alh* mutants, ORNs in the at4 sensilla, which normally express Or47b, Or88a, and Or65a, and connect to three glomeruli, all express OR47b without a change in general glomerular connectivity patterns with similar effects in a few other sensilla (Hueston et al. 2016). This suggests that independent molecular mechanisms govern olfactory receptor gene choice and glomerular selection programs during development.

There are a number of decisions a given ORN will make as it develops that determines which OR gene it will express. These include decisions regarding which olfactory appendage, sensilla subtype, and neuron within a subtype the ORN will form. Even though the exact molecular details of this choice for each of the 50 ORNs is not completely known, genetic studies suggest that olfactory receptor expression is not a single cellular decision made at the final stages of development among the possible olfactory receptor genes in the genome. Rather OR expression relies on multiple nested decisions that progressively restricts the OR possibilities that depends on the early lineage history of SOPs for each ORN through transcriptional and chromatin modulatory pathways (Fig. 16.2). Rn, for example, is a component of the early prepatterning gene regulatory network, and is expressed in half of all antennal sensilla subtype SOPs (Li et al. 2013). Loss of *rn* function is associated with loss of all ORNs in the *rn*-positive sensilla subtype, and an expansion of another subtype identity with its own unique ORN cluster. This is due to the conversion of *rn*-positive subtype identities at the precursor stages, as the *rn*-positive subtypes are now replaced with a single *rn*-negative subtype within each morphological type of sensilla. These conversions occur in the early precursor stages of ORN development, thus affect both sensory and wiring properties of ORNs. Similarly, in experiments modifying Notch signaling, the conversions occur not only at the level of olfactory receptor expression but also at their connectivity programs (Endo et al. 2007).

16.2.1 Decisions Regarding Sensory Appendage Selection for Olfactory Receptor Expression

Most of our understanding of decisions regarding sensory appendage expression of olfactory receptors comes from mutant analysis of factors identified through a biochemical screen for binding upstream of OR genes, and bioinformatic analysis of cis-regulatory elements upstream of olfactory receptor genes. A bioinformatic search for common elements upstream of olfactory receptors normally expressed in the maxillary palps, revealed two regulatory elements (Ray et al. 2007). One of these sequences, dyad-1, is required for the expression of these olfactory receptors in the maxillary palps, and elimination of this element in promoter reporters, results in a loss of OR-reporter expression. On the other hand, Oligo-1 element in these promoters functions to inhibit inappropriate expression of these olfactory receptors in the antenna. Elimination of Oligo-1 element from these promoters results in ectopic reporter expression in the antenna. It is plausible to imagine that tissue-specific transcription factors or chromatin modulatory proteins assemble at the sites to regulate OR expression. The identity of these factors still remains unknown.

Another regulatory insight into the olfactory appendage decisions impacting olfactory receptor expression comes from the regulation of CO_2 receptors, Gr21a and Gr63a, which are normally expressed in antennal ORNs. Interestingly, CO_2 receptors show a predisposition to be expressed in other chemosensory appendages among different insect species (Jones et al. 2007). For example, CO_2 ORNs in mosquitoes are in the maxillary palps, and mutations in *Drosophila mir-279* and *prospero* result in flies with ectopic CO_2 ORNs in the maxillary palps (Cayirlioglu et al. 2008; Hartl et al. 2011). The effect of these two genes on CO_2 receptor gene regulation is indirect, as both *mir-279* and *prospero* are expressed in maxillary palp precursors, and function to regulate the number of ORNs as the ectopic CO_2 ORNs are generated as extra neurons in maxillary palp sensilla.

Direct regulation and restriction of CO_2 receptors to antenna was shown to be under chromatin regulation. The Ray group has identified the DREAM complex, consisting of Myb, Mip120, Mip130, E2f2, and Rbf proteins that regulate the chromatin around CO_2 receptors that both facilitate its expression in specific antennal ORNs, and inhibition of their expression in other chemosensory appendages or the rest of the adult nervous system (Sim et al. 2012). Among these 5 components of the DREAM complex, E2f2 and Mip120, are required to repress inappropriate CO_2 receptor expression in other antennal ORNs or other chemosensory appendages, such as the second antennal segment, maxillary, and labial palps. On the other hand, Myb and Mip130 work together to facilitate expression of CO_2 receptors in the antenna. Assembly of DREAM complex components upstream of CO_2 receptor genes modifies chromatin marks and restricts their expression to antennal ORNs.

16.2.2 Neuron-Specific Regulation of Olfactory Receptor Expression

So far, only a few transcription factors were shown to directly regulate olfactory receptor expression. Lz, as one (Bai and Carlson 2010) of the prepatterning genes expressed early in the antennal disc, has binding sites upstream of olfactory receptor genes, yet direct binding of Lz protein to these sites has not been demonstrated (Ray et al. 2007). Other than Lz, two POU domain transcription factors, Acj6 and Pdm3, were shown to regulate olfactory receptor expression. Acj6, was shown by multiple groups to regulate gene expression of many antennal and maxillary palp ORs, and can directly interact with some OR promoters. Multiple splice isoforms of Acj6 exist, each with distinct functions to specify maxillary palp OR expression (Ayer and Carlson 1992). In addition to regulating olfactory receptor expression (Bai and Carlson 2010; Ayer and Carlson 1992; Bai et al. 2009; Jafari et al. 2012), Acj6 also regulates connectivity of some but not all ORN classes (Komiyama et al. 2004). It is possible that Acj6 regulates the expression of both olfactory receptors and cell surface proteins governing guidance and connectivity decisions of ORN axons. Pdm3, also regulates a small number of maxillary palp olfactory receptors—again with small effects on connectivity (Tichy et al. 2008). POU binding sites found upstream of olfactory receptors can provide regulatory functions to not only Acj6 but also Pdm3, supporting the combinatorial nature of olfactory receptor expression.

A systematic RNAi screen in postmitotic ORNs also identified six additional factors, Fer1, Xbp1, E93, onecut, sim, and zf30c as regulators of approximately 30 olfactory receptors (Jafari et al. 2012). Some of these transcription factors function as positive regulators of olfactory receptor expression in the antenna, while others work through repressing expression in inappropriate ORNs. This study provided an additional code for the combinations of transcription factors that are superimposed onto the prepatterning network to regulate olfactory receptor expression late in ORN development. In addition, chromatin modulators, such as Atrophin, work together with these factors for restricting olfactory receptor expression to specific sensilla subtypes (Alkhori et al. 2014). In *atrophin* mutants, in which histone H3 acetylation is misregulated, a number of olfactory receptors expand their expression to other ORNs in the antenna, however, further work is required to understand how Atrophin fits into the developmental progression of the chromatin and transcriptional networks driving specification of each ORN.

16.2.3 Cis-Regulatory Elements Upstream of Olfactory Receptor Genes

Cis-regulatory elements upstream of olfactory receptor gene promoters were analyzed for specific olfactory receptor genes to look for common elements that share developmental or lineage history that contributes to sensilla type/subtype/ORN-specific regulation. A promoter deletion analysis identified important regulatory elements upstream of olfactory receptors expressed in trichoid ORNs (Miller and Carlson 2010). Similar to the examples above, this study revealed cis-regulatory elements that are either required for the expression of trichoid OR genes, as well as repressor elements that inhibit ectopic expression not only in inappropriate ORNs, but also in non-neuronal cells of the same trichoid sensilla subtype.

Another study investigated POU domain protein binding sites upstream of two olfactory receptors and found that the spacing between the clusters of binding sites, as well as the chromatin context as important determinants of olfactory receptor gene expression (Jafari and Alenius 2015). Perturbations to the spacing and chromatin context of the clusters resulted in ectopic olfactory receptor expression and revealed constraints on olfactory receptor gene regulation that can be only fulfilled in a small number of neurons. However, almost all ORs have binding sites for these transcription factors yet are unaffected in *acj6* mutants, suggesting that transcription factors act in concert with the developmental context and history of ORNs, likely to be influenced by heritable differences in chromatin for different SOP populations that generate each ORN combination (Barish and Volkan 2015).

16.3 ORN Connectivity

Unlike spatial information, which is continuous, olfactory information is discrete rather due to the nature of the chemical space. This calls for a discrete odortopic map in the brain that encodes information about the complexity of odor stimuli. One organizational solution to simplify this task is forming circuits where ORNs detecting the same odor converges their axons to the same region in the brain to make connections with a few second order neurons. In *Drosophila*, ORNs expressing the same OR gene are dispersed through a broad region of the primary olfactory epithelium, converge onto a distinct and class-specific “glomerulus” in the antennal lobe to synapse with a small number of postsynaptic projection neurons (PNs). This organizational logic is also conserved in many species including mammals, yet these two organisms utilize different molecular and developmental mechanisms to arrive at this organizational configuration (Takeuchi and Sakano 2014).

Mice use ligand-independent olfactory receptor signaling during development for guidance and connectivity of ORNs. Unlike in mammals, *Drosophila* OR genes are not required for targeting of ORNs, as mutants in many OR genes were reported with no phenotypes in connectivity (Larsson et al. 2004; Kurtovic et al. 2007; Elmore and Smith 2001; Dobritsa et al. 2003). Flies that lack *orco*, the co-receptor essential for proper OR localization and function in ORNs, have normal ORN targeting (Larsson et al. 2004). Similarly, deletions of *Or22a/Or22b,* normally expressed by ORNs in ab3 sensilla, show no defects in Or22a ORN connectivity (Dobritsa et al. 2003). In addition, ectopic expression of another OR gene in these "empty" Or22a ORNs, does not reroute their axons to new glomerular locations. These results suggest that functional properties of the ORs are not required for appropriate axon targeting, and that programs for olfactory receptor expression and connectivity are independent, yet are linked at specific points in ORN development.

Drosophila ORN targeting, instead, seem to rely on a gradual restriction of target possibilities, much like programs that refine the OR choice itself. Early patterning cues coarsely sort different ORN axonal projections, which arrive at the developing antennal lobe around 16–18 h APF, to specific regions (Marin et al. 2005). Secondary cues then drive homophilic and repulsive interactions between the same classes of ORNs to drive fine-scale guidance. Finally, synapse—promoting factors ensure that the appropriate connections are made with class-specific projection neurons (PNs).

16.3.1 Early Patterning of the Future Antennal Lobe and ORN Axon Guidance

Prior to the arrival of adult ORNs developing in the antennal disc during early stages of metamorphosis, early born projection neuron dendrites occupy distinct zones in the future antennal lobe and pattern the tissue by laying down gradients of guidance proteins such as Sema-2a and Sema-2b. Semaphorin signaling among ORNs as well as between ORNs and PNs through cis-interactions or interactions of Sema proteins with their receptor PlexinB attract ORN fibers arriving through the antennal nerve to specific routes (Sweeney et al. 2011). ORNs originating from the same sensillum, send their axons through the antennal nerve to the lobe, and upon entering the antennal lobe segregate into two trajectories, ventromedial and dorsolateral (Joo et al. 2013; Jhaveri and Rodrigues 2002; Jhaveri et al. 2000) (Fig. 16.4). Notch-dependent and differential inhibition of *sema-2b,* restricts its expression to ventromedial ORNs (Notch OFF) and mediates the segregation and consolidation of the axon projections from dorsolateral ORN fibers (Notch ON). This phenomenon is reminiscent of the binary OR gene choice decisions mediated by Notch. In fact, perturbing Notch signaling shows that it not only segregates OR gene choice fates in a binary fashion but also affects axonal sorting of ORNs from the same sensillum. This occurs through the negative regulation of Sema-2b by Notch (Joo et al. 2013).

Mutants of Robo and Dscam (Down Syndrome Cell Adhesion Molecule) cause targeting defects during this process (Jhaveri et al. 2004; Hummel et al. 2003). Dscam is a repulsive hemophilic adhesion protein with forty thousand different variants that can be produced through alternative splicing (Schmucker et al. 2000; Zipursky and Sanes 2010; Zipursky et al. 2006; Hattori et al. 2008). This suggests a mechanism by which each set of ORN axons expressing the same OR gene would all express the same Dscam isoform and thus have a unique identifier (Hattori et al. 2007). However, *dscam* is required for correct targeting in an isoform

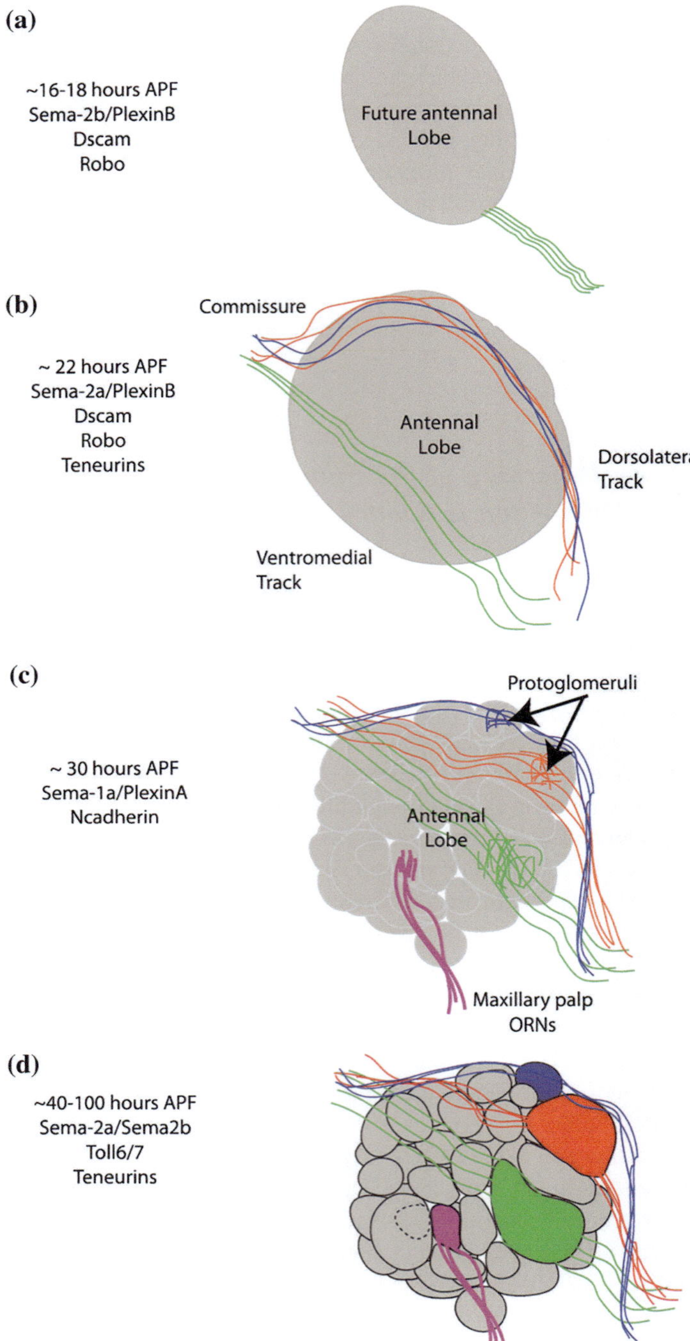
(a)
~16-18 hours APF
Sema-2b/PlexinB
Dscam
Robo
Future antennal Lobe
(b)
~ 22 hours APF
Sema-2a/PlexinB
Dscam
Robo
Teneurins
Commissure
Antennal Lobe
Dorsolateral Track
Ventromedial Track
(c)
~ 30 hours APF
Sema-1a/PlexinA
Ncadherin
Protoglomeruli
Antennal Lobe
Maxillary palp ORNs
(d)
~40-100 hours APF
Sema-2a/Sema2b
Toll6/7
Teneurins

◂**Fig. 16.4** Regulation of ORN connectivity. **a** ORN axons fasciculated within the antennal nerve reach the future antennal lobe around 16–18 h after puparium formation (*APF*). **b** Around 22 h, ORN axons segregate into two nerve tracks (ventromedial and dorsolateral) followed by sending out contralateral projections through the commissure. The stages in **a** and **b** are mostly governed by the function of Dscam, Robo as well as the Sema-2a/Sema-2b/PlexinB signaling between ORNs and the resident PNs. **c** At 30 h APF, ORNs start forming class-specific protoglomeruli, mostly driven by the hemophilic adhesive properties of Ncadherin, as well as the repulsive interactions between axons of different ORN classes mediated by Sema1a-PlexinA signaling. **d** By 40 h APF, which marks the onset of olfactory receptor gene expression in some ORNs, antennal lobe glomeruli start forming for some ORN classes, and ORN-PN matching induces formation of ORN class-specific synapses with distinct PN populations. Teneurins, Toll proteins, Sema2a/2b, and Dscam function to drive ORN-PN matching at these stages

diversity-independent manner (Hummel et al. 2003). The role of the different isoforms of *dscam* appears to be in the final convergence of ORNs into glomeruli in late targeting stages (Hattori et al. 2007, 2009).

16.3.2 Convergence and Maturation of Glomerular Connections

A critical process that occurs after groups of ORN axons reach distinct zones in the antennal lobe is selection of appropriate target projection neurons to make synapses within class-specific glomeruli. Class-specific and non-overlapping glomeruli formation requires both attraction between ORN terminals of the same class and repulsion between ORNs of different classes. Once ORN axons reach appropriate locations around 25–30 h APF in both ipsilateral and contralateral antennal lobes, they aggregate within protoglomeruli prior to establishing synapses with PNs (Fig. 16.4). This time also coincides with the arrival of ORN projections from the maxillary palps, guided to their appropriate target sites by Sema-1a signaling and the antennal ORNs already in the antennal lobe (Komiyama et al. 2007). At around approximately 40 h APF, around the onset of select olfactory receptor gene expression [i.e., Or47b (Hueston et al. 2016)], ORN axonal terminals start arborizing and synapsing with projection neurons as they define glomerular boundaries. During this process Ncadherin, a well-known homophilic adhesion molecule, acts as an ORN class-specific attractive cue within each ORN class. In *Ncadherin* mutants ORNs target to the appropriate regions of the antennal lobe but fail to form protoglomeruli and appropriate synapses within glomeruli (Hummel and Zipursky 2004). In contrast, interclass repulsive interactions among different ORN axons appear to be based on semaphorins. Sema1a expression in ORNs mediates repulsion between different ORN classes (Sweeney et al. 2007; Lattemann et al. 2007). For example, two ORN classes that target neighboring glomeruli, fail to segregate into distinct glomeruli in *sema-1a* mutants, and show an intermingling of axonal processes within a single glomerulus (Lattemann et al. 2007). Lack of axon–axon repulsion via Dscam isoforms also contributes to glomerular convergence. Dscam diversity is necessary for this process, as *dscam* mutant alleles, which encode only a single isoform, show severe loss of glomerular boundaries and ORN convergence onto multiple regions within the antennal lobe (Hattori et al. 2007).

16.3.3 ORN-PN Matching During Synapse Formation

Once ORN axons reach the target regions within the antennal lobe, how do they find the appropriate PN dendrites to initiate class-specific synapses? Given the large number of Dscam isoforms, the initial hypothesis was that combinations of Dscam isoforms in ORNs and PNs result in homophilic adhesion and regulate connection specificity. This hypothesis was swiftly eliminated once Dscam isoform-specific hemophilic adhesion was shown to lead to repulsion, and with the results showing overexpression of Dscam isoforms in projection neurons shifts the position of the glomerulus but not the ORN-PN matching (Zhan et al. 2004; Matthews et al. 2007; Zhu et al. 2006).

So far, two major cell surface molecule families have been shown to contribute to class-specific synaptic matching between ORNs and PNs. Two Teneurin proteins, Teneurin-m and Tenurin-a, encode cell surface proteins with hemophilic adhesive properties, and have been identified as synaptic matching molecules (Hong et al. 2012; Mosca and Luo 2014). ORN and PN classes express different levels of teneurins, and ORN-PN matching occurs once both express matching levels of either Teneurin-m or Tenurin-a. Genetic modulations that change the expression levels of these transmembrane proteins cause ORN-PN mismatching. However, the expression of Teneurin-m and Teneurin-a does overlap in some ORNs and PNs, indicating that combinatorial action of other factors together with Teneurins are required for precision of ORN-PN target matching.

Recently, members of the Toll family of receptors, long known for their involvement in the immune system and in embryonic development, have also been implicated in this process (Ward et al. 2015). Toll proteins encode cell surface molecules with leucine rich repeat domains (LRR) that mediate protein–protein interactions. Toll-7 is required for the correct targeting of ORNs in the anterolateral antennal lobe, while Toll-6 is required for their corresponding PNs. Interestingly, these effects do not require any of the downstream components of the Toll signaling pathway or even the cytoplasmic domains of the Toll receptors (Ward et al. 2015).

Despite the identification of cell surface molecules involved in ORN axon targeting in the *Drosophila* olfactory system, it is still unclear how axon targeting combines with the fate decisions that occur during SOP division to generate unique targeting locations for each ORN class. This is likely due to the combinatorial nature of the molecular regulation underlying ORN connection specificity. Although olfactory receptor genes are not required for appropriate targeting, the fact remains that the progressive restrictions of precursor fates that result in gene choice in an ORN also influence its possible axonal targets in the antennal lobe. Currently, it is unclear what mechanism links these two processes. Systems level developmental analysis of transcriptional programs from individual precursors and the ORNs they generate will likely help answer these questions.

References

Abuin L, Bargeton B, Ulbrich MH, Isacoff EY, Kellenberger S, Benton R (2011) Functional architecture of olfactory ionotropic glutamate receptors. Neuron 69(1):44–60

Ai M, Min S, Grosjean Y, Leblanc C, Bell R, Benton R et al (2010) Acid sensing by the Drosophila olfactory system. Nature 468(7324):691–695

Alkhori L, Ost A, Alenius M (2014) The corepressor Atrophin specifies odorant receptor expression in Drosophila. FASEB J. 28(3):1355–1364

Ayer RK Jr, Carlson J (1992) Olfactory physiology in the Drosophila antenna and maxillary palp: acj6 distinguishes two classes of odorant pathways. J Neurobiol 23(8):965–982

Bai L, Carlson JR (2010) Distinct functions of acj6 splice forms in odor receptor gene choice. J Neurosci 30(14):5028–5036

Bai L, Goldman AL, Carlson JR (2009) Positive and negative regulation of odor receptor gene choice in Drosophila by acj6. J Neurosci 29(41):12940–12947

Barish S, Volkan PC (2015) Mechanisms of olfactory receptor neuron specification in Drosophila. Wiley Interdiscip Rev Dev Biol

Benton R, Sachse S, Michnick SW, Vosshall LB (2006) Atypical membrane topology and heteromeric function of Drosophila odorant receptors in vivo. PLoS Biol 4(2):e20

Benton R, Vannice KS, Gomez-Diaz C, Vosshall LB (2009) Variant ionotropic glutamate receptors as chemosensory receptors in Drosophila. Cell 136(1):149–162

Cayirlioglu P, Kadow IG, Zhan X, Okamura K, Suh GS, Gunning D et al (2008) Hybrid neurons in a microRNA mutant are putative evolutionary intermediates in insect CO_2 sensory systems. Science 319(5867):1256–1260

Chou YH, Zheng X, Beachy PA, Luo L (2010) Patterning axon targeting of olfactory receptor neurons by coupled hedgehog signaling at two distinct steps. Cell 142(6):954–966

Clyne PJ, Warr CG, Freeman MR, Lessing D, Kim J, Carlson JR (1999) A novel family of divergent seven-transmembrane proteins: candidate odorant receptors in Drosophila. Neuron 22(2):327–338

Couto A, Alenius M, Dickson BJ (2005) Molecular, anatomical, and functional organization of the Drosophila olfactory system. Curr Biol 15(17):1535–1547

Dobritsa AA, van der Goes van Naters W, Warr CG, Steinbrecht RA, Carlson JR (2003) Integrating the molecular and cellular basis of odor coding in the Drosophila antenna. Neuron 37(5):827–41

Elmore T, Smith DP (2001) Putative Drosophila odor receptor OR43b localizes to dendrites of olfactory neurons. Insect Biochem Mol Biol 31(8):791–798

Endo K, Aoki T, Yoda Y, Kimura K, Hama C (2007) Notch signal organizes the Drosophila olfactory circuitry by diversifying the sensory neuronal lineages. Nat Neurosci 10(2):153–160

Endo K, Karim MR, Taniguchi H, Krejci A, Kinameri E, Siebert M et al (2011) Chromatin modification of Notch targets in olfactory receptor neuron diversification. Nat Neurosci 15(2):224–233

Enjin A, Zaharieva EE, Frank DD, Mansourian S, Suh GS, Gallio M et al (2016) Humidity Sensing in Drosophila. Curr Biol 26(10):1352–1358

Fishilevich E, Vosshall LB (2005) Genetic and functional subdivision of the Drosophila antennal lobe. Curr Biol 15(17):1548–1553

Goulding SE, zur Lage P, Jarman AP (2000) Amos, a proneural gene for Drosophila olfactory sense organs that is regulated by lozenge. Neuron 25(1):69–78

Gupta BP, Rodrigues V (1997) Atonal is a proneural gene for a subset of olfactory sense organs in Drosophila. Genes Cells 2(3):225–233

Gupta BP, Flores GV, Banerjee U, Rodrigues V (1998) Patterning an epidermal field: Drosophila lozenge, a member of the AML-1/Runt family of transcription factors, specifies olfactory sense organ type in a dose-dependent manner. Dev Biol 203(2):400–411

Hartl M, Loschek LF, Stephan D, Siju KP, Knappmeyer C, Kadow IC (2011) A new Prospero and microRNA-279 pathway restricts CO_2 receptor neuron formation. J Neurosci 31(44):15660–15673

Hattori D, Demir E, Kim HW, Viragh E, Zipursky SL, Dickson BJ (2007) Dscam diversity is essential for neuronal wiring and self-recognition. Nature 449(7159):223–227

Hattori D, Millard SS, Wojtowicz WM, Zipursky SL (2008) Dscam-mediated cell recognition regulates neural circuit formation. Annu Rev Cell Dev Biol 24:597–620

Hattori D, Chen Y, Matthews BJ, Salwinski L, Sabatti C, Grueber WB et al (2009) Robust discrimination between self and non-self neurites requires thousands of Dscam1 isoforms. Nature 461(7264):644–648

Henikoff S (2015) Epigenomic landscapes reflect neuronal diversity. Neuron 86(6):1319–1321

Hong W, Mosca TJ, Luo L (2012) Teneurins instruct synaptic partner matching in an olfactory map. Nature 484(7393):201–207

Hueston CE, Olsen D, Li Q, Okuwa S, Peng B, Wu J et al (2016) Chromatin modulatory proteins and olfactory receptor signaling in the refinement and maintenance of fruitless expression in olfactory receptor neurons. PLoS Biol 14(4):e1002443

Hummel T, Zipursky SL (2004) Afferent induction of olfactory glomeruli requires N-cadherin. Neuron 42(1):77–88

Hummel T, Vasconcelos ML, Clemens JC, Fishilevich Y, Vosshall LB, Zipursky SL (2003) Axonal targeting of olfactory receptor neurons in Drosophila is controlled by Dscam. Neuron 37(2):221–231

Jafari S, Alenius M (2015) Cis-regulatory mechanisms for robust olfactory sensory neuron class-restricted odorant receptor gene expression in Drosophila. PLoS Genet 11(3):e1005051

Jafari S, Alkhori L, Schleiffer A, Brochtrup A, Hummel T, Alenius M (2012) Combinatorial activation and repression by seven transcription factors specify Drosophila odorant receptor expression. PLoS Biol 10(3):e1001280

Jhaveri D, Rodrigues V (2002) Sensory neurons of the Atonal lineage pioneer the formation of glomeruli within the adult Drosophila olfactory lobe. Development 129(5):1251–1260

Jhaveri D, Sen A, Reddy GV, Rodrigues V (2000a) Sense organ identity in the Drosophila antenna is specified by the expression of the proneural gene atonal. Mech Dev 99(1–2):101–111

Jhaveri D, Sen A, Rodrigues V (2000b) Mechanisms underlying olfactory neuronal connectivity in Drosophila-the atonal lineage organizes the periphery while sensory neurons and glia pattern the olfactory lobe. Dev Biol 226(1):73–87

Jhaveri D, Saharan S, Sen A, Rodrigues V (2004) Positioning sensory terminals in the olfactory lobe of Drosophila by Robo signaling. Development 131(9):1903–1912

Jones WD, Cayirlioglu P, Kadow IG, Vosshall LB (2007) Two chemosensory receptors together mediate carbon dioxide detection in Drosophila. Nature 445(7123):86–90

Joo WJ, Sweeney LB, Liang L, Luo L (2013) Linking cell fate, trajectory choice, and target selection: genetic analysis of Sema-2b in olfactory axon targeting. Neuron 78(4):673–686

Komiyama T, Carlson JR, Luo L (2004) Olfactory receptor neuron axon targeting: intrinsic transcriptional control and hierarchical interactions. Nat Neurosci 7(8):819–825

Komiyama T, Sweeney LB, Schuldiner O, Garcia KC, Luo L (2007) Graded expression of semaphorin-1a cell-autonomously directs dendritic targeting of olfactory projection neurons. Cell 128(2):399–410

Kurtovic A, Widmer A, Dickson BJ (2007) A single class of olfactory neurons mediates behavioural responses to a Drosophila sex pheromone. Nature 446(7135):542–546

Larsson MC, Domingos AI, Jones WD, Chiappe ME, Amrein H, Vosshall LB (2004) Or83b encodes a broadly expressed odorant receptor essential for Drosophila olfaction. Neuron 43(5):703–714

Lattemann M, Zierau A, Schulte C, Seidl S, Kuhlmann B, Hummel T (2007) Semaphorin-1a controls receptor neuron-specific axonal convergence in the primary olfactory center of Drosophila. Neuron 53(2):169–184
Li Q, Ha TS, Okuwa S, Wang Y, Wang Q, Millard SS et al (2013) Combinatorial rules of precursor specification underlying olfactory neuron diversity. Curr Biol (CB) 23(24):2481–2490
Li Q, Barish S, Okuwa S, Maciejewski A, Brandt AT, Reinhold D et al (2016) A functionally conserved gene regulatory network module governing olfactory neuron diversity. PLoS Genet 12(1):e1005780
Marin EC, Watts RJ, Tanaka NK, Ito K, Luo L (2005) Developmentally programmed remodeling of the Drosophila olfactory circuit. Development 132(4):725–737
Matthews BJ, Kim ME, Flanagan JJ, Hattori D, Clemens JC, Zipursky SL et al (2007) Dendrite self-avoidance is controlled by Dscam. Cell 129(3):593–604
Miller CJ, Carlson JR (2010) Regulation of odor receptor genes in trichoid sensilla of the Drosophila antenna. Genetics 186(1):79–95
Mo A, Mukamel EA, Davis FP, Luo C, Henry GL, Picard S et al (2015) Epigenomic signatures of neuronal diversity in the mammalian brain. Neuron 86(6):1369–1384
Morata G (2001) How Drosophila appendages develop. Nat Rev Mol Cell Biol 2(2):89–97
Mosca TJ, Luo L (2014) Synaptic organization of the Drosophila antennal lobe and its regulation by the Teneurins. eLIFE 3:e03726
Ray A, van Naters WG, Shiraiwa T, Carlson JR (2007) Mechanisms of odor receptor gene choice in Drosophila. Neuron 53(3):353–369
Royet J, Finkelstein R (1997) Establishing primordia in the Drosophila eye-antennal imaginal disc: the roles of decapentaplegic, wingless and hedgehog. Development 124(23):4793–4800
Sakano H (2010) Neural map formation in the mouse olfactory system. Neuron 67(4):530–542
Sato K, Pellegrino M, Nakagawa T, Vosshall LB, Touhara K (2008) Insect olfactory receptors are heteromeric ligand-gated ion channels. Nature 452(7190):1002–1006
Schmucker D, Clemens JC, Shu H, Worby CA, Xiao J, Muda M et al (2000) Drosophila Dscam is an axon guidance receptor exhibiting extraordinary molecular diversity. Cell 101(6):671–684
Sen A, Reddy GV, Rodrigues V (2003) Combinatorial expression of prospero, seven-up, and Elav identifies progenitor cell types during sense-organ differentiation in the Drosophila antenna. Dev Biol 254(1):79–92
Sen A, Kuruvilla D, Pinto L, Sarin A, Rodrigues V (2004) Programmed cell death and context dependent activation of the EGF pathway regulate gliogenesis in the Drosophila olfactory system. Mech Dev 121(1):65–78
Silbering AF, Rytz R, Grosjean Y, Abuin L, Ramdya P, Jefferis GS et al (2011) Complementary function and integrated wiring of the evolutionarily distinct Drosophila olfactory subsystems. J Neurosci 31(38):13357–13375
Sim CK, Perry S, Tharadra SK, Lipsick JS, Ray A (2012) Epigenetic regulation of olfactory receptor gene expression by the Myb-MuvB/dREAM complex. Genes Dev 26(22):2483–2498
Song E, de Bivort B, Dan C, Kunes S (2012) Determinants of the Drosophila odorant receptor pattern. Dev Cell 22(2):363–376
Stocker RF (1994) The organization of the chemosensory system in *Drosophila melanogaster*: a review. Cell Tissue Res 275(1):3–26
Sweeney LB, Couto A, Chou YH, Berdnik D, Dickson BJ, Luo L et al (2007) Temporal target restriction of olfactory receptor neurons by Semaphorin-1a/PlexinA-mediated axon-axon interactions. Neuron 53(2):185–200
Sweeney LB, Chou YH, Wu Z, Joo W, Komiyama T, Potter CJ et al (2011) Secreted semaphorins from degenerating larval ORN axons direct adult projection neuron dendrite targeting. Neuron 72(5):734–747
Takeuchi H, Sakano H (2014) Neural map formation in the mouse olfactory system. Cell Mol Life Sci 71(16):3049–3057
Tichy AL, Ray A, Carlson JR (2008) A new Drosophila POU gene, pdm3, acts in odor receptor expression and axon targeting of olfactory neurons. J Neurosci 28(28):7121–7129
Van Bortle K, Corces VG (2013) Spinning the web of cell fate. Cell 152(6):1213–1217

Vosshall LB (2004) Olfaction: attracting both sperm and the nose. Curr Biol (CB) 14(21): R918–R920

Vosshall LB, Amrein H, Morozov PS, Rzhetsky A, Axel R (1999) A spatial map of olfactory receptor expression in the Drosophila antenna. Cell 96(5):725–736

Vosshall LB, Wong AM, Axel R (2000) An olfactory sensory map in the fly brain. Cell 102(2):147–159

Ward A, Hong W, Favaloro V, Luo L (2015) Toll receptors instruct axon and dendrite targeting and participate in synaptic partner matching in a Drosophila olfactory circuit. Neuron 85(5):1013–1028

Wicher D, Schafer R, Bauernfeind R, Stensmyr MC, Heller R, Heinemann SH et al (2008) Drosophila odorant receptors are both ligand-gated and cyclic-nucleotide-activated cation channels. Nature 452(7190):1007–1011

Zhan XL, Clemens JC, Neves G, Hattori D, Flanagan JJ, Hummel T et al (2004) Analysis of Dscam diversity in regulating axon guidance in Drosophila mushroom bodies. Neuron 43(5):673–686

Zhu H, Hummel T, Clemens JC, Berdnik D, Zipursky SL, Luo L (2006) Dendritic patterning by Dscam and synaptic partner matching in the Drosophila antennal lobe. Nat Neurosci 9(3):349–355

Zipursky SL, Sanes JR (2010) Chemoaffinity revisited: dscams, protocadherins, and neural circuit assembly. Cell 143(3):343–353

Zipursky SL, Wojtowicz WM, Hattori D (2006) Got diversity? Wiring the fly brain with Dscam. Trends Biochem Sci 31(10):581–588

Zur Lage PI, Prentice DR, Holohan EE, Jarman AP (2003) The Drosophila proneural gene amos promotes olfactory sensillum formation and suppresses bristle formation. Development 130(19):4683–4693

Chapter 17
The Developmental Origin of Cell Type Diversity in the *Drosophila* Visual System

Claire Bertet

Abstract The *Drosophila* visual system is composed of complex neural circuits involving an incredible variety of neurons, making it an excellent model system to study how cell type diversity is generated during development. Recent studies using new genetic tools and cellular markers have shown that this diversity is generated by at least four neurogenesis modes involving four different types of progenitors localized in distinct regions of the developing optic lobes. In this chapter, I will first describe the anatomical organization of the visual system and then review the different neurogenesis modes generating cell diversity in the four optic lobe neuropils.

17.1 Introduction

Understanding how embryonic progenitors generate the vast diversity of neural cells of the adult nervous system is a central question in neurobiology. The *Drosophila* visual system, which contains around 120,000 neurons comprising more than 110 subtypes (Hofbauer and Campos-Ortega 1990; Meinertzhagen 2014), is an excellent model to study this question. In this system, visual information is detected in the retina by photoreceptors, specialized neurons containing light-sensing Rhodopsin proteins. Photoreceptor signals are then transmitted to the optic lobes where they are processed (Behnia and Desplan 2015; Borst and Helmstaedter 2015) (Fig. 17.1). *Drosophila* optic lobes are divided into four neuropils, which together contain more than 60% of brain neurons. The overall organization of these neuropils and the vast diversity of neuronal subtypes forming them have been described in exquisite detail more than 25 years ago, first by Cajal and Sanchez and later by Fishbach and Dittrich (Fischbach 1989). However, until recently, only the very early stages of optic lobe formation were known and the

C. Bertet (✉)
Aix Marseille Univ, CNRS, IBDM, Marseille, France
e-mail: claire.BERTET@univ-amu.fr

A. Çelik and M.F. Wernet (eds.), *Decoding Neural Circuit Structure and Function*, DOI 10.1007/978-3-319-57363-2_17

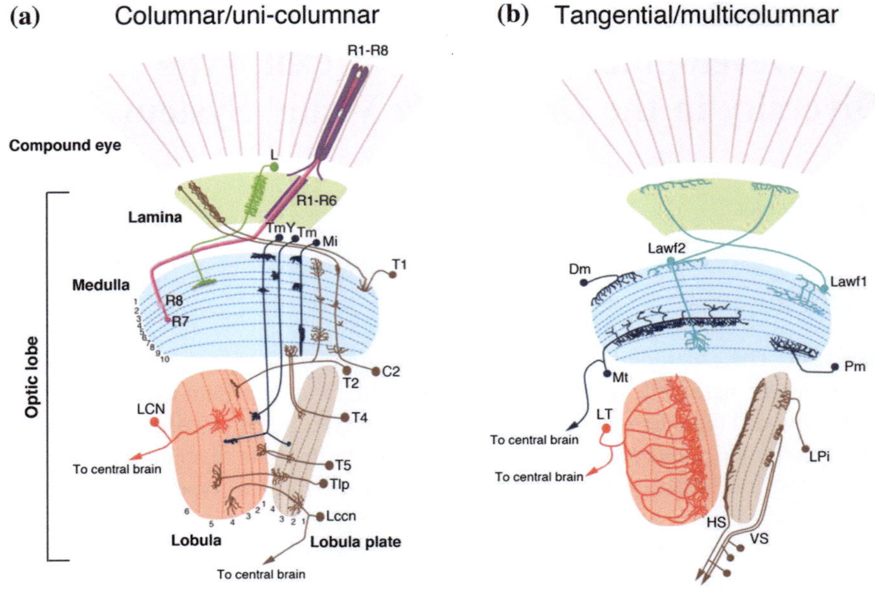

Fig. 17.1 The *Drosophila* visual system. **a** The compound eye transmits visual information to the optic lobes. The optic lobes comprise four neuropils: the lamina (*green*), medulla (*blue*), lobula (*red*), and lobula plate (*brown*). R1–R6 photoreceptors (*purple*) project into the lamina while R7 and R8 (*pink*) project into the medulla neuropil. Uni-columnar neurons of the optic lobes: Lamina neurons (*L*, *green*), T cells (*T1*, *brown*), C cells (*C2*, *brown*), Medulla intrinsic neurons (*Mi*, *blue*), Transmedullary neurons (*Tm*, *blue*), Transmedullary Y neurons (*TmY*, *blue*), Lobula Columnar neurons (*LCN*, *red*), Translobula plate neurons (*Tlp*, *brown*), T4 and T5 neurons (*brown*), and Lobula complex columnar neurons (*Lccn*, *brown*). **b** Tangential and multi-columnar neurons of the optic lobes: Lamina wide field neurons (*Lawf1* and *Lawf2*, *cyan*), Distal medulla neurons (*Dm*, *blue*), Proximal medulla neurons (*Pm*, *blue*), Medulla tangential neurons (*Mt*, *blue*), Lobula Tangential neurons (*LT*, *red*), Lobula plate tangential neurons (*HS* and *VS*, *brown*) and Lobula Plate intrinsic neurons (*LPi*). *Numbers* indicate layers of the different neuropils. Adapted from Apitz and Salecker (2014) and Neriec and Desplan (2016)

mechanisms generating the tremendous cell diversity in these structures remained elusive (Hofbauer and Campos-Ortega 1990). A series of new studies has addressed this problem and has uncovered at least four different modes of neurogenesis involving four different types of progenitors generating this cell diversity. In this chapter, I first describe the anatomical organization of the *Drosophila* visual system and outline the different neuronal subtypes found in each optic lobe neuropil. I then describe the different strategies by which these neurons are generated and specified during development.

17.2 Organization of the *Drosophila* Visual System

The *Drosophila* compound eye is made of 800 independent units called ommatidia. Each ommatidium looks in a different point in space and contains eight photoreceptor cells projecting to the optic lobes (Kumar 2012) (Fig. 17.1). Outer R1–R6 photoreceptors are involved in motion and dim light vision whereas inner R7 and R8 photoreceptors are sensory receptors for color and polarized light vision (Heisenberg 1977; Yamaguchi et al. 2008). *Drosophila* optic lobes are divided in 4 ganglions/neuropils—the lamina, medulla, lobula and lobula plate—that all have a columnar organization (Fig. 17.1). The number of columns in each neuropil ($\sim$800) corresponds to the number of ommatidia in the compound eye. Retinotopic connections between photoreceptors and neuropil columns ensure a coherent representation of the visual world (Fischbach 1989; Meinertzhagen and Sorra 2001; Zhu 2013; Morante and Desplan 2004).

The first ganglion, the lamina, consists of around 4000 neurons. Its neuropil is innervated by outer photoreceptors (R1–R6) and is involved in the early stages of motion computation (Douglass and Strausfeld 1995). The lamina neuropil is organized into an array of 800 retinotopic columns named cartridges, each of which corresponds to one pixel in the visual field (Meinertzhagen and Sorra 2001). There are 12 subtypes of lamina neurons that can be divided in two populations (Fischbach 1989; Tuthill et al. 2013) (Fig. 17.1): uni-columnar neurons contacting a single cartridge (Lamina monopolar cells L1–L5, T1, C2, and C3 cells) and multi-columnar neurons contacting several cartridges [Lamina intrinsic amacrine neurons (Lai, not shown), Lamina wide-field neurons (Lawf1, Lawf2) and lamina tangential neurons (Lat, not shown)]. Only L1–L5 neurons have their cell bodies in the lamina and are formed in response to induction by signals from photoreceptor axons (see below).

The second ganglion, the medulla, is innervated by inner photoreceptors (R7 and R8) and by L1–L5 lamina monopolar neurons. It processes both motion and color vision (Morante and Desplan 2008). The medulla has the largest and most complex neuropil. It alone contains around 40,000 neurons comprising >80 subtypes whose projections form 10 layers (Hofbauer and Campos-Ortega 1990; Fischbach 1989; Morante and Desplan 2008; Takemura et al. 2008; Hasegawa et al. 2011; Raghu and Borst 2011; Varija Raghu 2011; Raghu et al. 2013). Layers M1 to M6 are referred to as the 'distal medulla' while layers M7 to M10 are referred to as the 'proximal medulla'. The medulla neuropil is organized in repetitive columnar units orthogonal to the 10 layers that are comparable to lamina cartridges. As in the lamina, medulla neurons can be subdivided into two broad classes based on their projection pattern, uni-columnar and multi-columnar neurons (Fischbach 1989; Morante and Desplan 2008; Takemura et al. 2013) (Fig. 17.1). Uni-columnar neurons have arborizations limited to one medulla column and receive/process information from one point in space. Some of these neurons remain local and connect the distal with the proximal medulla (e.g. Mi neurons) while others project outside the medulla into the lobula and the lobula plate (Tm and TmY projection neurons). Multi-columnar neurons possess wider arborizations spreading over multiple columns, suggesting that they likely compare information covering a much larger receptive field. Their arborizations are often restricted to one layer (Dm and Pm neurons). Some of these neurons connect the medulla neuropil with the central brain (Mt neurons).

The deepest visual ganglions, the lobula and the lobula plate, form the lobula complex which consists of 15,000 neurons (Hofbauer and Campos-Ortega 1990). These ganglions are not innervated by photoreceptors and represent the main output of medulla columnar neurons (Tm, TmY neurons) (Fischbach 1989).

The lobula is thought to be involved in color vision processing, spectral preference, polarized light vision and feature detection (Behnia and Desplan 2015; Morante and Desplan 2008; Karuppudurai et al. 2014; Homberg et al. 2011). Its neuropil can be divided in 6 layers and, like the medulla, is made of repetitive columnar units arranged perpendicularly to the layers (Fischbach 1989). The lobula neuropil is the most intimately linked to the central brain (CB), where the next steps of visual processing take place after the optic lobes. Almost all lobula neurons are visual projection neurons, whose processes all converge and merge at the medial edge of the neuropil to form a single fiber tract connecting the lobula with higher visual processing centers within the CB. There are 12 different subtypes of lobula projection neurons that can be divided into 2 categories, based on their projection pattern (Otsuna and Ito 2006) (Fig. 17.1): Columnar neurons (LCN neurons, Fig. 17.1a), whose arborizations receive input from few ommatidia and Tangential and tree-like neurons (LT neurons, Fig. 17.1b), whose arborizations receive inputs from a very large portion of the visual field.

The lobula plate is the site of motion detection (Borst et al. 2010). Its neuropil can be divided in 4 layers, each processing motion information along one of the four cardinal directions, front-to-back, back-to-front, up and down (Fischbach 1989; Maisak et al. 2013). There are ~28 subtypes of lobula plate neurons that can be classified into 3 categories, columnar, tangential and intrinsic (Fischbach 1989) (Fig. 17.1). (i) Columnar: There are four types of lobula plate columnar neurons. T4 and T5 neurons connect each of the different layers of the lobula plate to the medulla (T4) or to the lobula (T5). Translobula plate neurons (Tlp) connect different layers of the lobula plate with the lobula. Y cells connect the lobula plate with the lobula and proximal medulla. T2 and T3 neurons specifically connect the lobula with the proximal medulla. T2 cells also project into the distal medulla (Meinertzhagen 2014). (ii) Tangential: Lobula plate tangential neurons (LPTCs) are grouped in 2 categories according to their preferred orientation, i.e., whether they respond primarily to horizontal (HS cells) or vertical (VS cells) motion (Maisak et al. 2013; Borst et al. 2010). They transmit direction-selective visual information into the CB. New types of LPTCs resembling those existing in larger flies have recently been discovered: Hx neurons receive and process local motion signals from the T4–T5 system (Wasserman et al. 2015) while loom sensitive neurons integrate specific motion cues and induce escape behavior (de Vries and Clandinin 2013). (iii) Intrinsic: Lobula plate intrinsic neuron (LPi) projections only innervate the lobula plate neuropil. Recently discovered LPi3-4 and LPi4-3 neurons, which exclusively innervate the two layers of the lobula plate vertical system, integrate opposing motion in the visual field (Mauss et al. 2015).

Finally, lobula complex columnar neurons connect the lobula plate and the lobula with the CB (Lccn) (Fig. 17.1). There are two types of Lccn that differ in the position of their cell body and in the routes their axons take towards the central brain (Fischbach 1989; Raghu et al. 2013). Their function remains to be discovered.

17.3 Generation of Neuronal Diversity in Drosophila Optic Lobes

How is the vast diversity of neurons found in *Drosophila* optic lobe neuropils generated? Early stages of optic lobe development have been characterized more than 20 years ago (Hofbauer and Campos-Ortega 1990; Green et al. 1993; Urbach and Technau 2003, 2004). The optic lobe derives from an optic placode located in the posterior part of the embryonic head (Fig. 17.2). This placode, which forms during embryonic stages 11–13, is composed of densely packed columnar shaped epithelial cells that remain mitotically quiescent until the end of embryogenesis. Optic placode cells start proliferating again after larval hatching and become subdivided into two primordia: the Outer Proliferation Center (OPC) and the Inner Proliferation Center (IPC) (Hofbauer and Campos-Ortega 1990; White and Kankel 1978). By the end of the second larval stage, these two primordia adopt a crescent shape and become separated by newly generated cells (Fig. 17.2). Thus, at the beginning of the third larval stage, two proliferative centers start producing optic lobe neurons: the superficially located OPC mainly generates lamina and medulla neurons while the centrally located IPC produces lobula and lobula plate neurons (Fig. 17.2) (Meinertzhagen 1993).

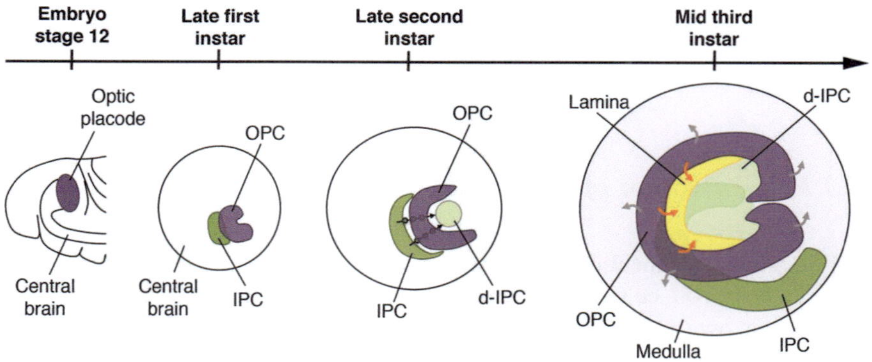

Fig. 17.2 Early optic lobe development. The optic lobe derives from an optic placode (*purple*) that forms during embryonic stages 11–13. At the end of the first larval instar, the optic placode becomes subdivided into the Outer Proliferation Center (*OPC*, *purple*) and the Inner Proliferation Center (*IPC*, *green*). By the end of the second larval stage, the OPC and the IPC adopt a crescent shape and become separated by newly generated cells (*black arrows*). These cells form the distal IPC (*d-IPC*, *light green*). During the third larval instar, the lateral side of the OPC (*orange arrows*) generates lamina neurons (*yellow*) while its medial side (*grey arrows*) generates medulla neurons (*light purple*). The IPC produces lobula and lobula plate neurons

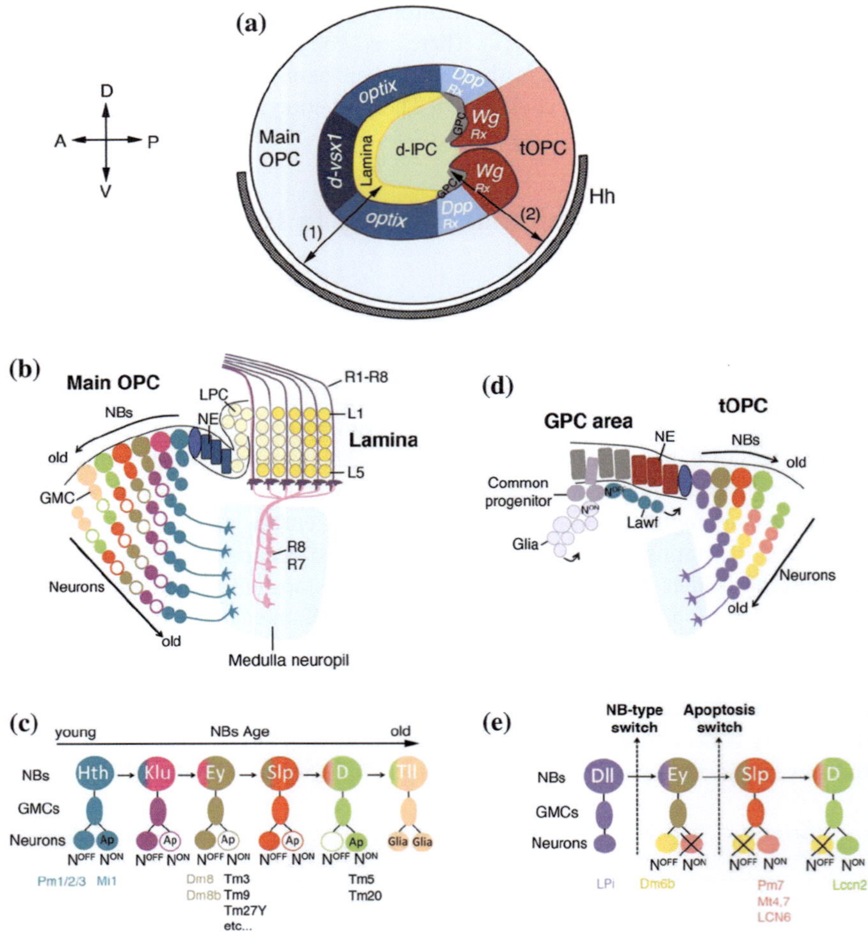

17.3.1 Generation of Neuronal Diversity in the OPC

At the beginning of the third larval stage, the OPC is composed of a single layer of approximately 700 columnar neuroepithelial cells that divide symmetrically to expand their own pool (Hofbauer and Campos-Ortega 1990; Nassif 2003). This neuroepithelium is subdivided into two domains having different modes of neurogenesis and generating different types of progenitors producing neurons for two distinct neuropils. The lateral OPC domain generates lamina neurons while the medial domain facing the central brain mainly produces medulla neurons (Figs. 17.2 and 17.3a).

◂**Fig. 17.3** The different neurogenesis modes generating cell diversity in the OPC. **a** Spatial patterning of the OPC (third larval stage, surface view). *d-VSX1* (*dark blue*), *optix* (*blue*), *dpp/Rx* (*light blue*), *wg* (*dark red*) and *Hh* (*black*) divide the OPC into 8 distinct spatial regions. The main OPC (*light blue*) is defined by *d-VSX1*, *optix* and *dpp/Rx* expression. The tOPC (*red*) is defined by *wg* expression. The lateral side of the OPC that is innervated by photoreceptors produces lamina neurons (*yellow*). The Glial Precursor Cell area (*GPC*, *grey*) is not innervated by photoreceptors. **b** Cross-section illustrating the lamina and the main OPC neurogenesis modes (*arrow* (*1*) in panel **a**). On the lateral side of the OPC, R1-R6 photoreceptors projections (*purple*) trigger lamina column formation by converting lamina precursor cells (*LPC*, *light yellow*) into lamina neurons (*L1-L5*, *yellow*). On the medial side, a pro-neuronal wave gradually converts main OPC neuroepithelial cells (*NE*, *dark blue*) into type I neuroblasts (*NBs*). These NBs divide to self-renew and generate ganglion mother cells (*GMCs*), which in turn produce neurons. **c** Main OPC NBs sequentially express Hth (*blue*), Klu (*pink*), Ey (*grey*), Slp (*red*), D (*green*), and Tll (*pale pink*) as they age. They generate GMCs that divide asymmetrically to produce one Notch^{ON} neuron (expressing Ap) and one Notch^{OFF} neuron. Each temporal window produces distinct neuronal subtypes (examples indicated at *bottom*). During the last temporal window (*Tll*) these NBs produce 2 glial cells. **d** Cross-sections illustrating the GPC area and tOPC neurogenesis modes (*arrow* (*2*) in panel **a**). In the GPC area, common progenitors (*purple*) delaminate from the NE and divide asymmetrically to generate Notch^{ON} proliferating glial precursor cells (*light purple*) and Notch^{OFF} proliferating Lawf precursor cells (*teal*). Glial cells and Lawf cells then migrate towards the lamina and the main OPC neuropil, respectively. tOPC NE produces NBs that divide asymmetrically and generate GMCs that produce neurons. Based on gene expression, only four classes of tOPC neurons can be identified (*purple, yellow, pink and green*). **e** tOPC neuroblasts sequentially express Dll (*purple*), Ey (*grey*), Slp (*red*), and D (*green*) as they age. They undergo 2 dramatic transitions (*vertical dashed arrows*), one in their division mode and the other in the systematic death of their neuronal progeny (Notch^{ON} first and then Notch^{OFF}, *black cross*). tOPC neuroblasts sequentially produce lobula plate (*Lpi*, *purple*), medulla (*Dm6b*, *yellow*; *Pm*, *pink*; *Mt*, *pink*), lobula (*LCN6*, *pink*) and lobula plate (*Lccn2*, *green*) neurons, indicated at *bottom*. Svp is in *yellow*, Toy is in *pink*. *A* anterior; *P* posterior; *D* dorsal and *V* ventral

17.3.1.1 Generation of Lamina Neurons

Lamina neurogenesis is directly coupled to innervation from the compound eye. During mid-third larval stage, photoreceptor axons from newly formed ommatidial cell clusters arrive in the optic lobes in a sequential order, from posterior to anterior (Fig. 17.3b). They fan out upon reaching the OPC to establish a retinotopic map along the anterior-posterior and dorso-ventral axes (Selleck and Steller 1991; Clandinin and Feldheim 2009). Simultaneously, the OPC neuroepithelium undergoes a morphological change leading to the formation of the lamina furrow at its lateral edge (Fig. 17.3b) (Hofbauer and Campos-Ortega 1990; Selleck and Steller 1991; Meinertzhagen and Hanson 1993). OPC neuroepithelial cells located posterior to this furrow gradually give rise to Lamina Precursors Cells (LPCs). These precursors express Tailless (Tll) and Dachshund (Dac) and remain arrested in G1 phase (Selleck et al. 1992; Huang and Kunes 1998; Guillermin et al. 2015). Upon arrival in the optic lobes, photoreceptor axon bundles trigger LPCs to complete a final symmetric division and induce neuronal differentiation by secreting Hedgehog (Hh), which controls LPC proliferation, and Spitz, which promotes LPC differentiation through Epidermal Growth Factor signaling (Selleck et al. 1992; Huang and Kunes 1996, 1998; Huang et al. 1998). Photoreceptor axons also induce migration and maturation of glial precursors into the lamina target field (Winberg et al. 1992; Chotard and Salecker 2007). Newly formed lamina neurons associate with glial cells and photoreceptor bundles to form lamina columns, in which they mature into neuronal subtypes L1–L5 (Fig. 17.3b). The molecular mechanisms specifying the identity of L1–L5 neurons are still unknown. The temporal control of lamina neurogenesis by incoming photoreceptor axons could be a key feature of this mechanism but this remains to be demonstrated (Huang and Kunes 1998). Interestingly, although they innervate the lamina neuropil, Lawf1 and Lawf2 neurons are generated by a restricted region of the lamina side of the OPC (Chen et al.2016; Suzuki et al. 2016), while C2, C3, and T1 cells are produced by the IPC (Meinertzhagen and Hanson 1993) (Hiesinger 2009) (see below).

17.3.1.2 Generation of Medulla Neurons

During the third larval stage, while lateral OPC cells generate lamina neurons, medial OPC cells are progressively converted into neuroblasts (NBs), the progenitors of medulla neurons (Fig. 17.3a–c). This conversion is sequential and is generated by a wave of expression of the pro-neural gene *lethal of scute* (*l'sc)* that sweeps from the medial edge toward the center of the OPC crescent over time (Yasugi et al. 2008). Progression of this pro-neural wave is controlled by at least four signaling pathways, Notch, JAK/STAT, EGFR and Hippo/Fat (Egger et al. 2010; Ngo et al. 2010; Yasugi et al. 2010; Reddy et al. 2010). OPC NBs express Deadpan (Dpn) and Asense (Ase) and undergo a type I division mode (Buescher et al. 1998; Egger et al. 2007): they divide asymmetrically to self-renew and generate intermediate precursors called Ganglion Mother Cells (GMCs) expressing Ase and Prospero (Pros). GMCs in turn divide once to produce medulla neurons (Fig. 17.3c).

Recent studies have shown that the vast diversity of cells found in the medulla is generated by a combination of spatial patterning, temporal patterning and binary cell fate decisions in the OPC (Hasegawa et al. 2011; Morante et al. 2011; Li et al. 2013; Suzuki et al. 2013; Bertet et al. 2014).

(i) *Spatial patterning*: The OPC is divided into different regions along the anterior-posterior axis defined by the expression of three transcription factors, *dVsx1* (Erclik et al. 2008), *optix* (Gold and Brand 2014), and *retinal homeobox (rx)* (Chen et al. 2016) (Fig. 17.3a). The *rx* region is itself subdivided into two subregions by the expression of the signaling molecules *decapentaplegic* (*dpp*) and *wingless* (*wg*) (Kaphingst and Kunes 1994). *Hedgehog* (*Hh*) further subdivides the OPC into dorsal and ventral halves (Evans et al. 2009; Chen et al. 2016). Each of the resulting eight spatially distinct regions generates different neuronal subtypes (Erclik et al. 2017).

(ii) *Temporal patterning*: NBs derived from the 'main OPC' (which includes the regions defined by *dVsx1*, *optix,* and *rx/dpp*) sequentially express 6 transcription factors—Homothorax (Hth), Eyeless (Ey), Klumpfuss (Klu), Sloppy paired 1 and 2 (Slp), Dichaete (D) and Tailless (Tll)—as they age

(Fig. 17.3a–c) (Morante et al. 2011; Li et al. 2013; Suzuki et al. 2013). These temporal factors and their overlap control expression of downstream transcription factors that mark the identities of the neuronal progeny. They define at least 12 NBs fates producing a large variety of distinct neuronal subtypes. Ey, Slp, and D cross-regulate each other: they are each required for turning ON the next transcription factor in the dividing NBs. Slp and D are also required for turning OFF the preceding transcription factor. Hth and Klu are not involved in these cross-regulations, suggesting that some key components of the temporal cascade remain to be discovered (Li et al. 2013; Suzuki et al. 2013).

(iii) *Notch-dependent binary cell fate decisions:* to further increase cell diversity, main OPC NBs produce GMCs that divide asymmetrically to generate two distinct neurons, one Notch^{OFF} and one Notch^{ON} neuron in which Notch signaling is active (Fig. 17.3b, c) (Li et al. 2013; Suzuki et al. 2013). Temporal gene expression is transmitted to GMCs and their neuronal progeny: Hth expression is transmitted to both Notch^{ON} and Notch^{OFF} neurons; Ey and Slp are transmitted to Notch^{OFF} neurons; D is transmitted to Notch^{ON} neurons (Fig. 17.3c). Temporal genes control expression of downstream transcription factors such as Brain specific homeobox (Bsh), Runt, Drifter (Drf), Lim-3, Twin of Eyeless (Toy), and Distalless (Dll). Apterous is specifically expressed in all Notch^{ON} neurons (Fig. 17.3c) (Li et al. 2013). Combinatorial expression of these factors confers a unique identity to each OPC neuron. As a result, the larval medulla is composed of several layers of Apterous-positive Notch^{ON} neurons intermingled with various types of Notch^{OFF} neurons (Hasegawa et al. 2011; Li et al. 2013). During the last temporal window, Tll-expressing NBs switch to glioblasts, and then undergo Pros-dependent cell cycle exit (Fig. 17.3b, c) (Li et al. 2013).

Efforts are now concentrated on determining the correspondence between larval transcriptional codes and adult neuron morphology. Loss-of-function approaches combined with neuronal subtype-specific Gal4 lines have allowed the identification of some of the neurons produced in each temporal window (Fig. 17.3c) (Hasegawa et al. 2011; Li et al. 2013). In the Hth temporal window, OPC NBs produces Hth$^+$ Bsh$^+$ Ap$^+$ uni-columnar Mi1 neurons as well as multi-columnar Hth$^+$ Lim-3$^+$ Svp$^+$ Pm1/2/3 neurons (Erclik et al. 2017). In the Ey temporal window, they produce Drf$^+$ Ap$^+$ Tm3, Tm3b, Tm9, Tm27, Tm27Y, and TmY3 uni-columnar neurons as well as Drf$^+$ Dm8 and Dm8b multi-columnar neurons. In the Slp temporal window, NBs produce uni-columnar Toy$^+$ Ap$^+$ Tm20 and Tm5 neurons. Although the precise transcriptional code defining each of these subtypes remains to be identified, these data suggest that all uni-columnar neurons are Ap$^+$ Notch^{ON} neurons whereas multi-columnar neurons are instead Notch^{OFF}. This raises the possibility that each GMC produces one Ap$^+$ Notch^{ON} uni-columnar neuron and one Notch^{OFF} multi-columnar neuron. This model would explain how the homogenous distribution of uni-columnar neurons throughout the medulla (and thus its columnar organization) is established. According to this model, optic lobe retinotopic map would be built progressively during larval development, by synchronizing the pro-neural wave that sequentially generates new rows of medulla uni-columnar neurons with arrival of each new row of photoreceptor projections in the OPC. This model, however, remains to be demonstrated.

17.3.1.3 Neurogenesis in the Posterior Tips of the OPC

Because they produce glial cells that migrate into the lamina, the posterior tips of the OPC (tOPC, defined by *wg* expression) were commonly described as the Glial Precursor Cell areas (GPC, Fig. 17.3a) (Perez and Steller 1996; Poeck et al. 2001; Dearborn and Kunes 2004; Chotard et al. 2005). However, a recent study has shown that this region also generates NBs that undergo complex neurogenesis to produce neurons for three different neuropils of the adult optic lobes (Bertet et al. 2014).

tOPC NBs sequentially express four transcription factors, Dll, Ey, Slp, and D as they age (Fig. 17.3d, e). Dll and D expression is transmitted to neurons whereas Ey and Slp are only expressed in NBs and GMCs. Ey, Slp and D cross-regulate each other. Although the tOPC and the main OPC have related temporal sequences, their modes of neurogenesis are very different (compare Fig. 17.3c, e). tOPC neurogenesis involves two dramatic transitions, one in the mode of GMC division and the other in systematic apoptosis of one of their neuronal progeny during Notch-mediated binary cell fate decisions (Fig. 17.3e). tOPC neuroblasts are initially specified as type 0 NBs (Karcavich and Doe 2005; Ulvklo et al. 2012) (Dll temporal window) and produce a single class of neurons. They then switch to type 1 NBs (Ey, Slp, and D temporal windows) but half of their progeny undergo apoptosis: Notch^{ON} progeny die first in the Ey temporal-window while Notch^{OFF} progeny die later, in the Slp and D temporal windows. As a result, tOPC NBs all produce hemi-lineages composed of four different neuronal classes, based on their transcriptional profile (one per temporal window, Fig. 17.3d, e) (Bertet et al. 2014).

In the tOPC, temporal patterning of neuroblasts generates cell diversity by controlling multiple aspects of neurogenesis. In addition to specifying the production of distinct neuronal subtypes over time, tOPC temporal factors also control Notch-mediated cell survival decisions. Dll and Ey determine Notch^{OFF} neuron survival and death, respectively, by controlling expression of the pro-apoptotic factor *hid.* Notch and Slp determine death and survival of Notch^{ON} neurons, respectively, by controlling *reaper.* Thus, sequential expression of Dll, Ey, and Slp dictates the switch in apoptosis (Bertet et al. 2014). The signals that modify the function of Ey and Slp in the tOPC are currently unknown. Genes expressed in this region such as *combgap* and *wg* are good candidates (Song et al. 2000). Whether the tOPC temporal series also controls the mode of intermediate precursor division remains to be determined.

A newly designed lineage tracing tool (FLEXAMP) has allowed the identification of the adult neuronal subtypes deriving from the four classes of larval tOPC neurons (Bertet et al. 2014). Previous work had established that the OPC specifically produces medulla neurons while the IPC generates lobula and lobula plate neurons (Meinertzhagen and Hanson 1993). Unexpectedly, clonal analyses with FLEXAMP have revealed that the tOPC produces medulla, lobula and lobula plate neurons in a sequential manner (Fig. 17.3e). When they are specified, Dll-expressing tOPC NBs produce lobula plate intrinsic neurons (LPi). They then switch to Ey expression and produce multi-columnar Distal medulla neurons (Dm6b). They next express Slp and produce Medulla tangential (Mt4, Mt7), Proximal medulla (Pm7), and Lobula Columnar (LCN6) neurons. They finally express D and produce Lobula complex columnar neurons (Lccn 2). The fact that the tOPC produces neurons for three distinct optic lobe neuropils could be due to the specific localization of this region in close contact with the IPC (Figs. 17.2 and 17.3a). The IPC could send non-autonomous signals to tOPC NBs that would instruct them to produce lobula and lobula plate neurons.

17.3.1.4 Neurogenesis in the Dpp-Domains of the Lamina Side of the OPC

The studies described above show that generation of lamina and medulla neurons involve two distinct modes of neurogenesis occurring in two different regions of the OPC. Lamina neurons are produced on the lateral side of the OPC (Lamina side) where neuroepithelial cells produce post-mitotic LPCs that express Dac and Tll and differentiate into neurons in response to signals from ingrowing photoreceptor axons. Medulla neurons are produced on the medial side of the OPC (medulla side) where neuroepithelial cells generate NBs that express Dpn and Ase and sequentially produce distinct neuronal subtypes over time. While investigating the origin of Lawf1 and Lawf2 neurons, two recent studies have identified a third mode of neurogenesis occurring in specific domains of the lamina side of the OPC (Fig. 17.3d) (Chen et al. 2016; Suzuki et al. 2016).

Lawf1 and Lawf2 are multi-columnar lamina neurons collecting signals from the medulla to modulate lamina neuron activity (Tuthill et al. 2014). Since they express Hth and have their cell body localized in the medulla, these neurons were assumed to be generated by medulla NBs during the Hth temporal window. Instead, the two new studies show that these neurons are in fact generated in the Rx-expressing domains of the lamina side of the OPC (Chen et al. 2016; Suzuki et al. 2016). These domains are not innervated by photoreceptors and are part of the GPC area that produces epithelial and marginal glia (eg/mg) (Fig. 17.3a, d) (Perez and Steller 1996; Poeck et al. 2001; Dearborn and Kunes 2004; Chotard et al. 2005). Lamina Rx-domains produce a unique class of 'common progenitors', which give rise to both eg/mg glial cells and to Lawf neurons. These progenitors do not express typical markers of OPC progenitors such as Dac and Tll (LPCs) or Dpn (NBs), but instead GMCs markers (Pros and Ase). However, unlike GMCs, they have a high mitotic potential (Fig. 17.3d).

Generation of eg/mg glia and LawF neurons occurs in 3 steps (Fig. 17.3d) (Chen et al. 2016): (i) during the third larval stage, common progenitors delaminate from the neuroepithelium of the Rx-expressing lamina side of the OPC. (ii) Common progenitors then undergo Notch-dependent asymmetric divisions to generate two populations of more restricted precursors, $Notch^{ON}$ gliogenic precursors and $Notch^{OFF}$ neuronal precursors. (iii) Restricted progenitors proliferate and migrate over long distances toward their target regions. They differentiate during the migration process. Eg/mg glial cells reach the lamina neuropil while Lawf neurons populate deep layers of the medulla cortex. Lawf migration is under the control of the Slit-Robo signaling system (Suzuki et al. 2016). Interestingly, production of Lawf1 and Lawf2 neurons is spatially controlled: Lawf1 neurons are specifically generated in the ventral arm of the OPC whereas Lawf2 are produced in the dorsal arm (Chen et al. 2016). This suggests that the Hh-dependent dorso-ventral patterning of the OPC specifies cell identities in both the medulla and lamina sides of the OPC (Fig. 17.3a).

17.3.2 Generation of Neuronal Diversity in the IPC

At the beginning of the third larval stage, the IPC is composed of a single layer of approximately 400 columnar neuroepithelial cells (Hofbauer and Campos-Ortega 1990) (Nassif et al. 2003). This neuroepithelium is subdivided into two domains, the proximal IPC (p-IPC) and the surface IPC (s-IPC), which together form an asymmetric horseshoe close to the central brain (Fig. 17.4a). A third proliferative domain, the distal IPC (d-IPC), derives from the p-IPC and forms a symmetric horseshoe adjacent to the lamina and OPC crescents (Fig. 17.4a, b) (Apitz and Salecker 2015). These three domains generate distal cells (including C2, C3, T2 and T3 neurons) as well as lobula neurons and lobula plate neurons (including T4 and T5 neurons) (Meinertzhagen and Hanson 1993; Hiesinger 2009).

Neurogenesis in the s-IPC, which is defined by *wg* expression (Fig. 17.4a, C.B. unpublished observations), is not fully understood. By contrast, the mechanisms underlying neurogenesis in the p-IPC and d-IPC have recently been uncovered. The d-IPC neuroepithelium is made of a central Brinker-expressing subdomain flanked by two Dpp-expressing subdomains (Apitz and Salecker 2015) (Fig. 17.4a). During the third larval stage, d-IPC neuroepithelial cells delaminate from these three subdomains by undergoing an epithelial-mesenchymal transition. They convert into progenitors that first proliferate and then migrate as four cell streams to generate a new proliferative zone named the distal IPC (d-IPC, Fig. 17.4a, arrows). The pro-neural gene L'sc, which is expressed at the inner edge of the p-IPC crescent, promotes neuroblast formation by controlling the rate of conversion of neuroepithelial cells and the progenitor supply (Apitz and Salecker 2015). During migration, progenitors do not divide and do not express characteristic NB or GMC markers (e.g.: Dpn, Ase, Pros). They acquire NB properties only when they reach the d-IPC. They start expressing Dpn and Ase and divide asymmetrically to self-renew and generate GMCs that produce neurons (Fig. 17.4b) (Apitz and Salecker 2015).

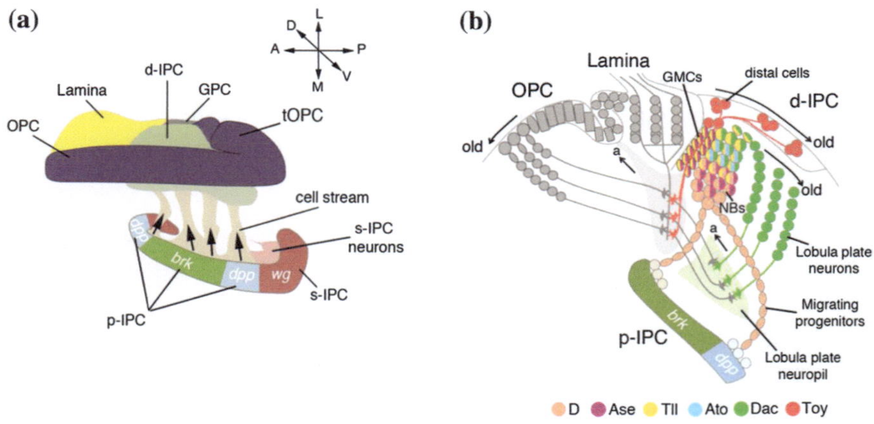

Fig. 17.4 Neurogenesis in the IPC. **a** *Side view* of an optic lobe during the third larval stage. The IPC is divided into 2 regions having different modes of neurogenesis: the surface IPC expresses *wg* (*s-IPC*, *red*) and produces lobula neurons. The posterior IPC (*p-IPC*) expresses *brk* (*green*) and *dpp* (*blue*) and generates progenitors that migrate (*cell stream*, *beige*) to form the distal IPC (*d-IPC*, *light green*). The d-IPC produces distal cells (C2, C3, T2, T3) and lobula plate neurons (*T4*, *T5*). OPC and tOPC are in *purple*, lamina in *yellow* and the GPC in *grey*. *A* anterior; *P* posterior; *D* dorsal; *V* ventral; *M* medial; *L* lateral. **b** Cross section of an optic lobe illustrating neurogenesis in the p-IPC and d-IPC. Neuroepithelial cells delaminate from the p-IPC following an epithelial-mesenchymal transition and convert into D-expressing progenitors (*light pink*). These progenitors migrate to form the d-IPC, were they convert into NBs. They first express D and Ase (*purple*) and produce GMCs that generate Toy-expressing distal cells (*pink*). They then express Ato (*cyan*), Tll (*yellow*) and Dac (*green*) and produce GMCs that generate lobula plate neurons (*green*). The oldest lobula plate neurons (*green*) and distal cells (*pink*) project near the oldest medulla neurons (*grey*), photoreceptors (*grey*), and lamina neurons (*grey*) in the posterior regions of the neuropils

The d-IPC NBs appear to transit through two stages during which they generate different neuronal populations (Fig. 17.4b) (Apitz and Salecker 2015). In the lower d-IPC, NBs express Dpn, D and Ase. They generate GMCs that give rise to Acj6 and Toy-distal cells (Fig. 17.4b). These neurons, which include C2, C3, T2, and T3 neurons, are born in the region between the lamina and the d-IPC and eventually move to form a layer above the d-IPC. In the upper d-IPC, NBs progressively downregulate D and Ase and upregulate Tll, Ato and Dac (Oliva et al. 2014; Apitz and Salecker 2015). They generate GMCs that produce Acj6 and Dac-expressing lobula plate neurons. These neurons, which include T4 and T5 neurons, form chains close to the d-IPC that get displaced centrally as they age (Fig. 17.4b) (Apitz and Salecker 2015). These data suggest that, like in the OPC, temporal patterning of progenitors is a key mechanism generating neuronal diversity in the IPC. Whether spatial patterning and binary cell fate decision are also involved in this process remains to be confirmed.

In all d-IPC neuronal populations, oldest neurons project in the posterior parts of the medulla and lobula complex neuropils while youngest neurons project to anterior neuropil regions. Since retina and lamina neurons processing information from the posterior eye are born first and innervate posterior neuropil areas (Hofbauer and Campos-Ortega 1990), these results reinforce the idea that the optic lobe retinotopic map is built progressively, from posterior to anterior, during larval development (Fig. 17.4b).

17.4 Conclusions

In summary, although the many neuronal subtypes forming the four neuropils of the *Drosophila* adult optic lobes have been described more than 25 years ago (Fischbach 1989), the mechanisms generating this diversity have only been identified recently. Studies performed in the past five years have substantially improved our understanding of this process by showing that optic lobe neurons are generated by at least four different modes of neurogenesis involving four different types of progenitors located in distinct regions of the larval optic lobes. These new studies have highlighted unanticipated similarities between neural circuit formation strategies in vertebrate and insect brains. Indeed, similarly to vertebrate central nervous system neurogenesis, optic lobe cell diversity is generated by a combination of spatial patterning, temporal patterning and binary cell fate decisions (Rowitch and Kriegstein 2010; Brand and Livesey 2011). These studies have also revealed that temporal patterning of neural progenitors generates cell diversity by controlling multiple aspects of neurogenesis including neuronal identity, Notch-mediated cell survival decisions, and probably the mode of intermediate precursor division. An outstanding question is why are such diverse modes of neurogenesis and progenitor types used to generate a single system? One possibility is that this reflects the gradual increase in complexity of the visual system that occurred throughout evolution and required the addition of new neurons with new functions. Another key challenge is now to understand how the optic lobe retinotopic map is established. This will involve identifying the mechanisms allowing columnar neurons of each neuropil to find and connect to each other as well as those controlling migration and targeting of multi-columnar neurons.

References

Apitz H, Salecker I (2014) A challenge of numbers and diversity: neurogenesis in the Drosophila optic lobe. J Neurogenet 28(3–4):233–249

Apitz H, Salecker I (2015) A region-specific neurogenesis mode requires migratory progenitors in the Drosophila visual system. Nat Neurosci 18(1):46–55

Behnia R, Desplan C (2015) Visual circuits in flies: beginning to see the whole picture. Curr Opin Neurobiol 34:125–132

Bertet C et al (2014) Temporal patterning of neuroblasts controls Notch-mediated cell survival through regulation of Hid or Reaper. Cell 158(5):1173–1186

Borst A, Helmstaedter M (2015) Common circuit design in fly and mammalian motion vision. Nat Neurosci 18(8):1067–1076

Borst A, Haag J, Reiff DF (2010) Fly motion vision. Annu Rev Neurosci 33:49–70

Brand AH, Livesey FJ (2011) Neural stem cell biology in vertebrates and invertebrates: more alike than different? Neuron 70(4):719–729

Buescher M et al (1998) Binary sibling neuronal cell fate decisions in the Drosophila embryonic central nervous system are nonstochastic and require inscuteable-mediated asymmetry of ganglion mother cells. Genes Dev 12(12):1858–1870

Chen Z et al (2016) A unique class of neural progenitors in the Drosophila optic lobe generates both migrating neurons and glia. Cell Rep

Chotard C, Salecker I (2007) Glial cell development and function in the Drosophila visual system. Neuron Glia Biol 3(1):17–25

Chotard C, Leung W, Salecker I (2005) glial cells missing and gcm2 cell autonomously regulate both glial and neuronal development in the visual system of Drosophila. Neuron 48(2): 237–251

Clandinin TR, Feldheim DA (2009) Making a visual map: mechanisms and molecules. Curr Opin Neurobiol 19(2):174–180

Dearborn R Jr, Kunes S (2004) An axon scaffold induced by retinal axons directs glia to destinations in the Drosophila optic lobe. Development 131(10):2291–2303

de Vries S.E, Clandinin T(2013) Optogenetic stimulation of escape behavior in *Drosophila melanogaster.* J Vis Exp (71)

Douglass JK, Strausfeld NJ (1995) Visual motion detection circuits in flies: peripheral motion computation by identified small-field retinotopic neurons. J Neurosci 15(8):5596–5611

Egger B et al (2007) Regulation of spindle orientation and neural stem cell fate in the Drosophila optic lobe. Neural Dev 2:1

Egger B, Gold KS, Brand AH (2010) Notch regulates the switch from symmetric to asymmetric neural stem cell division in the Drosophila optic lobe. Development 137(18):2981–2987

Erclik T et al (2008) Conserved role of the Vsx genes supports a monophyletic origin for bilaterian visual systems. Curr Biol 18(17):1278–1287

Erclik T et al (2017) Integration of temporal and spatial patterning generates neural diversity. Nature 541(7637):365–370

Evans CJ et al (2009) G-TRACE: rapid Gal4-based cell lineage analysis in Drosophila. Nat Methods 6(8):603–605

Fischbach KF, Dittrich AP (1989) The optic lobe of *Drosophila melanogaster*. I. A Golgi analysis of wild-type structure. Cell Tissue Res 258:441–475

Gold KS, Brand AH (2014) Optix defines a neuroepithelial compartment in the optic lobe of the Drosophila brain. Neural Dev 9:18

Green P, Hartenstein AY, Hartenstein V (1993) The embryonic development of the Drosophila visual system. Cell Tissue Res 273(3):583–598

Guillermin O et al (2015) Characterization of tailless functions during Drosophila optic lobe formation. Dev Biol 405(2):202–213

Hasegawa E et al (2011) Concentric zones, cell migration and neuronal circuits in the Drosophila visual center. Development 138(5):983–993

Heisenberg M, Buchner E (1977) The role of retinula cell types in visual behavior of *Drosophila melanogaster.* J Comp Physiol 117:127–162

Hiesinger PR, Meinertzhagen IA (2009) Visual system development: invertebrates. In: Squire LR (ed) Encyclopedia of neuroscience, vol 10. Academic Press, Oxford, pp 313–322

Hofbauer A, Campos-Ortega JA (1990) Proliferation and early differentiation of the optic lobes in *Drosophila melanogaster*. Roux's Arch Dev Biol 198:264–274

Homberg U et al (2011) Central neural coding of sky polarization in insects. Philos Trans R Soc Lond B Biol Sci 366(1565):680–687

Huang Z, Kunes S (1996) Hedgehog, transmitted along retinal axons, triggers neurogenesis in the developing visual centers of the Drosophila brain. Cell 86(3):411–422

Huang Z, Kunes S (1998) Signals transmitted along retinal axons in Drosophila: hedgehog signal reception and the cell circuitry of lamina cartridge assembly. Development 125(19):3753–3764

Huang Z, Shilo BZ, Kunes S (1998) A retinal axon fascicle uses spitz, an EGF receptor ligand, to construct a synaptic cartridge in the brain of Drosophila. Cell 95(5):693–703

Kaphingst K, Kunes S (1994) Pattern formation in the visual centers of the Drosophila brain: wingless acts via decapentaplegic to specify the dorsoventral axis. Cell 78(3):437–448

Karcavich R, Doe CQ (2005) Drosophila neuroblast 7-3 cell lineage: a model system for studying programmed cell death, Notch/Numb signaling, and sequential specification of ganglion mother cell identity. J Comp Neurol 481(3):240–251

Karuppudurai T et al (2014) A hard-wired glutamatergic circuit pools and relays UV signals to mediate spectral preference in Drosophila. Neuron 81(3):603–615

Kumar JP (2012) Building an ommatidium one cell at a time. Dev Dyn 241(1):136–149

Li X et al (2013) Temporal patterning of Drosophila medulla neuroblasts controls neural fates. Nature 498(7455):456–462

Maisak MS et al (2013) A directional tuning map of Drosophila elementary motion detectors. Nature 500(7461):212–216

Mauss AS et al (2015) Neural Circuit to Integrate Opposing Motions in the Visual Field. Cell 162(2):351–362

Meinertzhagen IA (2014) The anatomical organization of the compound eye's visual system. In: Josh Dubnau (ed) Behavioral genetics of the fly (*Drosophila Melanogaster*). Cold Spring Harbour Laboratory

Meinertzhagen IA, Hanson TE (1993) The development of the optic lobe. In: Bate M, Martinez Arias A (eds) The development of *Drosophila melanogaster*, vol II. Cold Spring Harbor Laboratory Press, pp 1363–1491

Meinertzhagen IA, Sorra KE (2001) Synaptic organization in the fly's optic lamina: few cells, many synapses and divergent microcircuits. Prog Brain Res 131:53–69

Morante J, Desplan C (2004) Building a projection map for photoreceptor neurons in the Drosophila optic lobes. Semin Cell Dev Biol 15(1):137–143

Morante J, Desplan C (2008) The color-vision circuit in the medulla of Drosophila. Curr Biol 18(8):553–565

Morante J, Erclik T, Desplan C (2011) Cell migration in Drosophila optic lobe neurons is controlled by eyeless/Pax6. Development 138(4):687–693

Nassif C, Noveen A and Hartenstein V (2003) Early development of the Drosophila brain: III. The pattern of neuropile founder tracts during the larval period. J Comp Neurol 455(4):417–434

Neriec N, Desplan C (2016) From the Eye to the Brain: development of the Drosophila Visual System. Curr Top Dev Biol 116:247–271

Ngo KT et al (2010) Concomitant requirement for Notch and Jak/Stat signaling during neuro-epithelial differentiation in the Drosophila optic lobe. Dev Biol 346(2):284–295

Oliva C et al (2014) Proper connectivity of Drosophila motion detector neurons requires Atonal function in progenitor cells. Neural Dev 9:4

Otsuna H, Ito K (2006) Systematic analysis of the visual projection neurons of *Drosophila melanogaster*. I. Lobula-specific pathways. J Comp Neurol 497(6):928–958

Perez SE, Steller H (1996) Migration of glial cells into retinal axon target field in *Drosophila melanogaster*. J Neurobiol 30(3):359–373

Poeck B et al (2001) Glial cells mediate target layer selection of retinal axons in the developing visual system of Drosophila. Neuron 29(1):99–113

Raghu SV, Borst A (2011) Candidate glutamatergic neurons in the visual system of Drosophila. PLoS ONE 6(5):e19472

Raghu SV, Claussen J, Borst A (2013) Neurons with GABAergic phenotype in the visual system of Drosophila. J Comp Neurol 521(1):252–265

Reddy BV, Rauskolb C, Irvine KD (2010) Influence of fat-hippo and notch signaling on the proliferation and differentiation of Drosophila optic neuroepithelia. Development 137(14): 2397–2408

Rowitch DH, Kriegstein AR (2010) Developmental genetics of vertebrate glial-cell specification. Nature 468(7321):214–222

Selleck SB, Steller H (1991) The influence of retinal innervation on neurogenesis in the first optic ganglion of Drosophila. Neuron 6(1):83–99

Selleck SB et al (1992) Regulation of the G1-S transition in postembryonic neuronal precursors by axon ingrowth. Nature 355(6357):253–255

Song Y, Chung S, Kunes S (2000) Combgap relays wingless signal reception to the determination of cortical cell fate in the Drosophila visual system. Mol Cell 6(5):1143–1154

Suzuki T et al (2013) A temporal mechanism that produces neuronal diversity in the Drosophila visual center. Dev Biol 380(1):12–24

Suzuki T et al (2016) Formation of neuronal circuits by interactions between neuronal populations derived from different origins in the Drosophila visual center. Cell Rep 15(3):499–509

Takemura SY, Lu Z, Meinertzhagen IA (2008) Synaptic circuits of the Drosophila optic lobe: the input terminals to the medulla. J Comp Neurol 509(5):493–513

Takemura SY et al (2013) A visual motion detection circuit suggested by Drosophila connectomics. Nature 500(7461):175–181

Tuthill JC et al (2013) Contributions of the 12 neuron classes in the fly lamina to motion vision. Neuron 79(1):128–140

Tuthill JC et al (2014) Wide-field feedback neurons dynamically tune early visual processing. Neuron 82(4):887–895

Ulvklo C et al (2012) Control of neuronal cell fate and number by integration of distinct daughter cell proliferation modes with temporal progression. Development 139(4):678–689

Urbach R, Technau GM (2003) Molecular markers for identified neuroblasts in the developing brain of Drosophila. Development 130(16):3621–3637

Urbach R, Technau GM (2004) Neuroblast formation and patterning during early brain development in Drosophila. BioEssays 26(7):739–751

Varija Raghu S, Reiff DF, Borst A (2011) Neurons with cholinergic phenotype in the visual system of Drosophila. J Comp Neurol 519(1):162–176

Wasserman SM et al (2015) Olfactory neuromodulation of motion vision circuitry in Drosophila. Curr Biol 25(4):467–472

White K, Kankel DR (1978) Patterns of cell division and cell movement in the formation of the imaginal nervous system in *Drosophila melanogaster*. Dev Biol 65(2):296–321

Winberg ML, Perez SE, Steller H (1992) Generation and early differentiation of glial cells in the first optic ganglion of *Drosophila melanogaster*. Development 115(4):903–911

Yamaguchi S et al (2008) Motion vision is independent of color in Drosophila. Proc Natl Acad Sci U S A 105(12):4910–4915

Yasugi T et al (2008) Drosophila optic lobe neuroblasts triggered by a wave of proneural gene expression that is negatively regulated by JAK/STAT. Development 135(8):1471–1480

Yasugi T et al (2010) Coordinated sequential action of EGFR and Notch signaling pathways regulates proneural wave progression in the Drosophila optic lobe. Development 137(19): 3193–3203

Zhu Y (2013) The Drosophila visual system: from neural circuits to behavior. Cell Adh Migr 7(4): 333–344

Chapter 18
Single-Cell Transcriptomic Characterization of Vertebrate Brain Composition, Development, and Function

Bosiljka Tasic, Boaz P. Levi and Vilas Menon

Abstract A fundamental effort in neuroscience is to identify and characterize the building blocks of the central nervous system. Starting from the early days of the field, researchers have classified brain cells into types based on cellular morphology, electrical properties, connectivity patterns and molecular characteristics. Recent advances in molecular techniques, DNA sequencing, and computational power have enabled high-throughput molecular characterization of individual cells through the use of single-cell RNA-sequencing. This chapter reviews the general notion of cell types in the brain, and then outlines methods to select, isolate, and profile individual cells using single-cell RNA-sequencing. Also included is an overview of analysis methods to define putative types from single-cell RNA-sequencing data, and additional methods to link these data to other modalities in order to obtain a comprehensive picture of the basic components of the central nervous system.

18.1 What Is a Cell Type?

The notion of discrete classes, or types, of neural[1] cells dates back to Ramon y Cajal's description of their diverse morphologies (i.e., shapes), combined with their specific locations within the nervous system. Over the past century, imaging, electrophysiological, and molecular techniques have examined a variety of neural cell characteristics and have used this information to classify cells into types (Ascoli et al. 2008; DeFelipe et al. 2013; Armananzas and Ascoli 2015). It is generally

The original version of this chapter was revised: New figure has been updated. The erratum to this chapter is available at https://doi.org/10.1007/978-3-319-57363-2_21

[1]Neural relates to the nervous system, and neuronal relates to the neuron. We use the term neural here, as Cajal described both neuronal and non-neuronal cells in his studies (Garcia-Marin et al. 2007).

B. Tasic (✉) · B.P. Levi · V. Menon
The Allen Institute for Brain Science, Seattle, USA
e-mail: bosiljkat@alleninstitute.org

A. Çelik and M.F. Wernet (eds.), *Decoding Neural Circuit Structure and Function*, DOI 10.1007/978-3-319-57363-2_18

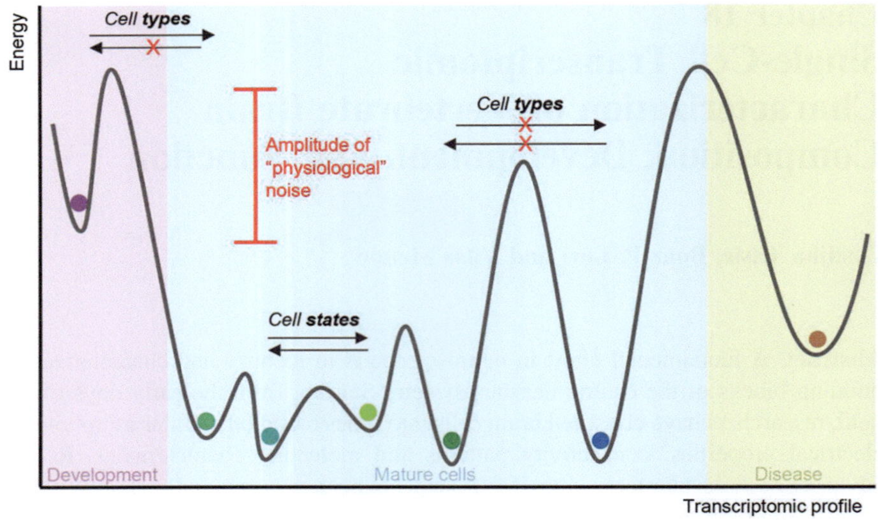

Fig. 18.1 A schematic of a putative transcriptomic landscape. This depiction of gene expression space (shown only in one dimension) shows "valleys" corresponding to the existence of cell types and states. Distinct cell types are separated by energy barriers that cannot be reversibly surpassed under "normal" or "physiological" conditions, whereas cell states are separated by smaller barriers that can be surmounted in either direction

agreed that a comprehensive and "ultimate" definition of a cell type should rely on a variety of cellular characteristics including its function (Armananzas and Ascoli 2015; Poulin et al. 2016). However, defining function is not straightforward, and a cell's observed function depends on how it is tested and analyzed. As cellular function is the result of other cellular properties, it is reasonable to assume that cells with similar properties across multiple data modalities are likely to share similar function. Therefore, more easily measurable cell characteristics (shapes, connection patterns, gene expression, responses to stimuli) have been used to categorize cells into groups and to help provide insights into their function.

Fundamentally, cell types can be defined as discrete regions within a multidimensional feature space where cells are more likely to reside. For example, in transcriptomic space, each feature is the expression of a gene or a set of genes, and cells do not populate the entire gene expression space uniformly, but rather tend to aggregate in "valleys" that correspond to more stable molecular profiles (Fig. 18.1). This concept is similar to a Waddington landscape[2] or an energy diagram in physics. In adulthood, two cell types can be conceived as distinct if the barrier between them is large enough to be effectively insurmountable under normal

[2]Waddington landscape is an illustrated example of the epigenetic landscape of cellular differentiation where one can imagine the process of differentiation as a marble rolling down a hill and traveling down the valleys or grooves. The ridges between the valleys represent epigenetic barriers between lineage trajectories or cell types.

conditions in vivo. In other words, a cell type is based on a unique combination of stably expressed genes, whose expression has been established through development and is mostly invariable to stimuli in adulthood. Conversely, if the barrier between valleys is small enough to be surmounted in both directions (i.e., reversibly) by a physiologically relevant mechanism (such as neuronal activity, stimulus-dependent signaling, or perhaps stimulus-independent coordinated fluctuation in the levels of key transcripts defining the valleys), then these two valleys are more accurately described as "cell states", rather than cell types.

The adult cell-type landscape takes its shape through orchestrated transitions between a series of temporally connected landscapes during development. Aging or disease can also be depicted as additional contours in the landscapes or additional landscapes, depending on the temporal window considered. If the transition between two valleys is irreversible, and occurs only through non-physiological stimuli (such as transdifferentiation or reprogramming) if at all, then the cells occupying each valley should also be termed distinct "cell types".

In addition to the difference between cell types and cell states, another important distinction is between discrete types or states and a phenotypic continuum. A discrete cell type in the transcriptomic landscape is one where the variation in gene expression among cells of that type is less than or approximately equal to biological noise. A transcriptomic continuum, however, is elongated in one or more gene expression dimensions, such that a range of types exists with no clear division or multimodal distribution along the axis of variation. In this case, cells at either end are clearly distinct, but there also exists a range of "intermediate" cells exhibiting gene expression profiles that span the space between the two ends. Not enough evidence exists to conclude if states are more likely to be continuous than types, but given lower energy barriers between states, that indeed may be the case.

Any description of a cell-type landscape based on experimental data is strongly influenced by cell sampling and methodological noise (Fig. 18.2). If cells belong to a single discrete type, which is defined by a unique combination of expressed genes with little biological variation within the type, they are unlikely to be misidentified as being part of a continuum. However, if a continuum exists and only portions of it are sampled, the analysis may incorrectly suggest the existence of one or more discrete classes. Alternatively, if two highly related but distinct cell types exist, technical noise may blur them into a false continuum.

In summary, any assessment of cell types as being discrete or belonging to a continuum has to take into account the sampling (if non-exhaustive) and the level of noise introduced by the profiling method.

18.2 Why Single-Cell Transcriptomics?

Historically, transcriptomic profiling has been applied to cell populations defined by fine dissections (Belgard et al. 2011; Miller et al. 2014), a common cell-surface marker, or expression of a transgene (Sugino et al. 2006; Heiman et al. 2008; Doyle et al. 2008). These studies have provided important advancements toward defining

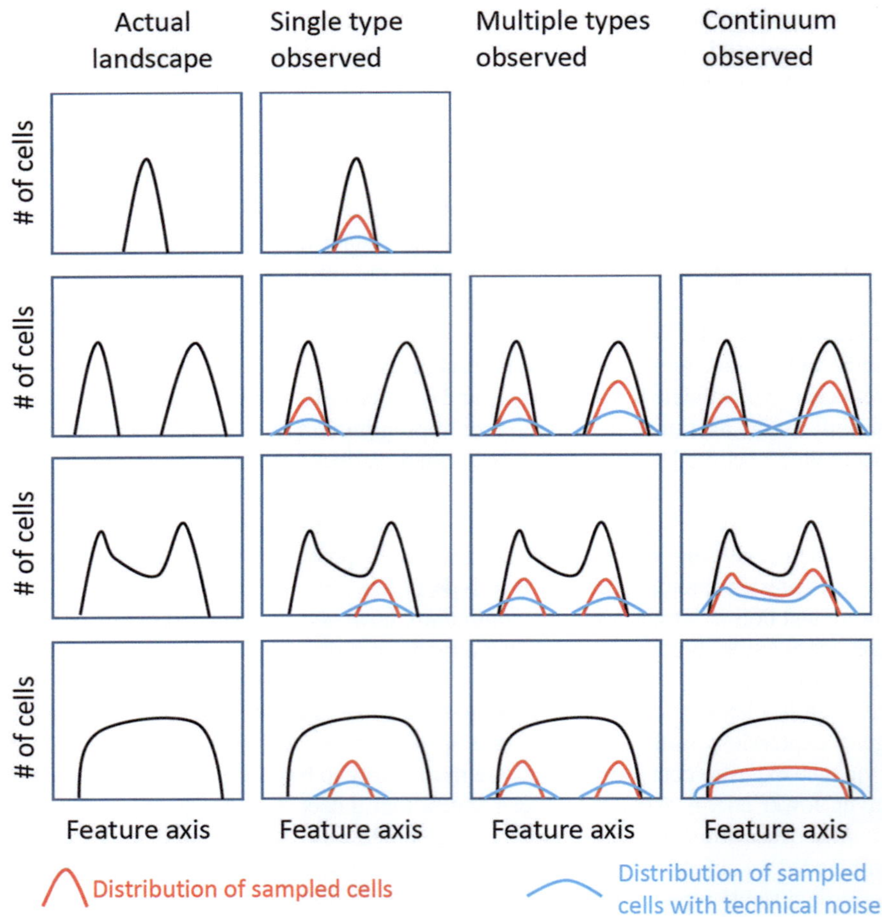

Fig. 18.2 Effect of sampling on identification of discrete and continuous cell types. The underlying "true" distribution of cells in continuous or discrete cell types is always subject to the sampling approach of a given study. This schematic depicts cells existing as discrete or continuous types (along a single gene expression axis, for simplicity). Sampling of cells along this axis in various cases, combined with technical noise, affects whether the true underlying distribution is correctly deduced from the sampled data

new cell class markers and developing methods for cell isolation. However, any approach that pools cells for profiling inevitably averages gene expression and may mask cellular heterogeneity. In contrast, non-genome-wide methods for assessing gene expression [e.g., immunocytochemistry or RNA in situ hybridization (ISH)] provide single-cell resolution, but generally permit examination of only one or a few molecular markers at a time (Lein et al. 2007; Zeng et al. 2012). Given that individual cells are the building blocks of multicellular organisms, and that each cell expresses thousands of genes, genome-wide gene expression profiling at the single-cell resolution is key to cell classification.

Transcriptomic single-cell profiling was demonstrated more than 20 years ago and applied to individual neurons, but was limited in scale (Eberwine et al. 1992; Tietjen et al. 2003; Trimarchi et al. 2007). Multiplexed single-cell RT-PCR has been used as an alternative to provide higher cell throughput, but it does not permit genome-wide gene expression analysis: one has to choose the genes to be examined —therefore prior knowledge of informative markers is necessary and unknown markers will be missed (Poulin et al. 2014; Johnson et al. 2015; Chiu et al. 2014). Developments in nucleic acid amplification protocols (Tang et al. 2009; Ramskold et al. 2012; Islam et al. 2011), DNA sequencing technology (Shendure and Ji 2008), single-cell isolation, computational power, and bioinformatics analysis have enabled transcriptomic characterization of large numbers of individual cells (Poulin et al. 2016; Junker and van Oudenaarden 2015). The most prominently used approach is single-cell RNA-sequencing (scRNA-seq), which in principle quantifies expression of all genes in a cell.[3] It is a multistep process that reverse transcribes RNA (usually mRNA) in each cell, amplifies the obtained cDNA, and then adapts the amplified cDNA libraries for sequencing on a next-generation sequencing instrument (e.g., Illumina HiSeq). Similarities in gene expression between individual cells can then be used to group cells into classes or types. Once the transcriptomic cell types are defined, individual markers or their combinations that are specific for each type can be revealed. Compared to other methods for single-cell characterization, scRNA-seq is a quantitative high-throughput method that generates high-dimensional data for cell classification. It has therefore become a method of choice for characterization of the "cell-type landscape" within many parts of the nervous system (Poulin et al. 2016).

Expression of specific genes in a cell type revealed by scRNA-seq suggests certain cellular functions, and it can be used to identify potential therapeutic targets, or as a readout to guide or improve procedures (for example derivation of cells in vitro). For example, expression of certain cell-surface receptors has been used to predict or evaluate responses to extracellular stimuli and these genes may be potential therapeutic targets (Pollen et al. 2015). Through the evaluation of $Lmx1a^{+}$ lineage cells, new markers were found that can be used to improve purification of a single-cell type from a mixture of cells (Kee et al. 2017). Similarly, an inferred lineage from scRNA-seq data helped guide interventions to bias cell-type production to the desired cell types (Kee et al. 2017). Such strategies will likely be clinically valuable as they enrich for cells with the highest therapeutic potential, while minimizing potential side effects from contaminating off-target cell types (Kirkeby et al. 2017). Finally, marker genes or their regulatory elements can be employed to build tools to isolate, access, or perturb specific transcriptomic cell types (Taniguchi et al. 2011; Harris et al. 2014; Madisen et al. 2012; Madisen et al. 2015). The genetic tools based on cell-type-specific markers, especially using intersectional genetic strategies, will facilitate careful assessment of

[3]We say "in principle" because the detection of the number of genes per cell depends on the efficiency of reverse transcription and cDNA amplification and sequencing depth (see below).

cell-type-specific phenotype and function. For example, a transcriptomic interneuron-type expressing somatostatin (Sst) and Nos1 in the cortex, named Sst-Chodl (Tasic et al. 2016), was labeled by intersectional genetic labeling (with Somatostatin-driven Flp and Nos1-driven Cre-recombinase transgenic lines, and an intersectional Cre/Flp reporter) to specifically examine morphology and physiology of only this type of cell (He et al. 2016).

Although this chapter focuses on single-cell characterization and classification based on transcriptomic profiling, this approach does not obviate the need for characterization of other cellular properties. In fact, the gene expression pattern of an adult neuron may not reflect all adult neuronal properties; for example, a neuron's axonal projection pattern in adulthood may depend on molecules expressed only transiently during development. Therefore, examination of other cellular properties is key to confirming the 'working hypotheses' resulting from cell classification based solely on cellular transcriptomes. Multimodal or integrative neuronal phenotyping that takes into account many cellular properties is the future of cell classification, and we will discuss recent methodological advances in this area at the end of the chapter.

In summary, single-cell transcriptomics has been used to profile and classify cells in various parts of the nervous system. Analysis of cells from a single time point can build a taxonomy, and analysis from one or more time points during development can be used to infer a developmental trajectory. Once the cell types are defined, specific genes become "handles" that can inform future investigation of specific cell types, including multimodal phenotypes and functional analysis.

18.3 Cell Classification

A summary of classification studies provided by single-cell transcriptomics is available in Fig. 18.3 and Table 18.1. Most were performed on live cells isolated from the mouse nervous system. They have employed a variety of tissue and cell isolation and sampling techniques, different cDNA amplification approaches, and orders of magnitude different cell numbers and sequencing depth. A few have sampled similar or the same tissues and provide a window into comparability of results obtained by different experimental and computational approaches. We discuss them here to provide a snapshot of the present state of the field, and to showcase the power of single-cell transcriptomics for cell-type classification.

18.3.1 Mouse Cortex and Hippocampus

Three studies have characterized the gene expression profiles from single cells of the mouse cortex and hippocampus (Fig. 18.3) (Tasic et al. 2016; Habib et al. 2016; Zeisel et al. 2015). Zeisel et al. (2015) employed microdissection with and without genetic

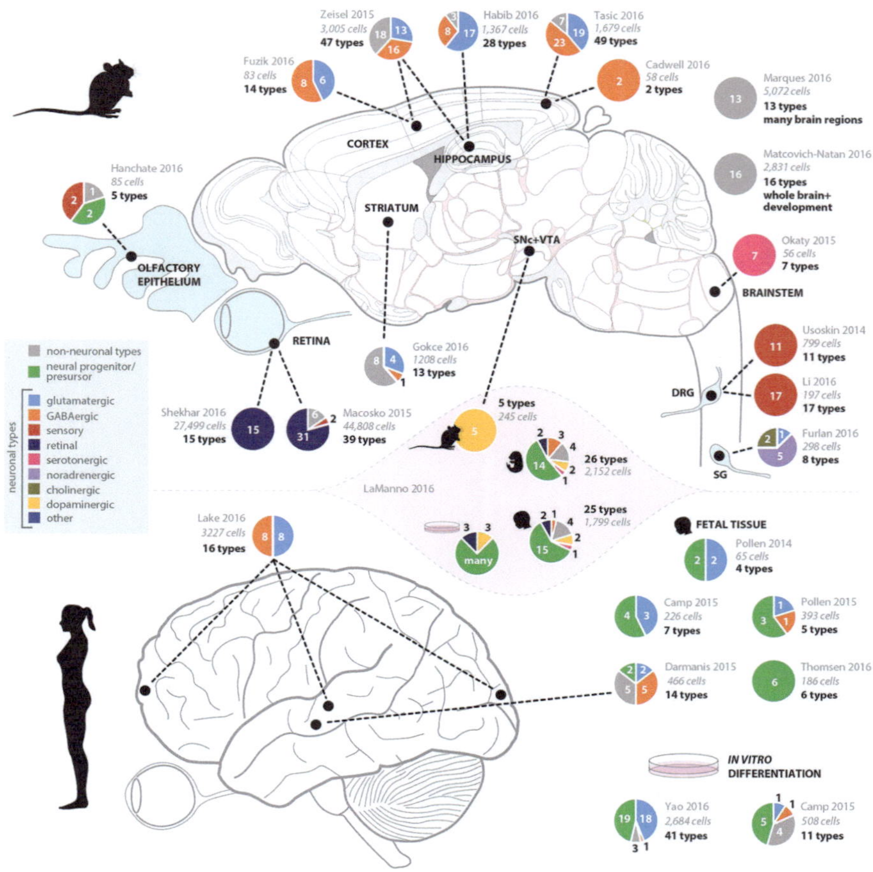

Fig. 18.3 Summary of transcriptomic cell types described so far from mouse and human brain. Each study is represented by a pie chart representing the number of cell types defined, and is pinned to approximate locations of tissue used for isolation from the mouse and human brain. We did not include population-based studies, like Chiu et al. (2014) for DRG and Cembrowski et al. (2016) for hippocampus, or studies that characterized general molecular diversity without specifying individual transcriptomically derived cell types. *DRG* dorsal root ganglion, *SG* Sympathetic ganglia. Image of human brain: (https://commons.wikimedia.org/wiki/File%3AHuman_Brain_sketch_with_eyes_and_cerebrellum.svg); file URL: https://upload.wikimedia.org/wikipedia/commons/0/04/Human_Brain_sketch_with_eyes_and_cerebrellum.svg; By Hankem (own work) (public domain or public domain), via Wikimedia commons)

labeling to identify cell types from the somatosensory cortex and the hippocampus of the juvenile mouse brain. Single-cell profiling of >3000 cells led to the identification of 47 cell types. Across these two brain regions, the authors identified 16 types of interneurons, 13 types of excitatory neurons, and 18 types of non-neuronal cells. Tasic et al. profiled live, adult single cells purified by fluorescence-activated cell sorting (FACS) after manual dissociation of the visual cortex (Tasic et al. 2016). This study relied on positive marking via Cre-recombinase transgenes to enrich for distinct

Table 18.1 Cell sampling, isolation, library preparation, and sequencing methods from recent single-cell transcriptomic studies of the brain

Publication	Tissues	Cell isolation	RNA harvest/cDNA AMP	Number of cells	Depth of sequencing
Usoskin et al. (2015)	Mouse dorsal root ganglionic sensory neurons	Manual dissection, automated cell picking	STRT	799 cells	1.12M reads/cell
Pollen et al. (2014)	Fetal human cortex	Whole cortex isolation/C1 capture	C1-SMARTer	65 cells	Subsampled to 50,000 reads/cell
Pollen et al. (2015)	Fetal human germinal zone	germinal zone microdissection/C1 capture	C1-SMARTer	393 cells	2.9M reads/cell
Okaty et al. (2015)	Serotonergic neurons	Manual dissection and single-cell isolation	Ovation RNA-seq System v2	56 cells	18M reads/cell
Zeisel et al. (2015)	Juvenile mouse somatosensory cortex and hippocampus	Manual dissection/C1 capture	C1-STRT	3005 cells	Median $\sim$12,900 UMIs (transcripts)/cell
Macosko et al. (2015)	Mouse retina	Manual dissociation	Drop-seq	44,808 cells	$\geq$900 genes detected per cell
Thomsen et al. (2016)	Fetal human cortex progenitors	Manual dissection/FACS enrichment	FRISCR/SmartSeq2	186 cells	$\sim$3500 genes/cell
Caldwell et al. (2016)	Mouse adult neocortex, layer 1	Patch pipet aspiration of cytoplasm	Patch-seq/SmartSeq2	58 cells	$\sim$7000 genes/cells
Li et al. (2016)	Mouse dorsal root ganglionic sensory neurons	Manual dissection and picking	SMARTer	203 cells	58.2M reads/cell

(continued)

Table 18.1 (continued)

Publication	Tissues	Cell isolation	RNA harvest/cDNA AMP	Number of cells	Depth of sequencing
Tasic et al. (2016)	Mouse adult visual cortex	Manual microdissection/FACS enrichment	SMARTer	1697 cells	~8.7M reads/cell
Lake et al. (2016)	Multiple adult human cortical areas	Manual microdissection/FACS sorted nuclei	C1-SMARTer	3227 nuclei	8.43M reads/nucleus
Gokce et al. (2016)	Mouse striatum	Manual dissection/FACS or magnetic enrichment	C1-SMARTer or FACS/SmartSeq2	1208 cells	1–5M reads/cell
Marques et al. (2016)	Oligodendrocyte lineage cells, multiple mouse brain areas	Manual dissection/FACS or magnetic enrichment	C1-STRT	5072 cells	Median ~7600 UMIs (transcripts)/cell
Shekhar et al. (2016)	Mouse adult retinal bipolar cells	Manual dissection/FACS enrichment	Drop-seq	27,499 cells	8200 reads/cell
Habib et al. (2016)	Mouse hippocampus	Manual dissection/FACS sorted nuclei	Nuc-seq/SmartSeq2	1367 nuclei	2000–8000 genes/cell
Camp et al. (2015)	Fetal human and PSC-derived neocortical cells	Microdissection/C1 capture	Smarter	734 cells	2–5M reads/cell

(continued)

Table 18.1 (continued)

Publication	Tissues	Cell isolation	RNA harvest/cDNA AMP	Number of cells	Depth of sequencing
Yao et al. (2017)	Fetal human and PSC-derived brain cells	Manual dissection/FACS enrichment	Cel-Seq or SmartSeq2 for PSC cells, FRISCR/SmartSeq2 for Fetal cells	2864 Cel-Seq or 1846 SmartSeq2 for PSC cells, 472 fetal human cells	Cel-Seq: Subsampled to 20,000 UMI (transcripts)/PSC cell, SmartSeq2: ~1.6M reads/PSC cell, 1.5M reads/fetal human cell

Table listing the methods used to select, prepare, and sequence single-cell libraries from various regions of the brain. *UMI* unique molecular identifier, a proxy for transcript count. *PSC* pluripotent stem cell neuronal cell populations and identified 49 discrete cell types from just over 1600 cells: 23 interneuron, 19 excitatory, and seven non-neuronal cell types. Habib et al. profiled 1367 single nuclei extracted from the adult mouse hippocampus and identified eight major cell types. The cell types identified in these studies showed partial overlap with each other, but the attempted comparison between the Zeisel et al. (2015) study and Tasic et al. (2016) noted that some ambiguous correspondences may be due to analysis of different cortical areas, mice of different age, and different cell isolation and cDNA amplification strategies between the two studies (Tasic et al. 2016). Two studies have exclusively focused on cortical and extracortical non-neuronal cells: oligodendrocytes (Marques et al. 2016) and microglia (Matcovitch-Natan O et al. 2016). They discovered the developmental trajectories of these cell types, but also profound involvement of these cells in nervous system development, plasticity, and homeostasis.

18.3.2 Human Cortex

Profiling of adult human cortex faced additional technical hurdles. Live human tissue can only be obtained infrequently from neurosurgical operations (usually epilepsy or brain tumors), the specimens may be compromised by the existing pathology, and tissue dissociation is challenging. One study profiled 466 dissociated live cells from adult and fetal human tissues captured by a microfluidic device, and identified a variety of cell types including an array of glial and neuronal cell types (Darmanis et al. 2015). Clustering of the scRNA-seq data revealed seven neuronal clusters: two excitatory and five inhibitory. The relatively low number of cell types identified compared with the mouse studies is likely due to the low number of neurons profiled (131 total). To circumvent the difficulties in obtaining live human neurons, single nucleus profiling from post-mortem human brains was developed as an alternative. This approach, applied to 3227 single nuclei from six human cortical regions from post-mortem tissue, identified eight excitatory and eight inhibitory neuronal cell types (Lake et al. 2016). Although most scRNA-seq studies characterize gene expression using exon reads, the authors found increased cell-type resolution by analyzing reads that mapped to both exons and introns (Lake et al. 2016).

18.3.3 Other Brain Areas

scRNA-seq studies have reported cell diversity in multiple regions of the nervous system, currently in a nonsystematic fashion, usually driven by specific brain regions of interest. The diversity of sensory neurons was characterized in the dorsal root ganglion (Li et al. 2016; Usoskin et al. 2015), and 11 and 17 different sensory neuron types were identified from each study, respectively. In both studies, the cell types identified by transcriptomic signatures were shown to have distinct physiological sensory roles. The sympathetic nervous system was investigated by scRNA-seq and eight neuronal cell types were identified from the sympathetic ganglia that innervate nipple- and pilo-erector muscles (Furlan et al. 2016). In this study, several neuronal types were shown to be born embryonically but remain unspecialized until the target organ matured postnatally. Serotonergic neurons were identified by microdissected tissue regions from genetically labeled cells and then profiled by population and scRNA-seq (Okaty et al. 2015). The use of scRNA-seq allowed further subdivision of functionally distinct serotonergic cell types beyond shared lineage and anatomy. Finally, 368 mouse striatal neurons were profiled by scRNA-seq (Gokce et al. 2016), and analysis identified heterogeneity of D1 and D2 medium spiny neurons that appear to be more accurately described as phenotypic continua.

The mouse retina is known to contain many morphologically, functionally, and molecularly distinct cell types (Siegert et al. 2012), and has now been evaluated by large-scale single-cell transcriptomics. Thirty-nine distinct cell types were identified from the mouse retina after profiling 44,804 cells using Drop-seq including eight bipolar cell types (Macosko et al. 2015). In a subsequent study of more than 25,000 bipolar cells genetically labeled and enriched by FACS, 15 distinct types of bipolar cells were identified using Drop-seq (Shekhar et al. 2016). Each type identified by transcriptomics was matched to a morphological type, and two new bipolar cell types were discovered, including one type with a unique monopolar morphology.

18.4 Neural Development In Vivo and In Vitro

In addition to identifying cell types in mature brains, scRNA-seq can be used to characterize cell types throughout development in vivo and in vitro, and to infer developmental lineage relationships among the defined cell types. Even profiling a single developmental time point can be informative, as the developing tissue usually contains a mix of cells at different stages of development, and progression among types can be inferred based on the overlap in gene expression (see computational methods below). However, a more stringent way of connecting cells is by using a pulse label that marks dividing progenitors and is retained in the progeny. Developmental progression can then be inferred based on both shared genes and label retention/dilution. The labels can be genetic, viral, or chemical.

For example, a fluorescent label ("FlashTag") was injected into mouse ventricle at embryonic day (E) 14.5 to pulse label the radial glial progenitors lining the ventricle. These progenitors gave rise to label-retaining cells including newborn neurons, which migrated away from the ventricle. scRNA-seq profiling of individual label-retaining cells at several time points within the 48 h revealed gene expression signatures of early neuronal differentiation in the mouse cortex (Telley et al. 2016), including sequential phases dominated by ribosome biogenesis/translation, DNA double-strand breaks, and chemotaxis. In the hippocampus, a genetically encoded label expressed in quiescent neural stem cells aided the molecular characterization of adult neurogenesis. Single-cell transcriptomic signatures defined progression from quiescent to activated stem cells and then to intermediate progenitors, and a new computational method was employed to infer a trajectory of neuronal maturation (Shin et al. 2015). In the mouse olfactory system, scRNA-seq revealed a transient immature state, in which olfactory receptor neurons express multiple odorant receptor genes prior to maturing and selecting a single odorant receptor gene per cell (Hanchate et al. 2015). These studies have provided a precise picture of the molecular changes that accompany the process of neuronal differentiation from multiple regions of the mouse brain.

Single-cell profiling has also helped characterize progenitor diversity in the developing fetal human cortex. The diversity of human neocortical progenitors has been of interest to the field since a greater diversity of morphological types is observed in human than in mice, but the corresponding molecular distinctions are not yet known (Lui et al. 2011). Initial studies profiled the most abundant cells recovered following tissue dissociation (Darmanis et al. 2015; Camp et al. 2015; Pollen et al. 2014) and resolved the transition from radial glial (RG) progenitors to immature and then more mature neurons. Subsequent studies profiled only cells selected by expression of intracellular molecular markers of radial glia (Thomsen et al. 2016) or by microdissection of the human cortical germinal zones (Pollen et al. 2015); these approaches revealed the distinct molecular identities of two different types of radial glia: ventricular zone-restricted RG (vRG) and outer RG (oRG). Outer RGs are particularly interesting because they are present in human neocortex but rarely observed in mice. These cells have been proposed to be partially responsible for the substantial increase in primate brain size and complexity during evolution.

ScRNA-seq has been used to characterize neuronal differentiation from pluripotent stem cells (PSCs) in vitro and to compare them to primary developing cells (Yao et al. 2017; Camp et al. 2015; La Manno et al. 2016). These PSC-derived neurons and neuronal progenitors have important clinical implications for cell-based therapies, drug discovery, and modeling of human development and disease (Lancaster and Knoblich 2014). Through comparison to primary human gene expression atlas data and scRNA-seq, Yao et al. showed that several neuron types derived from hESCs are similar to primary forebrain cells. By comparison with primary single cell from the forebrain (Camp et al. 2015) and midbrain (La Manno et al. 2016), human PSC differentiation models were shown to mimic aspects of fetal differentiation in vivo, but also exhibited transcriptional signatures specific to the hPSC-derived cells. Off-target cell types were often detected among hPSC-differentiated neurons: along with the desired forebrain cell types, ventral cell types were detected in one organoid study (Camp et al. 2015), and mid/hindbrain cell types in another (Yao et al. 2017).

In summary, scRNA-seq is becoming an important tool to benchmark brain cells made in vitro, and can help identify and mitigate off-target cell types, and guide the future development of in vitro differentiation protocols to mimic the in vivo counterparts.

18.5 Methods for Single-Cell Transcriptomics

In the following section, we review methods commonly used for (1) obtaining mRNA from single cell, (2) conversion of single-cell mRNA to an NGS library, (3) computational analysis of scRNA-seq data, and (4) strategies for combining single-cell transcriptomics with other phenotypic analyses.

18.5.1 Cell, Nucleus, or Cell Content Collection

To obtain mRNA from single cells, methods have been developed to harvest an entire cell, its nucleus, or a portion of its cytoplasm (Fig. 18.4). The majority of single-cell transcriptomic studies to date profiled live cells from dissociated tissues (Tasic et al. 2016; Zeisel et al. 2015; Usoskin et al. 2015; Okaty et al. 2015). Embryonic tissue is easy to dissociate, however, with age, cells in the brain associate into elaborate and stable structures and tissue dissociation becomes a major challenge. Dissociation is usually performed by treatment with proteases and subsequent trituration. These procedures introduce several confounding factors to the final data set: cells are removed from their natural environment, some cell types may be particularly sensitive to dissociation, and portions of cells, such as dendrites and axons, are severed and mostly lost. Therefore, even when cells are processed in an 'unbiased' manner from a single-cell suspension, the cell suspension may already be biased due to differential cell survival during tissue dissociation. Additional biases can be introduced by purification steps or by microfluidic capture (see below for Fluidigm C1). For example, the first unbiased profiling study of cortical cell types did not capture any oligodendrocyte precursor cells (OPCs) (Zeisel et al. 2015), and captured very few cortical parvalbumin-expressing GABAergic cells. Cell dissociation also generates a large amount of debris that can be difficult to distinguish from nucleated cells. The debris may contain cellular mRNA and can be attached to cells or captured with individual cells (Fig. 18.4). The degree of cell-to-cell contamination will depend on the exact method of dissociation and cell isolation, and may be substantial especially if one cell type is extremely abundant, like rods and cones in the retina (Macosko et al. 2015).

Single-cell isolation can be carried out with a number of techniques. The simplest with respect to required instrumentation is manual picking of single cell aided by a dissection microscope (Hempel et al. 2007). This approach allows direct observation of each cell, but is labor-intensive. Improved throughput of this technique was achieved by the aid of a robotic setup to collect 799 cells including 622 sensory neurons from dissociated dorsal root ganglia (Usoskin et al. 2015).

Fluorescence-activated cell sorting (FACS) allows rapid purification and collection of cells based on multiparameter fluorescence detection. Each fluorescently encoded parameter is tied to a particular cellular characteristic: shape, size, specific molecular components, etc. FACS can isolate single cells by sorting directly into individual wells/tubes with lysis buffer, and is compatible with very dilute cell suspensions. Although FACS can damage fragile cells, post-sorting viability is not a concern since cells are immediately lysed in the targeted wells. Dead cells and

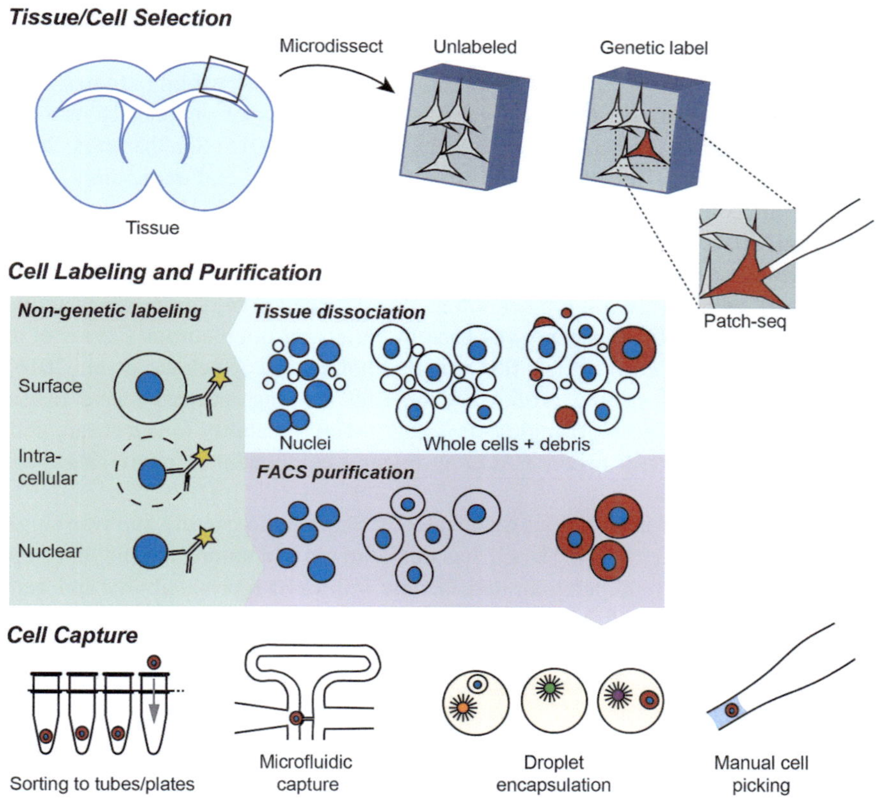

cDNA Amplification

Common methods	UMI	Read bias	Amp.	Batch amp.
SMARTer	No	Low 3'	PCR	No
Smart-seq2	No	None or low 3'	PCR	No
STRT	Yes	5'	PCR	Yes
CEL-seq	Yes	3'	IVT	Yes
Drop-seq	Yes	3'	PCR	Yes

Fig. 18.4 Processing steps involved in scRNA-seq studies in the brain. Diagram depicts steps in recovery of single cells or cell contents, labeling and purification strategies, cell or mRNA capture methods, and cDNA amplification strategies used in mouse single-cell transcriptomics studies

doublets can be avoided by careful gating, but it is always recommended to perform additional controls for sorting efficiency and accuracy (selection of single cell as opposed to doublets[4]), which can be achieved by test sorts into Terasaki plates for microscopic evaluation (Tasic et al. 2016). FACS can also be used to enrich for a

[4]Two cells in a droplet are referred to as a "doublet". These cells may be attached to each other (due to incomplete dissociation into single cell, or re-association of "sticky" cells) or may just have happened to be trapped in a single droplet.

certain cellular population or to remove debris (Zeisel et al. 2015; Shekhar et al. 2016), and single cells can subsequently be captured by another method. In that case, preservation of viability is important. In recent brain profiling studies, cells have been evaluated and purified by FACS on the basis of genetically encoded markers (Tasic et al. 2016; Zeisel et al. 2015 Okaty et al. 2015; Shekhar et al. 2016; Marques et al. 2016), dye accessibility (Florio et al. 2015), and expression of cell surface, intracellular or nuclear markers (Johnson et al. 2015; Habib et al. 2016; Lake et al. 2016; Li et al. 2016; Thomsen et al. 2016; Florio et al. 2015). FACS is often used as a purification method upstream of droplet- or microfluidic-based scRNA-seq methods for cell-type enrichment and depletion of debris. For example, GFP from *Htr3a-GFP* transgene was used to enrich for interneurons (Zeisel et al. 2015) and *Vsx2-GFP* to enrich for retinal bipolar cells (Shekhar et al. 2016). Although single-cell sorting proficiency requires training and expensive instrumentation, the robustness of single-cell isolation and the ability to select rare cells from dilute samples still makes FACS an integral component in many single-cell profiling strategies.

A microfluidic device, C1, was developed by Fluidigm to capture and lyse single cell, and prepare their cDNA. The C1 has miniaturized and automated the first steps in scRNA-seq library preparation, and thereby improved reproducibility and gene detection (Pollen et al. 2014; Wu et al. 2014). This device has been adopted by multiple groups to study cell types from the developing and mature brain (Pollen et al. 2015; Zeisel et al. 2015; Pollen et al. 2014; Marques et al. 2016). It automates the single-cell capture and cDNA preparation procedure within integrated fluidic circuits (IFCs) containing 96–800 cell capture sites that become enclosed into separate micro-chambers for subsequent biochemical reactions. Following cell capture, the IFC is imaged to document each captured cell, thus empty wells, dead cells, and wells with multiple cells can be omitted from sequencing library preparation and analysis. The limitations of this approach are that optimal capture efficiency requires cells to be loaded at a high concentration, and that cells of different sizes and shapes are not captured with equal efficiency due to fixed pore size on each IFC. The C1 has also been used to isolate nuclei, but it has proven difficult to distinguish between a single nucleus and doublet nuclei at the capture sites (Lake et al. 2016). Thus, computational approaches were needed to identify and remove the data that was likely to be derived from doublets (Lake et al. 2016).

Recently, nanodroplet-based cell-encapsulation approaches, Drop-seq and inDrop, have enabled dramatic increases in the numbers of cells analyzed by scRNA-seq (Macosko et al. 2015; Klein et al. 2015). The key process is encapsulation of single cells within individual aqueous droplets and barcoding of all cDNA molecules in the droplet phase prior to batch amplification. This technology has expanded the scale of single-cell transcriptomes generated and analyzed per experiment by almost 100-fold while substantially reducing the cost of library production per cell. Two studies have profiled retinal cell types by Drop-seq, and although there is a large variability in the number of reads per cell, and cells are sequenced at a low read depth, good cell-type resolution is conferred by the high number of cells profiled (Macosko et al. 2015; Shekhar et al. 2016). To date, most studies employing these techniques sequenced many cells to low read depth (usually <10,000 reads per cell). However, this, at least for Drop-seq, does not appear to be molecular limitation of the technique, but a limitation based on the cost of sequencing and the necessary read depth to resolve the desired cell types (Shekhar et al. 2016). The limitations include large numbers of cells that are required for input and the fact that many cells are discarded to ensure two cells are not encapsulated into one droplet. In addition, the focused 3′-end sequencing is necessary since cDNAs are amplified as a pool, and each transcript is given a cell-specific 3′-barcode that must be sequenced with every read to identify the cell of origin for each transcript. As rapidly as these methods are developed, commercial products follow with the promise to standardize and simplify their adoption. Droplet technologies are being commercialized by Dolomite Microfluidics, 1CELLBIO, and 10× Genomics, and high-throughput cell deposition, imaging and barcoding systems are being developed by Wafergen (now part of TakaraBio).

Acquisition of scRNA-seq information does not require live and intact cells, and can rely on the nucleus or cytoplasm from a single cell as the starting material. FACS and microfluidic isolation have been used with slight modifications for nuclei (Lake et al. 2016; Krishnaswami et al. 2016) and fixed cells (Thomsen et al. 2016). Isolation and profiling of nuclei has the advantage that cell viability is not a concern —a major advantage when processing adult human neural tissue that is refractory to dissociation. Nuclear isolation can also be applied to frozen archived tissue (Lake et al. 2016), and nuclei are unlikely to be as differentially sensitive to dissociation as live cells. Therefore, cell-type ratios obtained by nuclear profiling may be more similar to the ones in situ. Both fixed cell profiling and nuclear profiling have the advantage that intracellular and/or nuclear antigens can be used to perform FACS-based enrichment (Habib et al. 2016; Thomsen et al. 2016; Krishnaswami et al. 2016). This marking is particularly useful in tissues from humans and other non-genetically accessible organisms to enrich for defined populations of cells. Transcriptomic data from nuclei are not interchangeable with data from whole cells. Less mRNA is isolated from nuclei, and a high percentage of reads map outside the reference transcriptomes and to introns and intergenic regions (Lake et al. 2016).

Tissue dissociation can be completely avoided by isolating mRNA from cells using a membrane permeable oligonucleotide mRNA capture (Lovatt et al. 2014), or by aspirating the cytoplasm of a cell by patch pipette, such as in Patch-seq (Cadwell et al. 2016; Foldy et al. 2016; Fuzik et al. 2015). These techniques are much lower in throughput, but permit integration of transcriptomic data with other data modalities—an essential step toward integrative neuronal phenotyping.

The choice of the cell capture technique depends on a variety of factors, including tissue source, abundance of cells, doublet tolerance, full-length mRNA coverage, and the interest in phenotypes beyond transcriptomics. For example, doublet contamination varies dramatically among techniques for capturing single cells: 0–2.3% with FACS (Tasic et al. 2016; Jaitin et al. 2014), 0.36–11.3% with Drop-seq (Macosko et al. 2015), and 11–44% with Fluidigm C1 (Macosko et al. 2015; Shalek et al. 2014). Each technique requires care in experimental design and regular monitoring of controls to mitigate the impact of doublet contamination on cell-type identification. If tissue is abundant, and overview of cell-type landscape is desired, Drop-seq techniques are suitable. However, if cells are rare, but can be selected based on marker expression, FACS would be more appropriate. Combination of the two may be ideal in some cases (Shekhar et al. 2016).

18.5.2 From Single-Cell mRNA to an NGS Library

Multiple strategies exist to generate and amplify cDNA from single cells and generate RNA-seq libraries. Here, we provide an overview over the methods and discuss the variations in amplification strategy, mRNA coverage, sensitivity, scalability, and susceptibility to batch effects (reviewed in Grun and van Oudenaarden 2015; Kolodziejczyk et al. 2015).

All single-cell RNA-seq strategies rely on converting mRNA to cDNA through reverse transcription (RT) and subsequent cDNA amplification. Reverse transcription is usually performed with oligo-dT primers. Some methods use oligo-dT primers containing unique molecular identifiers (UMIs) composed of a random string of nucleotides (Islam et al. 2014; Kivioja et al. 2012; Shiroguchi et al. 2012). UMIs can mitigate the distortion of transcript abundance during cDNA amplification; however, they are limited to methods where only the 3′- or 5′-end of the transcript is analyzed. Once cDNA from a single cell is created, it is amplified exponentially by PCR, or linearly by in vitro transcription.

Most linear amplification methods require more molecular steps and have not been shown to be more sensitive than PCR-based methods in the detection of endogenous transcripts (Grun et al. 2014; Hashimshony et al. 2012; Picelli et al. 2013). Many PCR-based strategies rely on so-called "template switching" to incorporate primer binding sites at the 5′ end of a transcript (Tang et al. 2009; Islam et al. 2011). Template switching followed by PCR amplification (such as SMARTer or SMART-seq) is commonly used for full-length cDNA coverage, albeit with varied degree of 3′-end bias depending on the exact protocol (Ramskold et al. 2012; Picelli et al. 2013). Other methods focus on 5′-end [STRT (Islam et al. 2011)] or 3′-end of cDNAs [e.g., Drop-seq (Macosko et al. 2015), CEL-seq (Hashimshony et al. 2012), MARS-seq (Jaitin et al. 2014), and Quartz-seq (Sasagawa et al. 2013)]. Sensitivity and technical noise of scRNA-seq reactions can be monitored by the addition of External RNA Control Consortium (ERCC) (Consortium T.E.R.C. 2005) spike-in mRNAs to each sample. These polyadenylated mRNAs are synthetic or of bacterial origin and cannot be confused with mammalian mRNAs. ERCCs vary in concentration over six orders of magnitude, and span 250–2000 nucleotides in length. ERCC spike-in reads provide a measure of single-molecule detection sensitivity that can be used to model technical noise, and can thus establish genes that display higher variance than technical noise (Brennecke et al. 2013). The efficiency of single-molecule detection by different scRNA-seq methods is typically in the range of 5–20% (Yao et al. 2017; Tasic et al. 2016; Zeisel et al. 2015; Macosko et al. 2015; Thomsen et al. 2016; Grun et al. 2014).

Application of cell-specific barcodes added either in the first step or after cDNA amplification allows parallel processing of hundreds of cells at one time, a substantial increase in the scale of scRNA-seq experiments (Islam et al. 2011; Hashimshony et al. 2012). Microfluidic technology also provided a solution to increase throughput of cell capture and cDNA amplification. Each C1 chip can capture, reverse transcribe, and amplify the cDNA from up to 96 or 800 cells at once. The number of cells profiled by these scRNA-seq studies was dwarfed by the first Drop-seq study, which profiled >40,000 retinal cells (Macosko et al. 2015), although each individual cell was sampled at substantially lower read depth. Although droplet-based methods allow examination of a large number of cells, the sequencing depth per cell varies substantially, and the sequencing depth is usually limited. In addition, the current methods mostly read the 3′-ends of the expressed transcripts. Thus, if full-length mRNA coverage is desired, the droplet-based methods are not suitable in their current state.

The resolution of single-cell transcriptomic experiments is frequently limited by the number of cells profiled and/or the number of reads evaluated per single cell. Given a fixed total sequencing depth, the number of reads per cell decreases as the number of cells profiled increases. Although the ideal case is high-depth sequencing of large numbers of cells, until detection of new cell types and cell-type resolution reaches a plateau, this is usually not feasible due to budgetary constraints. As a result, this depth-vs-cell number trade-off is a practical consideration for most single-cell profiling experiments. For detection of overall transcriptomic cell-type landscape, if cells are abundant, Drop-seq technologies are appropriate. In the cortex, transcriptomic profiles of excitatory neurons can be very similar, and many genes that distinguish them are not expressed at a high level: therefore high read depth is informative to distinguish between these neuronal types (Tasic et al. 2016; Li et al. 2016).

After the amplification of cDNA, some of the procedures (e.g., Smart-seq) require additional steps for preparation of cDNA libraries for next-generation sequencing (NGS). Specific adapters are attached to the amplified cDNA fragments to accommodate sequencing on an NGS platform. This usually involves fragmentation by ultrasonic shearing and adapter ligation, or random transposon insertion and PCR amplification ("tagmentation" by Nextera procedures). After that the libraries are sequenced on an NGS platform (e.g., Illumina HiSeq), and the resulting NGS reads are stretches of nucleotides (usually 50–100 bases) that identify the expressed transcript, and additional short barcode reads that identify the cell of origin of that transcript.

18.5.3 Quality Control of scRNA-seq Data

scRNA-seq has proven to be a valuable neuroscience tool, and its rapid development has led to a variety of methods, each with its own strengths and weaknesses. Different techniques vary in their scale, signal-to-noise ratio, read bias, and susceptibility to batch effects and artifacts. Quality control is often ad hoc, and the field lacks a common set of standards. Thus, data interpretation and navigation can be challenging.

Researchers conducting studies currently often rely on multiple metrics to evaluate the quality of library preparation and scRNA-seq data. Successful cDNA amplification can be determined in some methods (such as Smart-seq) by analyzing the cDNA fragment distribution on a fragment analyzer. Quality control is also carried out after sequencing. Regular evaluation of the number and percentage of reads mapping to genes, introns, and intergenic regions can often identify outlier libraries that may have technical problems. After sequencing, outlier cells that have low numbers of unique genes or UMIs detected per cell are often omitted from further analysis. Cells with low reads could reflect the poor health of the input cells, or failed steps in the scRNA-seq process. Data from these basic quality control steps are usually reported in single-cell transcriptomic studies.

scRNA-seq results are sensitive to experimental variation and batch effects. Sufficient experimental replication is essential to ensure that any observed variability is not an artifact. Batch effects can be introduced at different stages of single-cell processing. Differences in sample handling, differentiation, and dissociation can lead to batch differences in gene expression. Standardization of sample derivation and handling is essential to minimize this type of batch effect. Inclusion of standards such as ERCCs or a standard control cell sample can help evaluate if the observed effect is of technical or biological origin. Batch effects can also arise from the scRNA-seq library preparation steps. Some methods are particularly prone to batch effects, which appear in the data as various degrees of cell-to-cell contamination (e.g., RNA from one cell is labeled by the barcode from another cell). For example, Cel-seq requires careful monitoring, and through analysis of the proper controls, some of this crosstalk can be computationally mitigated (Yao et al. 2017; Paul et al. 2015).

Because of the large variety of methods used in scRNA-seq studies, direct comparison of data between studies is often challenging. Differences in methodological read bias (e.g., full-length or 3′-biased), sequencing depth, sensitivity, and gene coverage all contribute to this problem. As the field of scRNA-seq matures, implementation of uniform quality control standards will be important. Adoption of only a few standard methods for scRNA-seq will aid in the comparison of scRNA-seq data between studies, but multiple methods will probably always be necessary, dictated by budget, expertise, and experimental objectives.

Finally, conclusions drawn from scRNA-seq should be validated with independent experiments. Co-expression or mutually exclusive expression of genes can be confirmed by qRT-PCR, RNA in situ hybridization (ISH), or immunohistochemistry (IHC) (Pollen et al. 2015; Yao et al. 2017; Zeisel et al. 2015; Usoskin et al. 2015; Thomsen et al. 2016; La Manno et al. 2016). Correlating gene expression-based cell types to orthogonal phenotypic properties reinforces the conclusion that certain cell types are distinct from each other by more than just gene expression. Therefore, it is always highly recommended to correlate expression of select marker genes, especially the newly discovered ones, with other cellular properties.

18.6 Computational Approaches to Identify Clusters in scRNA-seq Data

Historically, the existence of distinct neuronal types was implied using one or a handful of molecular markers, often correlated to a separate data modality such as location, electrophysiology, or connectivity patterns. However, individual or small numbers of genes are not sufficient to develop models of the transcriptomic space. With simultaneous measurement of thousands of genes in large numbers of individual cells, correlated gene expression patterns provide a more comprehensive depiction of the cell-type landscape. Various clustering and machine learning methods can then identify regions in gene expression space where cells are more likely to exist, thus defining "cell types" in a purely transcriptomic sense.

In general, the procedure to identify data-driven transcriptomic cell types follows standard approaches to clustering large data sets. This process includes the selection of genes/features, dimensionality reduction, establishment of a cell–cell distance metric, clustering, and assessment of cluster robustness. Although not all published methods explicitly include all of these steps, most incorporate variations of these procedures into the overall clustering algorithm (Table 18.2). The initial selection of genes for clustering is often based on an assessment of gene expression variance. To assess variance, first a dispersion model is fitted to expression of all genes or spike-in controls (ERCCs, see above). "High-variance" genes are then selected as those showing dispersion above the model fit. Alternatively, high-variance genes can also be selected based on a global variance measure, such as contribution to the first few principal components.

Once the initial set of high-variance genes has been selected, many published methods reduce the dimensionality of the gene expression space by collapsing the expression data onto a smaller number of features. Standard dimensionality reduction techniques such as principal component analysis (PCA) and weighted gene co-expression network analysis (WGCNA) have both been widely used for single-cell analysis in neuronal data sets. PCA identifies a set of orthogonal features, each a linear-weighted combination of all genes, which maximally account for the variance. WGCNA, by contrast, clusters genes by similarity of expression, and identifies modules (or groups) of genes that follow the same global patterns. Overall, these two methods are similar, but impose different constraints on the ultimate features—orthogonality—and support over all genes for PCA, and non-negativity and mutually exclusive module membership for WGCNA. Ultimately, both of these methods (along with others) provide a reduced set of features on which cells can be clustered.

Once the appropriate genes and/or features for clustering have been selected, the next step is to establish a dissimilarity metric among cells and to apply a clustering algorithm to identify groups of cells. Here, the published methods use a variety of dissimilarity metrics; in addition to standard correlation and Euclidean distance metrics, there are topological overlap measures, t-SNE embedding, and biclustering-based distance metrics. Once a metric has been chosen, it is

Table 18.2 Clustering methods from recent single-cell transcriptomic studies of the brain

Publication	Gene selection	Dimensionality reduction	Distance metric	Clustering algorithm	Assessment of robustness	Iterative
Usoskin et al. (2015)	Overdispersed genes	PCA	Distance in PC space	PCA	Range of variance filters and outgroups	Yes
Pollen et al. (2014)	Top genes loaded on principal components		Euclidean distance	Hierarchical clustering	Bootstrap aggregating	
Pollen et al. (2015)		PCA	Expectation Maximization			
		PCA	Euclidean distance	Hierarchical clustering	Bootstrap aggregating	
Okaty et al. (2015)	Overdispersed genes		Correlation	Hierarchical clustering	Differential gene expression	
Zeisel et al. (2015)	Overdispersed genes showing some degree of correlation		Biclustering using BackSPIN		Bootstrap aggregating	Yes
Macosko et al. (2015)	Overdispersed genes	PCA and t-SNE	t-SNE embedding	Density clustering	Differential gene expression	
Thomsen et al. (2016)	Overdispersed genes, benchmarked by spike-ins	WGCNA	Correlation	Hierarchical clustering	Differential gene expression	
Cadwell et al. (2016)	Overdispersed genes		Euclidean distance	Affinity propagation	Bootstrap aggregating	

(continued)

Table 18.2 (continued)

Publication	Gene selection	Dimensionality reduction	Distance metric	Clustering algorithm	Assessment of robustness	Iterative
Li et al. (2016)	Top genes loaded on principal components		Correlation	Hierarchical clustering	Intra- and inter-cluster correlation	
Tasic et al. (2016)	Overdispersed genes, benchmarked by spike-ins	PCA	Euclidean distance	Hierarchical clustering, binary splitting	Random forest validation	Yes
	Overdispersed genes, benchmarked by spike-ins	WGNCA	Topological overlap based on adjacency matrix	Hierarchical clustering	Random forest validation	Yes
Lake et al. (2016)	Overdispersed genes		Correlation	Hierarchical clustering, binary splitting, Reassignment of cells using Random Forest	Random forest out-of-bag error rate	Yes
Gokce et al. (2016)		t-SNE	t-SNE embedding	Density clustering	Differential gene expression	Yes
	Genes with correlation above a threshold	robust PCA	Distance in PC space	Examination of bimodality		
Marques et al. (2016)	Overdispersed genes showing some degree of correlation		Biclustering using BackSPIN		Differential gene expression	
Shekhar et al. (2016)		PCA	Jaccard overlap index	Louvain algorithm for graph clustering	Differential gene expression, bootstrap aggregating	
Habib et al. (2016)		PCA and t-SNE with cosine distance	t-SNE embedding	Density clustering	Differential gene expression	Yes

(continued)

Table 18.2 (continued)

Publication	Gene selection	Dimensionality reduction	Distance metric	Clustering algorithm	Assessment of robustness	Iterative
Camp et al. (2015)	Top genes loaded on principal components	t-SNE	t-SNE embedding	Density clustering	Differential gene expression	
Yao et al. (2017)	Overdispersed genes showing some degree of correlation	WGNCA	Topological overlap based on adjacency matrix	Hierarchical clustering	Bootstrap aggregating	Yes

Table listing the methods used to cluster single-cell data and identify cell types from different nervous system tissues. *PCA* principal component analysis, *WGCNA* weighted gene co-expression network analysis, *t-SNE* t-stochastic neighbor embedding incorporated into a clustering algorithm—this includes standard hierarchical clustering methods (most often with complete linkage or Ward's method), biclustering, affinity propagation, density-based clustering, and graph theory-based clustering methods. Although no clustering method has been shown to be superior in all instances, the use of a wide range of methods suggests that neuronal gene expression data is amenable to standard cluster analyses.

Given that the number of genes detected in a single cell is on the order of thousands, several published methods use an iterative approach to transcriptomic cell-type identification. Iterative clustering assumes that the optimal selection of genes and features depends on the group of cells being considered. For example, the ideal features to discriminate excitatory and inhibitory neurons may not be optimal to discriminate subtypes of excitatory neurons from each other. A key element to iterative clustering is a set of termination criteria that stop the iterations; these include limits on the number of cells, the number of variably expressed genes, or the inability to distinguish robust groupings of cells. For data sets comprising a wide variety of cell types, of which some are transcriptionally similar, iterative clustering has been regularly used, as it can subdivide broad classes of cells identified in the first pass into smaller, more refined groups in subsequent iterations (Tasic et al. 2016).

An important question arising in all clustering analyses is the robustness of the clusters. In general, there are three major ways to assess cluster robustness: cluster tightness, cluster reproducibility, and cluster validation using a different data modality. The first assessment—cluster tightness—is implicit in all methods, since clusters are generally selected to have high intra-cluster similarity and low inter-cluster similarity. Certain published methods have assessed cluster tightness explicitly, and this serves as a useful validation of the final clusters and their separation. The second assessment—cluster reproducibility—examines whether cells fall into the same clusters repeatedly, using subsets of the data or using different methods. Clustering different subsets of the data is a version of bootstrap aggregation, and is incorporated into many of the published methods as a way to characterize the reproducibility of the final results, especially when the clustering methods themselves have a degree of stochasticity (Yao et al. 2017). Less common, but very informative, is the use of multiple clustering methods to identify whether clusters can be reproducibly identified; Pollen et al. (2014), Tasic et al. (2016), and Shekhar et al. (2016) have used multiple clustering methods to show that the final clusters are reliably discovered from the data. The final way to assess cluster robustness is to compare the results against those obtained from a different data modality, such as electrophysiology, connectivity, or morphology (see below).

The methods above can also be applied to cells isolated during development to identify new markers and new developmental cell types. For example, the examination of clusters of cells with previously unobserved combinations of genes has resulted in the identification of novel markers for intermediate cell types such as different types of radial glia (Pollen et al. 2015; Thomsen et al. 2016). Computational methods have also been developed to infer lineage relationships among cell types identified using single-cell RNA-sequencing. The underlying idea behind these methods is that there is overlap in gene expression patterns among cells that share a common lineage. Identifying these links is challenging without a priori knowledge of the genes involved; thus, transcriptome characterization using scRNA-seq has afforded a major breakthrough in computational lineage reconstruction. One common approach for generating lineage-related hypotheses from single-cell data includes drawing minimum spanning trees after dimensionality reduction in gene expression space; these include Monocle (Trapnell et al. 2014), Waterfall (Shin et al. 2015), TSCAN (Ji and Ji 2016), and Eclair (Giecold et al. 2016). Alternative graph theory-based approaches to linking cells in a reduced dimensionality projection, such as Wanderlust (Bendall et al. 2014) and Wishbone (Setty et al. 2016) have been used to uncover lineage relationships using single-cell mass cytometry data. This class of methods has been applied specifically to single-cell neuronal RNA-seq data to elucidate cell trajectories in adult neurogenesis (Shin et al. 2015) as well as in vitro neuronal development (Camp et al. 2015). These minimum spanning tree methods enable the discovery of linear cell trajectories and simple branching patterns, which are apparent from the arrangement of cells in a reduced dimensionality representation. Alternatively, for systems with more complex lineage relationships, other methods seek to find transcription factor relationships key to branching and cell fate decisions (Yao et al. 2017). Therefore, computational methods have already begun to relate gene expression data to additional phenotypic characteristics such as developmental origin, and the incorporation of this type of data to additional modalities is crucial to an increasingly refined understanding of cell types and their function.

18.7 Combining Single-Cell Transcriptomics with Other Experimental Modalities

Initial transcriptional analyses of brain cell types show that there are likely hundreds of cell types present in the brain (Fig. 18.3). Ultimately, transcriptional classification into types generates predictions about groups of cells that may have functional differences. A given sampling and clustering method might yield distinct clusters in transcriptomic space, but further testing using alternative data modalities is necessary to determine the extent to which these clusters are biologically meaningful. As a result, it is crucial to correlate these transcriptomically defined cell types with other neuronal characteristics including electrophysiology, morphology, protein expression, location, axonal projection pattern, and function.

The notion of a cell type is not limited to transcriptomics: any modality that yields quantitative or binary data can be used to classify cells into putative types. If multiple modalities result in consistent segregation of the same class of cells, then that class is more likely to be a biologically relevant cell type. For example, neurons in layer 6a of the mouse cortex can be separated into at cortico-cortical and cortico-thalamic classes based on their projection patterns; these same classes of cells also have distinct transcriptional signatures, which are strong enough to segregate them by unsupervised clustering of transcriptomic data (Kee et al. 2017). The concordance between multiple data modalities is unlikely to be one-to-one; rather, a given data modality may show subdivisions or overlap among groups segregated by a different type of data. Overall, clustering across multiple modalities is necessary to compile a taxonomy of cell types with putative functional differences.

There are two broad approaches to link single-cell gene expression data to other modalities. The first class of methods constitutes a "forward" or prospective approach, where cells are labeled by some aspect of their molecular profile prior to acquiring data from other modalities. For example, cells can be labeled by one or more mouse transgenes. Unfortunately, for cells in the mouse cortex, all tested Cre/reporter mice label multiple transcriptomic cell types, often with only the ability to resolve broad cell classes (Tasic et al. 2016). Careful gene selection and intersectional strategies using Cre-Lox and related systems may allow generation of "Rosetta stone" lines, in which only a single specific transcriptomic cell type is labeled (Madisen et al. 2012). In organisms for which transgenic lines do not exist or are rare, antibody or viral labeling strategies offer potential avenues to achieve prospective cell-type marking (Liu et al. 2013).

The second class of methods constitutes a "post-hoc" or retrospective approach, where transcriptomic cell identity is determined at the same time or after other modalities. These methods include extracting cellular contents after electrophysiological recordings and cell-filling neurons for morphological reconstruction (Cadwell et al. 2016; Fuzik et al. 2015), as well as RNA ISH or IHC after electrophysiological recording, in vivo imaging, or projection labeling (Chen et al. 2015; Lee et al. 2014). The transcriptomic data from the retrospective approaches (e.g., Patch-seq) may be of lower quality due to contamination by genes from neighboring cells or incomplete collection of the cytoplasm (Fuzik et al. 2015). As a result, RNA-seq data collected is sometimes characterized by mapping (though machine learning classifiers) to already-defined transcriptomic cell types (Fuzik et al. 2015). ISH or IHC can also be used retrospectively, but they are not genome-wide, and careful marker selection based on scRNA-seq is important to provide unambiguous conclusions.

Both prospective and retrospective methods benefit from pre-existing transcriptionally defined cell types, and they are not mutually exclusive: when intersected they can be used to further refine the cell type examined. Ideally, tool selection used for prospective isolation is informed by transcriptomically defined cell types such that the cell-type enrichment is as high as possible.

Currently, single-cell transcriptomics data has been combined with several modalities, including electrophysiology, morphology, connectivity, and lineage. For linking to electrophysiology, the most successful method so far has been to extract cellular content and profile it using scRNA-seq methods after in vitro slice electrophysiological recording (Patch-seq). This approach yields information about gene expression and intrinsic electrical properties from the same cell (Cadwell et al. 2016; Fuzik et al. 2015). If Patch-seq is conducted with biocytin in the patch pipet, morphological neuronal reconstruction may be accomplished along with transcriptional profiling and electrophysiology, although collection of the whole cell body may interfere with acquisition of high-quality morphology. With respect to connectivity, retrograde labeling of cells using a fluorescent reporter can identify cells with selected projection targets; these cells can subsequently be isolated and processed for scRNA-seq (Tasic et al. 2016). Recent barcoding strategies infer connectivity through sequencing (Kebschull et al. 2016), and provide a promising avenue toward potentially linking transcriptomics and projection patterns using a sequencing-only approach. Viral sequence-based barcoding strategies can also be used to establish clonal relationships and derive a cell lineage. A viral barcode expressed as part of the reporter mRNA was used in one scRNA-seq study to establish the clonal relationship between transcriptomic cell types (Yao et al. 2017). Finally, with the advent of in situ-based methods where the expression of large numbers of genes can be read simultaneously, it is possible to obtain positional information about individual cells and assess their transcriptomes directly in the tissue. Methods such as highly multiplexed FISH (Chen et al. 2015; Lubeck et al. 2014) and fluorescent in situ sequencing (Lee et al. 2014) combine single-cell transcriptional profiling and anatomy and eliminate the single-cell dissociation step. As of yet, published sets of data combining transcriptomics with other modalities are relatively small, which allows only for de novo clustering on a small scale (Cadwell et al. 2016; Foldy et al. 2016; Fuzik et al. 2015).

18.8 Conclusions and Future Directions

Single-cell transcriptomic analysis is rapidly becoming integral to many neuroscience studies: in the past few years, it has created the most comprehensive set of cell-type identities to date for a few regions for the brain. Yet, it has a long way to go. The mouse brain is composed of $\sim$75 million neurons, but so far scRNA-seq has only profiled tens of thousands of neurons and support cells from the mouse and human brains. Further definition and exploration of transcriptomic cell types will be central to building a complete parts list for the mouse and human brains.

Though clearly successful, scRNA-seq experiments have limitations. Studies so far have uniformly relied on tissue dissociation prior to analysis, thus disrupting tissue architecture, morphology, and connectivity, and retaining spatial information only as fine as the microdissections employed. Especially sensitive, rare, or unmarked cells are frequently absent from the current taxonomies. Further developments of in situ technologies, which would permit molecular (and other) analyses of cells in their natural proportions and in their tissue context with better spatial resolution, are critical for complete understanding of the cell-type landscape. To build a complete atlas of brain cell types that can be useful across all of neuroscience, the transcriptomic cell types must also be evaluated for other phenotypes including morphology, physiology, connectivity, lineage, and function. Techniques and tools are being developed to allow multimodal analysis of neurons and will be essential for building a more complete taxonomy of brain cell types that incorporate many basic neuronal properties. The joint outcome will be a better understanding of the molecular processes at work in different cell types in development and adulthood, and their cell-type-specific involvement in nervous system disorders.

References

Armananzas R, Ascoli GA (2015) Towards the automatic classification of neurons. Trends Neurosci 38(5):307–318

Ascoli GA et al (2008) Petilla terminology: nomenclature of features of GABAergic interneurons of the cerebral cortex. Nat Rev Neurosci 9(7):557–568

Belgard TG et al (2011) A transcriptomic atlas of mouse neocortical layers. Neuron 71(4):605–616

Bendall SC et al (2014) Single-cell trajectory detection uncovers progression and regulatory coordination in human B cell development. Cell 157(3):714–725

Brennecke P et al (2013) Accounting for technical noise in single-cell RNA-seq experiments. Nat Methods 10(11):1093–1095

Cadwell CR et al (2016) Electrophysiological, transcriptomic and morphologic profiling of single neurons using patch-seq. Nat Biotechnol 34(2):199–203

Camp JG et al (2015) Human cerebral organoids recapitulate gene expression programs of fetal neocortex development. Proc Natl Acad Sci U S A 112(51):15672–15677

Cembrowski MS et al (2016) Hipposeq: a comprehensive RNA-seq database of gene expression in hippocampal principal neurons. Elife 5:e14997

Chen KH et al (2015) RNA imaging. Spatially resolved, highly multiplexed RNA profiling in single cells. Science 348(6233):aaa6090

Chiu IM et al (2014) Transcriptional profiling at whole population and single cell levels reveals somatosensory neuron molecular diversity. eLife, vol 3

Consortium T.E.R.C. (2005) The external RNA controls consortium: a progress report. Nat Meth 2(10):731–734

Darmanis S et al (2015) A survey of human brain transcriptome diversity at the single cell level. Proc Natl Acad Sci U S A 112(23):7285–7290

DeFelipe J et al (2013) New insights into the classification and nomenclature of cortical GABAergic interneurons. Nat Rev Neurosci 14(3):202–216

Doyle JP et al (2008) Application of a translational profiling approach for the comparative analysis of CNS cell types. Cell 135(19013282):749–762

Eberwine J et al (1992) Analysis of gene expression in single live neurons. Proc Natl Acad Sci 89(7):3010–3014

Florio M et al (2015) Human-specific gene ARHGAP11B promotes basal progenitor amplification and neocortex expansion. Science 347(6229):1465–1470

Foldy C et al (2016) Single-cell RNAseq reveals cell adhesion molecule profiles in electrophysiologically defined neurons. Proc Natl Acad Sci U S A 113(35):E5222–E5231

Furlan A et al (2016) Visceral motor neuron diversity delineates a cellular basis for nipple- and pilo-erection muscle control. Nat Neurosci 19(10):1331–1340

Fuzik J et al (2015) Integration of electrophysiological recordings with single-cell RNA-seq data identifies neuronal subtypes. Nat Biotechnol

Garcia-Marin V, Garcia-Lopez P, Freire M (2007) Cajal's contributions to glia research. Trends Neurosci 30(9):479–487

Giecold G et al (2016) Robust lineage reconstruction from high-dimensional single-cell data. Nucl Acids Res 44(14):e122

Gokce O et al (2016) Cellular taxonomy of the mouse striatum as revealed by single-cell RNA-Seq. Cell Rep 16(4):1126–1137

Grun D, van Oudenaarden A (2015) Design and analysis of single-cell sequencing experiments. Cell 163(4):799–810

Grun D, Kester L, van Oudenaarden A (2014) Validation of noise models for single-cell transcriptomics. Nat Methods 11(6):637–640

Habib N et al (2016) Div-seq: single-nucleus RNA-seq reveals dynamics of rare adult newborn neurons. Science 353(6302):925–928

Hanchate NK et al (2015) Single-cell transcriptomics reveals receptor transformations during olfactory neurogenesis. Science 350(6265):1251–1255

Harris JA et al (2014) Anatomical characterization of Cre driver mice for neural circuit mapping and manipulation. Front Neural Circuits 8:76

Hashimshony T et al (2012) CEL-seq: single-cell RNA-seq by multiplexed linear amplification. Cell Rep 2(3):666–673

He M et al (2016) Strategies and tools for combinatorial targeting of GABAergic neurons in mouse cerebral cortex. Neuron 92(2):555

Heiman M et al (2008) A translational profiling approach for the molecular characterization of CNS cell types. Cell 135(19013281):738–748

Hempel CM, Sugino K, Nelson SB (2007) A manual method for the purification of fluorescently labeled neurons from the mammalian brain. Nat Protoc 2(11):2924–2929

Islam S et al (2011) Characterization of the single-cell transcriptional landscape by highly multiplex RNA-seq. Genome Res 21(7):1160–1167

Islam S et al (2014) Quantitative single-cell RNA-seq with unique molecular identifiers. Nat Methods 11(2):163–166

Jaitin DA et al (2014) Massively parallel single-cell RNA-seq for marker-free decomposition of tissues into cell types. Science 343(6172):776–779

Ji Z, Ji H (2016) TSCAN: pseudo-time reconstruction and evaluation in single-cell RNA-seq analysis. Nucl Acids Res 44(13):e117

Johnson MB et al (2015) Single-cell analysis reveals transcriptional heterogeneity of neural progenitors in human cortex. Nat Neurosci 18(5):637–646

Junker JP, van Oudenaarden A (2015) Single-cell transcriptomics enters the age of mass production. Mol Cell 58(4):563–564

Kebschull JM et al (2016) High-throughput mapping of single-neuron projections by sequencing of barcoded RNA. Neuron 91(5):975–987

Kee N et al (2017) Single-cell analysis reveals a close relationship between differentiating dopamine and subthalamic nucleus neuronal lineages. Cell Stem Cell

Kirkeby A et al (2017) Predictive markers guide differentiation to improve graft outcome in clinical translation of hESC-based therapy for Parkinson's disease. Cell Stem Cell

Kivioja T et al (2012) Counting absolute numbers of molecules using unique molecular identifiers. Nat Methods 9(1):72–74

Klein AM et al (2015) Droplet barcoding for single-cell transcriptomics applied to embryonic stem cells. Cell 161(5):1187–1201

Kolodziejczyk AA et al (2015) The technology and biology of single-cell RNA sequencing. Mol Cell 58(4):610–620

Krishnaswami SR et al (2016) Using single nuclei for RNA-seq to capture the transcriptome of postmortem neurons. Nat Protoc 11(3):499–524

La Manno G et al (2016) Molecular diversity of midbrain development in mouse, human, and stem cells. Cell 167(2):566–580.e19

Lake BB et al (2016) Neuronal subtypes and diversity revealed by single-nucleus RNA sequencing of the human brain. Science 352(6293):1586–1590

Lancaster MA, Knoblich JA (2014) Organogenesis in a dish: modeling development and disease using organoid technologies. Science 345(6194):1247125

Lee JH et al (2014) Highly multiplexed subcellular RNA sequencing in situ. Science 343(6177):1360–1363

Lein ES et al (2007) Genome-wide atlas of gene expression in the adult mouse brain. Nature 445(7124):168–176

Li C-L et al (2016) Somatosensory neuron types identified by high-coverage single-cell RNA-sequencing and functional heterogeneity. Cell Res 26(1):83–102

Liu Y-J et al (2013) Tracing inputs to inhibitory or excitatory neurons of mouse and cat visual cortex with a targeted rabies virus. Curr Biol 23(18):1746–1755

Lovatt D et al (2014) Transcriptome in vivo analysis (TIVA) of spatially defined single cells in live tissue. Nat Methods 11(2):190–196

Lubeck E et al (2014) Single-cell in situ RNA profiling by sequential hybridization. Nat Methods 11(4):360–361

Lui JH, Hansen DV, Kriegstein AR (2011) Development and evolution of the human neocortex. Cell 146(1):18–36

Macosko EZ et al (2015) Highly parallel genome-wide expression profiling of individual cells using nanoliter droplets. Cell 161(5):1202–1214

Madisen L et al (2012) A toolbox of Cre-dependent optogenetic transgenic mice for light-induced activation and silencing. Nat Neurosci 15(5):793–802

Madisen L et al (2015) Transgenic mice for intersectional targeting of neural sensors and effectors with high specificity and performance. Neuron 85(5):942–958

Marques S et al (2016) Oligodendrocyte heterogeneity in the mouse juvenile and adult central nervous system. Science 352(6291):1326–1329

Matcovitch-Natan O et al (2016) Microglia development follows a stepwise program to regulate brain homeostasis. Science 353(6301):aad8670

Miller JA et al (2014) Transcriptional landscape of the prenatal human brain. Nature 508(7495):199–206

Okaty BW et al (2015) Multi-scale molecular deconstruction of the serotonin neuron system. Neuron 88(4):774–791

Paul F et al (2015) Transcriptional heterogeneity and lineage commitment in myeloid progenitors. Cell 163(7):1663–1677

Picelli S et al (2013) Smart-seq2 for sensitive full-length transcriptome profiling in single cells. Nat Methods 10(11):1096–1098

Pollen AA et al (2014) Low-coverage single-cell mRNA sequencing reveals cellular heterogeneity and activated signaling pathways in developing cerebral cortex. Nat Biotechnol 32(10):1053–1058

Pollen AA et al (2015) Molecular identity of human outer radial glia during cortical development. Cell 163(1):55–67

Poulin JF et al (2014) Defining midbrain dopaminergic neuron diversity by single-cell gene expression profiling. Cell Rep 9(3):930–943

Poulin JF et al (2016) Disentangling neural cell diversity using single-cell transcriptomics. Nat Neurosci 19(9):1131–1141

Ramskold D et al (2012) Full-length mRNA-Seq from single-cell levels of RNA and individual circulating tumor cells. Nat Biotechnol 30(8):777–782

Sasagawa Y et al (2013) Quartz-seq: a highly reproducible and sensitive single-cell RNA sequencing method, reveals non-genetic gene-expression heterogeneity. Genome Biol 14(4): R31

Setty M et al (2016) Wishbone identifies bifurcating developmental trajectories from single-cell data. Nat Biotechnol 34(6):637–645

Shalek AK et al (2014) Single-cell RNA-seq reveals dynamic paracrine control of cellular variation. Nature 510(7505):363–369

Shekhar K et al (2016) Comprehensive classification of retinal bipolar neurons by single-cell transcriptomics. Cell 166(5):1308–1323 (e30)

Shendure J, Ji H (2008) Next-generation DNA sequencing. Nat Biotechnol 26(10):1135–1145

Shin J et al (2015) Single-cell RNA-seq with Waterfall reveals molecular cascades underlying adult neurogenesis. Cell Stem Cell 17(3):360–372

Shiroguchi K et al (2012) Digital RNA sequencing minimizes sequence-dependent bias and amplification noise with optimized single-molecule barcodes. Proc Natl Acad Sci U S A 109(4):1347–1352

Siegert S et al (2012) Transcriptional code and disease map for adult retinal cell types. Nat Neurosci 15(3):487–495

Sugino K et al (2006) Molecular taxonomy of major neuronal classes in the adult mouse forebrain. Nat Neurosci 9(1):99–107

Tang F et al (2009) mRNA-Seq whole-transcriptome analysis of a single cell. Nat Meth 6(5):377–382

Taniguchi H et al (2011) A resource of Cre driver lines for genetic targeting of GABAergic neurons in cerebral cortex. Neuron 71(6):995–1013

Tasic B et al (2016) Adult mouse cortical cell taxonomy revealed by single cell transcriptomics. Nat Neurosci

Telley L et al (2016) Sequential transcriptional waves direct the differentiation of newborn neurons in the mouse neocortex. Science 351(6280):1443–1446

Thomsen ER et al (2016) Fixed single-cell transcriptomic characterization of human radial glial diversity. Nat Methods 13(1):87–93

Tietjen I et al (2003) Single-cell transcriptional analysis of neuronal progenitors. Neuron 38(2):161–175

Trapnell C et al (2014) The dynamics and regulators of cell fate decisions are revealed by pseudotemporal ordering of single cells. Nat Biotechnol 32(4):381–386

Trimarchi JM et al (2007) Molecular heterogeneity of developing retinal ganglion and amacrine cells revealed through single cell gene expression profiling. J Comp Neurol 502(6):1047–1065

Usoskin D et al (2015) Unbiased classification of sensory neuron types by large-scale single-cell RNA sequencing. Nat Neurosci 18(1):145–153

Wu AR et al (2014) Quantitative assessment of single-cell RNA-sequencing methods. Nat Methods 11(1):41–46

Yao Z et al (2017) A single-cell roadmap of lineage bifurcation in human esc models of embryonic brain development. Cell Stem Cell

Zeisel A et al (2015) Brain structure. Cell types in the mouse cortex and hippocampus revealed by single-cell RNA-seq. Science 347(6226):1138–1142

Zeng H et al (2012) Large-scale cellular-resolution gene profiling in human neocortex reveals species-specific molecular signatures. Cell 149(2):483–496

Chapter 19
Transcriptional Profiling of Identified Circuit Elements in Invertebrates

Marta Morey

Abstract One approach to understand how neural circuits contribute to behavior is to dissect the function of discrete neuronal components of the network. The transcriptional profile of a neuron is a starting point to infer morphological, biochemical, and physiological properties that determine its functionality. This chapter aims to provide an overview on the challenges and advances to gain genetic access to distinct neuronal cell types, and the transcriptional profiling methods used to query their gene expression. In addition, it also surveys the contribution of transcriptional profiling experiments to our knowledge on aspects of circuit structure and function, which include dendritic morphology, wiring specificity, synaptogenesis, remodeling, and physiological states and functional properties of neurons. Based on the limitations of the current transcriptional profiling approaches, this chapter also addresses the perspectives and new developments that are expected to push cell type-specific gene expression profiling to new frontiers.

19.1 Introduction

To understand how neural circuits generate behavior, it is necessary to identify the neuronal cell types within a circuit and determine their connectivity and function. One way to reveal the molecular basis of neural function is to characterize the gene expression blueprint that determines the highly specialized phenotype of different types of neurons. Neuronal phenotypes are determined by molecules that regulate the morphological, biochemical, and physiological properties of a cell. Thus, the unique phenotype of different types of neurons is expected to be the result of their differential gene expression. Hence, comparison of gene expression profiles between neurons should identify key molecular components that specify their distinct functions.

M. Morey (✉)
School of Biology, Department of Genetics, Institute of Biomedicine (IBUB), University of Barcelona, Barcelona, Spain
e-mail: mmorey@ub.edu

A. Çelik and M.F. Wernet (eds.), *Decoding Neural Circuit Structure and Function*, DOI 10.1007/978-3-319-57363-2_19

Neuroscientists have long desired to be able to measure cell type-specific gene expression. However, two main issues have slowed down progress in this direction: difficulties in genetically manipulating specific neuronal cell types and in obtaining their gene expression profiles.

A neuronal cell type can be defined as a group of neurons that carry out a distinct task. Most often the way to identify a neuronal cell type is through its shape, determined by the dendritic arborization and projection pattern of the axon. This approach is based on the fundamental premise that a neuron's shape is a direct reflection of its connectivity, and hence of its unique function. It is reasonable to imagine that the distinct spatial position of neurons classified as belonging to the same cell type could result in further subdivision of that cell type into distinct subpopulations. This could be due to unidentified subtle morphological changes or physiological differences that would go undetected. Thus, one could argue that each neuron is unique. While the scientific community is aware of the drawbacks of morphological classification, at this point, in most cases, morphology is the easiest feature to score.

In the case of *Caenorhabditis elegans* and *Drosophila*, early studies identified distinct cell types through their morphology. A complete reconstruction of the *C. elegans* nervous system was undertaken using EM serial sections (White et al. 1986), and Golgi staining was used in *Drosophila* to characterize cell types in the optic lobe (Fischbach and Dittrich 1989). This information served as a mere, although very informative, catalog until experimental tools were developed that provided genetic access to specific neuronal cell types.

The nervous system is characterized by containing numerous highly intermixed cell types with irregular morphology. Many neuronal cell types are found in small numbers and are frequently difficult to access manually. For these reasons, cell type-specific gene expression analysis has also been dependent on the development of techniques that enable isolation of transcripts in a cell type-specific fashion. Technical advances in high-throughput gene expression analysis platforms have also been crucial to the success of these approaches.

This chapter aims to review the efforts to gain genetic access to specific neuronal cell types, an essential step to then apply profiling technologies since these depend on the expression of transgenes in a cell-specific fashion. In addition, it intends to provide an overview of the different types of profiling techniques that have been applied in *C. elegans* and *Drosophila*, with emphasis on the neuronal cell types to which these different techniques have been applied. To conclude, examples are given of biological questions related to the function of neural circuits that have been addressed through gene expression profiling.

19.2 Labeling Specific Neuronal Cell Types

Most methods used for cell type-specific profiling rely on the expression of some sort of transgene that distinguishes the cell type of interest from the rest of the neurons in the tissue. The nature of the transgene expressed will differ depending on the profiling approach taken, and this will be addressed in the corresponding section. Transgenesis techniques employed in *C. elegans* and *Drosophila* are well established and beyond the scope of this section, thus they will not be discussed. Here, we present the genetic approaches to molecularly mark specific neuronal populations.

19.2.1 Genetic Toolkit for Labeling Neurons

The preferred genetic methods to label any cell type of interest, in this case neurons, can be divided in two main types: regulatory sequence/reporter fusions and binary systems.

19.2.1.1 Regulatory Sequences/Reporter Fusions

In this strategy, the regulatory sequence of a known gene that is highly expressed in the neuronal cell type of interest is placed upstream of the coding sequence of a marker.

Identification of the regulatory sequence of the gene of interest is not necessarily a simple feat. Complementary approaches such as in situ hybridization and/or immunohistochemistry, if an antibody against the protein is available, can determine the correlation between the expression of the regulatory sequences/reporter fusion transgene and the endogenous expression of the gene.

The 7.4 kb 5′ regulatory sequences of the *Drosophila choline acetyltransferase* (*ChAT*) gene, which labels the cholinergic neuronal population, was determined through the generation of fusions of different lengths of 5′ flanking sequences of *ChAT* to the *lacZ* reporter gene, and comparison to the distribution of endogenous ChAT protein. Smaller fragments directed the *lacZ* expression in selected subsets of cholinergic neurons (Kitamoto et al. 1992). For cell type-specific genes of sensory neurons, such as opsins and odorant receptors, fusions of regulatory sequences to reporters have been quite successful (Couto et al. 2005; Fortini and Rubin 1990; Tahayato et al. 2003), probably due to the smaller size of their cis-regulatory regions.

The compacted nature of the *C. elegans* genome, and hence the fact that regulatory regions might be smaller, has facilitated the widespread use and success of regulatory sequences/reporter fusion transgenes in the worm. A significant number of cell type-specific fusions are available, among which there are many examples of mechano- and chemosensory neurons (Zaslaver et al. 2015).

To use direct fusion transgenes in profiling experiments it is necessary that the transgene used is expressed at levels compatible to the profiling approach that will be used. In this front, another reason for the success of regulatory sequences/reporter fusion transgenes in the worm is the presence of multiple transgene copies when the transgenesis approach involves extrachromosomal arrays.

The ease of genome editing using CRISPR technology, available both in the fly and the worm (Li and Ou 2016; Paix et al. 2015; Xu et al. 2015), could facilitate the generation of reporter lines where the marker expression is under endogenous regulation. This could be achieved by substituting one copy of the gene by the marker of choice or introducing the marker upstream of the translational start of the gene. This approach would be useful as long as the level of expression of the marker is sufficient for the profiling approach to follow.

19.2.1.2 Binary Transactivator/Responder Systems

Binary systems consist of a transactivator that binds to a specific DNA sequence to promote the transcription of a downstream responder. Spatial control of the expression of the responder is dictated by the choice of regulatory sequences that control the transactivator expression. The main virtue of this approach is the ability to control temporal and/or level of expression of the responder. This is achieved, thanks to the existence of transactivator repressors and compounds that positively or negatively modulate transactivator or repressor activity. Another advantage of this system is amplification of responder expression levels.

The main binary systems used in *Drosophila* are:

1. GAL4-*UAS*: The yeast GAL4 transcription factor binds to the Upstream Activating Sequences (*UAS*) placed upstream of the responder (Brand and Perrimon 1993; Fischer et al. 1988). Additionally, the GAL4-*UAS* system is repressible by the GAL80 protein (Lee and Luo 1999; Ma and Ptashne 1987). The most widely used strategy to regulate temporal expression of GAL4 is to use the temperature sensitive GAL80 repressor (GAL80^{ts}) (McGuire et al. 2001). This mutant version of the protein represses GAL4 transcriptional activity at 17 °C and releases repression at 29 °C or higher temperatures.
2. LexA-*lexAop*: This system is based on the LexA bacterial repressor that binds to specific lexA operator (*lexAop*) sequences. LexA DNA binding domain (DBD) has been fused to several activation domains (AD). Fusions to GAL4 AD render the system sensitive to GAL80, conferring temporal control through the use of GAL80^{ts} (Lai and Lee 2006; Szüts and Bienz 2000). Fusions to viral VP16 and human p65 strong activation domains result in chimeric proteins that transcribe high levels of responder expression and are insensitive to GAL80 (Lai and Lee 2006; Pfeiffer et al. 2010). The lexA system has been optimized to obtain better inducible expression and reduce leakiness and toxicity (Pfeiffer et al. 2010; Yagi et al. 2010).
3. QF-*QUAS*: This recently developed system relies on components identified from the fungus *Neurospora crasa* (Potter et al. 2010). The QF transactivator binds to QF upstream activating sequences (*QFUAS*), triggering the transcription of downstream responders. The activity of the QF system can be temporally controlled through the presence of the suppressor QS in the genetic background and addition of quinic acid (QA) to the fly food. Interestingly, repressor activity can be titrated by varying the concentration of QA fed to the animal, adding an extra layer of regulation. Recent modifications of this system have generated

less toxic versions of QF AD that have been proven to function in GAL4-QF AD and LexA-QF AD chimeras (Riabinina et al. 2015), enriching the tools available for responder expression regulation.

Binary expression systems are starting to become available in *C. elegans* but are not yet widely used. One of these approaches is a binary system employing heat shock induction (Bacaj and Shaham 2007). This is based on cell type-specific rescue of mutants defective in the heat shock response. The heat shock response factor (HSF) is expressed under a cell type-specific promoter in the cell of interest. The HSF transactivator activity is regulated by heat shock stress, which results in the formation of transcriptionally active trimers. The presence of an additional transgene containing HSF binding sites upstream of a marker gene triggers its expression in a cell type-specific manner. Transient or sustained heat shock pulses allow for temporal control of marker expression. In addition, a repressible Q binary system has been developed (Wei et al. 2012). Efforts to adapt the GAL4-*UAS* system for its use in the worm have entailed the systematic comparison of the transcriptional efficacy of three major components of this system—the DNA-binding domain, the activation domain, and UAS copy number. The Sternberg laboratory has found that performance of GAL4 is heavily dependent on temperature, acting poorly at 20 °C or below. Through evolutionary analysis they have identified *Saccharomyces kudriavzevii* GAL4, which functions robustly across the 15–25 °C range. Their optimized GAL4 system is capable of driving expression in a variety of tissues, including neurons (Wang et al. 2017). Long desired by the community, the GAL4/*UAS* system is expected to become widely used in the near future.

Alternatively, a two-part system for conditional FLP-out of FRT-flanked sequences in the worm has been developed to control gene expression in a spatially and/or temporally regulated manner (Davis et al. 2008; Voutev and Hubbard 2008). In this system, transcription is blocked by the presence of an "off cassette", composed of a transcriptional terminator flanked by FLP recognition targets (FRT), between the promoter and the coding sequence of the desired product. FLP-mediated excision of the cassette brings together the promoter and coding sequence activating transcription. Temporal control of marker expression can be regulated through heat shock-mediated expression of FLP. In addition, this system could be used to spatially restrict expression in a subset of cells that can only be addressed as the intersection of two available promoters (Davis et al. 2008). In this context, FLP expression would be under a cell type specific promoter.

19.2.2 Endeavors to Gain Access to Neuronal Cell Types

19.2.2.1 Searching for Regulatory Sequences

A key factor in implementing the above approaches is to identify regulatory regions that label the cell type of interest. It is relatively easy to find regulatory regions that label a large population of neurons based on a molecular characteristic (e.g., neurotransmitter used). A gene expression analysis of such a population may reveal broad characteristics, but this knowledge will be obtained at the expense of understanding the diversity of cell types comprising the population. In consequence, concerted efforts have been made to gain access to smaller populations of neurons.

Over the years, the *Drosophila* community has made enormous progress in gaining genetic access to specific cell types. A first approach was based on the random insertion of transposable elements and their capacity to act as enhancer traps, enabling identification of genomic enhancers. Initial studies used P elements containing lacZ (O'Kane and Gehring 1987). The generation of P elements containing sequences coding for GAL4 (Brand and Perrimon 1993) paved the way for binary systems, and many GAL4 lines have been generated by this means (Brand and Perrimon 1993; Hayashi et al. 2002). Though not as extensive, similar collections have been made for GAL80 (Suster et al. 2004), and more recently for LexA (Miyazaki and Ito 2010). More recently, transposable element vectors have been designed that make it possible to swap DNA content through various methods. Thus, these new collections permit researchers to customize a pre-existing line according to their needs. MiMIC (*mi*nos-*m*ediated *i*ntegration *c*assette) lines (Venken et al. 2011) contain two inverted *attP* sites that allow DNA replacement using RMCE (*r*ecombinase-*m*ediated *c*assette *e*xchange) (Bateman et al. 2006). MiMIC lines inserted in the first noncoding intron can be replaced with the transactivator or suppressor of choice. G-MARET (*G*AL4-based *m*osaic-inducible *a*nd *r*eporter-exchangeable *e*nhancer *t*rap) (Yagi et al. 2010) and InSITE (*in*tegrase *s*wappable *i*n vivo *t*argeting *e*lement) (Gohl et al. 2011) insertion collections allow replacement of GAL4 with other transactivators. A recurrent finding with all these transposable element collections is that the expression patterns obtained often tend to be broad because the same gene can be expressed in more than one cell type. Since these lines often include different neural cell types, their usefulness for profiling is limited.

In an attempt to generate lines with more restricted expression patterns, Rubin and colleagues at Janelia Research Campus took the following approach. They selected a group of 925 genes for which available expression data or predicted function indicated expression in neurons in the adult brain. These genes included transcription factors, neuropeptides, receptors, and ion channels, among others. The approach consisted of cloning relatively small fragments of genomic DNA upstream of these genes to a promoter and the GAL4 coding sequence (Pfeiffer et al. 2008). These plasmids were integrated at a specific docking site in the genome using phiC31 integrase, yielding thousands of GAL4 lines (Jenett et al. 2012) and LexA lines (Bloomington FBrf0222940). These lines were then curated for expression in the embryonic, adult and larval CNS, providing an excellent resource for the community (http://flweb.janelia.org/cgi-bin/flew.cgi). The entire collection of lines covers most *Drosophila* neurons, and over half of the fragments drive unique expression in 10–200 cells in the brain (Pfeiffer et al. 2008). Plasmids are available to clone the identified enhancers for fusion to various transactivators or fluorescent proteins. A complementary collection of GAL4 and LexA lines has been generated by the Dickson and Stark research groups (Kvon et al. 2014), and their expression pattern in the nervous system has been cataloged (VDRC Vienna tiles http://brainbase.imp.ac.at/bbweb/#6?).

The *C. elegans* community has undertaken several genome-wide gene expression projects. Hope and colleagues pioneered these studies using lacZ reporters and later developed the "promoterome": a genome-wide resource of *C. elegans* promoters to generate transgenic animals expressing GFP (Dupuy et al. 2004; Hope 1991; Lynch et al. 1995). Together with other groups, a collection of over 2000 transgenic lines carrying promoter: GFP fusions have been created and their spatiotemporal expression patterns curated (350 TF and almost 1900 genes) (Dupuy et al. 2007; Hunt-Newbury et al. 2007; Reece-Hoyes et al. 2007). Transgenic *C. elegans* strains for studying miRNA expression have also been generated (Isik et al. 2010; Martinez et al. 2008). Expression patterns are compiled in several databases: the Hope Lab Expression Patten Database: http://bgypc059.leeds.ac.uk/~web/; *C. elegans* Promoter/Marker Database: http://www.grs.nig.ac.jp/c.elegans/promoter/index.jsp?lang=english; the Promoterome Database: http://worfdb.dfci.harvard.edu/promoteromedb/; the BC *C. elegans* Gene Expression Consortium: http://gfpweb.aecom.yu.edu; and the Localizome Project: http://localizome.dfci.harvard.edu/index.php?page=home. Together with lines generated by researchers for their specific studies, these collections have expanded the catalog of regulatory regions with characterized expression patterns.

A cautionary note on expression patterns derived from transgenic enhancer/promoter constructs in *C. elegans* and *Drosophila*: besides the difficulty of defining regulatory regions that recapitulate the endogenous expression pattern of the gene of choice, factors such as integration site and surrounding chromatin structure can affect transgene expression. Thus, it is advisable to verify that expression of the reporter matches the endogenous gene expression. It is worth noting that the approach taken by the Rubin group was aimed at identifying small fragments in the putative upstream regulatory sequences of neuronal genes that would label subsets of neurons. It is possible that some of these fragments label subsets of neurons where the gene is actually not expressed. This situation could occur if the identified fragment lacked repressor sequences that under normal conditions repressed expression of the gene in those cells. Provided that the identified fragment labels neurons of interest for the researcher, the reporter is a valid reagent to genetically manipulate those neurons.

In addition, the modENCODE project aims to identify all of the sequence-based functional elements in the *C. elegans* and *Drosophila melanogaster* genomes. The work of this consortium (Gerstein et al. 2010; modENCODE Consortium et al. 2010; Nègre et al. 2011) and other laboratories (Kvon et al. 2012, 2014; Shi et al. 2009) could, in principle, aid researchers in the search of regulatory regions functioning as enhancers for neurons or specific neuronal populations in their gene of interest.

19.2.2.2 Applying Intersectional Strategies

All these efforts have yielded an exceptional collection of transgenic lines and a catalog of expression patterns in the nervous system in *C. elegans*, and especially in *Drosophila*. However, while some cell type-specific lines exist, many still label several neuronal populations. To overcome this issue, intersectional strategies have been developed. These are aimed at defining an expression domain that is cell type-specific. When reporter expression cannot be restricted to the cell type of interest using one particular regulatory sequence, the combined use of two or more unrelated regulatory sequences is employed to define a cell type-specific expression domain.

In *Drosophila*, the ample collection of enhancer and binary factor lines available, together with the fact that binary systems are specific and do not cross talk, and can be combined either together or with other genetic techniques, renders intersectional strategies a useful approach to label specific neuronal cell types (del Valle Rodríguez et al. 2012). Through intersectional strategies, cell type-specific expression domains can be obtained as a result of addition, intersection, or subtraction of the expression domains of the combined binary systems and/or other elements used. Below, we describe some of the possible combinations used in intersectional strategies (Fig. 19.1).

When independent lines each label a different subset of cells of the same type, addition is the simplest strategy to label the entire cell-type population. This can be achieved by the combination of transactivators of the same or different type, for instance GAL4 + GAL4 or LexA + GAL4. In the latter case, the given responder transgenes for the two types of transactivators should be present in the background (Fig. 19.1A).

Regulatory sequences driving transactivators usually label more than one cell type; however, different intersectional strategies can restrict expression to the cell type of interest. When the cell type of interest falls within the expression domain common to the two regulatory sequences used, split binary systems are a useful option (Fig. 19.1B). This variation was pioneered by the split-GAL4 system (Luan et al. 2006), and has recently been developed for LexA (Ting et al. 2011). The transactivator is separated into two hemi-proteins, each of which is expressed from a different regulatory sequence. One hemi-protein contains the GAL4 DNA-binding domain (DBD) or LexA, while the other hemi-protein contains the activation domain (AD). The use of distinct ADs renders these split systems GAL80-sensitive or insensitive. A functional transactivator will only reconstitute and activate the transcription of the responder when expressed together in the same cell. One drawback of this approach is that it often requires the generation of new hemi-lines. A considerable improvement offered by the split-LexA system is that it can leverage the wealth of pre-existing GAL4 lines by placing the expression of one of the hemi-lines under UAS control (UAS-split-LexA or UAS-split-AD), while the other hemi-driver can be expressed from a direct fusion (Ting et al. 2011). Alternatively, based on the fact that binary systems do not cross talk, they can be combined. The use of binary system-specific responder transgenes encoding for different

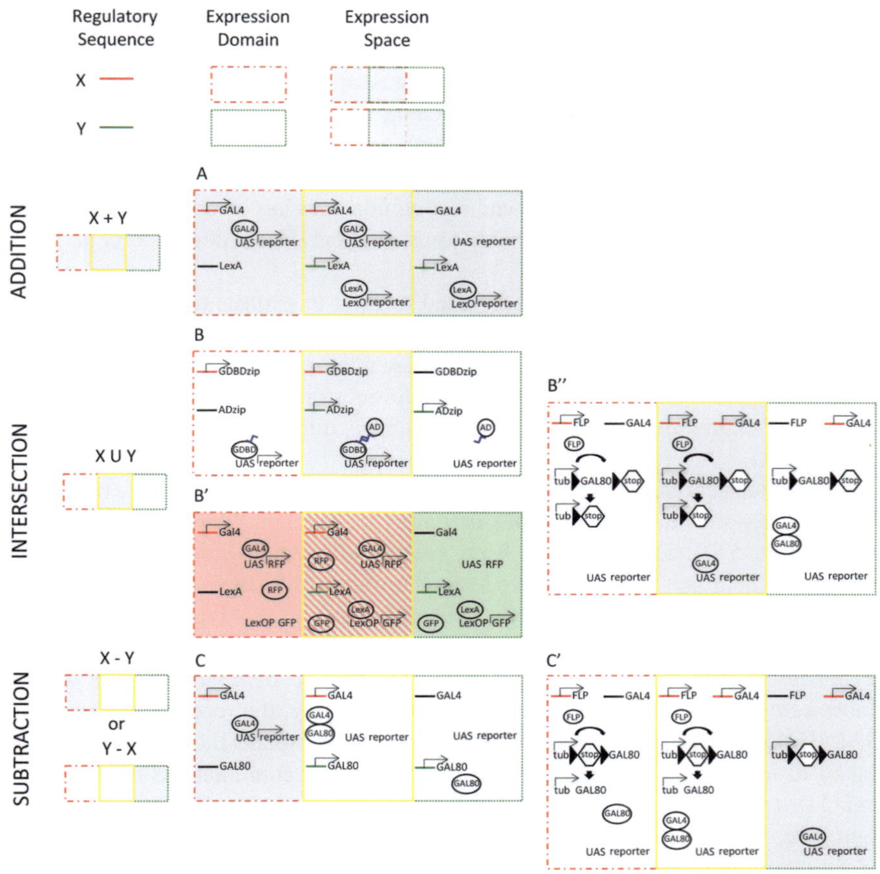

Fig. 19.1 Examples of intersectional strategies used to restrict expression to the cell type-specific neuronal population of interest in *Drosophila*. (*A*) Addition strategy combining both Gal4 and LexA binary systems and the use of the same reporter for both of them. (*B*) Intersection strategy based on the split GAL4 approach (*GDBD* GAL4 DNA binding domain; *AD* activation domain; *zip* zipper). (*B′*) Intersection strategy based on the combination of the GAL4 and LexA binary systems and system-specific fluorescent reporters. Cells in the common domain are identified by the coexpression of Gal4 and LexA reporters. (*B″*) Flip-based example of an intersection strategy where an FRT flanked ORF is eliminated. (*C*) Example of subtraction strategy using the GAL4 system and the GAL80 repressor. (*C′*) Flip-based example of a subtraction strategy where the FLP-out of an interruption cassette results in the expression of a downstream ORF

fluorescent proteins enables identification of cells in the common domain as the double-labeled cell type (Fig. 19.1B′).

When the cell type of interest falls within a specific expression subdomain driven by a regulatory sequence, it is possible to restrict expression by subtraction. The simplest method is by expression of a transactivator repressor, such as GAL80 when using the GAL4 or LexA GAL80 sensitive binary systems (Fig. 19.1C).

A combination of binary systems and FLP recombinase can be used to define intersecting domains and in subtraction strategies (Fig. 19.1B″, C′). In these scenarios, expression of the transactivator, repressor, or responder is regulated by recombinase activity removing an intervening FRT stop cassette. Many creative genetic designs have emerged from the combined use of the FLP recombinase and binary systems (for a review, see del Valle Rodríguez et al. 2012). The recent development of new recombinases and recognition sites has increased the numerous combinatorial options already available to researchers (Hadjieconomou et al. 2011; Nern et al. 2011).

In *C. elegans*, the most commonly used method to express transgenes is based on regulatory sequences/reporter fusions. By combining regulatory sequences yielding overlapping expression patterns, researchers can engineer worm strains that label specific subsets of neurons. Addition strategies can be pursued with transgenes expressing the same reporter under different regulatory sequences (Fig. 19.2A).

One intersectional strategy that can be also applied using regulatory sequences/reporter fusion transgenes is multicolor labeling. Triple color combinations (CFP, YFP, DSRed) have been successfully employed to label separate classes of neurons using cell type-specific regulatory regions (Hutter 2003). Similarly, one could use this strategy to identify the neuronal type of interest when distinct regulatory sequences/GFP variant fusions are combined in the same organism. In this scenario, specific cell types can be detected by their distinct fluorescent marker combination (Fig. 19.2B). Similarly, the recent addition of the GAL4/*UAS* system to the worm toolkit promises to expand the possibilities with regard to intersectional strategies. For example, doing combinations of the GAL4 and Q systems, or either of these systems with a regulatory sequence/reporter fusion transgene, where the two regulatory sequences label a common set of neurons.

Another intersectional strategy is based on the split approach. This approach has been applied to obtain cell type-specific GFP reconstituted expression (Fig. 19.2B′). Identified N-GFP and C-GFP peptides fused to leucine zippers can reconstitute GFP fluorescence in vivo when expressed in the same cell type (Zhang et al. 2004). Expression vectors have been constructed that are suitable for cloning regulatory sequences. Alternatively, the split Q system has been generated and used in worms to label neurons common to two distinct promoters (Fig. 19.2B″) (Wei et al. 2012).

The FLP-out system also offers the possibility of using intersectional strategies in the worm (Davis et al. 2008). This could be achieved when FLP expression and the FLP-out cassette are under regulatory sequences that label a set of common neurons (Fig. 19.2B‴).

Finally, the subtraction approach has been achieved by combining QF, QS, and two distinct fluorescent reporters, for example, mCherry and GFP. By means of this strategy, neurons can be distinguished based on their single or double reporter expression pattern (Fig. 19.2C).

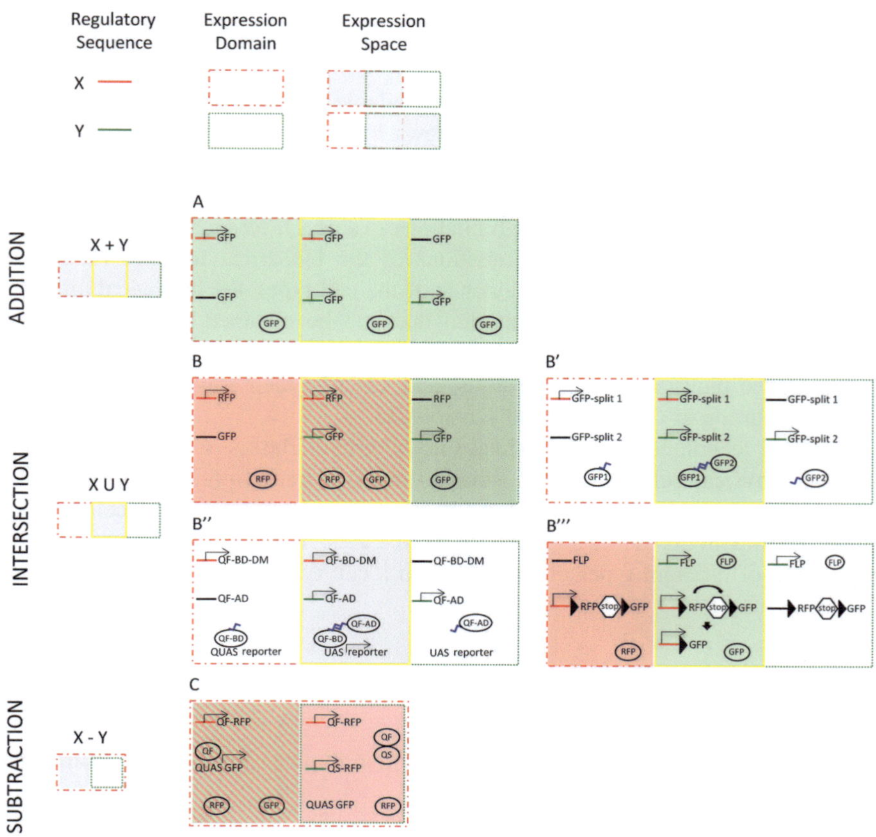

Fig. 19.2 Examples of intersectional strategies used to restrict expression to the cell type-specific neuronal population of interest in *C. elegans*. (*A*) Addition strategy. (*B*) Intersection strategy based on multicolor labeling. Cells in the common domain are identified by the coexpression of fluorescent proteins. (*B′*) Split GFP intersection strategy. Cells in the common domain are identified by reconstitution of GFP fluorescence. (*B″*) Split Q system intersection strategy. (*B‴*) FLP-based refinement of expression patterns. Excision of an FRT flanked cassette containing a fluorescent protein and stop sequence results in the expression of a downstream ORF, normally not expressed, that codes for a different fluorescent protein. (*C*) Example of subtraction strategy using the Q system and the QS repressor. This approach is useful to label a subset of neurons in the X expression domain for which there is no available regulatory sequence that labels them. This can be achieved if there is a promoter that labels the complementary subset of neurons and is used to express the QS repressor. The cells of interest for which there is no specific regulatory sequence are identified as double labeled

19.3 Methods to Profile Transcriptional Activity

A wide variety of techniques are currently available to profile the transcriptomes of specific cell types. Recent excellent reviews have discussed the key issues that influence their choice (McClure and Southall 2015; Otsuki et al. 2014). Yield, accuracy, technical difficulty, and cost are among the factors to consider. Since each method has its own strengths and limitations (Table 1), researchers must reach a decision based on the physical limitations of the biological material (ability to access the cell type of interest, abundance of the cell type), the biological question to address, and the type of information that can be obtained from the selected methodology. In this section, we will present these methods, discuss the nature of the transgene required to label neurons, and describe their use to profile specific types of neurons in *C. elegans* and *Drosophila*.

Profiling techniques can be divided into two main classes. One set of methods involves physical cellular/nuclear isolation prior to transcriptional profiling. The other techniques rely on capturing the transcriptional activity of the cell type of interest while in its tissue context. In both cases, it is necessary to drive the expression of different types of transgenes in a cell type-specific fashion.

19.3.1 Profiling Using Physical Cellular/Nuclear Isolation

In these techniques, physical isolation is used to minimize contamination from other cell types in the tissue sample.

19.3.1.1 Manual Isolation

Conceptually, manual isolation and identification of cells is the most straightforward technique. This procedure usually consists of dissecting the tissue containing the cells of interest, dissociating the cells and diluting the suspension to a concentration where cells can be individually viewed, and extracting them by aspiration with a micropipette. In principle, provided that the researcher can differentiate cells based on shape and/or size, this approach achieves very high purity. In general though, manual isolation is aided by the expression of fluorescent proteins in the cells of interest. In particular, when cells cannot be distinguished in any other way, it is essential that fluorescent protein expression is strong enough to allow for in vivo sorting under the microscope and that there is no leaky expression outside the neurons of interest.

This approach has been successfully used in *Drosophila* to address transcriptional changes in distinct types of larval and adult neurons in the circadian circuit (Abruzzi et al. 2017; Abruzzi et al. 2015; Kula-Eversole et al. 2010; Nagoshi et al. 2010). This methodology is highly suitable for such studies since collecting cells

Table 19.1 Overview of individual isolation methods and their respective advantages and limitations

Method	Isolated material	Profiling material	Technical description	Pros	Cons
Manual	Cells	Total RNA*	Manual aspiration of cells from cellular suspension	High purity	Time consuming
FACS	Cells	Total RNA*	Automated isolation of cells from cellular suspension by fluorescense sorting	High purity	Low yield
MACS	Cells	Total RNA*	Automated isolation of cells from cellular suspension by antibody coated magnetic beads	Quick procedure	Low yield
LMD	Cells	Total RNA*	Automated isolation of cells from tissue sections by laser dissection	More highthroughput than manual isolation for rare cell types	Contamination by neighbouring cells
INTACT	Nuclei	Nuclear RNA*	Automated isolation of nuclei from tissue homogenetes by antibody coated magnetic beads	Easy method	No cytoplasmic information
BITS-CHIP	Nuclei	Nuclear RNA*	Automated isolation of nuclei from tissue homogenetes by fluorescence sorting	Tissue fixed prior isolation	Expensive, so far only used for ChIP experiments
TU-tagging	Transcriptome	Nascent RNA including polyA and non-polyA transcripts	Isolation of 4-TU tagged RNA from a total RNA extraction by pull down	Precise spatiotemporal control	Reports of toxicity
Poly-A mRNA tagging	Transcriptome	PolyA mRNA	Isolation of poly-A RNA bound to PABP from a total RNA extraction by pull down	High sensitivity	Reports of toxicity
Ribosome tagging (TRAP and T-TRAP)	Translatome	PolyA mRNA	Isolation of poly-A mRNA bound to ribosomes from a total RNA extraction by pull down	Identifies translated mRNAs	Contamination by inspecific binding of RNA during pull down procedure
TaDa	Methylated DNA	DNA	Isolation of methylated DNA from a total DNA extraction by PCR amplification	Avoids handling RNA and pull down protocols	Limited quantitative resolution

[*] Type of RNA analyzed will depend on whether a poly-T or random primed amplification method is used.

from entrained brains at different circadian times requires rapid isolation protocols. In addition, given that some of these types of neurons are present in very reduced numbers, this approach reduces the signal-to-noise ratio and allows detection of mRNAs that would be masked by mRNA in the rest of the brain. Using this approach, the Rosbash group has profiled the transcriptomes of small and large PDF-expressing ventral lateral neurons (s-LNvs and l-LNvs) known to drive the morning activity period at different circadian times (Abruzzi et al. 2015; Kula-Eversole et al. 2010). These cells, 8 s-LN_vs and 10 l-LN_vs per brain, were labeled with GFP using a Pdf-GAL4 line and isolated by the size of their cell bodies. 100 cells obtained from around 100 brains provided sufficient material to perform microarray analysis (Kula-Eversole et al. 2010). The same researchers have recently developed an RNA amplification protocol that has enabled them to obtain enough mRNA to generate libraries for RNA deep sequencing (Abruzzi et al. 2015) and profile additional clock neurons as well as dopaminergic neurons (Abruzzi et al. 2017).

A recent study reports the harvesting of different types mushroom body neurons (a/b and g Kenyon cells (KC)) and mushroom body extrinsic neurons (V2, DAL, MBONa3, MBONg5b′2a, MBONb2b′2a) using GAL4 cell type-specific lines. In this case, GFP-labeled cell bodies were manually extracted in vivo, from intact brains, via patch clamp electrodes on an electrophysiology rig. RNA-seq was performed with material obtained from pooling approximately 100 cells from a single fly for each KC sample, and 4–14 neurons from one or two flies for each mushroom body extrinsic neuron sample (Crocker et al. 2016).

In theory, manual sorting could be performed for identifiable cells for which there are no cell type-specific lines available by filling them with fluorescent dyes. This strategy has been used to identify gene expression profiles through microarray analysis on single cells isolated from living embryos (Bossing et al. 2012).

19.3.1.2 Automated Isolation

Several alternative methods exist for automated isolation:

– *Fluorescence-Activated Cell Sorting (FACS)*

This flow cytometry isolation technique is based on sorting dissociated cells according to their fluorescent properties. In the case of *C. elegans* and *Drosophila* profiling, fluorescence is provided by genetic means in the cell type of interest. This fluorescent label must be sufficiently strong to be detected by the sorter in live cells.

FACS has been extensively used in *C. elegans*, especially in studies involving profiling of embryonic neurons since embryonic dissociated tissues can be cultured. An extensive collection of different types of neurons, including olfactory, thermosensory, and motor neurons, has been profiled with microarray experiments (Blacque et al. 2005; Cinar et al. 2005; Colosimo et al. 2004; Etchberger et al. 2007; Fox et al. 2005; Hallem et al. 2011; Von Stetina et al. 2007a; Zhang et al. 2002).

Recently, the development of culture protocols for larval tissues has facilitated the use of FACS to isolate larval neurons and perform RNA-seq analysis (Spencer et al. 2014). Starting material is not a limiting factor since large numbers of larvae are easily generated using standard culture conditions. Indeed, even neuronal cell types consisting of 2 neurons per worm have been profiled from worm cultures containing approximately 3 million larvae. The NSM serotonergic neurosecretory neurons (2 neurons/worm, 6 million neurons in 3 million larvae) have been purified with a yield of 0.85%; in other words, as few as 30,000–50,000 neurons have been isolated through FACS and used to generate sequencing libraries (Spencer et al. 2014).

In *Drosophila*, the use of FACS to isolate neuronal cell types has been more limited. One of the earliest instances was a study by Jasper and colleagues where they used SAGE to profile a subset of photoreceptor neurons in larval stages (Jasper et al. 2002). FACS has also been used to profile multidendritic neurons, wild type and mutant motor neurons (Parrish et al. 2014), and wild type and mutant LN_vs pacemaker neurons (Mizrak et al. 2012; Ruben et al. 2012). In the latter case, microarray analysis was performed with as few as 150–300 cells obtained from 50 brains. Recent publications have reported profiling cell type-specific neurons using RNA deep sequencing, including seven different neuronal cell types from the fly visual system, which have been used to create libraries from as few as 8000 sorted cells (Tan et al. 2015), and ultralow input RNA-seq data from 100 larval multidendritic neurons (Williams et al. 2016).

– *Magnetic Activated Cell Sorting (MACS)*

MACS is an affinity-based purification strategy. Magnetic particles coupled to antibodies are used to capture the cell of interest from a suspension of dissociated tissue. Sorting specificity is based on the use of antibodies against membrane-targeted antigens specific to the cell type of interest. After the incubation period, beads are recovered with a magnet and cells eluted for further processing. Given that cell type-specific membrane-targeted antigens are often not known or antibodies are not available, the use of this method in *Drosophila* has been facilitated by the GAL4-*UAS* system and the exogenous expression of UAS-mCD8-fluorescent protein transgenes in the cell of interest.

MACS has been adapted to isolate dendritic arborization (da) neurons from the larval peripheral nervous system (Hattori et al. 2013; Iyer and Cox 2010; Iyer et al. 2013a). Peripheral neurons are difficult to isolate due to their low numbers and difficult-to-reach location below the chitinous larval cuticle. This approach has enabled Iyer and colleagues to isolate 1500–2000 da neurons (classes I–IV), and 300–500 Class IV da neurons, from 30 to 40 larvae (Iyer and Cox 2010). Using an intersectional approach, Class I da neurons have also been isolated. Since the Class I driver faintly labeled Class IV neurons, Class I driver expression was restricted to Class I neurons using the regulatory sequence of the Class IV driver fused to GAL80 (Iyer et al. 2013b). With this reported amount of material, they performed transcriptional profiling on microarrays.

MACS has also been used to isolate the dopaminergic neuronal population in the adult brain (Iyer et al. 2013a).

– *Laser microdissection of cells*

Laser-based dissection enables isolation of single cells or single-cell clusters from complex tissue without the need for cell dissociation. Where the cell type of interest presents a recognizable shape, there is no need to use antibodies or genetic labels, either. Laser capture microscopy is the most common procedure and involves positioning a thermoplastic film over the frozen and/or fixed tissue sections. While cells are visualized under the microscope, a low-power infrared laser is used to locally melt the membrane around the cells of interest, binding them to the film. Lifting the membrane separates the cells from the rest of the sample.

This technique has been applied to profile the transcriptomes of *Drosophila* larval and pupal mushroom body neurons (Hoopfer et al. 2008), and larval insulin producing cells (IPCs) (Cao et al. 2014). Both these neuronal populations are characterized by the fact that their cell bodies form clusters, rendering LCM a useful isolation approach. Both studies relied on the use of cell type-specific lines and fluorescent reporters to visualize the cell bodies in the tissue sections. Mushroom body studies were performed with material pulled from 40 captures at a rate of 100 cells/capture per replicate (i.e., 4000 cells/replicate), and these were used to perform microarray analysis (Hoopfer et al. 2008). The number of IPCs per brain is just 14, distributed into two 7-cell clusters. Using both membrane (GFP) and nuclear (RFP) reporters to label these cells significantly accelerated the process and increased the reliability of their identification. The spatial resolution provided by LCM made it possible to use just 23 IPC cells/replicate, and improved amplification protocols enabled the construction of sequencing libraries for RNA-seq analysis (Cao et al. 2014).

A protocol for LCM isolation of da neurons has also been established (Iyer and Cox 2010). In this case, given the difficulty of capturing sparse da cell bodies in transversal sections of the whole larva, these researchers opted to isolate the cuticle from internal larval tissues and section the cuticle pellet. This modification enabled them to increase the number of cells accessible in the sample and to isolate single da cell bodies.

– *Isolation of nuclei*

When cells are hard to dissociate, nuclei isolation presents itself as an alternative. Nuclei may be gently released from tissue homogenates without the need for hard dissociation, and are relatively unaffected by changes in the cytoplasmic RNA and protein pool. Most importantly, microarray-based mRNA expression analysis using nuclear RNA samples yields results comparable to those obtained using total RNA (Barthelson et al. 2007; Zhang et al. 2008). Another advantage of nuclei isolation is that it can be used for other high-throughput genomic characterization protocols besides transcriptional profiling.

The *i*solation of *n*uclei *ta*gged in specific *c*ell *t*ypes (INTACT) method involves the coexpression of a nuclear envelope protein modified to sustain biotinylation, and a biotin ligase in the cell of interest. Incubation of the nuclear suspension with streptavidin-coated magnetic beads provides a rigorous method to isolate the nuclei of interest. Although this approach was developed in *Arabidopsis* (Deal and Henikoff 2010), it has been adapted to isolate nuclei from muscle of adult *C. elegans* and mesoderm from *Drosophila* embryos (Steiner et al. 2012). Shortly after the publication of the above studies, similar conceptual approaches and adaptations of INTACT were developed to isolate the nuclei of *C. elegans* and *Drosophila* neurons (Haenni et al. 2012; Henry et al. 2012; Ma and Weake 2014).

The difficulty in accessing post-embryonic tissues in *C. elegans*, mainly due to its tough cuticle, small size, and extremely complex tissue dissection, prompted the development of fluorescent activated nuclei sorting (FANS) (Haenni et al. 2012). Similar to INTACT, this procedure is based on cell type-specific nuclear labeling. However, it uses fluorescent labeling that does not need to be targeted to the nuclear envelope per se, since isolation is based on fluorescent sorting and not on antibody recognition. To gauge the scope of the technique, the method was tested on distinct cell types, including neurons. Although these studies focused on intestinal gene expression, it is expected that sequencing of cell type-specific neurons will also be feasible. This nuclear isolation protocol has been set up for large-scale worm cultures, and thus starting material should not present a problem in the case of small neural populations.

In *Drosophila*, various groups have developed procedures to isolate nuclei from cells in the adult brain and larval central nervous system following the INTACT rationale. Taking advantage of the GAL4-*UAS* system, these approaches are based on the cell type-specific expression of GFP-tagged nuclear envelope proteins and the use of anti-GFP antibody-coated magnetic beads for their isolation (Henry et al. 2012; Ma and Weake 2014). Nuclei of neuronal populations as small as 100–150 neurons per brain can be isolated from 600 tagged heads as starting material, without the need to dissect the brain, with high purity and around a 50% yield (Henry et al. 2012).

*B*atch *i*solate *t*issue-*s*pecific chromatin for immunoprecipitation (BiTS-ChIP) is an alternative nuclei isolation procedure developed in *Drosophila* (Bonn et al. 2012a, b). It is particularly suitable for ChIP experiments since the tissue is fixed before nuclei isolation, and the method is based on cell type-specific expression of an epitope-tagged histone protein, immunostaining against the tag, and fluorescent sorting of the nuclei. Alternatively, given the advances in extraction and quantitation of RNA from fixed sorted cells, as well as its integrity (Nilsson et al. 2014; Russell et al. 2013), one can envisage that nuclear RNA could be obtained from these nuclei, which would expand the use of this procedure beyond ChIP analysis.

19.3.2 Profiling Without Cellular/Nuclear Isolation

These techniques are aimed at minimizing possible acute transcriptional changes due to the stress caused by physical isolation procedures. Techniques developed to capture the transcriptional activity of the cell of interest are based on tagging the RNA, proteins interacting with the RNA, or proteins interacting with the DNA in a cell type-specific fashion. This enables distinction of the transcriptional activity of the cell type of interest from the rest of the cells in the sample.

19.3.2.1 Tagging RNA

The most prominent technique for tagging RNA is TU tagging. This technique is based on the properties of the *Toxoplasma gondii* uracil phosphoribosyltransferase (UPRT) enzyme, which when provided with 4-thiouracil (4-TU) inserts this analog in place of uracil in nascent RNA (Cleary et al. 2005). Subsequent biotinylation of thio-RNA enables affinity purification using streptavidin-coated magnetic beads.

The TU-tagging method was developed in *Drosophila* and introduced spatial regulation of RNA tagging through the GAL4-UAS system (Miller et al. 2009). This was achieved by the cell type-specific expression of UPRT. Thus, even if RNA is isolated from the whole animal, tagged RNA from the cells expressing UPRT can be selectively recovered. In addition, this technique allows for temporal control by timing and duration of 4-TU administration. 4-TU has been provided to embryos by immersion and fed to larvae and adult flies. Though no reports are available of TU tagging during pupal stages, 4-TU could be provided to pupae by injection, as is done in mouse. Exposure to 4-TU for up to 8 h has enabled detection of 4-TU-tagged RNA from whole animal RNA extraction in neural populations as small as mushroom body neurons in larval and adult brains. For smaller populations (250 cells), dissection of the brain was necessary (Miller et al. 2009). As few as 50 larvae per sample have been used to perform TU tagging and RNA-seq analysis of wild-type and mutant larval neuroblasts (Lai et al. 2012). There are reports that TU feeding can lead to background incorporation into mRNA and is toxic to flies (Thomas et al. 2012). Oxonic acid can be added to prevent a salvage pathway, which can use 4-TU without the presence of UTPR (Lai et al. 2012).

19.3.2.2 Tagging Proteins Interacting with RNA

Once again, different techniques exist in this area as well:

– *Poly-A Binding Protein tagging*

This approach uses the endogenous transcriptional machinery to isolate poly-adenylated mRNA. The cell type-specific expression of a FLAG-tagged poly (A) binding protein (PABP) enables isolation of poly-A mRNA from the cell of interest. Using FLAG antibodies, the cell type-specific mRNA can be immunoprecipitated from a total RNA lysate.

This technique was developed in *C. elegans* to overcome the difficulty of working with larval and adult worms where cell isolation was problematic. Initially developed for muscle cells (Roy et al. 2002), it was soon applied to the nervous system. In a first study of the nervous system, mRNA was isolated from ciliated sensory neurons, which comprise approximately 50 cells in the worm, confirming the applicability of this approach to small numbers of cells (Kunitomo et al. 2005). Subsequently, the procedure was successfully applied for profiling, using microarrays of different types of motor neurons (Petersen et al. 2011; Von Stetina et al. 2007a, b) and the two PVD multidendritic nociceptor neurons of *C. elegans* in wild type and mutant backgrounds (Chatzigeorgiou et al. 2010; Smith et al. 2010, 2013), and to identify differential gene expression between the gustatory neurons ASER and ASEL (Takayama et al. 2010).

Poly-A mRNA tagging has also been used in the fly to isolate mRNA from adult photoreceptors using *Drosophila* PABP (Yang et al. 2005). However, this study also reported toxicity effects upon expression of dPABP, depending on the spatiotemporal expression of the GAL4 lines used. This toxicity might be partially reduced by controlling temporal expression of GAL4. Additionally, toxicity caused by overexpression of dPABP could be due to deregulation of the translation initiation and mRNA stabilization/degradation roles that this protein might cause when interacting with other types of proteins.

The study showed that the use of hPABD, whose C-terminal interacting domain only shares 30% similarity to the fly, was an alternative to dPABP. In an attempt to use this technique in photoreceptor neurons with a different set of GAL4 lines, we detected developmental defects caused by dPABP overexpression (Morey and Zipursky, unpublished). Given the possible appearance of morphological defects and lethality when expressing dPABP, toxicity should be carefully assessed before opting for this approach.

– *Ribosome tagging*

*T*ranslating *r*ibosome *a*ffinity *p*urification (TRAP) (Heiman et al. 2008) and RiboTag (Sanz et al. 2009) were developed in mouse and are based on tagging a ribosomal subunit with a tag antigen. Ribosomes and their attached RNA can then be isolated through immunoprecipitation with magnetic beads coated with antibodies against the tag antigen. While this approach does not recover noncoding RNAs, it offers a snapshot of the putative translatome: transcripts being actively translated. Thus, it provides a more relevant insight into the cellular environment as a proxy for the cellular proteome.

Integration into the GAL-UAS system has provided the means to isolate mRNA associated with ribosomes in a cell type-specific fashion in *Drosophila*. UAS transgenic lines expressing the mouse or *Drosophila* RpL10 ribosomal subunit tagged with EGFP have been generated and successfully used with cell type-specific GAL4 lines to analyze transcriptomes by RNA-seq. One of these studies successfully isolated ribosome-bound RNA from adult neurons, and a small population of around 200 neurons of the *pars intercerebralis* of the brain from whole head extracts using 500–1000 heads (Thomas et al. 2012). Another study documented the rhythmic translatome of clock neurons (150 cells/brain). Using a GAL80-based intersectional strategy to restrict GAL4 expression in clock neurons, ribosome-bound mRNAs were profiled at six different time points of the circadian cycle (Huang et al. 2013). In this case, 200 heads (30,000 clock neurons) were lysed for each affinity purification experiment.

One factor influencing the success of any affinity purification method is the signal-to-noise ratio. To this end, Zhang et al. (2016) have recently developed Tandem-TRAP (T-TRAP), which includes a second tag to facilitate an additional purification step. They generated a UAS line where the N-terminus of *Drosophila* RpL10 was modified with two tandemly arranged epitopes, 3X FLAG and GFP, separated by the tobacco etch virus (TEV) protease site. They expressed TRAP and T-TRAP transgenes in photoreceptor neurons and purified ribosomal mRNA from dissected retina–optic lobe complexes. They next assessed enrichment of photoreceptor specific versus optic lobe transcripts comparing TRAP and T-TRAP samples to reference RNA obtained from the retina–optic lobe complexes. Using two sequential purification steps resulted in higher cell type-specific enrichment for T-TRAP (TRAP 1–10 times, T-TRAP 25–500 times). Although this enrichment came at the cost of a 30% decrease in the mRNA yield compared with TRAP, the amount of material obtained with T-TRAP was sufficient to perform RNA deep sequencing. They used 40 retina–optic lobe complexes per sample to perform T-TRAP, which represents a total of 240,000 photoreceptor neurons (6000 photoreceptors/retina–optic lobe complex). Attempts to isolate cell type-specific mRNA from populations of 750 cells/retina–optic lobe complexes showed non-specific mRNA presence. The transgene encoding T-TRAP has recently been modified to further reduce background noise by increasing expression via the inclusion of noncoding sequences enhancing translation (Pfeiffer et al. 2012), and by mitigating the effects of leaky expression of the UAS construct by inserting a transcriptional stop sequence flanked by FRT recombination sites (unpublished data). Intersectional strategies targeting FLP expression to the cell type of interest coupled with cell type-specific GAL4 expression will further increase the potential of this method.

– *RISC tagging*

This strategy is based on tagging specific proteins of the RNA-induced silencing complex (RISC) where the miRNA and its mRNA target interact. Thus, pull down of RISC permits the identification of associated miRNAs and their targets. Among the first reports of this approach were studies performed on *Drosophila* and *C. elegans* (Easow et al. 2007; Zhang et al. 2007). Tissue-specific identification of miRNA has been reported in the worm using intestine and muscle-specific enhancers driving the expression of tagged RISC proteins (Kudlow et al. 2012). Cell type-based analysis of miRNA profiles has been successfully performed for glutamatergic and GABAergic neurons and subtypes in the mouse brain (He et al. 2012). Thus, in principle, this technique could be used to profile miRNAs and their targets in *C. elegans* and *Drosophila* neurons by targeting expression of the tagged RISC complex in the neuron of interest.

19.3.2.3 Tagging Proteins Interacting with DNA

*Ta*rgeted *Da*mID (TaDa) is an adaptation of the original DamID technique (van Steensel and Henikoff 2000; van Steensel et al. 2001). The DamID system is based on identifying methylation footprints generated by the DNA adenine methyltransferase (Dam) enzyme from *Escherichia coli*. When this enzyme is fused to a protein that binds DNA, it methylates GATC sites in the vicinity of the binding site. These methylated sites can be conveniently digested with the methyl-sensitive restriction enzyme DpnI, and the fragments amplified by PCR for profiling with microarrays or deep sequencing. DamID has been used to study chromatin-associated protein interactions with DNA to understand transcriptional regulation (through transcription factors-Dam fusions) and chromatin states and dynamics (for a review, see Aughey and Southall 2016).

In order to use DamID in a cell type-specific manner, Southall and colleagues developed targeted DamID in *Drosophila* using the GAL4-UAS system (Southall et al. 2013). To this end, it was necessary to limit the expression levels of the Dam fusion protein, since its inherent high activity causes cell toxicity. This was achieved by leveraging ribosome reinitiation constructing a UAS transgene that carried a fluorescent protein followed by the Dam fusion protein, which is expressed at very low levels. This approach has been used to profile RNA Pol-II occupancy, thus giving a readout of transcription (Southall et al. 2013), and was applied to study neuroepithelial and neuroblast populations in the fly brain using between 100 and 300 brains depending on the developmental stage analyzed. Combining use of the GAL4-UAS system with $GAL80^{ts}$ allowed temporal restriction of Dam expression.

Efforts are being made to maximize the potential of TaDa as a cell type-specific transcriptional profiling approach, when it uses a fusion of the Dam enzyme to Pol-II. From a technical standpoint, TaDa has many advantages over other methods for cell type-specific profiling. It does not require cell isolation, avoiding any possible transcriptional responses to tissue dissociation protocols, nor is crosslinking or antisera use necessary, eliminating the noise caused by these procedures. Furthermore, fixation artifacts are avoided, since TaDa profiles protein binding in vivo. This protein binding and methylation is achieved with very low levels of enzyme-fused protein, limiting the impact of protein overexpression. It also uses DNA as readout, avoiding the technical complications of working with RNA.

Two main aspects of the initial protocol have been modified and improved. The publication describing the transcriptional profiling application of TaDa used tiling arrays as a means of mapping expression. The new protocol includes preparing the material for next generation sequencing and can be accomplished in 5 days from collection of the tissue samples to generation of the sequencing libraries. In addition, the number of targeted cells required for TaDa is very low. This new protocol has achieved RNA-seq transcriptional profiling with approximately 10,000 cells in total from 100 Drosophila heads (100 neurons/head). At >200,000 cells per head, this represents a 1:2000 ratio of methylated DNA to total DNA (Marshall et al. 2016).

The main limitations of TaDa are: (1) it does not provide direction of transcription, which can be an issue for nearby genes transcribed in opposite directions, and (2) it does not provide quantitative levels of RNA produced. However, in addition to embryonic and larval neural stem cells, this method has been successfully used to profile larval and adult neurons (Southall et al. 2013; A. Estancio-Gomez and T.D. Southall, unpublished). Furthermore, the TaDa protocol has been used to compare the transcriptional states of distinct sets of neurons, enabling the identification of differentially expressed genes (A. Estacio-Gomez and T.D. Southall, unpublished). Importantly, other laboratories have used this protocol successfully. In a recent publication, the laboratory of Dr. Edgar used TaDa to identify target genes of the Capicua (Cic) transcriptional repressor (Jin et al. 2015). They generated a UAS-cic-Dam construct that was expressed specifically in Drosophila intestinal stem cells (ISC) for 24 h using an ISC GAL4 line and GAL80ts. Taken together, the preliminary data in neurons, and the easy implementation of the protocol, suggest that TaDa could become a widely used approach for cell type-specific transcriptional profiling, in addition to its many other applications (Aughey and Southall 2016).

19.4 Contributions of Profiling Experiments to Circuit Structure and Function

Profiling experiments have yielded an insight into molecular underpinnings regulating distinct developmental processes involved in the assembly of functional circuits. In addition, they have revealed distinct physiological states and properties of different types of neurons, which explain their unique functionality in the circuit.

19.4.1 Neural Circuit Architecture

19.4.1.1 Dendritic Morphology

Dendritic architecture is a neuronal feature with important functional implications in circuit assembly, signal processing, and neural function. Profiling approaches have contributed to the identification of molecular strategies regulating dendrite branching both in *C. elegans* and *Drosophila.*

Studies on *C. elegans* have identified a set of transcription factors regulating the morphology of PVD dendrites. Distinct transcription factors appear to control discrete steps in PVD dendritic morphogenesis and either promote or limit PVD branching at specific developmental stages (Smith et al. 2010, 2013).

In the fly, analyses of transcriptional differences between arborizations of two classes of dendritic neurons with uniquely distinct dendritic morphologies have been conducted (Hattori et al. 2013; Iyer et al. 2013b). Class I da neurons exhibit selective innervations of dendritic territories and occupy relatively small receptive fields, whereas Class IV da neurons exhibit an elaborate space-filling network of dendrites that completely and nonredundantly tile the larval body wall. Protein synthesis and proteolysis gene classes appear differentially expressed and directly correlate with the complexity of the dendritic arbors of the two classes. In addition, genes associated with oxidation and mitochondria appear enriched in Class IV, suggesting underlying differences in their metabolic demands. Similarly, more transcription factors appeared differentially expressed and showed phenotypes in Class IV than in Class I (Iyer et al. 2013b). Transcription factors had already been shown to regulate dendrite morphogenesis (Parrish et al. 2006). Profiling experiments have provided an insight into their cell type-specific diversity, and have revealed their context-dependent functions. For example, some differentially expressed transcription factors showed phenotypes in both classes of da neurons, and in some cases showed opposing effects (Iyer et al. 2013b). Additionally, distinct transcription factors can regulate the expression of the same target gene in different cell types, but do so at different levels resulting in distinct dendritic arborization patterns (Hattori et al. 2013).

19.4.1.2 Wiring Specificity

In order to assemble a functional neural circuit, neurites need to discriminate between one another and form connections with their specific synaptic partners. Langley and Sperry proposed that molecular differences between neurons would account for their specific connectivity. These molecular differences can be readily identified through cell type-specific profiling experiments.

In *C. elegans*, the expression profiles of wild type and mutant motor neurons have been compared to address the specificity of motor circuit synapses. In wild type animals, VA and VB motor neurons arise as sister cells that adopt distinctive morphologies and synapse with separate sets of interneurons. In UNC-4 mutants, morphological differences are preserved; VAs, however, are miswired with inputs from interneurons normally restricted to their VB sisters. Thus UNC-4, together with the corepressor UNC-37 (Groucho), explicitly controls synaptic choice and not axonal growth or process placement, which could indirectly alter wiring specificity. Comparison of wild type versus UNC-4 mutant VA transcriptional profiles identified VB genes to be negatively regulated in VA motor neurons (Von Stetina et al. 2007a). Of these, CEH-12, an HB9 family member, functions downstream of UNC-4 to regulate synaptic choice. This study revealed a developmental switch in which motor neuron input is defined by the differential expression of transcription factors that select alternative presynaptic partners.

A recent approach to investigate wiring specificity in *Drosophila* has been to obtain the transcriptional profiles of developmentally related neurons with distinct connectivity patterns (Tan et al. 2015). This study characterized the cell surface membrane and secreted molecule complement of neuronal types with distinct connectivity patterns, and proposed a molecular strategy underlying the selection of synaptic partners. How many cell surface and secreted molecules a neuron expresses has been a long-standing question in the field. The relevance of this question resides in the fact that these types of molecules are the final effectors of cell–cell interactions, since they mediate contact-dependent recognition (through attraction/adhesion or repulsion events) and synapse assembly. Lamina neurons (L1-L5) and photoreceptors R7 and R8 all have a unique morphology, including layer-specific arborizations and connectivity patterns in the medulla neuropil. Their expression profiles were obtained at a developmental time point just prior to (R7, R8) or in the early stages of synapse formation (L1-L5). Using stringent settings, these neurons express between one-quarter to one-third (247 for R7 and 322 for L3) of the 976 genes encoding cell surface membrane and secreted molecules (CSMs) in the fly genome. While these neurons express roughly the same amount of CSMs, marked differences in the type of CSMs are observed between neurons. Classification of CSMs into families led to the detection of particular families with unique paralog combinations expressed in a cell type-specific fashion. One of these is the Dpr family, comprising 21 members. Detailed immunohistochemistry analysis of this family and the Dpr interacting protein (DIP) family (9 members) revealed colocalization of interacting Dpr and DIP members in layers where Dpr expressing lamina neurons and photoreceptors R7 and R8 establish synapses with medulla neurons. This suggests that Dpr–DIP interactions could regulate synaptic connections within a layer. Indeed, this study identified cell type-specific Dpr–DIP interactions between lamina neurons and the R7 photoreceptor and a subset of their synaptic partners. Supporting this notion, a recent study (Carrillo et al. 2015) has shown defects in a subset of R7 photoreceptors that make connections with DM8 neurons. Defects observed in R7 cells, when analyzing either mutations for the Dpr expressed in them or mutations in the DIP expressed in Dm8 cells, are consistent with synaptic defects.

The simplest interpretation is that the matching of Dpr and DIPs between synaptic partners specifies connections between them. It is possible that these interactions regulate other aspects of wiring specificity in DM8, such as viability through trophic support, given that a reduction in the number of DM8 cells is observed when the DIP expressed in these cells is mutated. More detailed genetic analysis will be required to definitively establish the precise function of Dpr–DIP ligand receptor interactions in circuit assembly.

19.4.1.3 Synaptogenesis

Neural circuit assembly requires coordination of recognition events between synaptic partners and the establishment of synaptic connections. Presynaptic development is a complex process, the study of which can be hindered by the complex temporal dynamics of neural development, with different types of neuron being born and establishing synaptic connections at different time points.

A recent study adopted a profiling approach to analyze the conversion of growth cones to synaptic terminals (Zhang et al. 2016), taking advantage of the synchronicity of this process in the *Drosophila* photoreceptor population. An analysis was conducted of mRNAs bound to ribosomes over time, thus reflecting protein rather than gene expression during this process. Consistent with the coordination of recognition events and presynaptic development, substantial changes were observed in many mRNAs encoding CSM, including those implicated in recognition and synapse formation. The pattern of expression suggests a massive restructuring of the neuron cell surface in closely spaced time points, with a downregulation of CSM preceding the transformation of growth cones to presynaptic terminals (35–40hrs after puparium formation), and a strong upregulation of CSM correlating with the first morphological manifestation of presynaptic differentiation (40–45hrs after puparium formation). Interestingly, changes in the levels of transcripts of synaptic molecules were modest. However, a doubling in the length of the 3′ UTRs for these transcripts was correlated with an increase in the number of binding sites for RNA binding proteins implicated in the regulation of mRNA localization, stability, and translation, which were expressed at constant levels. These findings suggest strong post-transcriptional regulation of presynaptic differentiation.

19.4.1.4 Remodeling

Neural circuits are remodeled by developmental signals and experience. This plasticity is embodied in structural changes that include dendrite and axon pruning and synapse relocation. The study of developmentally regulated plasticity through profiling experiments can uncover molecular components of remodeling programs.

Pruning of neuronal connections is a widely used mechanism in metazoan nervous systems to achieve a mature connectivity pattern. In *Drosophila*, early born mushroom body gamma neurons undergo axon pruning at the onset of metamorphosis in a process regulated by ecdysone. Comparison of wild type and ecdysone receptor mutants identified the upregulation of genes in the UPS (ubiquitin proteasome system), providing a mechanistic link to pruning. Unexpectedly, an RNA-binding protein promoting translation was identified as a negative regulator or developmental axon pruning, which suggests that post-transcriptional regulation might be an important mechanism regulating axon remodeling (Hoopfer et al. 2008).

C. elegans Dorsal D (DD) GABAergic motor neurons undergo stereotypical synaptic changes during development. Initially formed ventral DD synapses are relocated to the dorsal side with no evident changes in DD process morphology. Ventral D (VD) GABAergic motor neurons, which are functionally and structurally related to DDs, do not remodel due to the action of UNC-55, the COUP transcription factor homolog, which has been shown to function as a negative regulator of transcription. UNC-55 mutant VDs relocate synapses to the dorsal side, similar to DD developmental synapse relocation, and thus UNC-55 target genes would be enriched in this scenario compared to wild-type VDs. Profiling experiments identified the Iroquois homeodomain protein IRX-1 as both necessary and sufficient for synaptic remodeling (Petersen et al. 2011).

19.4.2 Physiological States and Functional Properties of Neurons

The link between gene expression and behavior is best exemplified in circadian rhythms, which result in cycling physiological states of neurons in the circuit. Pacemaker neurons possess molecular clocks that control gene expression. The circadian function of clock molecules is regulated by negative feedback loops of transcription and post-transcriptional modifications that modulate their stability and activity in a rhythmical fashion. The core clock then regulates transcription of other output molecules, which also accumulate rhythmically or have rhythmic activity. These output molecules regulate electrical activity rhythms to more directly generate overt circadian behavior.

In *Drosophila*, the circadian circuit is comprised of about 75 clock neurons on each side of the adult brain. Of these, two key groups of neurons control adult locomotor activity, which peaks twice a day in anticipation of dawn and dusk transitions. Genetic screens and microarray analysis from whole fly heads collected at different circadian times have identified many cycling mRNAs (100–200). However, given the existence of seven classes of neurons in the circuit, it is possible that mRNA cycling in only a small number of clock neurons is masked by non-cycling mRNAs in other neurons and head tissues. Furthermore, mRNAs that are only expressed in the clock neurons or in a subset of these should only comprise a tiny fraction of head RNA, and may therefore escape detection in both cycling and non-cycling analyses of head RNAs. Cell type-specific profiling has provided a means to address the above issues and has indeed identified genes that are expressed in subsets of distinct clock neurons and that affect distinct aspects of rhythms (Abruzzi et al. 2017; Nagoshi et al. 2010). Moreover, potent oscillations of gene expression have been observed in clock neurons, as well as enrichment of certain transcripts important for the neural function of clock neurons themselves, suggesting that some physiological aspects such as firing rhythms and/or electrical excitability may be rhythmically regulated (Flourakis and Allada 2015; Kula-Eversole et al. 2010; Ruben et al. 2012). Interestingly, altered electrical activity of clock neurons results in overt transcriptional changes involving a large set of circadian genes. This suggests a positive feedback loop between transcription and electrical activity, which would add robustness and precision to circadian behaviors (Mizrak et al. 2012).

An analysis of the circadian translatome of clock neurons has revealed that translation of most rhythmic transcripts coincides with behavioral quiescence, prior to initiation of locomotor activity, and thus protein synthesis may occur predominantly at circadian phases associated with reduced metabolic expenditure. In addition, the synchronized translation of functionally related mRNAs suggests a clock-orchestrated activation of biological processes (Huang et al. 2013).

Taken together, the knowledge gained from these profiling studies has revealed distinct mechanisms that regulate the rhythmic physiological state of distinct neuronal populations in the circadian circuit.

Functional specialization is a hallmark of sensory neurons. The *C. elegans* nervous system is richly endowed with sensory neurons. This organism navigates its environment by chemo-, thermo-, and aerotaxis, and thus exhibits behavioral responses to these types of stimuli. This is accomplished through 24 sensillar organs and some isolated sensory neurons. Most sensory neurons are characterized by the presence of ciliated endings. Many of the early studies focused on identifying chemotaxis mutants through genetic screens (Dusenbery 1974; Dusenbery et al. 1975; Ward 1973); however, this approach does not favor the detection of genes with redundant function or genes that give subtle phenotypes when mutated. Profiling complements genetic methods by providing a direct examination of genetic networks in a cell type-specific fashion (Blacque et al. 2005; Kunitomo et al. 2005; Zhang et al. 2002). Indeed, while it was through a genetic screen that the transcription factor DAF-19 was shown to regulate ciliated sensory neuron formation (Swoboda et al. 2000), gene expression analysis was necessary to obtain a transcriptome of ciliated neurons (Blacque et al. 2005; Kunitomo et al. 2005) and identify new ciliary components under the regulation of DAF-19 (Blacque et al. 2005). The genetic networks regulating the differentiation of touch receptor neurons and the ASE gustatory neuron have also been characterized. Cell type-specific profiling experiments combined with transcription factor motif discovery have started to unveil the regulatory logic behind sensory neuron differentiation programs (Etchberger et al. 2007; Zhang et al. 2002).

In addition, recent profiling studies have identified genes regulating the functional properties of particular types of sensory neurons. Using in vivo calcium imaging, Hallem and colleagues showed that CO_2 specifically activates BAG neurons, and using profiling unveiled that their CO_2-sensing function requires a particular type of cyclic nucleotide-gated ion channel and receptor-type guanylate cyclase (Hallem et al. 2011). Similarly, Chatzigeorgiou and colleagues have identified a distinct set of channels involved in responses to thermal and mechanical stimuli in polymodal nociceptor PVD neurons (Chatzigeorgiou et al. 2010). Thus, cell type-specific transcriptional analysis has shed light on the genetic programs regulating differentiation of sensory neurons and the molecular mechanisms that explain their physiological properties.

19.5 Perspectives and New Developments

Two of the main goals in the field are to achieve progress on the issue of neuronal classification and to work toward improving current profiling techniques.

The best functional classification of neurons would be one combining morphological and physiological data. Profiling based on cell types defined by morphology has revealed unknown physiological properties of the studied neurons. However, it has not given an overview of the physiological differences across morphologically defined cell types. In addition, cell type population profiling does not detect differences or variances among cells from a morphologically defined cell type. Two distinct complementary approaches are emerging as possible ways to address these issues: single-cell profiling, fluorescent in situ sequencing (FISSEQ) and Patch-seq.

In recent years, low input RNA-seq methods have been adapted to work in single cells (Tang et al. 2009). Single-cell RNA-seq (sc-RNA-seq) methods are now robust and economically practical, and are becoming a powerful tool for high-throughput, high-resolution transcriptome analysis (Liu and Trapnell 2016). Data analysis is not easy, since the low input material for scRNA-seq creates high levels of technical noise (Brennecke et al. 2013; Ding et al. 2015; Grün et al. 2014; Marinov et al. 2014). In addition, only around 10% of each cell's transcript complement is represented in the final sequencing libraries (Islam et al. 2014), and this technique is unable to reliably detect low-abundance transcripts (Deng et al. 2014; Islam et al. 2014; Saliba et al. 2014). Many of the genes detected are housekeeping genes such as ribosomal subunits, and thus uninformative; therefore, reads from multiple cells must be combined to detect biologically meaningful gene expression differences between groups of single cells (Grün et al. 2014). Nevertheless, scRNA-seq has revealed intrapopulation heterogeneity in various tissues, including the brain (see the many references in the following reviews Poulin et al. 2016; Johnson and Walsh 2017).

Recently developed fluorescent in situ hybridization (FISH) techniques such as single-molecule FISH (sm-FISH) (Raj et al. 2008), which allows visualization of bright fluorescent spots that can be counted to determine the copy number of the gene of interest and its cellular location in individual cells, are typically performed on one RNA species at a time. Efforts to massively multiplex the sm-FISH imaging method have culminated in the development of multiplexed error-robust fluorescence in situ hybridization (MERFISH) (Chen et al. 2015). This method achieves large-scale multiplexing by assigning error-robust barcodes to different RNA species and then reading out these barcodes through successive rounds of hybridization and imaging on the same sample, so far up to 1000 genes. This technique can extend the benefits of sm-FISH toward the transcriptome scale. However, in situ hybridization techniques rely on a defined set of probes. Church and colleagues developed an unbiased and transcriptome-wide sampling method for quantitative visualization of RNA in situ. This technique is called fluorescent in situ sequencing (FISSEQ) (Lee et al. 2014, 2015), and combines the benefits of in situ hybridization with RNA-seq. It is based on the generation of stably cross-linked complementary DNA (cDNA) amplicons, which are sequenced manually on a confocal microscope within the biological sample. FISSEQ enriches biologically active genes, enabling the discrimination of cell type-specific processes with a small number of reads. However, it is not clear how such enrichment occurs. It has been proposed that active RNA molecules are more accessible to FISSEQ than ribosomal transcripts trapped in ribonucleoproteins, spliceosomes, or stress granules. Further elimination of still remaining transcripts of this nature (i.e., using random priming with rRNA depletion) will increase the number of cell-specific reads and enable FISSEQ to generate single-cell gene expression profiling that is biologically meaningful (Lee et al. 2015). Alternatively, sc-RNAseq is starting to be combined with tissue reference maps (for examples see the review by Moor and Itzkovitz 2017).

Patch-seq is a method that combines whole cell electrophysiological recordings, sc-RNA-seq and morphological characterization. This technique has been used to characterize pyramidal cells and cortical interneurons (Fuzik et al. 2016; Cadwell et al. 2016). While the efficiency of mRNA capture in Patch-seq is lower than that of in sc-RNA-seq on dissociated tissues is still sufficient to sample genes with low expression. This allowed to make inferences on the specificity and heterogeneity of afferent inputs for different cell types (Fuzik et al. 2016) and the identification of genes associated to neurological disorders such as autism and schizophrenia in particular neuronal subtypes (Cadwell et al. 2016). These studies were also able to render associations between the expression of ion channels and synapse-related proteins and biophysical parameters of action potentials. Thus, Patch-seq has an enormous potential in the vertebrate brain to precisely map neuronal subtypes and predict their network contributions in the brain.

Both sc-RNAseq and FISSEQ could in principle be easily adapted to invertebrates. Patch-seq will depend on the development of electrophysiological probes suitable for the small size of Drosophila and C. elegans neurons. Importantly, especially in Drosophila, Patch-seq could be done in vivo for behaviors that can be assessed in tethered flies.

19.6 Concluding Remarks

Gene expression profiling approaches are making important contributions to the understanding of neural circuit structure and function. Gene expression profiling experiments can address various biological questions depending on their design. Initial experiments characterized broad neuronal populations by identifying enriched transcripts versus the whole animal or neural tissue reference sample. However, as a result of advances in technology and knowledge, researchers are shifting their focus to discrete neuronal cell types. Thus, they are now addressing questions such as what genes determine the unique morphology or physiology of related neuronal cell types, by comparing their gene expression patterns, or what are the genetic programs and downstream molecular determinants that drive these differences when comparing wild type versus genetically manipulated gene expression in a particular neuronal cell type. These types of profiling experiments have shown clear potential for discovery and are becoming increasingly popular.

Nevertheless, the qualitative and quantitative information obtained from gene expression analysis of a particular neuronal cell type will always be dependent on two factors: (1) the definition of cell type and (2) the specificity of the data obtained depending on the profiling method used.

Cell types are often arbitrarily defined by the expression of markers or their morphology. However, it is possible that definition by these criteria can include heterogeneous neurons, even in a small population. This is exemplified in a recent study analyzing R7 photoreceptors and their major postsynaptic partner DM8 neurons. A specific subset of DM8 neurons was identified, and based on genetic analysis of mutants, suggested to be selectively targeted by a subset of R7 cells (yR7) (Carrillo et al. 2015). In these scenarios, discerning between different types of neurons might require complementary knowledge such as the electrophysiological properties of discrete neurons in the population or detailed connectivity maps. Obtaining this type of data might not be feasible for certain neuronal cell types and/or in certain organisms.

All profiling techniques have their advantages and disadvantages. A major concern in profiling approaches based on cell/nuclei isolation is the potential transcriptional changes caused by the cellular stress associated with dissociation procedures. It is assumed that in experiments designed to pinpoint differential expression between cell types, these transcriptional responses will be equal in both cell types, and thus will not interfere in the bioinformatic identification of differentially expressed transcripts. However, it is possible that distinct neurons present different sensitivity to cellular stress. In addition, if the aim is simply to characterize the gene expression profile of a particular cell type, these techniques will not differentiate between naturally expressed genes versus gene expression caused by dissociation stress. The main issue with profiling techniques that do not involve cell/nuclei isolation is nonspecific contamination by RNA in the total sample. Considerable efforts are being made to improve protocols and strategies in order to minimize this type of contamination. However, the smaller the size of neuronal cell type population under study versus the tissue sample used, the lower the signal-to-noise ratio. Where there is sufficient knowledge about the cell type being studied, unspecific contamination can be identified. For example, if the neurotransmitter identity is known, the presence of other neurotransmitters can be a sign of the presence of contaminating transcripts. However, in cases where there is little knowledge about the cell type being studied, caution is required when interpreting the data obtained.

Single-cell transcriptional profiling using microarrays and specially RNA-seq is emerging as a technology that sheds light on cell type identification in the nervous system. While signal-to-noise ratio can be an issue (Brennecke et al. 2013; Grün et al. 2014; Wu et al. 2014), improvements to reduce noise and the development of microfluidic technology to perform parallel sequencing of large numbers of cells simultaneously will contribute greatly to unveiling neural heterogeneity. In addition, the pursuit of knowledge will lead to the development of techniques that minimize both dissociation stress and nonspecific RNA contamination. This has already commenced with the development of fluorescence in situ sequencing (FISSEQ) (Lee et al. 2014, 2015). This technique has achieved RNA sequencing in cells within their tissue by combining biochemical with fluorescence imaging processing steps. One can envisage that high-throughput parallel single-cell FISSEQ would address both neural heterogeneity and the technical issues associated with profiling techniques. Moreover, Patch-seq is arising as an approach to classify neurons based on their physiology as well as their morphology and gene expression pattern. The endless creativity of researchers and multidisciplinary collaborative efforts will certainly push technology toward such new frontiers.

Acknowledgements The author apologizes to those whose work was not cited due to space constraints, and thanks A. Ferrús, T. Southall, L. Broday, and A. Sapir for critical reading, thoughtful comments, and insightful suggestions on this manuscript. M. Morey is supported by a Ramón y Cajal contract (RYC-2011-09479) and a Ministerio de Economía y Competitividad grant (BFU2015-69689-P).

References

Abruzzi K, Chen X, Nagoshi E, Zadina A, Rosbash M (2015) RNA-seq profiling of small numbers of *Drosophila* neurons. Methods Enzymol 551:369–386

Abruzzi KC, Zadina A, Luo W, Wiyanto E, Rahman R, Guo F, Shafer O, Rosbash M(2017) RNA-seq analysis of *Drosophila* clock and non-clock neurons reveals neuron-specific cycling and novel candidate neuropeptides. PLoS Genet 13(2):e1006613. doi: 10.1371/journal.pgen.1006613 (eCollection 2017 Feb)

Aughey GN, Southall TD (2016) Dam it's good! DamID profiling of protein-DNA interactions. Wiley Interdiscip Rev Dev Biol 5:25–37

Bacaj T, Shaham S (2007) Temporal control of cell-specific transgene expression in *Caenorhabditis elegans*. Genetics 176:2651–2655

Barthelson RA, Lambert GM, Vanier C, Lynch RM, Galbraith DW (2007) Comparison of the contributions of the nuclear and cytoplasmic compartments to global gene expression in human cells. BMC Genom 8:340

Bateman JR, Lee AM, Wu C (2006) Site-specific transformation of *Drosophila* via phiC31 integrase-mediated cassette exchange. Genetics 173:769–777

Blacque OE, Perens EA, Boroevich KA, Inglis PN, Li C, Warner A, Khattra J, Holt RA, Ou G, Mah AK et al (2005) Functional genomics of the cilium, a sensory organelle. Curr Biol 15:935–941

Bonn S, Zinzen RP, Perez-Gonzalez A, Riddell A, Gavin A-C, Furlong EEM (2012a) Cell type-specific chromatin immunoprecipitation from multicellular complex samples using BiTS-ChIP. Nat Protoc 7:978–994

Bonn S, Zinzen RP, Girardot C, Gustafson EH, Perez-Gonzalez A, Delhomme N, Ghavi-Helm Y, Wilczyński B, Riddell A, Furlong EEM (2012b) Tissue-specific analysis of chromatin state identifies temporal signatures of enhancer activity during embryonic development. Nat Genet 44:148–156

Bossing T, Barros CS, Fischer B, Russell S, Shepherd D (2012) Disruption of microtubule integrity initiates mitosis during CNS repair. Dev Cell 23:433–440

Brand AH, Perrimon N (1993) Targeted gene expression as a means of altering cell fates and generating dominant phenotypes. Development 118:401–415

Brennecke P, Anders S, Kim JK, Kołodziejczyk AA, Zhang X, Proserpio V, Baying B, Benes V, Teichmann SA, Marioni JC et al (2013) Accounting for technical noise in single-cell RNA-seq experiments. Nat Methods 10:1093–1095

Cadwell CR, Palasantza A, Jiang X, Berens P, Deng Q, Yilmaz M, Reimer J, Shen S, Bethge M, Tolias KF, Sandberg R, Tolias AS (2016) Electrophysiological, transcriptomic and morphologic profiling of single neurons using Patch-seq. Nat Biotechnol 34:199–203. doi: 10.1038/nbt.3445

Cao J, Ni J, Ma W, Shiu V, Milla LA, Park S, Spletter ML, Tang S, Zhang J, Wei X et al (2014) Insight into insulin secretion from transcriptome and genetic analysis of insulin-producing cells of *Drosophila*. Genetics 197:175–192

Carrillo RA, Özkan E, Menon KP, Nagarkar-Jaiswal S, Lee P-T, Jeon M, Birnbaum ME, Bellen HJ, Garcia KC, Zinn K (2015) Control of synaptic connectivity by a network of *Drosophila* IgSF cell surface proteins. Cell 163:1770–1782

Chatzigeorgiou M, Yoo S, Watson JD, Lee W-H, Spencer WC, Kindt KS, Hwang SW, Miller DM, Treinin M, Driscoll M et al (2010) Specific roles for DEG/ENaC and TRP channels in touch and thermosensation in *C. elegans* nociceptors. Nat Neurosci 13:861–868

Chen KH, Boettiger AN, Moffitt JR, Wang S, Zhuang X (2015) RNA imaging. Spatially resolved, highly multiplexed RNA profiling in single cells. Science 348:aaa6090

Cinar H, Keles S, Jin Y (2005) Expression profiling of GABAergic motor neurons in *Caenorhabditis elegans*. Curr Biol 15:340–346

Cleary MD, Meiering CD, Jan E, Guymon R, Boothroyd JC (2005) Biosynthetic labeling of RNA with uracil phosphoribosyltransferase allows cell-specific microarray analysis of mRNA synthesis and decay. Nat Biotechnol 23:232–237

Colosimo ME, Brown A, Mukhopadhyay S, Gabel C, Lanjuin AE, Samuel ADT, Sengupta P (2004) Identification of thermosensory and olfactory neuron-specific genes via expression profiling of single neuron types. Curr Biol 14:2245–2251

Couto A, Alenius M, Dickson BJ (2005) Molecular, anatomical, and functional organization of the *Drosophila* olfactory system. Curr Biol 15:1535–1547

Crocker A, Guan X-J, Murphy CT, Murthy M (2016) Cell-type-specific transcriptome analysis in the *Drosophila* mushroom body reveals memory-related changes in gene expression. Cell Rep 1–17

Davis MW, Morton JJ, Carroll D, Jorgensen EM (2008) Gene activation using FLP recombinase in *C. elegans*. PLoS Genet 4:e1000028

Deal RB, Henikoff S (2010) A simple method for gene expression and chromatin profiling of individual cell types within a tissue. Dev Cell 18:1030–1040

del Valle Rodríguez A, Didiano D, Desplan C (2012) Power tools for gene expression and clonal analysis in *Drosophila*. Nat Methods 9:47–55

Deng Q, Ramsköld D, Reinius B, Sandberg R (2014) Single-cell RNA-seq reveals dynamic, random monoallelic gene expression in mammalian cells. Science 343:193–196

Ding B, Zheng L, Zhu Y, Li N, Jia H, Ai R, Wildberg A, Wang W (2015) Normalization and noise reduction for single cell RNA-seq experiments. Bioinformatics 31:2225–2227

Dupuy D, Li Q-R, Deplancke B, Boxem M, Hao T, Lamesch P, Sequerra R, Bosak S, Doucette-Stamm L, Hope IA et al (2004) A first version of the *Caenorhabditis elegans* promoterome. Genome Res 14:2169–2175

Dupuy D, Bertin N, Hidalgo CA, Venkatesan K, Tu D, Lee D, Rosenberg J, Svrzikapa N, Blanc A, Carnec A et al (2007) Genome-scale analysis of in vivo spatiotemporal promoter activity in *Caenorhabditis elegans*. Nat Biotechnol 25:663–668

Dusenbery DB (1974) Analysis of chemotaxis in the nematode *Caenorhabditis elegans* by countercurrent separation. J Exp Zool 188:41–47

Dusenbery DB, Sheridan RE, Russell RL (1975) Chemotaxis-defective mutants of the nematode *Caenorhabditis elegans*. Genetics 80:297–309

Easow G, Teleman AA, Cohen SM (2007) Isolation of microRNA targets by miRNP immunopurification. RNA 13:1198–1204

Etchberger JF, Lorch A, Sleumer MC, Zapf R, Jones SJ, Marra MA, Holt RA, Moerman DG, Hobert O (2007) The molecular signature and cis-regulatory architecture of a *C. elegans* gustatory neuron. Genes Dev 21:1653–1674

Fischbach K-F, Dittrich AP (1989) The optic lobe of *Drosophila melanogaster*. I: A. Golgi analysis of wild-type structure. Cell Tissue Res 258:441–475

Fischer JA, Giniger E, Maniatis T, Ptashne M (1988) GAL4 activates transcription in *Drosophila*. Nature 332:853–856

Flourakis M, Allada R (2015) Patch-clamp electrophysiology in *Drosophila*circadian pacemaker neurons. Methods Enzymol 552:23–44. doi: 10.1016/bs.mie.2014.10.005

Fortini ME, Rubin GM (1990) Analysis of cis-acting requirements of the Rh3 and Rh4 genes reveals a bipartite organization to rhodopsin promoters in *Drosophila melanogaster*. Genes Dev 4:444–463

Fox RM, Von Stetina SE, Barlow SJ, Shaffer C, Olszewski KL, Moore JH, Dupuy D, Vidal M, Miller DM (2005) A gene expression fingerprint of *C. elegans* embryonic motor neurons. BMC Genom 6:42

Fuzik J, Zeisel A, Máté Z, Calvigioni D, Yanagawa Y, Szabó G, Linnarsson S, Harkany T (2016) Integration of electrophysiological recordings with single-cell RNA-seq data identifies neuronal subtypes. Nat Biotechnol 34:175–183. doi: 10.1038/nbt.3443

Gerstein MB, Lu ZJ, Van Nostrand EL, Cheng C, Arshinoff BI, Liu T, Yip KY, Robilotto R, Rechtsteiner A, Ikegami K et al (2010) Integrative analysis of the *Caenorhabditis elegans* genome by the modENCODE project. Science 330:1775–1787

Gohl DM, Silies MA, Gao XJ, Bhalerao S, Luongo FJ, Lin C-C, Potter CJ, Clandinin TR (2011) A versatile in vivo system for directed dissection of gene expression patterns. Nat Methods 8:231–237

Grün D, Kester L, van Oudenaarden A (2014) Validation of noise models for single-cell transcriptomics. Nat Methods 11:637–640

Hadjieconomou D, Rotkopf S, Alexandre C, Bell DM, Dickson BJ, Salecker I (2011) Flybow: genetic multicolor cell labeling for neural circuit analysis in *Drosophila melanogaster*. Nat Methods 8:260–266

Haenni S, Ji Z, Hoque M, Rust N, Sharpe H, Eberhard R, Browne C, Hengartner MO, Mellor J, Tian B et al (2012) Analysis of *C. elegans* intestinal gene expression and polyadenylation by fluorescence-activated nuclei sorting and 3′-end-seq. Nucleic Acids Res 40:6304–6318

Hallem EA, Spencer WC, McWhirter RD, Zeller G, Henz SR, Rätsch G, Miller DM, Horvitz HR, Sternberg PW, Ringstad N (2011) Receptor-type guanylate cyclase is required for carbon dioxide sensation by *Caenorhabditis elegans*. Proc Natl Acad Sci U S A 108:254–259

Hattori Y, Usui T, Satoh D, Moriyama S, Shimono K, Itoh T, Shirahige K, Uemura T (2013) Sensory-neuron subtype-specific transcriptional programs controlling dendrite morphogenesis: genome-wide analysis of Abrupt and Knot/Collier. Dev Cell 27:530–544

Hayashi S, Ito K, Sado Y, Taniguchi M, Akimoto A, Takeuchi H, Aigaki T, Matsuzaki F, Nakagoshi H, Tanimura T et al (2002) GETDB, a database compiling expression patterns and molecular locations of a collection of Gal4 enhancer traps. Genesis 34:58–61

He M, Liu Y, Wang X, Zhang MQ, Hannon GJ, Huang ZJ (2012) Cell-type-based analysis of microRNA profiles in the mouse brain. Neuron 73:35–48

Heiman M, Schaefer A, Gong S, Peterson JD, Day M, Ramsey KE, Suárez-Fariñas M, Schwarz C, Stephan DA, Surmeier DJ et al (2008) A translational profiling approach for the molecular characterization of CNS cell types. Cell 135:738–748

Henry GL, Davis FP, Picard S, Eddy SR (2012) Cell type-specific genomics of *Drosophila* neurons. Nucleic Acids Res 40:9691–9704

Hoopfer ED, Penton A, Watts RJ, Luo L (2008) Genomic analysis of *Drosophila* neuronal remodeling: a role for the RNA-binding protein Boule as a negative regulator of axon pruning. J Neurosci 28:6092–6103

Hope IA (1991) "Promoter trapping" in *Caenorhabditis elegans*. Development 113:399–408

Huang Y, Ainsley JA, Reijmers LG, Jackson FR (2013) Translational profiling of clock cells reveals circadianly synchronized protein synthesis. PLoS Biol 11:e1001703

Hunt-Newbury R, Viveiros R, Johnsen R, Mah A, Anastas D, Fang L, Halfnight E, Lee D, Lin J, Lorch A et al (2007) High-throughput in vivo analysis of gene expression in *Caenorhabditis elegans*. PLoS Biol 5:e237

Hutter H (2003) Extracellular cues and pioneers act together to guide axons in the ventral cord of *C. elegans*. Development 130:5307–5318

Isik M, Korswagen HC, Berezikov E (2010) Expression patterns of intronic microRNAs in *Caenorhabditis elegans*. Silence 1:5

Islam S, Zeisel A, Joost S, La Manno G, Zajac P, Kasper M, Lönnerberg P, Linnarsson S (2014) Quantitative single-cell RNA-seq with unique molecular identifiers. Nat Methods 11:163–166

Iyer EPR, Cox DN (2010) Laser capture microdissection of *Drosophila* peripheral neurons. J Vis Exp

Iyer EPR, Iyer SC, Cox DN (2013a) Application of cell-specific isolation to the study of dopamine signaling in *Drosophila*. Methods Mol Biol 964:215–225

Iyer EPR, Iyer SC, Sullivan L, Wang D, Meduri R, Graybeal LL, Cox DN (2013b) Functional genomic analyses of two morphologically distinct classes of *Drosophila* sensory neurons: post-mitotic roles of transcription factors in dendritic patterning. PLoS ONE 8:e72434

Jasper H, Benes V, Atzberger A, Sauer S, Ansorge W, Bohmann D (2002) A genomic switch at the transition from cell proliferation to terminal differentiation in the *Drosophila* eye. Dev Cell 3:511–521

Jenett A, Rubin GM, Ngo TTB, Shepherd D, Murphy C, Dionne H, Pfeiffer BD, Cavallaro A, Hall D, Jeter J et al (2012) A GAL4-driver line resource for *Drosophila* neurobiology. Cell Rep 2:991–1001

Jin Y, Ha N, Forés M, Xiang J, Gläßer C, Maldera J, Jiménez G, Edgar BA (2015) EGFR/Ras signaling controls *Drosophila* intestinal stem cell proliferation via capicua-regulated genes. PLoS Genet 11:e1005634

Johnson MB, Walsh CA (2017) Cerebral cortical neuron diversity and development at single-cell resolution. Curr Opin Neurobiol 42:9–16. doi: 10.1016/j.conb.2016.11.001

Kitamoto T, Ikeda K, Salvaterra PM (1992) Analysis of cis-regulatory elements in the 5′ flanking region of the *Drosophila melanogaster* choline acetyltransferase gene. J Neurosci 12:1628–1639

Kudlow BA, Zhang L, Han M (2012) Systematic analysis of tissue-restricted miRISCs reveals a broad role for microRNAs in suppressing basal activity of the *C. elegans* pathogen response. Mol Cell 46:530–541

Kula-Eversole E, Nagoshi E, Shang Y, Rodriguez J, Allada R, Rosbash M (2010) Surprising gene expression patterns within and between PDF-containing circadian neurons in *Drosophila*. Proc Natl Acad Sci U S A 107:13497–13502

Kunitomo H, Uesugi H, Kohara Y, Iino Y (2005) Identification of ciliated sensory neuron-expressed genes in *Caenorhabditis elegans* using targeted pull-down of poly(A) tails. Genome Biol 6:R17

Kvon EZ, Stampfel G, Yáñez-Cuna JO, Dickson BJ, Stark A (2012) HOT regions function as patterned developmental enhancers and have a distinct cis-regulatory signature. Genes Dev 26:908–913

Kvon EZ, Kazmar T, Stampfel G, Yáñez-Cuna JO, Pagani M, Schernhuber K, Dickson BJ, Stark A (2014) Genome-scale functional characterization of *Drosophila* developmental enhancers in vivo. Nature

Lai S-L, Lee T (2006) Genetic mosaic with dual binary transcriptional systems in *Drosophila*. Nat Neurosci 9:703–709

Lai S-L, Miller MR, Robinson KJ, Doe CQ (2012) The Snail family member Worniu is continuously required in neuroblasts to prevent Elav-induced premature differentiation. Dev Cell 23:849–857

Lee T, Luo L (1999) Mosaic analysis with a repressible cell marker for studies of gene function in neuronal morphogenesis. Neuron 22:451–461

Lee JH, Daugharthy ER, Scheiman J, Kalhor R, Yang JL, Ferrante TC, Terry R, Jeanty SSF, Li C, Amamoto R et al (2014) Highly multiplexed subcellular RNA sequencing in situ. Science (80-) 343:1360–1363

Lee JH, Daugharthy ER, Scheiman J, Kalhor R, Ferrante TC, Terry R, Turczyk BM, Yang JL, Lee HS, Aach J et al (2015) Fluorescent in situ sequencing (FISSEQ) of RNA for gene expression profiling in intact cells and tissues. Nat Protoc 10:442–458

Li W, Ou G (2016) The application of somatic CRISPR-Cas9 to conditional genome editing in *Caenorhabditis elegans*. Genesis 54:170–181

Liu S, Trapnell C (2016) Single-cell transcriptome sequencing: recent advances and remaining challenges. F1000Research 5

Luan H, Peabody NC, Vinson CR, White BH (2006) Refined spatial manipulation of neuronal function by combinatorial restriction of transgene expression. Neuron 52:425–436

Lynch AS, Briggs D, Hope IA (1995) Developmental expression pattern screen for genes predicted in the *C. elegans* genome sequencing project. Nat Genet 11:309–313

Ma J, Ptashne M (1987) The carboxy-terminal 30 amino acids of GAL4 are recognized by GAL80. Cell 50:137–142

Ma J, Weake VM (2014) Affinity-based isolation of tagged nuclei from *Drosophila* tissues for gene expression analysis. J Vis Exp

Marinov GK, Williams BA, McCue K, Schroth GP, Gertz J, Myers RM, Wold BJ (2014) From single-cell to cell-pool transcriptomes: stochasticity in gene expression and RNA splicing. Genome Res 24:496–510

Marshall OJ, Southall TD, Cheetham SW, Brand AH (2016) Cell-type-specific profiling of protein-DNA interactions without cell isolation using targeted DamID with next-generation sequencing. Nat Protoc 11:1586–1598. doi: 10.1038/nprot.2016.084

Martinez NJ, Ow MC, Reece-Hoyes JS, Barrasa MI, Ambros VR, Walhout AJM (2008) Genome-scale spatiotemporal analysis of *Caenorhabditis elegans* microRNA promoter activity. Genome Res 18:2005–2015

McClure CD, Southall TD (2015) Getting down to specifics: profiling gene expression and protein-DNA interactions in a cell type-specific manner. Adv Genet 91:103–151

McGuire SE, Le PT, Davis RL (2001) The role of *Drosophila* mushroom body signaling in olfactory memory. Science 293:1330–1333

Miller MR, Robinson KJ, Cleary MD, Doe CQ (2009) TU-tagging: cell type-specific RNA isolation from intact complex tissues. Nat Methods 6:439–441

Miyazaki T, Ito K (2010) Neural architecture of the primary gustatory center of *Drosophila melanogaster* visualized with GAL4 and LexA enhancer-trap systems. J Comp Neurol 518:4147–4181

Mizrak D, Ruben M, Myers GN, Rhrissorrakrai K, Gunsalus KC, Blau J (2012) Electrical activity can impose time of day on the circadian transcriptome of pacemaker neurons. Curr Biol 22:1871–1880

modENCODE Consortium, Roy S, Ernst J, Kharchenko PV, Kheradpour P, Negre N, Eaton ML, Landolin JM, Bristow CA, Ma L et al (2010). Identification of functional elements and regulatory circuits by *Drosophila* modENCODE. Science 330:1787–1797

Moor AE, Itzkovitz S (2017) Spatial transcriptomics: paving the way for tissuelevel systems biology. Curr Opin Biotechnol 46:126–133. doi: 10.1016/j.copbio.2017.02.004

Nagoshi E, Sugino K, Kula E, Okazaki E, Tachibana T, Nelson S, Rosbash M (2010) Dissecting differential gene expression within the circadian neuronal circuit of *Drosophila*. Nat Neurosci 13:60–68

Nègre N, Brown CD, Ma L, Bristow CA, Miller SW, Wagner U, Kheradpour P, Eaton ML, Loriaux P, Sealfon R et al (2011) A cis-regulatory map of the *Drosophila* genome. Nature 471:527–531

Nern A, Pfeiffer BD, Svoboda K, Rubin GM (2011) Multiple new site-specific recombinases for use in manipulating animal genomes. Proc Natl Acad Sci U S A 108:14198–14203

Nilsson H, Krawczyk KM, Johansson ME (2014) High salt buffer improves integrity of RNA after fluorescence-activated cell sorting of intracellular labeled cells. J Biotechnol 192 Pt A:62–65

O'Kane CJ, Gehring WJ (1987) Detection in situ of genomic regulatory elements in *Drosophila*. Proc Natl Acad Sci U S A 84:9123–9127

Otsuki L, Cheetham SW, Brand AH (2014) Freedom of expression: cell-type-specific gene profiling. Wiley Interdiscip Rev Dev Biol 3:429–443

Paix A, Folkmann A, Rasoloson D, Seydoux G (2015) High efficiency, homology-directed genome editing in *Caenorhabditis elegans* using CRISPR-Cas9 ribonucleoprotein complexes. Genetics 201:47–54

Parrish JZ, Kim MD, Jan LY, Jan YN (2006) Genome-wide analyses identify transcription factors required for proper morphogenesis of *Drosophila* sensory neuron dendrites. Genes Dev 20:820–835

Parrish JZ, Kim CC, Tang L, Bergquist S, Wang T, Derisi JL, Jan LY, Jan YN, Davis GW (2014) Krüppel mediates the selective rebalancing of ion channel expression. Neuron 82:537–544

Petersen SC, Watson JD, Richmond JE, Sarov M, Walthall WW, Miller DM (2011) A transcriptional program promotes remodeling of GABAergic synapses in *Caenorhabditis elegans*. J Neurosci 31:15362–15375

Pfeiffer BD, Jenett A, Hammonds AS, Ngo T-TB, Misra S, Murphy C, Scully A, Carlson JW, Wan KH, Laverty TR et al (2008) Tools for neuroanatomy and neurogenetics in *Drosophila*. Proc Natl Acad Sci U S A 105:9715–9720

Pfeiffer BD, Ngo T-TB, Hibbard KL, Murphy C, Jenett A, Truman JW, Rubin GM (2010) Refinement of tools for targeted gene expression in *Drosophila*. Genetics 186:735–755

Pfeiffer BD, Truman JW, Rubin GM (2012) Using translational enhancers to increase transgene expression in *Drosophila*. Proc Natl Acad Sci U S A 109:6626–6631

Potter CJ, Tasic B, Russler EV, Liang L, Luo L (2010) The Q system: a repressible binary system for transgene expression, lineage tracing, and mosaic analysis. Cell 141:536–548

Poulin J-F, Tasic B, Hjerling-Leffler J, Trimarchi JM, Awatramani R (2016) Disentangling neural cell diversity using single-cell transcriptomics. Nat Neurosci 19:1131–1141. doi: 10.1038/nn.4366

Raj A, van den Bogaard P, Rifkin SA, van Oudenaarden A, Tyagi S (2008) Imaging individual mRNA molecules using multiple singly labeled probes. Nat Methods 5:877–879

Reece-Hoyes JS, Shingles J, Dupuy D, Grove CA, Walhout AJM, Vidal M, Hope IA (2007) Insight into transcription factor gene duplication from *Caenorhabditis elegans* Promoterome-driven expression patterns. BMC Genom 8:27

Riabinina O, Luginbuhl D, Marr E, Liu S, Wu MN, Luo L, Potter CJ (2015) Improved and expanded Q-system reagents for genetic manipulations. Nat Methods 12:219–222, 5 p following 222

Roy PJ, Stuart JM, Lund J, Kim SK (2002) Chromosomal clustering of muscle-expressed genes in *Caenorhabditis elegans*. Nature 418:975–979

Ruben M, Drapeau MD, Mizrak D, Blau J (2012) A mechanism for circadian control of pacemaker neuron excitability. J Biol Rhythms 27:353–364

Russell JN, Clements JE, Gama L (2013) Quantitation of gene expression in formaldehyde-fixed and fluorescence-activated sorted cells. PLoS ONE 8:e73849

Saliba A-E, Westermann AJ, Gorski SA, Vogel J (2014) Single-cell RNA-seq: advances and future challenges. Nucleic Acids Res 42:8845–8860

Sanz E, Yang L, Su T, Morris DR, McKnight GS, Amieux PS (2009) Cell-type-specific isolation of ribosome-associated mRNA from complex tissues. Proc Natl Acad Sci U S A 106:13939–13944

Shi B, Guo X, Wu T, Sheng S, Wang J, Skogerbø G, Zhu X, Chen R (2009) Genome-scale identification of *Caenorhabditis elegans* regulatory elements by tiling-array mapping of DNase I hypersensitive sites. BMC Genom 10:92

Smith CJ, Watson JD, Spencer WC, O'Brien T, Cha B, Albeg A, Treinin M, Miller DM (2010) Time-lapse imaging and cell-specific expression profiling reveal dynamic branching and molecular determinants of a multi-dendritic nociceptor in *C. elegans*. Dev Biol 345:18–33

Smith CJ, O'Brien T, Chatzigeorgiou M, Spencer WC, Feingold-Link E, Husson SJ, Hori S, Mitani S, Gottschalk A, Schafer WR et al (2013) Sensory neuron fates are distinguished by a transcriptional switch that regulates dendrite branch stabilization. Neuron 79:266–280

Southall TD, Gold KS, Egger B, Davidson CM, Caygill EE, Marshall OJ, Brand AH (2013) Cell-type-specific profiling of gene expression and chromatin binding without cell isolation: assaying RNA Pol II occupancy in neural stem cells. Dev Cell 26:101–112

Spencer WC, McWhirter R, Miller T, Strasbourger P, Thompson O, Hillier LW, Waterston RH, Miller DM (2014) Isolation of specific neurons from *C. elegans* larvae for gene expression profiling. PLoS ONE 9:e112102

Steiner FA, Talbert PB, Kasinathan S, Deal RB, Henikoff S (2012) Cell-type-specific nuclei purification from whole animals for genome-wide expression and chromatin profiling. Genome Res 22:766–777

Suster ML, Seugnet L, Bate M, Sokolowski MB (2004) Refining GAL4-driven transgene expression in *Drosophila* with a GAL80 enhancer-trap. Genesis 39:240–245

Swoboda P, Adler HT, Thomas JH (2000) The RFX-type transcription factor DAF-19 regulates sensory neuron cilium formation in *C. elegans*. Mol Cell 5:411–421

Szüts D, Bienz M (2000) LexA chimeras reveal the function of *Drosophila* Fos as a context-dependent transcriptional activator. Proc Natl Acad Sci U S A 97:5351–5356

Tahayato A, Sonneville R, Pichaud F, Wernet MF, Papatsenko D, Beaufils P, Cook T, Desplan C (2003) Otd/Crx, a dual regulator for the specification of ommatidia subtypes in the *Drosophila* retina. Dev Cell 5:391–402

Takayama J, Faumont S, Kunitomo H, Lockery SR, Iino Y (2010) Single-cell transcriptional analysis of taste sensory neuron pair in *Caenorhabditis elegans*. Nucleic Acids Res 38:131–142

Tan L, Zhang KX, Pecot MY, Nagarkar-Jaiswal S, Lee P-T, Takemura S-Y, McEwen JM, Nern A, Xu S, Tadros W et al (2015) Ig superfamily ligand and receptor pairs expressed in synaptic partners in *Drosophila*. Cell 163:1756–1769

Tang F, Barbacioru C, Wang Y, Nordman E, Lee C, Xu N, Wang X, Bodeau J, Tuch BB, Siddiqui A et al (2009) mRNA-Seq whole-transcriptome analysis of a single cell. Nat Methods 6:377–382

Thomas A, Lee P-J, Dalton JE, Nomie KJ, Stoica L, Costa-Mattioli M, Chang P, Nuzhdin S, Arbeitman MN, Dierick HA (2012) A versatile method for cell-specific profiling of translated mRNAs in *Drosophila*. PLoS ONE 7:e40276

Ting C-Y, Gu S, Guttikonda S, Lin T-Y, White BH, Lee C-H (2011) Focusing transgene expression in *Drosophila* by coupling Gal4 with a novel split-LexA expression system. Genetics 188:229–233

van Steensel B, Henikoff S (2000) Identification of in vivo DNA targets of chromatin proteins using tethered dam methyltransferase. Nat Biotechnol 18:424–428

van Steensel B, Delrow J, Henikoff S (2001) Chromatin profiling using targeted DNA adenine methyltransferase. Nat Genet 27:304–308

Venken KJT, Schulze KL, Haelterman NA, Pan H, He Y, Evans-Holm M, Carlson JW, Levis RW, Spradling AC, Hoskins RA et al (2011) MiMIC: a highly versatile transposon insertion resource for engineering *Drosophila melanogaster* genes. Nat Methods 8:737–743

Von Stetina SE, Fox RM, Watkins KL, Starich TA, Shaw JE, Miller DM (2007a) UNC-4 represses CEH-12/HB9 to specify synaptic inputs to VA motor neurons in *C. elegans*. Genes Dev 21:332–346

Von Stetina SE, Watson JD, Fox RM, Olszewski KL, Spencer WC, Roy PJ, Miller DM (2007b) Cell-specific microarray profiling experiments reveal a comprehensive picture of gene expression in the *C. elegans* nervous system. Genome Biol 8:R135

Voutev R, Hubbard EJA (2008) A "FLP-Out" system for controlled gene expression in *Caenorhabditis elegans*. Genetics 180:103–119

Wang H, Liu J, Gharib S, Chai CM, Schwarz EM, Pokala N, Sternberg PW (2017) cGAL, a temperature-robust GAL4-UAS system for *Caenorhabditis elegans*. Nat Methods 14:145–148. doi:10.1038/nmeth.4109 (Epub 2016 Dec 19)

Ward S (1973) Chemotaxis by the nematode *Caenorhabditis elegans*: identification of attractants and analysis of the response by use of mutants. Proc Natl Acad Sci U S A 70:817–821

Wei X, Potter CJ, Luo L, Shen K (2012) Controlling gene expression with the Q repressible binary expression system in *Caenorhabditis elegans*. Nat Methods 9:391–395

White JG, Southgate E, Thomson JN, Brenner S (1986) The structure of the nervous system of the nematode *Caenorhabditis elegans*. Philos Trans R Soc Lond B Biol Sci 314:1–340

Williams CR, Baccarella A, Parrish JZ, Kim CC (2016) Trimming of sequence reads alters RNA-Seq gene expression estimates. BMC Bioinformatics 17:103

Wu AR, Neff NF, Kalisky T, Dalerba P, Treutlein B, Rothenberg ME, Mburu FM, Mantalas GL, Sim S, Clarke MF et al (2014) Quantitative assessment of single-cell RNA-sequencing methods. Nat Methods 11:41–46

Xu J, Ren X, Sun J, Wang X, Qiao HH, Xu BW, Liu LP, Ni JQ (2015) A toolkit of CRISPR-based genome editing systems in *Drosophila*. J Genet Genomics 42:141–149

Yagi R, Mayer F, Basler K (2010) Refined LexA transactivators and their use in combination with the *Drosophila* Gal4 system. Proc Natl Acad Sci U S A 107:16166–16171

Yang Z, Edenberg HJ, Davis RL (2005) Isolation of mRNA from specific tissues of *Drosophila* by mRNA tagging. Nucleic Acids Res 33:e148

Zaslaver A, Liani I, Shtangel O, Ginzburg S, Yee L, Sternberg PW (2015) Hierarchical sparse coding in the sensory system of *Caenorhabditis elegans*. Proc Natl Acad Sci U S A 112:1185–1189

Zhang Y, Ma C, Delohery T, Nasipak B, Foat BC, Bounoutas A, Bussemaker HJ, Kim SK, Chalfie M (2002) Identification of genes expressed in *C. elegans* touch receptor neurons. Nature 418:331–335

Zhang S, Ma C, Chalfie M (2004) Combinatorial marking of cells and organelles with reconstituted fluorescent proteins. Cell 119:137–144

Zhang L, Ding L, Cheung TH, Dong M-Q, Chen J, Sewell AK, Liu X, Yates JR, Han M (2007) Systematic identification of *C. elegans* miRISC proteins, miRNAs, and mRNA targets by their interactions with GW182 proteins AIN-1 and AIN-2. Mol Cell 28:598–613

Zhang C, Barthelson RA, Lambert GM, Galbraith DW (2008) Global characterization of cell-specific gene expression through fluorescence-activated sorting of nuclei. Plant Physiol 147:30–40

Zhang KX, Tan L, Pellegrini M, Zipursky SL, McEwen JM (2016) Rapid changes in the translatome during the conversion of growth cones to synaptic terminals. Cell Rep 14:1258–1271

Part V
Epilogue: Future Outlook and Challenges in the Field of Circuit Science

Chapter 20
Perspective: A New Era of Comparative Connectomics

Ian A. Meinertzhagen

Abstract Morphological studies on brains have recently entered a new phase of circuit analysis identified under the newly designated area of connectomics, the study of brain wiring diagrams exact at synapse level that can now be produced by means of electron microscopy and automated reconstruction. The most comprehensive examples come from the brains of invertebrates with few neurons, which Nature provides in great abundance especially among marine larval invertebrates. Two complete examples, the nematode C. elegans and the larva of the ascidian Ciona intestinalis, are now published; others are in the pipeline. Each species has its advantages and champions, especially clearly so in Drosophila, which offers outstanding opportunities for functional analysis of complex behaviours using genetics-based methods. Collectively, these offer an ultimate prospect for the causal analysis of behaviour. In addition, the availability of multiple connectomes from behaviourally different species will reveal features of the network design that are common to all, and that enable comparison with networks from different levels of biological organization, as well as with those from networks that have evolved from human technologies.

Morphological studies on brains have recently entered a new phase of circuit analysis identified under the newly designated area of connectomics, the study of

I.A. Meinertzhagen
Department of Psychology and Neuroscience, Life Sciences Centre,
Dalhousie University, Halifax, NS, CanadaB3H 4R2

I.A. Meinertzhagen
Department of Biology, Life Sciences Centre, Dalhousie University,
Halifax, CanadaB3H 4R2

I.A. Meinertzhagen
Howard Hughes Medical Institute, Janelia Research Campus,
19700 Helix Drive, Ashburn, VA 20147-2408, USA

I.A. Meinertzhagen (✉)
Life Sciences Centre, Dalhousie University, Halifax, NS, CanadaB3H 4R2
e-mail: I.A.Meinertzhagen@Dal.Ca; iam@dal.ca

A. Çelik and M.F. Wernet (eds.), *Decoding Neural Circuit Structure and Function*, DOI 10.1007/978-3-319-57363-2_20

exact brain wiring diagrams (Lichtman and Sanes 2008). A connectome is of course no new idea, merely one now finally enabled in select systems of neurons by efficient digital imaging and computer-aided reconstruction tools. Additionally, it relies on an enhanced range of electron imaging methods, including scanning block-face microscopy, SBFM (Denk and Heinz 2004) and focused ion beam milling (FIB)-SEM (Knott et al. 2008, 2011) as well as more traditional serial-section EM, ssEM methods (e.g. Fahrenbach 1984; Hall 1995; Harris et al. 2006), that all have sufficient resolution to identify clearly the synaptic contacts formed between identified neurons. Neuron identification in those cases is therefore usually enabled by 3-D reconstruction of neuron shapes, and comparisons between these and profiles from light microscopy. Obvious as it may be to some, not all are convinced by the approach, even when the contra arguments of its detractors have been robustly countered (Morgan and Lichtman 2013).

The *history of connectome reports* is rather sparsely populated by studies, most of which have taken advantage of small brains with few neurons. This epilogue summarizes such studies, highlighting ways in which invertebrate brains serve as a foundation for connectome studies, in parallel with alternative approaches on vertebrate brains, especially in the retina (e.g. Helmstaedter et al. 2013; Ding et al. 2016). Without doubt, pride of place among invertebrates goes to the first complete connectome report from *Caenorhabditis elegans* (White et al. 1986). But even this had its own forerunner, in the subway maps of *Ascaris* neurons published by Richard Goldschmidt nearly 80 years earlier (1908, 1909), and cleverly avoided patterns of rewiring in species with entirely different feeding habits (Bumbarger et al. 2013). As for genomes, the importance attributed to the connectome by its chief architect, Sydney Brenner, lies in it being complete, accurate and permanent, insofar as these absolutes can be determined. The initial version of the connectome (White et al. 1986) has in fact undergone some steps in its completion but these are very minor (Varshney et al. 2011), while comparisons between the reconstructions of two animals have demonstrated features of its accuracy both in the hermaphrodite (Durbin 1987), and in another comparison, with the posterior CNS of the adult male (Jarrell et al. 2012). Major early contributions on different systems were also made using photographic imaging, especially by the group of Cyrus Levinthal and colleagues at Columbia University, who pioneered in the development of approaches to automate serial image capture using pre-digital methods (e.g. Ware and LoPresti 1975), and examined amongst other systems, the entire nervous system of the rotifer *Asplanchna brightwelli* (Ware 1971) and the visual system of the crustacean *Daphnia magna* (Macagno et al. 1973).

The rest of the gang. The diminutive brains of invertebrates, especially those of larval stages with few cells (Meinertzhagen 2016a), are part of Nature's bounty for connectomic analyses. Marine invertebrate larvae, many with essential virtues of small size, suitable for EM, and complete transparency, suitable for LM, are rich in opportunity, but generally poor in the functional studies they enable. The brains of larval polychaetes, which show considerable diversity, have some of the smallest numbers and are especially well suited to connectomic studies. For example, the 48-h trochophore of the polychaete *Spirobranchus* with only ~36 neurons (Lacalli 1984), compares to the CNS of dwarf male *Dinophilus gyrociliatus* with a total of 68 neurons (Windoffer and Westheide 1988), and the larva of *Platyneiris dumerilii,* with 71 neurons for visual navigation (Randel et al. 2014, 2015); while the rotifer CNS has 196 brain cells in *Asplanchna* (Ware 1971). These are to be compared with the tadpole larva of the tunicate *Ciona intestinalis*, a basal deuterostome with ~330 cells of which 177 are neurons with 6618 synapses (Ryan et al. 2016), and the nematode *C. elegans* with 302 neurons and 6393 synapses in the hermaphrodite (White et al. 1986), for both of which an entire connectome has been completed. In addition, there is the immediate prospect of the entire connectome for the CNS of the first-instar larva of the fruit fly *Drosophila melanogaster*, for which excerpts have appeared (e.g. Ohyama et al. 2015).

To identify behavioural function in a synaptic circuit, neuron by neuron, targeting transgenes that either disable neuron function or selectively restore it in null mutants is possible in *Drosophila* (Simpson 2009; Meinertzhagen and Lee 2012) and other genetic models, but targeted single-cell photoablations are also possible in some other species.

Three overlapping motivations have mostly driven work in this area: first, the analysis of the neural substrate for interesting behaviours; second, the availability of powerful genetic methods that enable the targeting and functional analyses of identified neurons; and third, the attraction of simple nervous systems having few, morphologically simple cells favouring the construction of a complete connectome. Some species may indulge us in two of these three, but none has them all. For example, *Drosophila* has both powerful genetic methods (e.g. Luan et al. 2006; Pfeiffer et al. 2010; Jenett et al. 2012) and interesting behaviours, especially for vision (e.g. Heisenberg and Wolf 1984; Borst 2009; Silies et al. 2014), but its neurons branch profusely to yield many slender neurites that require special methods to reconstruct at EM level (e.g. Feng et al. 2015), albeit still relatively slowly and somewhat incompletely. The nematode *C. elegans* has both powerful genetic methods and a landmark connectome, made possible by the simple tubular features of its neurons, but its behavioural repertoire is less dynamic, compelling and completely analysed, especially in the depth dimension. It is difficult to know how to weight these three requirements, and whether the relative ease of 3-D reconstruction in, for example *C. elegans* or *C. intestinalis*, might eventually outweigh the disadvantages of either. For the moment, the prize goes to the construction of a single connectome in each species, whereas the eventual need is to reconstruct many, in order to seek their common features.

The brain as a network of networks. In addition to whole-brain connectome studies, considerable progress has been made in documenting smaller brain regions containing behaviourally interesting pathways. Early work on the connectome of individual neuropiles started most obviously with work on the first optic neuropile, or lamina, in various arthropods—notably the branchiopod crustacean *D. magna* (Macagno et al. 1973), the horseshoe crab *Limulus polyphemus* (Fahrenbach 1985), the housefly *Musca domestica* (Strausfeld and Campos-Ortega 1977) and fruit fly *D. melanogaster* (Meinertzhagen and O'Neil 1991; Meinertzhagen and Sorra 2001). These all capitalized on the parallel pathways endowed by compound eyes, and captured what was possible with limited ssEM. More recently, Randel et al. (2014) have reported the four-eye, 71-neuron sensory-motor circuits for visual navigation in *Platyneiris* that are comprehensively organized into circuits with 1106 connections. These have a high level of stereotypy between individuals (Randel et al. 2015).

Drosophila is today's star. Contemporary approaches to connectomics may show few bounds. Yet, even the tiny brain of *Drosophila* is too complex for us to imagine fully. In parallel with EM studies above, light microscopy in flies provides accounts for the cells and circuits that underlie visual behaviour (*Musca*: Douglass and Strausfeld 2003; Strausfeld and Nässel 1980; and *Drosophila*: Meinertzhagen 2014), for which there is a large body of quantitative behavioural studies (Heisenberg and Wolf 1984; Borst and Egelhaaf 1989; Borst and Haag 2002). The objectives at EM level may even be powerfully aided by wiring networks derived from light microscopy in *Drosophila* (Chiang et al. 2011; Shih et al. 2015). These are enabled in turn by new tools for neuroanatomy and neurogenetics (Pfeiffer et al. 2008), and 3-D registration (Peng et al. 2011). Thus, light microscopy identifies 58 tracts between 41 neuropiles and 6 hubs in *Drosophila* and a total of 16,000 neuron classes that fasciculate consistently within these (Chiang et al. 2011), upgraded to 43 local processing units (Shih et al. 2015). The individual connectomes for each of these can only be assembled piecemeal, in an initial strategy of divide and conquer followed by one of unification. In their entirety these obviously contribute unimaginable complexity, and the whole is only more manageable when each is considered apart from the others. Examples include not only the optic neuropiles (Takemura et al. 2008, 2013, 2015, 2017a; Shinomiya et al. 2014), but also the antennal lobe (Rybak et al. 2016; Tobin et al. 2017), mushroom body—alpha lobe (Takemura et al. 2017b) and calyx (Butcher et al. 2012). The FlyEM team at the HHMI Janelia Research Campus in fact proposes to complete the *Drosophila* connectome in exactly this manner, neuropile by neuropile, from 20 μm slices of an entire brain cut on a hot diamond knife (Hayworth et al. 2015) each then individually imaged by FIB-SEM, and the consecutive slices merged.

Simpler yet, but nevertheless still daunting, the CNS of a first-instar larval *Drosophila* is another candidate system for which detailed synaptic networks from specific systems are now available. The functional features of various anatomical connectomic circuits have received attention, for example for action selection in rolling behaviour (Ohyama et al. 2015), or the competitive interactions among neurons that mediate choice behaviour. Thus, selecting one behaviour over another, behavioural choice is mediated by reciprocally connected feedforward inhibitory interneurons, while maintaining a chosen behaviour is mediated by feedback disinhibition; sequence transitions are mediated by lateral disinhibition (Jovanic et al. 2016).

All *Drosophila* brain regions reconstructed to date reveal an unsuspected complexity of synaptic circuits. These are obvious in the adult fly, for the visual pathways in the medulla (Takemura et al. 2013, 2017a, b) and especially for the alpha 2 and 3 output lobes of the mushroom body (Takemura et al. 2017a, b), the substrate for aversive olfactory conditioning. The latter incorporate: direct connections from dopaminergic neurons (DAN) to the output neurons; universal anatomical connection of Kenyon cells (KC) to output neurons (whereas electrophysiology had indicated that only 30% are functionally connected); and KC to DAN connections. Once revealed anatomically, the functions of these and the many other connections in the connectome can of course be investigated using genetic, behavioural, and electrophysiological approaches, and it is these that may be remembered by posterity. But the fact that, despite intense behavioural investigations in the field of olfactory learning, circuits that had been neither detected by light microscopy nor proposed in any model of learning were revealed only by EM (Takemura et al. 2017a), illustrates clearly the predictive value of EM connectomics as an enabling technology, in revealing the entire envelope of possible behaviours, each in need of functional validation.

No less powerful, an even simpler case had in fact already been revealed, for UV phototaxis pathways in *Drosophila* (Gao et al. 2008). The history of their discovery illustrates the predictive power of connectomics. Initially, Gal4 reagents identified a rather nondescript distal amacrine cell in the medulla, called Dm8 (Fischbach and Dittrich 1989), and early ssEM identified that it received synaptic inputs from the UV-signalling photoreceptor neuron R7. These synapses were identified largely because Dm8's neurites are somewhat coarse and could be reconstructed from EM with relative ease (Gao et al. 2008). Based only on this rather slender evidence for UV sensitivity, subsequent behavioural tests of UV phototaxis revealed that Dm8 neurons were in fact both necessary and sufficient for flies to exhibit phototaxis towards UV, in preference to green light (Gao et al. 2008). Thus, even though UV phototaxis is a low-level visual behaviour and Dm8 its rather nondescript substrate, this analysis provided a very clear precedent for the predictive power of connectomics in assigning an essential role to an identified neuron, in this case Dm8.

Comparing connectomes. The connectomes of different species and systems provide a rich data source for comparative computational studies. We need always to bear in mind that brains are the product of harsh behavioural selection over millions of years, and that evolutionary relationships, uncertain as these may be, are essential to interpreting connectome organization in ways not revealed by single-species connectomes (Hale 2014). Key to that organization is, of course, how the network functions.

The two comprehensive connectomes now available, the one in a widely analysed adult hermaphrodite *C. elegans* (White et al. 1986), a protostome, and the other in an entirely different deuterostome, the larva of *C. intestinalis* (Ryan et al. 2016), exhibit network similarities. Both comprise relatively few neurons and not surprisingly neither is a random nor a regular network. Instead, both constitute so-called 'small-world' networks, characterized by highly connected local sub-networks linked by fewer long-range connections (Watts and Strogatz 1998; Bassett and Bullmore 2006; Bezares-Calderón and Jékely 2016). These mirror many human-constructed networks, from power grids to social media (Watts and Strogatz 1998), and offer an intuitive appeal in their economy, supporting high levels of dynamic complexity while tending to minimize wiring costs (Bassett and Bullmore 2006). Those costs are rarely specified or quantified within a network, but have received treatment (Sterling and Laughlin 2015) in energetic rather than morphogenetic terms.

Network motifs, simple building blocks of connected elements, offer a further element of commonality among many forms of network. Recognition that network motifs can be identified not only in biological systems that range from food chains to genetic networks, but also in engineering and information-processing networks (Milo et al. 2002), provided an important insight to the interpretation of an entire connectome network. Thus, specific two-, three- and four-element motifs, especially—among the latter—bi-fan and bi-parallel patterns of interconnection, occur with a significantly higher frequency in the actual synaptic networks of the *C. elegans* connectome, than in matched randomized networks having the same single-node features (Milo et al. 2002). This preference suggests that such motifs confer special functional properties, yet the physiology of the motifs and their elements is rarely if ever known and unlikely to be the same in all cases. Of course, this lack merely emphasizes the need for functional studies that can indicate the operational features of individual motifs, in order to identify and define classes of networks and network homologies more closely. Connectomic data also highlight higher-order networks, for example, in the CNS of larval *Drosophila* (Jovanic et al. 2016; Schneider-Mizell et al. 2016), where they identify a particular role for disinhibition. Thus, interneurons involved in promoting distinct behaviours via disinhibition have extensive reciprocal connections (Jovanic et al. 2016). In a different example, among the visuomotor circuits of the polychaete *Platyneiris*, the reciprocal wiring reasoned to provide strong mutual inhibition between the circuits of the two eyes is interpreted to represent a network motif that enhances contrast detection (Randel et al. 2014). Essential knowledge of the neurotransmitter used by each neuron, and—even more important—the neurotransmitter receptors expressed by its postsynaptic partners, is required for all such interpretations, and represents the next stage in any functional analysis.

Opinions may differ on the relative utility of computational approaches to connectome networks, and few would deny the value of lower level information on the anatomy and physiology of synaptic circuits. The latter would at least help us identify synaptic pathway strengths. In some cases, synapse numbers may give a sufficient first impression, but where synapses vary in size the aggregate synapse size is also important in determining the area over which vesicle shedding and postsynaptic signalling can occur. Moreover, the processing depth of each element of a circuit is also crucial, because it identifies not only the synaptic gain resulting from the quantitative strength of its connection, but also the gain that the element inherits from, or is offset by, its upstream partners.

Finally, *no connectome is fixed*. Even when completed comprehensively, synaptic partnerships and their connection strengths may change. Thus the literature on morphological plasticity in fly synaptic circuits (Meinertzhagen 2001, 2016b) reveals that there are many forms of plasticity among circuits that may appear fixed, but are evidently plastic. Such changes often go unacknowledged in the nervous systems of flies, which are generally assumed to have a fixed structural phenotype. In fact the fixity of synaptic circuits is at least partly the outcome of fixity in the conditions under which flies are usually raised in laboratory culture. More than this, the boundaries of connectome data are in many cases set by the developmental stage of the individual, especially in larval forms that are undergoing morphogenetic change. Even so, to the extent that the overall network remains unchanged, especially in *Drosophila*, such data will have archival as well as heuristic value, provided they can be made available in some accessible way for future generations. A related problem is therefore how to store such data, which rely on multi-terabyte datasets and need tools for rapid access and manipulability, needs and considerations that will not be treated in this perspective.

Acknowledgments The author acknowledges various sources of support for his work summarized in this review, especially grant DIS-0000065 from the Natural Sciences and Engineering Research Council, for research on the larval nervous system of *Ciona*, and the FlyEM team at the Janelia Research Campus of HHMI for work on *Drosophila*. Dr. Kerrianne Ryan read an earlier version of the manuscript.

References

Bassett DS, Bullmore E (2006) Small-world brain networks. Neuroscientist 12:512–523

Bezares-Calderón LA, Jékely G (2016) Think small. eLIFE 5. pii: e22497. doi:10.7554/eLife.22497

Borst A (2009) *Drosophila*'s view on insect vision. Curr Biol 19:R36–R47

Borst A, Egelhaaf M (1989) Principles of visual motion detection. Trends Neurosci 12:297–306

Borst A, Haag J (2002) Neural networks in the cockpit of the fly. J Comp Physiol A Neuroethol Sens Neural Behav Physiol 188:419–437

Bumbarger DJ, Riebesell M, Rödelsperger C, Sommer RJ (2013) System-wide rewiring underlies behavioral differences in predatory and bacterial-feeding nematodes. Cell 152:109–119

Butcher NJ, Friedrich AB, Lu Z, Tanimoto H, Meinertzhagen IA (2012) Different classes of input and output neurons reveal new features in microglomeruli of the adult *Drosophila* mushroom body calyx. J Comp Neurol 520:2185–2201

Chiang AS, Lin CY, Chuang CC, Chang HM, Hsieh CH, Yeh CW, Shih CT, Wu JJ, Wang GT, Chen YC, Wu CC, Chen GY, Ching YT, Lee PC, Lin CY, Lin HH, Wu CC, Hsu HW, Huang YA, Chen JY, Chiang HJ, Lu CF, Ni RF, Yeh CY, Hwang JK (2011) Three-dimensional reconstruction of brain-wide wiring networks in *Drosophila* at single-cell resolution. Curr Biol 21:1–11. doi:10.1016/j.cub.2010.11.056

Denk W, Heinz H (2004) Serial block-face scanning electron microscopy to reconstruct three-dimensional tissue nanostructure. PLoS Biol 2:e329. doi:10.1371/journal.pbio.0020329

Ding H, Smith RG, Poleg-Polsky A, Diamond JS, Briggman KL (2016) Species-specific wiring for direction selectivity in the mammalian retina. Nature 535:105–110

Douglass JK, Strausfeld NJ (2003) Anatomical organization of retinotopic motion-sensitive pathways in the optic lobes of flies. Microsc Res Tech 62:132–150

Durbin RM (1987) Studies on the development and organisation of the nervous system of *Caernorhabditis elegans*. Doctoral thesis, University of Cambridge

Fahrenbach WH (1984) Continuous serial thin sectioning for electron microscopy. J Electron Microsc Techn 1:387–398

Fahrenbach WH (1985) Anatomical circuitry of lateral inhibition in the eye of the horseshoe crab, *Limulus polyphemus*. Proc R Soc Lond B 225:219–249

Feng L, Zhao T, Kim J (2015) neuTube 1.0: a new design for efficient neuron reconstruction software based on the SWC format. eNeuro 2(1). pii: ENEURO.0049-14.2014

Fischbach K-F, Dittrich APM (1989) The optic lobe of *Drosophila melanogaster*. I. A Golgi analysis of wild-type structure. Cell Tiss Res 258:441–475

Gao S, Takemura S-Y, Ting C-Y, Huang S, Lu Z, Luan H, Rister J, Yang M, Hong S-T, Wang JW, Odenwald W, White B, Meinertzhagen IA, Lee C-H (2008) Neural substrate of spectral discrimination in *Drosophila*. Neuron 60:328–342

Goldschmidt R (1908) Das Nervensystem von *Ascaris lumbricoides* und *megalocephala*, I. Z wissenschaftliche Zool 90:73–136

Goldschmidt R (1909) Das Nervensystem von *Ascaris lumbricoides* und *megalocephala*, II. Z wissenschaftliche Zool 92:306–357

Hale ME (2014) Mapping circuits beyond the models: integrating connectomics and comparative neuroscience. Neuron 83:1256–1258

Hall DH (1995) Electron microscopy and three-dimensional image reconstruction. Methods Cell Biol 48:395–436

Harris KM, Perry E, Bourne J, Feinberg M, Ostroff L, Hurlburt J (2006) Uniform serial sectioning for transmission electron microscopy. J Neurosci 26:12101–12103

Hayworth KJ, Xu CS, Lu Z, Knott GW, Fetter RD, Tapia JC, Lichtman JW, Hess HF (2015) Ultrastructurally smooth thick partitioning and volume stitching for large-scale connectomics. Nat Methods 12:319–322

Heisenberg M, Wolf R (1984) Vision in *Drosophila*. Springer, Berlin

Helmstaedter M, Briggman KL, Turaga SC, Jain V, Seung HS, Denk W (2013) Connectomic reconstruction of the inner plexiform layer in the mouse retina. Nature 500:168–174

Jarrell TA, Wang Y, Bloniarz AE, Brittin CA, Xu M, Thomson JN, Albertson DG, Hall DH, Emmons SW (2012) The connectome of a decision-making neural network. Science 337:437–444

Jenett A, Rubin GM, Ngo TT, Shepherd D, Murphy C, Dionne H, Pfeiffer BD, Cavallaro A, Hall D, Myers EW, Iwinski ZR, Aso Y, DePasquale GM, Enos A, Hulamm P, Lam SC, Li HH, Laverty TR, Long F, Qu L, Murphy SD, Rokicki K, Safford T, Shaw K, Simpson JH, Sowell A, Tae S, Yu Y, Zugates CT (2012) A GAL4-driver line resource for *Drosophila* neurobiology. Cell Rep. 2:991–1001

Jovanic T, Schneider-Mizell CM, Shao M, Masson JB, Denisov G, Fetter RD, Mensh BD, Truman JW, Cardona A, Zlatic M (2016) Competitive disinhibition mediates behavioral choice and sequences in *Drosophila*. Cell 167(858–870):e19. doi:10.1016/j.cell.2016.09.009

Knott G, Marchman H, Wall D, Lich B (2008) Serial section scanning electron microscopy of adult brain tissue using focused ion beam milling. J Neurosci 28:2959–2964

Knott G, Rosset S, Cantoni M (2011) Focussed ion beam milling and scanning electron microscopy of brain tissue. J Vis Exp 53:e2588. doi:10.3791/2588

Lacalli TC (1984) Structure and organization of the nervous system in the trochophore larva of *Spirobranchus*. Philos Trans R Soc Lond B Biol Sci 306:79–135

Lichtman JW, Sanes JR (2008) Ome sweet ome: what can the genome tell us about the connectome? Curr Opin Neurobiol 18:346–353

Luan H, Peabody NC, Vinson CR, White BH (2006) Refined spatial manipulation of neuronal function by combinatorial restriction of transgene expression. Neuron 52:425–436

Macagno ER, Lopresti V, Levinthal C (1973) Structure and development of neuronal connections in isogenic organisms: variations and similarities in the optic system of *Daphnia magna*. Proc Natl Acad Sci USA 70:57–61

Meinertzhagen IA (2001) Plasticity in the insect nervous system. Adv Insect Physiol 28:84–167

Meinertzhagen IA (2014) The anatomical organization of the compound eye visual system. In: Dubnau J (ed) Handbook of behavior genetics of *Drosophila melanogaster*, vol 1. University Press, Cambridge, pp 1–19

Meinertzhagen IA (2016a) Morphology of invertebrate neurons and synapses. In: Byrne JH (ed) Handbook of invertebrate neurobiology. Oxford University Press

Meinertzhagen IA (2016b) Connectome studies on *Drosophila*: a short perspective on a tiny brain. J Neurogenet 30:62–68

Meinertzhagen IA, Lee C-H (2012) The genetic analysis of functional connectomics in *Drosophila*. Adv Genet 80:99–151

Meinertzhagen IA, O'Neil SD (1991) Synaptic organization of columnar elements in the lamina of the wild type in *Drosophila melanogaster*. J Comp Neurol 305:232–263

Meinertzhagen IA, Sorra KE (2001) Synaptic organisation in the fly's optic lamina: few cells, many synapses and divergent microcircuits. Progr Brain Res 131:53–69

Milo R, Shen-Orr S, Itzkovitz S, Kashtan N, Chklovskii D, Alon U (2002) Network motifs: simple building blocks of complex networks. Science 298:824–827

Morgan JL, Lichtman JW (2013) Why not connectomics? Nat Methods 10:494–500

Ohyama T, Schneider-Mizell CM, Fetter RD, Aleman JV, Franconville R, Rivera-Alba M, Mensh BD, Branson KM, Simpson JH, Truman JW, Cardona A, Zlatic M (2015) A multilevel multimodal circuit enhances action selection in *Drosophila*. Nature 520:633–639

Peng H, Chung P, Long F, Qu L, Jenett A, Seeds AM, Myers EW, Simpson JH (2011) BrainAligner: 3D registration atlases of *Drosophila* brains. Nat Methods 8:493–500

Pfeiffer BD, Jenett A, Hammonds AS, Ngo TT, Misra S, Murphy C, Scully A, Carlson JW, Wan KH, Laverty TR, Mungall C, Svirskas R, Kadonaga JT, Doe CQ, Eisen MB, Celniker SE, Rubin GM (2008) Tools for neuroanatomy and neurogenetics in Drosophila. P Natl Acad Sci USA 105:9715-9720

Pfeiffer BD, Ngo TT, Hibbard KL, Murphy C, Jenett A, Truman JW, Rubin GM (2010) Refinement of tools for targeted gene expression in *Drosophila*. Genetics 186:735–755

Randel N, Asadulina A, Bezares-Calderón LA, Veraszto C, Williams EA, Conzelmann M, Shahidi R, Jékely G (2014) Neuronal connectome of a sensory-motor circuit for visual navigation. eLIFE 3. doi:10.7554/eLife.02730

Randel N, Shahidi R, Verasztó C, Bezares-Calderón LA, Schmidt S, Jékely G (2015) Inter-individual stereotypy of the *Platynereis* larval visual connectome. eLIFE 4:e08069. doi:10.7554/eLife.08069

Ryan K, Lu Z, Meinertzhagen IA (2016) The CNS connectome of a tadpole larva of *Ciona intestinalis* highlights sidedness in the brain of a chordate sibling. eLIFE 5:e16962

Rybak J, Talarico G, Ruiz S, Arnold C, Cantera R, Hansson BS (2016) Synaptic circuitry of identified neurons in the antennal lobe of *Drosophila melanogaster*. J Comp Neurol 524:1920–1956

Schneider-Mizell CM, Gerhard S, Longair M, Kazimiers T, Li F, Zwart MF, Champion A, Midgley FM, Fetter RD, Saalfeld S, Cardona A (2016) Quantitative neuroanatomy for connectomics in *Drosophila*. eLIFE 5. pii: e12059

Shih CT, Sporns O, Yuan SL, Su TS, Lin YJ, Chuang CC, Wang TY, Lo CC, Greenspan RJ, Chiang AS (2015) Connectomics-based analysis of information flow in the *Drosophila* brain. Curr Biol 25:1249–1258

Shinomiya K, Karuppudurai T, Lin T-Y, Lu Z, Lee C-H, Meinertzhagen IA (2014) Candidate neural substrates for off-edge motion detection in *Drosophila*. Curr Biol 24:1–9

Silies M, Gohl DM, Clandinin TR (2014) Motion-detecting circuits in flies: coming into view. Ann Rev Neurosci 37:307–327

Simpson JH (2009) Mapping and manipulating neural circuits in the fly brain. Adv Genet 65:79–143

Sterling P, Laughlin S (2015) Principles of neural design. The MIT Press, London

Strausfeld NJ, Campos-Ortega JA (1977) Vision in insects: pathways underlying neural adaptation and lateral inhibition. Science 195:894–897

Strausfeld NJ, Nässel DR (1980) Neuroarchitectures serving compound eyes of *Crustacea* and insects. In: H Autrum (ed) Handbook of sensory physiology, vol VII/6B. Comparative physiology and evolution of vision in invertebrates. Springer, Berlin, pp 1–132

Takemura S, Lu Z, Meinertzhagen IA (2008) Synaptic circuits of the *Drosophila* optic lobe: the input terminals to the medulla. J Comp Neurol 509:493–513

Takemura S, Bharioke A, Lu Z, Nern A, Vitaladevuni S, Rivlin PK, Katz WT, Olbris DJ, Plaza SM, Winston P, Zhao T, Horne JA, Fetter RD, Takemura S, Blazek K, Chang L-A, Ogundeyi O, Saunders MA, Shapiro V, Sigmund C, Rubin GM, Scheffer LK,

Meinertzhagen IA, Chklovskii DB (2013) A visual motion detection circuit suggested by *Drosophila* connectomics. Nature 500:175–181

Takemura S, Xu CS, Lu Z, Rivlin PK, Olbris DJ, Parag T, Plaza S, Zhao T, Katz WT, Umayam L, Weaver C,Hess H, Horne JA, Nunez J, Aniceto R, Chang L-A, Lauchie S, Nasca A, Ogundeyi O, Sigmund C, TakemuraS, Tran J, Langille C, Le Lacheur K, McLin S, Shinomiya A, Chklovskii DB, Meinertzhagen IA, Scheffer LK(2015) Multi-column synaptic circuits and an analysis of their variations in the visual system of Drosophila.Proc Natl Acad Sci USA 112:13711–13716

Takemura S et al (2017a) EM reconstruction of α2 and α3 lobes of the mushroom body in adult *Drosophila*. eLIFE (submitted)

Takemura S, Nern A, Plaza S, Chklovskii DB, Scheffer LK, Rubin GM, Meinertzhagen IA (2017b) The comprehensive connectome of a neural substrate for 'ON' motion detection in *Drosophila*. Elife (under review)

Tobin W, Wilson R, Lee W-C (2017) Wiring variations that enable and constrain neural computation in a sensory microcircuit. eLIFE (submitted)

Varshney LR, Chen BL, Paniagua E, Hall DH, Chklovskii DB (2011) Structural properties of the *Caenorhabditis elegans* neuronal network. PLoS Comput Biol 7:e1001066

Ware R (1971) Computer aided nerve tracing in the brain of the rotifier, *Asplanchna brightwelli*. Ph.D. thesis, Massachusetts Institute of Technology, Boston, 213 pp

Ware RW, LoPresti V (1975) Three-dimensional reconstruction from serial sections. Int Rev Cytol 40:325–440

Watts DJ, Strogatz SH (1998) Collective dynamics of 'small-world' networks. Nature 393:440–442

White JG, Southgate E, Thomson JN, Brenner S (1986) The structure of the nervous system of the nematode *Caenorhabditis elegans*. Philos Trans R Soc Lond B Biol Sci 314:1–340

Windoffer R, Westheide W (1988) The nervous system of the male *Dinophilus gyrociliatus* (Polychaeta, Dinophilidae): II. Electron microscopical reconstruction of nervous anatomy and effector cells. J Comp Neurol 272:475–488

Erratum to: Single-Cell Transcriptomic Characterization of Vertebrate Brain Composition, Development, and Function

Bosiljka Tasic, Boaz P. Levi and Vilas Menon

Erratum to:
Chapter 18 in: A. Celik and M.F. Wernet (eds.), *Decoding Neural Circuit Structure and Function*, https://doi.org/10.1007/978-3-319-57363-2_18

In the original version of the book, Figure 18.4 has to be updated with the new figure in Chapter 18, which is a belated correction. The erratum chapter and the book have been updated with the change.

The updated online version of this chapter can be found at
https://doi.org/10.1007/978-3-319-57363-2_18

A. Çelik and M.F. Wernet (eds.), *Decoding Neural Circuit Structure and Function*, https://doi.org/10.1007/978-3-319-57363-2_21

MIX
Papier aus verantwortungsvollen Quellen
Paper from responsible sources
FSC® C105338

If you have any concerns about our products,
you can contact us on
ProductSafety@springernature.com

In case Publisher is established outside the EU,
the EU authorized representative is:
Springer Nature Customer Service Center GmbH
Europaplatz 3, 69115 Heidelberg, Germany

Printed by Libri Plureos GmbH
in Hamburg, Germany